Engineering Software IV

Proceedings of the 4th International Conference,
Kensington Exhibition Centre, London, England,
June 1985

Editor:
R.A. Adey

Springer-Verlag Berlin Heidelberg GmbH

R.A. ADEY

Computational Mechanics Centre
Ashurst Lodge,
Ashurst, Southampton,
Hampshire SO4 2AA
U.K.

British Library Cataloguing in Publication Data

Engineering software IV : proceeding of the 4th
    International Conference, Kensington Exhibition
    Centre, London, England, June 1985.
    1. Engineering——Data processing
    I. Adey, R.A.   II. Computational Mechanics Centre
    620'.0028'5425   TA345

ISBN 978-3-662-21879-2   ISBN 978-3-662-21877-8 (eBook)
DOI 10.1007/978-3-662-21877-8

© Springer-Verlag Berlin Heidelberg 1985

Originally published by C.M. Ltd, Southampton in 1985.
Softcover reprint of the hardcover 1st edition 1985

PREFACE

Since the First International Conference on Engineering Software was held six years ago, the role of computers in engineering has increased dramatically. Advances in computer hardware have provided low cost powerful microcomputers, high performance graphical displays and super-computers which enable problems to be solved which were inconceivable just a few years ago. Looking back, similar changes have taken place in the development of engineering software. Sophisticated graphics-based modelling systems and advanced mathematical modelling have provided solutions to previously unsolvable problems and, together with microcomputers and powerful mini-computers, have made such solutions widely available.

These proceedings contain over seventy of the papers presented at ENGSOFT 85 held in London. The Conference sessions include Software Development, Construction, Numerical Methods, Engineering Application of Expert Systems, Micro Applications, Structural Engineering, Hydraulics, Geometric Modelling, Simulation and Databases in Engineering.

The emphasis of the Conference was on the new applications which is demonstrated by the number of papers on expert systems, engineering databases and micro applications. However, papers describing new developments in the more 'mature' application areas are well represented.

R.A. Adey
Editor

CONTENTS

Preface

1. OPENING SESSION

Teaching Civil Engineers to Write CADDD Software   1-3
J.L. Jorgenson, North Dakota State University,
U.S.A.

A Practical Approach to the Integration of   1-13
Engineering Software Packages
M. Milivojcevic and S. Simonovic, Institute for
Development of Water Resources, Yugoslavia

Specification and Evaluation of Computer Aided   1-25
Engineering (CAE) Systems
A.J. James, Fluor Ltd., London, UK

2. SOFTWARE DEVELOPMENT

Some Thoughts on Design, Development and   2-3
Maintenance of Engineering Software
K. Bell, The Norwegian Institute of Technology,
Norway

On Halstead's Language Level   2-21
N.C. Debnath, Iowa State University, U.S.A.

Management of Interactive Graphics Functionality   2-31
Beyond "GKS"
R. Chauhan, Honeywell Information Systems, U.S.A.

AIM-AN Adaptable Input Module for Engineering   2-47
Programs
Z. Grodzki and M. Winiarski, Technical University
of Warsaw, Poland

3. CONSTRUCTION

Computing in Construction   3-3
I. Hamilton, CICA, Cambridge, England

Finite Element Analysis of Deep Foundations   3-9
Subjected to Uplift Loading
C.F. Mahler and A.P. Ruffier**,*
*COPPE/UFRJ-Coordination of Postgraduate Programmes
in Engineering, **CEPEL-Electrical Energy Res.
Centre, Brazil

The Use of Computers for the Stability Analysis 3-19
of Natural Slopes and Earth Dams Under Earthquake Conditions
A. Calvaruso, G. Seller and R. Trizzino,
University of Bari, Italy

The Optimal Design of Road and Industrial Pave- 3-35
ments Using Microcomputers
B. Shackel, The University of New South Wales,
Australia

Analysis of Laterally Loaded Pile Groups 3-51
C.F. Leung and Y.K. Chow, National University of
Singapore, Singapore

GEOSHARE - A New Dimension in Geotechnical/ 3-61
Geological Databases
*L.A. Wood and D.E. Wainwright and E.V. Tucker***
Queen Mary College, London, UK

4. ENGINEERING APPLICATIONS OF EXPERT SYSTEMS

Engineering and Scientific Software Control 4-3
Standards Modelling: An Interactive Expert System
Using Telecommunications
A. Rushinek and S. Rushinek, University of Miami,
U.S.A.

Computer Package for Reservoir Yield Estimation 4-17
L.M. Miloradov and S.P. Simonovic, Institute for
the Development of Water Resources, Yugoslavia

An Expert System for Preliminary Numerical Design 4-33
Modelling
K.J. MacCallum and A. Duffy, University of
Strathclyde, Scotland, UK

A Non - Deterministic Approach to the Computer 4-49
Aided Design of the Electrical Motor
I. Fucik, V. Klevar and J. Pistelak, Research
Institute for Electrical Machines, CSSR

Venture Analysis with a Microcomputer 4-59
G. Singh, University of Leeds, UK

Expert Systems in the Selection of Process Equipment 4-73
P. Norman and Y.W. Voon, University of Newcastle-
upon-Tyne, UK

Reliability and Integrity of Computer Assisted 4-89
Decision Making
M. Hashim, Burroughs Machines Ltd. Scotland, UK

5. ENGINEERING APPLICATIONS OF EXPERT SYSTEMS POSTER SESSION

A Rule-Based Program to Construct an Influence   5-3
Line for a Statically-Determinate Multi-Span
Beam Structure
G.G. Haywood, Bolton Institute of Higher Education, UK

6. NUMERICAL METHODS

The Fundamentals of the Space-Time Finite Element   6-3
Method
M. Witkowski, Technical University of Warsaw, Poland

A Computer Program for Large Eigenvalue Problems in   6-13
Dynamic Analysis
*A. Vale e Azevedo, LNEC - National Laboratory of Civil
Engineering, Portugal*

Accuracy Assessment by Finite Element P-Version   6-35
Software
*P. Angeloni, R. Boccellato, E. Bonacina, A. Pasini
and A. Peano, ISMES, Italy*

A Powerful Finite Element Method for Low Capacity   6-49
Computers
*J. Jirousek, Swiss Federal Institute of Technology,
Switzerland*

An Explicit Finite Difference Solver by Parameter   6-65
Estimation
S.K. Dey and C. Dey**, *Eastern Illinois Univer-
sity and **Charleston Jr. High School, U.S.A.*

7. NUMERICAL METHODS POSTER SESSION

Microcomputer Application of Non-Rectangular Space-   7-3
Time Finite Elements in Vibration Analysis - Direct
Joint-by-Joint Procedure
C.I. Bajer, Engineering College, Podgorna 50, Poland

Multiextremal (GLOBAL) Optimiaation Algorithms for   7-17
Engineering Applications
*J. Pinter and J. Szabo, Research Centre for Water
Resources Development, Hungary*

Eigenvalue Calculations Using the Collocation Finite   7-27
Element Method
*N.G. Zamani and S.N. Sarwal, Technical University of
Nova Scotia, Canada*

8. MICRO-APPLICATIONS

Structural Analysis Using Microcomputers State of the   8-3
Art
D. Nardini, University of Zagreb, Yugoslavia

Application of Microcomputers in Thermal   8-11
Insulation Design and Analysis
S.N. Tay, Nanyang Technological Institute, Singapore

On the Development of a Finite Element Program for   8-25
Microcomputers
S.A. Brown and M.P. Kamat**, *Virginia Polytechnic
Institute and State University and **University of
Tennessee at Chattanooga, U.S.A.*

Development of Hardware and Software for Machine   8-41
Control Applications
G.D. Alford, Teesside Polytechnic, U.K.

FLAPS — A Fatigue Laboratory Applications Package   8-53
*S. Dharmavasan, D.R. Broome, M. Lugg and W.D. Dover,
University College London, UK*

9.   MICRO-APPLICATIONS POSTER SESSION

The Design of Non-linear Structures Using a Small   9-3
Micro-Computer
I. Davidson, Warrington, UK

PECI — Calculation of Building Structures   9-15
S.S. Kalmus, Centro de Informatica — FAAP, Brazil

Complex Waveforms from a PET and BASIC   9-25
D.A. Pirie, Glasgow University, Scotland, UK

The Analysis of Speckle Interferometric Records of   9-31
Strain by Micro Computer
I. Grant and G.H. Smith, Heriot-Watt University, UK

10.   STRUCTURAL ENGINEERING

VAPSIMM: A Computer Program for Vibration Analysis   10-3
of Piping Systems Incorporating a Moving Medium
C.W.S. To, The University of Calgary, Canada

Computer Aided Design of Shell Structures   10-19
H. Sardar A. Saleh, Katholic University Leuven

SSCAD on a Desktop Computer   10-33
*J.J. Jonatowski and D. Koutsoubis, Space Structures
Int. Corp., U.S.A.*

Computer Programme for Analysis of Reinforced   10-45
Concrete Structures
*M. Sekulovic and Z. Prascevic, University of
Belgrade, Yugoslavia*

Exploiting Parallel Computing with Limited Program   10-61
Changes Using a Network of Microcomputers
J.L. Rogers, Jr. and J. Sobieszczanski-Sobieski,
NASA Langley Research Center, U.S.A.

Through Thickness and Surface Stress Distribution for   10-75
Welded Tubular T-Joint Using Finite Element Analysis
G.S. Bhuyan, K. Munaswamy and M. Arockiasamy,
Memorial University of Newfoundland, Canada

11.   STRUCTURAL ENGINEERING POSTER SESSION

Dynamics of Multi-Storeyed Buildings by Means of   11-3
Shear Waves
A. Mioduchowski and M.G. Faulkner, The University
of Alberta, Canada

Computer Procedure for Optimal Design of Structures   11-13
Under Traffic Loads
Lj. R. Savic, University of Belgrade, Yugoslavia

Optimization of Structures with Random Parameters   11-21
S.F. Jozwiak, Politechnika Warszawska, Poland

Analysis of Plates by the Initial Value Method   11-31
H.A. Al-Khaiat, Kuwait University, Kuwait

Computer-Aided Design of Steel Floor Beam Framework   11-37
M.C. Thakkar and S.J. Shah, Elecon Eng. Co. Ltd. Gujarat

An Estimating Package for Steel Fabricated Structures   11-41
H.C. Ward, CAD Unit, Teesside Polytechnic, UK

12.   HYDRAULICS

Optimal Design for the Run-of-the-River Plant System   12-3
M. Petricec*, J. Margeta and N. Mladineo**
*Institut za Elektroprivredu and **V. Maslese bb,
Yugoslavia

The Influence of Engineering Software for the   12-15
Design and Analysis of Urban Storm Drainage
R.K. Price, Hydraulics Research, Wallingford, UK

Irrigation Network Modelling   12-23
S. Selvalingam, S.L. Ong and S.Y. Liong, National
University of Singapore, Singapore

Computerised Forecast System for River Flows   12-35
J.M. Dujardin, J.L. Rahuel and P. Sauvaget,
SOGREAH, France

Mathematical Modelling Software for River Management: 12-49
'CARIMA' and 'CONDOR' Systems
Ph. Belleudy, J.A. Cunge and J.L. Rahuel, SOGREAH, France

Operational Control of a Water Distribution System   12-63
Utilizing Micro Computers
G.W. Lackowitz, Consultant Engineer, U.S.A.

13.   HYDRAULICS POSTER SESSION

Unsteady Flow Computation for Planning Flood   13-3
Control Projects
*S. Bognar and L. Somlyody, Research Centre for
Water Resources Development, Hungary*

Modelling of an Unsteady Flow in a Complex Water   13-11
Resources System Applied to Management
*M. Baosic*and B. Djordjevic**, *Institute for the
Development of Water Resources and **University of
Belgrade, Yugoslavia*

Irrigation Water Requirement Model   13-23
S-Y Liong and O. Sutjahyo & B. Rasli**, * National
University of Singapore, Singapore and **PT Dacrea,
Jakarta, Indonesia*

14.   GEOMETRIC MODELLING

Solid Modelling : A Teenaged Art - When will it Mature?   14-3
M.J. Pratt, Cranfield Institute of Technology, UK

EUCLID(*): A Powerful Tool for Interactive CAD and   14-15
Specialist Software Development
*K.R. Colman, European Organization for Nuclear
Research, Geneva*

On Quadratic Splines and Their CAD-Application   14-37
*F. Fenyves and G. Kovacs, Hungarian Academy of
Sciences, Hungary*

15.   SIMULATION

CAMAS, A Computer Aided Modelling, Analysis and   15-3
Simulation Environment
*J.F. Broenink and G.D.N. Twilhaar, Twente
University of Technology, The Netherlands*

ST.EX.OM - A Statistical Package for the Pred-   15-17
iction of Extreme Values
J. Labeyrie, IFREMER, France

Interaction of Suspension Design with Vehicle Ride   15-33
P.J.H. Wormell, Royal Military College of Science, UK

Random Signals Analysis Applied to Vehicle-Track   15-47
Interaction
C. Esveld, Netherlands Railways

2D Finite Element Analysis Package for Sensibility   15-63
Analysis in Electrotechnique
J.L. Coulomb and A. Kouyoumdjian, Laboratoire
D'Electrotechnique de Grenoble, France

SYNTRA - An Interactive Program to Design Grashof   15-77
Planar Four Bar Mechanisms
C.R. Barker, University of Missouri, U.S.A.

The Use of the Fluid Flow Program Phoenics in   15-93
Engineering Design
N. Rhodes, S.A. Al-Sanea and K.A. Pericleous,
Concentration Heat and Momentum Ltd. London, UK

Simulation of Fuel Injection in Diesel Engines   15-107
G. Chiatti and R. Ruscitti, Universita degli
Studi di Roma, Italy

Computer Aided Design for the Selection Process   15-119
of Hazardous and Nuclear Wastes Sites
G.V. Abi-Ghanem and V. Nguyen, ARDI Corporation,
U.S.A.

16.   SIMULATION POSTER SESSION

COMPAMM - A Program for the Dynamic Analysis of   16-3
Multi-Rigid-Body Systems
J. Unda, A. Avello, J.M. Jimenez, J. Garcia de
Jalon, University of Navarra, Spain

Computer Aided Methods for the Synthesis of   16-17
Mechanism Kinematic Models: Planar and Spatial Cases
P. Fanghella and C. Galletti, University of Genoa

17.   DATABASES IN ENGINEERING

ESTSOL - A Subroutine Library for Establishment and   17-3
Solution of Linear Equation Systems
H. Tagnfors, N-E. Wiberg and L. Bernspang, Chalmers
University of Technology, Sweden

Management Method of a Database System for Scientific   17-15
Computer Codes
M. Paolillo, Electricite de France

Database Administrator Facilities for Engineering   17-31
Data Management
H. Steenbergen and F.J. Heerema, National Aerospace
Laboratory NLR, The Netherlands

A Data Base Manager for Engineering Computations   17-51
E. Backx, Catholic University of Louvain

Engineering Database - Weights and Centres 17-57
S.M. Fraser, University of Strathclyde, Scotland, UK

Database Used for Optimal Control of the DJERDAP 17-65
Hydropower System
S. Opricovic, and M. Miloradov**, *Faculty of Civil
Engineering, Belgrade, **Institute for Development
of Water Resources, Yugoslavia*

PAPERS RECEIVED AFTER 14th MARCH

Engineering with Software Prototypes 14-47
M.P. Williamson, APPLICON (UK) Ltd.

Comparative Analysis BEM-FEM for Stress Con- 6-77
centration Around the Holes in Elasto-Plastic
Infinite Medium
*V.F. Poterasu and N. Mihalache, Polytechnic
Institute of Jassy, Romania*

An Application of Systems Modelling in Steel 15-127
Production Scheduling
*J.R. Wright and M.H. Houck, Purdue University,
U.S.A.*

BEASY Boundary Element Analysis SYstem 15-141
*C.A. Brebbia, D.J. Danson, J.M.W. Baynham and R.A. Adey,
CM BEASY Ltd. Southampton, England*

DOLMEN: A Complete CAD System for Microcomputers 10-87
F. Biasioli, Politecnico di Torino, Italy

1. OPENING SESSION

TEACHING CIVIL ENGINEERS TO WRITE CADDD SOFTWARE

James L. Jorgenson

North Dakota State University, Fargo, N. D.

ABSTRACT

This paper describes a course on the principles and procedures of writing computer programs to be used for computer
aided design, drafting, and detailing (CADDD). The need
for the course developes from an individuals desire to be
more productive. With the proper computer software and hardware an engineer's productivity can be significantly
increased.

The principles covered in the course start with the benefits
and problems in using a CADDD system. Next, characteristics
of activities which can be effectively programmed are considered. Since many programs will use the same data it is advisable to establish a data base. The principles of program/
user interaction are also presented.

These principles are illustrated through the student preparation of a class project. The project consisted of the student preparation of programs to do the designing, detailing,
and drafting of a reinforced concrete foundation. Programs
written by the students include: structural design of a
wall footing, detailing the reinforcing for the foundation,
and the drafting for the foundation drawing. A program used
by the students which they did not write was the design
of single column spread footings.

At the end of the course the students were able to use their
software and the available computer and plotter to design,
detail, and draft a building foundation in less than one-
half hour, thereby significantly increasing their productivity.

INTRODUCTION

The Keyword for the 1980's is productivity. The word is most closely associated with industrial output. One component of industrial output is the productivity of the engineering personnel. Engineering productivity can be significantly improved if the engineers have the appropriate computers and computer software to assist them in their work. It has been common practice for many years for engineers to use computers in design and in fewer cases for computer driven plotters to be used in drafting. These changes have increased engineering productivity, however, a more significant increase in productivity results when the design, detailing, and drafting can be carried out by a single program.

Engineering has been described as an area where every problem is different. Even though most design problems are unique there are many categories of problems in which the design parameters are limited both in number and scope. The productivity of engineers can be significantly increased if they will, (1) determine those design areas which have similar solutions, (2) write computer programs to do the design, detailing, and drafting, and (3) assemble the programs with the appropriate user interaction to obtain a computer aided design, detailing and drafting (CADDD) package.

This paper describes a course on teaching engineers to write CADDD software.

THE COURSE

The course objective is that exiting students will be able to demonstrate an ability to plan, layout, program, test, and document projects in computer aided design, detailing, and drafting.

A description of the course follows: the course will begin with how to select a portion of an engineering design project on which one can use the computer to assist in the design, detailing, and drafting of construction drawings. This will be followed by studies on how to design, detail and prepare the construction drawings for the project. Next, students will prepare or modify, test, and document computer programs which will be used for designing, detailing, and drafting the selected project. The final step in the course will be for each student to demonstrate the application of programs by designing, detailing, and drafting construction drawings on three problems that are within the scope of the selected project.

The course prerequisites are a knowledge of Fortran programm-

ing and a knowledge of design of reinforced concrete. It
is being taught to first year graduate students. It could
be taught to undergraduates, however, the added experience
of the graduate students gives them a better perspective
of the overall process.

Computer hardware and software were selected on the basis
of what was available. The hardware was the university's
computer: IBM 370; plotter: Complot DP-8 drum plotter; graphics terminal: Tektronix 4050. There are a number of terminals available for the students to access the computer
and the computer driven plotter. The graphics terminal permitted one to view parts of the drawing during the development stage.

The computer software is the operating system available
to the students, the Complot drafting subroutines, and a
computer program for the design of single column spread
footings.

A microcomputer and attached plotter may well have been
used for the class, however, there wasn't any available
in sufficient number to provide easy access to the students.
The principles and procedures discussed in this paper are
equally applicable to micro computers.

No suitable textbook was available for the course. References appropriate to each of the subject areas were used.

COURSE OUTLINE

The course outline is shown in Table 1. The Table has three
headings: subject area, class periods, and topics.

Introduction to the course was covered in two periods. It
consisted of going over the course outline and then giving
a number of examples where engineering productivity has
been increased. A full illustration was given on the software being used by structural engineers in certain metal
building companies. It clearly illustrated how an engineers
productivity can be significantly improved. The emphasis
on improving ones productivity was used throughout the
course to motivate student interest in the subject.

Computer programming was a course prerequisite hence the
students came with at least one Fortran programming course.
With only two periods on the subject, the major emphasis
was on program debugging and documentation. Most students
are more interested in codeing a problem than they are in
documenting their program. However, in programs used by
others or used by the programmer at a later date, the pro-

TABLE 1
COURSE OUTLINE

SUBJECT AREA	CLASS PERIODS	TOPICS
INTRODUCTION	1	COURSE OUTLINE
	1	ENGINEERING PRODUCTIVITY
PROGRAMMING	2	COMPUTER PROGRAMMING: DESIGN, DEBUG, TESTING AND DOCUMENTATION
DESIGN OF WALL FOOTINGS	2	STRUCTURAL DESIGN WALL FOOTINGS
	2	COMPUTER PROGRAM FOR DESIGN OF WALL FOOTINGS
DESIGN OF SINGLE COLUMN FOOTINGS	1	CRITERIA FOR THE DESIGN OF SINGLE COLUMN SPREAD FOOTINGS
	2	COMPUTER PROGRAM FOR THE DESIGN OF SINGLE COLUMN SPREAD FOOTINGS AND EXAMPLES
DETAILING	2	DETAILING REINFORCING FOR REINFORCED CONCRETE FOOTINGS
DRAFTING	2	ESSENTIALS OF FOUNDATION DRAWINGS, DETAILS, DIVISIONS OF DRAWING
	1	INTRODUCTION TO COMPLOT PLOTTER AND TEKTRONIX TERMINAL
	1	COMPLOT SUBROUTINES
	6	WRITING PLOTTING PROGRAMS
	2	REPORT ON SUBROUTINES WRITTEN BY STUDENTS
	1	FOUNDATION TEST CASE
COMBINING PROGRAMS	3	USER/PROGRAM INTERACTION PROGRAM/PROGRAM INTERACTION
	2	CLIST AND OTHER CONTROL PROGRAMS
EXAMPLE	1	INDIVIDUAL CADDD ASSIGNMENT IN USING THE PROGRAM

gram documentation is the most important aspect. The program documentation should consist of the following: (1) objective; (2) problem analysis including assumptions, figures, diagrams, equations, and references; (3) sample problem with hand calculations; (4) description of program input; (5) description of progam output, (6) flow chart; (7) code with sufficient comments for one to understand the coding; (8) definition of all variables; (9) test cases; and (10) instructions to the program user.

The next four subject areas of the course outline cover designing, detailing, and drafting of the project material chosen for the course. The project chosen was to design and prepare plans for a reinforced concrete building foundation. The foundation was to be rectangular in plan and consist of wall footings and, if required, single column spread footings. Many other projects could have been used. It is a matter of selecting something that has the potential for productivity increase and is of interest to both the students and teacher.

Two parts of the concrete foundation were to be designed, they were the wall footings and the single column spread footings. Each subject was initiated by a discussion of the foundation and structural design criteria. This was followed by writing a computer program to do the design. In the case of the single column spread footing, an available program was used, however, the students had as an individual assignment to write the computer program to design wall footings. This was done in three stages. The first stage was the first three items in program documentation. The second stage contained documentation items 4 through 6. The third stage was to complete the documentation.

The next subject was how to detail the reinforcing and concrete in a reinforced concrete foundation. This consisted of preparing a list containing the shape and weight of all reinforcing bars as well as the volume of concrete for each pour in placing the concrete. The students wrote a computer program to do the detailing. Input information for the computer program was the output information from the foundation design, namely, the size and reinforcing for all foundation elements.

The subject area with the major time requirement (43%) was drafting. The objective was to have the plotter prepare the foundation drawing. To know how the final drawings should appear it was necessary to spend two class periods on the essentials of foundation drawings, divisions of the drawing, and details of the drawing.

The Fortran Plotting User's Manual was the reference used

in learning how to communicate with the plotter. It is necessary to instruct the plotter on each movement of the pen. This is accomplished with about ten complot subroutines. The complot subroutines are used as statements in a Fortran program. The most common complot subroutine was PLOT (X, Y, IPEN) where X and Y are the new coordinates for the pen and IPEN is a variable used to indicate if the pen is up or down. The other common subroutines were SYMBOL and NUMBER which permit the user to place letters and numbers at the desired locations on the drawings.

To gain experience and confidence in using the plotter the students completed a number of plotting assignments. It started with drawing a rectangle of a given size, then writing a rectangle subroutine, then using the rectangle subroutine to draw a foundation plan of variable size and location on the drawing.

Next, it was necessary to prepare a computer program that would direct the plotter to prepare the complete foundation drawing. This was accomplished by preparing the data input sheet in class and then assigning each student to write the program to do a portion of the drawing. These student assignments were written in subroutines. Each student reported to the class on how to use the subroutine. Each student could then prepare a program for the final drawing incorporating the subroutines prepared by the other students.

The class is now at the stage where they have four programs for the reinforced concrete foundation. They have two on design, one on detailing and one on drafting. They can be run individually where the user takes the output of one program to prepare the input of the next program. The manual operation can be eliminated if the programs can be operated by a control program.

The control program must have the capability of manipulating data files, calling programs, and providing the appropriate user/program interaction. A flow chart of the control program is shown on Figure 1. As can be seen it is a sequence of: preparing a data file, running the program, and then requesting the operator for approval in going to the next step.

To operate the control program, the user prepares the data file containing the job identification, building geometry, and column reactions. Then, the control program is called and instructed to read the prepared data file. From this point on the control program will complete the design, detailing, and drafting with the exception of the approvals required by the user after the completion of each step. The approval consists of the user being able to view the

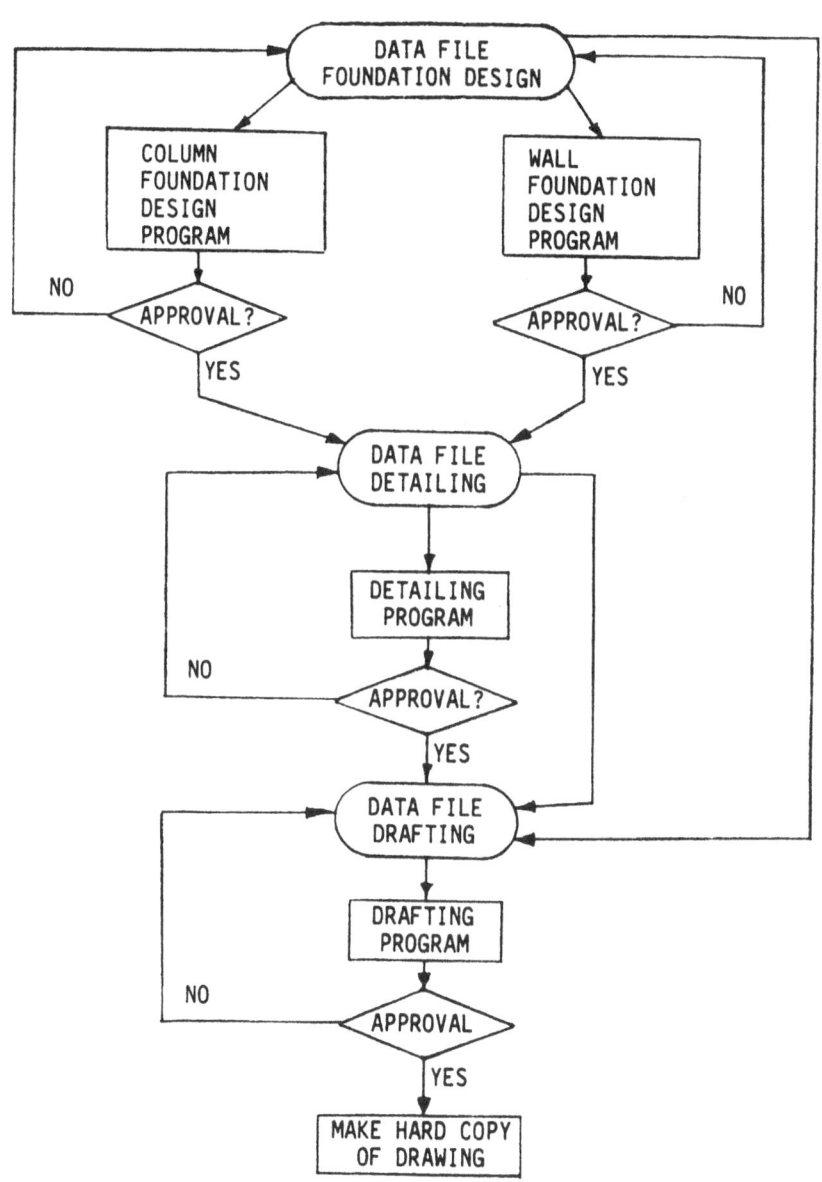

FIGURE 1

FLOW CHART FOR CONTROL PROGRAM

results of the design, detailing, or drafting and then either approve the results and go to the next step or revise the previous input file and redo that step.

The final topic in the course outline is the example. The assignment was a specific building for which they were to use their developed CADDD software. An example of the final drawing is shown on Figure 2. The actual drawing was 24 x 36 inches. The reduction process has made the drawing difficult to read. The foundation plan is shown on the upper left side of the drawing. Details showing anchor bolt locations are shown on the lower left. The foundation wall schedule and spread footing schedules are shown on the upper right side of the drawing. The title block, construction notes and volume of concrete are shown on the lower right side of the drawing. The detailing of the reinforcing bars was not completed to a state of being placed on the drawing.

SUMMARY

This paper has described a course in which the student experienced (1) selecting an area of engineering design which should be prepared for CADDD, (2) writing computer programs to do the design, detailing, and drafting, (3) assembling the programs with the appropriate user interaction so they operated as a CADDD package, and (4) using the CADDD package on typical design problems thereby significantly increasing their engineering productivity.

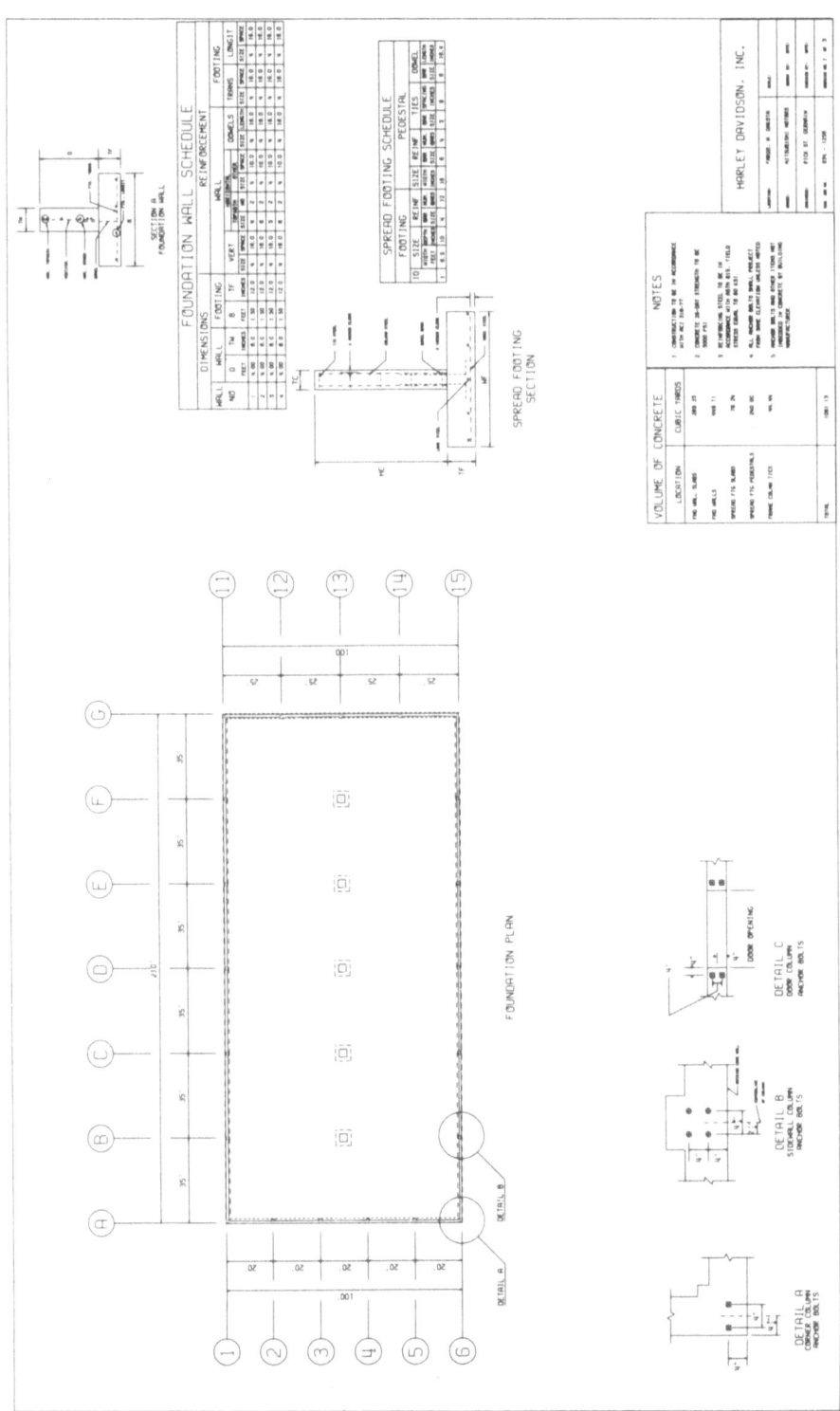

FIGURE 2   EXAMPLE OF FOUNDATION DRAWING

A PRACTICAL APPROACH TO THE INTEGRATION OF
ENGINEERING SOFTWARE PACKAGES

M. Milivojcevic and S. Simonovic

Institute for Development of Water Resources "Jaroslav Cerni"
P.O. Box 530, 11000 Beograd, Yugoslavia

1. INTRODUCTION

The aim of this paper is to present a convenient approach to
the functional integrating of large packages of engineering
modelling software. This approach is applied when solving the
problem of integrating a large hydraulic-hydrologic simulation
package HEC5, developed by the U.S. Army Corps of Engineers
(reference Anonymous, 1979) and a general-purpose optimization
program. It is based on the use of some very useful facilities
that are most often available on the operating systems of
modern computers.

These facilities can easily be used and it is therefore very
important to point them out in the hope that they will be widely used by engineers involved in computer applications in water
resources as well as in other areas of engineering.

Section 2 presents the background for the work, giving an overall description of the problem, as well as the problem statement and the set objectives. The solution for the integration
of the considered software packages is given in section 3. A
description of the general elements of the solution applicable
to various modern operating systems has first been given, this
being followed by a description of the specific implementation
on the VAX 11/780 having an operating system VAX/VMS Version
V.2.3.

An example of the application in water resources engineering
(section 4) serves to illustrate the usefulness of the entire
work in integrating these particular optimization and simulation
software packages.

2. BACKGROUND

Computer applications in water resources systems engineering are
becoming more and more complex. Numerous software packages that

have already been developed perform even more complicated tasks.
This is due to the fact that they have become an indispensable
tool, the use of which has resulted in quantitatively better,
more reliable and less expensive planning, design and management. The development of a series of more or less general software packages brings up a possibility to accomplish even more
complex tasks in a fully automated manner.

In the process of solving one type of practical problems of
water resources management (minimization of flood damages on
a river system) (Simonovic, Potic and Milivojcevic, 1984) it
was conjectured that an automatic coupling of optimization and
simulation techniques could bring us better management solutions.
It has already been demonstrated in the literature (Houck and
Datta, 1981, Jacoby and Louks, 1972, Simonovic, Potic, Milivojcevic, 1984) that the combined use of optimization and simulation models may be very useful in river basin planning and
management.

The very complex HEC5 hydrologic simulation package developed
by the Hydrologic Corps of Engineers, U.S.A. (Anonymous, 1979)
has already been installed at the computer center of the "Jaroslav Cerni" Institute and it has been used for the computer
simulation of some water resources systems. On the other hand,
we also had available the general purpose optimization program
BOX. Their combining in the form of a series of possibly many
isolated runs could prove to be very cumbersome, especially
when solving some real engineering problems when it is not known
 a priori where the optimal solution really lies. It was, therefore, our intention to try to integrate these packages to
function as one programming system.

A schematic presentation of the program linkage and data flow
for the BOX optimization program is shown in Fig.1. BOX is the
main program while the rest are subprograms and all were written
in standard FORTRAN '66. The scheme itself is self-explanatory.
The BOX program uses the algorithm of Box to perform the optimization task. This algorithm is generally recommended for non-
linear objective functions in a limited space of decision variables (Kuester and Mize, 1973).

Fig. 2 presents an overall schematic view of the flow of control
and of the data flow for the HEC5 simulation package. In fact,
the complex software of HEC5 is composed of two large programs
HEC5A and HEC5B. Their automatic serial execution is realized
simply through the use of operating system command language
statements in the form of a command procedure. The files and
corresponding data flow paths are depicted in a synoptic way.

The main problem that has here been considered consists of designing and implementing a functional integration of the optimization and simulation software packages, that will enable the

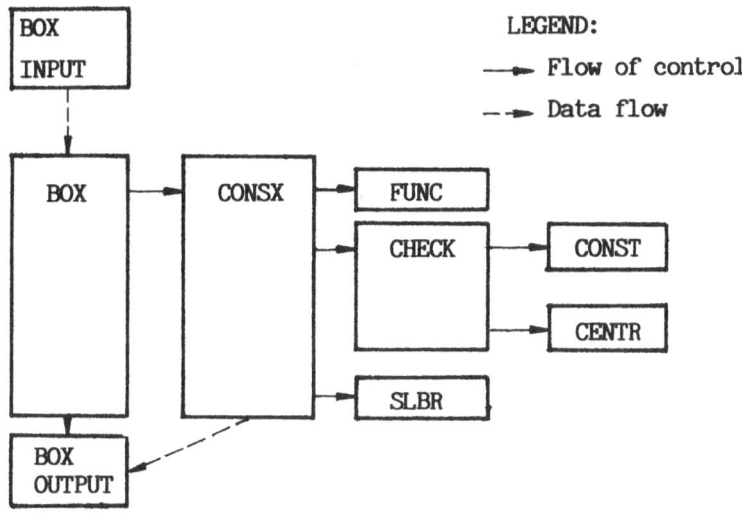

Figure 1. Program linkage and data flow for BOX

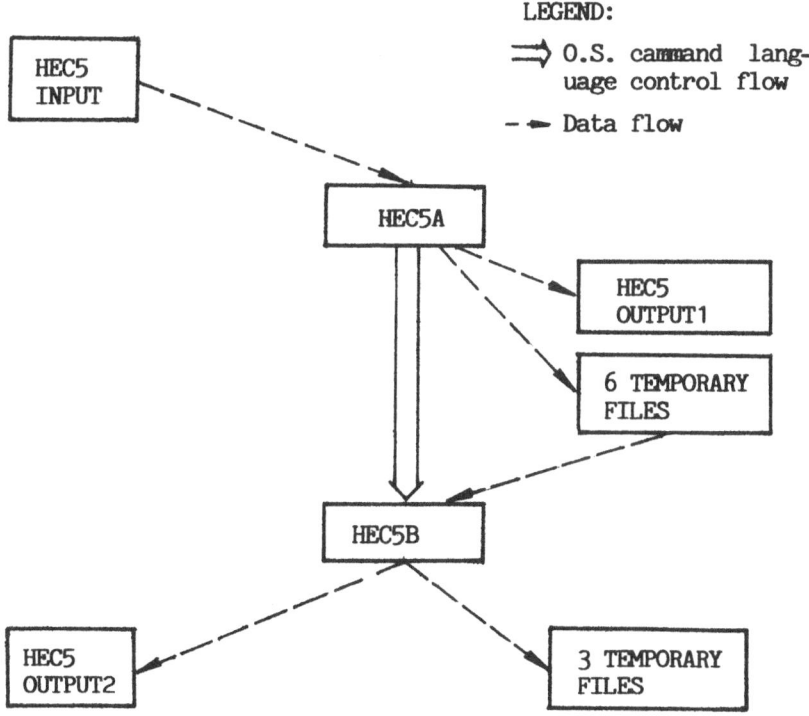

Figure 2. Flow of control and data flow for HEC5

accomplishment of a complex optimization-through-simulation task in a fully automatic manner. In fact, the question is how to perform an integration of these basically two FORTRAN packages in a manner that will enable the optimization package to call the simulation package of two large programs to get the results of the simulation for a new physical system state (generated by the optimization program); to take the control back with all the relevant results and to continue its optimization task in the same manner until it satisfies the specified criteria.

An additional desirable objective would consist of introducing only small number of changes of the software packages that are integrated, without any changes in their general structure if possible.

3. SOLUTION APPROACH

Considering the previously stated problem, it could be thought that a simple approach consisting of the use of the operating system command language statements in order to realize an appropriate coupling of the optimization and simulation package is possible. Unfortunately, although it may be possible in principle, this solution is not an acceptable one since the entire most significant part of the package BOX from FORTRAN would need to be rewritten into the operating system command language.

The integration of software packages frequently brings out the question of file sharing. The same question can also be expected to arise in the process of solving our problem since the simulation packages HEC5A and HEC5B use input and output statements that do not allow the use of "OPEN file" and "CLOSE file" statements.

Description of Solution
The solution that has been proposed is based on the following two elements:

1. use of interprocess communication facilities,
2. use of the technique of chaining two or more program images.

Fig. 3 is a schematic presentation of the main general points of the solution. The files and data flow have not been shown in order to simplify the scheme. The solution requires the generation of two processes on two interactive terminals. One process is devoted to the execution of the optimization image, using the main program MBOXHC that was made by a slight modification of the main program BOX (Fig.1) in order to integrate it with HEC. The second process should realize the ordered running of the images of the software packages MHEC5A and MHEC5B that were made by extremely simple modifications of HEC5A and HEC5B consisting of the addition of only one FORTRAN statement to each of them.

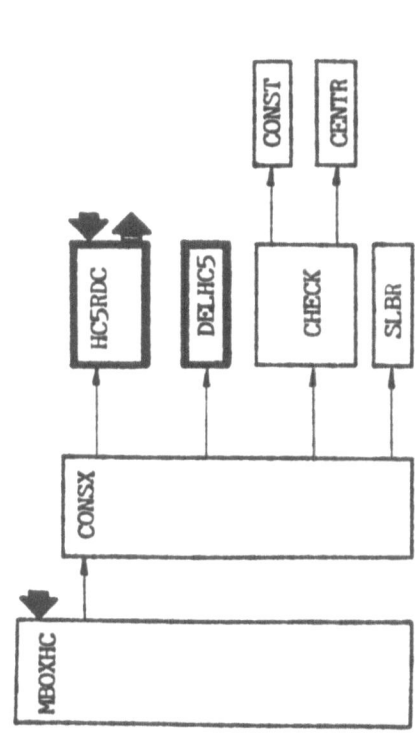

PROCESS 2
(Terminal T2)

PROCESS 1
(Terminal T1)

LEGEND :

☐ Unchanged or slightly changed modules

☐ New program modules

◆ Interprocess communication

◁— Program chaining

—— Subroutine call

Figure 3. Schematic presentation of the flow of control
for the integrated packages

In general terms, a process can be defined as the basic schedulable program entity executed by a computer processor. The transfer of control between these two processes is achieved by using some interprocess communication facilities that are available within a given computer system. The black arrows in Fig. 3 denote the places in the program where the control is transferred from (➡) and the places where the control is returned back (⬅) to a program. The shaded arrows depict the flow of control between several images within one process. This flow of control is realized by using some facilities for the chaining of images. In the proposed solution the chain is closed, i.e. it makes a cycle (or, a link).

This solution for the flow of control makes it possible for the optimization package to define (i.e. to generate according to its algorithm) the new state of the considered physical (hydrological) system, and to then activate two simulation packages to perform the simulation for the new state of the hydrological system, thus computing the new values of the objective function and returning the control back to the optimization program. This program can then use the new value of the objective function to automatically continue its optimization task until it reaches below an a priori defined optimization tolerance.

The files of data for the packages being integrated are exactly the same as those depicted in Figs. 1 and 2. The data flow paths are slightly different. It was necessary to add two new data paths. First, a two-way path from the new subroutine HC5RDC was added to the input file of HEC5, in order to store in this file any new (physical) system state at the appropriate place and in the right format. This was followed by another addition of a one-way path from the output file #2 of HEC5 to subroutine HC5RDC. In this way, HC5RDC retrieves the computed value of the objective function for the new system state.

In order to be able to implement the proposed solution on a computer system, it is necessary to use some operating systems facilities that are usually available. In fact, as far as the interprocess communication is concerned, only the following three functions need to be at one's disposal:

1. putting a process in a wait status,
2. sending a message to a process to resume execution,
3. clearing the indicator value (a flag) for the resumption of process execution.

The entire implementation of the interprocess communication is then very simple. In the solution given in Fig. 3, this is mainly localized in two small FORTRAN modules: the subroutine HC5RDC and the main program MIRHC5.

Since the simulation part consists of two software packages, it was also necessary to realize an appropriate transfer of the control between them. One straightforward way would seem to consist of having one process for each of these packages where the same type of interprocess communication would again be used. In some cases this may work fine, but for this particular case this was not applied. This is mainly due to the fact that this new scheme of the flow of control ends up with the file locking problem. Namely, the packages being coupled, should now share both the HEC5 input file and the HEC5 output file, as has already been explained. On the other hand, the conception of creators of packages HEC5A and HEC5B was to use input and output statements that do not allow the use of OPEN and CLOSE statements for the input and output files. They had their reasons for this, but in the integration of the considered packages, according to the three-processes scheme, this very fact resulted in the file locking problem.

Therefore, the solution for this part of the problem was to have two processes as in Fig. 3, and to use a facility to pass the control from one image to the other, but by completely getting out of the first image. The program chaining technique is therefore the right way to complete the solution. In this way, the control from HEC5A to HEC5B and further to MIRHC5 could then easily be transferred and it would be possible to completely exit from the preceding images, thus automatically closing all their files.

Brief Note on Specific Type of Implementation
The particular implementation of the proposed solution was done on the VAX 11/780 with the VAX/VMS virtual memory operating system version V2.3 installed. This presentation should be kept very short since it must be machine and operating system oriented. For this computing system, the process is defined as the basic entity scheduled by the system software that provides the context in which an image executes. It consists of an address space and of both the hardware and software context. The VAX/VMS offers various facilities for both the process control and interprocess communication.

In our work, we used the common event flags (see Anonymous,1978) as a means of implementing the interprocess control and communication as shown in Fig. 3. This is one of the convenient mechanisms that allows cooperating processes to coordinate their activity, communicate with each other and mutually transfer the control to each other if necessary. One process can create one or two so called common event flag clusters, each containing 32 flags. Once a cluster has been created, the other processes can then associate with the same cluster. Processes associated in this way can then easily mutually communicate using all the available event flags. All of this can be conveniently done on the VAX/VMS using a program unit written on a high level language, just as was done in this work in the FORTRAN programs.

For this purpose, specific system services, i.e. the so-called
event flag services, are always available. For example, in
order to make a process waiting for a common event flag number
n, one should use the following function given in a high language:

SYS$WAITFR (n)

In order to set an event flag m in a common flag cluster to 1,
the following event flag system service, for example in FOR-
TRAN, should be used:

SYS$SETEF (m)

Similarly, one may set an even flag 1, in a common event in
the following general high-level language format:

SYS$CLREF (1)

There are more additional event flag system services to perform
some other functions. Prefix SYS denotes that they belong to
the group of system services that the VAX/VMS offers.

The program chaining technique is also available in this operating system just as it is in many others of the older generation.
In the VAX/VMS this is available through a so called run-time
procedure. More specifically, if the general purpose run-time
procedure $RUN_PROGRAM is called in the program MHEC5A by re-
ference

LIB$RUN_PROGRAM ('MHEC5B')

it causes the image to exit at the point of the call and passes
the control to the image MHEC5B.

At the closure of this section it should be added that conceptually the same and practically a simple approach can be applied
in a similar manner for solving other forms of integration of
the engineering software. For example, in this way one can
easily perform the coupling of two or more packages in order to
realize the parallel execution of some of them in a parallel
algorithm. Whenever applicable, this could bring a great saving
in execution time.

4. PRACTICAL EXAMPLE OF APPLICATION

The presented general ideas relating to the integration of software and its implementation, have been finalized by the successful coupling of optimization and simulation packages. This section is an illustration of one possible application of such an
integrated composite package for the solution of a practical
engineering problem.

Fig. 4 depicts a river system in Yugoslavia, where the planning

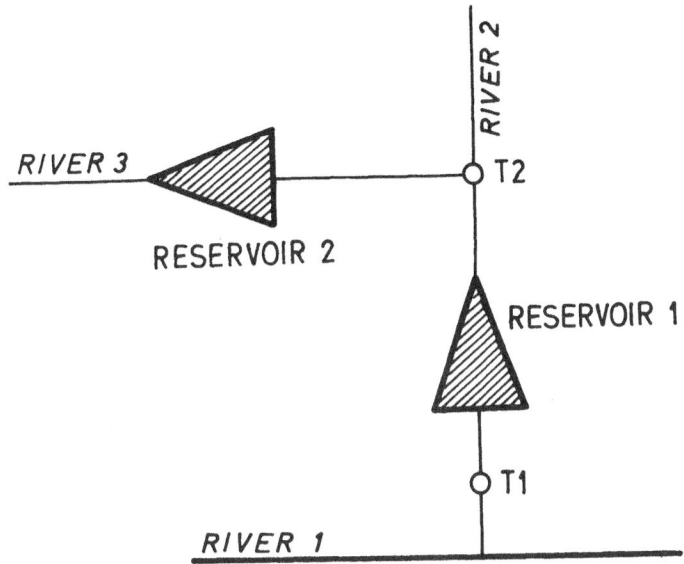

Fig. 4. Schematic presentation of a river system

and management of flood protection is a very present problem.
Since small towns T1 and T2 suffer the majority of damages, the
proposed solution was to build two reservoirs, RESERVOIR1 and
RESERVOIR2 as shown in the same Fig. The system management
problem consists of finding the optimal reservoir releases which
maximize the flood protection of the downstream region.

First the pure simulation approach was used employing the HEC5
package. This was followed by the combined optimization-simulation technique which was applied using integrated BOX-HEC software. The optimization package generates an appropriate set of
reservoir outflow hydrographs for both reservoirs and then calls
HEC5 (according to Fig. 3) to find the values of the reduced
flood damages. The optimization was actually done by maximizing
the reduced flood damages. This was done for three different
values of the optimization error tolerance (10%, 5%, 1%). The
results have been summarized in Table 1. Figures 5 and 6 show
the outflow hydrographs for the reservoirs.

The table shows that the optimization-simulation approach results in a very high flood damage reduction (more than 95% in
the last case), its performance being significantly better than
that of pure simulation.

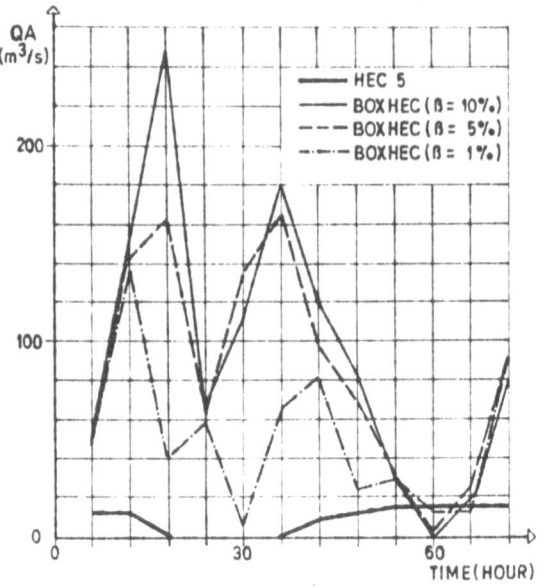

Figure 5. Output hydrographs of RESERVOIR1

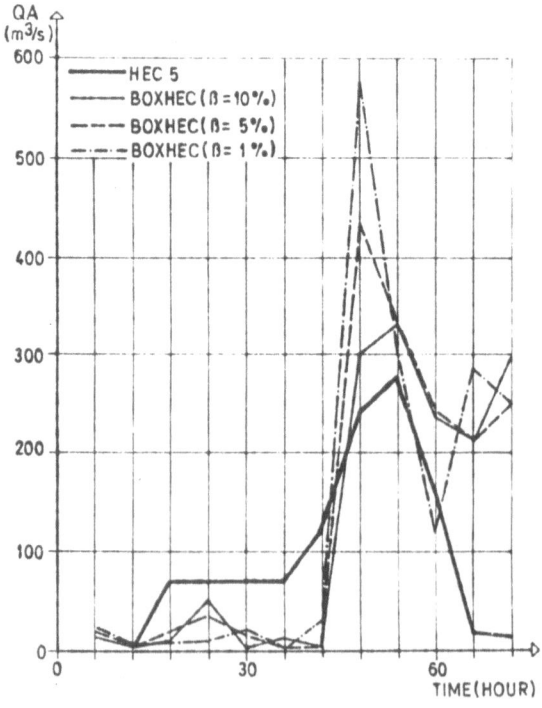

Figure 6. Output hydrographs of RESERVOIR2

Table 1. Final Results of Computing

Approach	Error tolerance	Expected Annual damages	Flood Damages reduction		Number of iterations
		$(10^3 din)$	$(10^3 din)$	(%)	
Simulation		18212.88	16017.59	87.95	−
Optimization-Simulation	10%	18212.88	16827.86	92.39	54
Optimization-Simulation	5%	18212.88	17105.10	93.92	93
Optimization-Simulation	1%	18212.88	17366.10	95.35	188

5. CONCLUSION

The paper presents a general but very practical approach to the functional integrating of large packages of modelling-industry software in order to accomplish some complex tasks in a fully automatic manner. This approach was implemented for the case of the integration of a well known large simulation package HEC-5 (widely available and frequently used in hydrological engineering community) and a general-purpose optimization program. Basically, the main problem concerns the realization of an integration of these FORTRAN packages in a way that enables the optimization package to act as a master program.

The proposed solution of this type of integration of two independently developed software packages employs the technique of interprocess communication of different processes. This facility is in various forms available in the operating systems of some modern computers. In addition, the program chaining technique is employed too. Our particular implementation was done on a VAX 11/780, using the VAX/VMS - V.2.3 operating system. A great flexibility and convenience of employment of these techniques that leads to a small number of very localized changes of the participating software, is recognized.

One possible application of such integrated software is illustrated on a practical engineering problem of flood damage minimization by means of the simulation of a reservoir system operations on a river in Yugoslavia. This application shows both technically significant reduction of annual flood damages and an important effort-and-time-saving in the process of solving.

It is important to add that the same solution approach could be used for the integration of engineering software for the purpose of the parallel work of several packages, whenever algorithm allows it, thus resulting in a great solution-time saving.

6. ACKNOWLEDGEMENT

One part of the research that initiated this work had been realized at the University of California, Davis within a project for Hydrologic Engineering Center of U.S. Corps of Engineers at Davis, California.

7. REFERENCES

Anonymous (1978) VAX/VMS System Services Reference Manual, Order No. AA-D018A-TE, Digital Equipment Corporation, Maynard, Massachusetts.

Anonymous (1979) U.S. Army Corps of Engineers, HEC-5 Simulation of Flood Control and Conservation Systems, Users Manual, Hydrologic Engineering Center, Davis, California.

Anonymous (1982a) VAX-11 Run-Time Library User's Guide, Order No. AA-L824A-TE, Digital Equipment Corporation, Maynard, Massachusetts.

Anonymous (1982b) VAX-11 Run-Time Library Reference Manual, Order No. AA-D036C-TE, Digital Equipment Corporation, Maynard, Massachusetts.

Houck, M.H. and Datta, B. (1981) Performance Evaluation of a Stochastic Optimization Model for Reservoir Design and Management with Explicit Reliability Criteria, Water Resources Research, 17,4:827-832.

Hufschmidt, M.F. and Fiering, M.B. (1966) Simulation Techniques for Design of Water Resources Systems, Harvard University Press.

Kuester, J.L. and Mize, I.H. (1973) Optimization Techniques with FORTRAN, McGraw-Hill, New York.

Jacoby, H.D. and Louks, D.P. (1972) The Combined Use of Optimization and Simulation Models in River Basin Planning, Water Resources Research, 8,6: 1401-1414.

Simonovic, S., Potic,O. and Milivojcevic, M. (1984) Optimization -Simulation Approach in Water Resources Systems Management, In: Proc.XI Yugoslav Operational Research Simposium, Herceg Novi, Yugoslavia, Oct.8-10,1984 (in Serbo-Croatian).

SPECIFICATION AND EVALUATION OF COMPUTER AIDED ENGINEERING
(CAE) SYSTEMS

A. J. James

Fluor (Great Britain) Limited, London

Summary

The functional activity of engineering as it is performed today
consists largely of a series of manual machinations supported
by a computer, within what is increasingly an information
management environment. However, current developments in data
processing technology clearly indicate a virtual revolution in
the way that such engineering activities will be performed in
the future. These developments include the emergence of
integrated Computer Aided Engineering (CAE) systems.

For the purposes of this paper, an integrated CAE system is
considered as the linking of stand-alone application software
packages directly into a suitable Data Base Management System
(DBMS), where the DBMS provides access to appropriate engineering and other reference data bases.

The potential benefits to be derived from such an integrated
CAE system are the ability to significantly improve both the
productivity and quality of the engineering service being
performed; hopefully at an economic cost. However, experience
with such systems to date have not proved an unqualified
success, and caution is prudent if Computer Aided Anarchy is
to be avoided.

This paper identifies those issues and criteria that are considered significant in the initial development of a specification for a CAE system; and in the subsequent evaluation
and selection of such a system; in order to maximise the
prospect of successful implementation. The content of this
paper is based upon experience gained during a recent CAE
evaluation carried out by Fluor (Great Britain) Limited, and
any opinions or conclusions expressed or implied are the
author's personal opinion.

Introduction

In order to develop a functional specification for a CAE
system, it is first necessary to examine present working
practices at the basic user level, in order to identify where
restraints on performance exist today, and likewise where
opportunities to improve performance occur. It is generally
accepted that the major part of engineering work today is concerned with the retrieval, analysis, computation or manipulation
of data - "information management".

Having identified the functional requirements at a user task
level, the next step is to define the functionality requirements of the total CAE system, i.e. the system boundary limits
and the functionality required within those limits. System
interface requirements are a natural fall-out from such a
system scope definition.

The effort that is necessary to consolidate the user and system
functionality requirements into a comprehensive CAE specification
is considerable, however, such a specification is a key document
in the subsequent evaluation of vendor offerings. Such a
specification will be dependent upon the nature of the
applications/disciplines/projects concerned, and this paper
reflects the functionality requirements typically experienced
by engineering/design contractors.

Integrated CAE systems consist of a number of component parts,
and any evaluation requires some appreciation of these various
components. Unfortunately, the computing world is burdened with
jargon which presents something of a mine-field to the
uninitiated. The concept of an integrated CAE system incorporating a DBMS is seen as the most likely method of achieving
the potential improvements in both productivity and quality of
the engineering project. The CAE system is further seen to be
independent of specific application software, this being a minor
part of the larger information management issue.

A review of the more important emergent data processing concepts
is presented, together with an assessment of their potential
impact of the functionality of future systems. An appreciation
of such concepts contributes towards the meaningful evaluation
of vendor offerings.

This paper is concluded with a perspective of the longer term
future, and the impact this will have on present working
practices.

User Functionality

Users and systems are increasingly inter-dependent, and the
effectiveness of the user/system interface has a major impact
upon the degree of success or otherwise. It is accordingly
necessary to define the required functionality at both the user
and system level.

At an engineering user level, functionality typically consists
of data retrieval/acquisition, manual calculations/sorting and
filing, with varying levels of computer support. Computer
calculations tend to be on a stand-alone application basis.
The acquisition of data, preparation, verification and transfer;
together with the inherent consistency/integrity checking;
represent a major part of the total engineering effort.

The above tasks are generally performed in a consecutive
sequence, with transfer of information to dependent engineering
disciplines by batch release of hard copy. This is generally
an inefficient process, and there is clearly much opportunity
for an improvement in performance.

A recent survey performed by an international oil company
suggested engineering time utilisation as follows:-

	%
Data Retrieval	20-25
Analysis/Calculation	20-30
Data Manipulation/Issue	35-40
Planning/Administration	15-20

Information management is seen to require some 55-65% of the
engineers available time. This finding is consistent with a
survey conducted by a U.K. chemical company, which concluded
that over 50% of engineering effort is concerned with information management. The study also highlighted:-

o   The extreme complexity of relationships that exist within
     a multi-discipline engineering environment.

o   The heavy dependence on hard-copy as a communication media,
     and the difficulties this poses for change management.

Engineering activities are generally seen to follow the same
generic task pattern, i.e:-

1.   Data Retrieval/Acquisition
2.   Select Parameters
3.   Perform Calculation/Analysis
4.   Review Results
5.   Document and Transmit Data.

Task (3) has received considerable attention, as is evidenced
by the abundance of available application software.

However, in an overall time utilisation context, this only accounts for some 20-30% of total effort. Tasks (1) and (5); collectively the major portion of the total effort; are pre-dominently performed manually with hard-copy media. These tasks present the major opportunity for improvement of per-formance, with benefits including:-

o   improved time utilisation of skilled engineers
o   improved quality and integrity of data
o   reduction in project duration
o   improved quality of design, through re-iterative develop-
     ment

The available computer support for these tasks at this time is perceived as:-

o   user interfaces are too complex for casual use
o   no general integration exists between CAE/PDS, or between
     the respective engineering application software programs
o   no generally acceptable DBMS is available for CAE

Tasks (2) and (4) are the necessary human interface to the system, where such intellectual input is necessary to assure application of:-

o   creativity
o   innovation
o   experience
o   compliance with good practices/codes
o   compliance with scientific principles

System Functionality

The next step is to define the boundary limits of the CAE system being considered. Such definition will identify the interfaces between the CAE system and the external environment, together with that functionality required internally within the system.

For this paper, CAE is considered to embrace all engineering disciplines normally associated with plant engineering and design. Figure 1 depicts the interfaces and functionality. Within such a definition, a major component is the 3-D Plant Design System (PDS), typically mature turn-key products which have received wide acceptance within the design engineering disciplines.

Within the integrated CAE system, data generated from cal-culations and analyses performed during the engineering definition is transferred under control to the PDS data base. Interfaces to material management and project controls permit controlled release of data from the CAE.

This presentation focuses upon those activities performed during engineering definition work, typically characterised by process, mechanical and control systems engineering. These disciplines will have a far higher level of interaction within a CAE environment than exists today. A number of vendors are now offering process simulation systems based upon data base technology, however, these are considered as restricted in the context of the definition above.

System Evaluation

Based upon the CAE system depicted in Figure 1, it is clear that there should be a multi-discipline interest in the system specification and evaluation. The approach recently adopted by Fluor (Great Britain) Ltd. was to establish a multi-discipline team including all relevant technical disciplines and d.p. personnel. This team identified over 50 system functionality requirements and over 70 engineering functions that the system should ideally be able to provide.

The system functionality conveniently divided into "universal" functionality which would be equally applicable to all activities and "dependent" functionality which would be activity specific.

Further consolidation into generic groupings provided the following breakdown:

Universal Functionality	o	Utility
	o	Hardware
	o	Data Base
	o	Modification
	o	Support
	o	Implementation
	o	Others
Dependent Functionality	o	User Friendliness
	o	Output Features
	o	Checking
	o	Characteristics
	o	Interfacing
Activities	o	Process
	o	Mechanical
	o	Control Systems
	o	Vessel
	o	Structural
	o	Electrical
	o	Piping
	o	Project

With the above specification as an objective target, an evaluation of vendor systems was performed.

The first conclusion was that a wide range of proven and excellent application software is available for general purposes and engineering use. Integration of such software within any CAE system is technically feasible with an appropriate inter- face, and engineering functionality is accordingly perceived as a minor part of a larger information management issue.

The second conclusion was that the successful CAE system of the future is going to be dependent upon emerging d.p. developments. These relevant d.p. concepts and functionality concepts are considered at this point.

Data Processing Concepts

It is evident that evolving d.p. developments will influence the performance of the future integrated CAE system. The author's perception of the direction and significance of the major developments is given below:

o   Architecture
     In terms of computer architecture, there is a clear
     evolving direction. Batch computing of the 1960's was
     characterised by very little or no interactivity, and very
     little or no sharing of peripherals and data files, see
     Figure II.

     Computer architecture evolved in two distinct directions
     during the 1970's, namely timesharing which provided large
     mainframe accessibility with poor interactivity; or ded-
     icated mini-computers which provided good interactivity and
     performance but sacrificed the sharing of data files and
     peripherals.

     Networking systems emerging in the 1980's offer the pros-
     pect of integrated sharing of peripherals and data files,
     together with the interactivity characteristic of dedicated
     mini-computers. Widespread adoption of network systems is
     anticipated.

o   Integration
     The term Integration merits consideration since it means
     different things to different people, including:

     -    Physical connection between workstations.

     -    Integration of various application programs.

     -    Integration of work of different engineers within a
          single project.

     -    Integration of work within corporate context.

     The common factor is the desire to minimise the number of
     times that information is entered into a system, while
     maintaining its integrity.

The successful integrated CAE system requires integration to an extent at all levels, and this should be recognised during specification development.

o   Data Base
The current concept of a data base derives from developments in the early 1970's when ANSI developed the concept of a Data Base Management System (DBMS), i.e. the separation of software utilities from data. The DBMS comprises those utilities which are used to define, access, manipulate and maintain large amounts of data. The actual data or data base is considered as an application of the DBMS, and may be any of the following conceptual structures, Figure III.

-    Hierarchical: This is a classic branches structure, where data is accessed through levels of increasing detail. Items of data have only a single access path; which is predetermined. In order to locate information, it is necessary to know where it is stored. Deviations from predetermined paths require schema redefinition.

-    Network: Data is accessed through paths among linked details/nodes/items, and accordingly data items can have multiple access paths, and is best suited to a complex data structure.

-    Relational: Data is accessed through rigorous operations on details (relations) which are organised as a two dimensional array, and is best suited to arbitrary queries of unknown arrival rate and content, essentially free form by category.

-    Inverted List: Data is accessed through searches conducted first through low level lists.

The structure of data within the data base may be completely transparent to the user, but the conceptual schema adopted for data organisation will have a major impact on the performance of specific application software, which is very sensitive to the schema employed.

Advantages of DBMS derived software include:

-    completeness/integrity
-    flexibility
-    data independence

Disadvantages include:

-    overhead of DBMS utilities
-    sensitivity of efficiency to choice of schema type.

In relative terms, hierarchical DBMS structures have minimum flexibility, while relation DBMS structures offer maximum flexibility.

Data base design is the key to the integrated CAE system performance and three desirable functional requirements can be identified:-

- Application - dependent data bases, for performance or other reason, but with computer based control analogous to manual control systems.

- Ability to share a common data base simultaneously between many users, not necessarily sharing the same CPU.

- Ability to overlay one data base above another.

Advanced DBMS structures based upon data "group" concepts appear to offer most promise for engineering applications.

Networks

The advent of Local Area Networks (LAN) provides an opportunity for:-

o   distributed work stations

o   network communications, i.e. sharing of files, printers, peripherals, etc.

o   integrated application software

o   effective interfaces

In the short term, the problems of data base management have increased, since software is lagging.

Two divergent network formats are available, either broadband (frequency division multiplexing), or baseband (time division multiplexing). It is important to recognise future requirements when specifying digital transfer rates, i.e.

o   Data/text   4m bit
o   Image/voice   32m bit
o   Video   100m bit

In order to facilitate interfacing with external networks, the International Standard Organisation (I.S.O.) Reference Model must be adopted. This classifies the communication process functions into a seven layer structure, which establishes a demarcation between transmission and applications, thereby permitting application programs which communicate via the transmission system to function largely independently of the transmission.

System Functionality Concepts

Engineering work at a system level is highly complex, and the appreciation of the level of complexity involved varies from good to poor within both computer system vendors and engineering companies alike. Engineering systems are characterised by:-

o   a high level of initial re-iteration, reducing as the design becomes definitive, i.e. transience and quality of data.
o   multiple case studies/design bases, evaluation of options and revisions.
o   discipline interdependence.

The following functionality issues should also be considered as part of the specification:-

o   concurrent access to data
o   consequence propogation as a result of change
o   data validation/audit trial
o   precedence ordered task execution
o   dynamic schema, i.e. the ability to modify the structure of information held in the system to match changing requirements
o   version control, with revision/changes identification
o   data ownership issues
o   control and authorisation of data access and transfer
o   security

The engineering company must specify the required functionality in the above areas, since a computer system vendor has a limited ability to anticipate the requirements inherent in a particular engineering company's working practices.

CAE System Evaluation

The methodology adopted by the Fluor evaluation team was to assign weighting factors to the CAE system functionality features, and to rank each system evaluated against these weighted factors. The functionality features were considered in the following order of decreasing significance:-

o   Data Base/DBMS
o   Integration
o   Interactivity, input/output
o   Design integrity
o   Data Verification
o   Utility Features
o   Hardware Features
o   Support
o   Implementation
o   Miscellaneous

Satisfactory functionality in DBMS and integration were seen to create an environment essentially independent of application software, and accordingly application software was not specifically part of the CAE system evaluation. Issues of particular interest include:-

o DBMS
   This subject is one of intense interest and development. Concepts to be considered include multiple data bases, particularly suited to assynchronous operations, and flexible schemas. Within such an environment, the management of data release and data base integrity comparitors become an offset overhead.

o Integration
   This is an equally rapidly developing area, particularly with the emergence of LAN's. International standards are in their infancy, however, adoption of the ISO Seven Layer standard provides a basis for inter-network communications.

o Interactivity
   Much effort is being directed at the human interface, with the objective of maximising interactivity for input and output. Graphical entry of data and sketches is possible, together with sophisticated window management systems permitting multiple activities to be run concurrently at a single work station.

o Design Integrity
   Emerging expert features within CAE systems will include checks against design rules, consequence propogation and design support/diagnostics.

Conclusions

Available CAE systems (excluding PDS) have limited functionality when evaluated against the Fluor specification, however, a range of new CAE systems are being developed based upon advanced DBMS concepts, which should offer improved functionality performance.

The integrated CAE system of the future includes:-

o the general availability of various intelligent work stations with interactive features
o network communications
o integrated and comprehensive application software
o effective interfaces

The key to success is dependent upon improved utilisation and availability of engineering data, particularly in a multi-discipline environment. This requires a DBMS which may be considered as an evolving collection of interrelated data representing a particular process plant.

In such a data base environment, although user will interact extensively with the data, he is free of the task of keeping trace of the data location within the computer since the data is handled by the DBMS.

In the longer term, the direction of computer technology is to develop systems where algorithms; or design rules; will increasingly substitute for individual judgement. The emergence of such expert systems will create an environment where the analytical and logical mode of working will be increasingly emphasised at the expense of individual experience. It is important that the intuitive and intellectual qualities of the mind are fully recognised, and that they have the opportunity to fulfil their proper role within the integrated CAE system of the future.

Glossary of Terms

ANSI   American National Standards Institute

CAE   Computer Aided Engineering

DBMS   Data Base Management System

d.p.   Data processing

ISO   International Standards Organisation

LAN   Local Area Network

mbit   mega bit/second

PDS   3-D Plant Design System

Figure I
CAE SYSTEM - COMPONENTS

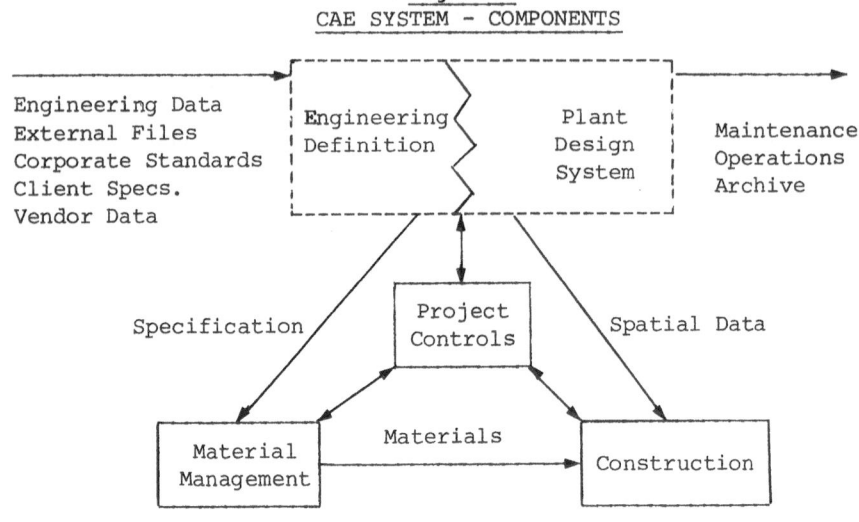

FIGURE II
COMPUTER ARCHITECTURE - EVOLUTION

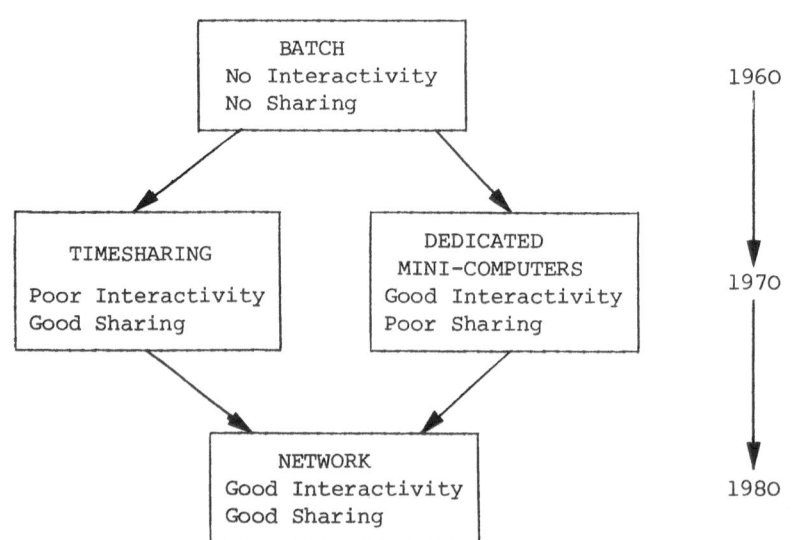

FIGURE III
DATA BASE STRUCTURES

Hierarchical:

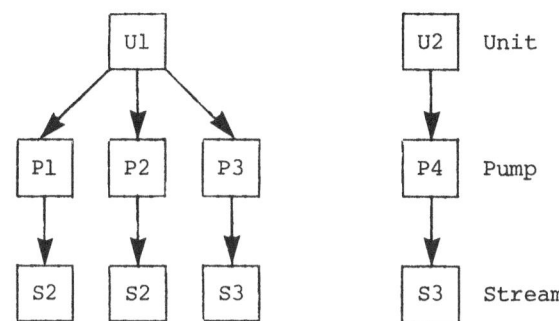

Network:

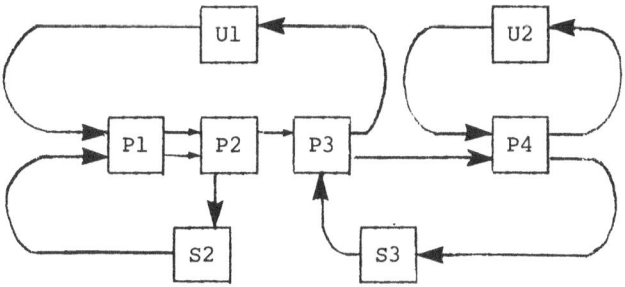

Relational:

Unit	Pump	Stream
U1	P1	S2
U1	P2	S2
U1	P3	S3
U2	P4	S3

2. SOFTWARE DEVELOPMENT

SOME THOUGHTS ON DESIGN, DEVELOPMENT AND
MAINTENANCE OF ENGINEERING SOFTWARE

Kolbein Bell

Division of Structural Mechanics
Department of Civil Engineering
The Norwegian Institute of Technology (NTH)

1. INTRODUCTION

The quality, or rather lack of quality of software in general
and engineering software in particular, is a problem attracting
more and more attention.

This paper is largely a personal view of the problems that exist
and of possible remedies that can be used to alleviate them.
I offer no magic formula, perhaps not even a single new idea,
and my approach to the issue is not a very academic one.
Although I am at present a full-time teacher, I consider myself
to be an engineer, and as such I am mostly concerned with practical solutions to problems.

Personal views are based on personal experiences, and a brief
account of my background is in order. I have been involved with
the design, development and maintenance of engineering software
for almost 20 years, in research (at SINTEF's division of structural engineering) and teaching (at NTH). Finite element techniques and FORTRAN are key words in these activities, and small
to medium-size (special purpose) programs dominate the scene.
I have had very little formal training in the art of computer
programming, and in spite of the many hours I have spent struggling with obstinate programs and difficult computers, I still
consider myself a structural engineer.

In the early seventies, before computer programming had gained
respectability (when anyone with a basic FORTRAN course was considered, particularly by him or herself, to be fully competent
to undertake even major programming tasks), we started a half-
hearted effort in SINTEF to derive internal guidelines and
standards for our program development and maintenance work.
However, adequate time and money were not allocated for this
project, and the outcome was insignificant.

It was not until the end of the seventies, when a major change of hardware made our problems very evident, that a new and more determined effort was made. Eventually we succeeded in producing a handbook [1] covering most aspects of the problem.

Our intention was to produce an internal reference manual. In the event, however, we decided to make some extra copies and to give the manual a proper binding, having a limited distribution in mind. The subtitle (*An internal reference manual*) still applies, and while some of the material has general interest, other sections of the book clearly apply to our local situation.

The handbook has been reviewed in *Engineering Computations* (Vol. 1, No. 2), and even if I have some reservations concerning a few (critical!) remarks made by the reviewer, it is a fair assessment.

The handbook is divided into 12 sections:
- Purpose and Scope
- The Digital Computer
- FORTRAN Syntax and Semantics
- Program Specification and Design
- Data Management
- Programming Style
- Errors, Testing and Debugging
- Documentation
- Maintenance
- Implementation and Distribution
- Standard Software
- Miscellaneous

It has now been in use for two years, and this paper is basically a brief review of it, section by section, emphasizing the most important points, in light of our experiences and current opinions.

Since the handbook is written for and by "ordinary" engineers, it is perhaps appropriate to pose the crucial question of who should design, develop and maintain engineering software? In the past practically all structural engineering programs have been designed and developed by structural engineers (by formal training) who have a varying understanding of and flair for computer programming (obtained on-the-job-training). Will this change? Personally I believe that engineering software must be the responsibility of the various engineering disciplines. But there is no need for every engineer to write intricate computer programs, and those who do should be properly trained for the task. This training must also give the engineer the ability and willingness to communicate with the "specialists" and to leave to the specialists what is rihgtly theirs.
Be this as it may, we can safely assume that for some time to come, engineers are going to continue to produce software which we shall have to live with, and which we may as well try to make the most of.

2. PURPOSE AND SCOPE

Our objective is to develop programs that are
reliable, *portable* and easy to
use, *modify* and *maintain*.

This can only be achieved through a combination of good planning,
personal skill, good management and proper quality assurance.

In an organization primarily concerned with small to medium size
projects (from ½ to 5 man-years) the largest gain would seem to
be achieved by improving the skill and attitude of the individual programmer, an assumption which is clearly reflected in the
choice, organization and presentation of the material in the
handbook.
Each section emphasizes its purpose, and we have tried to be
specific, using examples and indicating practical solutions as
much as possible.
On the whole we present guidelines and recommended techniques,
rather than rules. We do emphasize, however, the importance of
a standard and uniform approach within an organization, and point
out that individuality for its own sake has no place in a professional computer programming effort.
Programming is a serious logical business that requires concentration, discipline and precision, and we conclude this introductory section by subscribing to the ambitious goal that *the
time has come for programmers to write programs that work correctly the first time.*

3. THE DIGITAL COMPUTER

We have included some 30 pages containing a simple and schematic survey of the main components of a typical computer installation.
A basic understanding of how the computer (logically) represents,
stores and processes information is an essential prerequisite
for good programming.
It is also important to identify machine dependent features and
to establish a reference terminology from amongst the many varied
terms that have developed with the computer and its usage.

In order to develop *portable* programs it is important to have an
imaginary, "typical" computer installation as the *target* computer, rather than exploiting the capabilities of a particular
installation. A schematic configuration of our target computer
is shown in Figure 1. The transfer of data between primary and
secondary storage spaces is accomplished by a *data management
system*. This is, in its simplest form, a couple of FORTRAN subroutines; a more comprehensive system requires more routines, but
it can still be implemented by essentially *machine independent*
FORTRAN code.

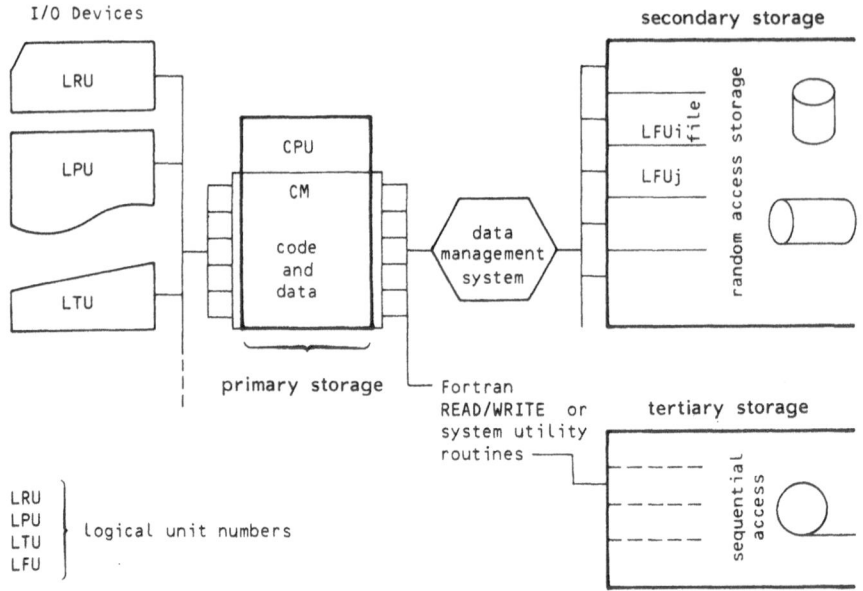

FIGURE 1 Configuration of target computer

The target computer can, in general, only compile a certain, well-defined level of the FORTRAN language, dealt with in the next section.

4. FORTRAN - SYNTAX AND SEMANTICS

Why FORTRAN? FORTRAN has become quite out of fashion with the "Academy", and it has also become a scapegoat for poor program quality.
To blame FORTRAN for the serious problems we are now faced with in development and maintenance of engineering software is a dangerous oversimplification of the issue and an erroneous diagnosis of the "software crisis".
We make no excuse for using FORTRAN. We realize and recognize its shortcomings, but we are convinced that if it is properly used it is quite possible to write excellent programs, even in a low level FORTRAN. And in real life FORTRAN has a few very important advantages: it is almost universally available, and it is potentially the most compatible (portable) programming langugage we have at our disposal. It is also worth keeping in mind that FORTRAN will dominate technical programming for some time to come, if for no other reason than that a very large investment has already been made in FORTRAN software.

A final argument in favour of FORTRAN, and one which I personally think is very important, is the vast amount of standard or library-type software, in the form of FORTRAN subprograms, that has accumulated over the years. Added to the programming language itself, this makes FORTRAN a powerful tool indeed in most areas of engineering.

Having said all this I would certainly not enter FORTRAN in a competition for the best high-level programming language. We all know that it is quite possible to write extremely bad FORTRAN programs. But why focus on the negative features? I cannot help thinking how much better off we would have been if some of the energy used in fighting FORTRAN had been used in a more constructive way, seeking acceptable solutions within the basic FORTRAN framework.

In my opinion a fundamental problem in computer programming is the lack of (primarily individual) discipline; the I-am-going-to-do-things-my-way attitude prevails. Hence if the compiler does not suit, we make a modification or, even better, a new language!

My motto is: accept FORTRAN, as it is, for what it is, and make the best of it. This obviously calls for constraints (on the capabilities/features of any particular FORTRAN compiler). In other words: we need a FORTRAN standard.

The two most widely accepted FORTRAN standards are defined by the American National Standards Institute in documents [2], FORTRAN 66, and [3], FORTRAN 77.

At the time we formed the basis for this section of the handbook (late 1980), FORTRAN 77 was not generally accepted, and some of the so-called FORTRAN 77 compilers in existence at that time were not impressive.

On the other hand FORTRAN 66 (or ANSI FORTRAN as it is also called) is, unfortunately, vague and open to different interpretations on several important points. A simpler and more restrictive, but above all, more precisely defined "dialect", called *Compatible Fortran* (CF), was suggested by Day [4], and with minor modifications we adopted CF as our FORTRAN standard. (FORTRAN 77 can be used with special permission!).

This choice was not the best, and the FORTRAN-section of the handbook is perhaps the one that is in most need for revision. I would still argue that CF, with minor modifications (primarily replacing Hollerith with Character), is quite adequate and provides for highly portable programs. With a general acceptance of FORTRAN 77 and highly improved compilers it has, however, been very difficult to enforce a strict use of CF.

We shall therefore have to rewrite this sections, but we will not adopt the full FORTRAN 77 language as our standard. With a view to both portability, simplicity and good coding practice we shall put clear restrictions on many statements, and we shall discourage altogether the use of some statements (such as ASSIGN, ENTRY and SAVE) and most use of others (such as COMMON and EQUIVALENCE). Other, potentially machine-dependent features (notably file opening and closing and direct access data transfer), should be isolated in a small number of standard subroutines.

Day [4] and Larmouth [5] provide recommended reading in matters related to FORTRAN and portability.

This section also includes a note on various forms of syntax checking, stressing the importance of manual checking (proof-reading), and we conclude with a strong recommendation to keep the code simple and to resist the temptation to be clever.

5. PROGRAM SPECIFICATION AND DESIGN

These are the most important phases in any programming project. The quality of the final program is decided here.

Our main objective in this section is to set out a *design procedure* that will ensure a "good" program, and to propose practical methods for analysing and describing this procedure.

A major difficulty is to establish measures of a "good" program. In the absence of quantitive measures we have identified characteristics of a good program. Listed in order of importance these are, in our opinion,

- reliability
- portability
- ease of use, and maintenance
- efficiency

These are fairly obvious criteria. A final judgement on how well we succeed in meeting them can only be passed after the program has been in use for some time ("the proof of the pudding is in the eating"), and then only quailtatively ("taste" is not an absolute measure). In spite of these obvious shortcomings it is important to recognize the relative importance of the main design criteria, and to relate more specific and measurable events and quantities to these main criteria.

The design procedure is a stepwise and partly iterative process comprising four distinct steps:

1. Functional specification

This is a precise and complete specification of what the program shall do, consisting of

- the purpose (aim) of the program
- the "boundary conditions"
- a functional survey (stating what the program shall do, shall not do and shall possibly do)
- specification of input and output data
- verification requirements (stating the test cases)
- type and amount of documentation.

The functional specification should always be kept as short as possible. But in order to serve as a stable basis for the further work, it must be complete, precise and consistent.

Far too many programs have evolved from a vague specification, and quite a few without any specification at all.

If you cannot decide what the program shall do, it is better left alone.

2. Data (flow) analysis

A formal data analysis, with the purpose of completely defining a convenient data representation (down to single variables), is better done independent of (and prior to the analysis of) the program logic. Due consideration must of course be given to the data requirements of existing (standard) software modules to be used.
Having chosen a method and an algorithm, and having identified the various data transformation processes (e.g. matrix operations) of the method, the data analysis is carried out "backwards" from each process, using a charting technique proposed by Gane and Sarson [6].

3. Program (logic) design

Here we adopt the *top-down* design method emphasizing *modularity* and *structured programming*. Through a procedure of successive refinement a (complex) process at a certain level is broken down into less complex sub-processes (modules), leading to a hierarchical tree-structure of the program. Again we adopt a charting technique similar to the one used by Gane and Sarson [6], and again it is important to incorporate the existing software modules (subroutines) we have at our disposal.

4. Module description

Before the actual coding starts, a detailed design of each module, including the flow of both control and data, should be worked out. To this end we propose the use of (structured) pseudo-code rather than the classical flow-chart.

In real life, steps 2, 3 and 4 will inevitably interfere with each other, and a certain amount of iteration must be expected. Even small adjustments of the specification must be tolerated.

This is an area of the problem in which we are still looking for better (and more precise) techniques. Most books and treatments dealing with program design seem to be concerned with really large scale programs, and they all seem to assume that one starts from scratch and that all code will be new code. This has hardly ever been our situation. Use of existing code is not only necessary (to meet time limits and limited budgets), but also desirable (it has already proved itself).

Nevertheless, a determined effort in these phases of program development, however insufficient it may seem to be, will pay handsome dividends.

6. DATA MANAGEMENT

The amount of data to be processed in a typical finite element program is large, often exceeding the primary memory and thus forcing us to use peripheral storage space (even on computers with virtual memory systems). Efficient use of both primary and secondary storage (disc), and efficient transfer of data between the two storage spaces, are therefore of considerable interest.

A major problem in FORTRAN is its "static" storage allocation (the initial specification of any FORTRAN array must be in terms of a fixed size storage requirement). "Dynamic" storage allocation can be simulated, and the problem thus reduced significantly, through the use of a single working space (in the form of a one-dimensional array), in which most of the individual data arrays are stored and kept track of through a system of pointers.
This section of the handbook shows how the working array concept can be implemented in an efficient and portable way, and it also indicates a very crude and simple "system" (consisting essentially of two subroutines) for allocation and deallocation of storage space in the working array. (COMMON is not used at all!)

Efficient transfer of data between primary and random access secondary storage is also dealt with in some detail. This can be accomplished by use of the direct access facilities defined in the FORTRAN 77 standard. It can, however, also be handled through direct use of the I/O-system (via assembly code or system software).
In any case the actual transfer should be performed by one or a few, well-defined subroutines. All machine dependency, such as the "optimal" size of the file addressing unit (sector length), should be hidden inside this or these subroutines, and they should preferably also provide the file address (or pointer) for the individual data records (which may of course have variable lengths).
Finally we deal with another area of potential machine dependence, namely file opening and closing, and suggest a portable solution.

7. PROGRAMMING STYLE

At the macro level the question of style is a philosophical one, and good style can be summarized in a few words: simplicity and clarity, perceivability and uniformity.

The first two need no further comments. By *perceivability* we mean a subjective measure of the complexity of a system. It is important that both software and its accompanying documentation are perceivable to the various persons involved at the various levels of development and use.

Uniformity, while not a goal in itself, is very useful for an organization engaged in program development and maintenance. How best to obtain uniformity is not an easy question. Too many and too detailed rules do not seem to produce the desired result. On the other hand, experience shows that recommendations alone are not enough. Hence we decided on a compromise, and in this section we present both guidelines and rules.

We strongly advise the use of one-entry-one-exit blocks of code, and that program logic is implemented, at all levels, by means of the basic constructs of structured programming.

We include detailed description of how the code of the individual program unit shall be organized. This ranges from the content and lay-out of a standard heading of comment lines, through a fairly rigorous ordering of the various statements, down to a series of minor points.

A little imagination and careful use of comments (including blank lines) and indentations may considerably improve code readability. (One can almost make good old FORTRAN appear to be a fairly well structured language).

Style is the sum of many details, and consistency is the key to good style.

A few comments on efficiency are also included. The importance here is to identify the real time-spenders and to make them efficient. (Normally the 10/90 rule, stating that 10 per cent of the code is responsible for 90 per cent of the execution time, applies). Chasing "microefficiency" throughout the code is at best a useless exercise; very often it also has the adverse effect of increased complexity (for instance, the many obscure structures resulting from the desire to re-use code).

The well-known slogans, *keep it simple to keep it fast* and *make it right before you make it fast* sum up our sentiments very well.

8. ERRORS, TESTING AND DEBUGGING

The cheapest errors are those never made! The most constructive approach to software reliability is therefore a design procedure, a coding technique and an attitude aimed at producing a program that works correctly the first time. This may sound unrealistic, but I am convinced that it is possible, if one really believes in it, and even if one is not one hundred per cent successful, it is a good (to my mind the best) basis for obtaining a high degree of program reliability.

Nevertheless, it is human to make mistakes, and it would be
foolhardy not to anticipate errors and to plan for their detection.
In general an error is more serious the earlier it is introduced
(or the longer it is allowed to remain), and it is therefore
important to have a continuous checking procedure throughout the
various stages of specification, design and development.

Through a combination of careful manual checking and reviewing
(of each others work), individual module testing and more integrated tests, errors should be weeded out before they can do
serious damage.
The program code should be instrumented to facilitate the
testing/debugging/verification procedure. Error traps (to
detect bad data), error reporting (both internally via error
flags and externally via printed messages) and selected print-
outs of intermediate data (controlled via print switches) are
all useful in detecting, not only program errors, but also input
(user) errors. The latter can also be prevented by including
data generation facilities, and they are more easily detected
through graphical display.

An important aspect of program testing and verification is to
choose a set of test problems that will ensure a thorough
check of all main paths. For a real, engineering program we
shall not be able to pass the verdict: guaranteed correct. A
program is wrong until proven correct, and such proof is hard
to come by.

We also include in this section some comments on debugging,
procedures and aid.

9. DOCUMENTATION

The following two statements form the starting point for our
discussion on documentation:
  - The quality of documentation is the best measure of
    the quality of the software.
  - Documentation is an integral part of any program
    development work (and not an independent activity
    added on as an afterthought).

The purpose of documentation is three-fold:
1) It is a means of communication amongst the people involved
   in *development* of software products.
2) Appropriate documentation should enable a person with the
   necessary qualifications to *use* the program correctly and
   efficiently without having any other knowledge about the
   program.

Document	Content	Target readership	Primary use	Format
General description	Description of most important system functions and features	Management Potential users	"Marketing"	Easily understood prose
User's manual	All information necessary for correct and efficient use of the program	Users	Use	Precise and complete text illustrated by examples. User's own notation
Functional specification	Complete and detailed definition of all system functions	Programmers Users	Development	Text, tables, charts
Design specification	Internal construction of the system	Programmers	Development Maintenance	Data flow, data dictionary, program structure, pseudo-code
Program description	Detailed description at code level	Programmers	Maintenance	Listings, variable lists, core pointers, file lay-out
Verification manual	Test examples	Users Programmers	Use Maintenance	Problem description, data input, printout, judgement of results

Table 1 Documentation of software products

3) The documentation should enable a person with the necessary qualifications, but who has no prior knowledge of the program, to *maintain* the program; i.e. to make corrections/changes/ modifications/extensions.

In order to serve its different purposes the documentation must be sufficiently comprehensive. However, too much documentation may be as bad as too little. Good documentation is *precise, to the point* and *complete*. It should be aimed at a typical member of the readership, and avoid longwinded explanations.

In order to meet the different needs, several pieces of documentation should be produced, as indicated in Table 1.

The handbook contains a fairly detailed description, including several examples, of the form, content and lay-out of the various documents, both for complete programs and general subprograms.

For small and medium-size programs we have had good experience with the PNB - Program Note Book - as a simple aid in development and maintenance. This is merely a binder (or perhaps a few binders) in which documents, listing and notes concerning a particular program are collected in an orderly manner as they are worked out.
The Program Note Book is divided into the following 10 sections:
- Program Log (operational status and important events)
- Specification
- Project Plan
- Theory
- Design
- Coding
- Testing
- Implementation
- Maintenance
- Miscellaneous

The PNB is not an additional document. It exists in only one copy and contains drafts and "working copies" of some of the documents listed in Table 1. It is supplemented and updated during development, use and maintenance, thus recording the history of the program.

10. MAINTENANCE

Maintenance is the collective term for all work necessary to keep a program successfully operating after it has been released for ordinary use. It consists of such activities as
- finding and correcting errors,
- adding new features or modifying existing ones,
- implementing the program on a new system,
- improving performance.

It is well-known that maintenance costs during the life of a reasonable "successful" program may easily exceed development costs by a factor of 2 or 3 or even more.

From a maintenance point of view the program consists of
- the program code,
- the user documentation,
- the program documentation, and
- the maintenance documentation.

These parts are in fact different descriptions of the same object; they are closely interrelated and it is of vital importance to maintain consistency between them.

Anyone who has been engaged in maintaining a computer program does not need to be told of the importance of good and cosistent documentation. But the organization of the work is equally important.
A common problem in maintenance is the strong interdependence between program and programmer. (More than one program has faded out after a key person resigned). It is therefore important to impose a certain amount of formalism, in order to secure that all changes are properly tested and recorded, and that all maintenance work is formally approved before it is undertaken.
Maintenance procedures will of course depend on the type, size and organization of the group or firm, and our solution, described in the handbook, may not be equally suitable for others. However, a few points concerning the source code may be of general interest. We maintain only *one* copy of the source code for each program.
A line of code that may differ due to the version of the program (e.g. single and double precision) or due to machine dependent characteristics (parameters), is duplicated and marked with a C in character position 1 and another (specific) letter in position 2. Depending on which version of the program is required, a specific combination of two letters (the first of which is always C, for comment) must be removed from all lines where it appears in positions 1 and 2, prior to compilation. (On some systems, for instance ND-computers , this is handled automatically by the compiler; so-called conditional compiling).

We have also found it useful to maintain a record of all changes in the source code itself. A line of code that needs to be changed is inactivated by a C in position 1 and marked with OLD plus the version number in positions 73 to 80. The new line is added and marked with NEW plus the version number in positions 73 to 80. An example is shown in Figure 2. Minor and obvious changes from one version to another may also be marked with "CHG x.y" instead of using OLD and NEW lines. Sometimes when there have been many changes, it may be necessary to clean up the code and start all over again. However, a listing of the source code before clean-up will be saved as part of the maintenance documentation.

```
      SUBROUTINE EXPL91(A,B,L,M,NA,LPU,IERR)
C
C ***********************************************************************
C
C     HANDBOOK OF
C     C O M P U T E R   P R O G R A M M I N G : SUBROUTINE EXPL91
C
C     PROGRAMMED BY : TOR G. SYVERTSEN, SINTEF DIV. 71
C     DATE/VERSION  : 1982-04-29 / 1.0
C                     1982-07-14 / 2.0  BY T.G.S.                    NEW 2.0
C
C     THIS ROUTINE DEMONSTRATES HOW MODIFICATIONS (MAINTENANCE) SHOULD
C     BE RECORDED IN THE SOURCE CODE.
C
C ***********************************************************************
C
      INTEGER        L, M, NA(M), LPU, IERR
      REAL           A(L,M), B(M)
C
C     INTEGER        I, J, K, IARR(4)                               OLD 2.0
      INTEGER        I, J, K, IARR(4), LLPU                         NEW 2.0
      CHARACTER      DATE*8, VERSN*3
      DATA           DATE, VERSN / '82-07-14', '2.0' /              CHG 2.0
      .....

      LLPU = IABS(LPU)                                              NEW 2.0
      .....

C --------------------------
C     PRINTOUT OF RESULTS
C --------------------------
C                                            SUPPRESS PRINT?        NEW 2.0
      IF (LPU.LE.0) GO TO 1000                                      NEW 2.0
      WRITE(LLPU,6000)                                              NEW 2.0
      WRITE(LLPU,6010) (B(I),I=1,M)                                 NEW 2.0
C                                                                   NEW 2.0
C     WRITE(LPU,6000)                                               OLD 2.0
C     WRITE(LPU,6010) (B(I),I=1,M)                                  OLD 2.0
      .....

C --------------------
C --- R E T U R N ---
C --------------------
 1000 CONTINUE                                                      NEW 2.0
      RETURN
      ...

      END
```

FIGURE 2   Example of maintenance recording in the source code

11.   IMPLEMENTATION AND DISTRIBUTION

This section outlines a procedure for program implementation
and points out some problems frequently encountered. General
solutions are not suggested, but a check-list is included.

The main problems of software distribution, including maintenance of distributed software, are discussed in some detail.

12.   STANDARD SOFTWARE

The term *standard software* denotes, in our terminology, software
modules of the "black-box" type designed to solve well-defined

problems or tasks which occur frequently. Characteristics of standard software are high reliability (obtained through rigorous testing and/or much use) and good documentation.

We distinguish between three categories of standard software:
- *Packages of independent subprograms* performing well-defined standard operations related to a certain class of problems (e.g. matrix operations). This type of standard software can be regarded as extensions of the programming language, i.e. similar to the FORTRAN library. The well-known NAG library [7] fits into this category. We have several in-house packages.
- *Programming systems* which are collections of subprograms designed around some common philosophy (conceptual model, data storage scheme etc.) for modelling and solving a general problem. The modules are (closely) interrelated, and less suited for independent use outside the system. NORSAM [8] is an example of this category.
- *Service programs.* These are executable (stand-alone) programs which perform subsidiary tasks and which can be connected to any relevant program via well-defined interfaces (data files). Examples are general pre- and postprocessors for finite element programs, e.g. FEMGEN [9].

The advantages of using standard software in program development are obvious:
- *economy;* it already exists.
- *reliability;* the modules are tested and verified.
- *portability;* standard software is normally available for a broad range of computer systems.
- *maintenance;* is provided by the supplier.
- *efficiency* ; critical parts are often optimized.
- *documentation* is (should be) good.

The disadvantages of using standard software are, in my opinion, few. The "black-box" behaviour of a standard module is sometimes considered a drawback; it cannot be modified to fit a particular application. I would argue that this is both a characteristic and a strength of standard software; if it could easily be changed to suit the individual programmer's wishes it would not be standard software and consequently would not possess the advantages of standard software. Perhaps a better solution would be to modify the algorithm/data representation to fit the standard software?
The only real disadvantage of using standard software that I can see, may arise if the program is to be implemented on a new system. This requires that the standard software is available on the recipient system. If it is not already available this may or may not cause a problem, depending on the supplier of the standard software. If one's organization has all rights to the standard software, including the source code, there are normally no problems. If this is not the case one must expect costs and possibly a time delay.

Any organization actively engaged in development of engineering software ought to have a considerable amount of "self-controlled" (and preferably "self-supplied") standard software, particularly in the area of numerical analysis (such as matrix operations, linear equations, eigenproblems, finite element properties), but also in the more problem dependent areas of input and output. These are central areas in which it is important to maintain some "local" expertise and also complete control of the software. This does not rule out the use of general libraries, such as NAG [7], which may be very useful for special purposes.

At SINTEF we have a mixture of standard software at our disposal, most of which we have produced ourselves and organized in "in-house" packages, ranging from "free-format-reading" and standard matrix operations to finite element libraries and a command processor for interactive programs. But joint efforts, such as NORSAM [8], and internationally well-known systems, such as FEMGEN [9] and FEMVIEW [10] are also available and briefly described (in terms of abstract sheets) in the handbook.

In spite of the standard software we currently possess, I cannot help thinking that more than 15 years of program development should have accumulated more and better standard software than is the case. The task of writing a typical, special-purpose finite element program today should have consisted of little more than organizing standard software.

In my opinion the best form of standard software is small packages of straightforward FORTRAN subroutines, and stand-alone programs with well-defined interfaces. The large programming systems, such as NORSAM, are hard to "sell", particularly in a medium to small size organization.

13. MISCELLANEOUS

This section of the handbook contains a sub-section on variable names in FORTRAN programs. A list of commonly used abbreviations is included along with a complete description of the ASCII character set.
We have also included key information (pertaining to program portability) for a number of the most common computer systems in the mini and super-mini range, plus a few main-frames. This information is recorded in a standard (one-sheet) format for each particular system.

14. CONCLUDING REMARKS

I think it is fair to state that engineering software, by and large, is still of low quality. I also think it is easy to identify some of the major problem areas.

However, when it comes to suggesting remedies, we are faced with some difficult decisions, most of which have to do with the attitudes and qualifications of the presonnel involved. To aim at perfection, suggesting and/or imposing procedures and techniques which are new and foreign to the programmer, may easily backlash and produce very little improvement, if any at all.

Our approach is a practical one. Our handbook addresses itself to the average programmer of engineering software who, at least in our country, still has little formal training in computer programming. We have tried to bridge the gap between the professional programming community and the engineering, or rather FORTRAN, community.
To say that we must also give our engineering students (or at least some of them) a better formal training in programming may seem to be stating the obvious. It is, however, not easy to squeeze more programming into already tight engineering curricula.

We emphasize simplicity, consistency and discipline, and although we cannot make do without some formalism, we have tried to keep it at a minimum.

Improving the skill of the individual programmer is, while extremely important, not enough to secure high quality. Effective project management is another essential element in both development and maintenance of any software product of some size. Good intentions are not sufficient to stop programmers from making short-cuts in an attempt to make up for lost time, and the result of improvisations in computer programming is more often than not disastrous.

15. REFERENCES

1. Bell, K., Carr, A.J. and Syvertsen, T.G.: "*Handbook of Computer Programming*; design, development and maintenance of engineering (Fortran) software", Report No. STF71 83005 from SINTEF (The Foundation of Scientific and Industrial Research at the Norwegian Institute of Technology), Trondheim, 1983.

2. *Programming Language* FORTRAN, X3.9-1966, American National Standards Institute, 1430 Broadway, New York 10018.

3. *Programming Language* FORTRAN, X3.9-1978, ANSI

4. Day, A.C.: *Compatible Fortran*, Cambridge University Press, 1978.

5. Larmouth, J.: "Fortran 77 Portability", *Software - Practice and Experience*, Vol. 11, 1071-1117 (1981).

6. Gane, C. and Sarson, T.: *Structured Systems Analysis*, Prentice-Hall Inc., New Jersey, 1979.

7. NAG - *Numerical Algorithms Group Limited*, 7 Bandbury Rd, Oxford OX2 6NN, U.K.

8. Bell, K., Hatlestad, B., Hansteen, O.E. and Araldsen, P.O.: "NORSAM - A programming system for the finite element method; User's Manual, Part I: General description", SINTEF, Trondheim, 1973.

9. FEMGEN - A general finite element pre-processor, FEGS Ltd, Cambridge, U.K.

10. FEMVIEW - A general finite element post-processor, FEGS Ltd, Cambridge, U.K.

ON HALSTEAD'S LANGUAGE LEVEL

N.C. Debnath

Computer Science Department, Iowa State University, Ames,
Iowa 50011, U.S.A.

1.   INTRODUCTION

The theory of Software Science, proposed by Halstead [1977]
deals with measurable properties of computer programs. In
software science, a computer program is considered to be a
string of tokens which are divided into "operators" and
"operands". Generally, any symbol or keyword group in a program that specifies an algorithmic action of the computer is
considered an operator, and any symbol used to represent data
is considered an operand. All software science measures are
functions of the counts of the operators and the operands.

The area of software science has been explicitly studied
by many independent research groups. Since many of the experimental results reported by Halstead [1977] and others
[Zweben et al. 1979, Shen et al. 1980] have been very en-
couraging, these metrics have received considerable attention
from the computer science community. Most of the work in
applying metrics of computer software using the methodology of
software science has concentrated on relatively few programming
languages such as Fortran, PL/I and Algol. COBOL has received
relatively little research attention with two notable excep-
tions. Zweben and Fung [1979] reported the results of a preliminary study of COBOL programs which were counted manually.
The work of Zweben and Fung [1979] initiated the writing of a
software science analyzer [Fung et al., 1983] for further study
in software science metrics in a COBOL environment. The analyzer provides a mechanical way of counting tokens (operators and
operands) of a COBOL program, and hence can produce all of the
software science statistics. Shen and Dunsmore [1980] have
done perhaps the most comprehensive study of COBOL programs
using software science. Recently, Debnath and Zweben [1984]
have reported the results of the analyses of a very large number of COBOL programs written by students at Ohio State University (OSU). This paper primarily discusses the results of a

study of software science metrics applied to a set of professionally-produced COBOL programs, which were collected from
the University systems computer center at the Ohio State University. Particular attention is directed at the behavior of
the Halstead language level metric and the length equation.
The results include support of the inclusion of Data Division
of a COBOL program in the software science counting strategy,
and nonsupport for the use of the software science language
level metric. In order to study the effects of different
counting methods on the properties of an algorithm, the same
set of production programs were run through two different
COBOL analyzers. The result of the comparison of the software
science statistics, obtained using two different counting
strategies, is also included. The present study provides a
reminder of how sensitive some of the metrics are and of how
careful researchers must be when drawing conclusions from software science measurements.

The relevant software science measures, which can be tabulated for any given program, are summarized below.[1]

n_1: the number of unique operators

n_2: the number of unique operands

N_1: the total number of operator occurrences

N_2: the total number of operand occurrences

n : the total number of types $(n = n_1 + n_2)$

N : the observed program length $(N = N_1 + N_2)$

$\hat{N}$: the estimated program length

$$(\hat{N} = n_1 \log_2 n_1 + n_2 \log_2 n_2)$$

V : the program volume $(V = N \log n)$

$\hat{L}$: the estimated program level $(\hat{L} = \frac{2}{n_1} * \frac{n_2}{N_2})$

D : the difficulty $(D = 1/\hat{L})$

$\hat{\lambda}$: the language level $(\hat{\lambda} = \hat{L}^2 * V)$

$\hat{E}$: the programming effort $(\hat{E} = V/\hat{L})$

A more detailed discussion and motivation of these metrics are
included in [Halstead 1977, Fitzsimmons et al. 1978].

2.   ANALYSIS OF COBOL PRODUCTION PROGRAMS

In order to study the behavior of software science metrics for
programs written in a production programming environment, ten
professionally-produced COBOL programs of various sizes were
obtained from the University Systems Computer Center at the
Ohio State University. These ten production programs were
much larger in size, and perform different kinds of functions

than the students' program considered by Debnath and Zweben [1984].

Each of the ten programs was run through the Software Science Analyzer [Fung et al., 1983] developed by the Software Metrics Research group at Ohio State University. Given a COBOL program, this analyzer can produce all of the software science statistics based on the counting rules proposed by this research group. The primary analyses will involve the Halstead length equation and the language level metric, so that we will be interested in N, $\hat{N}$, and $\hat{\lambda}$. The mean error, defined by $(N-\hat{N})/N$, was also computed for each program. The results of the analyses of these programs are shown in Table 1 and Table 2.

The analyses of these production programs show that for some programs the length equation works better for the procedure division alone, and for others the length equation gives a better estimate when the entire program is considered (i.e. when the data division is combined with the procedure division). It should be noted, however, that for almost all of these programs both the procedure division and the entire program give reasonable values of the error. A similar conclusion was drawn by Shen and Dunsmore [Shen et al., 1980] from the analysis of their COBOL analyzer program itself. Since the data division is a significant part of any COBOL program it still seems reasonable to include it in software science studies.

Software Science postulates that the language level ($\hat{\lambda}$) may be used to compare various programming languages. If $\hat{\lambda}$ is indeed a property of the programming language, we might expect that it is approximately a constant for all programs written in a given language. The present analyses, however, indicate that the language level is not constant. Its use in other software science relationships is therefore suspect, and it is not recommended as a useful metric to be applied to an individual program. It is also noticed that the $\hat{\lambda}$ for the data division is always very high compared to the $\hat{\lambda}$ of the procedure division and that of the program. This extremely large $\hat{\lambda}$ for the data division reflects the fact that COBOL provides for a compact representation of a good deal of information about type, size, structure and initial values of individual and group data items. Furthermore, it is important to note that the values of $\hat{\lambda}$ for this set of production programs are generally much lower than the $\hat{\lambda}$ values found for the students programs [Debnath et al., 1984]. Shen and Dunsmore [1980] observed that $\hat{\lambda}$ seems to fall as the program size increases. However, the two largest programs (OLD-VX-75 and NEW-VX-75) in this set have the largest values of $\hat{\lambda}$!

TABLE 1
Analysis of COBOL Production Programs

Program-ID	Div/Program	N	$\hat{N}$	$\dfrac{(N-\hat{N})}{N}$	$\hat{\lambda}$
OLD-SC-04	Data Div	1623	3406	-1.09	30.39
	Proc. Div	2567	3162	-0.23	0.80
	Program	4190	5755	-0.37	0.83
NEW-SC-04	Data Div	1696	3607	-1.13	32.43
	Proc. Div	2948	3506	-0.19	0.78
	Program	4644	6201	-0.34	0.81
OLD-AD-19	Data Div	1858	3165	-0.70	31.95
	Proc. Div	1916	2315	-0.21	0.91
	Program	3774	4420	-0.17	0.78
NEW-AD-19	Data Div	1870	3195	-0.70	32.34
	Proc. Div	2001	2397	-0.20	0.91
	Program	3871	4509	-0.16	0.80
OLD-AI-19	Data Div	1237	2143	-0.73	33.69
	Proc. Div	2191	1924	+0.12	0.45
	Program	3428	3033	+0.11	0.42
NEW-AI-19	Data Div	1249	2171	-0.74	34.09
	Proc. Div	1955	1949	+0.003	0.64
	Program	3204	3129	+0.02	0.53
OLD-YT-42	Data Div	956	1968	-1.05	35.33
	Proc. Div	2669	2830	-0.06	0.90
	Program	3625	4136	-0.14	0.78
NEW-YT-42	Data Div	1215	2497	-1.05	37.80
	Proc. Div	3127	3210	-0.02	0.88
	Program	4342	4821	-0.11	0.80
OLD-VX-75	Data Div	1782	3901	-1.19	61.00
	Proc. Div	5133	4680	+0.08	1.51
	Program	6915	7204	-0.04	1.47
NEW-VX-75	Data Div	2242	4755	-1.12	61.02
	Proc. Div	6835	5642	+0.17	1.23
	Program	9077	8552	+0.05	1.22

TABLE 2
Analysis of COBOL Production Programs

Program-ID	Div/Program	NOS	$\hat{L}$	$D=1/\hat{L}$	$\hat{E}$
	Data Div	268	0.0465	21.5	302322
OLD-SC-04	Proc. Div	590	0.0060	166.6	3704166
	Program	858	0.0046	217.4	8537391
	Data Div	278	0.0468	21.4	316410
NEW-SC-04	Proc. Div	653	0.0055	181.8	4706727
	Program	931	0.0043	232.5	10220465
	Data Div	322	0.0448	22.3	355401
OLD-AD-19	Proc. Div	385	0.0076	131.6	2087500
	Program	707	0.0048	208.3	7113125
	Data Div	324	0.0449	22.3	357371
NEW-AD-19	Proc. Div	401	0.0074	235.1	2250270
	Program	725	0.0048	208.3	7315625
	Data Div	208	0.0580	17.2	172706
OLD-AI-19	Proc. Div	501	0.0051	196.1	3467450
	Program	709	0.0038	263.2	7764473
	Data Div	211	0.0580	17.2	174724
NEW-AI-19	Proc. Div	437	0.0064	156.3	2468750
	Program	648	0.0044	227.3	6291590
	Data Div	162	0.0680	14.7	112382
OLD-YT-42	Proc. Div	642	0.0063	158.7	3609206
	Program	804	0.0049	204.1	6636938
	Data Div	209	0.0613	16.3	164127
NEW-YT-42	Proc. Div	718	0.0057	175.4	4755614
	Program	927	0.0045	222.2	8833333
	Data Div	326	0.0623	16.1	252279
OLD-VX-75	Proc. Div	1164	0.0057	175.4	8202456
	Program	1490	0.0047	212.7	14170000
	Data Div	412	0.0548	18.2	370802
NEW-VX-75	Proc. Div	1503	0.0044	227.3	14505000
	Program	1915	0.0037	270.3	24149459

Additional metrics, e.g. NOS (number of statements), $\hat{L}$, D and $\hat{E}$ have also been evaluated for each program. It should be observed that for this particular set of COBOL programs, the NOS and $\hat{E}$ metrics order the programs in the same way.

3. COMPARISON OF THE RESULTS BETWEEN THE OSU ANALYZER AND THE PURDUE ANALYZER

In order to observe the difference between the software science metrics values when using two different counting strategies, the programs collected from the University Systems were run through two different COBOL analyzers. One COBOL analyzer was developed at OSU, and the other analyzer was produced by the software metrics research group at Purdue University. The metrics values produced by the two analyzers are shown in Table 3 and Table 4. Since it was observed that the best software science estimates are generally achieved for the entire program rather than for the data or procedure division alone, this section includes the results only of the analysis of the entire program (i.e., combination of data and procedure division). The differences noticed in the values of the metrics are due to the differences in counting strategies of operators and operands as proposed by the two groups [Fung et al., 1983]. It appears that the Purdue analyzer was unable to analyze the largest program (NEW-VX-75) among all these ten programs obtained from the University Systems.

The results of both the analyzers on the University Systems production programs show that the length equation works well for almost all of the programs in the set. When the OSU analyzer is used, it is noticed that the length equation produces positive error for the two smallest (OLD-AI-19 and NEW-AI-19) and the largest program. For all other programs, the length equation produces negative errors. On the other hand, the use of Purdue's analyzer shows that the length equation produces negative errors for all the programs in the set except one of the smallest programs (OLD-AI-19). These results, as obtained for this particular set of production programs, contradicts the result observed for 11 AIRMICS production programs reported by Shen and Dunsmore [1980], namely that the length equation produces negative errors for small programs but positive errors for large programs.

It was also found [Shen et al. 1980, Smith 1980] that the range of program sizes for which the length equation appears to work best is $2000 \leq N < 4000$. It is of interest to note that the program length prediction for all the programs in that range are quite satisfactory.

Another result reported [Shen et al., 1980] for AIRMICS production programs is that language level $(\hat{\lambda})$ is affected by the size of the program. In particular, large N's are

TABLE 3
Comparison of the Software Science Metrics

	PURDUE				OSU			
Program-ID	n1	N1	n2	N2	n1	N1	n2	N2
OLD-SC-04	101	1486	329	1368	123	2286	540	1904
NEW-SC-04	105	1633	360	1562	127	2531	579	2113
OLD-AD-19	87	1324	268	1212	98	1970	431	1804
NEW-AD-19	88	1364	276	1250	99	2026	439	1845
OLD-AI-19	93	1303	200	1028	102	1958	288	1470
NEW-AI-19	87	1198	208	962	96	1800	303	1404
OLD-YT-42	87	1118	276	1281	103	2073	399	1552
NEW-YT-42	91	1323	326	1532	107	2442	463	1900
OLD-VX-75	89	2385	499	2243	100	3982	693	2933
NEW-VX-75	--	--	---	--	113	5259	806	3818

accompanied by smaller $\hat{\lambda}$'s. However, the $\hat{\lambda}$ values obtained for these 10 production programs, using two different analyzers, do not seem to support this particular result.

4. CONCLUSIONS

An investigation was made concerning the behavior of software science metrics for a set of COBOL production programs. Particular attention was given to Halstead's language level and the length equation. It appears that the COBOL studies provided mixed results. On the positive side, the length estimate was found generally satisfactory. This supports Halstead's first conjecture, namely, the observed length (N) and the estimated length ($\hat{N}$) of an algorithm are approximately equal. Furthermore, the results of the present analyses suggest that it is reasonable to include counts on data declarations and input/output statements (i.e. data division entries) in the software science counting strategy. It is appropriate to include them since they are a major portion of any COBOL program.

TABLE 4
Comparison of the Software Science Metrics

Program-ID	PURDUE				OSU			
	N	$\hat{N}$	$\frac{N-\hat{N}}{N}$	$\hat{\lambda}$	N	$\hat{N}$	$\frac{N-\hat{N}}{N}$	$\hat{\lambda}$
OLD-SC-04	2854	3423	-0.20	0.55	4190	5755	-0.37	0.83
NEW-SC-04	3195	3761	-0.18	0.55	4644	6201	-0.34	0.81
OLD-AD-19	2536	2722	-0.07	0.56	3774	4420	-0.17	0.78
NEW-AD-19	2614	2806	-0.07	0.55	3871	4509	-0.16	0.80
OLD-AI-19	2331	2137	+0.08	0.34	3428	3033	+0.11	0.42
NEW-AI-19	2160	2162	-0.001	0.44	3204	3129	+0.02	0.53
OLD-YT-42	2399	2798	-0.17	0.50	3625	4136	-0.14	0.78
NEW-YT-42	2855	3313	-0.16	0.53	4342	4821	-0.11	0.80
OLD-VX-75	4628	5048	-0.09	1.06	6915	7204	-0.04	1.47
NEW-VX-75	---	---	---	---	9077	8552	+0.05	1.22

On the negative side, further evidence against the utility of the language level measure was obtained. Large variances were observed, consistent with other studies, and conflicting evidence of the relationship between λ and $\hat{N}$ to that of Shen and Dunsmore [1980] was found. Contrary results to those of previous authors concerning the sign of the relative error in the length estimate was also noticed. The counting strategies for the OSU and Purdue analyzers appear sufficiently different that the actual values of several of the metrics changes dramatically. This has very serious implications if these metrics are to be used in an absolute sense, say as estimates of development time. It once again points out the need for taking great care in interpreting the results of software science studies, and in comparing these results with those of other researchers.

REFERENCES

Debnath, N.C., and Zweben, S.H. (1984) A Study of the Application of Software Metrics to COBOL, Tech. Rep. OSU-CISRC-TR-84-3, Computer and Information Science Research Center, Ohio State University.

Fitzsimmons, A., and Love, T. (1978) A Review and Evaluation of Software Science, ACM Computing Surveys, 10, 1:3-18.

Fung, K.C., Debnath, N.C., and Zweben, S.H. (1983) A Software Science Analyzer for COBOL, Tech. Rep. OSU-CISRC-TR-83-2, Computer and Information Science Research Center, Ohio State University.

Halstead, M. (1977) Elements of Software Science, Elsevier North-Holland, New York.

Shen, V.Y., and Dunsmore, H.E. (1980) A Software Science Analysis of COBOL Programs, Tech. Rep. CSD-TR-348, Department of Computer Science, Purdue University.

Smith, C.P. (1980) A Software Science Analysis of IBM Programming Products, IBM Santa Teresa Laboratory, Tr 03.081.

Zweben, S.H., and Fung, K.C. (1979) Exploring Software Science Relations in COBOL and APL, COMPSAC 79, 702-707.

MANAGEMENT OF INTERACTIVE GRAPHICS FUNCTIONALITY BEYOND "GKS"

Rohi Chauhan

Honeywell Information Systems, Phoenix, Arizona, U.S.A.

Engineering workstations are expected to offer an effective facility for integration of several functions in engineering design, drafting, machine tooling, manufacturing, and project management. The usefulness of these workstations is often compromised because of the industries' inability to rapidly develop and integrate complex applications software. With changing user needs, the ability to provide highly functional customizable interactive computer graphics applications is very significant to the effectiveness of workstations on a continuing basis.

Although the acceptance of Graphics Kernel Standard (GKS) has contributed to the standardization and portability of the applications, better tools are needed for speedy development and customization of interactive graphics intensive applications.

This paper presents a "Graphic Functions Manager" (GFM), which is intended to free an applications developer of repetitive graphics entity manipulation and display management related code, much beyond what can be expected from Siggraph Core and GKS packages. In order to facilitate applications development, the GFM provides a comprehensive system level support for the display, selection and management of menus, icons, creation and editing of objects, a custom user interface configuration facility, and general text editing functions with multiple user definable fonts. GFM offers an effective and localized way of managing several display details that are not germaine to an application's mainstream. The idea is to promote an environment for CAD/CAM applications specialists for generation of more-to-the-need applications, faster, with customized user interfaces, at lower cost, and with greater reliability.

CAD/CAM SYSTEMS

Due to the diversity of functions, an integrated configuration
of CAD/CAM systems generally involves several vendors. Also,
lacking detailed applications knowledge and with inadequate
ability to support end-users, the engineering workstation
vendors are, by and large, continuing to stay out of the
applications software market. The rapidly evolving technology
and changing user needs have been adding another component of
difficulty into the CAD/CAM integration process.   Inasmuch as
this scenario is likely to continue, in consideration of their
critical resources, costs, and limited know-how, most vendors
are likely to adopt the strategy of incremental product
development. The end-users, and applications software vendors
alike, therefore, will do better by evaluation of available
products in light of the following design criteria.

*   Applications programmers and systems integrators must look
    for consistent and standard ways for design and
    customization of user interfaces. The user interface must
    also fluently correlate with the displayed output in
    interactive sessions.

*   The interactive graphics support must be programmed in
    conformance to internationally accepted standards, such as
    GKS.   Its interface to applications software must be
    localized and separated from applications programs, to
    facilitate easier system upgrades as new programs become
    available.

*   In support of highly interactive design sessions,
    involving high resolution computer graphics, computational
    load on the sub-system supporting interactivity must be
    minimized.

*   Systems design must promote its upgrading to support an
    affordable level of parallel processing for maximization
    of the overall throughput.

*   Modularity of systems as well as applications software
    must accommodate segmentation of processing so that it is
    possible to position processes closer to the devices using
    them, particularly as special purpose processors become
    available at reasonable costs and as more and more
    functionality is fabricated into the same chip.

*   As long as no single vendor is offering the "complete
    solution," the system design must anticipate and allow for
    an easy integration of third party software.

This paper describes the Graphic Functions Manager (GFM) for
addressing the needs as stated above.   Portions of this

software have been implemented on a high performance interactive graphics workstation developed by Terak Corporation of Scottsdale, Arizona, USA.

TERAK GS/32 WORKSTATIONS

The graphics pipeline parallel processing architecture of the GS/32 series is shown in Figure 1. It consists of a Geometry

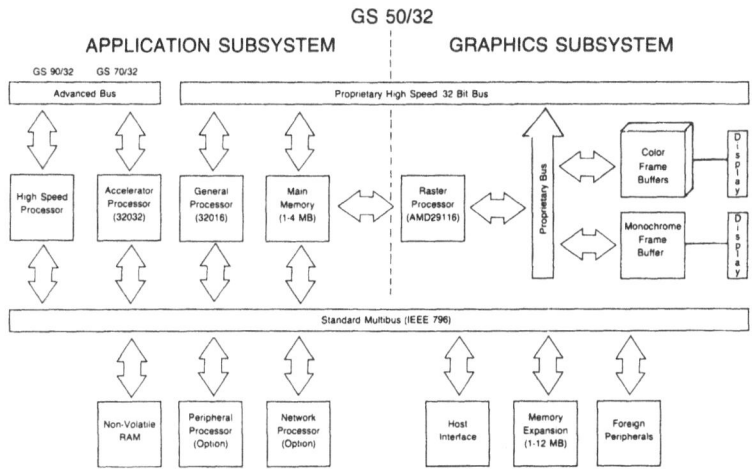

FIGURE 1:   Terak GS/32 Balanced Hardware Architecture

Engine, a Drawing Engine, accelerator processor, peripheral processor, local area network, high resolution 1280 x 1024 60 HZ non-interlaced raster graphic display, and UNIX 4.2BSD operating system.

The Graphics Processor (GP) is intended to be primarily used as a geometry engine. It is designed to furnish a high degree of interactivity required by display management processes related to display windows, cursor handling, object picks, image transformations, clipping and viewing. The GP is based upon the NS32016 32 bit chip set, with NS32081 for floating point operations.

Operating under the control of the Graphics Processor (GP), the Raster Processor (RP) makes the GS/32 drawing engine. It uses a AMD 29116 bit slice processor.

The Display Command Lists (DCL) are built and manipulated by
GP in its Main Memory (MM). Upon instructions from the GP,
the DCLs are executed by the RP and the resultant bit-map is
stored in the Frame Buffers (FB). The FBs are displayed
depending upon the visibility status of the bit-maps. While
the RP is drawing into a FB, the GP can, in parallel, continue
to build or edit DCLs in the MM for future executions by the
RP. Since the displays always take place from the bit-maps in
independent FB memory planes, the contention for the Main
Memory (MM) is substantially reduced. Taking advantage of
this facility, the GFM software needs to deal only with the
DCLs for manipulation of displayed objects.

For achieving high performance in the high resolution GS/32
drawing engines, the RP utilizes firmware assistance for many
drawing operations such as polylines, polygons, area fills,
text, and "bitblt" block move operations.

The GS/32 workstations are designed to use UNIX 4.2BSD. For
efficient interfaces between GS/32 functional modules, it is
essential that a GS/32 subsystem is not forced to wait on
another to complete a task. An effective pipelining of tasks
for inter-process communication and load regulation for the
GS/32 subsystems is required to achieve full utilization of
each subsystem's bandwidth.

In a typical GS/32 system, with a separate Applications
Processor (AP) or a host, a high level of interactivity can be
realized. The applications will reside on the AP. All user
interface, along with the Graphic Functions Manager (GFM),
will be available at all times from the GP. Full UNIX 4.2BSD
operating system will be resident in AP and a real-time UNIX
kernel (along with the Graphics Functions Manager) in GP will
support the interactive graphics functionality.

GRAPHIC FUNCTIONS MANAGER (GFM)

The primary intent of a GFM package is to free applications
software of repetitive graphic entity manipulation and display
management related code, by furnishing the high-level services
generally needed in an interactive graphics environment. An
applications program need not be concerned with the details of
how the information is displayed or with management of windows
on the screen. These are always handled by the GFM in GP.
All an applications program needs to do is to request access
to one or more display areas (which may be on the same or
different screens), just as it would access main storage
files.

The major GFM functions, as shown in Figure 2, can be
summarized as follows.

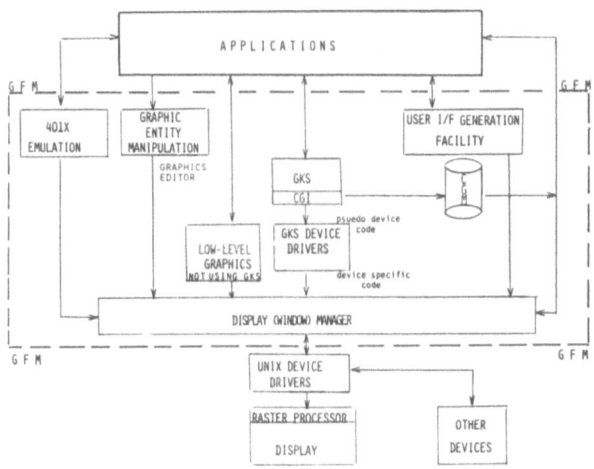

FIGURE 2: GFM - Directed to Easier Utilization of Graphics

*   Support for spawning and management of multiple
    independent process, and related icons, to be output on
    one or several devices, and comprehensive, management of
    displayed windows on screen.

*   Facility for an easy utilization of graphics standards
    such as GKS/CGI/CGM for processor and device independence.

*   Handling of menus, cursor, I/O, system prompts and screen
    house-keeping functions for an easier creation of user
    interfaces, customized to suit particular applications
    environments.

*   Monitoring of the system's cursor position and keeping it
    associated with the currently active display process.

*   Alphanumeric support, including a special display which
    emulates an ANSI X3.64 terminal.

*   Facility for creation of applications specific "help."

*   Facility for emulation of popular graphic terminals such
    as Tektronix 401X.

*   Special implements for graphic entity manipulation often
    used by applications.

*   Management of visual-aids, such as borders, partitions,
    highlighting, etc.

As a primary interface between application programs and interactive graphics, GFM must be capable of supporting a number of independent processes accessing one or more displays. Displays may be created by either the user to initiate a new independent and simultaneous process or by an application process for its own use.

GFM must also be capable of supporting programs which need direct access to the various devices as well as those programs which utilize graphic standards such as the Graphics Kernel Standard (GKS) or the SIGGRAPH GSPC Core System for device independence.

The major concepts used by GFM to provide the higher level functionality are the separation of pictures, transformations and displays and the use of various graphical objects. These concepts are described below.

Pictures
A "picture" is defined as an ordered list of graphical primitives and picture segments in model space (world coordinates). Once a picture has been constructed, it is not necessary to re-transmit the information in order to display the picture again. Also, since the picture is managed by GFM, it is possible to modify the picture simply by specifying the deletions and additions.

A picture may be manipulated and modified by any of a number of application processes other than the one which created it. This provides a significant capability for future systems using the full capabilities of processors such as the NS32000 or M68000 series, in UNIX environment which allows for mulitple cooperating processes. Protection from inadvertent access to a picture is provided by GFM generated picture names.

A picture may consist of either just graphical primitives or a number of hierarchical segments. A picture may be described as:

<picture> ::= <picture header>
                <picture item>...<picture item>
                <picture end>

where
    <picture item> ::=<graphical primitive>
                    : <picture segment>.

Graphical primitives are the basic building blocks of both pictures and picture segments. They include polylines, polymarkers, polygons, text, cell arrays, instance and symbol references,   and   attribute   settings.

A picture segment is a collection of graphical primitives and sub-pictures which may be referenced as an entity. The major value of picture segments is that they allow any picture to be copied into another picture or referenced as a "master" by an "instance."

Transformations

The "transformations" are constructs that convert pictures from model (world) space to the Normalized Device Coordinate (NDC) space which is equivalent to the coordinate space used by Display Command Lists (DCL). The raster processor executes the DCLs to make a picture bit-map for the frame buffers. This transformation is defined by specification of a window in the world coordinates and viewport in the NDC space. During the display process a transformation maps contents of NDC windows to viewports on logical output devices, with appropriate scaling.

A transformation may have one or more displays associated with it. If no display is currently associated with a given transformation the default is to the home display of the process, that allows a straight transformation of the model space to NDC space. The "home display" of a process is that display which provides standard I/O for the process. Usually it is a user created display from which the process was initiated.

Picture Processing

Pictures are displayed by associating a picture with a normalization transformation. The picture is then converted to NDC Space using the given transformation and is entered into the Display Command Lists of all the active displays currently associated with the given normalization transformation. When objects are added to an open picture, they are also converted to the NDC Space and posted to the Display Command Lists of the associated active displays.

GFM supports on-line storage of pictures. The application may direct storage to occur in a specific directory or on a specific device. A system picture storage facility is available for use as a default. The application process must provide a name under which the picture will be stored. Stored pictures may be recalled by GFM from specific on-line files as needed by an application process.

Text pictures are a special class of objects providing support for A/N displays similar to that provided by graphical objects for graphical displays. Text pictures may contain commands used to manipulate the contents of an A/N display and may also contain references to other text pictures.

GFM also supports retrieval from and storage of pictures in

the ANSI X3H3 Computer Graphics Metafile (CGM) format.

Display Management
A "display" is a logical output device consisting of a Display
Command List using Normalized Device Coordinates ranging from
[-1, -1] to [+1, +1]. A display window provides the user with
panning and limited zooming over the display. A display may
be accessed by outputting through the UNIX port associated
with the display or by using transformations associated with a
given display to display a picture.

The GFM provides the following display management functions.

* Display status management, including view priority and
  visibility status.

* Positive acknowledgement of user inputs.

* Paging of open displays, when exceeding memory.

* Creation, filing, retrieval, manipulation, and deletion of
  Display Command Lists (DCL) for logical displays. The DCL
  handler also allows for insertion of new DCL codes into
  the logical displays.

* Assignment of windows and viewports to different devices.

* General screen management by user (for local interaction)
  and by an application.

* ICON management, such as creation, deletion,
  identification with a process, etc.

* Window/viewport management, generally under local user
  control.

* Identification of an application (process) to an existing
  viewport or a new viewport. This implies spawning of a
  new UNIX shell.

* Special displays are supported for emulation of given
  devices such as the Tektronix 401X or an ANSI X3.64
  alphanumeric terminal.

* Control of systems cursor.

* Handling of systems prompts.

* Special displays are also used for supporting functions
  (usually interactive) such as menu selection, forms, help,
  etc. Management of menus involves primarily three types
  of menus, i.e.,

"pop-up menus":	at points of invocation
"applications menus":	in designated areas of the screen
"border menus":	along the screen edges

*   Handling of pre-defined screens for special objects (e.g., menus) under user or application control of

    -   Viewing priority
    -   Background colors
    -   Assignment of applications
        ICONS to different processes

*   Changing of input devices, e.g., mouse, keyboard, tablet.

*   Inclusion or exclusion of borders into display. The borders may contain explanatory legends and/or desired menus.

MENU MANAGEMENT

GFM supports special menu objects for the purpose of allowing an application to display available options and to allow a user to make a selection from these options. Menus operate as displays that are not associated with any I/O port. They will always be child displays of the home display of the process that activated them.

There are three basic kinds of menus; border menus, pop-up menus, and application menus. Each of these menus may be either alphanumeric or iconic. Alphanumeric menus display the options available as a strings of text. Iconic menus display the options available as symbols. An application menu has the additional aspect of allowing graphical primitives to be used in construction of the menu items.

GFM supports facility for interactive design and storage of menus, allowing menus to be stored on-line and recalled as needed.

Border Menus

Border menus are special objects that may be associated with displays. A display with a border menu associated with it will have a fixed minimum screen size extent according to where the border is placed. This will not restrict a display window from panning or zooming about the logical display or the positioning of the display viewport on the screen.

Border menus provide a set of choices associated with the display. These may be alphanumeric or iconic and may be pick, button, or keyboard selectable. In general, iconic menus should be restricted to pick selections.

Pop-up Menus

Pop-up menus are supported by the GFM for the purpose of allowing quick selection from a list of available choices. Pop-up menus will always consist of annotative text or symbols and will always require selection by a "request pick." While a pop-up menu is active, system cursor movement will be constrained to snap from one popped menu to another. If a pop-up menu is not active, it will not be visible.

Pop-up menus is generally enabled by the application program, invoked by the user, and managed by GFM. A uniform method for invoking a pop-up menu and selecting from it or canceling a selection is provided by GFM.

Only one pop-up menu may be enabled at any one time.

Pop-up menu selection is an event type interaction. A cancellation of a pop-up menu is not recognized as an event, requiring a selection (of RETURN for cancellation) to proceed further.

Application Menus

Application menus are special displays from which the user may make a selection and are positioned on a user given area of the screen. An application menu is expected to be constantly visible as part of a given display so that the user is aware of the functions currently available. A menu, when displayed, is provided with a viewing priority higher than the home display from which it was generated and higher than any previously specified child display of the home display. An application menu item may contain any graphic primitive in its description.

Menus may be defined separately from their display. When they are displayed, an instance name is returned by GFM so that the application process may identify the menu from which a selection was made.

Items may be selected from menus through the use of buttons, picks, or text input. The type of selection that will be allowed is determined by specifications at the time the menu was created.

GRAPHICAL OUTPUT PRIMITIVES

GFM supports both 2D and 3D graphics. Ultimately, all pictures and picture segments are defined in terms of the basic graphical primitives. There are two types of graphical output primitives, "effective" primitives which cause actual generation of display code and "attribute" primitives which set characteristics (e.g., line style and color) of the output commands. Polyline, Polymarker, Polygon, both filled and blank, and cell array primitives are supported, as defined by GKS. Also, GKS defines three types of attributes, namely, Geometric, Non-Geometric, and Indentification, which are supported by the GFM. The identification attribute may be expanded in a manner consistent with GKS, supporting the hierarchical nature of picture segments.

GFM supports two types of text, annotative text and picture text. The annotative text is not considered as an integral part of a picture. Only the position of annotative text is transformed to NDC Space, not the text itself. Annotative text may well be considered as "hardware" characters. The size, direction, and character path allowed for annotative text, which is always 2D, is provided by the Raster Processor.

Picture text is considered to be a part of the picture and is constructed from strokes which undergo full conversion to NDC Space using the normalization transforms applied to the picture. Picture text may be 2D or 3D. Although picture text is 2D in nature, it could consist of strokes that may be defined in a 3D space. Picture text is supported by all the text attributes defined in GKS.

A graphical primitive will be provided that will allow invocation of another picture as an instance of a current picture, without recursion for simplicity.

ADVANCED GRAPHICAL PRIMITIVES AND FUNCTIONS

GFM should be capable of supporting higher level graphic primitives and functions, such as Grids and Axes with tick marks, single end and double end arrow lines, double lines, splines, conics, and other functions useful in design of a drafting package. All specifications for these functions are to be given by applications in word coordinations. The GFM will be responsible for the transformations to NDC space and appropriate clipping operations.

Additional functionality to support surfaces, solids and shading will be welcome for an easier and standardized utilization of interactive computer graphics in solids modeling packages. Surfaces may be defined as planes or as objects of revolution. Solids can be defined both via boundary representation or as constructive solids geometry primitives, with appropriate display of patterns to indicate the solid aspect of the object in cut-away sections. Perspective creation and viewing, hidden line removal, and shading must also be supported.

Utility functions for an easier incorporation of drawing headings, Engineering Change Control process, revision and page numbers, current time and date will also be useful.

GRAPHICAL INPUT

All input, not determined to be local input, is assumed to be
input to the current working process. GFM is responsible for
directing and queueing input to the appropriate process. The
input devices are classified according to their normal
operating modes as keyboards, button, valuators, locators, or
picks. In addition, processes are available to transform
certain devices from one type into another and for support of
GKS input processes. For example, transformation of a locator
into a pick with specification of a set of keys or buttons
used to trigger the pick is allowed. The application process
can specify size of the "pick window" or may use a default
size.

A logical device is considered active for a given process if
the device has been initialized by the process and the process
is the current working process. Special input services, such
as dragging and menu selections, are provided as higher level
GFM functions. While many of these services provide input in
the same form and type as that provided by a corresponding
logical input device, the calls and references to these
services are separate and independent from the logical
devices.

GFM provides management of the physical input devices to
insure that a physical device is not multiply assigned either
to high level input functions or logical device functions.
Assignment of a physical input device is dependent upon the
current working display.

When a text input is formally requested by an application
program, the GFM will identify an area on the chosen display
in terms of lower left corner, number of characters per line
and number of lines. GFM creates a temporary alphanumeric
display in the identified area to manage the text input in
that area with limited text editing. Any text without a
formal request will be sent to a standard UNIX designated port
for an appropriate display.

The GKS provides for Locator, Request Pick, Buttons, Stroke,
Sample, and Event type inputs. All of these are supported by
GFM. When a locator input is chosen, the GFM allows an
applications process to specify which of many echo types is to
be used.

Picture Editing
GFM will support a stand-alone Graphics and Text Editor that
may be invoked either by a user directly or by a host
process. If invoked by a host process, the Display Command
List may be input to the host process as a set of graphical
primitives in world coordinates defined by a specified

transformation. Pictures created with the Editor may also be stored in on-line storage for future recall by the user or by a host process.

The GFM allows for maintenance, and management of special graphical objects. Some of these special objects will be strictly display oriented while others will be used for interaction. Symbols (graphical objects), with attributes to include color, reverse video, high-lighting, and blinking will be very useful in applications such as editing of multi-layered drawings which can be looked upon as a set of related complex graphical objects. GFM will manage a library of symbols. Any of the objects can be utilized as special "drag objects" to function as a "cursor" for positioning of the objects, with restrictions for position, scaling and rotating.

Rubber Lines, Rubber Boxes, and Snap Grids are examples of other useful special objects. The "Rubber Lines" are generally used as a cursor for placing a line on a display, while the "Rubber Boxes" allow specification of a rectangle of user chosen size and aspect ratio on a display which can be useful in an easy selection of a "pick-window." The "Snap Grids," containing spaced points, are useful for positioning of lines, icons, or other objects.

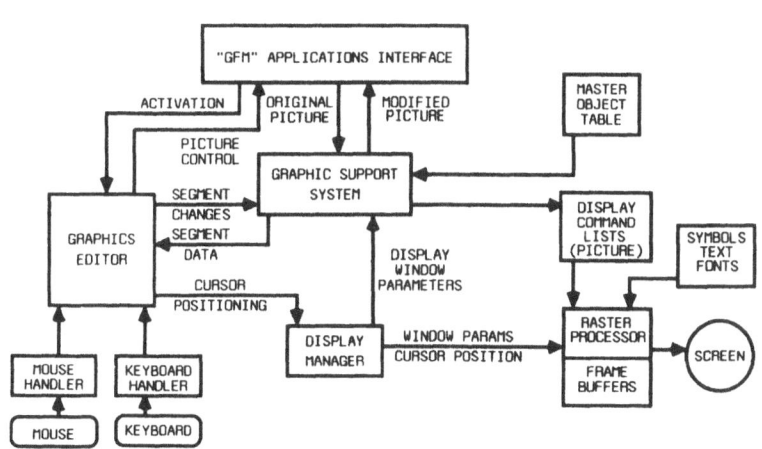

FIGURE 3: Graphics Editing with "GFM"

Figure 3 depicts the graphics editing operation. For a picture editor to be truly useful, some complex picture selection, modification, and filing operations will also be supported by the GFM. For example, it will be helpful to applications if GFM could provide for selection, modification, and filing of one or more picture segments within one or multiple "pick-windows" or throughout an entire drawing which may consist of many layers. Upon selection of picture segments the GFM would automatically issue positive acknowledgements, by temporarily modifying the segment attributes such as color, as given by the applications program.

USER INTERFACE CUSTOMIZATION

With facilities for creation, deletion, combination, association, modification, and storage of special objects, GFM furnishes an effective means of customization of user interfaces. Definition and application oriented organization of various types of menus with appropriate borders and user acknowledgements, contributes most to user friendliness in the customization process.

Once defined, a menu may be displayed and activated with a command, which will have arguments to contain the menu name and the screen area in which it should be displayed. A GFM transformation will cause the entire menu to be placed within the designated screen area. Annotative text and symbols will be clipped as in a normal picture. Zoom and panning are not permitted for menus. Association of specific function buttons or text strings with the menu items is done at the time of menu item definition. If two menu items are associated with the same button or a text string, then the last specified item will be selectable.

An interactive menu editor is provided for the modification of defined menus, to allow changes with user needs. This facility allows for creation, modification, and deletion of menus, associated icons and action references. It is possible to concatenate frequently used commands and associate them with a single new menu with user or application chosen attributes, such as background color and menu icon, and positioning.

SUMMARY

With the help of a comprehensive Graphic Functions Manager (GFM), the graphic's independence of applications can be carried to much beyond what is intended by use of the "GKS." This paper has conceptually presented such a GFM package, which is being implemented by Terak Corporation of Scottsdale, Arizona, U.S.A. It is expected that a higher than "GKS" level of standard for management of graphics functionality will emerge as more experience is gained with GFM type software.

Also, with careful modularization, the graphics functionality can be expected to move even closer to the end-devices and into silicon.

To the extent the stated objectives of a GFM are implemented, graphics independence in applications packages can be realized for portability. And, programmers can be expected to produce more to-the-need applications faster, with customized user interfaces, at lower costs and greater reliability.

ACKNOWLEDGEMENTS

Most of the concepts presented in this paper were developed during author's employment with Terak Corporation. The author is grateful to Terak for its contribution to the interactive graphics, by allowing publication of the information contained in this paper, and Honeywell Information Systems for supporting preparation and presentation of this paper. Callie Fish of Honeywell deserves all credits for preparation of this manuscript on a very short notice.

AIM-AN ADAPTABLE INPUT MODULE FOR ENGINEERING
PROGRAMS
Z.Grodzki, M.Winiarski

Technical University of Warsaw

INTRODUCTION

The importance of providing user-oriented, efficient input to engineering programs cannot be over-
emphasized. Most potential users are basing their
decision - to use or not to use a program - on
their impression of documentation and input only.
Over the past two decades, or more, a series of input solutions was devised, beginning with the printed data sheets using fixed format, through program
packages with unified input sheets, suites of inter-
acting programs in which the results of the preceding programs were providing the input to the following ones, to problem-oriented languages with free
format entry.
Each input solution was an improvement on the preceding one but each had its own drawbacks.
The integrated systems, set new high standards for
input facilities, but failed to fulfill the expectations in other areas because of their size and
cost, voluminous user manuals and-perhaps - chiefly
because of the existance of vast numbers of FORTRAN
application programs, thoroughly tested and well
established within the design environment that
could not be incorporated into these systems.
Input facilities provided by the integrated systems
were - on one hand - a great attraction to the
users, but - on the other - uneconomic to develop
for smaller programs. What more - they were outside
the competence and interest of engineer-programmers
who were the authors of most application programs.
With the introduction of alpha-numeric and graphics
terminals, the interactive dialogue is now accepted
as the most convenient form of man-machine communication.

An interesting attempt to develop program-indepen-
dent input language and facilities for engineering
programs was presented some time ago at the Institution of Civil Engineers. It was ADL - "A Data
Language" (Lim, 1981) with its Flexible Input Package. This was an important step in software design
philosophy as it marks the beginning of input module independence of the host program.
Another interesting solution of this kind is MIGS,
Menu Input Generating System for FORTRAN Programs,
presented at the ENGSOFT III Conference (Kovacic,
1983) . AIM, the Adaptable Input Module, presented
here, follows in the same direction, though using
different techniques.

THE BASIC IDEAS

The main problems facing the authors and developers
of engineering application programs are as follows:
how to write-at economically acceptable costs-effi-
cient dialogue input modules, for the following
classes of problems :
- existing smaller batch input programs without
  problem-oriented language,
- existing programs and systems with problem-oriented language,
- newly developed problem modules, or "academic"
  programs with provisional input,
- small micro-computer programs.
The general requirements for friendly input dialogue have been formulated by Jones in his "Four
principles of man-computer dialogue" (Jones, 1978).
Interactive design programs present additional requirements. Design is a step-wise, hierarchical
process, carried at several levels of detail. Accordingly, the input module structure should provide the possibility of dialogue at different levels. This means the provision of interrupts and
returns from one level to another, in addition to
echo generation, menu handling and input processing.
From the programming point of view, an Interactive
Dialogue Input Module should be - as far as possible - independent of the host program structure and
should communicate with this program by sets of
data and decisions. In fact, the task of the input
module is to supply the host program with data and
decisions. This enables the input Module to be developed and tested separately from the main program, which is particularly important when parts of
a larger system are being developed in parallel.
All the above requirements result in fairly compli-

cated software.
The Interactive Data Input Model, developed at the
Civil Engineering Faculty of the Technical University of Warsaw and presented at the C.A.D. 84 Conference (Grodzki 1984) has as its aim:
- to create a software tool, allowing the programmer to develop cost-efficient dialogue input modules, mainly by pre-developing program instructions that are repeated many times in various
  sequences, depending on the purposes of the programs,
- to enable engineer-programmers develop dialogue
  input modules without outside help.
The idea of the Model goes far in the direction of
input module independence from the host program.

SHORT DESCRIPTION OF THE MODEL

The idea of the Interactive Data Input Model is based on the theory of finite automata and on observation that only a few basic types of functions are
used to manipulate the input data, e.g. read text,
read integer, read real number check the record
just read in, etc.
Furthermore, these basic functions occur mostly in
a few sequential combinations.
The General Scheme of the Model is shown in Fig. 1
and the Organisation Diagram - in Fig. 2.
The main parts of the Model are:
THE PROGRAM STATE VECTOR. This is a very important
part of the model, as it defines the sequence of
progress throughout the entire input. The counter
indicates the number of an element in the vector,
which can contain either the number of the function
to be activated or an argument of this function.
The counter value is increased only when a given
dialogue step has passed all the checks and has
been released.
THE BLOCK OF FUNCTIONS - is the most important and
most labour-consuming part of the model to develop.
Once the elementary functions are written and tested, further work is much simpler.
THE DATA FILES, which serve to store both the input
data and organisation management data.
The functioning of the model in course of the dialogue input consists in execution of functions, indicated by the PROGRAM STATE VECTOR.
The functions contained in the BLOCK OF FUNCTIONS
(see Fig. 2) can be grouped as follows:
- System functions to help the programmer in the
  implementation of the input module to a given
  application program or system. They are mostly

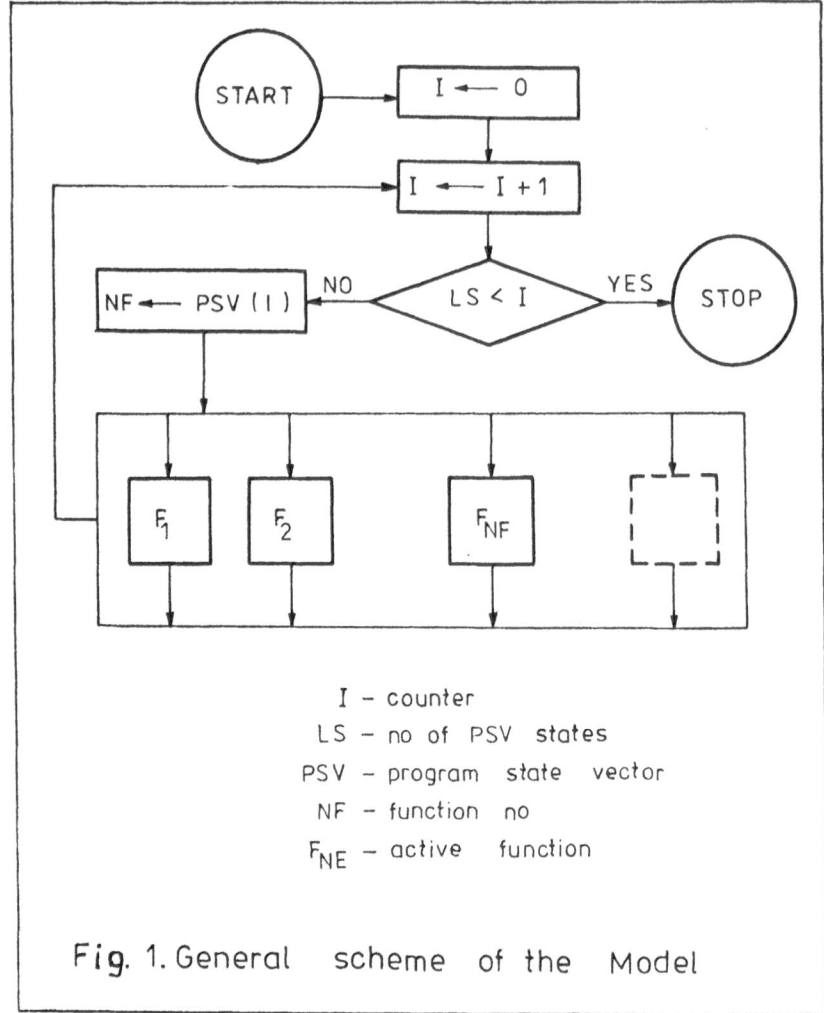

I — counter
LS — no of PSV states
PSV — program state vector
NF — function no
F_{NE} — active function

Fig. 1. General scheme of the Model

of control or information kind: the control pertains to the programmer's input to given data files (see Fig. 2) and the information consists in listing the texts and the data.
- System functions to help the user during the inputting of the data of informative character,
- Data manipulation functions, i.e. model functions at the disposal at the programmer, by means of which he organises the input dialogue, and
- problem functions
The data files consist of:
- relay record - containing information input by the user or that to be displayed to the user,
- working record - serving the purpose of storing

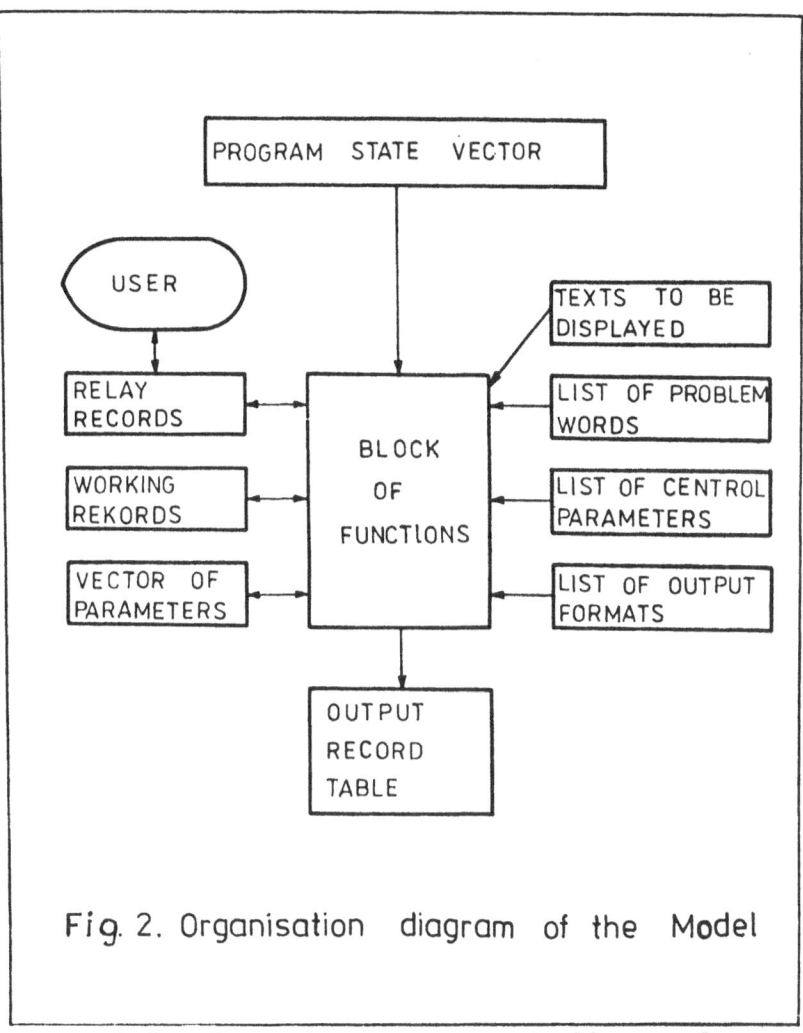

Fig. 2. Organisation diagram of the Model

partial information or automatically completing
the input to the host program in cases this information needs completing,
- vector of parameters - storing the values of
the decision variables on which the type of the
record to be read depends, i.e. which branch of
the dialogue to follow,
- list of texts - to be displayed on the screen.
- list of output records - containing data and
decisions ready for transmission to the host
program,
- list of problem words - occuring only in cases
when the host program uses a P.O.L.,

- list of controlling values - e.g. maximum or minimum magnitudes of numbers which are input.
In order to adapt the MODEL to form a DIALOGUE MO-
DULE, the application programmer has first to organise the files and prepare the library of the
functions mentioned above before starting to prepare the adjustments for a given application problem.
This necessary preparatory work has been done and
is presented in the next paragraph.

THE KERNEL MODULE

The implementation of the MODEL, discussed in the
preceding paragraph, in the form of adaptable IN-
PUT MODULES to engineering programs required the
development of the KERNEL MODULE, which contains
the part of software common to all application input modules. This software includes:
- organisation of data files
- a set of basic functions.

Data files
The KERNEL MODULE includes the following data
files:
- program state vector - a vector containing element numbers, the counter, function numbers and
  their parameters,
- list of texts to be displayed on the screen - a
  matrix storing the texts column - wise, as character files,
- list of constraints - two vectors containing the
  number of constraints together with their upper
  and lower values,
- list of output formats - contains the number of
  formats and the list of formats column-wise,
- list of problem words - a vector containing the
  total number of problem words and their list
  (when no problem words are input, this list is
  empty) ,
- vector of parameters - contains the values of
  parameters; it is automatically filled when the
  data is input by the user,
- parameter counter - a vector to store the actual
  loop counters, controlled by the parameter values,
- table of output records - a matrix to store the
  output in character form,
- relay records - input and output vectors,
- working records - vectors used to store and to
  complete intermediate information.
Since the output table is filled using full-length

(80 - byte) records, the information is also transmitted in full - length records.

Functions

In accordance with the INTERACTIVE DATA INPUT MO-
DEL, the KERNEL MODULE is fitted with the following function types:
- basic functions - functions of fundamental character such as: read number, copy, compare
  texts , which assure machine and compiler independence to the remaining functions of the MODEL.
  These are the only functions that need changing
  when implementing the module on different types
  of computers..

System functions to help the programmer include:
- listing of the programmer s options,
- listing of functions included in the module,
- listing of constraint values,
- listing of texts to be displayed to the user,
- listing of problem words,
- listing of output formats,
- listing of the program state vector in explicit
  form i.e. giving the name of the function, its
  number and parameters,
- the control of the input information.

System functions to help the user. They include:
- HELP function, giving additional information
  when requested,
- listing of the output records,
- control of data at the system level (depending
  on the type of the computer) ; this includes the
  check of type (INTEGER or REAL) , allowable magnitude and the number of digits,
- listing of the actual parameter values,
- repeating the last message,
- jump to data blocks, input previously,
- the way of signalling the end of a data block by
  the user, where no other means of control are
  possible.

Model functions include the following groups of
functions:
- reading of the data (texts or numbers) ,
- displaying the messages, questions or other information on the screen,
- control of data by setting of the constraint values and parameters,
- copying from one set of data to another,
- modifications to program state vector counter,
  defining which function is actually active.

The KERNEL MODULE is organised in such a way that
an application programmer can insert functions connected with the data structure of the host program.

If the structure of the data is subdivided into
blocks, each block can be given its own, local state vector.
The program state vector (global) treats then the
data blocks as macro-functions.
To adapt AIM to application programs, the programmer has at his disposal:
- the KERNEL MODULE, which enables the programmer
  to manage and organize the input,
- the list of basic functions and subroutines serving the manipulation of the flow of data,
- the prescriptions how to adapt the module to the
  host program or system at hand.
Using the documentation of the required input for
a given new or existing program he prepares the
input flow chart and - based on this - the necessary data for the module:
- the list of texts which will appear on the screen
  and prompt the user
- the list of constraints max. and min. values of
  numbers
- if the program has a problem-oriented language -
  the list of problem words
- if the host program input is in fixed format - a
  list of formats
- the program state vector containing the pointers
  to the graph of the data input.
All this work is equivalent to no more than supplying the data to a program.
In cases the binary form of program is only available, the module produces a complete data file
(graphic type) for this binary program.
The module is fully portable as it is written in
standard FORTRAN.

CONCLUSIONS

AIM - An Adaptable Input Module is the implementation of the Interactive Data Input Model.
It provides facilities to fit an application dialogue input module into existing or new programs
and systems. It is independent of the computer and
terminal devices. A fairly simple conversion only
is needed to adjust it to any engineering-type program. The costs of writing a new dialogue input
module are thus drastically reduced together with
the time needed for this task.
Adaptable Input Module is a direct help to the programmers and - particularly to engineer-programmers.
It is always advantageous if engineers take part in
program development-apart from writing the brief
and checking the results. The proposed model makes

it possible for engineer-programmers to undertake
themselves the development of efficient dialogue
input modules.
The efficiency of the implemented input for a Finite Element Method program used by the students
of our Civil Engineering Faculty - based on a sample of 2000 input cases - was found to by above
75% correct input at the first try on alpha - numerical terminals. After introducing simple graphics control this percentage went up to nearly 90%.
As the block of functions developed so far is oriented towards engineering applications, the module
is at present engineer-oriented. The MODEL is however more general and it is also possible to develop a module oriented towards other applications
by adding a new set of functions.

EXAMPLE

The example, presented below shows the beginning
of the input to FEAP, a well known Finite Element
Program.
The constraints at the level of the KERNEL MODULE
have been set depending on the type at computer:

        max. integer value: 32767
        min. integer value: -32767
        max no of digits in a real number: 8.
At the KERNEL level, the warning is also displayed
when a real number has been input instead of an
integer.
The following constraints were imposed for FEAP
input only:

No of nodes	2	NN	500	INTEGER
No of elements	1	NEL	500	"
No of materiales	1	NOM	20	"
No of dimensions	1	ND	3	"
No of degs.of				
freedom node etc.	1	NDOF	6	"

All kinds of mistaken numbers were input to show
the reactions of the module.

```
        —*F E A P   R E A D Y*-
NAME OF PROBLEM (MAX. 75 CHAR.)
SPACE FRAME 2
    -* INITIAL DATA *-
NO. OF NODES
1
*** NUMBER TOO SMALL,  (    2 ***
551
*** NUMBER TOO LARGE,  )   500 ***
32
```

```
NO. OF ELEMENTS
3.5
 *** THIS IS NOT AN INTEGER                          ***
 *** PLEASE REPEAT LAST ANSWER                       ***
35
NO. OF MATERIALS
0
*** NUMBER TOO SMALL, <      1 ***
1
NO. OF DIMENSIONS OF THE PROBLEM
33000
 *** INTEGER TO LARGE, GT. +32767                    ***
 *** PLEASE REPEAT LAST ANSWER                       ***
3
MAX. NO. OF DOF PER NODE
7
*** NUMBER TOO LARGE, >      6 ***
6
MAX. NO. OF NODES PER ELEMENT
1
*** NUMBER TOO SMALL, <      2 ***
2
DOES ELEMENT NO. 5 EXIST?  ? (0 - NO, 1 - YES)


HELP
  USERS'S HELP LIST :
  ON   - PRINT MESSAGES
  OFF  - SWITCH OFF MESSAGES
  LIST - PRINT OUTPUT FILE
  STOP - STOP THE PROGRAM
  WAIT - END OF A DATA BLOCK
  END  - END OF DATA
  GO   - RESUME
  ?    - PRINT LAST MESSAGE

STOP
```

REFERENCES

Grodzki Z., Winiarski M. (1984) IDIM An Interactive Data Input Model, CAD 84 Conference, Brighton.
Jones, P.F. (1978) Four Principles of man-computer dialogue. Computer Aided Design, Vol.10, No 3.
Kovacic, I. (1983) Menu Generating System for the FORTRAN programs. ENGSOFT III Conference, London.
Lim, P.T.K., Moffat, K.R. (1981) ADL - A Data Language for Engineers, Procedings I.C.E. Part 2.Vol. 71 March .

3. CONSTRUCTION

COMPUTING IN CONSTRUCTION

Ian Hamilton, BSc CEng MICE

CICA, Cambridge, U.K.

INTRODUCTION

The construction industry has been using computers since the arrival of the early valve driven machines in the 1950s. Most of the larger organisations had purchased their own mainframes by the end of the 1960s and many others were beginning to use remote terminals via GPO lines to timesharing bureaux. By the mid 1970s a considerable number of good, well tried and tested programs had been developed by a variety of different sources for a variety of different applications. The industry, however, is highly fragmented consisting of a number of different professional and other groups. Each distinct group comprises of firms varying in size from one employee (plus dog) to several thousand. This paper attempts to outline the progress that has been made by the construction industry in using computers to meet its own many and varied requirements.

CONSTRUCTION AND DESIGN

To the outsider, construction is often seen as being carried out by men in donkey jackets, hard hats and muddy wellington boots. While this is a large part of the process, there is an equally significant amount of work carried out 'at the drawing board' before the first spadeful of earth is excavated. This first stage, usually called the design stage and accounting for some 10% of the total cost, is where many of the significant decisions are taken. This can be put another way: the satisfaction or otherwise with the finished 'product' is very dependent on how well the various professionals involved in design do their job. The end product of the design stage consists essentially of a collection of drawings, bills of quantities and schedules which allow the contractor to submit a tender and, if successful, to build the project as planned.

DESIGN

It was at the design stage that the earliest applications of computers were made. Most of these were in the area of structural analysis. The computer's ability to perform large numbers of calculations rapidly enabled methods of analysis to be exploited whose mathematics had previously made them of academic interest only. It has been the development of these advanced methods of structural analysis that has made possible the design of complex structures, such as many of today's large bridges and oil rigs.

The numerical nature of structural engineering made the development of programs relatively straightforward and attractive. Thus, most of the organisations big enough to be involved in using analysis programs developed other programs, particularly in the areas of concrete and structural steel design.

The motorway programme started in the late sixties and, reaching its height in the mid seventies, gave further impetus to the development of civil engineering software. The calculations involved in planning mile after mile of motorway, including working out earthworks quantities, are tedious but repetitive. An ideal situation to let the computer do the work.

GOVERMENT SUPPORT

The public funding of motorway and other road schemes meant that considerable public resource could be devoted to both the develo pment of software and to insuring that it was developed to a high standard and also to meet the needs of all users. Perhaps the best known piece of software to emerge from this effort was the BIPS (British Integrated Programme Suite) suite of programs. This was a Government funded co-operative development carried out under the guidance of a team representing a variety of user organisations.

The joint initiative of the then Ministry of Transport and parti cipating County Councils did much, not only to create a library of software, but to create expertise and generally further the use of computers in engineering. It should be noted that most of this work was carried out using mainframe computers, usually 'owned' by a County Treasurer. The days of 'stand alone' mini computers, let alone today's micros, were still to come. It is not always appreciated that considerable use had been made of computers by a large number of organisations long before such names as Apple, Commodore, Sinclair and Acorn appeared on the scene.

COMPUTER GRAPHICS AND CAD

The advent of minis and micros has brought computing within the reach of a large number of organisations. An equally important development has been the increasing availability of computer graphics.

The ability of the computer to draw has been available since the late 1960's. Indeed, many of the highway design programs referred to earlier have been enhanced by the addition of graphics. To be able to 'view' a three-dimensional model of a highway or bridge greatly improved the understanding of what could previously only be reliably represented by two-dimensional drawings. To the public at large, computer graphics is probably best known through computer games, e.g. 'Space Invaders', and its use in the artwork for many of the adverts appearing on television.

However, a whole industry has grown up around the use of the computer to draw. From its origins in Electronics and Aerospaces, computer drawing has expanded to be of considerable importance to all industries, including construction. This is now known as CAD, standing for Computer-Aided Design (or Draughting). In manufacturing, this is linked with CAM, or Computer-Aided Manufacture. Thus, this now major area if computing is usually referred to as CADCAM. Most industrialised nations recognised CADCAM as being essential to the competitiveness of their manufacturing base. As yet, however, numerically controlled machines and robotics have few applications in construction, so most of the activity here is in CAD alone.

From the early caveman's wall paintings to the present day, drawings have been an essential vehicle for the communication of ideas between different individuals. To the Victorian engineers building railways and canals, the drawing was the legal contract document with both client's and engineer's signatures appearing on it. As with many other aspects of progress, the volume of paper associated with a construction project has increased many times since the Victorian days. Thus, much of the work in the design of any project is in the production of drawings. In building work, as distinct from civil engineering, most of the drawings are produced by the architect. Thus, it is perhaps not too surprising that architects have become the first major group in building to invest heavily in CAD. At the moment, they have overtaken the civil engineers in making use of sophisticated computer techniques. The nature of much of the architects' drawing work means that the facilities of CAD can be used to good effect. Such things as repetition of detail and the frequent changes that occur when a project is 'on the drawing board' can be easily and speedily dealt with by the computer.

Now most large and medium sized architectural practices have seriously considered using CAD, and many have invested in it. With the start-up price in the region of £ 100,000, this represents a considerable departure for a group previouslly unused to investing in capital equipment. It has been said that some clients now consider the possession of a CAD system as part of an architect's suitability to be given work.

While one or two CAD systems have been developed specially to meet the needs of building design, the majority on the market have their origins in other industries. It is interesting to note that a few of the latter group of vendors have taken Construction very seriously as a market place, and at least one has achieved some very good results. This is very much a growth area and one of continuing and interesting development.

To the Building Services Engineer, CAD holds great interest, traditionally dependent upon the architect's drawings on which to superimpose his layouts of pipes, etc.,. Any change on the architect's part has to be taken into consideration; CAD provides a very powerful system of overlays which allows changes by either party to be checked quickly. The major pay-back from services work will, however, come with three-dimensional service modelling. Much of the use of CAD until very recently was of a two-dimensional nature. As the technique has gained acceptance, however, considerable advances are now being made in the more difficult, but extremely valuable, area of three-dimensional service modelling.In buildings such as hospitals, services can account for 40-50% of the construction costs; thus a system which enables the designer to produce optimum design will show considerable savings.

The drawings having been produced (with or without CAD), the task of producing detailed bills of quantities and schedules falls to the Quantity Surveyors. Usually operating as separate firms, quantity surveyors are very much the cost accountants of building work. Much of their task requires the extraction of information which should appear on the drawings. Considerable use is now made of computers to do the word processing and computational side of document production. However, any direct, useful, link with CAD systems has still to be developed. An effective half-way stage using the electronic drawing board part of CAD systems is now in use. This allows information on drawings to be measured and input directly into a computer program for the production of bills of quantities.

CONSTRUCTION

The Contractor is on the receiving end of the drawings and other documents produced by architects, engineers and quantity surveyors. He has to submit a price for the job and, if he wins the contract, carry it out and make a profit. Most of the major firms of contractors have been using computers to handle the financial side of their activities for some time. As with design, the advent of the mini and the micro has stimulated the interest of medium and small firms. Accounts, and in particular integrated accounts, is usually the first priority. The need to keep track of labour, materials and plant is vital to the con-tinuing well-being of the firm. Much of the contractor's computing activity is office based, yet great benefit could be obtained by recording infor-mation on site when and as events take place, rather than days or weeks later, when records are processed at head office. The immediate demands of happenings on site are not always compatible with sitting down to record them at a computer terminal; the environ-ment is totally different from the design office.

However, interesting possibilities are beginning to arise. The use of sophisticated input devices using techniques such as video cameras coupled with pattern recognition can, by focussing on a tower crane, record directly materials handled by the day. While the use of these techniques is still at the research stage, the basic methods themselves are becoming increasingly robust.

ARTIFICIAL INTELLIGENCE

Those who follow computing in general will be aware
of the reaction caused by the Japanese Government's
announcement of their 'Fifth Generation' computer
programme. One of the main outcomes of this will be
computers that have some forms of reasoning or
'Artificial Intelligence'. Our own UK Government has
responded by launching the Alvey Programme, guided by
the Department of Industry. Using existing computers,
it is possible to develop 'Expert Systems'. The basic
function of these is to encapsulated knowledge in
such a way that it can guide non-experts in solving a
problem. One of the more publicised areas of
application is in the field of medical diagnosis.
Here, the knowledge of experts has been embedded in a
computer in such a way that qualified, but less
specialised practitioners can draw on this easily in
helping them to make their own diagnosis. Many
similar, yet perhaps simpler, applications could be
found in construction, for example fault diagnosis of
mechanical plant, or the interpretation of the
Building Regulations and other rules.

CONCLUSION

In many respects we are now entering a period of
reality in the use of computing. Many previous dreams
are now capable of fulfilment, or at least the
reasons why they must remain dreams are better
understood. Computing techniques and industrial
awareness and acceptance are now well matched.
Inevitably, there will still be 'overselling' of some
techniques, but there is a growing body of experience
to counteract this. The computer is now established
as a useful tool for the Construction Industry. The
future success of the Industry, both at home and
overseas, may be, at least partially, dependent on
its ability to continue to develop its use of
computing.

FINITE ELEMENT ANALYSIS OF DEEP FOUNDATIONS SUBJECTED TO UPLIFT
LOADING

Claudio F. Mahler and Aureo P. Ruffier

COPPE/UFRJ-Coordination of Postgraduate Programmes in Engineering, Rio de Janeiro, Brazil
CEPEL-Electrical Energy Research Centre, Rio de Janeiro, Brazil

1. INTRODUCTION

The study of foundations subjected to pull-out forces has been
developed very much in the last years in Brazil. The necessity
of transmission towers, self-supporting and guyed, with greater
dimensions, is basically a product of a possible future growing
demand of electrical energy in the great urban centres. Allied
to this, there is more potency transmitted per line. As this
power, in Brazil, comes from far locations, like Itaipu for example, a secure and economical design for each tower will avoid
extra costs.

The usual procedure for estimating pull-out resistance of deep
foundations was developed for sedimentary soils. However, in
Brazil the great majority of transmission towers are founded on
residual soils. The experience has shown that there is a lack
of knowledge about such phenomena in residual soils. In this
paper, there is an attempt to advance a little more in the understanding of the phenomena and to furnish a good tool to make
new investigations, on the behaviour of pier foundations under
pull-out forces.

Several cases of full-scale tests which had been performed (Barata et alii, 1978) in a residual soil near Rio de Janeiro are
examined.

The traditional methods have been investigated but only results
obtained using the formulation proposed by Martin (1973) are presented. Only this method was really developed for pier foundations pull-out resistance.

However, the intention of this study is not only to verify the
failure load, but to observe the behaviour of the whole soil-
structure system. So it was necessary to adopt another method,
different from that usually used in designs. The F.E. Method

was adopted. The developed programme uses an incremental-iter-
active formulation to simulate the pull-out load tests. Joint
elements were used for the soil-structure interaction and axi-
symmetric elements for the soil and the pier foundation. Soil
non-linearity and plastification were taken into account. Interesting results have been obtained about the behaviour of the
whole system. The comparison of the results obtained with field
measurements suggests that the proposed solution can be used in
predictions of the pull-out resistance of similar foundations.

2. METHOD OF ANALYSIS

Incremental-Iteractive Approach

Consider a solid body surrounded by a non-linear soil mass. The
model idealized to represent this described system will present
two different components: the soil and the structure. Interface will be called the contact between both components.

When load increment is applied to the soil-structure system the
tendency of separation between soil and structure is expected
due to their different stiffness characteristics.

Finite deformation should be included into the formulation in
order to accommodate the relative large displacement involved,
however infinitesimal displacement and joint elements were
adopted as an alternative approach. Based on the fundamental
energy equation and following the usual finite element procedure one reads

$$\underset{\sim}{K}\underset{\sim}{u} - \underset{\sim}{R} = \underset{\sim}{0} \qquad \qquad \text{Equation 1}$$

the well known finite element equation which represents the
equilibrium equations for static boundary conditions where
$\underset{\sim}{K}$ - the stiffness matrix
$\underset{\sim}{u}$ - the displacement vector
$\underset{\sim}{R}$ - vector of nodal forces

Material non-linear effect was incorporated by adopting the in-
cremental-iteractive Newton-Raphson procedure. For each increment (or general iteraction) equation (1) may be transformed to

$$^{m}\underset{\sim}{K} \; ^{m+1}\underset{\sim}{u} - \; ^{m}\underset{\sim}{F} = \underset{\sim}{0} \qquad \qquad \text{Equation 2}$$

where
$^{m}\underset{\sim}{K}$ - tangent stiffness matrix
$^{m+1}\underset{\sim}{u}$ - vector of displacement increment
$^{m}\underset{\sim}{F}$ - out of balance force
m - represents the m increment or general iteraction.

The initial stress state necessary to initialize ^{m}K is evaluated using the "gravity turn on" process where the calculated displacements are neglected.

As the pier foundation is surrounded by a cohesive over-consol-
idated soil, the excavation effect was not taken into account.
Also the simulation of the pier foundation concreting was not
considered.

Re-writing equation (2) in a more convenient form leads to

$$m_K^{i-1} \; u^{i-1} - m_R + m_F^{i-1} = 0 \qquad\qquad \text{Equation 3}$$

where

m   - m increment
i   - i interaction
m_R - m - external load increment
m_F^{i-1} - i-1 - balanced load for the m-increment

$$m_F^{i-1} \; = \; -\int_v {}_o^T B \; \Delta \sigma \; d \text{ volume} \qquad\qquad \text{Equation 4}$$

and

${}_o^T B$ - initial linear strain displacement transformation
        matrix
$\Delta \sigma$ - vector of incremental stresses

The difference $m_R - m_F^{i-1}$ represents the out of balance force
for the incremert m and interaction i-1.

Convergence Criterion
The convergence criterion is defined based on the displacement
criterion, and is represented by the following equation

$$\frac{1}{n} \; \frac{\| \; u^i \; \|_2}{\| \; m_u^i \; \|_2} \leqslant \varepsilon_d \qquad\qquad \text{Equation 5}$$

where

n   - number of degrees of freedom
ε_d - represent the tolerance and is defined by the program
        user
u^i and m_u^i - are as already defined in this report.

In this report, the analysed problems (presented in future sections) required five average itercations to converge to the
adopted tolerance ($\varepsilon_d = 10^{-4}$), where a VAX 11/780 computer was
used.

Adopted Elements
The 4CST element was used to simulate the soil and the structure. The good performance of this element for axisymmetric
problems was checked again.

The joint element (Goodman, 1976) was used for the interface.
In spite of this element having been developed for jointed
rock, it has shown a very good performance in the present study.

Two kinds of movements are presented in the joint element (Figure 1). These movements can be simultaneous.

Material Non-linearity
The non-linear, stress dependent stress-strain characteristics of the soil were considered. For this an hyperbolic stress-strain relationship (Duncan & Chang, 1970) was adopted and used to obtain the tangent value of Young's modulus. The tangent value of Poisson's ratio was obtained through an exponencial formulation (Lade, 1972), although, in most cases, it was maintained constant.

For the interface, the value of k_s, tangential stiffness of the joint changes with stress and strain in the following manner:

$$k_s = \frac{c + \sigma_n \tan\emptyset}{\varepsilon_\varsigma} \qquad \text{Equation 6}$$

where
   c - cohesion or tangential stress for $\sigma_n = 0$
   $\emptyset$ - friction angle
   σ_n - normal stress
   ε_ς - tangential strain

The value of k_n is maintained constant in this study.

Failure/Plastification
For both elements two failure criteria were adopted:
a) Mohr-Coulomb
b) Traction limits.

For the 4-CST element, in both cases, once it fails, the programme adopts a really low value for the tangent elasticity modulus. For the joint element k_s and k_n are put equal zero.

3. PARAMETERS DETERMINATION

The field tests were done in a plateau situated on a hill near Rio de Janeiro. The site consists of a layer of tropical/residual soil, mature, originated from a gneissic rock with something like 2.5 m depth, superimposed to a less weathered layer, very thick. The water table was not reached by sounding until 15 m in depth.

The in-situ tests consisted of a series of pull-out essays in pier foundations and footings (Barata et alii, 1978). For this study only five pier foundations are reported, two being with enlarged base (P1 and P3) and three without enlarged base (P4, P5 and P6). The least distance between the piers is about 8.5 m. The dimensions from them and the depth of the first layer in each case are shown in Figure 2.

The adopted parameters were obtained from laboratory tests (Ba-

rata et alii, 1978) and from backwards analysis of plate loading tests (Werneck et alii, 1979) and from pull-out load tests
(Barata et alii, 1978). Their determination is well described
by Ruffier (1985). The parameters here used are presented in
Tables 1, 2 and 3, for the interface, soil and pier foundation,
respectively. The method of determination of the hyperbolic
parameters can be seen by Duncan & Chang (1970).

The value adopted for k_n has the finality to avoid an interpe-
netration of two neighbouring elements.

TABLE 1 - PARAMETERS ADOPTED FOR THE INTERFACE

	Upper Layer	Inferior Layer
k_n (MPa/m) (normal stiffness)	1000000	1000000
k_s (MPa/m) (tangential stiffness)	500	500
c (MPa) (cohesion)	0.044	0.035
$\emptyset$ (degree) (friction angle)	18.9	20.3
σadm (MPa) (admissible tensile)	0.1	0.1

TABLE 2 -   SOIL PARAMETERS

		Upper Layer	Inferior Layer
γ (kN/m3) (unit weight)		16.5	18.0
c (MPa) (cohesion)		0.029	0.023
$\emptyset$ (degree) (angle of internal friction)		27	29
Hyperbolic parameters	K	600	350
	n	0.52	0.50
	R_f	0.75	0.85
	K_{ur}	900	590
ν (Poisson's ratio)		0.4	0.4
E_R (MPa) (residual Young's modulus)		0.1	0.1
σadm (MPa) (tensile stress)		0.1	0.1

TABLE 3 -   CONCRETE PARAMETERS

γ (kN/m^3) (unit weight)	25
E (MPa) (elasticity modulus)	2.1×10^4
ν (Poisson's ratio)	0.20

4. NUMERICAL EXAMPLES

Five cases of pier foundations subjected to uplift loading have
been simulated by the described method.

Two meshes have been used in the Finite Element Analysis and
are presented in Figure 3. One represents a pier foundation
with enlarged base and the other without enlarged base.

One difficulty in analysing the infinite boundary surface problem by F.E.M. is an adequate choice of the limit boundary surfaces. A study about the boundary influence has been made.
The only ones presented are the final satisfactory meshes.

To simulate the up-lift process, a 25   kN   constant increment
load has been chosen.

The plastification process initiates at the base of the pier
foundation and propogates, element by element, along the shaft
towards the surface. For the pier foundation with enlarged
base, the soil plastification initiates at the extreme lateral
point of the base and propagates until a certain vertical portion on the soil mass.

For both kinds of pier foundation, the plastified surfaces (defined by the same level of failure displacement) have been noticed to develop along the shaft of the pier foundation. This
prediction is also in agreement with the field observations.

The comparison between the predicted and observed ultimate resistance is displayed in Table (4). Good agreement can be noticed.

Finally, the predicted and observed displacement versus up-lift
force are presented in Figures 4 and 5. Again, reasonable agreement can be seen.

TABLE 4 - PULL-OUT RESISTANCE AT DIFFERENT PIER FOUNDATIONS

Pier	P1	P3	P4	P5	P6
Field Test	217.5	106.0	151.0	97.5	95.5
Martin (1973)	209.5 (96.3%)	102.5 (96.7%)	85.7 (56.8%)	85.8 (88.0%)	32.3 (71.0%)
F.E.M.	185.0 (85.1%)	97.5 (92.0%)	115.0 (76.2%)	110.0 (112.8%)	50.0 (109.9%)

Units: N x 10^5

5. CONCLUSIONS

The finite element method has been applied to analyse the behaviour of pier foundation subjected to up-lift force. Two
kinds of pier foundation, one with enlarged base and another
without enlarged base, have been considered to investigate the
displacement field, the failure load and the shape of failure
surfaces.

The displacement field and the failure load predicted by finite
element have shown a good agreement with the measured results.
Also the average failure load has been shown to be equivalent
to the one obtained by Martin (1973).

The predicted shape of the failure surface which develops along
the pier foundation shaft has agreed very well with the observed one. It is shown clearly that the failure process starts
around the pier foundation base and moves, element by element,
towards the surface.

The performance of the joint element to analyse axisymmetric
deep foundation by means of F.E.M. has been shown to be quite
adequate. The main influence on the ultimate up-lift force has
been shown to be provenient from the lateral friction on the
pier foundation.

Finally, the finite element programme developed in this report
has proved to be a very strong computation option to be used as
an instrument for research and engineering design.

REFERENCES

Barata, F.E.; Pacheco, M.P.; Danziger, F.A.B. and Pinto, C.P. (1979) Foundations under pulling loads in residual soils - Analysis and application of the results of load tests.
Proc. 6th Panam. Conf. SMFE, Lima, Peru.

Duncan, J.M. and Chang, C.-Y. (1970) Non-linear Analysis of Stress and Strain in Soils.
Journal of the Soil Mechanics and Foundations Division, Proc. ASCE, Vol. 96, No. SM5, September, pp.495-498.

Goodman, R.E. (1976) Methods of Geological Engineering in Discontinuous Rocks.
West Publishing Co.

Lade, P.V. (1972) The Stress-Strain and Strength Characteristics of Cohesionless Soils.
Thesis presented to the University of California, Berkeley in partial fulfillment of the requirements for the degree of Doctor of Philosophy.

Martin, D. (1973) Calcul des Pieux et Fondations a Dalle des Pylones de Transport D'Énergie Électrique - Étude Théorique et Resultats d'Essais en Laboratoire et in-situ.
Annales de L'Institut Technique du Batiment et des Travaux Publics, No. 307-308, Juillet-Août, pp. 106-131.

Ruffier, A.P. (1985) Analysis of Foundations submitted to pullout forces by the Finite Element Method.
Thesis presented to COPPE at Federal University of Rio de Janeiro in partial fulfillment of the requirements for the degree of Master of Sciences (in Portuguese).

Werneck, M.L.G.; Jardim, W.F.D. and Almeida, M.S.S. (1979) Deformation Modulus of a Gneissic Residual Soil Determined from Plate Loading Tests.
Solos e Rochas, Vol. 2, December, pp. 3-17.

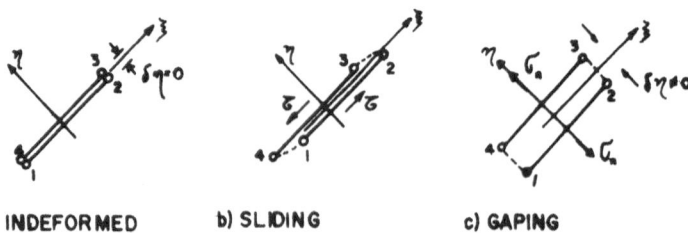

a) INDEFORMED   b) SLIDING   c) GAPING

Fig. I - JOINT ELEMENT

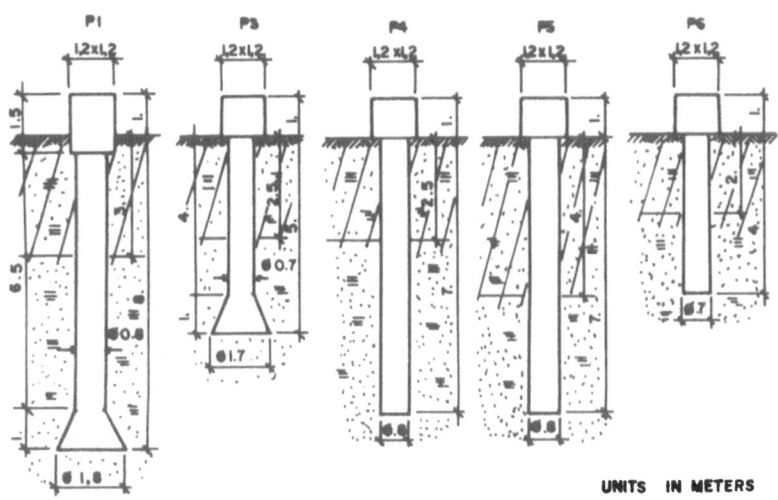

Fig.2 - PIER FOUNDATION'S GEOMETRY AND SOIL LAYERS

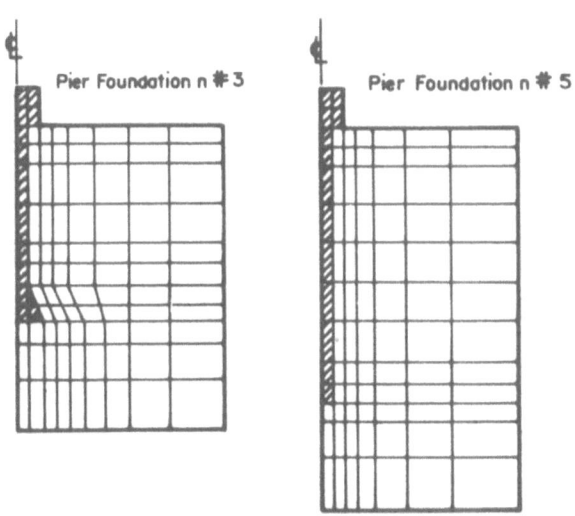

Fig.3 - FINITE ELEMENT MESHES (TWO EXAMPLES).

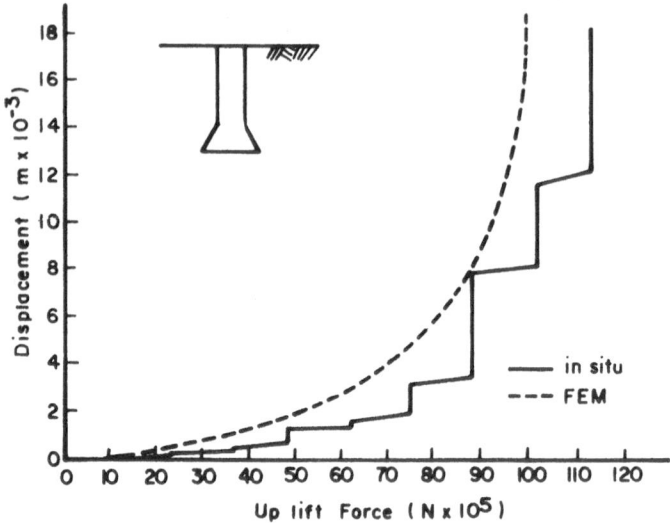

Fig. 4 — DISPLACEMENT VERSUS UP LIFT FORCE FOR PIER FOUNDATION N# 3.

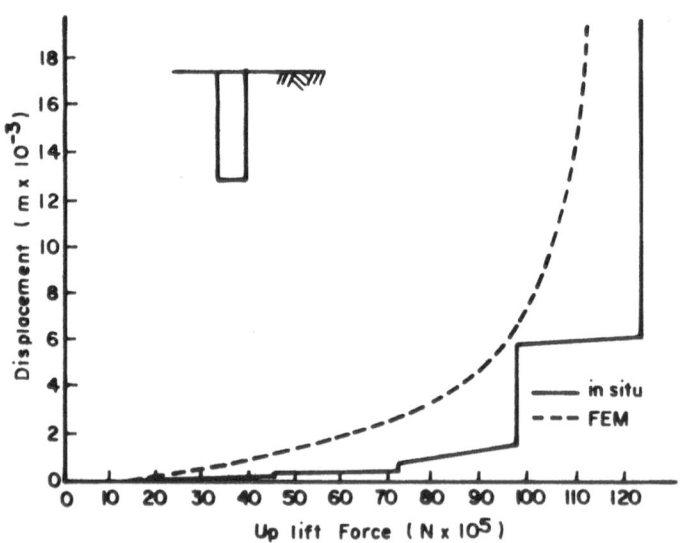

Fig. 5 — DISPLACEMENT VERSUS UP LIFT FORCE FOR PIER FOUNDATION N# 5.

THE USE OF COMPUTERS FOR THE STABILITY ANALYSIS OF NATURAL
SLOPES AND EARTH DAMS UNDER EARTHQUAKE CONDITIONS

A. Calvaruso, G. Seller and R. Trizzino

Engineering Faculty, University of Bari, Italy

INTRODUCTION

Although increased use of electronic computers for stability
analysis of slopes, both natural and artificial, has been
made over the last 20 years, calculation techniques more
accurate than in the past are still very rare in soil engineering practice. The methods of slope stability analysis usually
utilized (for example, the Fellenius method and others
similar) are greatly simplified by assuming a regular shaped
slope and circular failure surfaces, and neglecting the
effective interslice stresses. Under earthquake conditions
the evaluation of the stability of natural slopes and earth
dams is particularly arduous, so that the aid of an electronic
computer becomes essential.

Nowadays, as sophisticated and cost effective computer
hardware back up the soil engineer, the main problem is
to develop a suitable base software. The method presented
here provides a reliable design technique for simple and
rapid application to various safety problems of embankments
and earth dams located in highly seismic areas, computing
the critical seismic acceleration of the slope.

PRINCIPLES OF THE METHOD

The calculation method is based on the stability analysis
of embankments and slopes proposed by Sarma (1973). A pseudostatic analysis is therefore carried out. As a matter of
fact, a seismic force produces on the slope a set of stresses
and strains that vary from point to point and from time
to time; therefore the most complete and correct approach

to the problem undoubtedly consists in a dynamic analysis
which, starting from a design accelerogram and taking into
account the non-linear soil behaviour, provides continuously
the actual distribution of accelerations and shear stresses
for each particular point of the soil mass. However, the
application of the Finite Element Method (FEM) proves to
be very arduous and expensive even with the aid of an electronic
computer, as it requires soil characteristics to be more
exactly known, that is often very difficult to obtain.
Hence, increased use has been made of the methods known
as "pseudostatic analyses"; they simulate the seismic stresses
– varying with time and space – through a set of horizontal
forces constant with time and having the same value in each
point of the soil mass.

So, as reported by Seed (1979), it is assumed according to
Terzaghi (1950) that: "An earthquake with an acceleration equivalent K produces a mass force acting in a horizontal direction
of intensity K per unit of weight of the earth". The seismic
force is thus simulated by a horizontal force acting on the centre of gravity of the sliding mass towards the instability:

$$S=KW \tag{1}$$

where W is the total weight of the mass. Starting from the princi
ple of limiting equilibrium, the "critical acceleration" is defined as the minimum value of the equivalent seismic acceleration
that is required to bring the soil mass, bounded by a potential
slip surface of any shape, and the ground surface to a state of
limiting equilibrium.

Formulation

It seems useful to report here the formulation given by Sarma (1973) for a better comprehension of the calculation program
presented here. With reference to Figure 1, the potential sliding mass is divided into N vertical slices. The forces
acting on the ith slice are shown in the figure. Since just before failure the whole mass must be in equilibrium, considering
the vertical and horizontal equilibrium of the ith slice one obtains:

$$Ni \cdot \cos(Ai) + Ti \cdot \sin(Ai) = Wi - DXi \tag{2}$$

$$Ti \cdot \cos(Ai) - Ni \cdot \sin(Ai) = K \cdot Wi + DEi \tag{3}$$

If we assume that there are no other external forces acting
on the free surface, it must be:

$$\sum DEi = 0 \tag{4}$$

$$\sum DXi = 0 \tag{5}$$

According to the limiting equilibrium concepts, under the
action of the force S=KW the complete shear strength of the soil

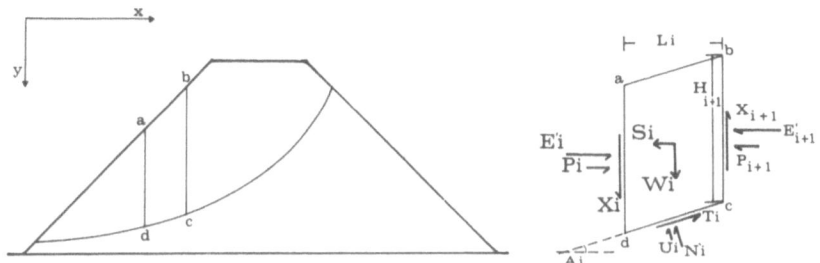

Figure 1: Forces acting on a single slice.

Hi	=Height of sliding mass at section i
Li	=Breadth of slice i
Ai	=Angle made by slip line d-c with the horizontal
Ami	=Average inclination of the slip surface at section i
xi,yi	=Coordinates of mid point of base of slice i
XG,YG	=Coordinates of centre of gravity of sliding mass
K	=Seismic acceleration
Si	=Seismic force
Ni	=Normal force at base of slice i
Ti	=Shear force at base of slice i
Pi	=Resultant water pressure on section i
Pmi	=Resultant water pressure on the middle section of slice i
Ui	=Resultant pore pressure at base of slice i
Xi	=Shear force on section i
$Ei=E'i+Pi$	=Lateral thrust on section i
Wi	=Weight of slice i
W	=Total weight of the sliding mass
Gi,Ci,Fi	=Geotechnical parameters of single soil stratum at section i
Gmi,Cmi,Fmi	=Average geotechnical parameters along section i
Fsi,Csi	=Average geotechnical parameters at base of slice i
Rui	=Pore pressure ratio on section i
Rumi	=Pore pressure ratio at the middle of slice i
FL	=Local safety factor

along the potential sliding surface is mobilized. Thus the factor of safety becomes equal to 1 and the coefficient K will represent the requested critical acceleration Kc, expressed as a fraction of the acceleration of gravity. According to the Mohr-Coulomb failure criterion, in terms of effective stresses the shear force acting at the base of the slice is:

$$Ti=(Ni-Ui)\cdot\tan(Fsi)+Csi\cdot Li\cdot\sec(Ai) \qquad (6)$$

The total pore pressure U at the base of the slice may be expressed by the pore pressure ratio Ru (Bishop and Morgenstern, 1960):

$$Ui=Rumi\cdot Wi\cdot\sec(Ai) \qquad (7)$$

Considering the complete equilibrium of the whole mass and satisfying the moment equilibrium about the centre of gravity of the sliding mass, one obtains the resolutive equations:

$$\sum DXi\cdot\tan(Fsi-Ai)+\sum Kc\cdot Wi=\sum Di \qquad (8)$$

$$\sum DXi\cdot((yi-yg)\cdot\tan(Fsi-Ai)+(xi-xg))=$$
$$=\sum Wi\cdot(xi-xg)+\sum Di\cdot(yi-yg) \qquad (9)$$

where

$$Di=Wi\cdot\tan(Fsi-Ai)+(Csi\cdot Li\cdot\cos(Fsi)-$$
$$-Rumi\cdot Wi\cdot\sin(Fsi))\sec(Ai)\ /\cos(Fsi-Ai) \qquad (10)$$

The quantities at the right-hand sides of equations (8) and (9) are known. If it is possible to find a set of X forces that satisfy those equations, the only unknown remains the critical acceleration Kc and the problem is completely defined. It has been found (Sarma, 1973) that the following definition gives satisfactory results:

$$Xi=\lambda fi((Mi-Rui)\cdot Gmi\cdot Hi^{2}\cdot\tan(Fmi)/2+Cmi\cdot Hi) \qquad (11)$$

$$Ei=Mi\cdot Gmi\cdot Hi^{2}/2 \qquad (12)$$

where Rui,Fmi,Cmi and Gmi are the weighted average values along the vertical section i, by means of which the non-homogeneity of the mass can be taken into account; Mi is a coefficient depending on the geometric and physical pattern of the sliding mass and expressing the relationship between the horizontal and vertical stresses; fi is a number to be selected depending on the degree of mobilization of the shear strength along the vertical section i. It is usually assumed to be equal to 1. Then it follows:

$$DXi=\lambda(Xi+1-Xi) \qquad (13)$$

By substituting equation (13) in equations (8) and (9), one obtains two simultaneous equations with two unknown: λ and Kc; thus a unique solution is obtained. Such a kind of analysis is very general and applies to the most varied situations, as no assumptions have been made about the shape of the failure surface nor about the mechanical and geometrical features of the soil mass.

THE COMPUTER PROGRAM

The program allows a stability analysis of any kind of slope to
be performed by means of simple algorithms; the method presented above provides a complete analysis, even if more sophisticated
finite element techniques are not utilized; among other things,
these require a storage capacity not always available on personal
computers. The program has been drawn up with BASIC on a HP-86B
personal computer.

The main effort was to obtain a very versatile tool that could
be easily used avoiding every possible error. Therefore, a set
of error messages has been introduced, in order both to avoid
obtaining wrong results, having wrongly interpreted the situations to be analysed, and to lead the operator to the step in
the program where the error is located. Figure 2 shows the flowchart of the program.

Geometrical characteristics of the soil mass are described
through a series of lines that are the projections of the deli-
miting surfaces on the plane, i.e.:
- ground surface
- piezometric surface
- slip surfaces
- surfaces separating soil portions having different physical
  characteristics.

Non-homogeneity of the soil above the slip surface is thus
taken into account by singling out well defined surfaces delimi-
ting areas which are innerly homogeneous, according to a simplified model that is often assumed in all engineering branches.

The program approximates each of the above-mentioned lines
to a piecewise-linear. The chosen representation has been seen
to best fit in describing both anyhow complex situations and
localized non-uniformities; moreover, it is very easy to be treated as input and more controllable than higher order curves are.
Therefore, the geometrical characteristics of surfaces are introduced as data files, by giving, for each of them, the number of
the points of the piecewise-linear and the x,y-coordinates of
each point in a fixed reference system. Although it is evident
that a slip can take place only downwards, the computer notices
an ambiguity about the acting direction of instabilizing forces;
therefore, it has been necessary to provide it with this information, by fixing the positive y-axis downwards in the reference
system and letting the slip take place between the two extremes
of the failure surface towards the one of higher y-coordinate.
No condition is required for x-axis. Number of points assigned
for each surface can be various, but not smaller than zero (for

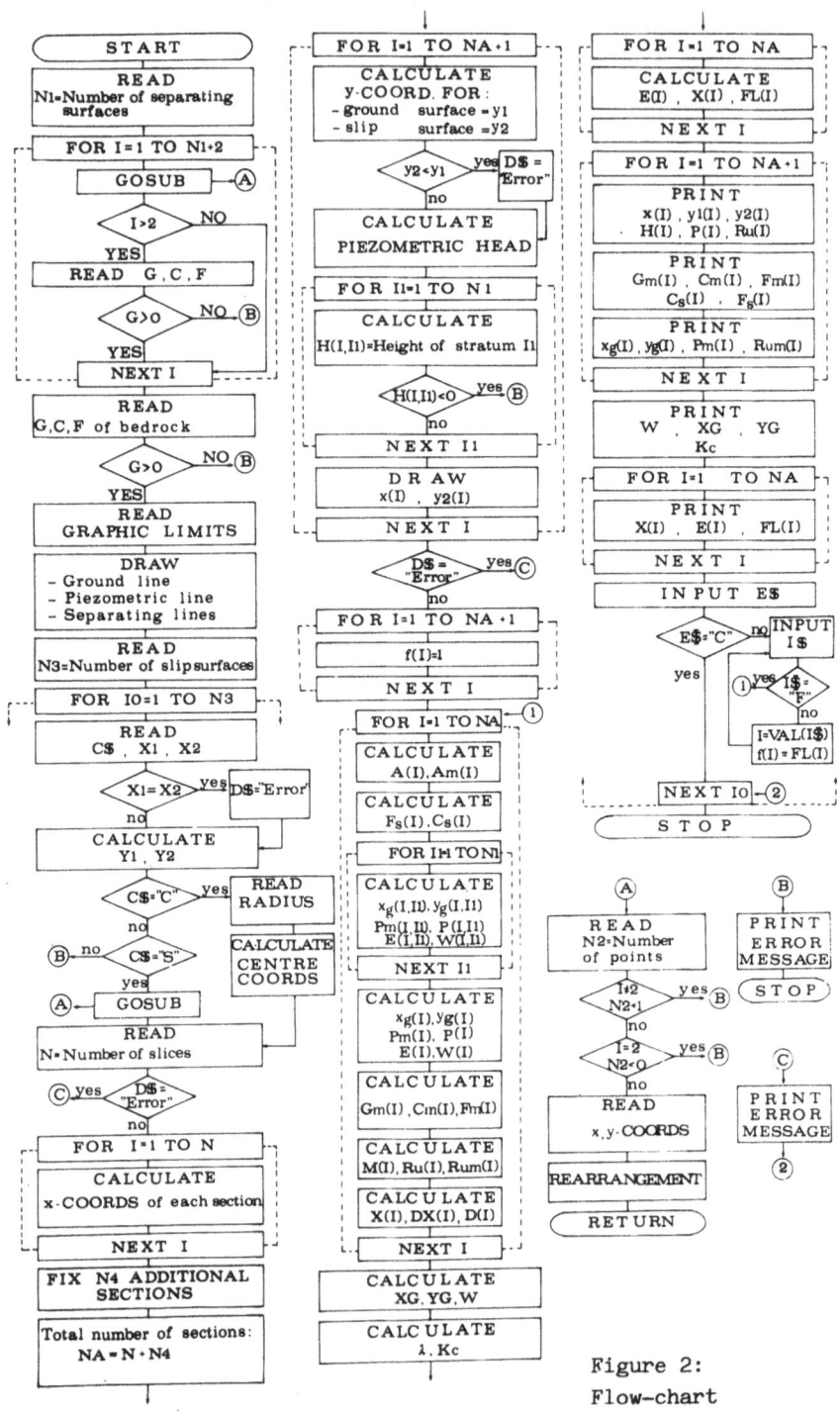

Figure 2:
Flow-chart

example, this may happen when data are wrongly entered). For piezometric surface only, a number of points equal to zero means absence of water. For all other surfaces such an assignment generates an error message. When only one point is given, the piecewise-linear is assumed to be conventionally coincident with the horizontal line through that point.

A suitable subroutine rearranges the given points in accordance with increasing x-coordinates; the requested surface will be identified by the piecewise-linear joining the points in the order specified. The two extreme portions are extrapolated to infinity beyond the first and last given points. The result of these operations may be regarded as a x-function, where only one y-value corresponds to each x-value. It would thus be impossible to represent surfaces that are vertical or reentrant and having two or more y-values for one x-value (Figures 3 and 4). This can be avoided by a simple trick in data assignment. To represent a vertical surface, two points are given having x-coordinates very little different from each other. In order to obtain an exact representation, it must be remembered that the points are re-ordered by the program according to increasing x-coordinates; therefore, a different statement might produce a wrong representation, as shown in Figure 3. Another trick can also be used when folded layers or suspended lenses are present inside strata with different characteristics. For example, the situation shown in Figure 4 can be solved by regarding the two parts of the folded layer as two distinct layers with the same geotechnical characteristics; as it can be observed from the figure, the two real layers become four dummy ones, having the same properties two by two. As the separating surfaces must anyway be defined along the whole length of the section that is analysed, the dummy surfaces -where useless- can be kept coincident, so characterizing layers of height equal to zero: the program, infact, excludes such strata from processing. The same method may be used when a suspended lens has to be represented (Figure 5).

In the representation of coincident surfaces it is necessary to take care to do it in such a way that they coincide without ever intersecting each other; when this occurs there is an error message. To avoid this, it is best to represent such surfaces assigning them exactly the same points. The geotechnical characteristics of the various soil strata can be represented simply by giving only three parameters, that can easily be obtained in a soil mechanics laboratory:

- unit weight (G) expressed in kN/cubic meters
- cohesion (C) expressed in kPa
- internal friction angle (F) expressed in sexagesimal degrees.

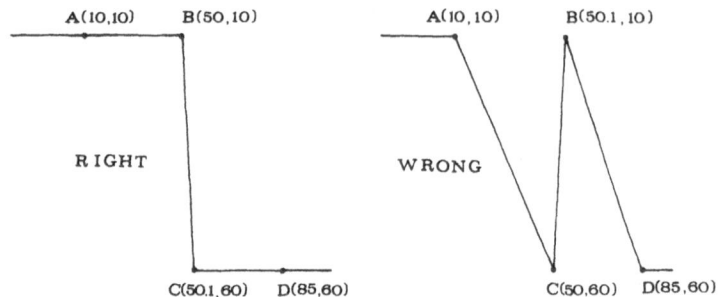

Figure 3: Vertical surface representation.

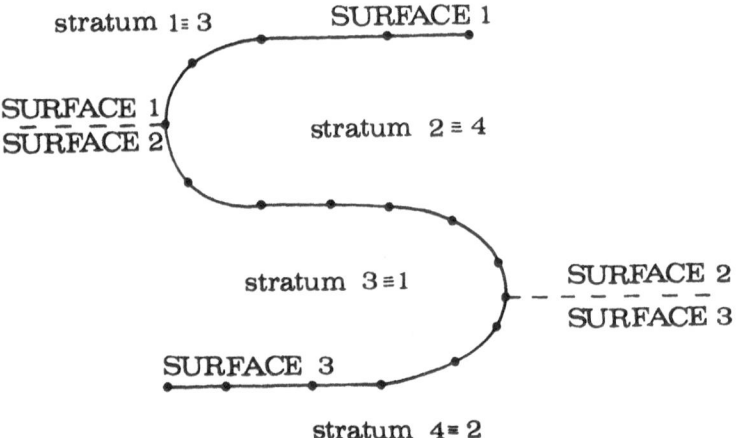

Figure 4: Folded stratum representation.

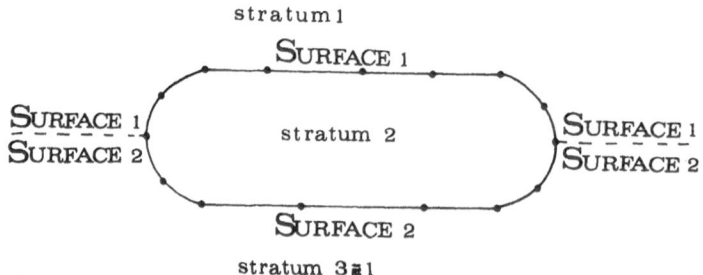

Figure 5: Suspended lens representation.

Although the International System (SI) has been adopted, it
is possible also to use other units slightly modifying the program. After reading the geometrical and geotechnical characteristics of the various strata, the computer reads data relative
to the bedrock, where this may be located. For each separate
slice in which the soil mass has been subdivided, the soil strength
along the slip surface is expressed as a function of the geotechnical parameters of the strata which contain that surface . If
in a single slice different strata are intersected, the computer
calculates the average values of the effective shear strength
parameters along the slice, producing a weighted average propor-
tionally to the length of each segment intersected. In such a
scheme it is also possible to take account of the presence of
joints, representing them as very thin strata, having geotechnical
characteristics which are average in those of the materials constituting the joint. If the slip surface lies exactly along a
joint, it is necessary to be careful not to give it a zero thick
ness, but it is necessary to assign a minimal thickness doing
it in such a way that the slip surface will be at each point
completely contained inside the joint, without reaching its edges.

As is often the case in the normal practice, there is a need
to consider cylindrical circular slip surfaces (for example,
searching for the critical circle); to be able to assign such
a surface, in the computer program a double possibility has been
anticipated by means of an alphanumeric flag: the piecewise-
linear will be assigned by inserting in the data the alphanumeric
value "S" or, alternatively, one could assign a circle giving
the value "C" to the flag. In this case the input parameters
more suitable both for the operator and the computer are the
two points at which the slip surface intersect the ground surface and the radius, the latter as form factor. The centre coor
dinates will be automatically calculated. It is so possible
to analyse an infinite number of surfaces, both circular and
non-circular, by introducing the required number of surfaces as
a datum followed by the chosen alphanumeric. This allows us to
isolate the critical slip surface under earthquake conditions,
that is ,the surface in which the minimal value of the critical
acceleration is found. For both types of surface the fundamental
parameters are the initial and final point, that is,the only two
points in which the surface meets the ground surface. To identify
them it is sufficient to assign their x-coordinate, because the
y-coordinate is automatically fixed by the program at the ground
surface. After this the other parameters of the surface are specified: the radius for the circle, the coordinates of other points
for the piecewise-linear.

Finally, for each particular slip surface the number of slices
in which one intends to subdivide the soil mass has to be assigned
The computer automatically adds these the sections corresponding
to each particular point assigned for the slip line. Therefore,
if there are very irregular local conditions -in the slip
surface as well as in the ground surface- to carry out a correct
analysis it is enough to increase in those zones the number of
subdivisions simply by assigning the coordinates of the projection
of these points on the slip surface.

At this point the program goes on calculating the quantities
in equations (8) and (9), taking into account the different characteristics of the layers concerned, as shown in the flow-chart
(Figure 2). By solving the two simultaneous equations, the value
of critical acceleration for each slip surface will be obtained;
then the computer automatically repeats calculations for the
next surface. Moreover, the local factor of safety (FL) is calculated for each section. It is defined as:

$$FL = ((E_i - P_i) \cdot \tan(F_{mi}) + C_{mi} \cdot H_i)/X_i \qquad (14)$$

The solution is acceptable if $FL \geqslant 1$ is obtained in most sections.
Otherwise the operator can enter the program by modifying the
value of fi in equation (11) through an input; then the calculations are repeated and a new value of Kc is obtained. As already
stated, the critical surface is that for which the minimum value
of Kc is obtained. Reducing by a known quantity the strength
parameters of the material along the slip surface so recognized
and calculating the corresponding Kc, it is then possible to
obtain the value of the static factor of safety, as that which
gives Kc=0, i.e. absence of earthquake force.

Error messages

As we have just seen, the most important thing is to outline
with the maximum accuracy the actual features of the slope. To
this end some error messages have been introduced into the computer program to avoid obtaining unreliable results due to a
wrong data assignment without knowing it. The list of the error
messages is as follows:

- NUMBER OF POINTS FOR GROUND SURFACE IS WRONG. This means that
  the operator has assigned a number of points smaller than one
  for the piecewise-linear of the ground surface; so it is undefined. There are similar messages for the surfaces separating different strata, with the indication of the exact surface,
  that allows a quick discovery of the error. Only for the piezometric surface a number of points equal to zero means water
  table absence. There is an error message only when the number
  of points is smaller than zero. The same happens for the slip

surface:because having anyway assigned the first and last point, only a number of picewise-linear points smaller than zero leads to wrong results.

- UNIT WEIGHT OF LEVEL I1 IS WRONG. It means that the computer has read for the unit weight of a stratum a number equal or smaller than zero. As this variable is often put in the denominator of a function, this condition could lead to a "warning" from the computer without knowing where and why. For the internal friction angle F and the cohesion C there is no error message even if they are smaller than zero, because this does not lead to a stop of the program. Of course, also in this case negative values will lead to unrealistic results, so it is necessary to be very careful with data entering.
- AMBIGUITY. FOR x=...: y=... AND y=.... This means that for a single value of x-coordinate more than one value of y-coordinate have been assigned. It is also pointed out the surface for which this occurs. As above shown, this fact is unacceptable for the computer program. It is allowable only for the initial and final point of the slip surface, because calculating the true y-coordinate of such points the computer ignores any other assignment.
- FIRST AND LAST x-COORDINATES OF SLIP SURFACE no.... ARE THE SAME: SURFACE IS UNDEFINED. In fact, it would be reduced to a single point.
- SLIP SURFACE no.... TYPE IS WRONGLY ASSIGNED ($\neq$ C,S). The alpha numeric datum identifying the slip surface type is other than "C" or "S". This kind of error could also arise from a wrong data arrangement that leads the computer to read a numeric value instead of the expected alphanumeric one. If this occurs, the error message avoids waste of time and yields a very easy discovery of subsequent computer errors. So it warrants a true data ordering.
- SLIP SURFACE no.... IS ABOVE GROUND SURFACE AT x=...: NOT ACCEPTABLE. In fact, for the continuity of the soil mass it is unacceptable a slip surface intersecting the ground surface at points other than the assigned initial and final points. If this occurs, the surface should be considered as two different slip surfaces.
- BOUNDARY SURFACES OF LEVELS I1 AND I1+1 ARE WRONG: THEY INTERSECT AT x=.... This means that two successive boundary surfaces intersect, giving rise to a situation meaningless for the computer program. There is no error message if either the surfaces only touch or coincide without passing beyond each other or the same occurs above the ground surface due to a meaningless extrapolation, since this does not interfere and

lets all that must be taken into account be exactly defined. The error messages listed above help in error discovering where both to do this is difficult and the error itself arises from the specific inner structure of the program without the operator intuitively realizing it. Some errors lead to unreliable results (as those deriving from cohesion or internal friction angle negative values), some others lead to error messages recognized by the computer itself (for example, an alphanumeric instruction in a wrong position or a radius negative or smaller than half the distance between the two extreme points of the circular slip surface). In general, such errors can easily be both realized by the operator and then corrected through a data inspection. Possibilities of error are obviously innumerable, neither they all can be foreseen; nevertheless, the present status of the program provides more than enough tools in order to get a prompt error diagnostic. In any case, the whole program has been tested by blocks and operation of each part of it has been verified.

Graphics

A very simple subroutine has also been inserted in the program in order to graphically represent the assigned surfaces. It consists in subdividing the space available in the "GRAPH" mode into 100 parts and calculating for each corresponding x-coordinate the y-coordinate of each surface, i.e. ground surface, piezometric surface and separating surfaces. Slip surfaces are represented in a manner that looks more suitable, by means of the coordinates of the points of each section fixed as above. Points so obtained are jointed together so that a suitable and quite exact representation will be achieved, although this procedure is rather slow. However, representation of a very complicated situation -a slope with 20 separating surfaces- will be drawn in about 15 minutes. Finally, portions of the piecewise-linear situated above the ground surface are not drawn according to the calculation method that has been used.

APPLICATION

Through the pattern described above, soil masses of various kinds -from very simple homogeneous soils with regular surfaces to the most complex ones with many stratifications- can be analysed. Only few examples are reported here. In order to verify the correspondence between the computer program and the calculation method that has been chosen, it seemed useful to analyse the examples reported by Sarma (1973) in his formulation (Figures 6, 7 and 8). The scheme of a very variously stratified slope, with a lens suspended inside a homogeneous layer, that is report-

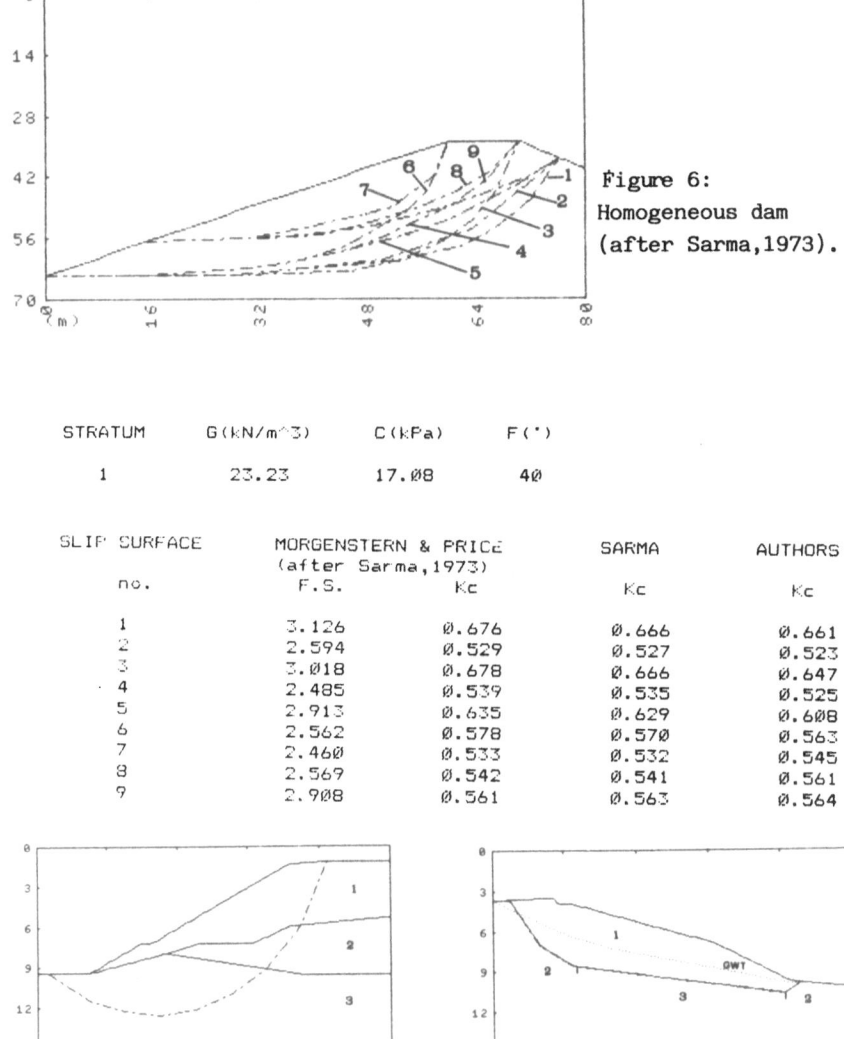

Figure 6:
Homogeneous dam
(after Sarma,1973).

STRATUM	G(kN/m^3)	C(kPa)	F(°)
1	23.23	17.08	40

SLIP SURFACE	MORGENSTERN & PRICE (after Sarma,1973)		SARMA	AUTHORS
no.	F.S.	Kc	Kc	Kc
1	3.126	0.676	0.666	0.661
2	2.594	0.529	0.527	0.523
3	3.018	0.678	0.666	0.647
4	2.485	0.539	0.535	0.525
5	2.913	0.635	0.629	0.608
6	2.562	0.578	0.570	0.563
7	2.460	0.533	0.532	0.545
8	2.569	0.542	0.541	0.561
9	2.908	0.561	0.563	0.564

STRATUM	G(kN/m^3)	C(kPa)	F(°)
1	20.50	0	35
2	20.02	0	33
3	20.50	7.32	20

MORGENSTERN & PRICE (after Sarma,1973) F.S.	SARMA Kc	AUTHORS Kc
1.557	0.224	0.262

Figure 7: Non-homogeneous slope
(after Whitman and Bailey. In:
Sarma, 1973).

STRATUM	G(kN/m^3)	C(kPa)	F(°)
1	18.74	0	23
2	18.74	0	19
3	18.74	0	16

MORGENSTERN & PRICE (after Sarma,1973) F.S.	SARMA Kc	AUTHORS Kc
1.014	-0.005	0.025

Figure 8: Non-homogeneous
slope (after Sarma,1973).

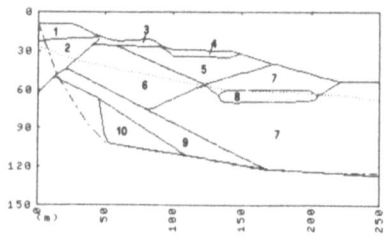

Figure 9: Output of a hypothetic slope scheme.

SLIP SURFACE no. 1

Slip surface first point coordinates (m): x= 0   y= 9
Slip surface last point coordinates (m): x= 500   y= 54

NUMBER OF SLICES= 9

SECTION	x (m)	y1 (m)	y2 (m)	H (m)	P (kPa)	Ru
1	0	9	9	0	-	-
2	4.5	9	22.5	13.5	911.25	.55
3	14	9	48.5	39.5	7801.25	.53
4	37.5	14.75	87.5	72.75	26462.813	.5
5	51	20.7	101.5	80.8	32643.2	.51
6	110	29.62	111	81.38	33114.297	.51
7	168	38.37	122	83.63	34972.103	.51
8	220	53.71	125	71.29	25412.509	.5
9	250	54	125	71	25205	.5
10	500	54	54	0	-	..

SECTION	Gm (kN/m^3)	Cm (kPa)	Fm (°)	Cs (kPa)	Fs (°)
1	--	-	-	.194	36.75
2	18.03	.207	36.384	7	28
3	18.7	4.902	28.505	9.073	34.062
4	19.877	7.79	31.269	10	35
5	19.586	4.614	30.093	10	35
6	19.751	6.454	24.24	10	35
7	19.729	6.38	24.165	10	35
8	20	7	23	7	23
9	20	7	23	7	23
10	--	-	-	-	-

SLICE	xg (m)	yg (m)	Pm (kPa)	Rum
1	3	13.51	227.813	.55
2	10.05	23.52	3511.25	.54
3	27.02	41.67	15750.078	.51
4	44.35	56.28	29472.003	.51
5	80.58	65.73	32878.327	.51
6	139.13	75.6	34036.862	.51
7	193.37	84.84	30001.931	.5
8	234.99	89.43	25308.648	.5
9	333.33	77.67	6301.25	.5

TOTAL WEIGHT (kN)   : W = 540053.083
CENTRE OF GRAVITY COORDINATES (m): XG= 198.19   YG= 74.16

CRITICAL ACCELERATION   : Kc= .147077415485 g

INTERSLICE STRESSES:

SECTION	X (kN)	E (kN)	FL
1	-	-	-
2	245.747	1329.601	6.70753292993
3	2393.43	12533.243	4.53308159104
4	8496.187	46433.437	5.14358632424
5	22281.301	64682.983	2.51461412785
6	35575.276	56825.588	1.12353498994
7	42202.785	55270.188	.9467921465
8	37385.501	47317.573	.812428492369
9	32929.546	46217.496	.905571266892
10	-	-	-

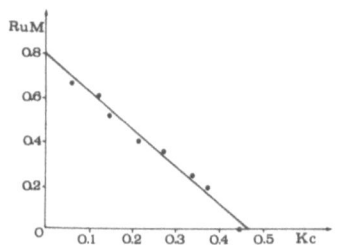

Figure 10: Critical acceleration against average pore pressure ratio.

ed in Figure 9, gives an idea of the program capacity. On the same slope a parametric analysis has been carried out in order to study the influence of groundwater level fluctuations on the critical acceleration values (Figure 10).

CONCLUSIONS

By applying the bi-dimensional model above presented, it has been possible to carry out a stability analysis under earthquake conditions, also in the presence of anyway complex situations. Data assignment is very simple and does not require the user to have a specific knowledge in the geotechnical field. Utilization of the program is made even easier as error messages have been introduced; this allows wrong data entries to be immediately discovered. Obtaining the critical seismic acceleration of either natural slopes or earth dams and embankments provides a sound tool in both designing and analysing these structures when located in a seismic area. In this case, in fact, if average ground acceleration values that are expected for the service life of the designed structure are higher than the critical acceleration value that has been obtained, design parameters must be changed and analysis must be repeated until Kc values higher than those expected are obtained, depending on the safety degree requested.

Finally, by changing the boundary conditions it will be possible both to analyse how stability conditions will vary with time and to forecast eventual triggering of instabilities due, for example, to variations of the piezometric levels.

ACKNOWLEDGEMENTS

The work described in this paper was carried out equally by each author. They are grateful to Dr. S.K. Sarma of the Imperial College of Science and Technology, London for his precious suggestions.

REFERENCES

BISHOP, A.W. and MORGENSTERN, N.R. (1960). Stability coefficients for earth slopes. Géotechnique 10, 4:129–150.
SARMA, S.K. (1973). Stability analysis of embankments and slopes. Géotechnique 23, 3:423–433.
SEED, H.B. (1979). Considerations in the earthquake-resistant design of earth and rockfill dams. Géotechnique 29, 3:215–263.
TERZAGHI, K. (1950). Mechanism of landslides. Application of Geology to Engineering Practice, Berkey Vol. Geological Society of America, 123 p.

THE OPTIMAL DESIGN OF ROAD AND INDUSTRIAL PAVEMENTS USING MICROCOMPUTERS

B. Shackel

School of Civil Engineering, The University of New South Wales, Australia.

1. INTRODUCTION:

The construction of roads and industrial pavements consistently represents one of the major areas of expenditure in Civil Engineering. Despite this the methods used routinely in pavement design are often crude and unsophisticated and rely more on precedent than on objective analysis for their successful implementation. This reflects two factors. The first is that pavement failures are seldom catastrophic and, therefore, escape public scrutiny even though their economic consequences may be of major importance. Secondly, the design and analysis of pavements poses a number of seemingly intractable problems which often require a level of design effort which is disproportionate to the professional risks and liabilities involved. For these reasons pavement design methodology has tended to lag behind the substantial progress made in such fields as structural engineering.

This paper demonstrates that, by the use of microcomputers, it becomes feasible to make sophisticated pavement design procedures available to practising engineers. The paper begins by summarising the advantages of a computer-based methodology and lists the practical requirements for implementing such an approach. The nature of the pavement design problem is then outlined and computer based solutions are described. Finally, the implementation of such solutions is illustrated using design procedures developed by the author as examples.

2. ADVANTAGES OF COMPUTER-BASED DESIGN PROCEDURES:

Several advantages accrue from adopting computer-orientated design methods. These may be summarised as follows:

(a)   The methods are systemmatic. Normally, the methods lead the user through a systematic design procedure which ensures that all essential design requirements are considered and that the designer's attention is drawn to the consequences of poor or incorrect decision. For example, it is possible to incorporate messages warning against the use of materials which do not meet the requirements implicit in a particular design method. In this way it is possible to assist the tyro designer.

(b) The methods concentrate on essentials. By isolating the designer from the minutia of complex design calculations he is left free to concentrate on the important objective of optimising his design. For example, he can readily adopt a "what if ?" approach to comparing different combinations, qualities and costs of materials.

(c) The methods provide a rapid design solution. By yielding a fast means of obtaining design solutions computer-based methodology encourages a designer to examine the relevance and importance of the design criteria and to refine progressively the design assumptions. Moreover, it becomes easy to quickly adopt a design to unforseen changes in design requirements.

(d) The methods can automatically maintain design records. The output of all design studies can be systemmatically stored as either computer files or as hard copy.

(e) The methods are suitable for automation. It is relatively easy to ensure that the design procedures automatically both assemble and print the specification necessary to implement a particular pavement design. Moreover, subject to the choice of hardware it is possible to plot pavement cross-sections etc.

3. PRACTICAL CONSIDERATIONS FOR IMPLEMENTING COMPUTER BASED METHODS:

In implementing computer based design procedures, the following factors need to be considered :

(a) Accessibility. Many engineers need on occasion to have access to pavement design procedures. It is therefore desirable to ensure that the method can be implemented on a wide range of computers.

(b) Ease of Use. Relatively few engineers have had specialist training in pavement design. Accordingly, the program should seek to present expert procedures and methodology in a form accessible to non-specialist engineers without compromising the design requirements.

(c) <u>Interaction</u>.   As noted above, one of the main benefits of computer based design is to allow the designer to consider and compare a range of alternative solutions. This requires interaction between man and machine. Such interaction is difficult to establish unless the computer provides solutions at a sufficiently fast rate to warrant the presence of the designer at the computer console.   Thus program speed represents one of the prime criteria for successfully applying computer-based procedures.

These requirements have implications for the choice of hardware and software.

Two considerations influence the choice of hardware.   The first of these is the amount of memory (RAM) available.   The use of 16-bit microcomputers capable of addressing up to 1 Megabyte of memory are obviously to be preferred to 8-bit machines, which normally do not address more than 64 K-bytes and which, therefore, may require time-consuming memory overlay techniques to accommodate sophisticated design programs.   The second consideration is that of operating speed.   Here math coprocessors should be utilised wherever possible because these can increase the speed of programs heavily based on mathematical operations by a factor of 10 or more.   In the author's experience the use of a 8088 CPU operating at 4.7 MHz in conjunction with an 8087 coprocessor combined with suitable software is capable of providing adequate speed to ensure effective interactive use of the pavement design programs described below. However, by comparison the use of 8-bit CPUs such as the widely available Z80 is much less effective in achieving truly interactive program operation.

In the case of software, consideration should again be given to speed.   This implies the use of compact, effective program code, the use of optimising compilers and the provision of software support for such hardware features as the presence of a math coprocessor.

4.   THE PAVEMENT DESIGN PROBLEM:

Mechanistic pavement design procedures involve the determination of the stresses and strains at critical locations throughout the pavement.   These are then compared with the values predicted to cause fatigue failure under traffic.   Typically, the critical locations lie on or near the vertical load axis at the bottom of all bound (brittle) layers and at the top of the subgrade.   The fatigue life of the bound layers can be related to the repeated tensile strains or stresses whilst the permanent deformation or rutting of the pavement is normally assumed to be related to the repeated vertical compressive strains at the top of the subgrade.   Thus mechanistic pavement design involves the formulation and solution of an exceptionally complex boundary value problem. The practical difficulties and their means of solution are now considered with respect to the three principal components of the boundary value problem. These comprise:

(a) Delineation and modelling of the boundary conditions.

(b) The characterisation of the properties of the pavement materials.

(c) The solution of the three-dimensional stress-strain conditions within the pavement consistent with (a) and (b) above.

4.1 Boundary Conditions:

For a pavement the significant boundary conditions comprise, firstly, a complex loading history normally involving a wide spectrum of axle loads, vehicle configurations and load repetitions and secondly, the effects of environmental or climatic factors of which temperature is the most important. Each of these areas is now considered in more detail.

Traditionally, engineers have sought to reduce the complexity of the actual loading history by expressing the design traffic in terms of an equivalent number of standard axle loads or ESA. The ESA concept supposes that the damaging effect of an axle is proportional to the ratio, raised to some power n, of the axle load to that of the standard load. The power n is commonly assumed to have a value close to 4.0 but accelerated trafficking tests conducted by Freeme and others [1981] have shown that n may range between about 2 and 8, depending on the type of pavement and the choice of materials. This means that, in practice, it is often difficult to model accurately the effects of actual traffic in terms of ESAs, particularly, where the pavement incorporates brittle materials such as concrete or cement-stabilised layers where values of n frequently exceed 6.

Because of the uncertainties associated with equivalent load concepts, pavement engineers are increasingly tending to design for the actual spectrum of expected design loads rather than merely for an equivalent number of ESAs. Indeed this is now routine procedure for the design of rigid pavements [PCA, 1966; Packard, 1973].   Similar approaches are now beginning to be applied to flexible pavements. For example, in Australia, the idealised loading spectrum given in Table 1 has recently been recommended by Potter and Donald [1984] as the basis for road pavement design.

As well as loading spectra , such as that given in Table 1, it is necessary to consider environmental effects. Effectively, the effects of environment and climate can be expressed in terms of temperature changes. Temperature affects the stiffness of viscoelastic pavement materials such as asphaltic concrete

[e.g. vide Table 2] and also causes warping and expansion in
rigid pavements. Consequently, seasonal and daily fluctuations
in the temperatures within a pavement need to be considered at
the design stage. For most developed countries the variations
in temperature throughout a pavement can be readily predicted
from climatic indices [e.g. Croney, 1977; Dickinson, 1984]
It then becomes possible to predict the proportions of time for
which a pavement will experience in any given range of temperatures. For example, Table 2 illustrates an idealised distribution of temperatures and asphalt moduli, typical of parts of
eastern Australia.

TABLE 1

LOADING SPECTRUM

Axle	Single	Single	Tandem	Triaxle
No. of Tyres	Single	Dual	Dual	Dual
Proportion of all commercial vehicles	0.370	0.253	0.285	0.092
Axle load tonnes	Proportion of each axle type			
1	0.00	0.016	0.004	0.00
2	0.029	0.087	0.092	0.186
3	0.092	0.108	0.152	0.128
4	0.271	0.124	0.078	0.055
5	0.533	0.078	0.068	0.090
6	0.071	0.095	0.124	0.442
7	0.001	0.135	0.274	0.099
8	0.001	0.187	0.198	0
9	0.001	0.136	0.010	0
10	0	0.025	0	0
11	0	0.001	0	0
12	0	0.001	0	0
13	0	0.001	0	0
14	0	0.002	0	0
15	0	0.002	0	0
16	0	0.001	0	0

The implications of using distributions of load and temperature,
such as those given in Tables 1 and 2 to model the boundary conditions in the pavement design problem, are twofold. First, it
becomes necessary to perform a large number of calculations.
For example, combining the data of Tables 1 and 2 implies that
no less than 234 separate determinations of stress and strain
will be needed at each critical location in the pavement.
Clearly, this could not be practically achieved using manual
methods but is ideally suited to computer-based methodology.
The second implication of modelling the boundary conditions by
loading and temperature distribution is that some means of
characterising each of the pavement material must be found so
that the cumulative effects of each individual combination of
temperature and load can be assessed. This is discussed further
in the next section of the paper.

4.2 Materials Characterisation:

Two problems concerning materials characterisation can be identified. The first concerns the type of deformation law adopted,
whilst the second involves the formulation of some form of

TABLE 2

TEMPERATURE EFFECTS

TEMPERATURE °C	ASPHALT MODULUS MPA	PROPORTION OF TIME
15	2750	0.16
20	1850	0.16
25	1200	0.16
30	700	0.18
35	350	0.18
40	240	0.16

fatigue or cumulative damage law.

As noted in Section 4.3 below the exigencies of achieving rapid stress analyses by microcomputers makes it highly desirable to select linear isotropic elastic deformation laws to characterise the pavement materials. This approach has been widely adopted in pavement analysis. Such a procedure provides acceptable modelling of stiff materials, such as concrete or stabilised layers but is open to objection when modelling unbound materials, such as crushed rocks and gravels. However, the non-linear response of such material can be modelled, at least in part, by dividing the layers of such material into a number of sublayers to which moduli etc. are then assigned on the basis of the stress conditions generated by the traffic loads (Barker, et al 1982). Because this is normally an iterative procedure it is not suited to manual calculations but is easily incorporated into a computer-based method. This is discussed further below.

Having chosen the deformation law it becomes necessary to find some means to characterise the effects of the loading and temperature spectra. This can be accomplished by adopting Miner's linear cumulative damage hypothesis. In this respect the methodology is similar to that already routinely used in the design of rigid concrete pavements [PCA, 1966; Packard,1973]

The Miner hypothesis states that, irrespective of the magnitude of the stress, each stress repetition is responsible for a certain amount of fatigue damage. It is assumed that there is a linear rate of fatigue damage irrespective of the order of load application and that fatigue failure occurs when the sum of the damage increments at each level of stress accumulates to unity. The law can be expressed in the form

$$\sum_{i=1}^{n} \frac{n_i}{N_i} = 1 \qquad [1]$$

Where N_i is the number of cycles to failure at stress level

i and n_i is the number of cycles actually adopted at stress
level i.

Having decided to adopt Miner's hypothesis it remains only to
postulate the mechanisms of failure in order to predict the
life of the pavement. Two criteria of failure can be adopted
for unbound and cement-stabilised bases and subbases respectively. In the case of unbound materials the pavement is
assumed to fail by the gradual accumulation of permanent,
rutting deformation. In flexible pavement design, it has been
commonly accepted that the rutting deformation can be related
to the magnitude of the vertical compressive strain in the top
of the subgrade. Various criteria have been proposed to relate the magnitude of these strains to the number of strain
repetitions that the pavement can carry before developing
unacceptable rutting. These have been reviewed elsewhere by
Barker et al, [2]. In the design method proposed here the
subgrade strain criterion developed by Claessen et al [1977] and
which forms the basis of the Shell design procedure for flexible asphalt pavements has been adopted.

In the case of cement-bound base or subbase materials a fatigue
failure criterion based on the load-induced curvature of the
lower boundary of the base or subbase has been adopted. The
relationships between the numbers of stress repetitions needed
to cause failure and the curvature of soil-cement beam specimens
developed at the PCA by Larsen et al, [1969] have been adopted
to suit the three-dimensional stress and strain conditions in
pavements. Separate relationships are used to characterise
either coarse-grained or fine-grained stabilised materials.

By combining the failure and damage criteria described above
with the Miner hypothesis, it becomes possible by trial and
error to determine the thicknesses of base and/or subbase needed to ensure that the cumulative damage factor will not exceed
unity. This forms the basis of the design methodology described
here.

4.3 Stress/Strain Analysis:

Prior to the advent of high speed digital computers in the
1960's, little progress had been made in the analysis of layered systems. However, since that time a variety of elastic
solutions have been published including such widely known programs as CHEVRON, ELSYM, BISTRO and CIRCLY. Some of the
simpler programs such as CHEV 4 - published by the National
Institute for Transport and Road Research [1977] can be adapted
to run on microcomputers. Such programs are usually restricted
to single rather than multiple loads and to classical isotropic
elastic material properties. Programs such as CIRCLY which
permit the analysis of anisotropic systems are available but
are too large to run on microcomputers. The alternative to
elastic solutions has been to use finite element programs.

Generally, these have not been suitable for microcomputers although, recently a powerful alternative to the use of class-ical finite element methods has been described byBooker et al, [1984] for the analysis of layered systems using small micro-computers.

Where existing layered system programs have been adopted from a main-frame computing environment to run on microcomputers they are generally too slow and too profligate of memory to be satisfactorily incorporated into a mechanistic design procedure requiring large numbers of iterative analyses (vide Section 4.1 above). For this reason it is, at present, necessary to adopt approximate methods of solution.

At least two approximate methods for the analysis of layered systems deserve consideration as the basis for an iterative design procedure. The first of these is to use existing main-frame programs to derive a series of statistically based reg-ression equations to link the critical stresses and strains in a pavement to such factors as layer thickness, layer moduli, fatigue relationships, applied loads and the subgrade properties. Although Hadley et al [1977]have shown that the method is capable of yielding equations of high predictive value, it usually requires a major computing and analysis effort to derive the design equation. However, the method yields a very compact module of computer code for determining the stresses and strains. This advantage is offset by the need to ensue that the design conditions fall wholly within the normally very limited domain of the regression analyses. This severely limits the usefulness of the regression approach.

An alternative to the regression method is to adopt approximate methods of stress analysis. Of these the Method of Equivalent Thicknesses (MET) originally developed by Odemark [1949] and further developed by Ullidtz and others [19, 20, 21] offers a number of practical advantages. Firstly, subject to certain restrictions [19, 20] on the thicknesses and modular ratios of successive layers it can be used to analyse a wide range of practical pavement situations. Secondly, it yields very compact computer code. Here the amount of code needed is typically about 3 to 5 times that needed for the regressions method described above but this only represents a fraction (typically less than 10%) of the code needed for more exact programs such as CHEV 4 [1977].This means that it is ideally suited to iterative design procedures implemented on microcomputers.

Details of the method of equivalent thicknesses have been given elsewhere (Odemark [1949] and Ullidtz [1981]. Essentially, the method involves replacing the actual thickness of each layer by equivalent thickness so that the pavement may be progress-ively modelled in terms of an equivalent elastic half space. Because the method is approximate, Odemark [1949] originally noted the need to incorporate correction factors in the calculation of the equivalent thicknesses. Typically such factors range in value between about 0.75 and 1.0 [e.g.19]. Although some authorities such as the British Ports Association [1982] have apparently ignored the need for correction factors, in general it is prudent to incorporate them into the design analysis. The factors vary depending on the materials used but can be determined by trial by comparing MET results with those obtained by more exact analyses such as, for example, CHEVRON. Provided suitable calibration factors are determined, then good correspondence can be obtained between the MET analysis and other methods. This is illustrated, using typical data obtained by the author in Figure 1 for the critical strain components in a 3-layer system incorporating a surfacing and bound base overlaying an unbound subgrade.

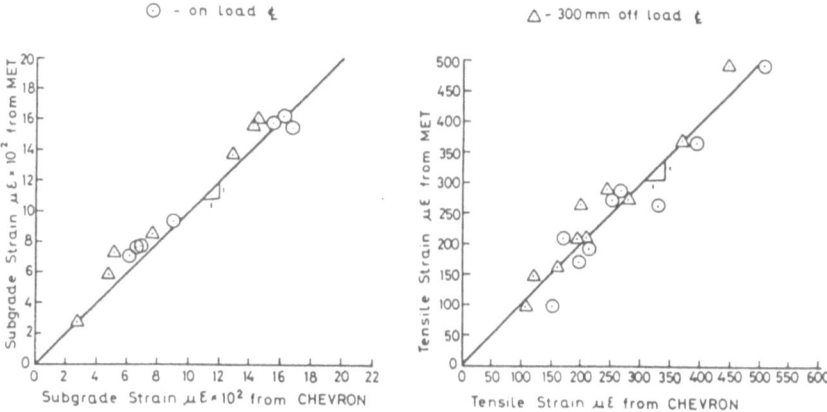

Figure 1: Comparison of Results from CHEVRON and MET programs.

The method of equivalent thicknesses has been successfully applied in the analysis both of flexible pavements [21] and of interlocking concrete block pavements [4, 18]. To date, such analyses have assumed that the pavement materials are isotropic. However, it is worth noting that the MET procedures can, if required, be extended to embrace limited cases of anisotropy (e.g. Ullidtz [1981].

5. EXAMPLES OF PAVEMENT DESIGN USING MICROCOMPUTERS:

The application of the principles outlined above are now illustrated, first for a flexible road pavement and then for a rigid industrial pavement.

5.1 Flexible Road Pavement:
Here, as an example, a series of pavement design curves has been produced to meet the requirements of 10^7 commercial vehicle movements distributed in accordance with the idealised loading dis-

tribution given in Table 1 for a region whose temperature distribution is as described in Table 2. It is assumed that the wearing surface of the pavement comprises a structural layer of dense-graded asphaltic concrete whose modulus is temperature dependent in accordance with Table 2. Two different types of base have been considered. The first comprises a layer of cement-treated crushed rock. This is assumed to have a modulus of 1000 MPa. The second type of base comprises an unbound crushed rock. Here the modulus is assumed to be stress-dependent in accordance with relationships recommended by Potter and Donald [1984] for road pavement design in Australia. The subgrade is assumed to be an unbound material whose modulus following accepted practise [5] is equal to 10 times the CBR value. In all cases the Poisson's ratio of the various material is assumed to be 0.35.

Using an IBM.PC microcomputer, the necessary thicknesses of base have been determined with a computer program that implements the following steps :

(a) A trial thickness of base is assumed.

(b) A trial thickness (> 50 mm) of asphaltic concrete surfacing is assumed.

(c) For each of the 234 combinations of load and temperature implied by Tables 1 and 2 the tensile strains are calculated at the bottom of the asphaltic layer. Where the loads are assumed to be applied by multiple wheels (vide Table 1) separate calculation are needed for each wheel position. The combined effects of the wheel group may then be determined using the principle of super-position.

(d) If the cumulative damage exceeds unity, a new trial thickness of asphalt is assumed (using increments of 5 mm) and then steps (b) to (e) are repeated.

(f) Once a satisfactory thickness of asphaltic concrete is determined then the thickness of base is progressively varied until the fatigue requirements for a cement-stabilised base or the rutting requirements for a crushed rock base, as detailed above, are satisfied.

(g) If the thickness of base determined in step (f) exceeds that assumed in step (a) then steps (a) to (f) are repeated until satisfactory agreement is obtained.

Consideration of the procedure outlined above shows that, depending on the proximity of the assumptions of surface and base thickness in steps (a) and (b) above to the true values, many thousands of calculations are involved. Typically, not less than 10,000 iterations will be necessary to achieve a design. This

emphasises the needs, discussed above, to use compact program
code, to employ floating point math co-processors (where available) and to use optimising compilers in the preparation of the
design program.

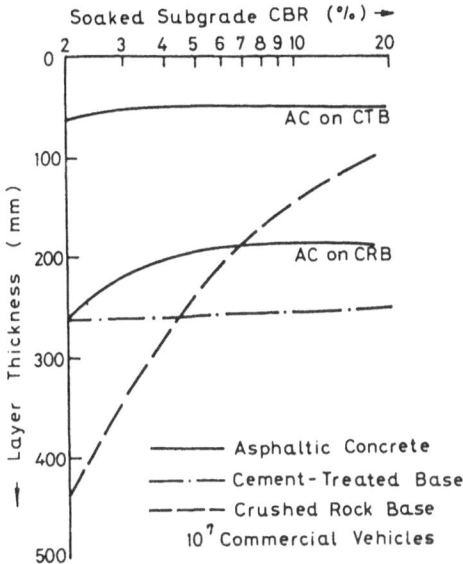

Figure 2:   Results of the Road Pavement Analysis Program

Typical results for the hypothetical design case are shown in
Figure 2. From this, it may be seen that there are substantial
differences in the thicknesses of both surface and base depending on whether or not a cement-stabilised base is selected.
This illustrates one of the fundamental advantages of using
computer-based design methodology, i.e. that it permits rapid
comparison between different materials. In this respect it is
possible and indeed desirable to incorporate cost as a design
factor in the program. For example, by computing the present
worth-of-cost for eachdesign it becomes easy to select the
optimal, i.e. most cost effective alternative. Indeed this
is already a feature of some design methods. [Maree, 1980].

5.2 Rigid Industrial Pavement:
In this section of the paper the design of an unreinforced concrete pavement is described. Here computer-based design methodology recently developed by the author for the Cement and Concrete Association of Australia has been used. For the purposes
of illustration the effects of temperature variations are not
considered here because they are reflected merely in terms of
warping reinforcement rather than as changes in slab thickness.

In the example it is assumed that the pavement is to withstand
10^6 movements of a forklift having the characteristics summar-

ised in Table 3. The forklift is assumed to be used to move
containers comprising 50% ISO 40 and 50% ISO 20 boxes. The
weights of the containers are assumed to follow a skewed distribution determined for British ports by Barber and Knapton
[1, 4]. For design purposes the front and rear axles of the
forklift are treated as separate axle groups. The concrete
has been assumed to have a modulus of 30,000 MPa with a Poisson's
ratio of 0.15. The 90 day tensile strength has been assumed to
be 4.5 MPa and a load safety factor of 1.1 has been chosen.

TABLE 3

FORKLIFT CHARACTERISTICS

A. PEAK LOAD

AXLE	TRACK WIDTH (MM)	WHEEL SPACING (MM)	TYRE PRESSURE (MPA)	L O A D S TARE (TONNES)	LADEN (TONNES)
FRONT	2750	450	0.7	56.28	56.28
REAR	2500	-	0.65	18.65	5.31

B. DISTRIBUTED LOADS

REPETITIONS x 100

AXLE LOAD (TONNES)	9.7	628	894	1284	1523	2433	2401	605	96	38	4
FRONT	23.04	26.74	30.43	34.12	37.82	41.51	45.20	48.89	52.59	56.28	59.97
REAR	17.32	15.98	14.65	13.3	11.98	10.65	9.31	7.98	6.64	5.31	3.98

Here the design procedures are slightly different from those
described above for the flexible pavement in that the method
of equivalent thickness is not used as the basis for the strain
and stress calculations. Rather an approximate model of the
Pickett and Ray influence charts for edge loading on a slab
resting on a Winkler foundation [15] has been used. This model
was developed within the Australian Department of Housing and
Construction and permits the computation of tensile stresses
caused by any designated vehicle geometry. The placement of
the vehicle relative to the edge of the slab is automatically
varied until the most severe stresses are encountered. In
other respects the design procedures are similar to those
described in Section 5.1 above in that the slab thicknesses
are progressively incremented until the Miner's law damage
criterion represented by Eqn. 1 is satisfied.

The results of the analysis are shown in Fig. 3. Here comparisons are displayed between the effect of dowelled and undow-
elled joints and between the fork-lift carrying containers whose

laden weights are assumed to be distributed and the fork-lift
always running at the rated maximum container loads . As might
be expected intuitively the designs, based on the assumption
that all movements of the design vehicle are at the maximum
rated load capacity, lead to the selection of substantially
greater pavement thicknesses than where designs using practical
load spectra are employed.

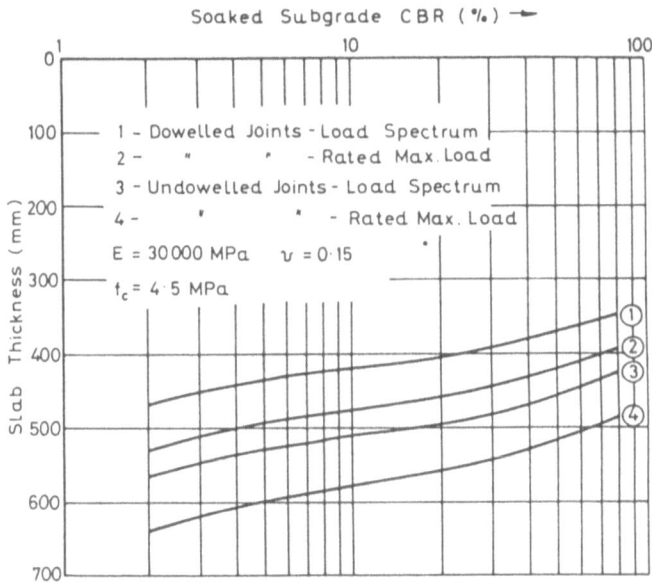

Figure 3: Results of the Industrial Concrete Pavement
           Design Program.

The data plotted in Fig. 3 can be obtained in about 30 minutes
using an IBM PC micro-computer fitted with a 8087 math coprocessor. This illustrates the speed with which a designer can compare a range of design alternatives without the need to make
unrealistic simplifying assumptions.

6. CONCLUDING COMMENTS:

This paper has shown that the boundary conditions applicable to
the mechanistic design of both roads and industrial pavements
realistically require the consideration both of complex loading
spectra and of a range of temperatures. Typically, to provide
even an idealised model of these inputs requires that several
hundred load/temperature combinations be studied. For each of
these combinations it is necessary to compute the stresses and
strains at a number of critical locations throughout the pavement. To accomplish this at reasonable speeds requires the use
of simplified and, therefore, approximate methods of stress

analysis. Nevertheless, it has been demonstrated that subject to careful calibration of the models, procedures suitable for rapid iteration on microcomputers can be developed. These procedures can be made to yield results than, on average, closely approximate the solutions obtained from more exact and elaborate main-frame computer analyses.

As illustrations of the potential of microcomputer pavement design methodology selected results obtained using procedures developed by the author have been reported. Early versions of these methods have already been successfully applied around the world in a number of major industrial pavement designs. These include the pavements at the Massey Coal Terminal, Newport News, Virginia, and at the Canadian-Pacific Intermodal Facility, Edmonton, Alberta. Experience shows that it is practical to rapidly examine a range of design alternatives. Consequently, a designer can concentrate on the choice of an optimal combination of materials and thicknesses rather than to become distracted by the complexities of the loading conditions and material characteristics. This coupled with the speed and other advantages of a computer-based design methodology is likely to promote a much wider use of microcomputers in pavement design and, thereby, to complement the advances common elsewhere in transport technology.

REFERENCES:

1. Barber, S.D. and Knapton, J. (1979) Port Pavement Loading. The Dock and Harbour Authority, April.

2. Barker, W.R., Brabston, W.N. and Chou, Y.T. (1982) A General System for the Structural Design of Flexible Pavements. Proc. 5th Int. Conf. on Stl. Design of Asphalt Pavements, Delft.

3. Booker, J.R., Small, J.C. and Balaam, N.P. (1984) Application of Microcomputers to the Analysis of Three-Dimensional Problems in Geomechanics. Engg Software for Microcomputers (Ed.Schrefler, Lewis and Odorizzi) Pineridge Press, Swansea.

4. British Ports Assn. (1982) The Structural Design of Heavy Duty Pavements for Ports and Other Industries. BPA.

5. Claessen, A.I.M., Edwards, J.M., Sonimer, P. and Uge, P. (1977) Asphalt Pavement Design - The Shell Method. Proc. 4th Int. Conf. on Stl. Design of Asphalt Pavements, Ann Arbor.

6. Croney, D. (1977) The Design and Performance of Road Pavements, D.E. Dot HMSO.

7. Dickinson, E.J. (1984) Bituminous Roads in Australia, Austn. Road Research Board.

8. Freeme, C.R., Meyer, R.G. and Shackel, B. (1981) A Method for Assessing the Adequacy of Road Pavement Structures using a Heavy Vehicle Simulator. Proc. IRF, Stockholm.

9. Hadley, W.O., Hudson, W.R. and Kennedy, T.W. (1972) A Comprehensive Structural Design for Stabilised Pavement Layers. Res. Report 98-13 Centre for Hwy. Res., University of Texas, Austin.

10. Larsen, T.J., Nussbaum, P.J. and Colley, B.E. (1969) Research on Thickness Design for Soil-Cement Pavements P.C.A. Dev. Dept., Bulletin D142.

11. Maree, J.H. and Freeme, C.R. (1980) Structural Design Methods and Pavement Types, NITRR, CSIR, Pretoria.

12. National Institute for Transport and Road Research (1977) Stresses and Strains in Layered Systems, CHEV 4. Computer Prog. P4, NITRR, CSIR, Pretoria.

13. Odemark, N. (1949) Investigation as to the Elastic Properties of Soils and Design of Pavements According to the Theory of Elasticity, State Road Institute, Stockholm.

14. Packard, R.G. (1973) Design of Concrete Airport Pavement. PCA Skokie.

15. Pickett, G. and Ray, G.K. (1951) Influence Charts for Concrete Pavements. Trans. ASCE, Vol.116.

16. Portland Cement Assn. (1966) Thickness Design for Concrete Pavements, PCA Skokie.

17. Potter, D.W. and Donald, G.S. (1984) Revision of NAASRA Interim Guide to Pavement Thickness Design. Proc. 12th Conf. Aust. Road Research Bd., Hobart.

18. Shackel, B. (1985) Evaluation, Design and Application of Concrete Block Pavements. Proc. 3rd Int. Conf. on Concrete Pavement Design and Rehabilitation, Purdue University.

19. Ullidtz, P. (1974) Some Simple Methods for Determining the Critical Strains in Road Structures. Doctoral Thesis, Tech. Univ. of Denmark.

20. Ullidtz, P. (1981) Prediction of Pavement Response Using Non-Classical Theories of Elasticity. Inst. for Roads and Traffic, Denmark.

21. Verstraeten, J., Veverka, A.V. and Francken, L. (1982) Rational and Practical Designs of Asphalt Pavements to Avoid Cracking and Rutting. Proc. 5th Int. Conf. on Stl. Design of Asphalt Pavements.

ANALYSIS OF LATERALLY LOADED PILE GROUPS

C F Leung and Y K Chow

Department of Civil Engineering
National University of Sing~pore, Singapore.

INTRODUCTION

Various theoretical methods are available for the analysis of single laterally loaded piles, namely the modulus of subgrade reaction approach (or p-y method), finite element method and the boundary integral method. However, many practical applications require the use of piles arranged closely in a group rather than in isolation (for capacities and/or deformation considerations). Examples include foundations for bridge piers, abutments and offshore steel platforms where the pile groups are subjected to a substantial amount of laterally forces in addition to vertical loads. The modulus of subgrade reaction approach, by defination is unable to deal with interaction problems. The pile-soil-pile interaction problem may be dealt with using the finite element method, but the three-dimensional nature of the problem makes the method expensive, even for the assumption of linear elastic soil behaviour. The boundary integral approach comes into its own when dealing with three-dimensional linear problems involving semi-infinite domains. Extensions of the approach to deal with layered soil and soil nonlinearity increase the complexity and cost of the analysis, making it less attractive for routine applications, particularly at the preliminary design stage.

In this paper, a "mixed" approach is described for the analysis of laterally loaded pile groups; in which the individual pile response is modelled using the modulus of subgrade reaction approach (or p-y method) while the pile-soil-pile interaction is modelled using a simplified boundary integral method based on Mindlin's solution. This approach is an extension of the method proposed by Chow (1985) for the analysis of vertically loaded pile groups, and the principles involved are very similar. Although the theoretical model necessarily involves a number of simplifying assumptions, it provides a cost effective method for the analysis of pile groups. Practical applications of the method are illustrated by studies on field tests on pile group.

METHOD OF ANALYSIS

Single Pile

The conventional approach of "p-y" curves is employed to analyse the lateral response of single pile. These curves are based on bending moments, deflections and rotations measured along the pile at different load levels from tests on instrumented piles in various types of soil by Matlock (1970) and Reese et. al. (1974 and 1975). Empirical relationships between pile-soil pressure p and deflection y were hence proposed for various soil types and the procedure has been summarised by Reese and Desai (1977).

In the present work, the pile is modelled by means of a number of one-dimensional discrete beam elements while the soil resistance is represented by non-linear "spring" at the nodal point of the element mesh. Values of soil spring stiffness at each nodes are evaluated from the appropriate p-y curves for each pile-head displacement increments.

Pile Group

The pile-soil-pile interaction problem may be represented schematically in Figure 1 for a pile group embedded in a homogeneous, isotropic elastic half-space. The interaction is modelled using a simplified boundary integral method based on Mindlin's (1936) solution. This technique has been described by Chow (1985) in detail for the analysis of vertically loaded pile groups. The procedure is summarised as follows. As the loading of any pile will induce additional displacements of other piles in the group, a general displacement-load relationship for all the pile nodes in the group may be expressed by the following matrix equation

$$\{w\} = [F] \{P\} \tag{1}$$

where $\{w\}$ = displacement vector for the n nodes in the pile group; $[F]$ = flexibility matrix; and $\{P\}$ = load vector. Equation 1 is inverted to give the load-displacement relationship of the soil for the pile group

$$\{P\} = [K_s] \{w\} \tag{2}$$

where $[K_s] = [F]^{-1}$ = soil stiffness matrix. The soil stiffness matrix is then assembled together with the individual pile stiffness matrices to yield the total stiffness matrix.

$$[K] = [K_s] + [K_p] \tag{3}$$

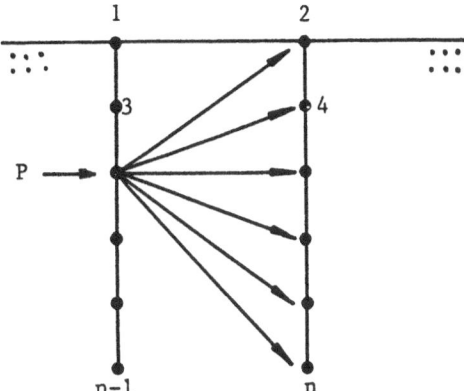

Figure 1 Model of Pile-Soil-Pile Interaction

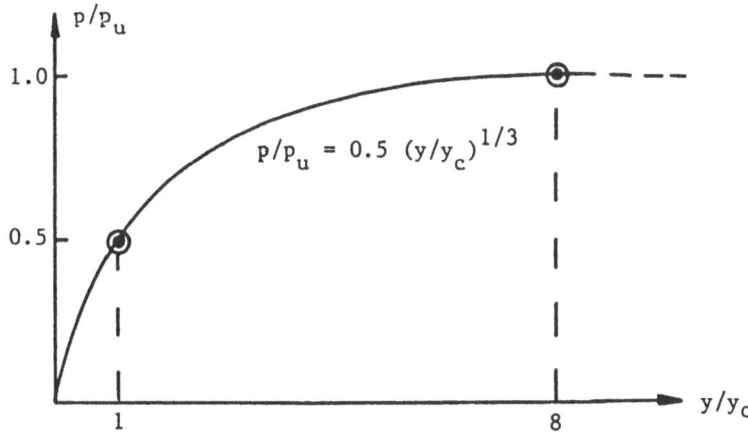

Figure 2 "p-y" Curve for Single Pile under
Short Term Static Lateral Loading
in Soft Clay (after Matlock, 1970)

where $[K_p]$ = assembled stiffness matrix of the piles. The complete load-displacement relationship for the pile group system is

$$\{P\} = [K] \{w\} \tag{4}$$

For each displacement increment, equation 4 can be solved to obtain the incremental load of individual pile in the group. In the present approach, the manner in which the soil inhomogeneity and soil nonlinearity is taken into account follows closely the work of Chow (1985) for vertically loaded pile groups.

COMPARISON WITH FIELD TESTS

Tests of Matlock, et al (1980)

The approach described in this paper was used to analyse field loading tests conducted on groups of piles embedded in soft clay. The field experiments were reported by Matlock, et. al. Part of the work include static loading tests on a single pile and on a five-pile circular group. Deflections were enforced at two elevations by a special loading device to simulate pile-head restraints typical of offshore structures. Measurements were made of the total load and deflection of the groups, and the load, shear and bending moment in individual piles. All piles had an external diameter of 168 mm, with a wall thickness of 7.1 mm, and they were driven to a penetration depth of 12.2 m below the bottom of an excavated pit. The centre-to-centre spacing was 3.4 pile diameter for the five-pile group. Based on Bogard and Matlock's (1983) reported values, the undrained shear strength of the soil at the mudline was taken as 10.4 kPa, increasing to 20.7 kPa at a depth of 0.3 m, thereafter increasing with depth at a rate of 1.743 kPa per m. An empirical correlation factor of E_s/C_u = 250 for Young's Modulus in soft clay (see for example Poulos and Randolph, 1983) was used in the present analysis. An approximate linearly increasing shear modulus profile was hence deduced with G = 875 kPa at the mudline, increasing at a rate of 210 kPa per m with depth. The Poisson's ratio of the soil was taken as 0.5.

Single Pile

The generation of p-y curves along the pile length for soft clay is based on the recommendations suggested by the American Petroleum Institute (1982). The guidelines follow closely with those postulated by Matlock (1970) who performed tests on instrumented full-scale pipe piles. The characteristic shape of the p-y curve, as shown in Fig. 2, is defined by a power function that was fitted to the shape of the experimental p-y curves:

$$p = 0.5 \, p_u \left(\frac{y}{y_c}\right)^{1/3} \tag{5}$$

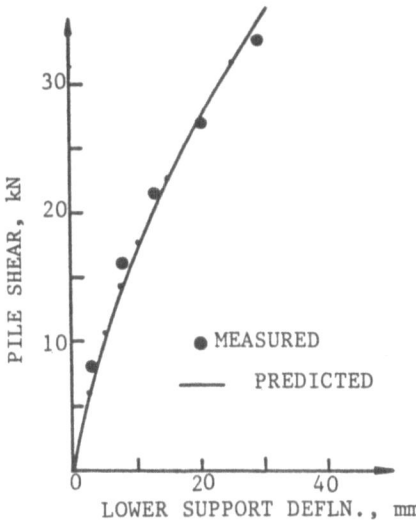

Figure 3   Pile Head Load-deflection Curve
for Single Pile

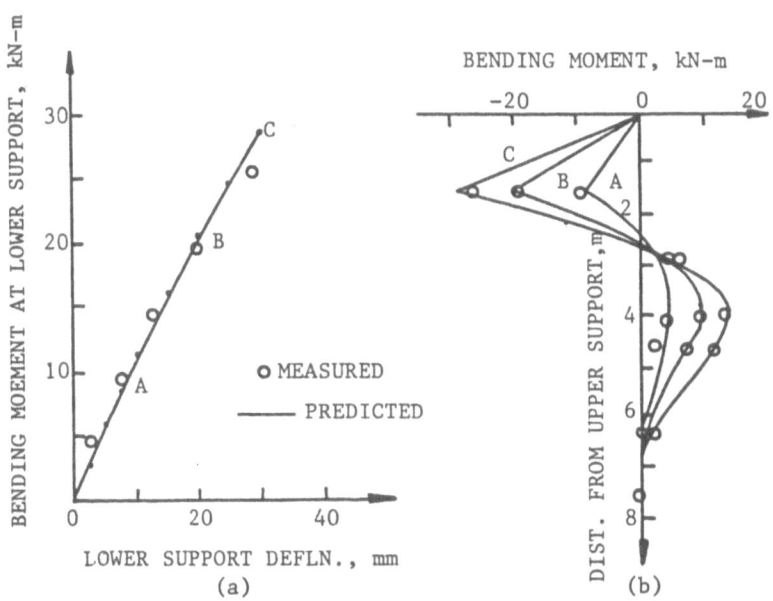

Figure 4   Development of Bending Moment in Single Pile

where p_u = ultimate soil resistance; and y_c = reference deflection of the pile. The variation of the ultimate soil resistance with depth is given by

$$p_u = N_p \, c_u \, D \qquad (6)$$

where c_u = undrained shear strength of soil at depth x; D = pile diameter; and N_p = ultimate lateral soil resistance coefficient = $3 + \sigma'/c_u + J \, x \, /D \leqslant 9$.

It is noted that σ' is the effective overburden pressure at depth x below ground surface and J is an empirical constant with an approximate value of 0.5 for soft clay.

The reference pile deflection is given as

$$y_c = 2.5 \, \varepsilon_c \, D \qquad (7)$$

where ε_c = major principal strain at one half the maximum deviator stress in a UU triaxial compression test (an approximate value of 0.01 was used in the present analysis).

The behaviour predicted for the single pile under static lateral loading is compared with the experimental data in Figures 3 and 4. There is generally reasonable good agreement between the two results. It is observed that the predicted initial pile-load response is slightly smaller than the measured response. It is believed that the agreement may further be improved if the shear strength of the soil is allowed to vary slightly. The good agreement between the computed and measured variations in bending moment along the pile (Figure 4b) suggests that both the magnitude and distribution of lateral resistance with depth was satisfactory. In addition, the shapes of the resistance-deflection curves should also be reasonably close to those measured during the field tests.

5-Pile Group
The predicted behaviour of the 5-pile group is compared to that measured in the field tests in Figures 5 and 6. Again reasonable good agreement was obtained between the two results for pile-head shears and bending moments. Similar agreement was also noted for the variation of bending moment along the pile.

A comparison between the behaviour of a 5-pile group and a single pile can be obtained by means of the group deflection ratio R_p which is defined as the ratio of the average group lateral deflection to the single pile lateral deflection at the same load per pile. Figure 7 shows that the predicted R_p is in remarkably good agreement with the corresponding values developed during the experiments.

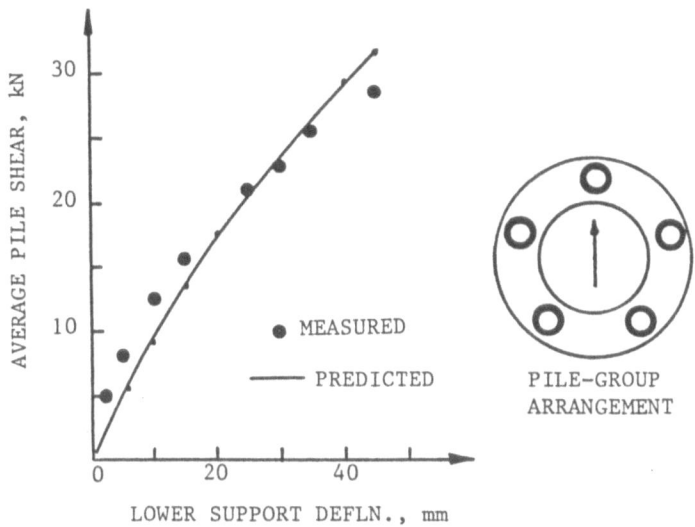

Figure 5   Average Pile Head Load-Deflection
Curve for 5-Pile Group

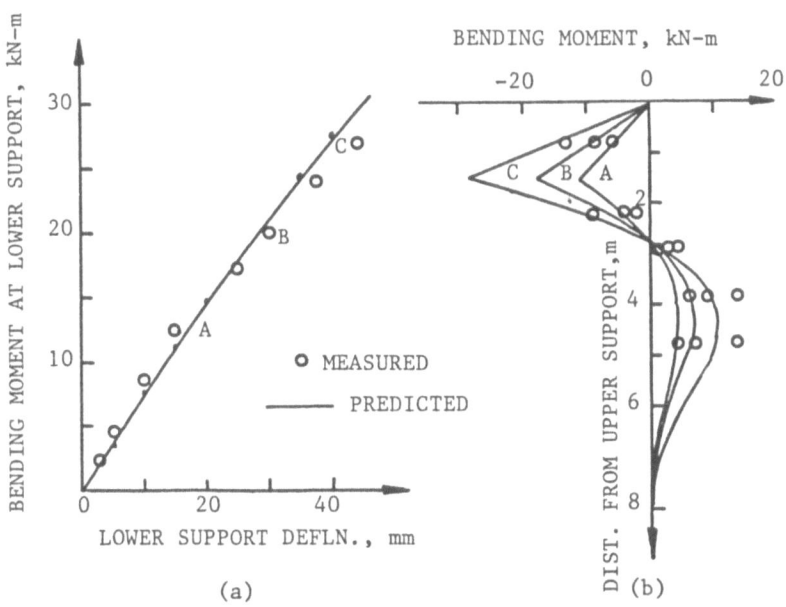

Figure 6   Development of Bending Moment in 5-Pile Group

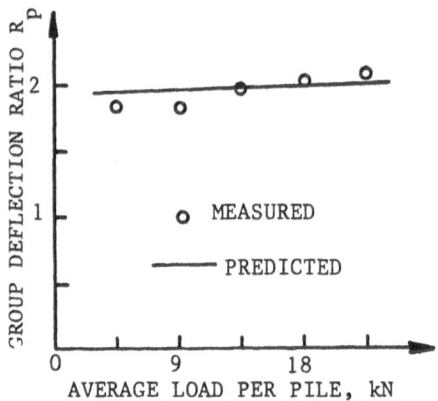

Figure 7   Comparison between Measured and
           Predicted Group Deflection Ratio

CONSLUSIONS

A semi-theoretical approach has been described for the
computation of load-deformation behaviour of laterally loaded
pile groups. This approach which was previously used for the
study on vertically loaded pile groups (Chow, 1985) is
demonstrated to be equally applicable to pile groups subjected
to lateral loads. Reasonably good agreement was observed when
comparing the theoretical solutions with existing field data
on laterally loaded pile group in soft clay. In general, the
lateral pile head shears and moments, and the resistance-
deflection shape along the pile may be obtained within
reasonable accuracy and economically using the proposed method
which basically employs one-dimensional discrete elements.
The cost of computing time is hence far smaller than those
employing three-dimensional finite elements. It is believed
that the solution procedure can be extended to deal with
cyclic loading condition provided the appropriate p-y curves
are used in each individual case.

REFERENCES

American Petroleum Institute (1982) Recommended Practice for
Planning, Designing, and Constructing Fixed Offshore
Platforms. API RP 2A, Washington, D.C., Thirteeth Edition.

Bogard, D., and Matlock H. (1983) Procedures for Analysis of Laterally loaded Pile Groups in Soft Clay. Proc. of Conference on Geotechnical Practice in Offshore Engineering, Austin, Texas, 499-535.

Chow, Y.K. (1985) Analysis of Vertically Loaded Pile Groups. Int. J. Numerical and Analytical Meth. in Geomechanics, (Paper in print).

Matlock, H. (1970) Correlation for Design of Laterally Loaded Piles in Soft Clay. Proc. 2nd Annual Offshore Technology Conference, OTC 1204, Honston, Texas, 577-594.

Matlock, H., Ingram, W.B., Kelley, A.E., and Bogard, D. (1980) Field Tests of the Lateral Load Behaviour of Pile Groups in Soft Clay. Proc. 12th Annual Offshore Technology Conference, OTC 3871, Houston, Texas, 163-174.

Mindlin, R.D. (1936) Force at a point in the interior of a semi-infinite solid. Physics, 7, 195-202.

Poulos, H.G., and Randolph, M.F. (1983) Pile Group Anslysis : A Study of Two Methods. J. of Geotech. Engr., 109, 355-372.

Reese, L.c., Cox, W.R. and Koop, F.D. (1974) Analysis of Laterally Loaded Piles in Sand. Proc. 6th Annual offshore Technology Conference OTC 2080, Houston, Texas, 473-483.

Reese, L.C., Cox, W.R. and Koop, F.D. (1975) Field Testing and Analysis of Laterally Loaded Piles in Stiff Clay. Proc. 7th Annual Offshore Technology Conference, OTC 2312, Houston, Texas, 671-680.

Reese, L.C., and Desai, C.S. (1977) Laterally Loaded Piles. Ch. 9 of Num. Meths. in Geot. Engr. Ed. C.S. Desai and J.T. Christian, McGraw Hill, New York.

GEOSHARE - A NEW DIMENSION IN GEOTECHNICAL/GEOLOGICAL DATABASES.

L.A. Wood, D.E. Wainwright, Dept. of Civil Engineering, Queen Mary College,
E.V. Tucker, Dept. of Geography and Earth Science, Queen Mary College, London.

1.0 INTRODUCTION

The Geotechnical/Geological database system "GEOSHARE" has been progressively developed over the last five years at Queen Mary College (Wood, et al 1981). The advantages of flexible computer manipulation of such data has been shown to be of advantage in the areas of
(i) Documentation
(ii) Stratum analysis and identification of formation boundaries (Day, et. al)
(iii) contouring and 3-D modelling (Halcrow et al, 1973);
(iv) the identification of channel and interfluve succession (Wood,et al, 1983).
To achieve clearer definitions, and to gain a better overall impression of the ground it was decided that more "data types" should be captured even though in some areas the data would be missing. After a thorough appraisal of the existing GEOSHARE, Fortran software, it was decided rather than attempting to upgrade the system further that it would be better to make a fresh approach taking account of recent advances in commercial database technology.

2.0 OBJECTIVES

Over the past two decades, numerous attempts to implement geotechnical/geological database's have been undertaken and on the whole each implementation can be judged a success if it achieved its design objectives. However, once the objectives are altered, a measure of successful implementation can be measured by the ease or difficulty encountered in altering the database to meet the new objectives. For one reason or another it has been seen how quite successful database systems have been abandoned and redesigned from scratch (Stauft, 1972) once it has been realised that for perhaps efficiency, security or consistency measures a different approach would prove better. Several of the failures can be accountered for by the systems being developed around a few known applications leaving the user with no choice but to juggle around application dedicated files to extract the required information (Sutterlin, 1973). Other failures can be accountered for by consideration of the systems friendliness. Systems where large amounts of data coding, data validation, use of mnemonics etc have been required are usually only exploited by the keen and dedicated scientists (Cripps, 1978). Systems have also fallen short of being useful databases for other reasons such as unrealiable data (Stauft 1972), poor documentation, uneconomic to use and so on.

At Queen Mary College, the development of the Geoshare system has proceeded with the ultimate goal of five major objectives broadly described as follows

   (i)   Information
   (ii) Efficiency
   (iii) User-Friendliness
   (iv) Documentation
   (v)   Security.

As far as it is possible each of these objectives is being developed simultaneously, since all functions of a data system are interdependent, and grave mishaps could be encountered from consideration of each item in isolation.

2.1 Information

The information stored within any database system must be considered as the most valuable attribute, the whole justification for its existence. To gain the confidence of its users and to be acceptable, the stored data must be a complete valid reflection of the information in its original form. Geotechnical/Geological borehole records and the associated testing details are regarded as unique, peculiar and in some cases too diverse to take account of computer technology, due to the fact that the information is a mixture of numerically quantifiable data, textual descriptions, observations and to some extent opinions. Most previous attempts to establish a database have tried to some extent, to reduce all of this information to numerical form by pre-coding or even omittance of textual data (Halcrow et al, 1973) (Cripps, 1979). The reason for numerical form, has been because most systems have evolved around the usual work horse applications which have generally been written in efficient scientific languages (Jeffrey et. al.) to relieve the geologist/engineer of large quantities of calculations. However, to make the best use of all available information we must be able to manipulate both numerical and textual data in such a way as to gain full advantage from each type.

2.2 Efficiency

Efficiency is another required objective of any data system. This can be measured by several criteria and the criteria considered for the GEOSHARE system are broadly described as follows

(i)   Response
(ii)   Storage
(iii) Flexibility
(iv)   Recovery

Obviously most of these criteria are interdependent and a balance has to be drawn between cost and the necessity for all of these requirements to be met..

2.2.1 Response   For any Information system there must come a point where the delay in waiting for the required information renders the system unworthwhile or uneconomic. Depending on the use of the Information system this delay factor may vary from seconds to perhaps weeks. With GEOSHARE the delay factor at which the system would be considered a failure is perceived to be of the order of a few minutes. To be of any use as an information base the response time of the system realistically wants to be a small as possible. However other criteria have to be examined before we can decide upon an economic response time.

Response times can drastically increase as the amount of information within a system becomes increasingly voluminous. Also, the way that the data is physically stored can also affect response times especially if complex selection requirements are asked of the database. These considerations have previously led to the downfall of several Information systems.

2.2.2 Storage The advantages of any information system could be completely out-weighed by actual physical storage requirements. Operational costs and maintenance overheads could all undermine the advantages, if vast quantities of data are stored in an inefficient way.

2.2.3 Flexibility Once a database has been established the requirements of the users may change with time and different views of the data will then be required. A database system must be resilient to these changing applications and should demonstrate relative ease in accommodating even major restructuring.

2.2.4 Recovery: At any particular time during a systems use, a hardware or software failure may cause the system to "crash", leaving the database in an inconsistent state. This inconsistency may take the form of corrupt data, lost data or even deletion of parts of the database. To ensure the validity of the data at all time it is then essential that procedures to interrogate and verify the data after such instances are employed.

2.3 User-Friendliness
Normally, when vast amounts of data are being input as their main process, the task is performed by secretaries, data processing clerks, bank clerks etc. Therefore to increase the input rate of the data, and relieve the trained engineer/geologist of a mundane task, the system must guide the user through the process, validating the data, correcting errors or bringing them to the users attention (Buller, 1972, Heer and Bie, 1973). Also, the commands given should be simple enough to navigate around the various applications with no knowledge of detailed computer programming.

Another aspect of user-friendliness is 'multi-user' application. It must be possible for more than one user to access and use the database, even the same data, at any time for a system to be classified as friendly.

2.4 Documentation

For any computer application to function for any length of time, it is essential that some form of documentation is produced. With a database system where there are a large number of data types, record types and so on, it is easy for data formats to become out of date as changes are required. The consequences of data getting out of control are becoming all too clear after two decades of data processing within the commercial data processing environment with problems arising from incompatible coding systems, suspect results and so on.

2.5 Security

With the development of large information systems attention has been brought to the problem of computer security due to numerous cases of breach of privacy, fraud and general abuse by non-authorised users. Any viable information system must demonstrate a degree of security to gain its users confidence, in terms of safe-guarding data from corruption and in offering protection for privacy and confidentiality.

3.0 INFORMATION TECHNOLOGY

3.1 Database Management Systems

Data-processing technology has been developing over the years at a remarkable rate and has come of age with the advent of Information Technology. With the advances in Information Technology both upon software and hardware, data processing systems have moved away from the techniques widely used in the late sixties where information was stored upon sequential storage media (magnetic tapes). It was through the introduction of direct access media (disc), that database management theory first made an impact. Database management theory first came about to alleviate problems that had started to arise when vast amounts of information were being collected. Basically problems arose in the following areas:

    (i)   Waste of file store
   (ii)   Duplicated data
  (iii)   Differing data structures

(iv) Waste of programmer effort
(v)   Loss of data accuracy

Database technology basically solves these problems by capturing each item of data only once and then making it available to different users as they require it. Subsets of the data in the central database are made available to meet user requirements and can be controlled to give overlapping subsets of data to different users at the same time.

Within the computing world ´Database´ is heralded as the new technology, the fourth generation in computing, and is supported by a great deal of research. Basically there are three different approaches to database management at present: the Relational Approach, the Network Approach (CODASYL) and the Hierachy approach. Of the three, the Relational approach is potentially the most efficient and powerful database management system. However, the problem with the concept is to implement its efficiency on current hardware. The approach relies heavily upon an extensive duplication of ´keys´, involving overheads of storage and use, which it appears can only be reduced by better hardware, cleverer disc controllers than are presently available on the average commercial system. It is very much a concept for the future.

The Network approach has however received a great deal of attention, and of the commercially available Database Management Systems (DBMS´s) by far the most popular systems adhere to this approach. In 1967, a committee of top researchers in the field of Database theory was established, the objectives of which were the further development of data processing technology. This committee was later to become CODASYL, the product of which became the CODASYL approach to Database management, as recommended in mumerous reports between 1968 and 1975 (CODASYL COBOL 1976). There are many commercial implementations of this approach available, best known of which are Univacs DMS 1100, Honeywells IDS/2, Siemens UDS, and available on IBM and ICL machines Cullinanes IDMS (Olle, 1980).

3.2 Data Dictionary Systems (DDS)
A data dictionary system is defined as a

"repository" of data about data", also known as meta-data. The Data dictionary system stores and handles a computer systems data definitions. Instead of holding the actual data item occurrences, the DDS software will create and maintain descriptions of the data and can produce them in response to various interrogation demands. The advantages of using a DDS include: system conversion, validation, documentation, consistency checking, operational performance evaluation, the design of systems for security and privacy, software evaluation, automatic program generation and hardware planning.

3.3 Data Analysis

To get full advantage from any information system a correct and full Data Analysis should be performed. The purpose of Data Analysis is to provide a way of structuring the data, then testing its use before design, removing inconsistencies, planning confidentiality controls, in preparation for designing the logical database. Data Analysis methodology is centred on two fundamental principles (Rock-Evans, 1981). First, analysis must be undertaken of the study area separately from any considerations of the technology which may eventually be used to implement the system. Second, to design systems intended to share data between a variety of users, it is essential to analyse the data independently of the various applications which may use it. This avoids the trap of designing application dedicated files, tied to the specific problem the system is orientated towards solving.

4.0 GEOSHARE SYSTEM

To realise many of our objectives, the importance of using a powerful DBMS was only too clear. By choosing to use a CODASYL style DBMS, we not only have the advantage of portability but also the fact that the system is designed to a recognised standard (Olle, 1980). Also, many of the known CODASYL approach implementations support the DBMS with compatible data dictionary systems, demonstrating many high powered useful facilities. The DBMS and DDS together form the heart of the information system. By exploiting both their facilities the GEOSHARE system can boast the following features from their use alone.

   (i)   Multi-user
   (ii) Efficient storage
 (iii) Full recovery
  (iv) Flexibility
   (v) Documentation
  (vi) Security

After a comprehensive data-analysis of the information in the geotechnical/geological data field the variety of information capable of being stored is illustrated in Fig. 1. It has been essential to provide flexible input routines to account for the varying styles of Borehole records, and in the presentation of test results. Some test results may be presented as a single result value, whereas others may go partly or fully towards a complete set of measurements and readings. Also, results may be presented in a summarised statistical manner, giving result ranges means, etc. All of these styles of presentation as recorded on manual media have been provided for in the input routines. On input the system automatically validates most numeric data and verifies all textual information against an in-system self updating vocabulary. In this way any spelling mistakes are picked out and corrected before storage. Of the numeric data not validated prompts are given frequently for the input user to correct any mistakes. Once input, the data is stored as read with no coding or abbreviations. By the method of data analysis, storing the data as read, checking and validating it before storage, we have realized our Information Objectives. Once stored we are now in a position to manipulate both numeric and textual data to any required degree. The system interfaces with a graphics package, utilising a refresh graphics screen. On input, screen formats are produced to simulate blank borehole records, test charts etc. The screen prompt automatically moves to each input position, and the input procedure time is greatly decreased. Error messages and comments automatically appear in a reserved screen position for user-replies and corrections. To enable users with little experience of computers to use the system responsibly, and gain confidence, the system is menu-driven. It is possible to retreive any data from the system by navigating the menus and using perhaps four or five numerical replies. By combining the graphics interface with the menu-driven style, the automatic validations and verification, the system is considered to have achieved the objective of user-friendliness.

Sizing of the GEOSHARE system requirements has been undertaken, and for one borehole record with all associated information (water, sample and test details) has resulted in a need for 1.3 k bytes of storage. This requirement was estimated using a 60% packing density, meaning the database is assumed full when 60% full. However, after tests monitoring performance have been undertaken, it may be possible to increase the packing density to 80%, reducing a single borehore's requirements to 0.9 kbytes. Also this figure of 1.3 kbytes allows 2/3 of the storage space to accommodate the results from perhaps 20 full scale triaxial tests with pore water measurement and consolidation. As this is perhaps the extreme, the figure of 1.3 Kbytes/boreholes is an over estimation and in practice less space may be required than actually designed for, therefore using less space than estimated. However taking the figure of 1.3kb/borehole, it has been estimated that 45,000 BH records and associated data can be accommodated on one inter-changeable disc.

To increase efficiencies in response time the DBMS takes advantage of a recent development in hardware technology, that being the computer addressable file store facility. This facility has been developed to improve computer response times when very large files of data are being searched. Basically, the facility replaces central mainframe software activity by peripheral disc-channel hardware activity. Searching of the data is essentially executed by an intelligent hardware device and therefore relieves the computers main software processors of most of the work. Field trials have demonstrated that a reduction load on the main processor in the order of 90% to 99.74% has generally been experienced when the device was utilised. Response times have also been demonstrated to become faster by an order of magnitude of 60 times. Due to these developments in computer performance, it is now possible for new computer applications to be developed that previously would not have been either cost effective or timely. To give an example it has been estimated that the device is capable of searching through and selecting specified data from 45,000 borehole records in a little less than a minute. By utilising these recent developments in computer hardware technology the GEOSHARE system can offer the following facilities.

    (i)   Fast Response
   (ii)   Economic Storage capability.

5.0 APPLICATIONS

The applications developed to date fall into two categories.

    (i)   Reporting facilities
   (ii)   Analytical facilities

5.1 Reporting
The reporting facilities basically give listings of the data stored. Options include

    (i)   Boreholes in a defined area
   (ii)   Boreholes from a particular contract
  (iii)   Samples from a borehole
   (iv)   Test details of samples

5.2 Analytical
Sort routines have been developed that will retrieve data on three main options, these being

    (i)   Strata descriptions
   (ii)   Depth
  (iii)   Parametric details

5.2.1 Strata
Within a defined area the system will retreive certain strata detials specified by the user. The user specifies the main lithological unit, and then has the option to specify (for example) certain colour or colours that are required, colour(s) not required, and colours whose presence does not matter. The user can then go on to specify the same options for the following strata attributes.

    (i)   Hardness
   (ii)   Lithological modifiers
  (iii)   Fracturing
   (iv)   Stratification
   (v)   Texture

The details returned to the user will depend on his option choice but can be Top and/or bottom levels, and/or certain parametric details chosen by the user. For example, by specifing the general description of London Clay

GREY BROWN BLUE
FIRM STIFF
FISSURED
SILTY
CLAY

it is possible to output the London Clay surface, in an area or say obtain all undrained shear strengths for a 3-D model of the London Clay.

5.2.2 Depth By specifying a certain level either strata details or parametric values can be obtained providing contours for perhaps the moisture content or strength over a site at a specific level (Hanna, 1966).

5.2.3 Parametric details If one is only interested in specific parameter variations over a site, by specifying the area, details are obtained for 3-D models of various attributes of all the strata in an area.

5.3 Application Usage
At present the ´analytical´ applications are being used as tools in two research projects. First the ´Natural Environmental Research Council´ is funding work to investigate the theoretical land form systems derived from Glaciation as compared to real life data. Second, work funded by the ´ Science and Engineering Research Council´ is trying to establish a soft clay test site with optimum conditions and have made use of the system.

6.0 COST

The GEOSHARE system runs upon the ICL 2980 mainframe computer at Q.M.C. interfaced to Tektronix screens. Outside of the University research environment rental of the software would come to approx. £30,000 per annum. However, many organisations already possess both the hardware and software requirements to run such a system, of which Southern Water Authority and Severn Trent Water Authority both have financial accounting systems utilising the technology. With the recent advances in communications it is feasible though to establish one main database, and because of the security measures offered, allow each user to manage their own sub-set of data within the main database.

7.0   FUTURE DEVELOPMENTS

At present, the system is running in a trial environment.   Once fully established it is intended to refine and develop the system, in order that future user requirements may be met. In the longer term, it is intended to create a ´hybrid´ system (Worden, 1984), by the inclusion of an expert shell system designed to be compatible with and to interact with the database system (Keen, 1984).   It is intended to use the expert system to interpret the textual data representing descriptions and remarks, giving expertly   interpreted   three   dimensional cross-sections and enhance, the ability to recognise discontinuities; to make statistical judgement and pattern recognition over vast quantities of data (Benson, 1984).   Before that however, it is intended to make full use of the Ordnance Survey´s map digitizing program (Ornance Survey, 1982).   These digitized maps, held on magnetic tape will be used as overlays for such requirements as plotting the positions of boreholes in an area.   It is also hoped that developments in the field of computer optical character recognition will dramatically improve data input.   One company has already started development of a system to handle the image of an A4 sheet of paper with typed, handwriten or printed text, tables and line diagrams.

REFERENCES

(1) Benson, I. (1984)   PROSPECTOR: An expert system   for   mineral   exploration.   SRI International.   Proc.   seminar   on   Computer Assisted Decision Making, Tara Hotel, 2-4 April ´84.

(2) Buller, J.V. 1972   Development of the Saskatchewan Computerised Well Information System 1964-1971.   24th Int. Geol. Congr. Section 16, 97-102.

(3) Cripps, J.C. 1978 Computer Storage of Site Investigation Data for Tunnelling.   Computer

methods in tunnel design. I.C.E. London, (a) 183.

(4) Cripps, J.C. 1979 Computer Geotechnical Data Handling for Urban Development, Bull. Int. Ass. Eng. Geol., No 19 290.

(5) CODASYL COBOL 1976 Journal of Development.

(6) Day, R., Tucker, E.V., Wood, L.A., 1983 The Computer as An Interactive Geotechnical Data Bank and Analytical Tool. Proc. Geol. Ass. 94(2) 123-132

(7) Hanna, T.H. 1966 Engineering Properties of Glacial-Lake Clays near Sarnia, Ontario, Ontario Hydro Research Quarterly Vol. 18, No. 2.

(8) de Heer. T. and Bie, S.W. 1973 Particular Requirements for the Dutch WIA Earth Science Information System in ´Computer-based systems for Geological Field Data´, Geological Survey ·of Canada paper 74/663 pp 78-81.

(9) Jeffrey, K.G. and Gill, K.M. 1973 G-EXEC : A Generalised FORTRAN System for Data Handling in ´Computer-based systems for Geological Field Data´, Geological Survey of Canada paper 74-663 pp 89-61.

(10)Keen, M.J.R. Expert systems in ICL. Proc. Seminar on Computer Assisted Decision Making, Tara Hotel 2-4 April ´84

(11)Olle, T.W. 1980 The CODASYL approach to Databse Management, John Wiley and Sons, 1980.

(12)Ornance Survey 1982 Information Leaflet - Digital Mapping, No. 48 August 82

(13)Rock-Evans, R. 1981 Data Analysis, IPC Business Press, 1981

(14)Stauft, D.L. 1972 Evolution of Data Systems in Oil Exploration, 24th Int. Geol. Congr. Section 16, 181-188

(15)Sutterlin, P.G. 1973 U.W.O. Safras System in ´Computer-based systems for Geological Field Data´, Geological Survey of Canada paper

74/663   pp 62-67

(16 Sir William Halcrow and Partners and Mott, Hay and Anderson (1973), Roads in Tunnel, GLC, London, 1973.

(17)Worden Logica, R. Blackboard Systems. Proc. Seminar on Computer Assisted Decision Making, Tara Hotel, 2-4 April ´84.

(18)Wood, L.A. Tucker, E. V. and Day, R.B. 1982 Geoshare - The development of a Databank of Geological Records. Engineering Software II, Pentech Press, in Advances in Eng. Software 1981, 4, 13, 6.

(19)Wood, L.A., Tucker, E.V. and Day, R.B. 1983 The Further Development of a Geotechnical Geological Database. Adv. Eng. Software, 1983, Vol. 5, No. 2 pp 81-85.

INFORMATION STORED

ORDNANCE SURVEY MAP DATA:

VOCABULARY: Self updating vocabulary with a Capability of holding any number of words

USER INFO: Information of users containing name, address and password

CONTRACT DETAILS: Details of contract number, client, contractor and testing laboratory

BOREHOLE DETAILS: (1)

Details contain grid-reference; ground-level-drilling methiod; location- starting; finish date; diameter: *remarks

TRIAL PIT DETAILS: (1)

Details contain grid-reference; ground level; excavation method; starting date; location; pit dimensions: * remarks

INSITU TEST DETAILS: (1)

Details contain grid; reference; ground level; location; starting date; name of tests: *remarks

GROUND WATER DETAILS: (1) Details contain casing depth; date and time of recording; depth to groundwater: *remarks

STRATA DETAILS: (1) Details contain strata thickness; depth; geological formation: *strata description:

INSITU TESTING DETAILS: (1)

Details contain full test details for 6 types of insitu tests: *remarks:

SAMPLE DETAILS: (1)

Details contain sample reference-no.; type of sample; top and bottom depths; *sample description

INSITU BOREHOLE TESTING DETAIL (1) (2) (3)

Details contain name of insitu test; depth of test; measured results (3):   * remarks

LABORATORY TESTING DETAILS: (1)

Detains contain name of test; full test details for 20 types of result or laboratory tests: * remarks

* There is no limit to the length of any Remarks and Descriptions:

(1) There is no restriction on the number of strata within a B/H or similarly, the number of samples from a particle strata etc.

(2) Suitable for continuous Well;logging;

(3) Data from sequential media (paper tape, magnetic tape) would ideally be fed in as batch imput

Figure 1.

4. ENGINEERING APPLICATIONS OF
   EXPERT SYSTEMS

ENGINEERING AND SCIENTIFIC SOFTWARE CONTROL STANDARDS
MODELLING: AN INTERACTIVE EXPERT SYSTEM USING
TELECOMMUNICATIONS

Avi Rushinek
Sara Rushinek

University of Miami, Coral Gables, FL U.S.A.

ABSTRACT

Selecting an Engineering and Scientific software (ES) system
is an arduous process. The overall satisfaction derived from
this type of system depends on many variables. This study
analyzes the influence of a number of ESS predictor variables
on overall satisfaction. This study confirms the theories
that suggest that ESS ease of operation, computer reliability
and tech support - trouble shooting are the major determinants
of overall computer user satisfaction.

INTRODUCTION AND OVERVIEW

The information obtained by user ratings of Engineering and
Scientific Software (ESS) systems, could be very useful to
ESS buyers and sellers who would like to see some type of
rating scale about these systems before deciding which type
of system to buy or sell (Datapro, 1984).

Traditionally, buyers or sellers who are interested in
evaluating overall user satisfaction of a potential new ESS
system have two options. One option is to study the technical specifications of the different ESS systems and their
respective user satisfaction reports. The disadvantages of
this option are that the buyer or seller may not have the time
nor the technical expertise to understand the specifications.
Moreover, many user satisfaction studies of ESS systems are
often incomplete, vague, inaccurate, subjective, ambiguous,
non-quantitative, and/or most importantly, too narrow to be
statistically generalizable (Bilbrey and House, 1981; Turney
and Laitala, 1976).

Another option that is available to buyers and sellers

of ESS systems is to hire consultants who can understand the technical specifications, discount inaccuracies and subjective judgments of trade publications, and most importantly use generic information to make suggestions custom-tailored to a particular installation. The main disadvantage of this option is that such experts are hard to come by, disruptive to the normal operation and rather expensive (Grueberger, 1981).

Traditionally, there has been a clear distinction among micro, mini, and mainframe computers. The power and capabilities of ESS systems have improved over the years; memory costs have gone down and performance distinctions between different systems have blurred (Sample, 1981). This challenges the traditional size distinction and its implications for evaluations. Thus, it is important to investigate if the traditional characteristics of size (whether it is a micro, mini, or mainframe) actually do have an effect on user satisfaction.

Choosing the right ESS from the bewildering array of systems, manufacturers and different configurations of components can be a frustrating and expensive experience for buyers and vendors alike (Barcus and Boer, 1981). If the parameters of ESS users satisfaction were known, then users' satisfaction could be maximized and the frustration level could be reduced, or at least controlled.

Buyers of ESS systems should look at advantages and/or disadvantages in cost, ease of operation, system reliability, and vendor reliability, such as established vending firms vs. newer and smaller firms (Cheney, 1979). Accordingly, the cost should be one of the most important determinants of user satisfaction. This may also suggest the inclusion of criterion variables that indicate the popularity of the vendor. Other such criterion variables include the number of systems, their average useful life, and the number of users that are using these systems.

In addition, due to the rapidly changing technology, management must be willing to commit time to the conversion of an outdated system. Thus the ease of conversion should be included in the study as an important determinant of user satisfaction. One can hypothesize that the easier the conversion process, the more satisfied the ESS users should be. Comparison of ESS systems should be done in the areas of support, service, ease of operation, compatibility and reliability of the computer, peripherals, compilers, and assemblers, as well as the cost of purchase and operation (Farmer, 1981).

The importance of having a written contract with the vendor has been discussed in the literature (Brandon, 1980). A thorough contract should cover reliability, performance, operating system compatibility, effectiveness, training, costs, and trouble shooting. Some studies cite maintenance, service, education, and documentation as the top concerns of ESS system users. Applications availability and reliability have been next highly rated with price being the most important criterion after that (Rosenfeld, 1980).

Some research reports that user support in terms of education and documentation seems to affect user satisfaction. Scannell (1982) has cited that users find software and support to be major problems. However, complaints that the computer industry does not provide adequate training, documentation, and manuals for users have been rebuffed by industry representatives (Lean et.al., 1983).

Questions have been raised about the effectiveness and responsiveness of traditional system maintenance services (Allerton, 1983; Howard, 1983). The issue of centralization vs. decentralization concerning maintenance contracts has been addressed (Linzey, 1983).

Another issue that may affect ESS user satisfaction is the method of acquisition. According to Kelly (1980) the impact of buying or leasing may be substantial.

ENGINEERING AND SCIENTIFIC SOFTWARE PRACTICE AND THEORY

Both practitioners and theorists have been struggling with various aspects of ESS systems. Some practitioners dealing with computers have raised perplexing questions (Computerworld, 1983; Bresnen, 1982; Business Week, 1982).

Choosing an appropriate ESS system and recognizing its limitations have been among the most difficult and confusing tasks for the Engineering Profession (Kull, 1984; Shoor, 1980) A better understanding of the determinants of ESS user satisfaction could not only partially answer the above questions, but could also facilitate the task of choosing a system and recognizing its limitations.

Theorists have wrestled with the problems from a more scientific point of view. They have examined the demise of help for utility planning engineers (Public Utilities Fortnightly, 1981), evaluated computer graphics in central office engineering (Smith and Strand, 1982), and investigated critical success factors in engineering firms (Scott, 1983).

Meanwhile, strides have been made in developing expert systems and artificial intelligence (ESAI) as they relate to ESS (Green, 1983). However, as in previous studies, the issue of user satisfaction has not been fully addressed. Therefore, the principal objective of this study is to supplement previous studies, and to integrate the issue of user satisfaction into the ESAI model, while building upon prior theoretical work.

EXPERT SYSTEM BASED ON ARTIFICIAL INTELLIGENCE (ESAI) DESIGN

A less traditional option is to use an expert system to aid in the evaluation of overall user satisfaction. Expert systems (ES) are specialized decision aids, which provide quantitative and qualitative analysis, probabilistic estimates and their respective explanations (Stefik et.al., 1982). Artificial intelligence (AI) has facilitated the use of ES through computerized data bases, which are frequently being updated (Duda and Gasching, 1981; Wong and Mylopoulos, 1977). Although AI has been applied to the configuration of computer systems (McDermott, 1982), it has not yet been applied to evaluating user satisfaction of ESS systems. The stated objective of the present study is to apply ES and AI to user satisfaction of ESS systems.

Using AI combined with a computerized expert system can help resolve some of the aforementioned problems. Accordingly a summary of industry statistics is placed in a computerized database, which is accessible to users, buyers and sellers by telecommunication networks (Kleinrock, 1982). An integrated telecommunication network can be composed of multiple local area networks (LAN) connected to each other via hard-wired, radio, microwave, terrestrial, and satellite communication.

A diagnostics program interrogates the users about the status of their ESS system. For each installation, the system accumulates user responses and saves them in a data base. Subsequently, this diagnostics program weighs the responses of the local users, and compares them to the industry standards (stored in the data base). Based on these comparisons, the diagnostics program can make specific technical and non-technical recommendations, which are based on industry statistical standards, but are also custom-tailored to a specific ESS installation.

The design of such an expert system basdd on artificial intelligence (ESAI) is one of the stated objectives of the present study. To accomplish this task, several steps are taken. First, the theories of consumer satisfaction are reviewed to identify and define the determinants of user satisfaction. These determinants along with the overall user satisfaction are quantified through a survey questionnaire. Then these determinant scores are regressed against the satisfaction scores to compute their respective weights and the industry averages in the model. Subsequently, the model is tested statistically to decide whether it is generalizable to the population of ESS users. Finally, if it is generalizable, it can then be incorporated into an interactive online ESAI model

Such an ESAI model compares industry standards (averages) to the averages of an individual ESS installation. It accumulates the responses of all the users in an installation. It then identifies the ABS system weaknesses, rank-orders, and proposes priorities and recommendations for improvements in an audit trail report. In addition, it also updates the data base, measures the adherence to previous recommendations and issues a progress report. This report indicates the gains or losses in overall ESS user satisfaction compared to the entire industry.

The above literature review sheds some light on the importance of different criterion variables and their consideration in the ESAI for ESS systems. System rating information could be a very useful tool to managers who are designing the acquisition of ESS systems, as well as to vendors, who must decide which systems to develop, market, and/or support.

Measurement of system ratings is quite complex and requires a selective of various criterion (independent) variables. It also requires an analysis of these variables to determine how they are related to one another. This paper describes the results of a system rating study in which the users were asked to respond to many questions. These questions (independendent variables), based on the literature, are the primary determinants of overall user satisfaction (dependent variable).

The overall user satisfaction is related to these ESS variables with the use of multiple regression analysis. This analysis is the basis for the design of an expert system (ESAI) for forecasting user satisfaction in a specific computer installation. ESAI can compare the current user satisfaction to industry standards, past levels of satisfaction and desirable future levels of user satisfaction.

SURVEY METHODOLOGY AND DATA COLLECTION

This survey was based on results received from questionnaires mailed to a very carefully controlled nth sampling from randomly drawn subsets of computer user lists. A total of 15,218 questionnaires were sent to computer users. The specific subsets were identified and qualified by a panel of experts. In an effort to improve the response rate, and thereby increase the statistical validity, the users were contacted twice; a first request was followed weeks later by a second request. The response rate was 32%, representing 4,597 users, who responded to 4,870 questionnaires (some users evaluated more than one computer model).

Judges invalidated 379 responses, including 179 users who rated two different computers at the same time; another 43

users rated more than two different systems simultaneously.
Datapro (1984) batched the remaining 4,448 valid returns by
vendor, model, users, and computer types [mainframes or plug
compatible mainframe computers (maxis), minicomputers and
small business computers (minis), and desk-top personal and
microcomputers (micros)] as follows:

	Maxis	Minis	Micros	Total
Users -------------	1,919	2,192	337	4,448
Computers ---------	67	93	19	179
Vendors -----------	10	28	17	55

Each questionnaire allowed the user to rate one system.
The recipient was enocuraged to reproduce the form if he/she
wished to rate more than one system. For each system the responses were averaged and recorded. Labels were used as initial validation vehicles and for identification and elimination of duplicate returns. Recipients were asked to summarize
their experiences with the systems currently being used and to
answer questions about them.

When returns were received, they were audited by an expert panel. Duplicate responses were invalidated. Also eliminated were all forws which failed on any of the following
points: did not identify the manufacturer or model; did not
withstand a "reasonableness" test; evaluated different systems
on one form; were forgeries; lacked system ratings; rated non-
computer systems, or revealed a vested interest on the system
being rated.

METHODS AND PROCEDURES

A total of 179 computer systems were represented in the survey. The present authors coded and stored the responses to 20
questions (variables) on the computer (see variable legend).
The data were tested for validity and consistency. For example, the percentage values were checked for the range between
0 to 100. Nonresponse bias was evaluated with an F-test and
found to be insignificant.

Multiple regression analysis determined the relationship
between the overall satisfaction (dependent variable) and the
independent criterion variables. Some of the assumptions for
multiple regression are: 1. The sample is drawn randomly;
2. The relationships among the variables are linear and additive; 3. The regression assumes a multivariate normal distribution of variables, equal variance, no multicollinearity, and
no autocorrelation (Kerlinger and Pedhazer, 1973; Overall and
Klett, 1973).

RESULTS AND DISCUSSION

The predicted variable (overall satisfaction) is regressed over the criterion (independent) variables. This is done by a forward stepwise inclusion procedure, in a manner which provides considerable control over the inclusion of independent variables in the regression equation (Theil, 1971; Nie et.al, 1975).

Table 1 presents the statistics used for the overall test for goodness of fit for the regression equation. This table shows the multiple R, R squared, the standard error and an analysis of variance (ANOVA) for the regression model. This step was selected because each additional variable added to the model increased the multiple R of the model while having an overall F value statistically significant at the .01 level.

According to these tests we can conclude that the sample R square of .818 indicates that 82% of the variation in over-all satisfaction is explained by these independent variables. The standard error of the estimate at the 19th step is 4.22. This means that on the average, the predicted overall satis-faction will deviate from the actual scores by 4.22 units on the overall satisfaction scale.

TABLE 1

MULTIPLE REGRESSION

Multiple R	.904	Analysis of variance	DF	Sum of Square	Mean Square
R Square	.818	Regression	19	12659.042	666.265
Adjusted R square	.796	Residual	159	2825.575	17.771
Standard error	4.216	Critical F 1.88		Calculated F 37.492	

The relative importance of each of the predictor or inde-pendent variables on the predicted or dependent variable is described in Table 2. This relative importance is described by the BETA, the change in satisfaction, due to one standard deviation change in the predictor criterion variable value. These variables and their coefficients are the basis for the ESAI model, used in an interactive on-line questionnaire.

According to Table 2, the model describes overall satis-faction as a function of criterion predictor variables, in descending order of their BETA values (relative importance). The ranking of the independent variables affecting the overall

TABLE 2

VARIABLES IN THE EQUATION

Significance Test for Specific Coefficients in the Regression

Step No.	Variable Name	B	BETA	RANK	Std. Error of B	F
1	Operation	.298+0	.260	1	.067	19.865
2	Trouble Shoot	.118+0	.137	3	.051	5.281
3	Computer	.214+0	.194	2	.056	14.494
4	Programming	.125+0	.128	4	.059	4.419
5	Peripherals	.107+0	.101	5	.054	3.970
6	Compilers	.633-1	.084	7	.034	3.472
7	Education	.559-1	.060	10	.057	.972
8	Expectations	.441-1	.075	9	.023	3.652
9	Op. System	.873-1	.091	6	.054	2.611
10	Mainframe	.391-1	.048	11	.030	1.557
11	Effectiveness	.740-1	.077	8	.052	2.055
12	Microcomputer	.509-1	.042	12	.050	1.050
13	Minicomputer	.222-1	.027	17	.032	.479
14	Life in Nos.	.517-1	.034	15	.054	.914
15	Rental	.966-2	.025	18	.015	.402
16	Applications	.201-1	.029	16	.027	.529
17	Conversion	.277-1	.035	13	.040	.481
18	Documentation	.311-1	.035	14	.051	.374
19	Lease	.385-2	.006	19	.025	.024

satisfaction of ESS users reveals some interesting results.
On the one hand, it appears that one standard deviation change
in the ease of operation and the technical support trouble
shooting (ranked 1 and 2) have the largest effects on the dependent variable. On the other hand, Lease From Third Party
and Documentation have the smallest effects on the overall
satisfaction with the ESS system.

The majority of variables have expanding or positive
effects on the overall satisfaction except that the variables
mainframes, life of computer, and rental from manufacturer,
had a contracting (negative) effect. Interestingly, the negative coefficient of mainframes is substantially smaller than
minis. This may be explained by the greater user control over
the microcomputer, thus the lesser aversion to them as compared to mainframes (Bates, 1982).

EXPERT SYSTEM FOR ENGINEERING AND SCIENTIFIC
SOFTWARE DIAGNOSTICS

Traditionally, experts have been using survey-questionnaires
to evaluate ESS. Such a questionnaire would usually be

administered manually through an interview. These manual
software evaluation methods have numerous disadvantages.

In contrast, an on-line interactive data collection
offers many advantages. Most importantly, selective clarifi-
cations are provided by help files and immediate feedback
becomes plausible. In fact, a computerized expert system
analyzes the data, immediately after it has been entered, providing immediate feedback and diagnostics to the user.

The ESS ESAI interactively interrogates the user, about
the system. User responses are underlined and recorded anony-
mously in a data-base. Subsequently, the ESAI generates the
ESS diagnostics audit trail. This report rails after the
interactive questionnaire, providing immediate feedback.
Later, it may also be used by an internal or external auditor,
manager, or user for system development. This audit trail is
self-explanatory. It compares the user's installation to
industry standards, based on the frequently updated data-base
information.

This ESS diagnostics audit trail sorts the report items
in ascending order of the current deviates, which reflect the
relative weaknesses (-) or strengths (+) of this installation
relative to the industry. It generates a current overall user
satisfaction score, compares it to the prior score, and computes the gain or loss in overall satisfaction.

This ESS ESAI decomposes the change in overall satisfaction, and it identifies the sources of the change. Based upon
that, it also generates prioritized recommendations for further improvements. The responses of the user, along with the
diagnostics audit trail are stored in a transaction file, and
eventually merged with the old data-base master file to form
the updated master file.

SUMMARY, CONCLUSION AND IMPLICATIONS

In summary, the multiple regression has been used to study the
dependence of overall satisfaction of an ESS system with many
ESS variables. The overall significance tests of the goodness of fit of the model have been conducted. The multiple
correlation coefficient was 0.904 thus the null hypothesis
that the correlation coefficient was zero was rejected.

In conclusion, many independent variables had regression
coefficients which were significantly different from zero.
The variables were rank ordered according to their BETA values. Ease of operation was ranked the single most important
factor for determining satisfaction. Other variables which
contributed overwhelmingly were as follows: trouble shooting,
computer reliability, ease of programming, and reliability of

peripherals. ESS application invariably has a negative effect on satisfaction. However, this effect diminishes as the applications are down loaded from mainframes into minis and micros. This may indicate that more attention should be devoted to ESS user satisfaction, especially for the mainframe computers.

It appears that the satisfaction depends on ease of operation, and trouble shooting while whether a computer is leased from a third party, or whether the technical support documentation is adequate had a minimal effect on overall satisfaction. The variable, acquisition method, did not contribute to any major extent to the overall satisfaction of the computer system. Therefore, it was completely excluded from the model.

The implications of the present study are many. The overall satisfaction of system users can be measured by answering certain questions and these results can be very useful to system users as well as buyers and vendors. ESS system buyers can compare different variables and thus can calculate the overall satisfaction they would derive by buying the system. The vendors and designers can build ESS systems based on the criteria which are important to users. Thus, they will maximize user satisfaction and eventually increase their sales

ESS vendors of the computer systems can determine the variables which would increase the overall satisfaction of their products. Thus, they would be more likely to incorporate some of these features in their systems. This could lead to better ESS systems as well as increase research and development. Vendors could also use these data as a marketing tool for their products. If their ESS systems have the features, which were highly ranked, they could advertise them and attract additional customers. These kinds of studies could promote vendors who are concerned with user satisfaction, and provide them an advantage over the competition.

Most importantly, this ESAI for ESS provides the buyer or user with an effective tool for system selection and upgrade. Buyers can evaluate potential ESS systems based on their user satisfaction scores, and eventually choose a system that will yield the highest satisfaction compared to other systems. Current users can evaluate the satisfaction at their installation and compare it to market standards, identifying weaknesses and strengths. Moreover, they can apply remedial action to improve their satisfaction and gauge their progress by running the ESAI on a regular basis.

--- VARIABLE LEGEND ---

1. Average life of computer systems in months(Life in Mos)
2. Rented from the manufacturer of the computer(Rental)

3. Leased from a third party (Lease)
4. Micro computer based (ESS systems) (Microcomputer)
5. Mini computer based (ESS systems) (Minicomputer)
6. Mainframe computer based (ESS systems) (Mainframe)
7. Ease of operation (Operation)
8. Reliability of the computer (Computer)
9. Reliability of peripherals (Peripherals)
10. Maintenance service effectiveness (Effective)
11. Technical support trouble-shooting (Trouble-Shoot)
12. Technical support education (Education)
13. Technical support documentation (Documentation)
14. Manufacturer's software operating system (Op. System)
15. Compilers and assemblers (Compilers)
16. Applications programs (Applications)
17. Ease of programming (Programming)
18. Ease of conversion (Conversion)
19. Systems meeting user expectations (Expectations)
20. Overall system satisfaction (Satisfaction)

REFERENCES

Allerton, J.L. (December, 1983) "Controlling Maintenance Costs via a Computer," Cost & Management, pp. 45-49.

Anderson, R.E. (February, 1982) "Consumer Dissatisfaction: The Effect of Disconfirmed Expectancy on Perceived Product Performance," Journal of Marketing Research, 10, pp. 38-44.

Anonymous (April 19, 1982) "Electronics Engineers Get a New Design Tool," Business Week (Industrial Edition), pp. 92H-92L.

Anonymous (July 25, 1983) "Firm Mohes Graphics Tools More Accessible to Engineers," Computerworld, pp. 43-44.

Anonymous(March 12, 1981) "Interactive Computing Helps Ability Planning Engineers," Public Abilities Fortnightly, pp. 61-64.

Barcus, S.W. and Boer, G.B. (March, 1981) "How a Small Company Evaluates Acquisition of a Mini-Computer," Management Accounting, pp. 13-23.

Bates, A. (February, 1982) "Choosing a Micro? Here's a Cautionary Tale," Accountancy, pp. 98-101.

Bresnen, Edward J., "Desk Top Computer Applications For Industrial Engineers," Management Journal of Methods Time Measurement, 10, No. 31983, pp. 17-20.

Bilbrey, C.P. and House, W.C. (July, 1981) "Mini-Computer Selection," Journal of Systems Management, pp. 36-39.

Brandon, D. (February, 1980) "Staying out of Court," Mini-Micro-Systems, pp. 127-129.

Cheney, P.H. (January, 1979) "Selecting, Acquiring and Coping with Your First Computer," Journal of Small Business Management pp. 43-50.

Duda, R.O., and Gaschnig, J.G. (1981) "Knowledge-Based Expert Systems Come of Age." Byte, 6, 9:238-281.

Farmer, D.F. (February, 1981) "Comparing the 4341 and M80.42," Computerworld, pp. 9-20.

Gevarter, Lillian B. (November 1983) "The Languages and Computers of Artifician Intelligence," Computers in Mechanical Engineering, pp. 33-38.

Green, Richard (August, 1983) "One Company's Answer: Computer Assisted Engineering," Interface Use, pp. 50-52.

Grueberger, F. (January, 1981) "Making Friends with User-Friendly," Datamation, pp. 108-112.

Hansen, J.V., and Messier, W. (October, 1982) "Expert Systems for Decision Support in EDP Auditing,"International Journal of Computer and Information Sciences, 11, 5:357-379.

Holmes, J.R. (December, 1979) "Microcomputer Limitations," Journal of Accountancy, pp. 48-50.

Howard, D. (November, 1983) "What to Look for When Buying Computer Maintenance," Canadian Datasystems, 15, 11:34-35.

Kerlinger, F.N. and Pedhazer, E. (1973) Multiple Regression in Behavioral Research. Holt, Rinehart and Winston, New York.

Kelly, N. (May, 1980) "The Impact of Buying or Leasing," Infosystems, p.78.

Kleinrock, L. (Spring, 1982) "A Decade of Network Development," Journal of Telecommunication Networks, 1, 5:1-11.

Kull, David (January 1984) "Personal Computers Let Engineer Retire Their Old Tools," Computer Decisions, pp. 136-145.

Lean, E., Stern, R., and Monds, T., (May, 1983) "Byting Back: The Industry Responds/Computers Made Easy," Training and Development Journal, 37, 5:13-23.

Linzey, H.T. (November, 1983) "Computerized Centralized Maintenance: An Independent's View," Telephony, 205, 22:150-155.

McDermott, J. (September, 1982) "Rl: A rule-based Configurer of Computer Systems," Artificial Intelligence, 19, pp. 39-88.

Nie, H.N., Hull, C.H., Jenkins, J.F., Steinbrenner, K., and Bentl, D.H. Statistical Package for the Social Sciences, McGraw-Hill, New York.

Olshavsky, R.W. and Miller, J.A. (February, 1972) "Consumer Expectations, Product Performance, and Perceived Product Quality," Journal of Marketing Research, 9, pp. 12-21.

Olson, J.C., and Dover, P. (April, 1979) "Disconfirmation of Consumer Expectations Through Product Trial," Journal of Applied Psychology, 64, pp. 179-189.

Overall, J.E., and Klett, C. (1973) Applied Multivariate Analysis. McGraw-Hill, New York.

Rosenfeld, K.E. (August, 1980) "Small Users Value Maintenance, Ease of Use," Computerworld, pp. 3-4.

Sample, R.L. (September, 1981) "Minis-Moving Beyond the Small Business User," Administrative Management, pp. 58-64.

Scannell, T. (June 7, 1982) "IPL, Singer Rated Tops in Main-frame Survey/Users Find Software, Support, Problem Areas," Computerworld, pp. 56-63.

Scott, Kim L. (Autumn, 1983) "Critical Success Factors in Architectural - Engineering Firms," Today's Executive, pp. 8-13

Shoor, Rita (June 2, 1980) "Panel Socks Human - Engineered Software Tools," Computerworld, pp. 36.

Smith, Harold and Straud, David (Nov/Dec., 1983) "The Hole of Computer Graphics is Central Office Engineering," GTE Automatic Electric Worldwide Communications Journal, pp. 169-176.

Stefik, M. (March, 1982) "The Organization of Expert Systems," Artificial Intelligence, 18, pp. 135-173.

Swan, J. and Martin, W.S. (1981) "Testing Comparison Level and Predictive Expectations Models of Satisfaction," Advances in Consumer Research, 8, pp. 77-82.

Swan, J.E. and Trawick, I.F. (Fall, 1981) "Disconfirmation of Expectations and Satisfaction with a Retail Service," Journal of Marketing, 57, pp. 49-67.

Theil, H. (1971) Principles of Econometrics. Wiley, New York.

Turney, P.B. and Laitala, P.H. (November-December, 1976) "A Strategy for Computer Selection by Small Companies," Managerial Planning, pp. 24-29.

User Ratings of Computer Systems, Datapro Research Corporation (1984).

Wong, H.K. and Mylopoulos (1977) "Two Views of Data Semantics: Data Model in Artificial Intelligence and Database Management," Infor, 15, 3.

COMPUTER PACKAGE FOR RESERVOIR YIELD ESTIMATION

Ljubiša M. Miloradov and Slobodan P. Simonović

"Jaroslav Černi", Institute for the Development of
Water Resources, Belgrade, Yugoslavia

SYNOPSIS

The computer package for the single multi-purpose
reservoir yield estimation based on the implicit
stochastic model is presented in the paper. A three
level algorithm is proposed for the reservoir yield
computation at the first level, the simulation approach is used for computing the objective function.
The second level gives computation of the seasonal
reservoir operating rules. The approach used for
deriving the reservoir operating rules is based on
the nonlinear unconstrained multivariable programming
method of M.J.D. Powell. The third level is used for
estimating the single multi-purpose reservoir yield
based on the predefined relative level of supply. It
is the Fibonacci search procedure that is used for
the optimization of the reservoir yield at this level.
The computer package described in this paper has been
used for estimating the reservoir yield as the first
stage in the cration of the Water Master Plan for the
Republic of Serbia.

1. INTRODUCTION

The Water Master Plan in Yugoslavia is a specific
planning document needed for the creation of a water
resources development strategy for an analysed region.
The time horizon for this kind of document in Yugoslavia is 30 years, and under the existing conditions its
creation consists of the following four major phases:

    i) water demand analysis (for different purposes
- with an analysis of the present and future demand);

    ii) analysis of the available water resources

(including surface water use, groundwater use and reservoirs);

        iii) generation of alternative solutions for satisfying the water demand with the available water (considering different technical solutions and other economic, quality and environmental aspects); and

        iv) optimal ranking of the alternative solutions on the basis of the chosen criteria (economic, social, environmental, etc.).

The proposed concept for the creation of a Water Master Plan used for the development of three major plans for Yugoslavia, uses the optimization approach in two major phases: optimization is first used in the second phase when the reservoir yield needs to be estimated. Secondly, the multi-objective optimization approach is used for the realisation of the fourth phase.

The computer package described in this paper is used for the optimal reservoir yield estimation in the second phase of the Water Master Plan creation. The specific character of the economic, social and political system in Yugoslavia requires that the approach to this problem be stated in a manner different from the one known from literature /2, 3, 7, 8/. The lack of strong and reliable economic criteria, which is typical for developing countries all over the world, is in this approach replaced by optimizing the reliability of the satisfying water demand from the reservoir. Since our main aim in this phase of the Water Master Plan creation is to determine the quantity of water available from the reservoir for different purposes (municipal and industrial water supply, irrigation, hydropower production, etc.), the specific demand is not known. The optimization procedure is therefore based on the use of the relative demand coefficient which describes the character of each demand during the planning horizon.

2.   MATHEMATICAL MODEL FOR RESERVOIR YIELD ESTIMATION

The approach used for estimating the reservoir yield belongs to the group of implicit stochastic techniques. The classical implicit stochastic approach /7, 9/ includes: i) 1 stochastic streamflow generation; ii) deterministic optimization; and iii) regression analysis. The final part i.e. the regression analysis is usually used for determining the reservoir operating rules (releases from the reservoir in terms of storage levels and previous inflow rates). Karamouz and Houck /5/ recently proposed an algorithm based on the deterministic dynamic programming, as well as on the regression analysis and on the simulation for the planning of reservoir operation.

The computer package for the reservoir yield estimation as presented in this paper, is based on an algorithm which only requires simulation for determining the reservoir operating rules. In order to solve the problem of reservoir yield estimation, a three level algorithm is proposed.

Simulation Model - First Level
The simulation model is based on the continuity constraints and the reservoir operating policy derived at the second level. If the following notation is introduced:

Q_t — reservoir inflow during time period t (month) as a result of the stochastic generation model;
R_t — reservoir release during time period t;
SMIN — minimum allowed reservoir storage;
SMAX — maximum allowed reservoir storage;

the continuity equation can be written as follows:

$$S_t = S_{t-1} + Q_t - R_t \quad \text{for every t} \tag{1}$$

where

$$SMIN \leq S_t \leq SMAX \tag{2}$$

$$0 \leq R_t \tag{3}$$

A number of authors /1, 5,9/ have tested both linear and more complex nonlinear forms of the monthly reservoir operating rules. However, Bhaskar and Whitlach /1/ proved that the quadratic operating rules are in many cases as good or even better than the more complex rules. The form of quadratic rules used in our work is:

$$\phi_t = A_m D_t + B_m (S_{t-1} - SMIN + Q_t)^2 / (SMAX - SMIN + \bar{Q}_m)^2 \bar{D} \tag{4}$$

$$\text{for every } t, m = 1, 2, \ldots, 12$$

where:

Q_m — average monthly inflow in month m;
$\bar{D}$ — relative water demand in month m;
A_m, B_m — coefficients (derived on the second level)
m — index of calendar month.

So, the reservoir operation could be defined as follows:

i) $R_t = \max(0, \Phi_t)$ (5)

$S_t = S_{t-1} + Q_t - \max(0, \Phi_t)$ (6)

if

$\text{SMIN} \leq S_{t-1} + Q_t - \max(0, \Phi_t) \leq \text{SMAX}$ (7)

ii) $R_t = S_{t-1} + Q_t - \text{SMAX}$ (8)

$S_t = \text{SMAX}$ (9)

if

$S_{t-1} + Q_t - \max(0, \Phi_t) > \text{SMAX}$ (10)

iii) $R_t = S_{t-1} + Q_t - \text{SMIN}$ (11)

$S_t = \text{SMIN}$ (12)

if

$S_{t-1} + Q_t - \max(0, \Phi_t) < \text{SMIN}$ (13)

The assumption used in the formulation of the reservoir operation was that the reservoir is a multipurpose one (municipal and industrial water supply, irrigation, hydropower production, etc.). Therefore, the total water demand in month t (D_t) could be expressed as the sum of the particular water demands $(d_{i,t})$.

It is obvious that different reservoir inflows could produce some nonideal reservoir operations this resulting in water shortages when $R_t < D_t$. The problem of water allocation in these cases is solved by applying the so called principle of uniform treatment which can be expressed as:

$$L_{1,t} = L_{2,t} = \ldots = L_{i,t} = \ldots = L_{n,t} = L_{e,t}$$ (14)

where:

$L_{i,t}$ - water shortage (loss) for user i in time period t; and

$L_{e,t}$ - equivalent shortage (loss) in period t.

Loss functions are assumed to be linear /4/:

$$L_{i,t} = \frac{1}{1 - P_i} - r_{i,t} \frac{1}{d_{i,t}(1-P_i)}$$ (15)

where:

$r_{i,t}$ - reservoir releases for user i in period t;

P_i - target supply level for user i in period

t $(P_i = p_{i,t}/d_{i,t})$; and

$p_{i,t}$ — reservoir releases which produce allow-able expected losses for user i.

Assuming that $L_{i,t}(p_{i,t}) = 1$ and that the total amount of the released water is the sum of the releases for every user, and some transformations of equations (14) and (15) we can conclude that:

$$L_{e,t} = (1 - \frac{R_t}{D_t}) \; / \; \sum_{i=1}^{N} (1 - \frac{p_{i,t}}{d_{i,t}}) \frac{d_{i,t}}{D_t} \qquad (16)$$

and

$$\frac{r_{i,t}}{d_{i,t}} = 1 - (1 - \frac{p_{i,t}}{d_{i,t}}) L_{e,t} \qquad (17)$$

The flow chart in Figure 1 describes the computation procedure at the first level.

Reservoir Operating Rules Derivation - Second Level
The main goal at this level is the optimization of the reservoir operating rule Φ. The objective function is the minimization of the total water shortages. The optimization problem is nonlinear and can be stated in the following form:

minimize LOSS   for $m = 1, 2, \ldots, 12$.   (18)
$/A_m, B_m/$

The variable LOSS can be expressed as:

$$LOSS = \frac{1}{T} \sum_{t=1}^{T} \sum_{i=1}^{N} L_{i,t} \qquad (19)$$

where $L_{i,t}$ are computed at the first level applying equations (1) through (17) subject to the boundary condition:

$$S_0 = (SMAX + SMIN)/2 \qquad (20)$$

The flow chart expressing the computation procedure at the second level is shown in Figure 2.

The Reservoir Yield Estimation - Third Level

At this level, the optimization task is to estimate the reservoir yield. The assumptions used in the pro-cedure of optimizing the yield are: i) known reservoir capacity, and ii) known relative demand coefficients for every reservoir purpose.

The relative reservoir yield is expressed as:

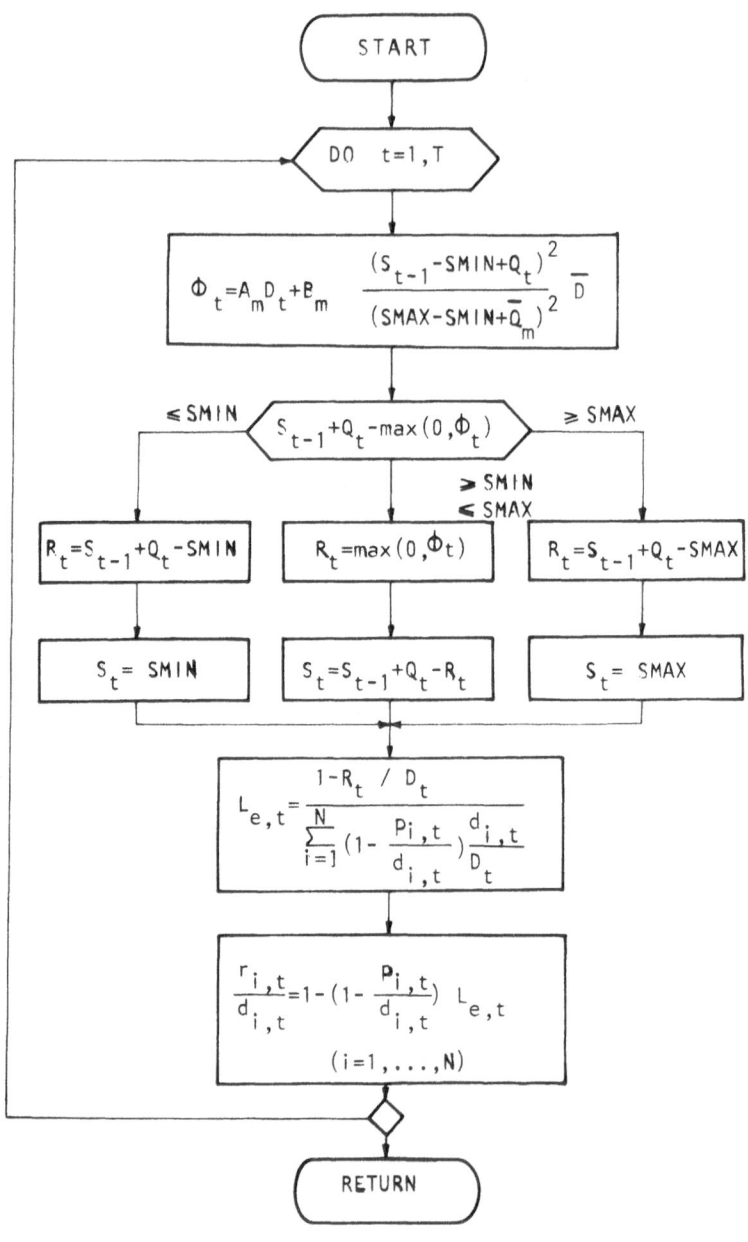

Figure 1. - Flow chart of the first level

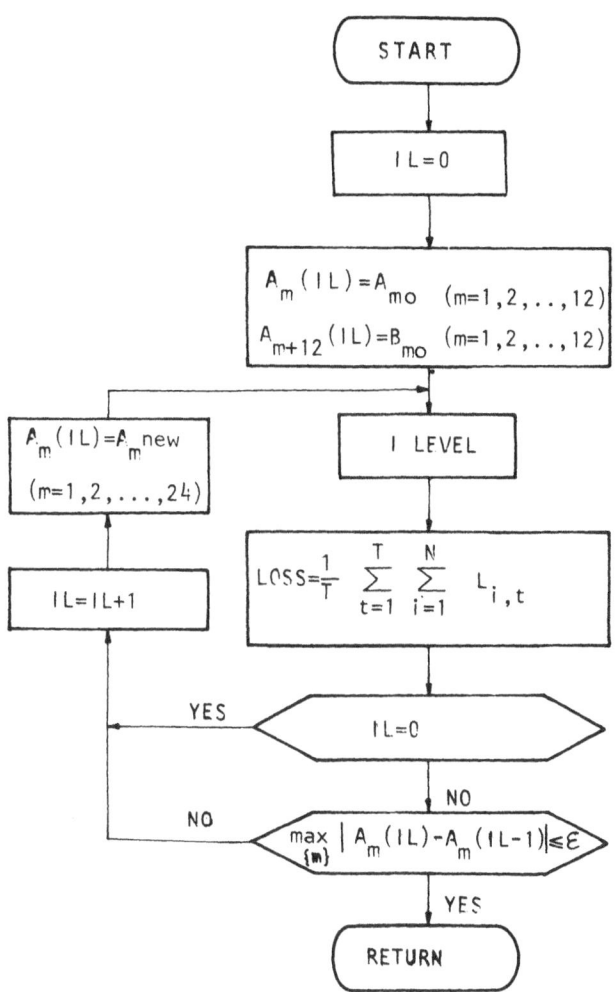

Figure 2 . - Flow chart of the second level

$$\alpha = (\frac{1}{T} \sum_{t=1}^{T} \sum_{i=1}^{N} d_{i,t}) / (\frac{1}{T} \sum_{t=1}^{T} Q_t) = (\sum_{i=1}^{N} W_i \bar{D}) /$$

$$/\bar{Q} = \bar{D}/\bar{Q} \tag{21}$$

where:

α — relative reservoir yield;
W_i — demand coefficents (weights);
$\bar{D}$ — average monthly water demand; and
$\bar{Q}$ — average monthly reservoir inflow

The optimization criterion is given as:

$$\text{minimize } (\sum_{i=1}^{N} \sum_{t=1}^{T} \frac{r_{i,t}}{D_{i,t} P_i} - N)^2 \tag{22}$$
$$\{\alpha\}$$

subject to

$$\alpha \geq 0 \tag{23}$$

The average monthly water demand $(\bar{D}_i)$ is also computed by estimating the yield value (α). The computation procedure at the third level is shown on the flow chart in Figure 3.

3. PROGRAM PACKAGE

The described computation algorithm is realised as a computer program package in the FORTRAN IV language for the VAX 11/780 computer. The program consists of a main program and of subroutines FIBON, POWELL, BOTM, FUNC and LINEAR.

The input data bases are divided into the following two groups:

i) Hydrological data base consisting of several generated inflow sequences of fifty years. These data are obtained by external ARMA model based on the historical sequence of data.

ii) Reservoir data base consisting of the reservoir capacity (SMAX and SMIN), reservoir volume curve, relative water demand coefficients for every reservoir user, and target yield reliability.

The final computation results are printed from the main program. The main result is the optimal reservoir yield (α). The additional results are contained in the table showing the available water per month for every user in each time period (month), and their

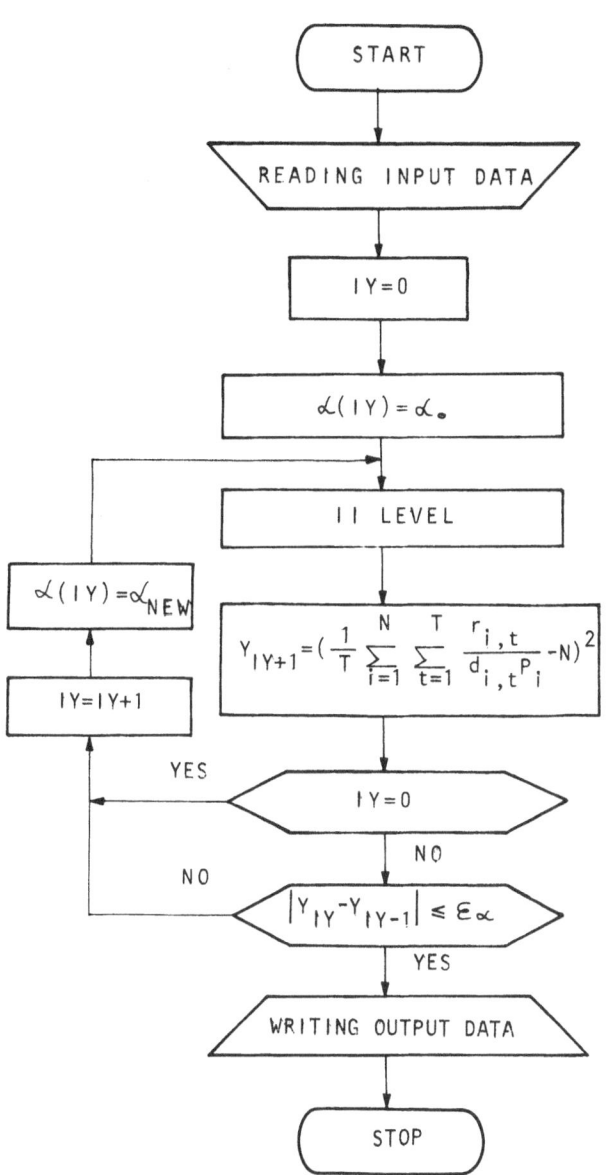

Fiqure 3. - Flow chart of the third level

corresponding average values computed by simulation model at first level. The final results also contained in the reliability table for each user and the shortages are shown together with the probability of occurrence. An example of the output tables is shown in Appendix 1.

The Subroutines Description

FIBON is the main subroutine for the reservoir yield optimization (third level). The objective function and the constraints are described in the previous chapter by equations (22) and (23). The procedure for optimizing the one dimensional constrained non-linear function is based on the Fibonacci search /6/. This subroutine is called from the main program. All the iterations can be printed in a special output file.

The POWELL subroutine is used as a tool for optimizing the reservoir operating rules (second level). The value of the objective function at this level is computed by using the simulation procedure. The mathematical form of the optimization problem solved in this routine is given by equations (18) and (19). This unconstrained multivariable nonlinear problem is solved using M.J.D.Powell's feasible direction method /6/. The routine POWELL is called from the FIBON subroutine and its intermediate results can be printed in a separate file too.

The BOTM subroutine is called from POWELL and it is used for the single variable search optimization of the reservoir operating rules.

The FUNC subroutine is called from BOTM and it is used for the specification of the objective function for the Powell method.

The LINEAR function subroutine is used for the linear interpolation of the reservoir storage curve. This subroutine is used only if energy production is one of the reservoir users.

CONCLUSIONS

The paper presents the computer program used for the optimal reservoir yield estimation and it describes the mathematical model applied for computation purposes. This program has been created for satisfying specific optimization demands within the Water Master Plan.

The program is very efficient although it uses the iterative procedure at two levels. It contains about 900 instructions and takes about 11 CPU minutes on the VAX 11/780 for the reservoir with three users.

The obtained results within the program package of one application have proved to be very useful in the further steps of the Water Master Plan creation. Up to now, the program package has been used for the optimization of the reservoir yield for about 100 reservoirs in Yugoslavia.

ACKNOWLEDGEMENTS

The authors highly appreciate the technical help of Dragica Teofilović, Rudolf Župančič, Vida Marjanović and Branka Jelić in preparing this manuscript. The work described in this paper was done under contract with the Serbian Water Resources Community of Interest.

REFERENCES

1   Bhaskar, N.R. and Whitlach (1980), Derivation of Monthly Reservoir Release Policies, Water Resources Research, 16 (6), 987-993.

2   Biswas,A.K. (1976) Editor. Systems Approach to Water Management, McGraw-Hill Book Company.

3   Haimes, Y.Y. (1977) Hierarchical Analysis of Water Resources Systems, McGraw-Hill Book Co.

4   Jettmar R.U. and G.K.Young (1975) Hydrologic Estimation and Economic Regret, Water Resources Research, 11 (5), 648-656.

5   Karamouz, M. and M.H.Houck (1982) Annual and Monthly Reservoir Operation Rules Generated by Deterministic Optimization, Water Resources Research, 18 (5), 1337-1344.

6   Kuester, J.L. and J.H.Mize (1973) Optimization Techniques with Fortran, McGraw-Hill Book Co.

7   Loucks,D.P. et all. (1981) Water Resource Systems Planning and Analysis, Prentice-Hall Inc., New York.

8   Texas Water Development Board (1974) Analytical Techniques for Planning Complex Water Resource Systems, Report 183.

9   Young G.K. (1966) Techniques for Finding Reservoir
     Operating Rules, Ph.D. dissertation, Harvard
     University, Cambridge, Mass.

APPENDIX 1

FINAL RESULTS

ALPHA = 0.92
BETA = 0.52

USER :

PERIOD	AV.INFLOW	M & I		HYD.POW.	
		DEMAND	SIMUL.	DEMAND	SIMUL.
1	5.625	1.715	1.639	3.118	2.290
2	7.708	1.549	1.480	2.816	2.068
3	10.205	2.053	2.029	3.118	2.902
4	7.898	2.414	2.386	2.012	1.872
5	6.817	2.988	2.883	2.072	1.639
6	5.171	3.344	3.226	2.072	1.586
7	3.945	3.638	3.550	2.079	1.777
8	2.250	3.638	3.550	2.079	1.777
9	2.825	3.017	2.932	2.012	1.866
10	3.383	2.598	2.525	3.118	2.892
11	4.668	1.986	1.929	3.017	2.895
12	6.144	1.715	1.665	3.118	2.991
SUM	66.638	30.593	29.794	30.593	26.556
	66.638	30.593	29.794	2.274	1.974
DEMAND RATIO :		0.500	0.447	0.500	0.399
SUPPLY LEVEL :		0.975	0.974	0.850	0.868

USER 1   TYPE M & I

FAILURE NUMBER AND RELIABILITY TABLE : PERIOD VERSUS REDUCTION(%)

REDUCTION: PERIOD	1	5	10	20	30	40	50	70
1	202	123	37	0	0	0	0	0
	0.327	0.590	0.877	1.000	1.000	1.000	1.000	1.000
2	59	30	0	0	0	0	0	0
	0.803	0.900	1.000	1.000	1.000	1.000	1.000	1.000
3	158	105	30	0	0	0	0	0
	0.473	0.650	0.900	1.000	1.000	1.000	1.000	1.000
4	143	43	23	0	0	0	0	0
	0.523	0.857	0.923	1.000	1.000	1.000	1.000	1.000
5	137	9	6	2	2	2	2	2
	0.543	0.970	0.980	0.993	0.993	0.993	0.993	0.993
6	12	12	10	2	2	2	2	2
	0.960	0.960	0.967	0.993	0.993	0.993	0.993	0.993

USER 2   TYPE   HYD.POW.

FAILURE NUMBER AND RELIABILITY TABLE : PERIOD VERSUS REDUCTION(%)

REDUCTION: PERIOD	1	5	10	20	30	40	50	70
1	207	204	190	158	123	98	55	21
	0.310	0.320	0.367	0.473	0.590	0.673	0.817	0.930
2	73	65	53	45	30	24	18	0
	0.757	0.783	0.823	0.850	0.900	0.920	0.940	1.000
3	168	160	152	146	105	57	35	21
	0.440	0.467	0.493	0.513	0.650	0.810	0.883	0.930
4	169	143	125	82	43	30	25	16
	0.437	0.523	0.583	0.727	0.857	0.900	0.917	0.947
5	215	153	54	9	9	9	6	6
	0.283	0.490	0.820	0.970	0.970	0.970	0.980	0.980
6	91	16	12	12	12	10	10	10
	0.697	0.947	0.960	0.960	0.960	0.967	0.967	0.967

AN EXPERT SYSTEM FOR PRELIMINARY NUMERICAL DESIGN MODELLING

K.J. MacCallum and A. Duffy

Department of Ship and Marine Technology, University of Strathclyde, Glasgow

1. INTRODUCTION

From the earliest days of Computer Aided Design (CAD) there has been a clear recognition that establishing effective communication between the computer and the designer is crucial to the development of a productive design system. Early papers on the philosophy and concepts of CAD stated quite clearly that man and machine had to work together in a co-operative environment; and a few systems were developed to show the potential of this kind of approach. A fine example of the appreciation of this potential is given in a paper by Robert Mann and Steve Coons in 1965 (1). In this paper, they state: "It is clear that what is needed if the computer is to be of greater use in the creative process, is a more intimate and continuous interchange between man and machine. This interchange must be of such a nature that all forms of thought that are congenial to man, whether verbal, symbolic, numerical, or even graphical are also understood by the machine and are acted upon by the machine in ways that are appropriate to man's purpose." The emphasis on the computer's understanding of man's purpose and on communication at the level of thought are significant.

Despite the early suggestions, an examination of developments in CAD systems leads to the conclusion that the greatest advances in the application of computers to engineering design have been the assistance in "number crunching" for design analysis, and the development of specialised systems for limited design tasks. Parallel improvements in man-machine communication have been achieved through improved availability and accessibility of computing power and significant advances in computer graphics, providing a more acceptable medium for exchange of information. However, even those advances have not made significant inroads into the problems of advanced communication. Only recently with increasing emphasis on product modelling has there been a gradual awareness that CAD systems require a much deeper level of user knowledge and problem knowledge than is currently normal. It would seem, therefore, that much of the early potential of CAD has not yet been realised.

The work described in this paper addresses itself to this goal. It is argued that progress towards the goal can only be made by building into design systems a greater understanding of man's purpose. The key to this is to build systems which have explicit knowledge and are able to manipulate that knowledge and reason with it. The concept of "expert systems" is built on this approach. This paper describes a system called DESIGNER which is an expert system concerned with building and handling knowledge of numerical relationships in preliminary design. The system is illustrated and evaluated using examples from preliminary ship design.

2.   THE NATURE OF THE DESIGN PROCESS

Before looking at ways of representing design knowledge, it is worthwhile examining the design process itself in order to understand what contribution we can expect computers to make. A key characteristic of much of engineering design is the complexity of the objects or systems of interest. Typically, a system will have many components, each of which will be related in different ways to other components through their characteristics. A designer's task is to create a specification for such a system, given a set of required functional objectives to be achieved in a given environment. The designer will rely on measures of performance, both objective and subjective, to select the most promising concepts for evaluation. However, complexity prevents the designer evaluating all concepts in detail; instead the design is broken down into parts and each part is tackled in a number of stages corresponding to levels of detail. Earlier stages have the least detail and use only the parameters which have the greatest influence on the overall design performance, whereas later stages operate within the constraints of previously defined parameters. A crucial feature of this approach is that individual stages are more tractable because the number of independent parameters and their interactions are reduced.

A conclusion which can be drawn from this brief description of the design process is that a designer's first expressions of concepts are in terms of objects, their characteristics, and the relationships which exist among them. One way of viewing design is as a process of modelling in which the above expressions constitute the model. Thus in every situation the designer creates some kind of abstract model which simulates some aspect of the behaviour of the thing being designed. In fact it is likely that, for each concept, the designer handles a variety of models simultaneously, each one representing a different abstraction, but being consistent with the other. The models are essentially mental models but will be formalised through graphical, numerical, logical, and physical means. The creation of a model, which is a process of synthesis, is difficult to formalise. Evaluation of a particular model, however, is a process of analysis. It requires effort and sometimes ingenuity; but in most cases the procedures of evaluation are well defined. The overall design process involves establishing and collecting a variety of models, interchanging synthesis and analysis, and allowing interaction between design objectives and model specifications (2).

In summary it is useful to identify some important characteristics of the design process:

(a) creative - it requires imagination and inventiveness to build conceptual models. As a result of creative activities, the form and structure of these models may change or develop as the design proceeds.

(b) multiple solutions - there can be many answers to a given design problem, all of which may achieve the objectives, and may thus be technically and economically feasible. Thus the design process is not deterministic.

(c) empirical - the process of creating and evaluating a model does not always follow well formalised rules with good theoretical bases. Very often relationships are of an empirical nature.

(d) approximate - because design is a modelling process which uses empirical relationships, the results obtained are generally approximate. Accuracy increases as the design proceeds and greater levels of detail are included. However, the concepts of expected and acceptable accuracy are important to the designer.

(e) requires expertise - the designer uses his expertise in many different ways during design, with respect to relationships used, the actual design process, and even in the judgement of the acceptability of proposed solutions.

These characteristics are most in evidence at the creative or preliminary stages of design during which basic concepts are being developed. However, the same characteristics are the ones which in many senses are the most difficult to computerise, involving intuition, experience,

approximation and empiricism. It is hardly surprising, therefore that conventional approaches of CAD to the creative stage
of design have had limited success.

3. EXPERT SYSTEMS APPROACH

One of the most promising developments in the use of computers
in recent years has been the work on expert systems (3). Its
significance is that it addresses itself to providing computer
systems which are able to make a "knowledgeable" contribution
to complex problems in a specific domain or field of interest,
that is, to act as an "expert". A human expert is someone
who has a specialised body of knowledge and is able to apply
it to solve problems, to advise, to act as a consultant,
and to communicate his knowledge with others. An expert
system is a computer system which is able to enact a similar
role. The major advance of expert systems compared to more
conventional software systems is the explicit representation
and manipulation of a body of knowledge. The "knowledge base"
can be used by the system in solving problems, in its own area
of relevance, and can be added to directly by the human
expert or by the system itself.

There are several important features in expert systems
which make them capable of tackling problems of great
complexity. The first of these is a description and
representation of knowledge in some formalised language.
This immediately makes the knowledge base available and
understandable to its users, and allows experts to examine
and modify the system's knowledge as new situations are
encountered. The second important feature is the system's
ability to reason using a combination of known facts and
generalised relationships in the knowledge base. Because
the reasoning mechanisms and their control can be structured
separately from the knowledge base it is easier to express
problems for solution. A third feature is the system's
ability to provide explanations of the steps taken to reach
a conclusion. This follows naturally from the formalised
representation of the knowledge.

While many simple expert systems have used the knowledge
representation and control techniques adopted in earlier
successful research work (4), there are still many areas of
concern for the longer term development of expert systems.
Some of these are the representations for new types of knowledge such as common-sense knowledge and uncertainty,
required depth of knowledge, control over the use of knowledge,
and the use of logic systems.

Michie (3) has identified three different user modes for
an expert system in contrast to the single mode (getting
answers to a problem) typical of the more familiar type of
computing:

a) user as a client - system acts as a consultant from whom the user wishes to get answers to problems.

b) user as a tutor - system accepts instructions from a domain specialist to improve or refine its knowledge.

c) user as a pupil - system can use its expert knowledge to instruct users in certain approaches.

To these three modes it is probably useful to add a fourth:

d) user as an assistant - system interacts with user to encourage user to find a solution to a problem with guidance advice and stimulation from the system.

It is this last mode which is most relevant to the design situation. The work described in this paper is based on the contention that the approaches being taken in expert systems provides a key to realising the potential of CAD. For many years now we have been building complex CAD systems which contain increasing amounts of knowledge. However, that know-ledge has been highly constrained to particular methods, and has been implicitly rather than explicitly available. Systems based on ideas of explicit knowledge representation and reasoning, offer the possibility of greater productivity in the contribution they make to man-machine communication.

4. A DESIGN ASSISTANT FOR NUMERICAL DESIGN

From the discussion of the previous section it is concluded that any system which intends to take a more active role in a design dialogue must have the following features:

- a powerful semantically rich interface
- a highly flexible design modelling and modification system
- an understanding (at least superficially) of design concepts and goals
- a capability for abstraction
- a method of capturing and using expertise in a useful way
- an ability to explain its own reasoning processes.

While we are still some way from achieving all these features, they indicate the direction of recent design system trends.

This section describes an approach to creating a design assistant for numerical design based on this philosophy. The system, called DESIGNER was constructed to meet the following requirements:

a) flexible representation of user's design models;

(b) ability to modify models, adding new design
    parameters or relationships, during the design
    process;

(c) ability to represent designer's expertise with
    respect to numerical modelling;

(d) provide feedback to the designer on the nature of
    relationships implied by a model;

and (e) variety of levels of control of the design process
    by the designer.

The requirements place emphasis on providing adequate
knowledge representations and control rather than on the user
interface. Thus the system will be described in terms of its
knowledge structure before providing methods and examples of
use.

4.1   A Network Model of Design

To define and illustrate the role of relationships in the
numerical design process, it is valuable to have a more
formalised way of presenting models. DESIGNER presents a
model as a directed network. In such a network, the various
nodes represent the characteristics of interest, and a link
between two nodes represents a dependency as contained in
some relationship, the direction of the dependency being
shown by an arrowhead. Thus Fig. 1 shows that the volume
of a box depends on its length, its breadth, and its depth.
The three dependencies together are contained in a single
relationship:

Volume = Length x Breadth x Depth

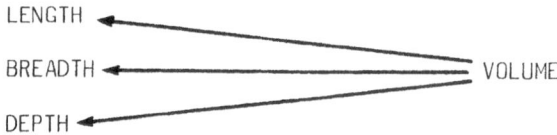

Fig. 1.   Dependency Network

which allows volume to be calculated from these three characteristics. A more realistic example, which represents a simplified
preliminary stage of ship design is shown in Fig. 2.

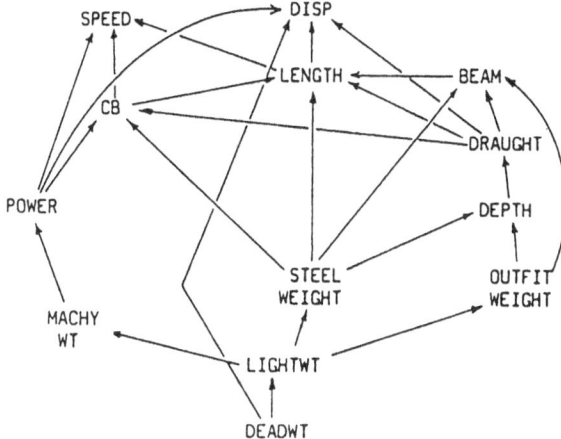

Fig. 2. Network for Preliminary Design

Several important points can be made in general about the approach of network modelling:

a) While some relationships are fixed in their form, either because they follow physical laws or they are defined by legislative requirements, many design relationships are of an empirical nature. Thus the results of using such relationships can only be considered to be approximately correct.

b) Again because of the empirical nature of numerical design, a particular characteristic may have a number of valid relationships for estimating a value. The choice of which relationship to use in a particular situation will be related to particular contexts, degree of detail being considered, availability of other information, and other expertise.

c) The network model can be considered to represent the network of currently active relationships. Thus the model can change with the progress of design.

d) In Fig. 1, if the Volume depends on Length, then Length can be considered to influence Volume, i.e. a change in the value of Length will affect the value of Volume. Thus the inverse of the dependency network is an influence network. The effect of any characteristic on another can be determined by tracing through all the intermediate paths in the network.

e) Some relationships may need to be used in a number of different forms depending on a designer's approach to a problem. For example, in Fig. 1 it would be just as correct in some situations to redraw it so that Length is dependent on Volume, Breadth and Depth.

In conclusion, the directed network is a useful way of modelling numerical design. However, it rapidly becomes complex to visualise, and can only illustrate an active network within a much more detailed network.

4.2 The Characteristic Frame

Knowledge about a characteristic is encapsulated in a "frame" of knowledge, that is a formalised structure containing relevant information in "slots" (Fig. 3). The main items of knowledge which are available to the user through his actions are:

a) a name - by which the characteristic can be referenced;

b) an explanation - an extended name which can be used for fuller explanations or for more complex referencing;

c) a current value - representing the latest answer calculated for this characteristic truncated to a value consistent with the known accuracy of the calculated value. A value may be "unknown";

d) accuracy - a measure of the accuracy of the value calculated,expressed as a likely error in the value;

e) units - the units in which the value has been measured;

f) relationships - a set of numerical relationships, any of which could be valid for estimating a value. Each relationship, in turn contains a list of dependent characteristics, the actual relationship, a list of conditions which have to be satisfied before the relationship is considered valid, and a list of conditions which have to be satisfied before the characteristic is updated with the generated value. Each relationship also has associated with it a reliability.

g) influences - a list of characteristics which this characteristic is known to influence with the currently active network. Associated with these influences is a "strength" indicating the degree of influence. Strength is equivalent to a numerical derivative.

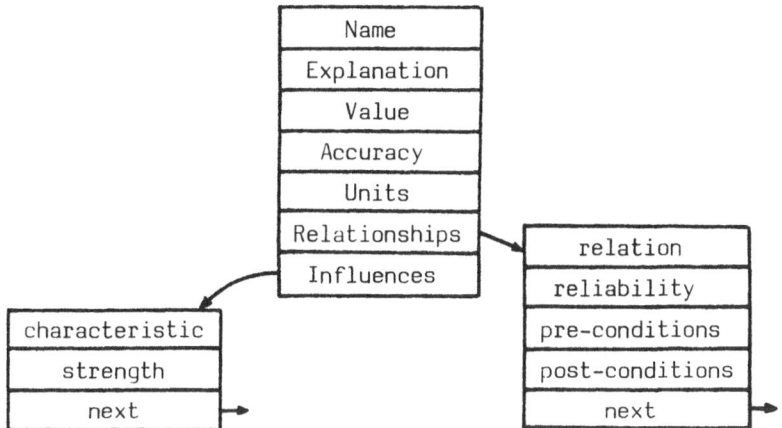

Fig. 3.   Characteristic Frame

4.3   The DESIGNER System

The network model and the characteristic frame together describe the underlying structure of DESIGNER. Superimposed on
that structure are mechanisms to control the use and
propagation of characteristic values and relationships, and a
user interface which allows the designer to express his model,
his expertise, his goals and his control over the design
process.

A user would normally expect to start a particular problem
with a set of characteristics and for each, a list of relationships available. He may then decide to provide some parameter
values he knows, or simply to ask for an "estimate" of values
for some performance variables. As he proceeds through
evaluations of the model, he may add new relationships together
with conditions under which it is appropriate to use them. At
any point, the system will provide information on the strength
of influence between any pair of characteristics. Eventually
the user will feel that he has reached a situation which
meets his goals to some acceptable degree of confidence and
will move to a more detailed stage of design.

Two basic actions taken when using DESIGNER are ESTIMATE
and UPDATE. ESTIMATE is used to obtain a value for a characteristic from known values of other characteristics using a
valid relationship. Where a relationship requires a characteristic value which is not available, one of two things
happen; either the system asks the user to provide a value,
or if an automatic search mode has been set, the system calls
ESTIMATE recursively until all values are resolved. There is
an important side-effect of using ESTIMATE. Every time a
relationship is used, each of the input characteristics can
"learn" something about its own role in the network. In

particular it learns of one other characteristic which it
directly influences and the strength of that influence. This
learning process is the basis of the system creating an
influence network.

Each time ESTIMATE is used, the system has to find a valid
relationship. In practice a single characteristic may have a
number of relationships which are valid in a particular context.
In this case the system selects the relationship which has the
highest reliability measure. It is open to the user when asked
for a value for a characteristic to respond "UNKNOWN". The
implications of this is that the recursive list of relationships which are depending on this value fail and some alternative relationships, perhaps of lower reliability, need to be
tried.

The UPDATE operation is used to allocate a known value to
a particular characteristic. In addition to updating a value,
it uses its knowledge of influences to do two things; first
it traces through the influence network to mark all influenced
values as inconsistent, and secondly it warns the user of the
immediate effects of his action. If the system is in automatic
propagation mode, then it will follow through all the
implications of these influences, calling UPDATE recursively.

The system is always able to provide the user with information about the current model structure, and a characteristic
knowledge including its influences. At any stage the user can
determine the strength of influence of one parameter on
another; the system using its knowledge of immediate
influence strengths to derive these. In addition, the system
can explain how values have been derived, identifying the
relationships which were successfully selected, and the
characteristic values used in their evaluation.

Throughout the evaluation process of a network model,
the system maintains a concept of approximation of derived
values. Inaccuracy in a derived value is expressed to the user
as a value ± tolerance. In general, inaccuracy in a value
depends on the reliability of a relationship used, and the
tolerances which already exist on the dependent characteristic
values. Those inaccuracies can be propagated through the network in a manner similar to strengths, providing the user
with soundly based measures of the degree of approximation
in results obtained.

5.   EXAMPLES OF USE

To illustrate DESIGNER and the features just described, a
simplified preliminary ship design problem will be considered.
The design model is similar to that shown in Fig. 2, but to be
more realistic, it includes a number of multiple relationships.
Table 1 summarises the main parameters included in the model,

and the number of relationships which have been defined for each model. The objective of the design process is to select a set of values for the main dimensions (L, B, T, D and CB) such that it meets requirements of deadweight (DWT) and speed (V). Simple empirical relationships have been taken from published sources to create this basic model. For the purposes of the example, the requirements are taken as a deadweight (or payload) of 22,300 tonnes and a trial speed of 14 knots. The designer as a first step assumes that a reasonable value of displacement is 26,000 tonnes and of beam/draught ratio is 2.4. He then inputs to the system:

Update V with 14; Update BT with 2.4;
Update DISP with 26,000; Estimate DWT

SYMBOL	EXPLANATION	NO. OF RELATIONSHIPS
B	moulded beam	4
BT	beam/draft ratio	1
CB	block coefficient	2
D	moulded depth	3
DISP	moulded displacement	1
DWT	full deadweight	1
GM	initial stability	1
L	L_{BP}	4
POWER	shaft power	1
T	draught	3
V	trial speed	0
WEIGHTS	lightship mass	1

Table 1. Key Parameters and Number of Relationships

The system responds to the Estimate by recursively searching for valid relationships for which information is known or can be supplied by the user, producing as it goes the following set of values:

L	143.35 m
CB	0.80
T	9.60 m
B	23.05 m
D	13.54 m
POWER	7351 Kw
WEIGHTS	5201 tonnes
DWT	20799 tonnes

The system's explanation of how these values were derived shows the specific relationships which were used. Alternatively, the system could be required to print out its search process as it goes along. An extract from this is given below:

```
B   ESTIMATE
    PHYS BEING TRIED
        B DEPENDS ON [DISP L T CB]

T   ESTIMATE
    PHYS BEING TRIED
        T DEPENDS ON [DISP L B CB]
        B CURRENTLY BEING ESTIMATED
        THIS RELATIONSHIP NOT SUITABLE
    SOLVE BEING TRIED
        T DEPENDS ON [DISP L BT CB]

CB  ESTIMATE
    EMP BEING TRIED
        CB DEPENDS ON [V        L]

ESTIMATE    OF   CB   IS    0.80

            OF   T    IS    9.60

            OF   B    IS    23.05
```

It is worthwhile noting in this case the failure of valid relationships because of lack of information, followed by an attempt with an alternative relationship. The names of the relationships being tried have been provided during the definition of the model.

The overall result from this first estimate is that the deadweight is too low by about 1500 tonnes. Using the strengths feature, the system reveals that an increase in deadweight of about 1500 tonnes could be achieved by increasing DISP by about 1750 tonnes. The effect of changing beam/draught ratio is negligible. Thus DISP can be "UPDATED" to 27700. Normally the system will respond by warning the user of the immediate influences of this update; in this case:

The following will be affected by DISP

```
            Deadweight
            B
            T
            L
```

If the automatic propagation mode was set then these immediate influences would be followed through making the remainder of the network consistent. Alternatively, the user can again ask for deadweight to be estimated. In this case the overall effect will be the same and a value of deadweight of 22,258 tonnes is produced.

As yet no unreliabilities have been mentioned, even though they should be associated with the relationships being used. If these are included using data from a range of available vessels then the latest set of values printed out would have read:

L	146.41 ± 4.39m	
CB	0.80 ± 0.02	
T	9.79 ± 0.21m	
B	23.50 ± 1.11m	
D	13.81 ± 0.75m	
POWER	7652 ± 1056 Kw	
WEIGHTS	5442 ± 668 tonnes	
DWT	22258 ± 668 tonnes	

With this level of approximation in the relationships, the deadweight requirement is well within the bounds of the answer.

The design process continues with an estimate of stability (GM). The value of 1.69 ± 0.93m is unsatisfactory and so further changes are required. The designer finds from the system that beam/draught ratio, and beam and draught them-selves are the parameters which most influence GM. However, he feels that beam should not increase and thus decides to achieve his requirements of GM by changing draught, and keep his deadweight requirement by changing Length. The final set of values is:

L	151.32m
CB	0.81 ± 0.02
T	9.50m
B	23.50m
D	13.39 ± 0.67m
POWER	7731 ± 1033 Kw
WEIGHTS	5629 ± 658 tonnes
DWT	22288 ± 658 tonnes
GM	1.95 ± 0.92

It is of interest to note that this last sequence of operations changed the independent variables from beam/draught ratio and displacement to length, beam and draught. Thus the relation-ships used and consequently the network of dependencies and influences has changed. A second point of note is that a change now to one of the independent variables in automatic propagation mode would result in both deadweight and GM being updated. This is now a different effect from just "estimating" deadweight, which would leave GM as inconsistent.

6.   DISCUSSION

This paper has been concerned with presenting an overall

knowledge-based approach to handling numerical relationships
in the preliminary stages of design. The last section
illustrated the system by stepping through a simple ship
design example. A number of features of the system were not
presented since the aim was simply to convey the "flavour" of
the system and to illustrate the overall concepts involved.
However, it is worth emphasising a number of points with respect
to the DESIGNER system and the examples used:

a) there is complete flexibility to set up large
   varieties of design models; i.e. there is
   nothing in the DESIGNER system which is related
   specifically to ship design;

b) the user of the system has a large degree of
   control over the process of design, both in
   terms of the sequence of steps and the rate of
   progress;

c) in defining the model for the example, there
   were no assumptions made about what information
   would be available; neither were there
   assumptions about the way the user would approach
   the design process;

d) the system uses built-in expertise to look for
   suitable relationships. Thus, only a part of
   the total model definition is active at one time;

e) the system learns about influences and strengths
   of influence through the use of the model. The
   results of this can be used to provide guidance
   to the user about possible solution areas.

Despite the achievements of the system in taking a step
towards a more knowledgeable design assistant, there are a
number of areas in which research is still required:

a) the present system deals only with numerical
   relationships. The other major aspect in any
   early design considerations is spatial
   arrangement. Although the present system
   allows graphical output of design trends or
   geometric representations, it cannot be considered to "know" about spatial arrangements
   in the same way that it knows about characteristics;

b) the major focus of attention has been to provide a
   flexible and rich set of facilities for handling
   numerical design problems. Little attention has
   been given to providing the most effective interface for handling this knowledge-based system;

c) because of the complexity of the system careful study needs to be given to providing the correct level of communication between the human designer and the DESIGNER system. This is more than just an interface problem; it implies a deeper modelling of the user and the system;

d) any long term accumulation of expertise by the system relies on relationships provided by designers. There should be links to design data bases which would allow relationships to be extracted and added to the system.

7. CONCLUSIONS

It has been stated that CAD still has considerable potential for advancing the process of design, and that recent developments in expert systems and knowledge-based systems currently hold the key to releasing some of that potential.

A system, called DESIGNER which is based on these developments has been presented. This demonstrates a number of features which are normally difficult to achieve in more conventional CAD systems. However, there are clearly many areas which require further research and development.

REFERENCES

1. Mann, R.W. & Coons, S.A. (1965). "Computer Aided Design". McGraw-Hill Yearbook - Science and Technology, McGraw-Hill.

2. Malhotra, A., Thomas, J.C., Carroll, J.M. & Miller, L.A. (1980). "Cognitive Processes in Design", Int. Journal of Man-Machine Studies, No. 12.

3. Michie, D. (1980). "Expert Systems", Computer Journal, Vol. 23, No. 4.

4. Alty, J.L. & Coombs, M.J. (1984). "Expert Systems - Concepts and Examples", N.C.C. Publications.

A NON - DETERMINISTIC APPROACH TO THE COMPUTER AIDED DESIGN OF THE ELECTRICAL MOTOR

I. Fučík, V. Klevar, J. Pištělák

Research Institute for Electrical Machines, Brno, CSSR

1. INTRODUCTION

One of the characteristic trends we witness in the design and production environment today is the ever increasing urge for accelerating the innovarion rate. Electrical machines represent the classical product of electrical engineering industry. The design methods are thoroughly understood. Most of the computations have been automated in the form of batch mode computer programs. When designing a modern machine a vast number of limitations, national and international parameters and dimensions standards as well as complex optimum criteria have to be taken into account. At the same time, however, the number of special machines or special modifications increases. Under these circumstances the application of CAD systems has been an obvious choice.

The first attempts to use CAD systems for electrical machines revealed somewhat special peculiarity of their design - an unusually high share of diverse computations needed to define the basic geometry of the machine. Without a substantial aid of purpose - oriented program modules the general and universal tools - the drawing or modelling systems - are rather inefficient and cumbersome to use in practice.

The traditional deterministic software packages for the electrical machines have been based on modelling the manual design process and copying its phases into a rather rigid succession of interactive program modules. The system of this kind has been used successfully for the design of unified series of induction motors. The basic conception of all series members must be the same and known well before the system is created and - together with all possible variants - it must be incarnated into

the program. In practice, the low flexibility of such a
purpose - written system proved to be its main bottleneck:

- a new conception of the machine results in costly
  modification of the system
- the user is forced to follow strictly the system
  requirements
- the system is not generally usable and expandable
  for other types of machines
- the system cannot "learn" without extensive
  programming costs.

To overcome the drawbacks a kind of simplified knowledge
- based system described in the contribution has been
suggested as a possible improvement. The system is non-
- deterministic from the user's point of view.

The user specifies his problem and the system offers a
possible solution. The system data-base stores the results
of successful solutions as well as the information
concerning the methods used.

2. STRUCTURE OF THE SYSTEM

The structure of the system comprises five levels (Fig.1).

The first level is represented by the user who specifies
the problem to be solved.

The second level comprises a monitor module. The monitor
is the basic tool for communication between the user and
the system. The monitor includes a user command interpreter
with an interface for various input/output devices. The
monitor analyses the commands and exercisses control over
the rest of the system accordingly.

The third level contains two fundamental subsystems - the
information system and the supervisor. The information
system provides the tools for passive information
retrieval from the system when no new data are created.
The supervisor can be divided into three purpose-oriented
modules: the predictor, the data-manager and the program-
- manager. The predictor looks for the system tools
suitable for solving a particular problem according to
the user specification, the data in the model and the
information stored in the knowledge-base. Generally, the
task of the predictor is to search through the relation
base until a terminal relation is encountered (see chapter
2.1). The control is passed to the data-manager module
then. This module manipulates the data for theforthcoming
active process (chapter 2.3). According to the input
descriptor table of the active process it draws data from

the model and - if missing - from the user through the
monitor. When the input data file is complete and checked
the control is passed further on to the program-manager.
The program-manager module controls the run of the active
process and takes care of non-standard or non-regular
situations.

The fourth level comprises the data access tools for the
heterogeneous data structures down in the fifth level.
The data access tools are represented by the data control
system consisting of two subsystems - the model control
and the data-base control. The model control subsystem
provides the data path to the dynamic data structure in the
model. The data-base control system enables the data flow
from the static data-base CAD/CAM and the knowledge-base.

The fifth level represents the data level divided into
three separate parts. The machine model is a dynamic data
structure. When starting the design of a machine this
model is empty. As the design proceeds more and more data
are generated and stored in the model. In the end the
model represents a complete data model of the new machine.
The CAD/CAM data-base is a traditional static data-base.
It stores information on successfully designed machines,
the production and material parameters as well as other
relevant data providing the general data support for the
design process.

The remaining module in the fifth level is the knowledge-
- base, the conceptual core of the system. The "knowledge"
in the previous systems has not been separately defined;
rather, it has been reflected only in the rigid algorithm
of the solution program.

The knowledge-base of the system suggested can be divided
into two parts - the method - base and the relation - base.

The method - base stores all the methods of solutions for
those problems known to the system. A "method" within the
terms of the system represents a program for solving
particular technical problem together with its input/output
descriptor tables which (running as the active process)
generate new data. In a way, the method - base is a program
library of a special form. The method is considered to be
the final tool for solving a problem.

The relation - base stores the information concerning the
solution of various problems in the form of relations. The
relations describe the relationship between the problem to
be solved and the method at hand. The relation-base
contains a hierarchy of knowledge accumulated in the form
of logical successions of necessary steps for solving the

design problems.

2.1 Relation - base

The relations in the relation - base represent a set of logical triads. A triad consists of three categories:

(PROBLEM) - (SIGNIFICANT - DATA) - (SOLUTION)

The (PROBLEM) category defines the name of a particular problem. This information enables the predictor to find whether the system knows this problem or not. More than one relation with the same name in the (PROBLEM) category are allowed as more solutions can be acceptable for the given problem. The suitable solution is chosen with the help of the next category.

The (SIGNIFICANT - DATA) category contains a list of parameters relevant for the decision whether the solution is acceptable for the concrete problem. The list is accompanied by the table of limits for each parameter.

The (SOLUTION) category comprises the information about the method leading to the solution of the problem. There are two types of relations according to the information in this category:

- the non-terminal relation
- the terminal relation.

The non-terminal relation contains a list of references to other relations. In this case, the "solution" offered is a decomposition of the problem into a succession of lower-level problems.

The terminal relation contains a single reference to a method (an active process).

When the user specifies his problem (by its "name") the predictor retrieves all the relations with the same name in the (PROBLEM) category. If the search is successful (the system knows the problem) the predictor calls to retrieve the real values of variables according to the parameters listed in the (SIGNIFICANT - DATA) category. The values are looked for the model or - if not present yet - asked from the user. After comparing the real values of significent data drawn from the model with the limits allowed in (SIGNIFICANT - DATA) category the system chooses the permissible relations and offers the solutions listed in their (SOLUTION) category to the user. The user is to make the final choice - if there is any left. After this decision the system either starts the appropriate active process if the relation was a terminal one or repeate the procedure again to find the solutions of the

new lower grade problems one after another if the relation
was a non-terminal one.

The relation-base structure does not permit recursion.
After a finite number of steps a terminal relation must
be encountered (even an empty one) to prevent the system
from a deadlock.

A simple example of a non-terminal relation:

(MOTOR-DESIGN) - (P/1,100/,...) - (ELMAG-DESIGN, CON-DES)
(CON-DES)   - (P/1,10/, ...) - (STATOR-DES,ROTOR-DES)

The first parameter P/1,100/ in the (SIGNIFICANT-DATA)
category list represents the performance of the machine
permitted within the limits one to one hundred kilowatts.

An example of a terminal relation:

(ELMAG-DESIGN) - (P/1,500/,)...) - (* ELMGDES 1)
- where ELMGDES 1 is the electromagnetic design program
in the method-base.

2.2 Machine model

The dynamic machine-model data structure provides a tool
for the unified data flow through the whole succession of
design steps. It is there, where the system turns to be
purpose-oriented and application dependent. The creation
of such a model suitable in the real practice is one of
the decisive and fairly non-trivial tasks. The model has
to be a single logical data structure composed from many
incompatible and heterogeneous data groups. The data have
to be sorted out into an unambiguously defined structure
to avoid redundancy and mismatch when retrieved for an
active process. Furthermore, the data have to be
accessible and distinguishable by their semantics.

Physically, the data describe the electrical and mecha -
nical parameters of the machine, the parts geometry, the
production data etc.To connect the heterogeneous data
logicaly the chain data structure has been used due to
its flexibility (Fig. 2).

2.3 Active process

The system based on the concepts described in chapter 2.
results in necessity to apply certain rules for the form
as well as the input/output structure of the system
programs.

It is useful to define three logical terms according to
the functional and implementation aspects of the programs.

A program - a coded algorithm for solving particular design problem. A tool to be run and generate data.

A method - a program in the method - base together with its input/output descriptors. A tool of the system to solve a specific problem.

An active process - a state of the system when a method is applied. New data are to be generated.

All the conversation between the user and the system runs under the control of the monitor. The programs in the method-base should be kept short of own conversation as much as possible.

The programs possess a unified input/output facilities. Each of the programs is accompanied by its input descriptor and output descriptor. The descriptors are implemented in the form of tables. The tables contain the data identifiers, the limits of values and the formats. This information is used by the data-manager to prepare the input data file before starting a program and to transfer the data from the output data file back into the model. The input/output operations of the programs are kept as simple as possible limiting themselves to a sequential file read/write or a passive CAD data-base access.

3. CLASSES OF ACTIVITY

At the first (the user) level, there are five basic classes of system activity. The classes are closely related to the function of lower-level modules.

INFORM — the passive information retrieval from the system. Typically, the data are read from the CAD/CAM data-base or relation-base (e.g. a list of probleme known to the system). Information system active.

PREPARE — the activation of the system. The "empty" machine model is created. The monitor's task.

PROBLEM-DEFINITION — the specification of the problem. A search through the relation- - base to arrive at a solution. The predictor active.

PROBLEM-SOLUTION   - the process of applying a method to the problem. An active process running. The data and program -manager active.

PROBLEM-TERMINATION   - the deactivation of the system. Clearing up the model. Storing data and relations. The monitor's task.

The classes are structured further into their subclasses.

The activity of the system can be seen in a simplified way as a four-state finite automaton (Fig.3). The first stage accords with the PREPARE class. From this, there is only one path to the second stage - the PROBLEM-DEFINITION class. When a non-terminal relation is encountered the system crosses into the stage two again to deal with the lower-grade problem. When a terminal relation is encountered the system moves into the stage three - the PROBLEM-SOLUTION class. After finishing the active process it returns back to the stage two if there is another non-terminal relation to be solved. If the terminal relation was the last one the stage four is reached - the PROBLEM-TERMINATION class.

4. DESIGN PROCESS

To make the system realistic in practice the design and production problems of electrical machines are considered as the primary application determining the underlying data structures.

The design process of an electrical machine generally follows a typical succession of phases:

- specifying the machine required
- check if such a machine is not a one in production
- electromagnetic design
- mechanical design and checks
- production drawings
- NC-programs
- CAP
... etc.

All the phases up into detailed problems can be described by means of a hierarchy of relations within the scope of the system described. However, a direct solution of lower - grade problems can be started if desired (e.g. the electromagnetic design only) without a necessary walk- - through along the whole preceeding phases. Hence, for example, the INDUCTION-MOTOR-DESIGN is not a rigid and complex one way program sequence as it has been so far. Rather, it is a flexible description of logical connections among the system tools to yield the desired solution.

5. CONCLUDING REMARKS

A non-deterministic system for computer aided design and production of electrical machines is suggested. Its main features and novelty with respect to the old one are:

- the three categories (triad) relation and relation-
  - base concept
- the active process and method-base concept
- the inherent flexibility
- the more general usability with postponing the application dependent elements down to the data-structure level.

The system can be viewed upon as a specialized operational system. It will be gradually implemented on an in-house high speed local minicomputer network which enables an effective access to the distant data or transactions in the distributed data-bases.

Currently, the first experiments have been carried out successfully to justify the concepts and the next effort. Further experiments are in progress to verify the effectiveness in the real industrial environment as well as the ability to handle non-trivial problems of the design practice.

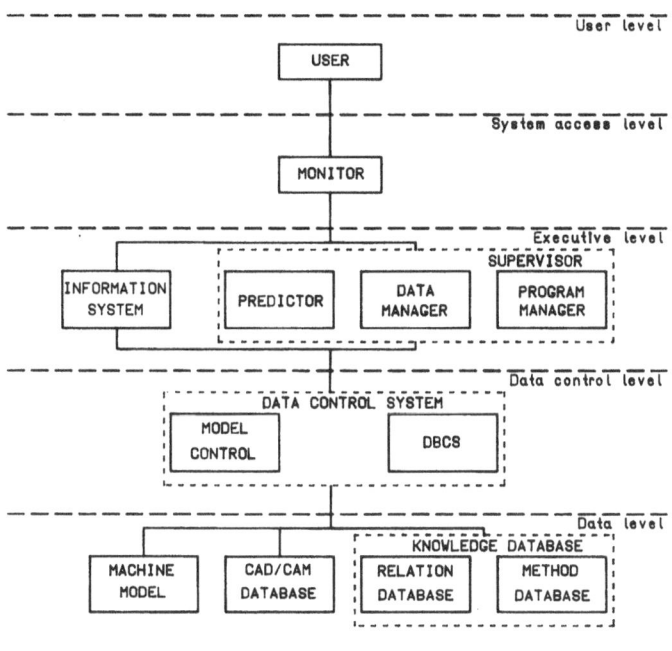

Fig. 1   The structure of the system

Machine data model

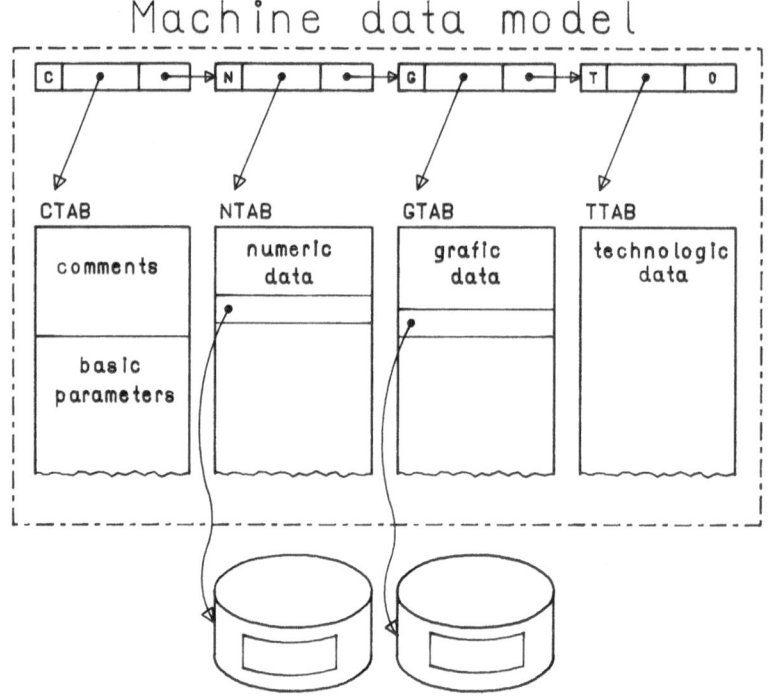

Fig. 2 The machine model data structure
(simplified sketch)

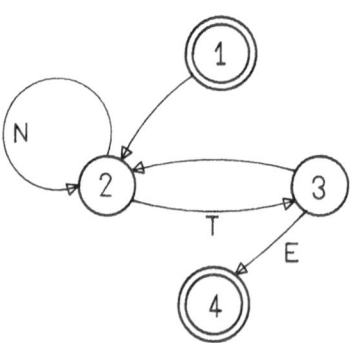

Fig. 3 The system activity graph

VENTURE ANALYSIS WITH A MICROCOMPUTER

G. Singh

Department of Civil Engineering, University of Leeds, LS2 9JT

ABSTRACT

Most engineering and business decisions are made in the environment of risk. The prevalent approach of calculating single values of objectives, based on likely, pessimistic or optimistic values of variables in the decision model, is incapable of distinguishing between high and low risk ventures.

VENTURER, an interactive and tolerant program, based on Monte Carlo simulation, has been developed on a microcomputer to aid decision making incorporating analysis of the risk and that of the sensitivity of the objective to the influencing factors. Application of the program is illustrated with an example of a problem in a civil engineering contracting firm.

INTRODUCTION

As Murphy would put it, "the future is uncertain; you can count on it". Allocation of resources for future gains is, at best, a venture. For the most part managers/engineers concentrate on single values of objectives or outcomes, such as net present value and/or internal rate of return or plant capacity, which are calculated from single value estimates of factors or variables involved in their projects. This deterministic approach is made usually to calculate the likely values of outcomes. In only some of these cases sensitivity analysis is performed on the basis of the estimates of only the optimistic and the pessimistic values of the variables.

A manager/engineer may or may not be in a position to control uncertainties but he should and can obtain a quantitative measure of the risk involved in a venture. Although optimistic and pessimistic estimates of variables do provide an indication of the uncertainty surrounding the estimates of the likely values of the variables, they fail to provide a complete description of that uncertainty. A full probabilistic approach

on the other hand, leading to a probability distribution of the
variable, provides a good indication of the relative
likelyhoods of different values of the variable concerned. In
this paper this distribution is termed UNCERTAINTY PROFILE.
Accepting the reality that the variables and the
objective/outcome are stochastic in nature, incorporation of
the uncertainty profiles into the venture model can lead to
determination of the probability distribution of the values of
the objective/outcome. This probability distribution is termed
VENTURE PROFILE.

Figure 1 shows that if only the likely values of the objectives
of two ventures, A and B, were made available to a decision-
maker then venture B would be prefered. Availability of the
venture profiles, on the other hand, not only provides insight
into the nature of the two projects but also throws a very
different light on their relative merits.

For a complete venture analysis it is also necessary to perform
analysis of the sensitivity of the venture profile to changes
in the uncertainty profiles. The primary purpose of this
analysis is to identify variables which contribute most to the
risk in a venture and to asses the effects of errors in the
estimation of uncertainty profiles. Efforts in obtaining
uncertainty profiles can then be made in proportion to the
sensitivity coefficients of the respective variables. Obviously
the results are also very valuable for directing management
attention and efforts towards reducing or controlling
uncertainties. It has often been found that risk analyses
improve estimates of the likely values of objectives.

From the above it can be seen that the analysis of the risk
sharpens thinking and puts importance of various uncertainties
into proper perspective.

METHODS OF PERFORMING VENTURE ANALYSIS

There are, broadly, two approaches to determination of the
venture profiles, analytical and Monte Carlo simulation. The
analytical method uses only the means and standard deviations
of the uncertainty profiles along with the coefficients of
corelations between the variables to obtain the mean and the
standard deviation of the venture profile (Hillier:1969)
(Wagle:1967) . This method does not provide a complete profile
and it relies on simplifying assumptions which may lead to bad
decisions.

Simulation
Monte Carlo simulation is the best and most powerful systems
approach available for venture analysis. In general, this
method is not only best used with a computer but is also
particularly suitable for programming. There is virtually no
constraint as far as the complexity of the venture model is

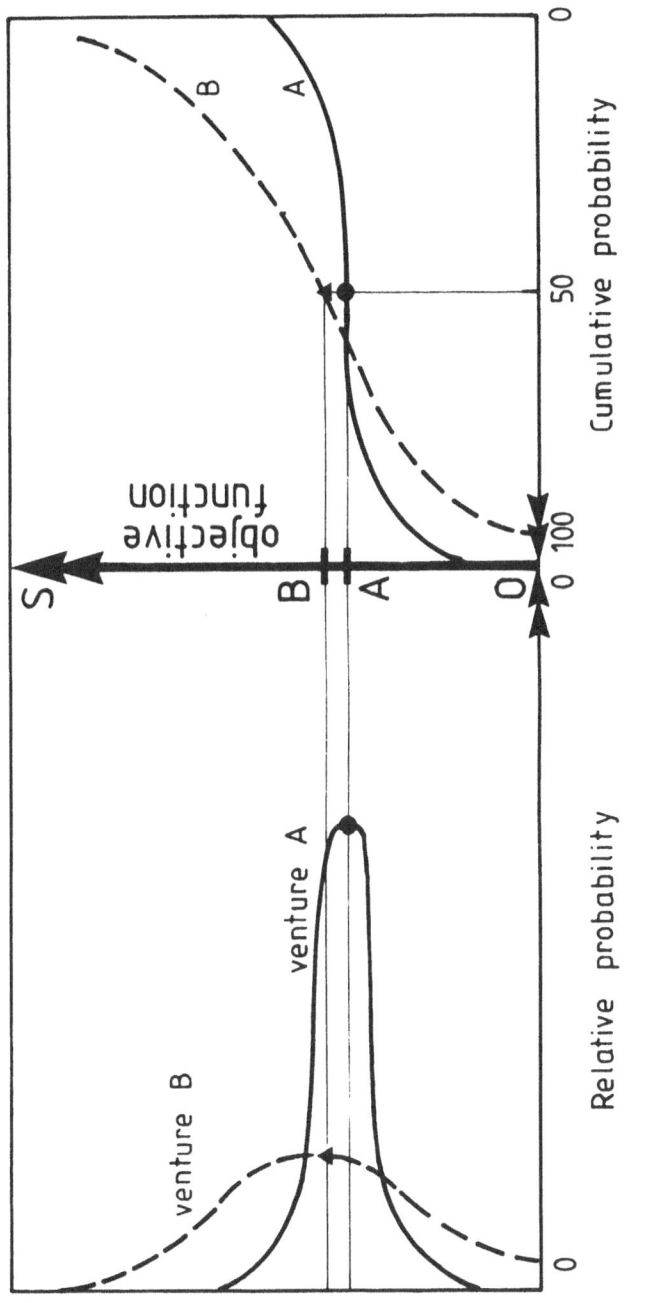

Figure 1   Risk profiles of two ventures

concerned. Despite the revolution in the availability and power
of computers, micros in particular, use of this approach is not
yet widespread in management, but is expected to grow rapidly
(Wagner:1975) (Pouliquen:1970) (Carter:1972) (Jones:1972). Lack
of this widespread use is generally thought to be due to the
large number of uncertainty profiles which the management are
required to estimate. A tolerant and interactive program on a
micro can be used to elicit this information from a single or a
group of managers. It is worth noting that managers are often
more willing to provide uncertainty estimates than they are to
make commitments in the form of single value estimates
(Woods:1966). Figure 2 depicts a typical methodology used in
this approach.

APPROACHES TO MAKING UNCERTAINTY PROFILES

For any type of analysis to be of value it is of upmost
importance that the input data be reliable. A reliable venture
profile can result only from a set of carefully estimated
uncertainty profiles. It should be stated that uncertainty
profiles are mostly subjective in nature. When estimates are
based on historical data it is important to be aware of the
changing environments and to give due weight to the expert
opinion of the management (Hussey:1974). To avoid an individuals motivational and cognitive biases it is often desirable to
use judgement of a group of individuals, but it is necessary to
avoid the possibility of the group being dominated by forceful
individuals. Delphi technique is considered to be useful for
estimating the profiles. The most important feature of this
technique is that it provides feedbacks and opportunities for
managers to improve their estimates (Dalkey & Helmar:1963).

Of the various methods designed to facilitate estimation of
profiles perhaps that suggested by Spetzler and Stael Von
Holstein(1975) is most useful. It comprises of five phases:
motivating, structuring, conditioning, encoding and verifying.
According to them, both fixed and variable methods should be
used for encoding. Studies by Hull(1978) show that fixed
interval methods are more accurate but the variable interval
methods may be found to be easier by managers. Spetzler and
Slael Von Holstein suggest that extremes of the profiles should
be elicited first to avoid "anchoring".

It is important to note that, when identifying variables and
building a model for the objective, independent sources of
uncertainty should be isolated as far as possible.

THE PROGRAM: VENTURER

The previous sections described the motivation for using
simulation and the methodology for venture analysis. The method
of providing input and the roles of feedback and sensitivity
analysis were also described.

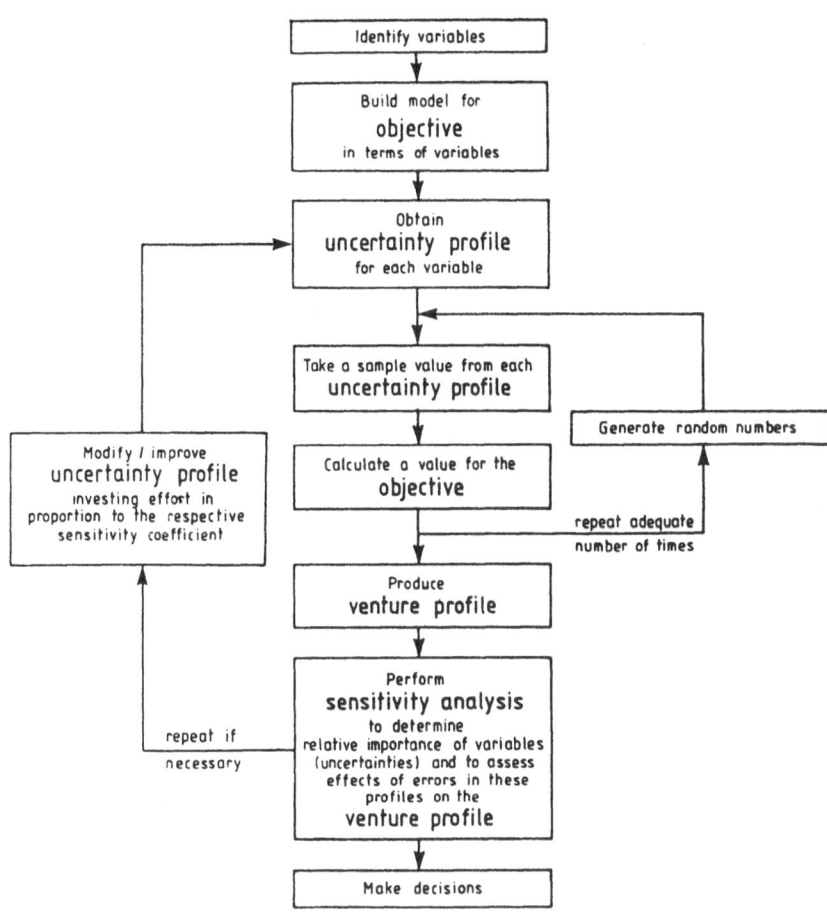

Figure 2   Typical methodology for venture analysis using Monte Carlo Simulation

In the view of the author, microcomputers are the best systems
for carrying out this analysis because of the special needs for
eliciting uncertainty profiles from managers. A single or a
network of micros can be used very efficiently to apply the
Delphi technique and the method suggested by Spetzler and Stael
Von Holstein at uncertainty and at venture profile levels. The
latter level can provide multi-venture analyses.

Planning of the program
In the development of the software attention has been paid to
the aspects of planning of input data, manager interaction,
data storage/retrieval, graphics and output. Particular effort
has been made to make the program user-friendly, tolerant and
completely interactive. Menu driven VENTURER takes the user
step by step, right through the analysis. A colour monitor can
be used for greater effect.

It is assumed that the user is aware of the various variables
(uncertainties) and their relationships, in the form of an
ordinary mathematical equation. Such a relationship or model is
shown in Figure 3.

Input and operation
A special feature of the input is the ease with which the model
can be entered directly and interactively without any need to
access the codes, and with full use of all the arithmatic
operations including exponentiation. Up to ten sets of
parentheses can be used. Meaningful variable names, up to ten
characters long, can be adopted. Input is prompted from the
screen and inbuilt ckecks prevent illogical or erroneous
entries. A previously saved model with data can be loaded in.
All inputs can be easily and quickly modified.

To illustrate the use of the program a contractor's venture,
named "Leeds Civil Contract" is considered. Extreme values of
the uncertainty profiles are elicited first (Figure 4),
followed by three intermediate values which can be chosen at
intervals to suit the needs and/or inclinations of the user
(Figure 5).The user can immediately see her/his inputs in the
graphical form such as Figures 6 to 8 and can revise them any
number of times. These revisions are facilitated through
displays of previous data sets (Figure 10). The number of
simulation runs (minimum 3) and the number of samples per run
(minimum 20) can be chosen to suit the model and the level of
confidence desired (Leeming:1963). For sensitivity analysis the
facility is available for inputting seed value for the random
number generator. This ensures that the effect of a changed
uncertainty profile is not disguised by the probabilistic
nature of the solution.

An algorithm has been adapted to give increased efficiency and
accuracy in simulations.

Venture Analysis

First of all you must build your own venture model.
For example, your model may take the following form:-
objective value=[sel.price-(manuf.cost+sel.cost)]*total
market*share of market-fixed cost.
But MUST be entered as:
    [selprice-(manufcost+selcost)]*totmarket*sharemarkt-
                                                    fixedcost

    so that all variable names consist of ONLY lower
case letters and are no more than TEN characters long.

            press ENTER to continue

FIGURE 3: A typical mathematical model.

Your model is:

  Objective value=(bid+claims)-(labour
cost+matercost+plantcost)

The total number of variables is 5

What value is your variable claims certainly likely to
      exceed ?100000
What value is your variable claims not likely to exceed
 ?
300000

Therefore you have chosen a possible range of 100000 to
 300000

Press ENTER

FIGURE 4: Input format for extreme values of uncertainty.

```
Range for variable claims : 100000 to 300000

Now choose three more values of your variable , one at
a time,and  assign for  each value the  probability (%)
that it will be exceeded.
The values must be entered in ascending order

Enter value no. 1       150000
What is the probability (%) that this will be exceeded?
                  50

Enter value no. 2       175000
What is the probability (%) that this will be exceeded?
                  30

Enter value no. 3       225000
What is the probability (%) that this will be exceeded?
                  10
PRESS c for hard copy OR press ENTER to continue
```

FIGURE 5: Elicitation of uncertainty estimates.

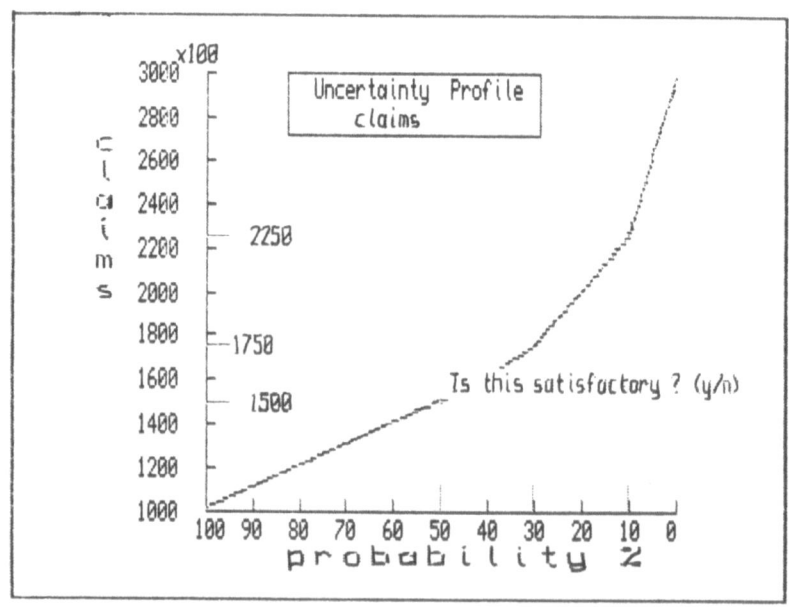

FIGURE 6: Uncertainty profile of CLAIMS: cumulative
probability distribution.

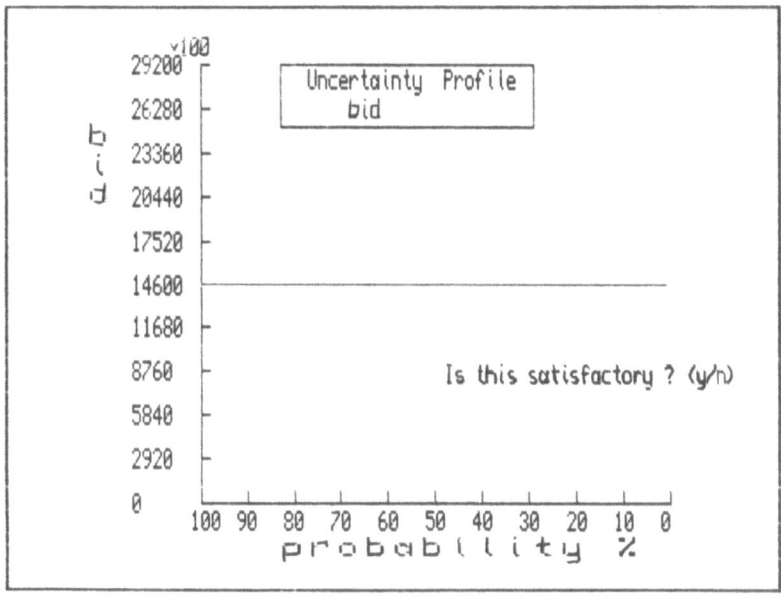

FIGURE 7: Uncertainty profile of BID: cumulative
probability distribution.

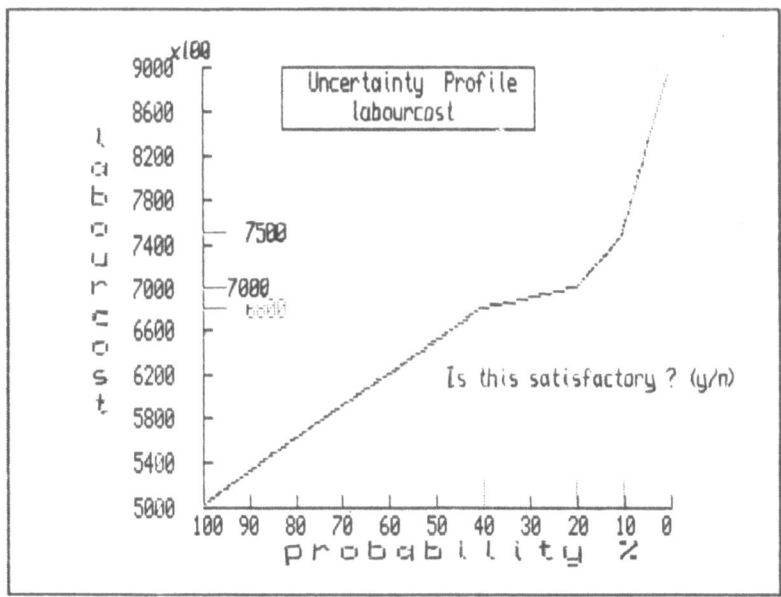

FIGURE 8: Uncertainty profile of LABOUR COSTS: cumulative
probability distribution.

Output
Any part of the input data(including the uncertainty profiles),
the model and the venture profile can be stored or printed out.
Figure 9 shows prints of the results of the simulation using
uncertainty profiles such as those shown in Figures 6 to 8. The
sequence of analysis including that of sensitivity, is recorded
by date and time printouts. A complete file including the cover
page incorporating information such as name of project and date
can be produced without the use of a pen.

Statistical results shown in Figure 9(a) give, for each run,
the values of the averages and the standard deviations of the
objective; as well as the standard deviation of the averages
and the overall average. It can be reasonably assumed that the
averages of the runs are themselves normally distributed.
Therefore one has the confidence that for about two-thirds of
the runs the average will lie within the limits 166916 $\pm$ 9576.
This represents a 5.7 percent error. If desired, this error can
best be reduced by increasing the number of simulations per
run.

Sensitivity analysis
The traditional method of studying the effect of an uncertainty
on the objective value is based on inputs of optimistic and
pessimistic (single) values of that uncertainty. The program
can be used to do this and/or to study the effects of complete
optimistic and pessimistic PROFILES.

Figures 10 to 12 illustrate the use of VENTURER for studying
the sensitivity of the venture profile to the uncertainty
profile of "claims" in the fictitious "Leeds Civil Contract".
This analysis is done through a hierarchy of menus. The
benefits of the analysis are obvious. This analysis can be
repeated for all the variables and their relative importance
established.

CONCLUSIONS

Assessment of ventures based on the uncertainty profiles is not
only more meaningful than that based on single value estimates,
but it also sharpens thinking and puts the relative importances
of various uncertainties into proper perspective.

Microcomputers are most suitable for venture analysis. A user-
friendly and completely interactive program, VENTURER, has been
developed for venture and multi-venture analysis. Its usefulness is illustrated with an example of an engineering contract.

ACKNOWLEDGEMENTS

The author is deeply grateful to Messrs. Ashok Singh and Ranjit
Singh, software consultants, SAR Investment Properties, Leeds,
for their collaboration.

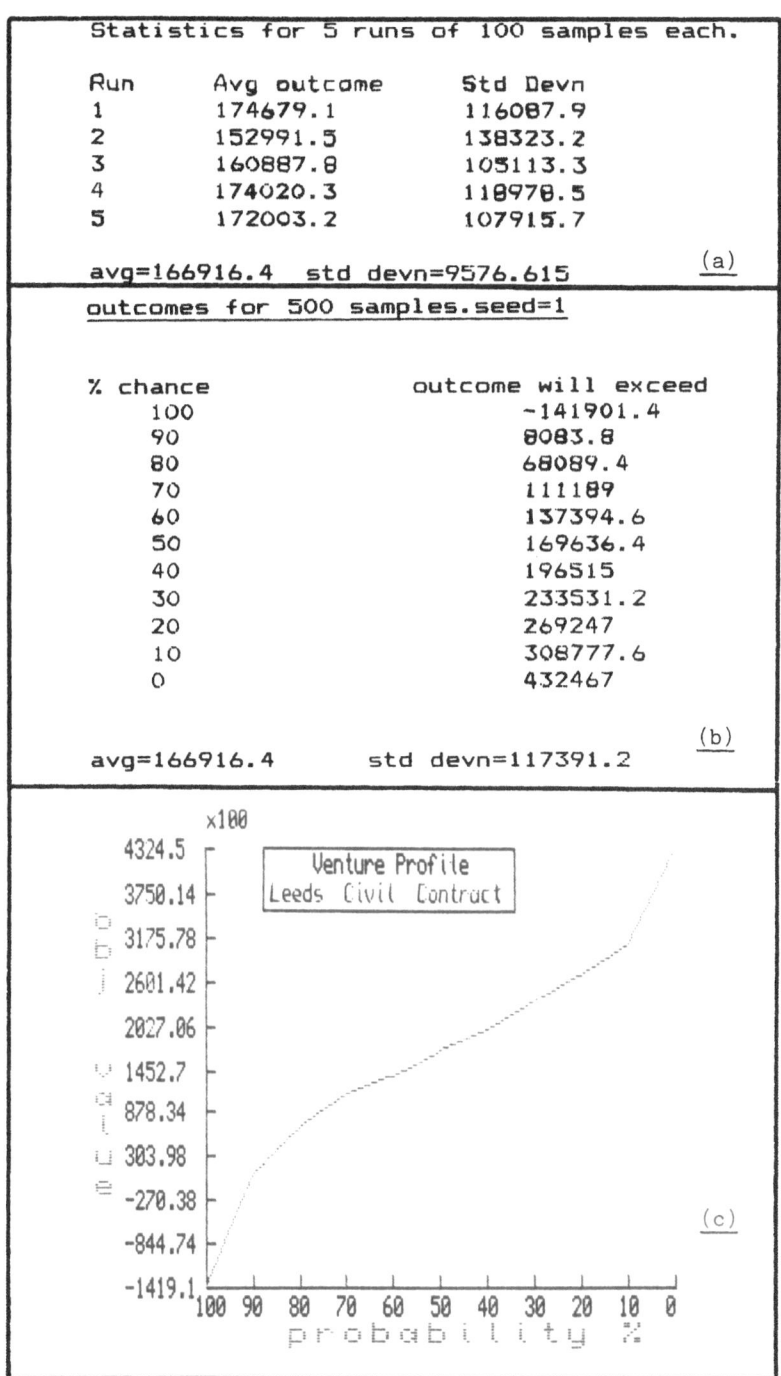

```
      Statistics for 5 runs of 100 samples each.

      Run        Avg outcome        Std Devn
      1          174679.1           116087.9
      2          152991.5           138323.2
      3          160887.8           105113.3
      4          174020.3           118978.5
      5          172003.2           107915.7

      avg=166916.4   std devn=9576.615                    (a)
```

```
      outcomes for 500 samples.seed=1


      % chance                    outcome will exceed
         100                         -141901.4
          90                          8083.8
          80                          68089.4
          70                          111189
          60                          137394.6
          50                          169636.4
          40                          196515
          30                          233531.2
          20                          269247
          10                          308777.6
           0                          432467

                                                          (b)
      avg=166916.4         std devn=117391.2
```

FIGURE 9: Statistical results and VENTURE PROFILE.

```
Range for variable claims : 100000 to 300000


Value no. 1=150000                                  (Old value: 150000)
Probability that this will be exceeded =90          (Old value: 50)


Value no. 2=175000                                  (Old value: 17500 )
Probability that this will be exceeded =80          (Old value: 30)


Value no. 3=225000                                  (Old value: 225000)
Probability that this will be exceeded =50          (Old value: 10)
```

FIGURE 10: Input format for sensitivity analysis.

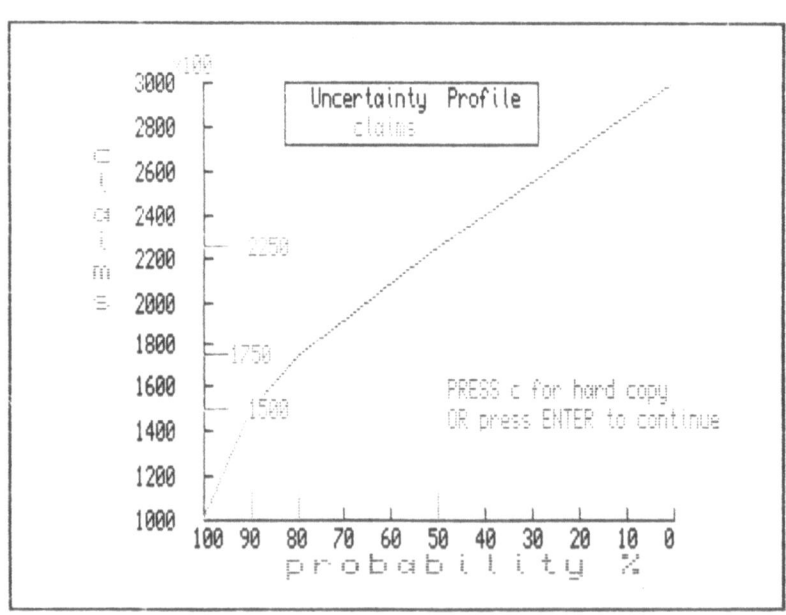

FIGURE 11: Revised uncertainty profile of CLAIMS for
sensitivity analysis.

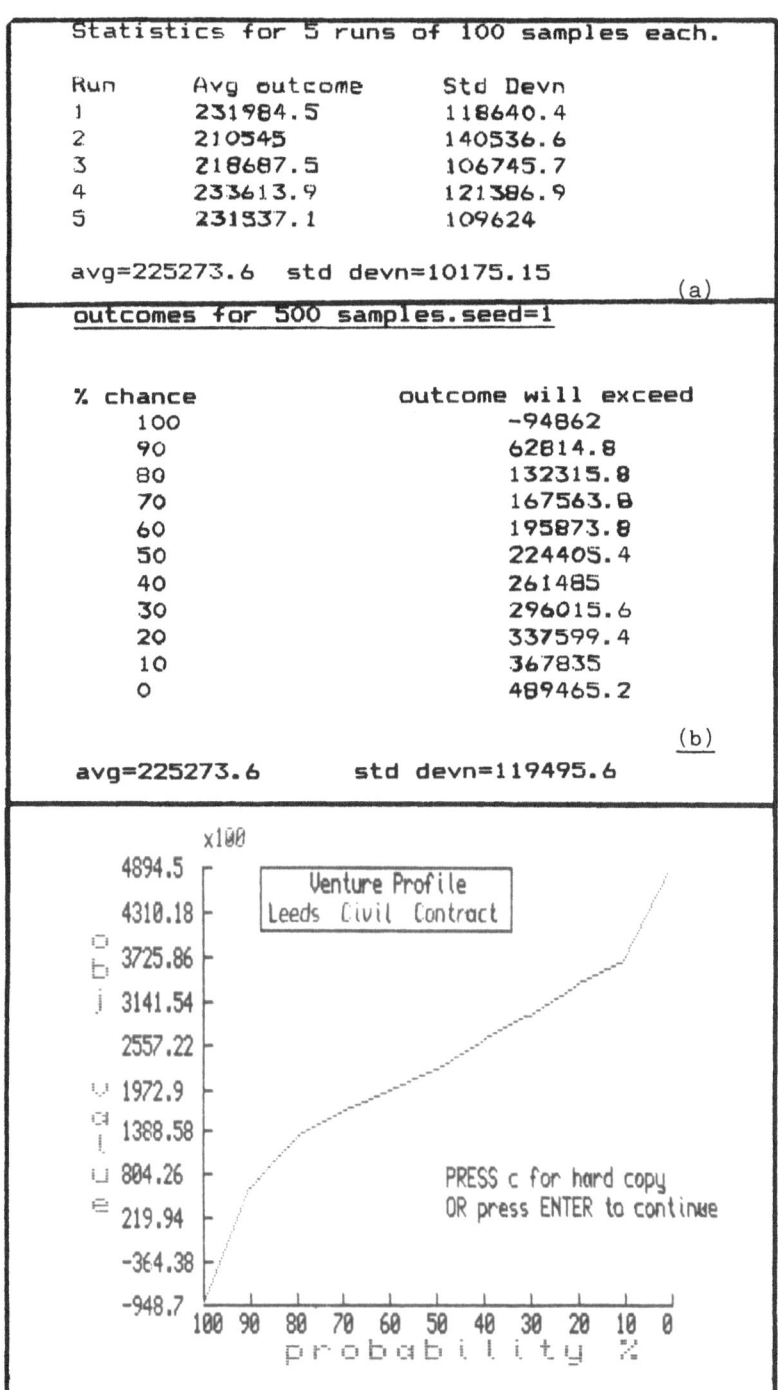

```
Statistics for 5 runs of 100 samples each.

Run      Avg outcome      Std Devn
1        231984.5         118640.4
2        210545           140536.6
3        218687.5         106745.7
4        233613.9         121386.9
5        231537.1         109624

avg=225273.6   std devn=10175.15
                                            (a)

outcomes for 500 samples.seed=1


% chance                    outcome will exceed
   100                         -94862
    90                         62814.8
    80                        132315.8
    70                        167563.8
    60                        195873.8
    50                        224405.4
    40                        261485
    30                        296015.6
    20                        337599.4
    10                        367835
     0                        489465.2

                                            (b)

avg=225273.6          std devn=119495.6
```

FIGURE 12: Revised VENTURE PROFILE revealing its sensitivity
to CLAIMS.

REFERENCES

Carter, E.E. (1972) What are the Risks in Risk Analysis. Harvard Business Review,July-August: 72-82.

Dalkey, N. and Helmar, O. (1963) An Experimental Application of the Delphi Methods to the Use of Experts. Management Science, Vol.9: 458-467.

Hillier, F.S. (1969) The Evaluation of Risky Interrelated Investments. North-Holland.

Hull, J.C. (1978) The Accuracy of the Means and Standard Deviations of Subjective Probability Distributions. Journal of R.S.S., Vol. 141, No. 1: 79-85.

Hussey,D.E. (1974) Corporate Planning: Theory and Practice. Pergamon Press, Oxford.

Jones, G.T. (1972) Simulation and Business Decisions. Penguin.

Leeming, J.J. (1963) Statistical Methods for Engineers. Blackie and sons.

Pouliquen, L.W. (1970) Risk Analysis in Project Appraisal. World Bank Staff Occasional Paper No. 11. Johns Hopkins Press for the International Bank for Reconstruction and Development.

Spetzler, C.S. and Stael Von Holstein, C.S. (1975) Probability Encoding in Decision Analysis. Management Science, Vol. 22: 340-358.

Wagle, B. (1967) A Statistical Analysis of Risk in Capital Investment Projects. Operational Research Quarterly, Vol. 18, No. 1:13-33.

Wagner, H.M. (1975) Priciples of Operations Research. Prentice-Hall.

Woods, D.H.(1966) Improving Estimates that Involve Uncertainty. Harvard Business Review, July-August: 91-98.

EXPERT SYSTEMS IN THE SELECTION OF PROCESS EQUIPMENT

P. Norman & Y.W. Voon

Department of Chemical Engineering,
University of Newcastle-upon-Tyne.

ABSTRACT

Process equipment is specified at the design stage by well established technical considerations and such calculations have been the subject of intensive committment to solution by computer. However, **selection** of items of equipment is largely dependent on the skill and experience of specialist procurement engineers and access to large unco-ordinated databases such as manufacturers' technical literature.
The latter area is amenable to the application of expert systems where judgement and experience are important criteria. This must be combined with conventional numerical treatment of the design and specification stage. Many existing expert system 'shells' do not incorporate a sufficiently powerful calculation capability.
In this paper, we examine the development of consultation programmes for the selection of process valves, as a prototype of equipment selection, under three different programming environments, i.e. PASCAL, LISP and PROLOG. We examine the advantages and disadvantages of these three different environments. We also discuss efficient ways of making decisions for process valve selection.
PASCAL, LISP and PROLOG are all suitable for this task and we suggest appropriate decision making techniques within each environment.

1.0 SYSTEM REVIEW

We investigate the development of Expert Systems in the Selection of Process Equipment (ESSPE) by examining the development of Expert Systems for Valve Selection (ESVS). Valve selection is chosen for the investigation because valves are sufficiently varied yet relatively simple pieces of equipment.

The fundamental task of ESVS is to provide consultative advice for the user in the course of selecting a suitable valve type or design, from the various types and designs of valves manufactured for the process industry.

1.1 Goals
1. To investigate the suitability of PASCAL, LISP and PROLOG for the development of ESVS.
2. To investigate the suitability of various decision techniques as the reasoning of ESVS.
3. One of our goals in the development of ESVS is to implement such a system on small machines, for example microcomputers, so that they may be made available for a wide spectrum of users.

The goals were examined by constructing the various versions of ESVS with these programming languages and with different types of decision making implemented in each version.

1.2 Constraints of the System Knowledge
The present system selects suitable materials of construction and valve-types; the selection of designs within types is not considered.
A material of construction is first selected, followed by the selection of a valve-type.
The two most important factors, i.e. temperature and corrosion resistance, are considered towards the selection of suitable materials of construction.
Ten examples of materials of construction are considered, as shown in Table 1.

Table 1: Some Common Materials of construction
[BVMA 1980; Perry & Chilton 1982]

Material of Construction	Temp. range oC	
	From	To
Iron	ambient	220
Steel	-200	600
Bronze	-200	280
Type 304 stainless steel	-255	800
Type 316 stainless steel	-255	800
'20' alloy	-255	800
Monel alloy (Ni/Cu alloy)	-255	800
Hastelloy alloy B	-255	800
Hastelloy alloy C	-255	1100
Aluminium and alloy	-255	205

The types of fluid media and their corrosion resistances at different temperature and concentration are based on the information in the

corrosion tables shown in [BVMA 1980].
Six valve-types, i.e. globe, ball, butterfly, plug,
gate and diaphragm valves are considered.
18 different working conditions are considered, and
are shown in Table 2.
Table 2 : Some Working Conditions with Valve Choices
[Zappe 1981]
1=Globe,2=Ball,3=Butterfly,4=Plug,5=Gate,6=Diaphragm.

WORKING CONDITIONS	VALVE CHOICES
Starting & stopping flow	1 2 3 4 5 6
Controlling flow	1 3 6
Moderate Throttling	2 4
Flow diversion	2 4
Frequent valve operation	1 2 3 4 6
Gases	2 3 4 5 6
Gases essentially free of solids	1 2 3 4 5 6
Liquids	2 3 4 5 6
Liquids essentially free of solids	1 2 3 4 5 6
Abrasive slurries	3 4 5 6
Non-abrasive slurries	2 3 4 5 6
Powder	3 5
Granules	3 5
Sticky fluids	4
Pharmaceuticals & food products	3 4 6
Fibres	5
Vacuum	1 2 3 4 5 6
Cryogenic	1 2 5

1.3 Valve Selection Criteria

The selection procedure is as follows:
1. Select suitable material of construction: first select suitable material(s) with respect to temperature, then select the material with the best corrosion-resistance.
2. Select suitable valve-type for the required working condition(s).

Selection criteria evaluated from a literature survey on valve selection are implemented on ESVS. These criteria may well not be complete or 100% accurate, because such knowledge may be continually updated. An expert system should be capable of accommodating this, and ESVS is expected to possess such capability.

1.3.1 Material-Temperature Selection.

A material is assumed to have an upper working temperature limit (Tu) and a lower working temperature limit (Tl). It is selected if its Tu is greater than the maximum working temperature (Tmax) required, and its Tl is smaller than the minimum working temperature (Tmin) required. For example, if the required Tmax and Tmin are 100oC and 0oC respectively, then Bronze may be selected since its Tu=280oC and Tl=-200oC.

1.3.2 Material-Corrosion Selection. Corrosion resistances of the materials selected by temperature are compared. Materials with higher corrosion resistances have higher preference than those with lower corrosion resistances. The corrosion resistance data are based on the corrosion resistance table shown in [BVMA 1980].

1.4 Programming Environment

Three high level programming languages are used to construct versions of ESVS in order to investigate their suitability for the task. The languages chosen for study are PASCAL, LISP and PROLOG.

PASCAL is chosen as a representative of algebraic languages, because numerical calculations may be required in the course of equipment selection. It is better than LISP and PROLOG in terms of calculations involving real numbers. It is available on most microcomputers and its syntax is clear.

LISP is chosen because it is widely used as an Artificial Intelligence (AI) programming language. It is a LIST processing language, lists being important data structures for the representation of knowledge as opposed to numerical data. One of its special features is that programmes and data share a common format (There is no distinction made between programmes and data).

PROLOG is also widely used as an AI programming language. Unlike PASCAL and LISP, which are PRESCRIPTIVE or ALGORITHMIC languages, PROLOG is a DESCRIPTIVE language [Clocksin & Mellish 1981; Swinson, Pereira & Biji 1983]. Its decisions are based on predicate logic (PROLOG stands for PROgramming in LOGic). It also manipulates list and symbols.

1.5 Decision Making

Decision making is the mechanism for using knowledge and is sometimes referred to as the 'inference engine'. Several decision techniques are used in the construction of separate versions of ESVS in order to investigate their suitability for the reasoning of this system. The most relevant techniques are: flowcharts, rules, intersection of sets, catalogue searching and probability [Voon 1985].

The flowchart is chosen for investigation because it is a common decision making technique used in pro-gramming with conventional languages. It is the algorithm of a programme in arriving at a decision.

Rule-based decision is chosen for investigation because expert systems, known as PRODUCTION SYSTEMs [Barr & Feigenbaum 1980], have been developed with rules as decision making tools.
Intersection of sets is chosen for investigation because the SET is a data structure in PASCAL. Sets can also be represented by lists, and hence can be manipulated in LISP or PROLOG.
Catalogue Searching is chosen for investigation because this is how equipment is selected manually, i.e. by refering to the manufacturers' catalogues. A valve catalogue can be constructed by means of data structures in PASCAL, LISP and PROLOG.
Probabilistic decision making is included since it is a widely used technique in expert system construction. Decision by flowchart, intersection of sets, catalogue searching, or rules is described as CATEGORICAL reasoning. This and PROBABILISTIC reasoning may be considered as the two extremes of decision making [Szolovits,Pauker, 1978]. Categorical decision may produce choices of valves without any preference. Under this condition, a probabilistic decision making mechanism may prove useful in reaching the final decision. The categorical decision techniques may be used as preliminary selection tools or as aids towards the final selection.

1.6 Knowledge Representation

The way that the system knowledge is represented depends on what decision technique is used. This is demonstrated as follows:

1.6.1 Decision By Flowcharts.

The knowledge required to decide on a valve type (the selection criteria) is represented by a flowchart or decision tree. Each node of the tree represents a decision point, and a question would be asked at this point. The answer input by the user would decide which branch of the tree should be followed next. In order to simplify the problem, a binary decision tree is used for valve-selection, and the user input is restricted to either YES or NO, for example:

Figure 1 : Flowchart for Valve Selection

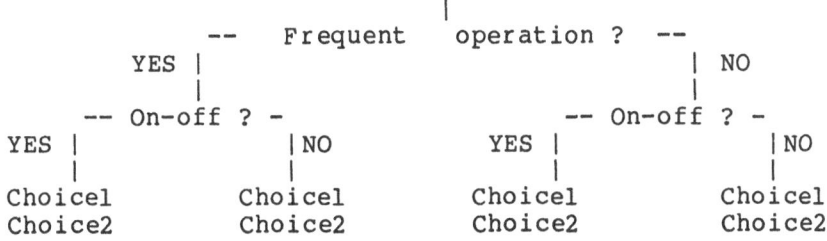

1.6.2 Decision By Rules. The knowledge is represented by rules of the form:

IF <condition> THEN <action>.
For example,
IF
(1) (minimum working temperature > 0) AND
(2) (maximum working temperature < 220)
THEN
(1) Iron may be selected.
Another example is:
IF
(1) the valve is to be used on a gas,
    including steam and air, AND
(2) the valve is to be used frequently, AND
'3) the valve is to be used for regulating flows
!HEN
(1) suitable choices are:
    obturating valve and rotating valve. [BVMA 1980]

1.6.3 Decision By Intersection of Sets. In this method, the knowledge is represented in terms of sets. The complete set of valve-types considered is: {globe,ball,butterfly,plug,diaphragm,gate}.
The choice of valve-type for any particular service is a subset of this. For example, the set of valve-type for flow control is {globe,butterfly,diaphragm}. The 18 different types of service considered are shown in figure 2, a set of valve-type is associated with each of the services.

1.6.4 Decision By Catalogue Searching. The system knowledge is arranged in the form of a catalogue; information of each material or valve-type is kept in separate records, for example:
Name : Diaphragm valve.
services: gases, liquids, on-off, flow-control, etc.
manufacturer: Saunders-valve-Co-Ltd, etc.

1.6.5 Probabilistic Decision Making. The certainty that a valve-type is selected is indicated by probability (real number range from 0 to 1), or likelihood (integer number of any range, eg. -5 to +5 or 1 to 10). For example, the following table [Saunders 1981] shows some likelihood values that the valve-types are selected for control application:

Table 3: Valves Selected for Control Application

Valve-type	likelihood	
Diaphragm	4	Higher value of
Ball	2	likelihood
Butterfly	3	indicates
Globe	5	higher
Gate	1	priority.
Lubricated plug	2	

2.0 SYSTEM CONSTRUCTION WITH PASCAL

There is no difficulty in programming **flowcharts** in
PASCAL. The decision at the node is achieved with an
'IF' statement. In order to facilitate updating of
the programme, as additional knowledge is acquired,
the main flowchart is broken down into simpler
flowcharts each performing a particular task. This is
accomplished by procedures in PASCAL. Modification of
the system at a particular location only requires
updating of a procedure.
For **Decision By Rules,** the programming is exactly the
same as that used for flowcharts. This is because the
syntax of a decision rule is similar to that of IF
statement. In fact, the decision rules for material-
temperature and valve-type selections are derived
from the decision flowcharts.
Intersection of sets presents a very quick way of
categorical decision making. Since the SET is a data
structure in PASCAL, the programming is easier to
handle than the IF-THEN-ELSE statement of flowcharts.
The resulting programme can be much simpler than that
with decision by flowcharts or rules. The
intersection of sets is performed by the operator
'*', which is the multiplication symbol.
Searching a valve-catalogue, in the form of a
database, can be effective. The main difficulty is
database design, but with the data structures of
RECORD and ARRAY in PASCAL, this is not a difficult
problem. Sorting procedures are used to sort out the
properties of valves or materials in order to reach a
decision. In particular, this can be used as an
initial filter to remove inapplicable data.
Finally, PASCAL is particularly effective for the
programming of **probabilistic** calculations involving
real number computation. This is a major advantage
over most versions of LISP and PROLOG.
The key point in the system design in PASCAL is to
store as much of the system knowledge as possible in
a knowledge-base (database) rather than in the

programme itself. This avoids updating of the programme when changes in the system knowledge are needed which would require recompilation of the programme. The kind of knowledge that can be stored as data is listed below:

1. Questions asked during query sessions.
2. Data, such as materials of construction, their names, upper and lower working temperature limits, corrosion-resistance data for different fluid media, and the valve-types, their applications, limitations, manufacturers, etc.
3. The valve choices for all the combinations of working conditions.
4. Selection criteria, or knowledge, about how to select a valve.

The different kinds of data are stored in separate datafiles as different data structures are used to access them. For example, the following demonstrates how the selection rules may be stored and accessed. Six questions would be asked, during a valve-type selection session, and the user input is restricted to either YES(1), or NO(0). (This could also be viewed as: YES means a probability of 1 and NO a probability of 0.) The user input is matched with the rules, which are combinations of numbers, in the database. Hence, the rules are stored as follows:

Response to dialogue	Suitable valve choices
0 0 0 0 0 0	4 5 6
0 0 0 0 0 1	2 3 4 5 6
0 0 0 0 1 0	4 6

The combinations of numbers, in the first six columns, are the rules, while the numbers in the rest of the columns are code numbers corresponding to the choices of valve type. Two arrays are used to access the rules and the choices within the programme.

3.0 SYSTEM CONSTRUCTION WITH LISP

The **decision** flowchart may be represented by a list expression. For example, the flowchart shown in Figure 1 may be represented by the following list:

```
((FREQUENT   (ON-OFF   choice1 choice2)
             NOT-ON-OFF  choice1 choice2)
  INFREQUENT (ON-OFF   choice1 choice2)
             NOT-ON-OFF  choice1 choice2)
```

The technique could be applied to other tree structures which are not binary.

The main disadvantage is that as the decision flow-chart gets larger, the list-expression representing the tree structure gets larger. However, the trans-formation of the flowchart into the list expression would not present any problem if the list structure is constructed "piece by piece" using the function CONS. For example, the small branches are first cons-tructed and then combined to form a larger structure:

```
(SETQ X1 (CONS 'ON-OFF '(CHOICE1 CHOICE2)))
   ==> (ON-OFF CHOICE1 CHOICE2)
(SETQ X2 (CONS 'NOT-ON-OFF '(CHOICE1 CHOICE2)))
   ==> (NOT-ON-OFF CHOICE1 CHOICE2)
(SETQ X3 (CONS X1 X2))
   ==> ((ON-OFF CHOICE1 CHOICE2)
        NOT-ON-OFF CHOICE1 CHOICE2)
(SETQ X4 (CONS 'FREQUENT X3))
   ==> (FREQUENT (ON-OFF CHOICE1 CHOICE2)
                 NOT-ON-OFF CHOICE1 CHOICE2)   etc.
```

Because LISP makes no distinction between programme and data, there is no problem in constructing decision **rules.** New rules may simply be appended to the rulebase.

There is no such data structure as **SET,** but LIST, which is a collection of elements (or no element) enclosed by parentheses, may be treated as SET. If lists are treated as sets, then a function is required to perform intersection, and this can be easily defined in LISP.

For decision by **catalogue searching,** the Property List is a very powerful data structure. For example, the expression (PUTPROP 'DIAPHRAGM 'MANUFACTURER SAUNDERS) assigns the property (MANUFACTURER SAUNDERS) to the atom DIAPHRAGM, where PUTPROP is a LISP function. Similarly, other properties may be assigned. The result is a property list.

Depending on the implementation, LISP can cope with simple arithmetic operations and certain logical operations. Normally, these operations are performed on integer numbers. Therefore, instead of **probabili-ties** (which are real numbers range from 0 to 1) likelihoods(which are integer numbers range, say, from 1 to 10) are used [Voon 1985]. This may be a restriction where real number data must be manipulated to assign probability data.

The major advantage of using LISP is that the decision rules may be represented as data which can carry out actions as a function or programme. Therefore, the generation of new rules creates a dynamic system.

4.0   SYSTEM CONSTRUCTION WITH PROLOG

Since logical decisions can be achieved with PROLOG
itself, the method of decision by **Flowcharts** turns
out to be irrelevant with PROLOG.
Decision **rules** derived from flowcharts can be easily
defined in PROLOG. For example the following rule is
defined for material-temperature selection:

```
suitable_material(X,Tmin,Tmax)
        :- temp_limit(X,Tl,Tu), Tmin>=Tl, Tmax=<Tu.
```

where X is the material.
Decision by **Intersection of Sets** is inefficient if
used in PROLOG. The reason is twofold. Firstly, rules
in PROLOG have to be defined to perform the
intersection (whereas in PASCAL, this is done by the
multiplication operation "*"). Secondly, intersection
of sets is similar to logical decision which can be
done by backtracking in PROLOG using simple recursive
rules. For example, if the sets of choice for "flow-
control" and "cryogenic" are {butterfly, diaphragm,
globe} and {ball,gate,globe} respectively, then
intersection gives {globe}. Similar results can be
obtained in PROLOG as follows:

```
?- application(globe,flow_control).
?- application(butterfly,flow_control).
?- application(diaphragm,flow_control).
?- application(globe,cryogenic).
?- application(ball,cryogenic).
?- application(gate,cryogenic).
                [the following is the query]
?- application(X,flow_control),
                application(X,cryogenic).
X = globe ;     [ this is PROLOG'S answer ]
no              [more answers]
```

A valve **catalogue** or material **catalogue** can be
constructed by merely using facts in PROLOG;
searching the catalogue can be done by backtracking.
Rules can also be constructed to sort out the data,
for example the names of manufacturers in alpha-
betical order [Clocksin & Mellish 1981].
Rules constructed in PROLOG are general decision
rules, i.e. the rules can be applied generally to any
data stored in the facts; while in PASCAL and LISP,
the data is stored rigidly in the rule itself, this
increasing the number of rules very considerably as
the amount of data increases. For example, the rule
shown in 1.6.2 has specific data (gas, frequently,
regulating, and valve choices) embedded in it, and is
applicable to only one combination of working
conditions. The following rule defined in PROLOG

contains only variables, and is applicable to any combination of working conditions, as long as sufficient facts concerning "application" are defined:

suitable_type(X,[]).
suitable_type(X,[A|B] :- application(X,A),
                         suitable_type(X,B).

A similar effect can also be obtained in PASCAL and LISP, but extra procedures or functions have to be defined to match the data input by the user with the data stored in the database.
Depending on the implementation, PROLOG can cope with simple arithmetic operations and certain logical operations like the comparison of numbers. Normally, these operations are performed on integer numbers. Therefore, instead of probabilities (which are real numbers ranging from 0.0 to 1.0) likelihoods (which are integer numbers ranging, say, from 1 to 10) are used.
Decision in PROLOG is based on logical deductions from the knowledge stored as facts and rules in the database. PROLOG can be viewed as a general rule-based system and if given some facts and rules, and a question, it would try to deduce some logical solutions to the question. Therefore, a system constructed under PROLOG can be viewed as being a **Rule-based Database System.** The searching of the appropriate rules and facts, followed by logical deduction is done by PROLOG itself, no extra sub-programmes are needed (whereas in PASCAL or LISP, this must be provided in the form a seperate procedure or function).
A major disadvantage of the language for system construction is that of arithmetic facilities.
Due to the large database of facts and rules required, PROLOG is also not very suitable for the development of ESVS on small machines.

5.0 DECISION MAKING FOR VALVE SELECTION

5.1 Material-Temperature Selection
In the selection of material of construction with respect to temperature, the decision is quite rigid, and is best be done by Flowcharts. However, if more factors (e.g. pressure) are involved in the selection then it is better to use a probabilistic decision making technique, e.g. Bayes' Theorem.

5.2 Material-Corrosion Selection
In the selection of material of construction with respect to corrosion-resistance, only one fluid

medium is normally involved at a time, and the corrosion data is simply read off from a corrosion resistance table. Selection here is best done by simple information retrieval. However, if more than one fluid is involved, for example, 50% of the time water and 50% of the time acid, then a probabilistic decision making technique should be used in addition. If the method of selection by scores [Voon 1985] is used, then the score table is similar to a corrosion-resistance table.

5.3 Valve-type Selection

Valve-type selection may be done by any of the decision techniques considered. Probabilistic decision is the most appropriate for final selection, because decision by rules, flowcharts, intersection of sets or searching a catalogue produce categorical results, with the possibility of several choices of equal status.
Either Bayes' Theorem or the second law of probability may be used for the selection of valve-types for a combination of working conditions. For the selection under a single condition, Bayes' Theorem is recommended. Likelihoods, instead of probabilities, may be used for each of the working conditions. For example, instead of using the probability of, say, 0.7 and 0.3, likelihoods(or scores) of 7 and 3 may be used. If the second law of probability is used for the calculation, the results would be the same, if the "products" are divided by the sum of the products. This is the basis of Selection By Score [Voon 1985], which is used in the system construction with probabilistic decision making. The following example demonstrates the selection of valve type for a combination of working conditions: "high-purity" and "control-application", by the method of selection by scores:

Table 4:   Example of Selection by Score.
condition: a = high-purity, b = control-application,

Condition -->	Score a	b	Combined Score(axb)	Probability (axb/total)
Globe	1	5	5	0.106
Ball	2	2	4	0.085
Butterfly	5	3	15	0.319
Plug	1	2	2	0.043
Gate	1	1	1	0.021
Diaphragm	5	4	20	0.426
		total = 47		1.000

One of the advantages of probabilistic decision making is that decision under uncertainty can be achieved numerically. The major problem is that it is difficult to obtain reasonable estimates of the probability that the valve-types would be selected for each of the working conditions.

5.4 Catalogue Searching

In addition to a valve catalogue, a material catalogue, containing information of all the relevant materials of construction, is also essential. Since the two important factors considered in the material selection are temperature and corrosion, each of the material records must contain information about its temperature limits and corrosion property. A decision by catalogue searching requires two elements:
1. A catalogue of materials of construction and a catalogue of valves.
2. Software tools, such as a Data Base Management System, to search, update, examine and sort the data in the database.

Catalogue searching is an important aid to the consultation system, regardless of what technique the system is constructed from. In other words, a DBMS is an important component of this type of expert system. Therefore, the programming environment within which the consultation system is constructed, must be able to provide facilities for the DBMS construction.

6.0 CONCLUDING REMARKS

6.1 Decision making with PASCAL, LISP, and PROLOG

Since, decisions in PROLOG are made by logical deduction from the knowledge stored as facts and rules in the database, it produces the same effect as the decisions made by flowchart, intersection of sets, searching databases and rules. Furthermore, searching a database in PASCAL and LISP resembles backtracking in PROLOG. The searching is done by procedures and functions in PASCAL and LISP, while backtracking in PROLOG is a built-in feature of the system.

PROLOG lends itself naturally to the construction of expert systems for the type of application considered here and is probably easier to use by those who are not programmers. However, some tasks require more than logical decision and the ability to perform mathematical computation, perhaps for manipulating probabilistic data, as well as database construction, recommends the use of PASCAL. A PROLOG written in PASCAL is a good compromise and we have had some success in using the ISO-Standard York Portable

PROLOG which has built-in facilities to 'escape' to a PASCAL environment when required. [Spivey, 1984].
One of our goals was to investigate construction of ESVS on a small workstation. The original target of a 64K machine was only achievable with PASCAL. Both LISP and PROLOG require substantial work areas for efficient performance, although small versions of LISP are available (such as the Cromemco LISP which we tested). The PASCAL employed was PRO-PASCAL. In this, the passing of control to other programmes is known as CHAINing. The system may be controlled totally by a small central instruction programme. The control may then be passed to appropriate procedural programmes to perform special tasks, and then returned to the central programme upon completion. By this means, an expert system of considerable size can be constructed.
The cost and availability of 16-bit workstations is now such that the 64K memory limit is pessimistic. The York PROLOG has been made to run in both an IBM-PC (256K, 16-bit PRO-PASCAL) and a Cromemco workstation (256K, 16-bit D-series PASCAL). We therefore conclude that a PASCAL/PROLOG environment running on current small workstations is a viable way of accommodating an expert system such as ESVS.

6.2 Valve selection
If either PASCAL or LISP is chosen for the system construction, we would recommend the following decision modules:
1.   Decision Rules obtained from flowcharts,
2.   Score values, calculated by a probabilistic method, assigned to each rule.

If PROLOG alone is chosen for the system construction, decision by rules is the obvious choice. PROLOG can be treated as a general rule-based system. The system builder requires only to construct the facts, rules and relevant questions. [Clocksin and Mellish, 1981]. Values, for example, calculated by the method of Selection By Score, may be attached to the facts or rules in order to achieve a probabilistic decision.

References:

Barr, A., Feigenbaum, E.A., (Ed.)   (1981)
The Handbook of Artificial Intelligence Vol. 1,2,
William Kaufmann, Inc., L.A., California.

British Valve Manufacturers' Association (BVMA),
Valve Users Manual,
British Valve Manufacturers' Association Ltd.,
(1980).

B.V.M.A.,
Valves and Actuators from Britain (Buyers Guide),
British Valve Manufacturers' Association Ltd.,
(1981).

Clocksin, W.F. and Mellish, C.S., (1981)
Programming in Prolog,
Springer-Verlag.

Perry, R.H., Chilton, C.H., (1982)
Chemical Engineer's Handbook, 5th Edition,
McGraw Hill Inc.

Saunders Valve Company Ltd.,
Diaphragm Valve Data Manual (G1/E/9/81),
Saunders Valve Company Ltd, (1981).

Spivey, J.M., (1984)
University of York Portable Prolog System User's
Guide,
Software Technology Research Centre,
University of York.

Swinson, P.S.G., Pereira, F.C.N., & Biji, A., (1983)
A fact dependency system for the logic programmer,
Computer-aided design, vol 15 no. 4 July 1983,
Butterworth & Co (Publishers) Ltd.

Szolovits, P., Pauker, S.G., (1978)
Categorical and Probabilistic Reasoning in Medical
Diagnosis,
Artificial Intelligence 11, 115-144.

Voon, Y.W., (1985)
Ph.D. Thesis (in preparation),
University of Newcastle-upon-Tyne.

Zappe, R.W., (1981)
Valve Selection Handbook,
Gulf Publishing Company, Texas.

RELIABILITY AND INTEGRITY OF COMPUTER ASSISTED
DECISION MAKING

Mohammad Hashim

Burroughs Machines Ltd.,Cumbernauld,Scotland,u.k.

1.INTRODUCTION

The bewildering rapid advances of our computer age have
brought us to a cross road of physical and intellectual
illusions where 'real intelligence'(human brain) is being
replaced by 'artificial intelligence' and may be real people
replaced by artificial people(robotics). We also see the
wizardly machines called computers not only assisting us in
our decision making process in our daily life but often
making actually decisions for us even if we dont like them and
have no option to change them either. The noble, refined and
joyful experience of listening, pondering, thinking,judging and
then deciding upon to do or say something seems to have been
taken over by concepts such as expert systems, AI etc.

One of the most important element of computer assisted decision
making process(CADMP) is the set of those unseen commands and
instructions to the machines that do the so called magic work
and what we term as software. The software forms the core of
all Artificial Intelligence(AI) activities with unique requirement. Though substantial efforts are being spent on developing
the heaps of hardware and piles of software for CADMP, little
attention seems to have been paid to the fundamental requirement of integrity and reliability of software particularly
used in decision making process. The impact of integrity and
reliability of software on the society has not been fully
appreciated yet.

This paper discusses the important elements of CADMP in terms
of its socio-technical behaviour with emphasis on software. It
also proposes a reliability prediction model based on complexity and particularly applicable to CADMP and similar software.
The paper also highlights the unique features of computer
assisted decision making process and compares it with the
conventional(manual) decision making process.

2. COMPUTER ASSISTED DECISION MAKING PROCESS

Fig.1 shows a conventional decision making process without the use of computer where the know-how resides in the human brain. If this know-how(the body of knowledge as the experts call it) is transfered from the human brain to a computer memory, the process of thinking also transfers from one location(brain) to another(memory). This sounds simple and we want to leave it to such simplicity for the purpose of our understanding.

Fig. 2 shows the above mentioned location transfer of know-how where apparently nothing else seem to have changed except that human brain is now behaving like a robot. Unfortunately what seems simple is not what it is. The fundamental concept behind AI is the reproduction of 'thinking' mechanism which inspite of being imperfect in many respects, is still able to perform at least as good as a human being in at least one isolated area of knowledge. Computer Assisted Decision Making(CADM) is a complex process of: (i) presentation of a body of knowledge with its rules and guidelines to use it appropriately (ii) the retrievlal of this knowledge when called upon to do so with full reliability (iii) stepwise and logical scanning through the whole body of knowledge(information) and (iv) most difficult of all to make a decision, give a verdict or provide a sensible and authorised answer to the problem posed by the user. These decisions may be,for example, if a person is suffering from a heart disease (medicine); why a large sum(say £50K) should be spent on the construction of a warehouse(engineering); is Mr. XYZ a criminal?(law) and so on.

3. NATURE AND ELEMENTS OF CADMP

Computer assisted decision making is finding its use more and more both in engineering and service industries. Its particular use is seen recently in the dignosis of medical ailments from symptoms. If we look at the entire process of CADM, we find it very democratic. It starts with people writing the intelli-gent software for other people to use it, where the former utilize their 'thinking and intelligence' to spare the latter from using their 'thinking and intelligence'. Hardware or the machines only play a mere mechanical part in this whole process. That is why concepts like artificial intelligence and expert systems cannot be regarded as purely technical in nature. When Aleksander(1) wrote a book on Artificial Intelligence, he named it 'The Human Machine' and very appropriately so.

The CADMP has four basic elements namely, Hardware, Software, Man (as user and as well as manufacturer) and his environment or society.The hardware has already taken a giant step forward in the reliability field. Man as user is being asked to keep his real intelligence away from the process so he is behaving just like robotics in the expert system or AI environment. The

society is receiving the impact of CADMP in the form of ripe
fruit and is not directly involved in the actual process of
computer assisted decision making. We must,however, remember
that man as a manufacturer of CADMP is very much involved. The
fourth component, the software is the most crucial of all and
results from the efforts and interaction of the other three
elements to the maximum ,as it is established that CADMP would
become an impossible task without software explosion(2).

The reliability and integrity of each of the above mentioned
four components is not merely a technical concept or a known
methematical number. It is on the contrary a state of confidence and trustworthiness of each component that adds up to
give the overall integrity and reliability of CADMP.

4. THE UNIQUE FEATURES OF COMPUTER ASSISTED DECISION PROCESS

It seems appropriate to highlight some of the unique features
of CADMP before discussing the integrity and reliability of
this concept. These features are summarised as follows;

(a) In a conventional deceision making process(Fig. 1) man is
always 'on-line' with the problem, solving system and his
environment. In CADMP man as a user is not directly involved
to the high degree of thinking.

(b) CADMP can only house a limited amount of info. (no matter
how large computer storage is)from a body of knowledge which
is very commonly available and used in practice but may not
possess such rules and information that may be of critical
importance under special circumstances or in the solution of
boundary line cases. Decisions in these cases are almost always
urgent requiring high degree of integrity and accuracy as the
balance lies in the fact that a small piece of information is
present or not. Man in conventional process has a ready access
to such information or is known to him, computer can't think
what it requires to fill the gap or what is missing at the
first place.

(c) Rational behaviour, constraints, prejudices, sympathy,
considerations, change of mind etc all play important part in
the overall process of a conventional decision making process.
These factors,however,cannot be programmed into a computerised
decision process hence lacks the integrity and reliability of
decision to be of immense use to those concerned and affected
by the decision.

(d) Before a man makes a final decision he is at liberty to
iterate and reiterate his findings in view of social, economical and behavioural aspects going in favour or against his
decision . This gives him the freedom and opportunity to judge
the overall impact of the decision. He is not reluctant to a
change if need be. During the process of thinking a decision a

man gathers more and more information, wisdom, screening logic and balance between emotions and needs. These factors contribute immensely to the integrity and reliability of decision. A 'computer brain' is, unfortunately unaware of these scenarios and cannot, therefore, be relied upon completely with decisions made by it.

(e) There is always a small sub-set of correct decisions to a problem out of a universal solution set. Man is capable of and has access to this sub-set in entirity. Computer is bonded and can reach only a part of the above mentioned sub-set. This limitation is one of the greatest source of unreliability and lack of integrity of computer assisted decision process.

(f) CADMP exhibits a lack of freedom of thought thus reducing the integrity of the process of decision making.

The above features clearly indicate the need for each element (particularly software being the major component of CADMP) to be designed to revoke these limitations and intentional efforts be made to achieve the desired reliability and integrity of the whole system. Without this the due benefits of CADMP cannot be appreciated.

5. INTEGRITY AND RELIABILITY OF CADMP

Let us go back to the fundamental question of defining the terms 'integrity' and 'reliability' when used in relation to computer assisted decision making process and its components.

Integrity of CADMP

Literary meaning of word 'integrity' is the state of being entire; wholeness; honesty and uprightness. But I think it is more than that. In terms of CADMP it means providing a decision which is 'Timely' with respect to nature, type and requirement of the problem.(Functioning timely in physical time frame is called reliability/availablity).

Let us illustrate this by a well known example. A six minute alert in USA nuclear base following a fault in the system software that almost took the world at the doorsteps of a third world war may be an absolutely reliable from technical or operational view point but it certainly had no integrity and respect attached to it. It was indeed not a honest manifest-ation of computer assisted decision making process.

Integrity also means to respond quickly and strongly to the wrong input to the process thus avoiding unwanted or undesired decisions. The bulk of the responsibility to inculcate the integrity into the computer assisted decision process falls upon the decision process designers i.e. software expertise.

Reliability of CADMP

Reliability is often thought in technical terms, a number
associated to the system, confidence or trustworthiness. It
should, however, be looked into broader framework of social and
environmental scenarios too. The high mathematical probability
that a system will 'decided' rightly to stop functioning if
there occurs a malfunction in one of the sub-systems or components without causing a faulut or a failure is considered to
be the reliability of the decision process. This becomes a
crucial parameter if such decision has to be taken under a
correct function and time frame at the same time. A fault in
the system hardware or software could be more deadly and fatal
than input or data induced faults. This in turn throws the whole
burden on designing-in the reliability for the system components
like software and hardware.

The total effectiveness of a CADMP is therefore a function of
socio-technical reliability component of each element. In very
simple notation it can be expressed as follows:

$$R_p = r_h + r_s + r_m + r_e \quad , \text{ where,}$$

R_p , r_n, r_s , r_m , r_e are the reliabilities of the decision
process, hardware, software, man and environment. With the
advances in technology the values of most of these parameters
have already been achieved as high as possible. The software,
however has not yet achieved the required reliability. This is
the component that shares most of the burden of unreliability
and disrespect. One of the reason for this is the lack of
suitable models that take into consideration the true nature
of software development in relation to its socio-technical
behaviour particularly in the design phase. We shall now try
to concentrate on this component.

6. THE CADMP 'BUG' : THE SOFTWARE

Complexity plays a very significant part in the reliability and
integrity of a software specially if the software is eventually
to be used in complex and strategic dicision making situation.
Many works exist in the literature that provide models of one
type or the other for software reliability but few take the
social and technical complexity look together. One such work
described in (3) provides a suitable working model and is given
in APPENDIX A along with model equations. It uses a complexity
factor 'C' for prediction of number of errors at various stages
of software life cycle and development cycle. This allows the
management to foresee the trend and thus to mobilize the various
resources to the best use accordingly. The model symbols are
also listed in the Appendix A. The above model has been applied
to data obtained from many large modular software developments
in telecommunication, medical diagonosis software, and from

real time programmes developmed for the flexible manufacturing systems(FMS). The results on one of the software modules are shown in Fig. 3.

7. CONCLUSIONS

The overall objectives of computer assisted decision making is to provide accuracy, efficiency, speed and reliability of decision. The efficiency in terms of nano-seconds may be surely achieved through hardware but other parameters mentioned above and the integrity can not be hammered into as it depends on the software of CADMP that resides inside the machine as a human brain with limited memory. The professionals in the field have a heavy responsibility on their shoulders to make the software as reliable and respectful as possible within the human limits. The manufacturer, user and environment of CADMP is all socio-technical and its impact in case of failure to provide a right decision can be enormous and catastrophic. This component of CADMP is most critical and needs most attention.

This paper has suggested a complexity based error prediction model for the software during its various phases of development (including early phases) for the management and designers alike. No doubt corporate planning in engineering and service industry are looking forward to computer assisted decision making facilities made available to them which are atleast as reliable as their people.

REFERENCES

(1)   Aleksander, I. (1977), The Human Machine, Georgia Publishing Company, USA.

(2)   Hayes, J.E. and Michie, D (1983), Intelligent Systems: The Unprecedented Opportunity, Ellis Howard Ltd., U.K.

(3)   Hashim, M. (1984) 'A Socio-technical Model for the Planning and Prediction of Software Reliability', National Conference on Quality(NCQR'84), Bombay, India.

APPENDIX A

Symbols:

Q_O ,Estimated initial faults
N_f ,Faults found to time t
N_r ,Faults removed in j
N_m ,Maintenance induced faults
N_R ,Total faults removed
N_p ,Faults remaining at t
C ,Complexity factor
K ,Residual factor
L ,Programme(S/W) size
T_O ,Mean time to failure
f ,Fix factor
Z ,Complexity shadow
p ,Patching factor
P ,Programmability factor
s ,Team profile index
m ,S/W modularity
U ,Utility factor
E_d ,Error density
V .Severity factor
j ,Maintenance number
$MTTR$,Mean time to repair

Model Equations

$$C = \frac{U \times E_d \times MTTR}{P \times s \times m} \qquad 1$$

$$Z = \sqrt{C / Q_O} \qquad 2$$

$$f = \sqrt{1-C} \qquad 3$$

$$p = 2\sqrt{C} \qquad 4$$

$$K = \left[\text{Log}((L \times N_f)/(T_O \times \eta)) \right]^{-1} \qquad 5$$

$$Q_O = N_f \times K \qquad 6$$

$$V = j^2 \qquad 7$$

$$N_r = (N_f \times f) \qquad 8$$

$$N_m = (N_r \times V)^p \qquad 9$$

$$N_p = Q_O - N_r + N_m \qquad 10$$

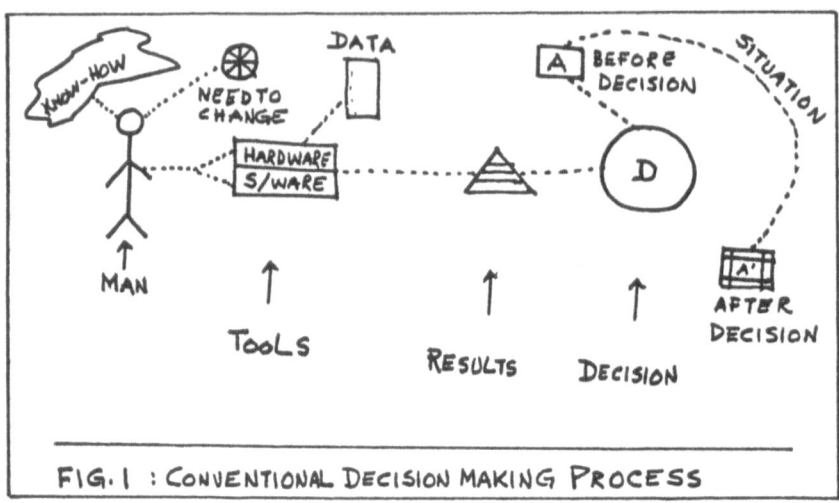

FIG.1 : CONVENTIONAL DECISION MAKING PROCESS

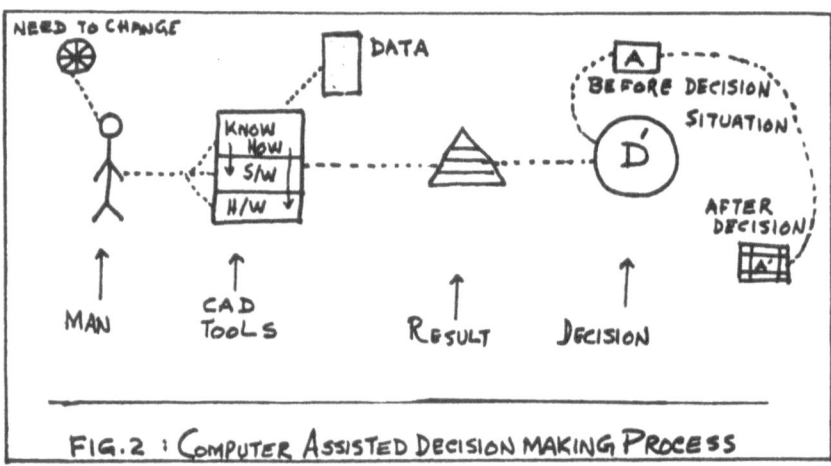

FIG.2 : COMPUTER ASSISTED DECISION MAKING PROCESS

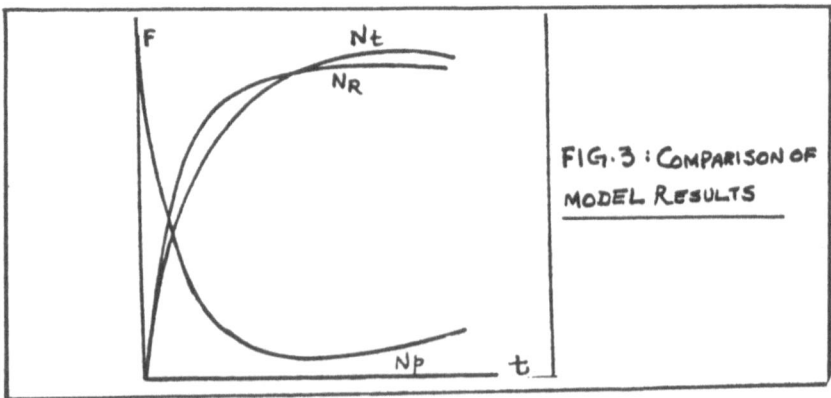

FIG.3 : COMPARISON OF MODEL RESULTS

5. ENGINEERING APPLICATIONS OF EXPERT
   SYSTEMS POSTER SESSION

A RULE-BASED PROGRAM TO CONSTRUCT AN INFLUENCE LINE FOR A
STATICALLY-DETERMINATE MULTI-SPAN BEAM STRUCTURE

G.G.Haywood

Bolton Institute of Higher Education, Bolton, U.K.

1. ABSTRACT

The form of any influence line for an internal or external
force present in a statically-determinate multi-span beam
structure is based entirely on straight lines. An engineer
can readily draw such influence lines by following a few
simple rules of procedure. The program presented in this
paper emulates this procedure by using a Production System, a
classic technique of Artificial Intelligence programming.

The structure is represented by a string of symbols, each
symbol corresponding to a certain type of node; this string
forms the data-base. Rules of the form, LHS implies RHS are
present in the knowledge-base, and have their analogue in the
geometrical properties of the influence line. The LHS of each
rule is matched, if possible, by the interpreter against the
literals of the data-base; if the match succeeds, then the
RHS of the successful rule deposits the y ordinate(s) of the
influence line into the data-base.

At the end of the matching process, the data-base contains
information about the (x,y) co-ordinates of the change points
of the influence line, which may then be plotted by connecting
each pair of (x,y) co-ordinates.

2. INTRODUCTION

An influence line is a means of evaluating an external reaction,
or internal force at a particular point in a structure, when
the structure is loaded by a moving unit load. If the value of
the required force is plotted against the position of the moving load, the graph obtained is the influence line for that
force.

The Principle of Virtual Work can be used to generate a
required influence line directly using the technique given

below, the procedure for which is encapsulated in the following
rules:-

Rule 1. Insert a release into the structure at the position of
the required force; the type of release is to correspond
to the force itself. Therefore, insert a moment release,
or pin, if the force is a bending moment; insert a
roller release to a support, if for a reaction, etc.

The structure resulting from the inserted release is called the
"reduced structure".

Rule 2. Give an appropriate unit displacement to the reduced
structure at the point of release, i.e. give a unit
rotation to a moment release, a linear displacement to a
support release, etc.

Rule 3. Ensure that the deformations of the reduced structure
comply with all the actual support conditions.

It can be shown by Virtual Work (2), that the shape of the
deformed reduced structure is also the influence line for the
desired force in the actual structure. The influence line obtained for a statically indeterminate structure will, of necessity be curved, because the reduced structure will be at the very
least statically determinate; however, if the structure is
itself statically determinate, then the insertion of a release
would render it a mechanism, whose deformation pattern would be
comprised of straight lines only.

The three rules given above are applicable to any class of
stable structure. This paper seeks to show how such rules may
be incorporated into a computer program to construct influence
lines. The technique used is that of Production Systems, a
major technique of Artificial Intelligence programming.

3. PRODUCTION SYSTEMS (PS)

A 'pure' PS consists of three components: a knowledge or rule-
base, a data-base which is operated upon by the knowledge-base,
and an interpreter (1). The knowledge-base typically contains
an ordered set of rules of the form: LHS implies RHS; the
data-base is an ordered set of symbols which models the domain
over which the knowledge-base is to be applied.

The interpreter scans the data-base in an attempt to match
any part of it to similar symbols contained in the LHS of the
knowledge-base rule under consideration; if the match succeeds,
then the symbols in the RHS of the rule replace those corresponding symbols in the data-base.

e.g. data-base contains the string: S = ABCD

knowledge-base contains the rules:

rule (a):   A -: EB   i.e. if S contains the literal "A", then
                         replace it by the RHS literals "EB".
rule (b): FCD -: G
rule (c): BB -: F

An application of the rules to the data-base will cause S to
change in the following manner:

$$
\begin{aligned}
S &= \underline{A}BCD \\
S &= \overline{EB}\underline{B}CD \quad \text{by rule (a)} \\
S &= E\overline{FCD} \quad \text{by rule (c)} \\
S &= E\overline{G} \quad \text{by rule (b)}
\end{aligned}
$$

Because the data-base models a domain, then it is important to
define and describe accurately the limits of the domain if the
knowledge-base is to be applied to it in a meaningful manner.

4.   ESTABLISHMENT OF THE DOMAIN

Rules of the PS type are incapable of describing the actions
necessary to determine the ordinates of the (typically curved)
influence lines for indeterminate structures, because the
values of the ordinates are calculated from a set of procedures which need only a few items of data to be effective,
e.g. span of beam, I-value of cross-section, and stress distribution throughout the structure. The procedures themselves are
derived from a limited set of laws such as: $M/EI=1/R$, the
knowledge for which is implicitly embedded in it, and is
inaccessible to a single rule formulation.

    PS rules are, however, perfectly capable of embodying the
knowledge of the geometrical properties needed to construct
the linear influence lines of a statically determinate
structure. Therefore, the domain of the data-base will be confined to that of the class of determinate structures. In addition, if the one-dimensional type of structure such as a multi-
span beam is chosen, in which the position of each node may be
specified by a single measurement along the structure, this
nodal information may be modelled completely by the ordered
symbolic string as used in a PS data-base.

    Therefore, the extent of the domain will cover only
statically determinate multi-span beam structures.

5.   ESTABLISHMENT OF THE DATA-BASE

The specification of the data-base is that of a list of ordered
3-tuples; the first element, or head, of each 3-tuple being a
symbol which corresponds to the type of node being represented;
the second element gives the distance, x along the structure;
the third element gives the y-ordinate of the influence line

at that node.

The five types of node that may be identified are:
(i) a fixed or cantilever node (C), (ii) a free node (F),
(iii) a support node with the members continuous over it,
(iv) a support node with the members on either side being pinned
to it (H), and (v) an internal pin (P).

e.g. the multi-span structure shown below in Figure 1. would be
represented as:

$$B = (H,0,*)(P,x_1,*)(S,x_2,*)(S,x_3,*)(P,x_4,*)(P,x_5,*)(C,x_6,*)$$

Note: the third item in each 3-tuple is unspecified until
operated upon by the knowledge-base.

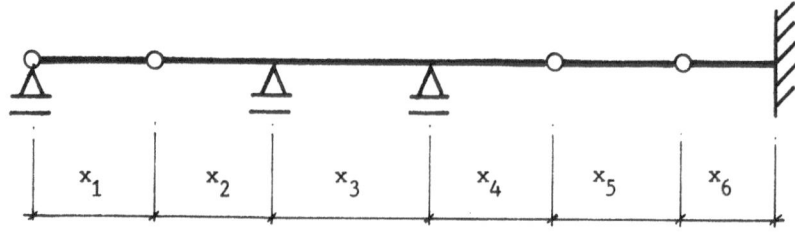

Figure 1.

6. ESTABLISHMENT OF THE KNOWLEDGE-BASE

The three rules given in the Introduction are not in the form
found in a standard PS; they are a series of procedural instructions, which, if followed, would produce the required influence lines. The knowledge they contain is implicit in the
instructions, and it is this knowledge which must be extracted
from these rules, and presented explicitly in PS format. Therefore, the three rules may be regarded as meta-rules or guides
to producing the actual PS rules to be used in the knowledge-
base.

There are two ways of framing the production rules:

(a) **as relevant to a single node**
If the node is a support node, e.g. S,H or C, then the y-ordinates will be defined by Rule 3, which insists on the reduced
structure complying with the support constraints. Therefore,
a PS rule will be of the form:

if the node is "S"
and the required force effect is a vertical reaction
at a support,
and the position of the required force is that of S
then the y-ordinate at S is equal to 1.

This may be expressed as: $(S,-,*) -: (S,-,1)$

where the dash "-" indicates an item in the data-base which does not require matching.

(b) <u>as relevant to a group of nodes</u>
This group is considered to be a single unit, which will appear as an ordered subset of the string B, and will be analogous to a continuous structural element.

Influence lines generated in statically determinate structures consist solely of straight lines which change their slope only at the position of an internal pin. Therefore, the type of node which bounds the specific internal group of nodes forming the unit, will be either a pin (P), or a hinged support (H); if the group is defining the situation found at the ends of the structure, then a fixed support (C), or a free node (F) could also be a boundary node.

An exhaustive survey shows that the total number of possible nodal configurations which comply with the determinate nature of the structure is limited to twenty-two, nine of which are mirror images;   these are shown in Appendix A, and together completely describe any statically determinate beam structure. All other combinations are inadmissible because they would describe implicitly a local region which would be statically indeterminate, or be a mechanism to more than one degree.

The purpose of this complex type of rule is to calculate the, as yet, unspecified y-ordinates of any nodes listed in the group. However, this calculation is difficult to achieve immediately without expressing the possible movements of the defining nodes graphically.

<u>Knowledge-base rules cast in graphical form</u> Figure 2. shows how the y-ordinates of a typical internal unit, (-PSSP-), may be calculated. The y-ordinates of each of the "S" nodes have been inserted previously into the data-base by the application of one of the single node rules.

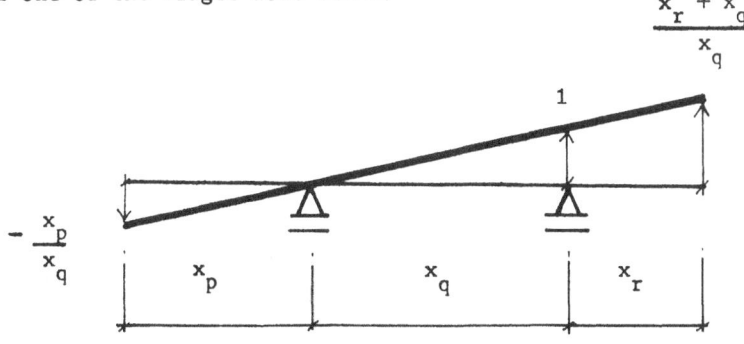

Figure 2.

Note that the linear property of the influence line is expressed implicitly in the calculation of the two y-ordinates.

<u>Knowledge-base rules cast in PS form</u>   Each unit of nodes can
now be expressed as a string e.g. (-PSSP-) which then forms
the LHS of the rule. The RHS replaces the corresponding undefined y-ordinates of the data-base 3-tuples by the calculated
values.

Therefore, the PS rule for the above (-PSSP-) rule is
given by:

$$(PSSP) \; -: \; (P,x_0, -\frac{x_p}{x_q} \;) \; (S,x_p,0) \; (S,x_q,1) \; (P,x_r, \frac{x_r+x_q}{x_q} \;)$$

operations on data-base

This type of PS rule is seen as a direct analogue representation of the pattern undergone by the equivalent beam
element as the reduced structure moves during the application of
the unit displacement to the release.

7.   OPERATION OF THE PROGRAM

The information regarding the nature and position of each node
defining the structure is input to the program using the cursor
control keys and function keys found on a modern microcomputer.
The actual dimensions between each pair of nodes may be given
explicitly so that a quantified influence line may be obtained;
alternatively, if all that is required is a qualitative appraisal, these dimensions need not be given, the resulting influence
line being merely a sketch. In either case, the line is shown
plotted graphically on the VDU screen.

The data-base for the actual structure is then constructed
using this information. The position and type of the required
force effect is next given, and this information is inserted
into the data-base in the correct position; thus the database now contains all information regarding the position and
type of each node and force effect stored as 3-tuples. Each
undefined y-ordinate is given the "*" symbol initially to show
that it is undefined for the matching process.

The y-ordinates at the release which correspond with the
unit displacement to the release are calculated and inserted
into the data-base by overwriting the "*" symbols in the
relevant 3-tuple. All instances of "H","S",or"C" are matched
by the interpreter against the single-node rules, and the
relevant y-ordinates replace the "*" symbols in the tail of the
3-tuples. The only unspecified y-ordinates left are those of
the internal pins or possible free nodes at the ends of the
structure. The interpreter uses the group-nodes rules as

detailed in the Appendix A to calculate and insert the remaining y-ordinate values.

Once the matching process is complete, the data-base contains all instances of the (x,y) co-ordinates of the influence line at all the nodes; the line is then plotted by connecting each pair of (x,y) co-ordinates.

8. DISCUSSION

A pure PS has the following characteristics:

(i) its fundamental mode of behaviour is the matching of the literals found on the LHS of the rule-base. with similar symbols in the data-base, and the subsequent replacement of those same symbols with those given in the RHS of the invoked rule,

(ii) it will not pass control from one rule to another via tags or markers deposited in the data-base, or use such devices to communicate between rules,

(iii) it will exhibit high modularity if rules may be added, deleted or replaced without affecting the action of the other rules in the knowledge-base. This modularity arises from the fact that the successful invokation of a rule is determined by the current state of the data-base,

(iv) there will be no conflict in the order of precedence of the rules; no two rules will compete to match the same set of literals in the data-base.

The PS described in this paper exhibits most of these characteristics, although the matching of literals and replacement of symbols in the data-base is more complex than that of a pure PS. The RHS of the rule uses information contained in the data-base to perform a simple procedural algorithm to determine the actual value of the symbol to be replaced (i.e. the y-ordinate of the 3-tuple of the relevant node). This complex action violates the spirit of a pure PS, but, even so, once the substitution has taken place, there are no devices used to call any other rule directly. A rule is only fired by the current state of the data-base.

The PS exhibits a high modularity because the rules used are exhaustive and unique; each rule matches a particular symbol or set of symbols whose equivalent isomorphic form is to be found in the actual structure. This modularity can be demonstrated by removing one or more rules, and observing the behaviour of the PS; it will continue to behave "reasonably" if the rules are modular. In this case, if the rule given by the LHS configuration, ((C P -) -:) is withdrawn from the knowledge-base, the PS is still able to construct any influence line for the structure, so long as the structure does not have its left-most node fully fixed.

Problems often encountered in other Systems such as resolution of conflicting rules have not arisen, possibly because of the nature of the domain being explored. A statically determinate structure is, by its nature, determined, and the order of application of the rules does not affect the final state of the data-base. There is a direct one-to-one relationship between the LHS of the rules used and the nodes of the actual structure. Because the structural elements are discrete and non-overlapping, then so are the rules. It may be difficult to display this degree of disconnectiveness in a more complex, two dimensional structure such as a plane frame.

An exhaustive pattern search over a wide data-base is generally very time-consuming; some PS programs run into the problem of the combinatorial explosion. This problem does not occur here, because the domain has been deliberately limited to that of statically determinate structures, mainly because of the difficulty, or even the impossibility, of devising suitable rules to define deflected shapes which exhibit curvatures, which occur in the influence lines for indeterminate structures. Therefore, there are only twenty or so rules needed to operate on a data-base which theoretically is limitless in its extent, but, practically, has an upper limit to its length, because it is modelling an actual multi-span structure.

9.   FURTHER DEVELOPMENTS

The program has been written in BBC BASIC to take advantage of the graphics facilities offered by the micro-computer. The language itself, being designed to operate on numerical equations, is not particularly good at manipulating symbolic strings. Therefore, a natural development would be to rewrite the program in LISP.

This would have the further benefit of formulating the rules as chunks of knowledge which may then be extracted and presented as part of a student tutoring program. The student would input the information for a multi-span beam structure, together with his estimate of how the required influence line should be drawn. The program would then be able to use the chunks of knowledge to assess and guide the student towards the correct solution.

10.   REFERENCES

1.   Davis, R. and King, A.   "An Overview of Production
                               Systems"
                               Machine Intelligence Ed. Elcock
                               & Michie (Wiley, 1976)

2.   Thompson, F. and Haywood, G.G. "Structural Analysis
                                       Using Virtual Work"
                                       (Chapman & Hall 1985)

11.   APPENDIX A

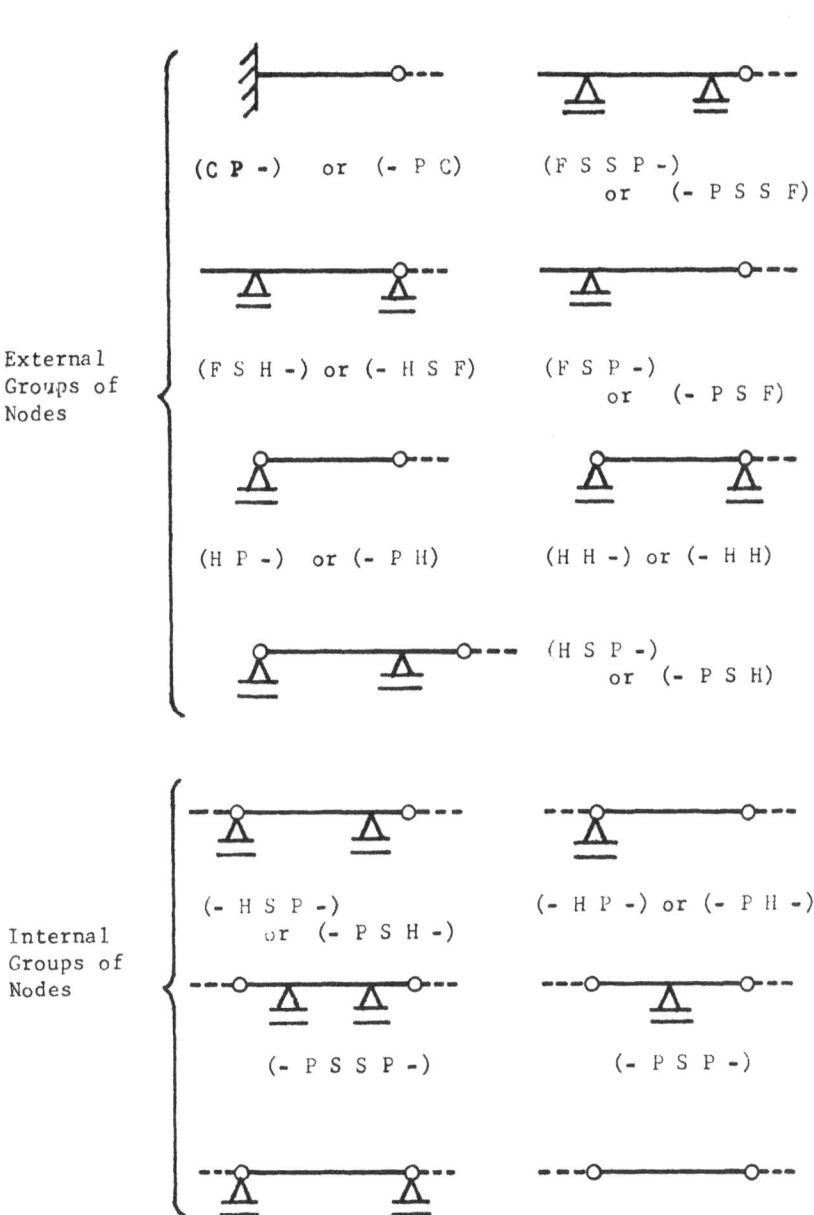

External
Groups of
Nodes

(C P -) or (- P C)   (F S S P -)
                                or   (- P S S F)

(F S H -) or (- H S F)   (F S P -)
                                or   (- P S F)

(H P -) or (- P H)   (H H -) or (- H H)

                             (H S P -)
                                or (- P S H)

Internal
Groups of
Nodes

(- H S P -)   (- H P -) or (- P H -)
  or (- P S H -)

(- P S S P -)   (- P S P -)

(- H H -)   (- P P -)

6.   NUMERICAL METHODS

THE FUNDAMENTALS OF THE SPACE-TIME FINITE ELEMENT METHOD

M. Witkowski

Technical University of Warsaw, Poland

1. INTRODUCTION

The problems of structural dynamics, described by partial differential equations, are usually solved in two phases. Firstly the partial equations system becomes the ordinary differential equations as a result of introducing the generalized coordinates $q(t)$, that are the time functions. The system has a well-known form

$$\underline{M}\,\ddot{\underline{q}} + \underline{C}\,\dot{\underline{q}} + \underline{K}\,\underline{q} = \underline{Q}\,, \qquad (1)$$

where $\underline{M}$, $\underline{C}$ and $\underline{K}$ are the mass, dumping and stiffness matrices, $\underline{Q}$ is the external load vector. All of them could be constants, variables in time or dependent on q, what influences on solving of the equations (1). Introducing to these equations some difference formulae for example central difference or Newmark's patterns we obtain the algebraic equations system. This reveals that space-coordinates are treated differently then time.

We are given also an other method of analysis that consists in direct discretization of four-dimensional space-time and that is a one-phase solving procedure. This process has been described since some time, for example in the works of Argyris, Scharpf, Chan [2], [1] and Oden [8]. In spite of interesting conclusions, particularly in the last paper, these conceptions were not developed.

It was Kączkowski [5] who returned to using the space-time in dynamics of structures. In the next years in Poland, particularly in Technical University of Warsaw, many works of the following authors were published: Kączkowski [6], Kacprzyk and Lewiński [4] and others who wrote in Polish. Moreover, a series of the computing programs based on the space-time element method was also written. In the presented paper the fundamentals of the method, the explanation of some characteristic conception and the bases of construction of the algorithms are introduced.

2. THE SPACE-TIME AND MOTION

Let us consider the motion of the particle in three-dimensional space. In order to describe the motion we usually introduce some coordinates, treating them next as some time functions (Fig. 1). If any coordinate depends

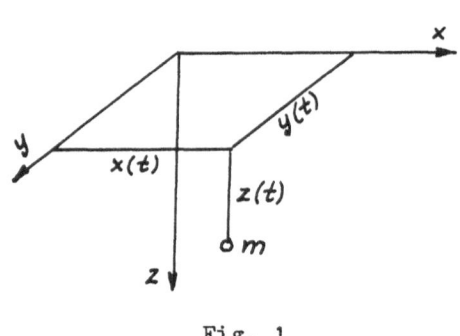

Fig. 1

on time, we admit that the mass m stays at rest, in opposite case it is in a state of motion. The criterions of the motion change when the time is not a parameter but a fourth independent coordinate. First of all we must state that the conception of the particle does not exist in the space-time. It is the line, called the life-line in relativistic mechanics that coresponds with the mass. The properties of the life-line are responsible for a motion or a state of rest. Let us show these properties refering to the two-dimensional space--time (Fig. 2).

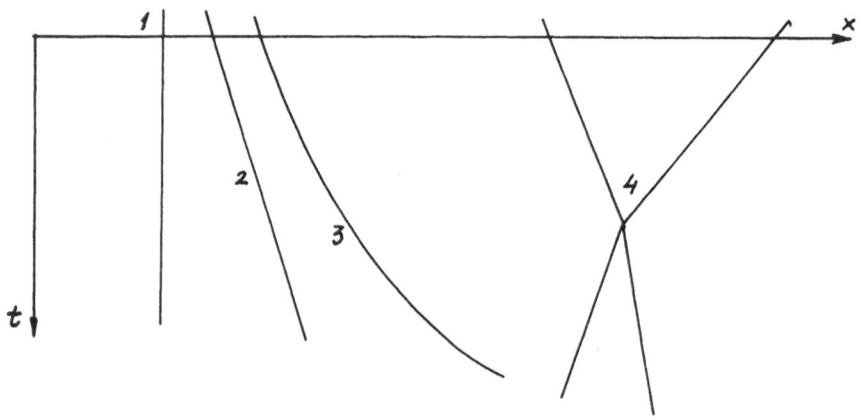

Fig. 2

The life-line 1 describes the immobility of the mass, the straight line 2 reflects the motion with the constant velocity, the curve 3 illustrates the motion with the variable velocity. The intersection of the two life-lines corresponds to the collision of the masses and their bounds (lines 4).

We must remark that any of the lines cannot deviate from the axis t at angle bigger than the relativistic mechanics permets. This condition is rather formal because the applied mechanics deals with very small velocities comparatively to the light speed and only for such velocities the Euclidean space can be the model of the space-time. Sometimes it is easier to construct the space-time as a metric space by multiplying the time by scale speed. In the relativistic mechanics it is the light speed; in the structural dynamics this must be a number much more smaller, for instance the vawe speed in the problems of vawes propagation. This operation brings one serious disadventage; the analized quantities loose their physical sense. It make us refer to the unmetric space-time while considering the problem.

Returning to the problem of the motion, the basic one in the dynamics, we state that in the space-time it looses its sense. The points of the motionless space-time are the sets of the events replacing the classic meaning of the motion. In this case the identic approach we may apply as well to the problems of statics as to dynamics, particularly we may consider the equilibrium equations. In this way for instance we may interpret the life-line of the mass as a space-time bending beam. The conditions of equilibrium for the infinitesimal element are prezented in the figure 3.

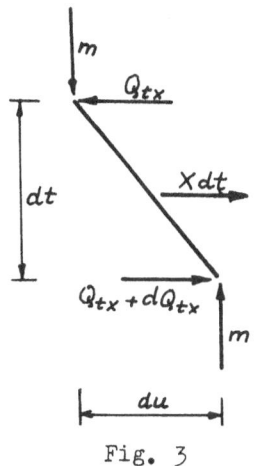

Fig. 3

The euquilibrium equations can be given the following form:

$$\frac{dQ_{tx}}{dt} + X = 0 \quad , \tag{2}$$

$$Q_{tx} = -m \frac{du}{dt} \quad .$$

The second equation indicate that we can interprete the internal force Q_{tx} as a momentum. If we can treat the velocity $\frac{du}{dt}$ as a angle of side γ_{tx} , the equation is a physical law

$$Q_{tx} = -m \, \gamma_{tx} \quad . \tag{3}$$

We remark that the second equation substituted to the first one leads to well-known equation of the motion for the mass with the acting force X

$$X - m \frac{d^2 u}{dt^2} = 0 \quad . \tag{4}$$

Similarly a straight bar is represented in the space-time

as a plate strip (Fig. 4a).

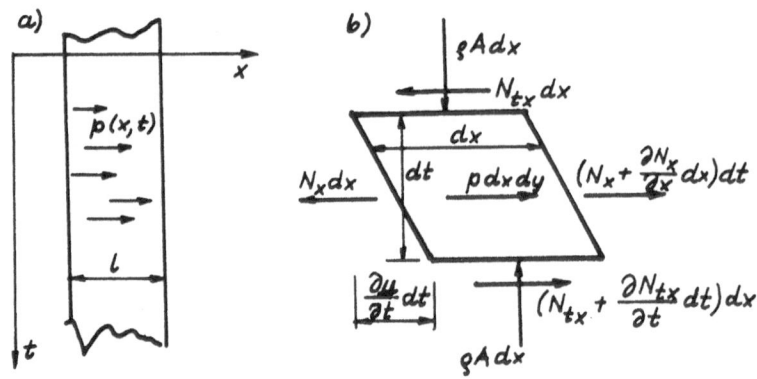

Fig. 4

The properties of the bar are following:
E - the Young's modulus, ρ - the density of the mass,
l - the lenght, A - the area of cross-section.
The conditions of the infinitesimal element of the space-
-time plate (Fig. 4b) lead to the equations

$$\frac{\partial N_x}{\partial x} + \frac{\partial N_{tx}}{\partial t} + p = 0 \quad ,$$

$$N_{tx} = -\rho A \frac{\partial u}{\partial t} \quad .$$

(5)

If we still regard the well-known physical relation

$$N_x = EA \frac{\partial u}{\partial x} \quad , \tag{6}$$

the Eqs.(5) and (6) can be written in the form

$$\underset{\sim}{\varepsilon} = \underset{\sim}{\partial} u \quad , \quad \underset{\sim}{\sigma} = \underline{E} \underset{\sim}{\varepsilon} \quad , \quad \underset{\sim}{\partial}^T \underset{\sim}{\sigma} + p = 0 \quad , \tag{7}$$

where

$$\underset{\sim}{\varepsilon} = \{ \varepsilon_x , \gamma_{tx} \} \quad , \quad \underset{\sim}{\sigma} = \{ N_x , N_{tx} \} \quad ,$$

$$\underset{\sim}{\partial} = \{ \frac{\partial}{\partial x} , \frac{\partial}{\partial t} \} \quad , \quad \underline{E} = \lceil EA , -\rho A \rfloor \quad .$$

(8)

The signs $\{\ \}$ indicate the vector and $\lceil\ \rfloor$ the diagonal matrix.
The above examples show that in the space-time we can ap-
ply a classic method for generating of the equations as
we do in the theory of elasticity. However the vectors
of stresses and strains must be increased by the momenta
and velocities. As a result the dimension of the constitutive matrix changes respectively too.

This reasoning can be repeated for the other structures
for instance the plates, shells and solids. The above way
of generating the space-time domain is used for the linear problems of the dynamics. If we want to consider the
large displacement we can generate the space-time domain
like this in a Fig. 5a. The cross-section of the domain,
that are perpendicular to the axis t, determine the variable configuration of the body.

We can also consider the space-time domains represented in the Fig. 5b. This last case does not mean generating of the matter but its appearance before the observer.
This situation occurs in some boundary-initial problems
descripted by differential hyperbolic equations. The direct solution of the domain Ω requires the generalization of the basic principles of statics for the space-time.

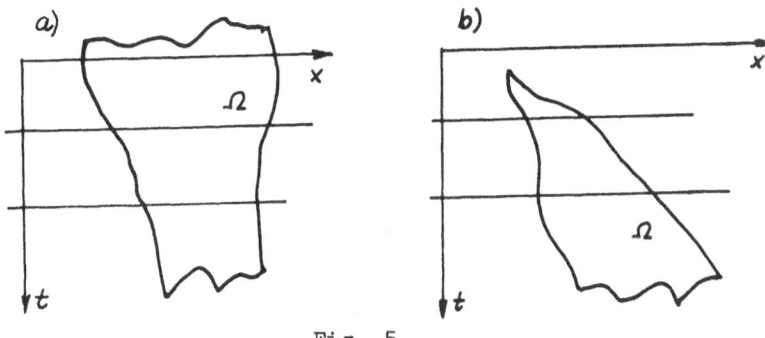

Fig. 5

3. THE VIRTUAL QUADRIWORK PRINCIPLE

The principle similar to the virtual work principle,
well-known in the structural mechanics is presented without derivation for the several important cases. Let us
take into consideration the space-time cylindrical domain.
The axis of the cylinder is parallel to the axis t and
the bases are given the equations $t=t_j=$const $(j=i,k)$.
For the two-dimensional space-time this body is a rectangle; for three dimensions we have the cylinder with the
volume Ω and area S (Fig. 6). In the four-dimensional
space-time we don't imagine this body and we can genera-
lize the conclusions of the principle only in the way of
induction.

We can write the equilibrium equations for the infinitesimal space-time element as follows:

$$\underline{\partial}^T \underline{\sigma} + \underline{p} = \underline{0} \quad , \tag{9}$$

where the matrices $\underline{\partial}$, $\underline{\sigma}$, $\underline{p}$ depend on the considering
problem and for the truss element they have a form of
the Eqs. (8). If the space-time domain represented in the

Fig. 6 is generated by the plane stress element the matrices have the forms, respectively

$$\underset{\sim}{6} = \{ N_x, N_y, N_{xy}, N_{tx}, N_{ty} \} \quad , \quad \underset{\sim}{p} = \{ p_x, p_y \} \quad ,$$

$$\underline{\partial}^T = \begin{bmatrix} \dfrac{\partial}{\partial x} & 0 & \dfrac{\partial}{\partial y} & \dfrac{\partial}{\partial t} & 0 \\[2ex] 0 & \dfrac{\partial}{\partial y} & \dfrac{\partial}{\partial x} & 0 & \dfrac{\partial}{\partial t} \end{bmatrix} \quad . \tag{10}$$

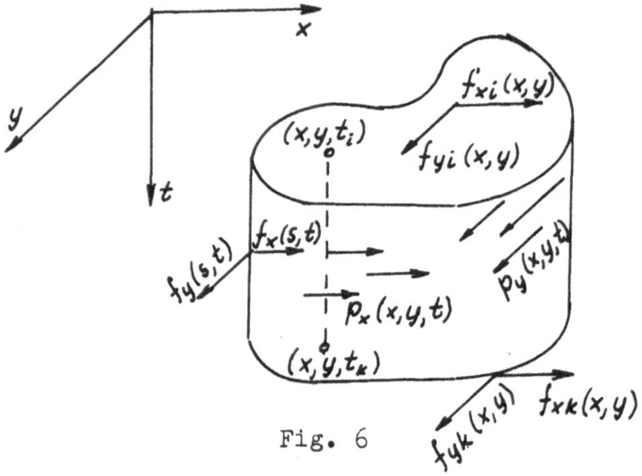

Fig. 6

Hence, the displacements and strains vectors are

$$\underset{\sim}{u} = \{ u, v \} \quad , \quad \underset{\sim}{\varepsilon} = \{ \varepsilon_x, \varepsilon_y, \gamma_{xy}, \gamma_{tx}, \gamma_{ty} \} . \tag{11}$$

Let us give to the space-time body virtual displacements

$$\delta \underset{\sim}{u} = \{ \delta u, \delta v \} \tag{12}$$

and after multiplying the Eqs.(9) by $\delta \underset{\sim}{u}$ we integrate them in the space-time domain Ω

$$\iiint\limits_{\Omega} \delta \underset{\sim}{u}^T (\underline{\partial}^T \underset{\sim}{6} + \underset{\sim}{p}) d\Omega = 0 \quad . \tag{13}$$

Basing on the principle of Gauss-Ostrogradski the above formula can be written

$$\iiint\limits_{\Omega} \delta \mathbf{u}^T \underset{\sim}{p} \, d\Omega + \iint\limits_{S} \delta \underset{\sim}{u}^T \underset{\sim}{f} \, dS = \iiint\limits_{\Omega} \delta \underset{\sim}{\varepsilon}^T \underset{\sim}{6} \, d\Omega , \tag{14}$$

where the surface integral spread over the whole domain limiting the body Ω .

The left-hand member of the Eq.(14) means the scalar products of the impulses by the virtual displacements. The right-hand member is a scalar product of the generalized stresses by the virtual strains. Therefore the scalar products are not work but an other physical quantity so called quadriwork.

In the analysis of the space-time domain represented in the Fig. 5 with the nonlinear geometric relations we act in a different way. we introduce a metric space-time and we derive the principle as for the elastic body in the way showed in the paper of Washizu [9] . The definite form of the quadriwork principle is identic with the Eq. (14).

It would be advicable to mention about the equivalent of this principle in the thermoelasticity problems. The principle of the virtual energy in thermoelasticity has been formulated by Biot [3] . After some modification it has become the basis of the virtual quadrienergy principle derived by Kączkowski [7] . This principle permets the analysis of the thermoelasticity equations with the use of the space-time description.

4. THE SPACE-TIME FINITE ELEMENT METHOD

The space-time domain represented in the Figs.4 and 5 can be direct discretisating by the finite elements. The discretisation of the space-time body is done by choice of finite set of the generalized coordinates $\underset{\sim}{q}$ independent on time and the shape functions $\underline{N}$. Also, the displacements are approximated as follows:

$$\underset{\sim}{u} = \underline{N}(x,y,z,t) \cdot \underset{\sim}{q} \quad . \tag{15}$$

We note that this approximation is different than the classic one, where we assume

$$\underset{\sim}{u} = \underline{N}(x,y,z) \cdot \underset{\sim}{q}(t) \quad . \tag{16}$$

Now, we describe some typical cases of the space-time elements nets.

The rectangular space-time elements
These elements have arbitrary shape in space and their net is invariable in time. It means that any section of the space-time body by the plane t=const gives the identic net of the space elements (Fig. 7). This net also calls the stationary net. The boundaries of the space-time element are always perpendicular or parallel to the time axis. In this case the matrix of the shape functions can be written as a tensor product

$$\underline{N}(x,y,z,t) = \underline{N}_t(t) \otimes \underline{N}_x(x,y,z) \quad . \tag{17}$$

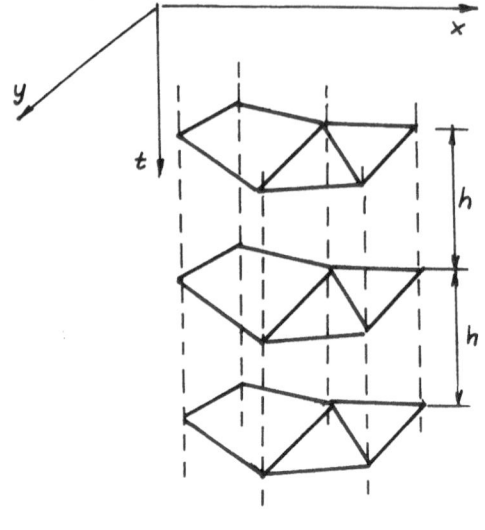

Fig. 7

This assumption with the use of the virtual quadriwork principle enables the conversion of the equations set to the form

$$
\begin{bmatrix}
\underline{A} & \underline{B} & & & \\
\underline{C} & \underline{H} & \underline{B} & & \\
& \underline{C} & \underline{H} & \underline{B} & \\
& & \cdot & \cdot & \cdot & \cdot & \cdot & \cdot
\end{bmatrix}
\begin{bmatrix}
\underline{q}_0 \\
\underline{q}_1 \\
\underline{q}_2 \\
\cdot \cdot
\end{bmatrix}
=
\begin{bmatrix}
\underline{Q}_0 \\
\underline{Q}_1 \\
\underline{Q}_2 \\
\cdot \cdot
\end{bmatrix}
\quad ,
\tag{18}
$$

where

$$
\underline{A} = \frac{h}{3} \underline{K} - \frac{1}{h} \underline{M} - \frac{1}{2} \underline{T} \quad ,
$$

$$
\underline{B}, \underline{C} = \frac{h}{6} \underline{K} + \frac{1}{h} \underline{M} \overset{+}{_-} \frac{1}{2} \underline{T} \quad ,
\tag{19}
$$

$$
\underline{H} = \frac{h}{3} \underline{K} - \frac{1}{h} \underline{M} \quad .
$$

In the formulae (19) h means time step, the matrices $\underline{K}$, $\underline{M}$, $\underline{T}$ are stiffness, mass and dumping matrices for whole space structure, respectively. The Eq.(18) is a difference, conditional stable equation. We can transform it to the unconditional stable form by the suitable choice of the virtual displacements.

The triangular space-time elements

The triangular elements family is a set of the elements which have a triangular, tetrahedronal or "fivesolids" shape.Certainly, we cannot imagine the last shape.

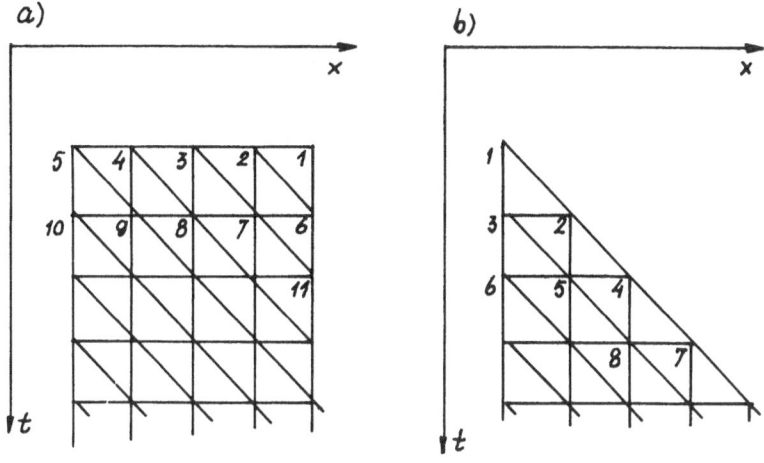

Fig. 8

The idea of these elements is exampled by the triangular
elements (Fig. 8). We take into consideration a net of
space-time elements represented in the Fig. 8a. We can
determine the stiffness matrix and load vector of the
space-time triangles in the standard way as it is in a
finite element method. We must stil remember that the integration refers to the space-time elements. In the case
of the Fig. 8a for the line 1-5 the initial conditions are
known. It permets a very simple construction of the equilibrium conditions. Namely, we generate the equilibrium
equations in the nodal numbers order. From the Eq.1 we
determine the displacement in the point 6, from the Eq.2,
displacement of the node 7 etc. This algorithms enables
us to determine one displacement from one equation.
It seems, it was Oden [8] who first descripted this property of triangular nets. This algorithm can be applied
to the other boundary-initial conditions for instance in
the hyperbolic equations with the characteristics 1-7
(Fig. 8b). We call this net a quasistationary net because
the position of the nodes becomes invariable in space.

The unstationary elements

In the case of the large displacements and of the contact
problems or the other nonlinear analysis we can apply the
space-time unstationary elements. They have a triangular
or tetragonal shapes (in the
two-dimensional space-time)
and the positions of the nodes
are variable in space (Fig. 9).
The configuration of nodes is
determined after solving the
suitable nonlinear problem
step by step.

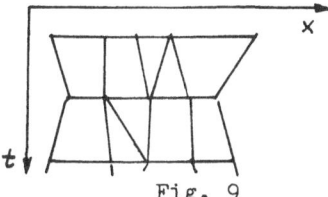

Fig. 9

REFERENCES

1. Argyris, J.H., Chan, A.S.L. (1972) Applications of Finite Elements in Space and Time, Ing.Archiv, 41, p.235-257.
2. Argyris, J.H., Scharpf, D.W. (1969) Finite Elements in Time and Space, Aero.J.of the RAS, 73, p.1041-1044.
3. Biot, M.A. (1959) New Thermoelastical Reciprocity Relations with Application to Thermal Stresses, J.Aero/Space Sciences, 26, p.401-408.
4. Kacprzyk, Z., Lewiński, T. (1983) Comparison of some Numerical Integration Methods for the Equations of Motion of System with a Finite Number of Degrees of Freedom, Engineering Transactions, 31, 2, p.213-240.
5. Kączkowski, Z. (1975) The Method of Finite Space-Time Elements in Dynamics of Structures, J.Techn.Phys., 16, 1, p.69-84.
6. Kączkowski, Z. (1979) General Formulation of the Stiffness Matrix for the Space-Time Finite Elements, Arch.Inżynierii Lądowej, 25, 3, p.351-357.
7. Kączkowski, Z. (1982) On Variational Principles in Thermoelasticity, Bull. de l'Acad.Pol. des Scientes, Appl.Mech., 30, 5-6, p.81-86.
8. Oden, J.T. (1969) A General Theory of Finite Elements. II. Applications, Int.J.Num.Meth.Eng., 1, 3, p.247-259.
9. Washizu, K. (1974) Variational Methods in Elasticity and Plasticity, Pergamon Press, Inc., Elmsford, N.Y.

A COMPUTER PROGRAM FOR LARGE EIGENVALUE PROBLEMS IN DYNAMIC
ANALYSIS

A. Vale e Azevedo

LNEC - National Laboratory of Civil Engineering, Lisbon, Portu
gal

ABSTRACT

This paper describes an efficient implementation of the
Subspace Iteration method to calculate the eigenvalues and eigenvectors of structures with a large number of degrees of free
dom.
A careful study has been made to optimize the use of
computer central memory, minimizing the input/output operations,
and to minimize the number of arithmetical operations.

The computer program has the possibility of generating
automatically the starting iteration vectors or, otherwise, the
user can choose a manual selection.

The overrelaxation of the iteration vectors to accele-
rate the convergence is included with a detailed study on the
choice of the overrelaxation factors.

The Sturm sequence property, to verify if the required
eigenvalues and eigenvectors have been calculated, is also implemented.

Some numerical examples show the most efficient computational implementations.

1 - INTRODUCTION

In civil engineering we often have to solve eigenvalue problems. In the case of Structural Dynamics, frequencies of vibration are associated with eigenvalues and vibration modes shapes are associated with eigenvectors.

In the dynamic analysis of structures under non-damped free vibrations, the problem of the eigenvalues is written as:

$$\underline{S} \cdot \underline{\phi} = \underline{M} \cdot \underline{\phi} \cdot \underline{\lambda} \qquad (1)$$

where $\underline{S}$ and $\underline{M}$ are respectively the stiffness matrix and the mass matrix of the structure;

$$\underline{\lambda} = \begin{bmatrix} \lambda_1 & 0 & \cdots\cdots & 0 & \cdots\cdots & 0 \\ 0 & \lambda_2 & & \vdots & & \vdots \\ \vdots & & \ddots & \vdots & & \vdots \\ 0 & \cdots\cdots\cdots\cdots & \lambda_i & & \vdots \\ \vdots & & & \ddots & & \vdots \\ 0 & \cdots\cdots\cdots\cdots\cdots\cdots & \lambda_n \end{bmatrix}$$

is the diagonal matrix of the eigenvalues, where $\lambda_i = \omega_i^2$ (frequencies of vibration)

and $\underline{\phi} = \begin{bmatrix} \phi_1, & \phi_2, & \dots \phi_i & \dots \phi_n \end{bmatrix}$ is the vibration modes shapes matrix, ϕ_1, ϕ_2, $\dots \phi_i \dots \phi_n$ being an orthonormed set of the vibration modes.

The computer time involved in the resolution of this problem is rather long, large storage areas are used and sometimes numerical precision and stability are important factors.

When the stiffness and mass matrices of the structure are of a high order (some hundreds or thousands of degrees of freedom) and are very sparse, one of the most efficient methods for solving the eigenvalue problem is the Subspace Iteration method developed by Bathe [2].

This paper shows an efficient implementation of the Subspace Iteration method, in wich:

- The computer central processor unity time spent was optimized;
- A balanced use between allocated positions of central memory and transferences to auxiliary memory (disk or magnetic tape) was adopted;
- The problems of numerical precision and stability regarding matrix factorization were taken into account.

2 - BRIEF DESCRIPTION OF THE SUBSPACE ITERATION METHOD

The object of the Subspace Iteration method is to calculate the "p" first eigenvalues and eigenvectors that fulfill equation (1).

All vectors ϕ_i are mutually orthogonal and the normali_zation condition

$$\underline{\phi}^T \cdot \underline{M} \cdot \underline{\phi} = \underline{I} \qquad (2)$$

can be introduced for them, $\underline{I}$ being the identity matrix.

Substituting (2) in (1) we get

$$\underline{\phi}^T \cdot \underline{S} \cdot \underline{\phi} = \underline{\lambda} \qquad (3)$$

The Subspace Iteration method can be divided into three basic steps:

a) Establishment of a set of "q" starting vectors, linearly independent and not orthogonal to the desired eigenvectors, to start the iterative process. If "p" is the desired number of eigenvalues and eigenvectors it is possible to impro_ve the convergence rate using "q" iteration vectors with q > p. The increase of the number of iteration vectors brings about an increase in the computer time of each iterative cycle, so that a balanced solution has, of course, to be found. Reference [2] suggests that "q" be defined as:

$$q = \min \{2p, p + 8\} \qquad (4)$$

b) Use simultaneous inverse iteration on the "q"vectors and Rayleight-Ritz analysis to extract the best approximated eigenvalues and eigenvectors from the "q" iteration vectors. The main idea in the Subspace Iteration method is that the eigenvectors in (1) form an $\underline{M}$ orthonormal basis of the "p" dimen_sional subspace of the operators $\underline{S}$ and $\underline{M}$, wich we will call E_∞. The starting iteration vectors generate an E_1 subspace and iteration proceeds until E_∞ is reached. The total number of iterations will depend on how close E_1 is to E_∞ and not on how close each iteration vector is to the eigenvector it will allow to calculate.

For $\kappa = 1, 2, 3, \ldots$ one goes from subspace E_κ to $E_{\kappa+1}$ using a simultaneous inverse iteration on the "q" vectors of $\underline{X}_\kappa$

$$\underline{Y}_\kappa = \underline{M} \cdot \underline{X}_\kappa \qquad (5)$$

$$\underline{S} \cdot \underline{X}_{\kappa+1} = \underline{Y}_\kappa \qquad (6)$$

Operators $\underline{S}$ and $\underline{M}$ are projected onto subspace $E_{\kappa+1}$ using the Rayleight-Ritz method in the form

$$\underline{S}_{\kappa+1} = \underline{X}_{\kappa+1}^T \underline{S} \cdot \underline{X}_{\kappa+1} \qquad (7)$$

$$\underline{M}_{\kappa+1} = \underline{X}_{\kappa+1}^T \underline{Y}_{\kappa+1} \qquad (8)$$

where

$$Y_{K+1} = \underline{M} \cdot \underline{X}_{K+1} \qquad (9)$$

With the projected operators $\underline{S}_{K+1}$ and $\underline{M}_{K+1}$ the problem of the eigenvalues is solved completely

$$\underline{S}_{K+1} \cdot \phi_{K+1} = \underline{M}_{K+1} \cdot \phi_{K+1} \cdot \lambda_{K+1} \qquad (10)$$

Convergence regarding the eigenvalues is checked. If it has not yet been achieved the following transformation is made to return to the "n" dimension space going back to (6) until convergence is reached

$$\underline{Y}_{K+1} = \overline{\underline{Y}}_{K+1} \cdot \phi_{K+1} \qquad (11)$$

If convergence is achieved in the "p" first eigenvalues one goes back to the "n" dimension space by means of the following transformation, wich corresponds to the last step of the Rayleight-Ritz method:

$$\underline{X}_{K+1} = \overline{\underline{X}}_{K+1} \cdot \phi_{K+1} \qquad (12)$$

c) Verify if the required "p" eigenvalues and corresponding eigenvectors have been calculated, equation (1) being fulfilled for each eigenvalue and eigenvector. The property of the Sturm sequence described later on in this paper should be used for this purpose.

3 - COMPUTER STORAGE OF THE MATRICES

It is sometimes impossible to solve the eigenvalues problem referred to using only computer central memory. This is so because the solution of equations (6) to (12) requires an amount of storage that may exceed that available in the computer central memory or reach the maximum limits allowed in the arrays for the computer equipment used. However, by making transfers of parts of the program (arrays) between the central memory and the auxiliary memory (disk or magnetic tape), under the control of the program itself, these transfers may be made following the logic of the resolution of the problem under study. Minimization of transfers and better use of the central memory may thus be achieved.

There are three major working areas in the Subspace Iteration method, namely the storage of the stiffness matrix $\underline{S}$, of the mass matrix $\underline{M}$ and of the iteration vectors $\underline{Y}_K$. The program developed stores matrices $\underline{S}$ and $\underline{M}$ in skyline (one-dimensional array) and the iteration vectors matrix $\underline{Y}_K$ in a two-dimensonal array.

The stiffness matrix $\underline{S}$ and the mass matrix $\underline{M}$ can be partitioned in blocks (groups of columns), as can be seen in fig.1. The iteration vectors matrix $\underline{Y}_K$ may also be partitioned in blocks, wich we shall designate by vector groups, as can be seen in fig. 2.

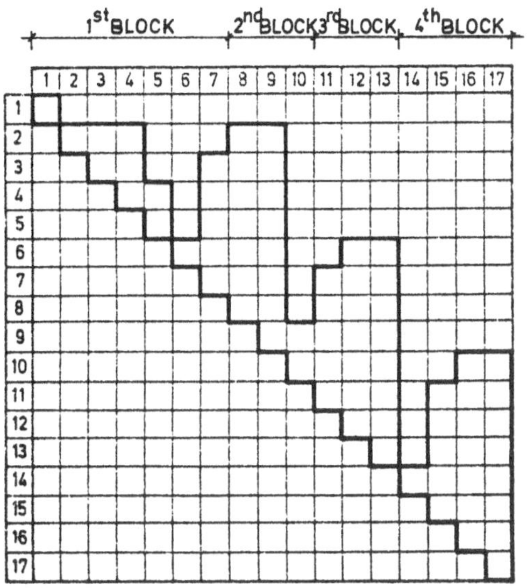

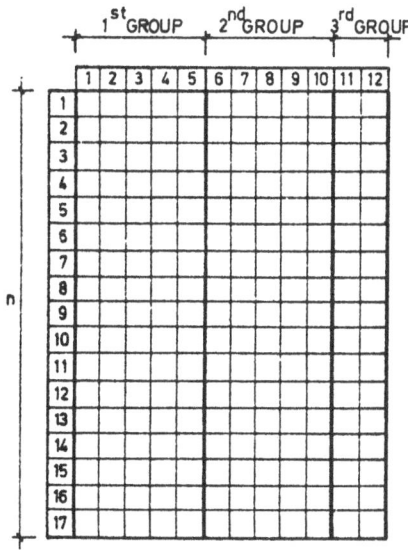

Figure 1 - Partitioning of stiffness matrix $\underline{S}$ and of mass matrix $\underline{M}$ in blocks (groups of columns)

Figure 2 - Partitioning of the iteration vectors matrix $\underline{Y}_K$ in groups of vectors

It should be noted that the computer program has been developed in such a way that the need of partitioning of matrices $\underline{S}$, $\underline{M}$ and $\underline{Y}_K$ is done automatically according to the problem that is to be analysed and to the dimensions of the work vectors.

The transfers to auxiliary memory are made by reading or writing matrices, without index control, adopting whenever possible sequential access.

As the time spent by the central processor unity of the computer may be important for solving the eigenvalue problem of a given structure, the program was prepared so as to allow restarting calculations of a previous problem.

4 - STARTING ITERATION VECTORS GENERATION

The choice of the starting vectors is essential to start the Subspace Iteration method. If the starting vectors are already close to the desired eigenvectors, convergence may be obtained after a few iterations. If, however, the initial subspace is a poor approximation of the final "p" dimension subspace, many iterations may be needed and the algorythm may become very expensive. Note that starting vectors cannot be orthogonal to the eigenvectors that are to be calculated.

Several algorythms have been proposed and studied for the choice of the starting iteration vectors (references [1] and [9]. The program developed considers the two hypothesis of choice described below.

4.1- Manual choice of the starting vectors
The user of the program should in this case indicate , for each of the "q" starting vectors, the nodal points of the structure, the directions and the values of the forces which he thinks will excite the vibration modes that are to be calculated. The "q" vectors thus chosen will be called "initial masters". Fig. 3 shows an example of the selection of initial masters for a structure.

4.2- Automatic choice of the starting vectors
The aim in this case is to establish automatically the starting vectors, based on the stiffness matrix and on the mass matrix [1]. The chosen vectors $\underline{Y}_1$ constitute the second member of equation (6) and proved efficient in the numerical examples analysed. The first column of matrix $\underline{Y}_1$ is made equal to the diagonal of the mass matrix $\underline{M}$. The last column of $\underline{Y}_1$ is a random vector and the remaining columns are unit vectors with the non-zero element equal to +1 and corresponding to those degrees of freedom that have the smallest ratios S_{ii}/M_{ii}. The reason for the choice of these vectors was firstly, that all degrees of freedom which have mass be excited in the first vector; secondly, the remaining vectors should be linearly independent

and should excite the degrees of freedom to which correspond
high mass and flexibility values; thirdly, the final random vec
tor should ensure the participation of all the modes.

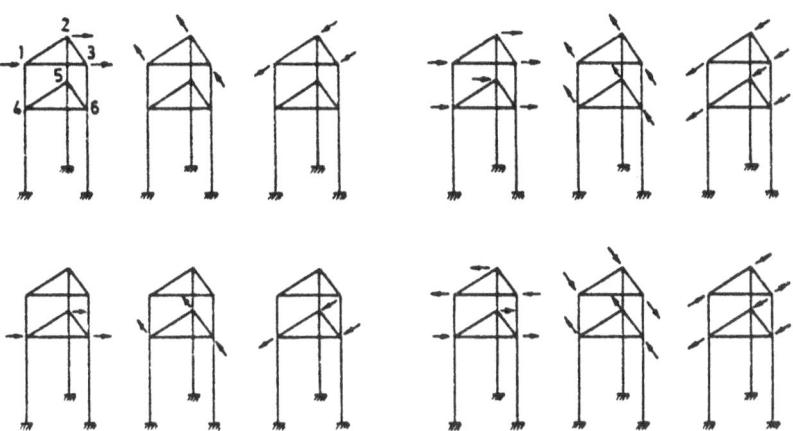

a) hypothesis 1 - 6 iteration vectors
b) hypothesis 2 - alternative to a)

Figure 3 - Choice of the initial masters for a tower
structure ⌐18⌐

5 - CONVERGENCE ACCELERATION BY OVERRELAXATION

The development of acceleration procedures regarding
the Subspace Iteration method has been the subject of various
papers (references ⌐ 5, 6, 19⌐).In principle, a number of
techniques such as Aitken acceleration, overrelaxation,the use
of Chebyshev polynomials and shifting can be used. The difficulty lies in working out a method of resolution that is reliable and at the same time significantly more efficient, and
this has so far not been achieved. For this reason, only the
technique of overrelaxation of the iteration vectors was implemented in this program and the acceleration factors to be
considered were carefully analysed.

All that is needed to implement this technique is to
change, in each iteration, the new vectors $\underline{Y}_{K+1}$ of equation
(11), which becomes

$$\underline{Y}_{K+1} = \underline{Y}_K + (\overline{\underline{Y}}_{K+1} \cdot \phi_{K+1} \cdot \underline{Y}_K) \cdot \underline{\alpha} \quad (13)$$

$\underline{\alpha}$ being the diagonal matrix of the overrelaxation factors
α_i, $i = 1, 2,\ldots, q$ of the different iteration vectors.

The use of overrelaxation of an iteration vector assu

mes that the vector has settled down and reached its asymptotic convergence rate. The overrelaxation α_i of each vector is obtained as a function of this rate of convergence and is calculated based on λ_{q+1}

$$\alpha_i = \frac{1}{1 - \dfrac{\lambda_i}{\lambda_{q+1}}} \qquad (14)$$

since the convergence rate of an iteration vector to the eigen vector originated by it is proportional to λ_i/λ_{q+1}. The value of λ_{q+1} must therefore be estimated in order to be able to use expression (14).

Supposing that some of the iteration vectors have reached their asymptotic convergence rate and that there is fairly good approximation to the corresponding eigenvectors, we can estimate values for λ_{q+1}. Let us consider the convergence rate for λ_i to be:

$$rc_i^{(\kappa+1)} = \frac{\lambda_i^{(\kappa+1)} - \lambda_i^{(\kappa)}}{\lambda_i^{(\kappa)} - \lambda_i^{(\kappa-1)}} \qquad i=1, 2, \ldots, q \qquad (15)$$

Depending on the iteration number κ the convergence rate estimates in equation (15) can be grossly in error and will be meaningless at or near convergence, owing to the arithmetic precision of the computer used. Nevertheless, if the two following conditions are satisfied

$$\frac{rc_i^{(\kappa+1)} - rc_i^{(\kappa)}}{rc_i^{(\kappa+1)}} \leq TOLR \qquad (16)$$

and

$$VLI \leq \frac{\lambda_i^{(\kappa+1)} - \lambda_i^{(\kappa)}}{\lambda_i^{(\kappa+1)}} \leq VLS \qquad (17)$$

The values estimated for $rc_i^{(\kappa+1)}$ are fairly reliable. The values of TOLR, VLI and VLS depend on the precision of the computer used. In the present case, in which a DEC 10 computer with 256 K words of 36 bits of central memory was used we have:

- Single precision corresponding to an 8 digit arithmetic
$$TOLR = 0,35; \quad VLI = 10^{-2}; \quad VLS = 10^{-6}$$

- Double precision corresponding to a 16 digit arithmetic
$$TOLR = 0,2 \text{ to } 0,35; \quad VLI = 10^{-3}; \quad VLS = 10^{-10}$$

Considering the convergence rate for λ_i expressed in

(15), bearing in mind that

$$\lim \ rc_i^{(\kappa+1)} = \left(\frac{\lambda_i}{\lambda_{q+1}} \right)^2 \tag{18}$$

and supposing that in iteration κ there are some eigenvalues estimates, of the "q" eigenvalues to be calculated, that pass in the tests indicated in (16) and (17), we may calculate an approximation for λ_{q+1}, using each one of these eigenvalues es timates, by expression

$$\lambda_{q+1} = \frac{\lambda_i^{(\kappa+1)}}{\sqrt{rc_i^{(\kappa+1)}}} \tag{19}$$

and use as value of λ_{q+1} the means $\overline{\lambda}_{q+1}$ of all the estimates calculated in this iteration.

It may also happen that the value estimated for λ_{q+1} is lower than some of the calculated eigenvalues estimates, thus leading the direct application of equation (14) to the broken line curve of figure 4. It can be seen that for the vectors be tween $\phi_{\ell+1}$ and ϕ_q the overrelaxation factor would be negative, which would not make sense. A second curve is therefore considered (dotted-line) for α_i, based on a value

$$\underline{\overline{\lambda}}_{q+1} = 1,1 \ . \ \lambda_q^{(\kappa+1)} \tag{20}$$

It is also assumed that no factor α_i, can be lower than α_{i-1}, which is easy to understand since the values of λ_i are in ascending order. The curve indicated by a full line in figure 4 is thus obtained for the values of the different overrelaxation factors α_i.

6 - STURM SEQUENCE

The third phase of the Subspace Iteration method consists in checking whether all the desired "p" eigenvalues and eigenvectors have been obtained. This can be done using the Sturm Sequence property. In the program developed this check is optional.
Let us construct the transformed matrix

$$\overline{\underline{S}} = \underline{S} - \mu\underline{M} \tag{21}$$

which corresponds to shifting the origin of the eigenvalue spe ctrum of the initial problem, expressed by equation (11) , in the positive direction of the axis of the eigenvalues, by a va lue μ as can be seen in figure 5.

Let us factorize $\overline{\underline{S}}$ as:

$$\overline{\underline{S}} = \underline{L} \ . \ \underline{D} \ . \ \underline{L}^T \tag{22}$$

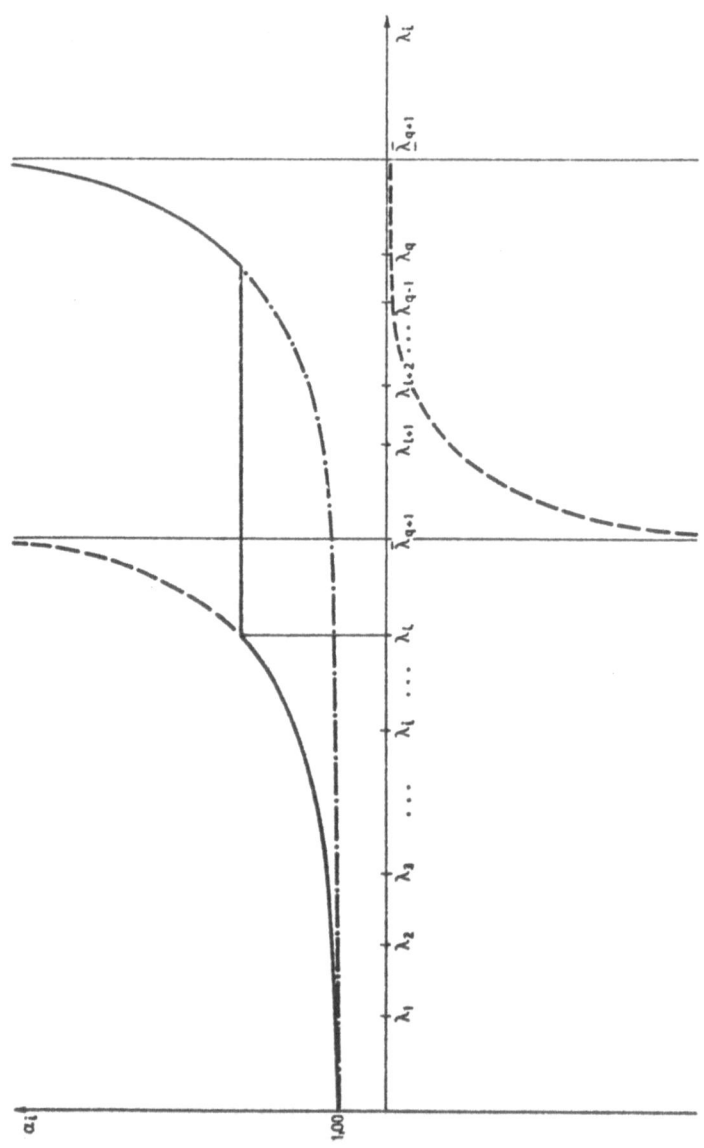

Figure 4 - Determination of the overrelaxation factors α_i

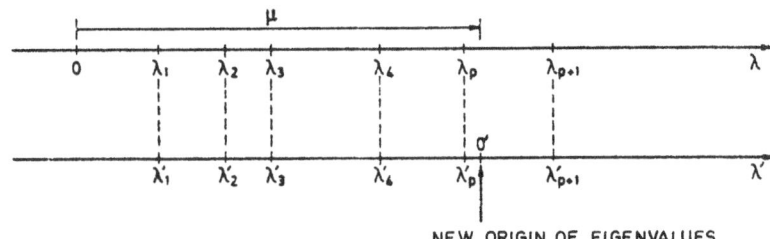

Figure 5 - Displacement of the origin of the eigenvalue spectrum

where L is an inferior triangular matrix with elements that are equal to 1 along the main diagonal and D is a diagonal matrix . The number of negative coefficients in $\overline{D}$ is equal to the number of negative eigenvalues λ_j, that is, equal to the number of eigenvalues λ_j that are smaller than μ.

To choose adequate values of μ it is necessary to establish bounds for the exact eigenvalues starting from the calculated approximate eigenvalues.

As an estimate of region in which the exact eigenvalue lies, the interval

$$0,99\ \lambda_i^K < \lambda_i < 1,01\lambda_i^K \qquad i = 1, 2,..., p \qquad (23)$$

is adopted, where λ_i represents the exact eigenvalues, assuming that matrices $\underline{S}$ and $\underline{M}$ have been truncated in the computer floating point representation and λ_i^K represents the calculated eigenvalues.

When there are grouped eigenvalues, the corresponding upper bound $(1,01\ \lambda_i^K)$ and lower bound $(0,99\ \lambda_{i+1}^K)$ may be superposed and the concept of bound of the group can be adopted as shown in figure 6.

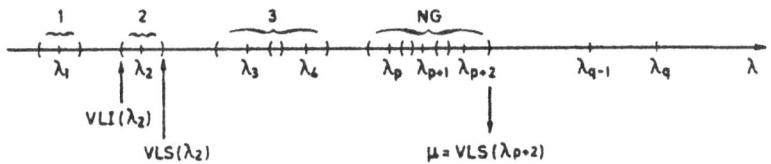

NG - NUMBER OF EIGENVALUES GROUPS

Figure 6 - Determination of value μ taking into account the closeness of eigenvalues

The number of groups of eigenvalues should thus be calculated, as well as the number of eigenvalues and the bounds of
each group. The number of groups is reached when the accumulated sum of eigenvalues in each group is greater or equal to the
number "p" of desired eigenvalues.

The value of μ should lie at the upper bound of the last
desired eigenvalue λ_p, so that it will be possible to check whether all "p" eigenvalues have been calculated when the factorization indicated in (21) is made.

7 - NUMBER OF ARITHMETICAL OPERATIONS

This part of the study concerns the estimation of the
number of arithmetical operations made in the main phases of the
program under consideration. Designating by "n" the number of
equations, by "f" the medium heigh of the stiffness matrix skyline columns, and by "q" the number of iteration vectors we have:

a) For the factorization of the stiffness matrix, in
which the Cholesky method is used, are carried out [14]

$$\frac{n(f^2 + f - 2)}{2} - \frac{f^3 - f}{3} \quad \text{arithmetical operations}$$

in addition to n square roots inherent in this method.

b) In the Subspace Iteration method we have for each
iteration:

Computing of	Number of operations
$\underline{S} \cdot \overline{\underline{X}}_{K+1} = \underline{Y}_K$ --------------------	$(2f - 1)\, nq$
$\underline{S}_{K+1} = \overline{\underline{X}}_{K+1} \cdot \underline{Y}_K$ ----------------	$\frac{1}{2} nq^2 + \frac{1}{2} nq$
$\overline{\underline{Y}}_{K+1} = \underline{M} \cdot \overline{\underline{X}}_{K+1}$ -----------------	$(2f - 1)\, nq$
$\underline{M}_{K+1} = \overline{\underline{X}}_{K+1} \cdot \overline{\underline{Y}}_{K+1}$ -------------	$\frac{1}{2} nq^2 + \frac{1}{2} nq$
$\underline{S}_{K+1} \cdot \underline{\phi}_{K+1} = \underline{M}_{K+1} \bar{+} \underline{\phi}_{K+1} \cdot \underline{\lambda}_{K+1}$ ----------	neglected
$\underline{Y}_{K+1} = \overline{\underline{Y}}_{K+1} \cdot \underline{\phi}_{K+1}$ -------------	nq^2

number of operations per iteration $= 4fnq - 2nq^2 - nq$

It may be assumed that the number of operations carried out in each iteration is about $4fnq + 2nq^2$.

c) To apply the Sturm sequence check we have:

Construction of the transformed matrix $\overline{\underline{S}} = \underline{S} - \mu \cdot \underline{M}$ ----	Number of operations
	nf

$$\text{Factorization of } \underline{\overline{S}} \text{ ------------ } \frac{n(f^2 + f - 2)}{2} - \frac{f^3 - f}{3}$$

and η square roots

8 - NUMERICAL EXAMPLES

In order to study the performance of the computer program dealt with in this paper the resolution of the problem of
the eigenvalues in 4 structures was considered.

The examples presented were processed in a DEC 10 compu
ter, with 256 K words (36 bits) of central memory. The computer
time presented concerns the Central Processor Unity (C. P. U.)
and includes the input and output time owing to the peculiari-
ties of the information of the processor used.

The 4 examples were thouroughly analysed so as to stress
the more significant aspects of the computer program. The 4
structures were thus calculated for a variable number (10, 20 ,
30, 40, 50) of iteration vectors without and with application
of acceleration by overrelaxation. Table I gives the characte -
ristics of each one of the analysed structures as well as the
C. P. U. time spent for the factorization of the stiffness matrix and for the Sturm sequence check.

TABLE I

Structure	Number of equations n	Mean height of the sky line columns f	CPU time (sec) for factoriza tion of the stiffness matrix	CPU time (sec) for the Sturm sequence check
I	450	10	0,84	1,25
II	918	54	35,55	58,76
III	600	96	76,11	118,52
IV	1 296	131	295,26	466,68

Table II shows, for each one of the structures analysed, the number of iterations needed to reach convergence, the
mean CPU time spent for each iteration and the total CPU time
spent for the complete solution of the problem, without and with
application of acceleration by overrelaxation on the iteration
vectors.

TABLE II

STRUCTURE	NUMBER OF VECTORS P/Q	WITHOUT ACCELERATION			ACCELERATION BY OVERRELAXATION			
		NUMBER OF ITERATIONS	C.P.U. TIME PER ITERATION (sec.)	TOTAL C.P.U. TIME (h : m : s)	NUMBER OF ITERATIONS	C.P.U. TIME PER ITERATION (sec,)	TOTAL C.P.U. TIME (h : m : s)	TOTAL C.P.U. TIME DECREASE (%)
I	5/10	23	9,84	3 : 56,12	21	10,00	3 : 39,48	7
	12/20	31	27,57	14 : 23,41	25	28,22	11 : 53,83	17
	22/30	18	49,00	14 : 50,25	18	49,23	14 : 54,42	0
	32/40	31	76,46	39 : 38,81	29	77,78	37 : 44,25	5
	42/50	54	104,13	1 : 33 : 51,61	47	105,92	1 : 23 : 6,78	11
II	5/10	7	51,90	9 : 22,89	7	61,94	9 : 23,88	0
	12/20	14	138,31	34 : 27,00	11	138,11	27 : 29,64	20
	22/30	15	246,04	1 : 3 : 35,10	15	243,89	1 : 3 : 2,68	0
	32/40	20	310,94	1 : 45 : 42,67	20	312,93	1 : 46 : 22,43	0
	42/50	52	451,66	6 : 33 : 39,57	39	452,08	4 : 55 : 57,77	25
III	5/10	7	67,02	11 : 29,71	7	66,42	11 : 25,55	0
	12/20	8	144,90	23 : 1,74	8	141,62	22 : 32,19	0
	22/30	28	215,52	1 : 44 : 11,59	19	217,73	1 : 12 : 34,05	30
	32/40	25	380,88	2 : 42 : 19,75	21	380,70	2 : 16 : 52,22	16
	42/50	29	484,34	3 : 58 : 1,41	20	459,84	2 : 36 : 54,17	34
IV	5/10	7	190,72	36 : 13,29	7	189,97	36 : 7,66	0
	12/20	13	470,39	1 : 55 : 38,27	13	472,12	1 : 56 : 1,22	0
	22/30	18	604,90	3 : 15 : 16,85	18	613,83	3 : 19 : 10,39	0
	32/40	28	887,72	7 : 8 : 51,43	26	881,02	6 : 35 : 31,19	8
	42/50	40	1 183,86	13 : 22 : 58,07	29	1 183,92	9 : 45 : 57,45	27

Figure 7 shows a graph with the number of iteration vec
tors in abscissae and, in ordinates, the total CPU time spent
in the resolution of the problem for each one of the structures
analysed. The points of this graph were joined by straight lines. The full-line curves concern the analysis without accelera
tion and the broken-line concerns the analysis with acceleration by overrelaxation.

The following figures show some vibration modes shapes
of the analysed structures.

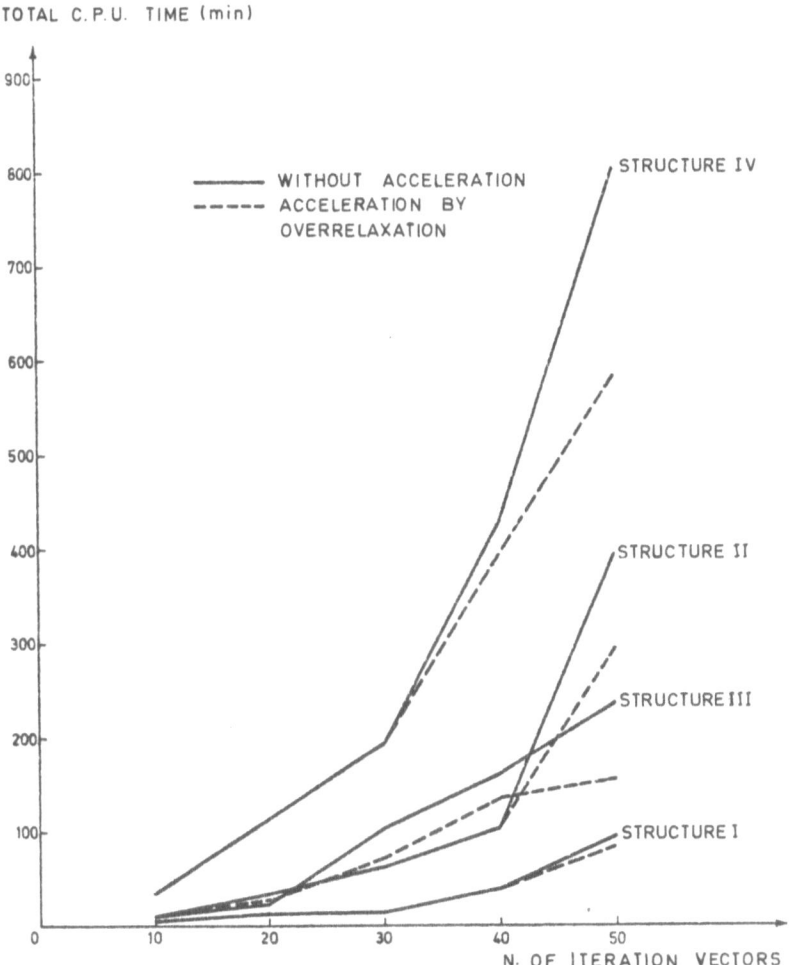

Figure 7 - Graph of the total CPU time for the resolution of the problem as a function of the num‐
ber of iteration vectors

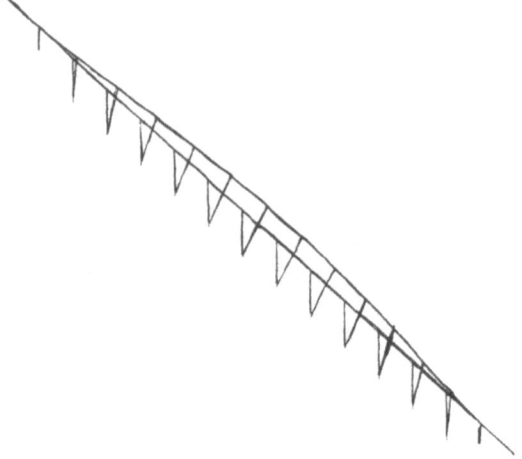

f = 1,94 Hz

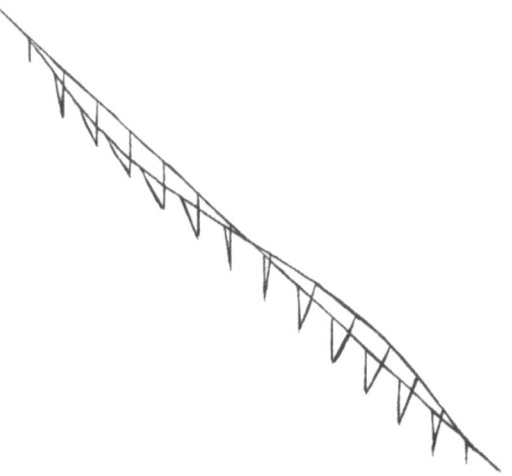

f = 2,03 Hz

Figure 8 - Structure I - 1st and 2nd vibration mode sha_
pe

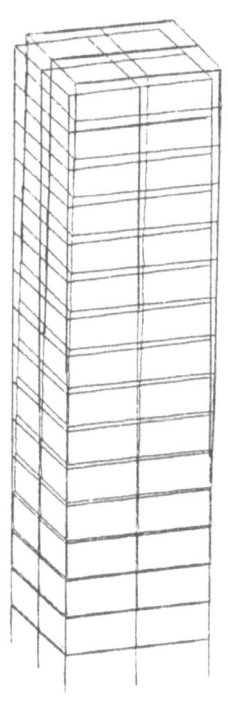 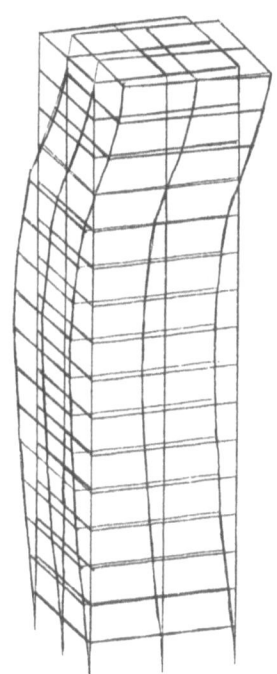

f = 0,57 Hz   f = 1,49 Hz

Figure 9 - Structure II - 1st and 5th vibration mode
shape

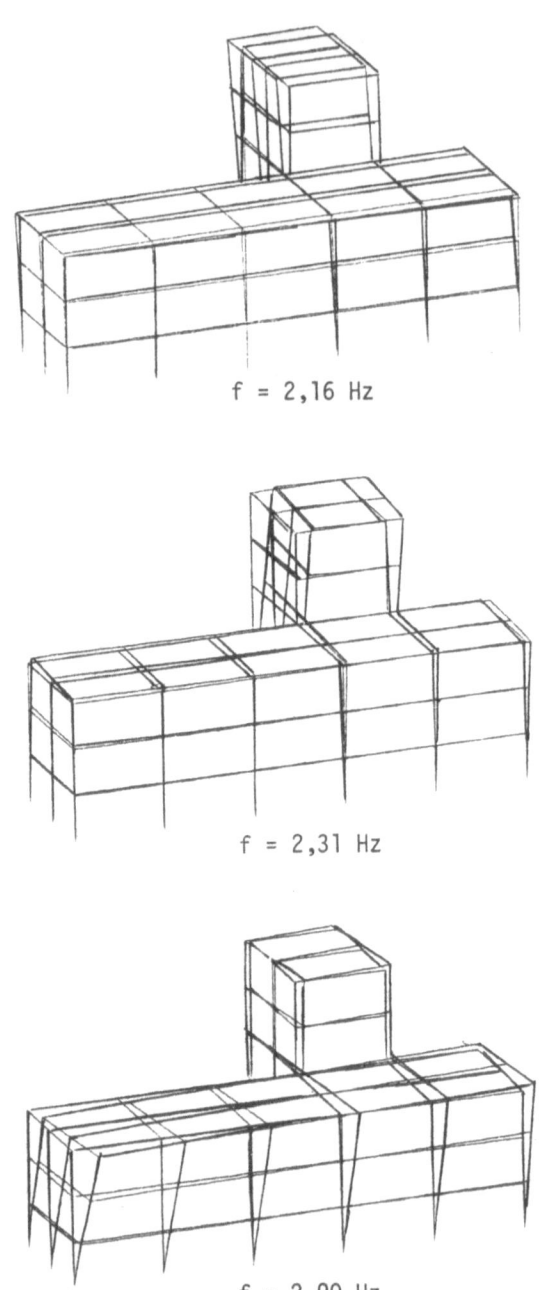

f = 2,16 Hz

f = 2,31 Hz

f = 2,90 Hz

Figure 10 - Structure III - 1st, 2nd and 3rd vibration
mode shape

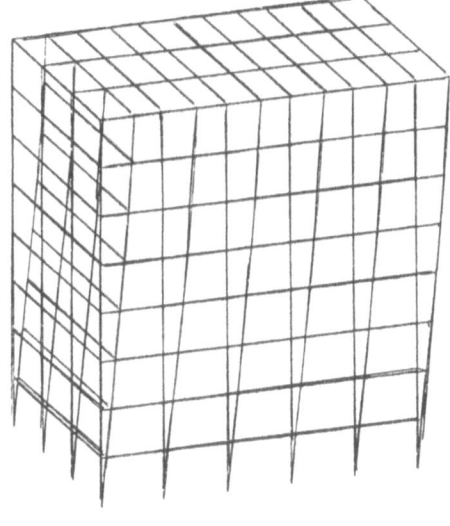

f = 0,86 Hz

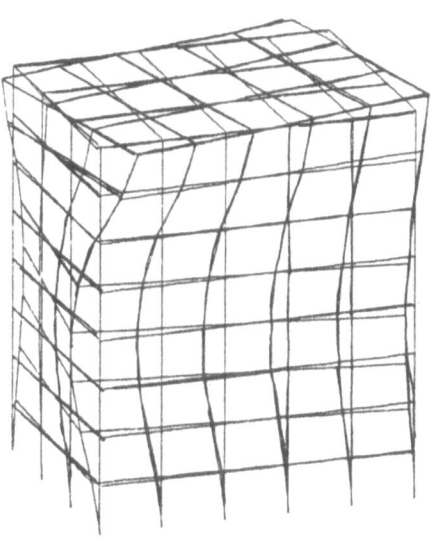

f = 3,13 Hz

Figure 11 - Structure IV - 1st and 6th vibration mode
shape

9 - CONCLUSIONS

The following conclusions can be drawn from the analysis of this paper.

a) The computer program developed includes data input with consistency tests for space structures (6 DOF), construction of the stiffness matrix and consistent mass matrix of the structure, and the calculation of eigenvalues and eigenvectors by the Subspace Iteration method.

b) The skyline technique, used to store the stiffness and mass matrices, proved very efficient as regards minimization of stored and operated coefficients and automatic partitioning into blocks with the corresponding transferences to auxiliary memory.

c) The Cholesky method was adopted for the factorization of the stiffness matrix and of the transformed matrix in the Sturm sequence, owing to its adaptability to storage in skyline partitioned into blocks.

d) The generation of the starting iteration vectors may be done automatically or based on initial masters defined by the user. The use of automatic generation, in the examples analysed, proved very efficient and supplied convergence for the required eigenvalues.

e) The computer program carries out the automatic partition of the matrix of the iteration vectors, with the corresponding transferences to auxiliary memory.

f) The partitions mentioned in b) and e) are closely related with the three work arrays defined in the beginning of the program, wich may be modified for computers with different memory capacities. Numerical examples showed that the use of smaller work arrays with the corresponding increase of partitions and transferences to auxiliary memory, does not bring about a significant increase in processing time.

g) The overrelaxation technique of the iteration vectors adopted proved efficient, particularly as the number of eigenvalues increases. Numerical examples showed a reduction in the total CPU time of resolution of the problem, which in some cases reached 34%. The use of this technique does not bring about a significant increase in numerical operations, but its use is nevertheless optional. The criterion adopted for the overrelaxation factors leads to great numerical stability and automatic use for any one of the "q" iteration vectors when necessary.

h) The implementation of the Sturm sequence, whose use in the program is optional showed:

- numerical stability in all the examples analysed.

- that convergence for the required "p" eigenvalues
was reached with the adoption of the automatic generation of the starting iteration vectors.

- that the aditional processing time is about 1,6 ti
mes that for the factorization of the stiffness ma
trix, which in turn is irrelevant as regards the
total time for the resolution of the problem.

    i) The computer program allows to restart calculations
using the results of previous problems. This technique of
restarting calculations makes it possible to calculate more ei
genvalues then those previously defined, and the generation of
the other starting vectors can be made either manually or automatically, no matter the procedure adopted in the initial
calculation.

REFERENCES

1 - Bathe, K.J. (1971) "Solution Methods for Large Generalized
Eigenvalues Problems in Structural Engineering".
University of California, Berkeley, California.

2 - Bathe, K.J. and Wilson, E.L. (1976) "Numerical Methods in
Finite Element Analysis".
Prentice-Hall, Inc., Englewood Cliffs, New Jersey.

3 - Bathe, K.J. and Wilson, E.L. (1972) "Large Eigenvalue Problems in Dynamic Analysis".
ASCE, Mechanics Division.

4 - Bathe, K.J. and Ramaswany, S. (1980) "An Accelerated Subspace Iteration Method".
Computer Methods in Applied Mechanics and Engineering 23.

5 - Bathe, K.J. (1977) "Convergence of Subspace Iteration".
in K.J. Bathe, J.T. Oden and W. Wunderlich (eds.), Formulations and Numerical Algorithms in Finite Element Analysis (MIT Press, Cambridge, MA).

6 - Bathe, K.J. and Ramaswamy, S. (1977) "Subspace Iteration
with Shifting for Solution of Large Eigensystems".
Report AVL 82448 - 7 (Dept. Mech. Eng. MIT, Cambridge, MA).

7 - Chee Hoong Loh (1984) "Chebyshev Filtering and Lanczos Pro
cess in the Subspace Iteration Method".
Int. Journal Numerical Methods in Engineering, Vol. 20,
nº 1, Jan.

8 - Clough, R.W. and Penzien, J. (1975) "Dynamics of Structures".
Mc Graw-Hill Inc.

9 - Gere, J.M. and Weaver, W.Jr. (1965) "Matrix Algebra for Engineers".
Van Nostram Reinhold Company.

10 - Meirovitch, L. and Baruth, H. (1981) "On the Cholesky Algo rithm with Shifts for the Eigensolution of Real Symmetric Matrices".
International Journal for Numerical Methods in Engineering, Vol - 17, 923 - 930.

11 - Mills, A.B. and Wood, L.A. (1981) "Cray - 1: A powerful De livery System for Engineering Software".
Advanced Engineeriong Software, Vol. 3, nº 2.

12 - Paige, C.C. (1972) "Computational Variants of the Lanczos Method for the EigenProblem".
J. Inst. Math Appl. 10, 373 - 381.

13 - Schwarz, H.R. (1974) "The Eigenvalue Problem $(\underline{A} - \lambda\underline{B})\underline{X} = \underline{0}$ for Symmetric Matrices of High Order".
Computer Methods in Applied Mechanics and Engineering 3.

14 - Soriano, H.L. (1981) "Sistemas de Equações Algébricas Li-neares em Problemas Estruturais".
Seminário 280, LNEC.

15 - Soriano, H.L. (1982) "Reduction of Degrees of Freedom at Substructural Level in Dynamics".
University Of Southampton.

16 - Soriano, H.L. and Azevedo, A.V. (1982) "Implementação do Método de Iteração por Subespaços utilizando Memória Auxi liar de Computador".
Revista Protuguesa de Engenharia de Estruturas, R.P.E.E., nº 14, Ano IV.

17 - Wilkinson, J.M. (1965) "The Algebraic Eigenvalue Problem".
Clarendon Press, Oxford.

18 - Wilson, E.L. (1978) "Numerical Methods for Dynamic Analysis".
Numerical Methods in Offshore Engineering, John Wiley & Sons.

19 - Yamamoto, Y. and Ohtsubo, H. (1976) "Subspace Iteration Acceleration by Using Chebyshev Polynomials for Eigenvalue Problems with Symmetric Matrices".
Int. J. Numer. Meths. Eng. 10, 935 - 944.

ACCURACY ASSESSMENT BY FINITE ELEMENT P-VERSION SOFTWARE

P. Angeloni, R. Boccellato, E. Bonacina, A. Pasini, A. Peano

ISMES, Bergamo, Italy

ABSTRACT

FIESTA is the only commercially available software based on the most recent developments of the finite element method: p-version to reduce the number of degrees of freedom, hierarchic elements for computational saving and better numerical stability as well as error indicators for optimizing the distribution of the degrees of freedom over the problem domain.

1.0 INTRODUCTION

Stress analysis of solid continua by conventional general purpose programs is a very time consuming and expensive process. Unless geometry is simple enough to lend itself to automated mesh generation, model preparation is generally a long and tedious task, even with advanced graphic capabilities. Once created, conventional 3D meshes are difficult to verify by inspection. Modification and mesh refinement add further burden on the analyst.
The time and cost required to prepare a conventional solid model often exceeds estimates significantly. Consequently, analysts and managers rarely attempt detailed three-dimensional analyses. Instead they simplify the model and are forced to accept an inherent loss of solution quality as a result. An additional drawback of 3D analyses is caused by the difficulty to interpret the computed results with confidence.

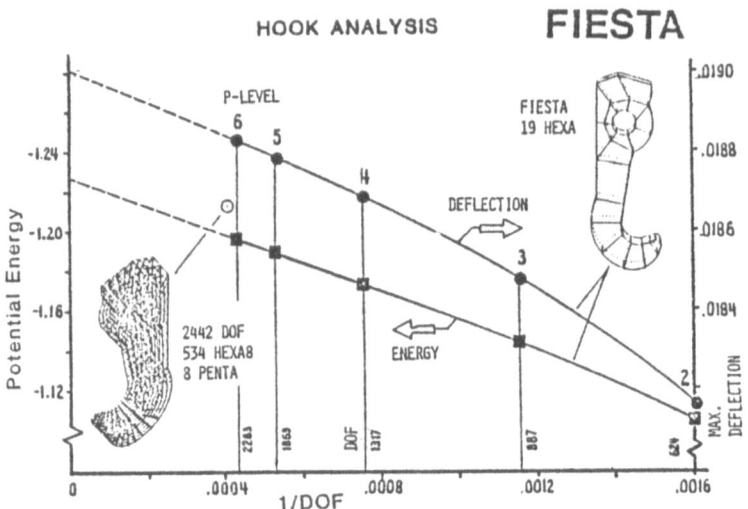

Figure 1 - Results of a hook analysis with
the h- and the p-version of the
F.E.M.

Two extreme situations often occur. If the modeling mesh is too coarse, the user may discover severe discontinuities occurring at element interfaces. When the model is sufficiently refined to
avoid discontinuities the user may be overwhelmed
by the huge amount of output data.
In neither case are error estimates available. A
result from a conventional analysis represents infact a single point on the convergence curve which
is not sufficient to assess the status of convergence (see Figure 1). On the other hand, successive analyses with increasing mesh refinement
would require prohibitive expenditure of both time
and money. Therefore convergence evaluation with
the conventional general purpose systems based on
the h-version of the finite element method is not
pratical.

The p-version of the finite element method offers
instead an efficient convergence process that allows the user to overcome the above mentioned
problems. Development of hierarchic p-version
finite element formulations for both C^0 and
C^1 problems was initiated in [Peano, 1976].
The distinguishing feature of this new convergence
process is that the number and distribution of

finite elements is fixed, while the number of
shape functions, which are complete polynomials of
order "p", are progressively increased over some
or all elements.
This means that the mesh can be designed to represent of model geometry only. The error of approximation is infact controlled separately through
the addition of polynomial shape functions. The
error control may be automated by criteria developed and extensively tested for 2D applications
several years ago [Peano A., Fanelli M. , Riccioni
R., Sardella L., 1978], [Peano A., Pasini A., Riccioni R., Sardella L., 1979], [Peano A., 1978],
[Peano A., Riccioni R., 1978]. This paper demonstrates application of the same concepts for 3D
applications in a commercially available program.

2.0 THE FIESTA CODE

The most recent theoretical advances concerning
the p-version of the finite element method, have
been implemented in the FIESTA code [ISMES, 1984].
The program, available for elastic and field analysis, makes it possible to analyze a single geometric configuration to yield multiple solutions,
each of which are associated with a particular
level of precision. This series of solutions may
be used to verify the trend toward convergence to
the true solution (see Figure 1). In addition,
the total project cost is substantially below that
of conventional methods because the FIESTA finite
element meshes are extremely simple to begin with,
and further, need not be refined when greater precision is desired. The model simplicity in FIESTA
provides the flexibility to make rapid configuration changes to the model so that it is possible
to evaluate a larger number of alternate designs
within given time and resource constraints. The
p-level grading is a distinguishing feature of the
code that gives the user the freedom to select the
order of approximation over one or more elements,
without any modification of the stored input data.
Each level of approximation is obtained by adding
higher order shape functions to the ones used in
the previous level. Six increasingly sophisticated levels of polynomial approximation (p-levels)
may be used in FIESTA. The maximum order of polynomial interpolation is fourth.
Typical shape functions corresponding to the various levels of approximation are illustrated in
Figure 2 with reference to a hexahedron element.

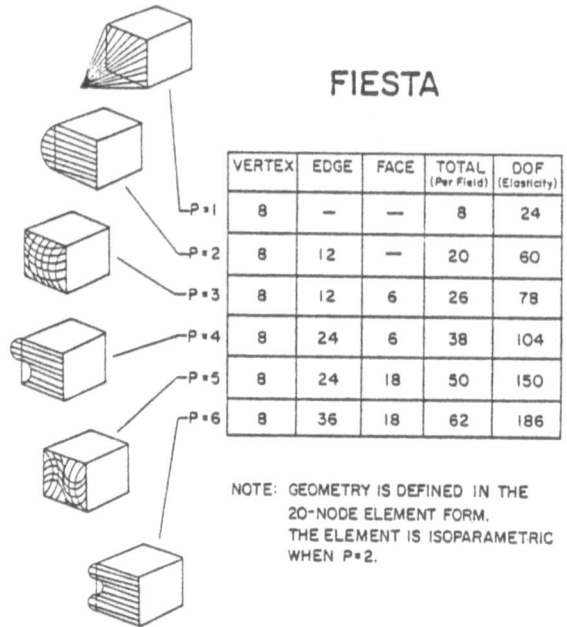

FIESTA

VERTEX	EDGE	FACE	TOTAL (Per Field)	DOF (Elasticity)
P=1	8	—	—	8	24
P=2	8	12	—	20	60
P=3	8	12	6	26	78
P=4	8	24	6	38	104
P=5	8	24	18	50	150
P=6	8	36	18	62	186

NOTE: GEOMETRY IS DEFINED IN THE
20-NODE ELEMENT FORM.
THE ELEMENT IS ISOPARAMETRIC
WHEN P=2.

Figure 2 - Number and kind of shape
functions in one hexahedron
element

The other element types available in the library
of the program are shown in Figure 3. All element
types, that can be either with curved or straight
edges, can be used within the same model.

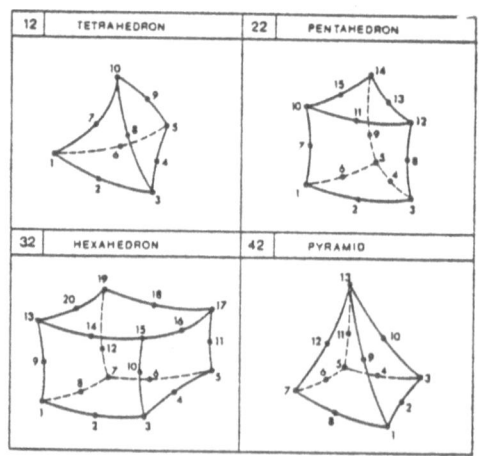

Figure 3 - Element types available in FIESTA

Complete interelement compatibility is maintained
even when adjacent elements are of different polynomial order. The interface boundary is reas-
signed to the lower of the occurring p-levels and,
in spite to the loss of a few high order terms in
the polynomial expansion of the element with
higher p-level, the remaining degrees of freedom
still provide minimal and exact conformity between
elements.

3.0 SELECTION OF THE POLYNOMIAL DISTRIBUTION

FIESTA is the only commercially available finite
element code which permits the user to vary the
polynomial order of the finite elements for the
purpose of controlling the quality of approximations.
A required level of precision is reached simply by
either uniformly or selectively increasing the polynomial order of the elements. Two options are
available to define the polynomial distribution
that is requested for computation.
In the first option the level of approximation to
be used within each element is specified directly
by the user. The selection is performed through
the execution of processor *PLEVEL and the user is
expected to apply his own "engineering judgement"
and use high polynomial orders in the same areas
where he would have used a more refined mesh.

The second option available in FIESTA [Peano,
1978] drastically reduces the need of user's interaction. The optimal polynomial distribution
compatible with the user's accuracy and cost requirements can be infact automatically defined by
the program.
This is achieved through the execution of processor *AUTOPL and the computation of an error indicator that orders all the available degrees of
freedom on the basis of the related contribution
to the quality of the solution. Each automatically selected polynomial distribution is comprised
of all degrees of freedom that contribute for more
than a given percentage to the improvement of the
solution.
The process that automatically selects of the various p-levels is only governed by a single parameter that specifies the maximum number of degrees
of freedom allowable for the solution of the problem being analyzed. This means, in other words,
that the user has only to establish a bound that

is primarily ruled by economic and/or operational considerations. All other decisions concerning the accuracy of the computed solution are instead directly managed by the program.

4.0 EFFICIENT SOLUTION STRATEGIES

Taking advantage of the capabilities given by the p-version approach of the finite element method, suitable solution strategies can be developed and applied for an efficient solution of engineering problems characterized by stringent safety requirements.

4.1 <u>Solution with direct user interaction</u>
    The following guidelines apply to the solution procedure based on the manual selection of the polynomial level of interpolation within each element:

A1) Solve for p-level=1
    This solution is inexpensive and provides the best check that input data are correct. Usually it is not worthwhile computing the stresses because they may be too inaccurate, assuming the model has a very coarse mesh. Displacements should have at least the correct order of magnitude and will be helpful for convergence analysis. Of course displacements will be too small in case a plate is modelled with only one element through the thickness.

A2) Solve for p-level=2
    This will usually provide accurate displacements. To assess it, compare the displacements for p-level=1 and p-level=2 and/or plot the total potential energy versus the inverse of number of degrees of freedom, and inspect the trend. For a preliminary assessment of the stress level, and to select areas where printed and graphic output are desired, stresses may be also computed. Since the mesh is coarse,it is not possible to expect convergence of all stress components yet. Moreover the stress components that are more important for the equilibrium of the structure will be surprising accurate already. For instance, if a plate in bending has been modelled with only one element through the thickness, the in-plane normal stress components will be more accurate than the shear stress components.

A3) Solve for p-level=4
Either uniform or graded distributions can be used, according to the degree of refinement of the mesh. For instance, it could be helpful to increase the polynomial order of the elements where stress output is needed and in the next element layer.

A4) Compare results for p-level=2 and p-level=4
Decide whether and where to further improve the polynomial approximation up to p-level = 6.

4.2 Solutions based on automatic polynomial distributions

The use of the capability offered by the program for an automatic selection of the optimal polynomial distribution, makes it possible to get the same results as above, by following a procedure that reduces the need of direct user's interaction.
The recommended solution strategy, in this case, is as follows:

B1) Solve for p-level=1
This step besides allowing a complete and inexpensive check of the correctness of the input data, prepares the initial solution that is needed as first requirement for the execution of processor AUTOPL. In case bending effects are significant, it may be better to solve for p-level=2 rather than p-level=1.

B2) Execute processor AUTOPL for the computation of the optimal polynomial distribution that complies with the assumed limits of cost and solution accuracy. During this execution it is also possible to get a forecast of the solution improvement that will be reached with the addition of all available degrees of freedom. The estimated solution accessed by the output processors, when compared with results of the initial solution gives, in general, an interesting idea of the most stressed areas of the model.

B3) Solve for the automatically defined polynomial distribution and produce the required output.

An additional analysis tool is available for the user in conjunction with the two above mentioned procedures.

Once two solutions corresponding to different numbers of degrees of freedom (NDOF1 and NDOF2) have been computed, it is usually beneficial to extrapolate to get the solution that could be computed with infinite degrees of freedom [Szabo B., Metha A., 1978], [Peano A., Szabo B., Metha A., 1978].

Indicating with SOL1, SOL2 the computed quantities (energies, displacements, stresses, etc.) the two results may be plotted in a diagram SOL versus increase of the number of degrees of freedom.

A line may be fitted between two or more solution points to get extrapolated solutions.

In practice the extrapolated solution (SOLEX) may be also directly computed by the formula:

$$SOLEX = \frac{SOL1*NDOF1 - SOL2*NDOF2}{NDOF1 - NDOF2} \qquad (1)$$

Although not necessarily corresponding to the true solution, the extrapolated results are generally better than the computed ones. With the exception of cases of very poor modeling (e.g. distorted elements and low order of interpolation), or presence of singularities (reentrant corners, concentrated forces, etc.) the method is proved to give accurate estimates of the asymptotic solution, and is of particularly easy application.

5.0 EXAMPLES OF APPLICATION OF THE FIESTA CODE

Two examples are presented to show engineering applications of the solution strategies made available from the p-version approach of the finite element method.

Focus is made in particular on the procedure of section 4.2. The reason of this choice is that the procedure based on the automatic selection of the polynomial distribution allows the computation of accurate solutions without requiring any particular care or known-how from the user. This is of course of great importance expecially in view of the generalized application of the finite element method that is now occurring.

The first example is the well-known problem of the semi-elliptical surface crack in a plate under remote tension. The problem is particularly interesting as the numerical solution can be easily

compared with results available in   the   technical
literature,   as well as for the fact that it highlights the extreme accuracy that can   be   obtained
by FIESTA.   The problem and the mesh used for computation are shown in Figure 4.

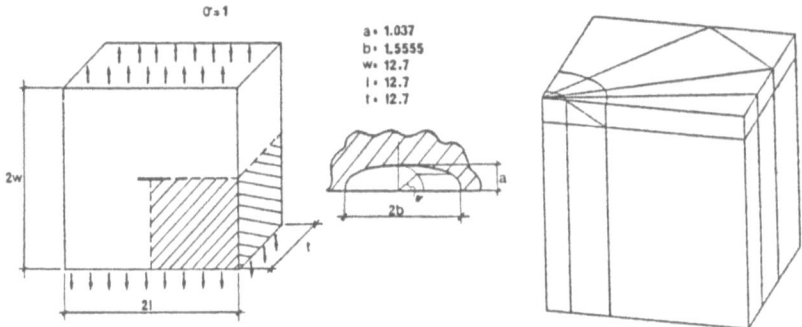

Figure 4 - F.E. model used to solve the
            problem   of the part through
            crack

Using all element at the same level of   interpolation,   a number of degrees of freedom ranging from
153 to 2241 can be defined over the model of   Figure   4.   According to the procedure of Section 4.2
the problem has been initially solved with a   uniform   distribution   of   elements at p-level=1.   As
only 153 degrees of freedom are used, the   initial
solution   at   p-level=1 can be effectively used as
an inexpensive check of all input data.

After the initial solution, processor *AUTOPL   has
been executed to compute the two optimal polynomial distributions bounded by 600 and   1200   degrees
of freedom, respectively.
These distributions are presented in Figure 5.   As
expected,   higher   order   elements are used around
the crack front, while lower   order   elements   are
progressively   used   going   away from the critical
area.   Note that fourth order polynomials are   selected   in both cases in the elements on the crack
surface.
One or two orders distinguish instead the level of
approximation in the surrounding elements.

The values of the stress intensity factor computed
along   the crack front with the polynomial distributions of Figure 5 are presented in Figure 6.

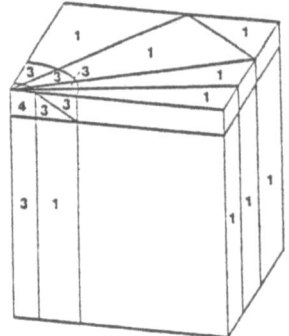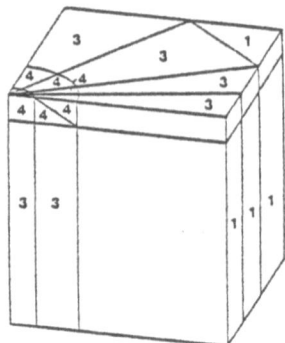

Figure 5 - Polynomial distributions selected
by the program with an allowable
number of 600 and 1200 degrees of
freedom

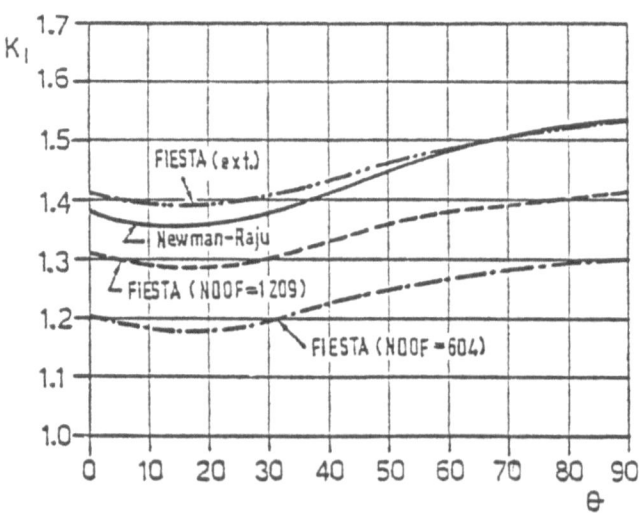

Figure 6 - FIESTA results for the
part-through crack problem

In the same Figure results of the reference solution [Newman et al., 1979], as well as results extrapolated using Equation 1 of section 4.0 are
also presented. The stress intensity values computed with 600 and 1200 degrees of freedom exactly
follow the trend of the reference solution with a

constant discrepancy of 15 and 7.5 percent along
the whole crack front.
As expected, extrapolated results are much more
accurate and are within a maximum discrepancy of
2.5 percent. Taking into account that the reference analytical solution is considered to be within 5 percent of the exact solution, the FIESTA solution may be regarded as equally accurate as (and
probably more accurate than) the reference solution.

The second example of application of the solution
strategy based on the error indicator capability
of FIESTA, is the analysis of a typical boiling
reactor feedwater nozzle.

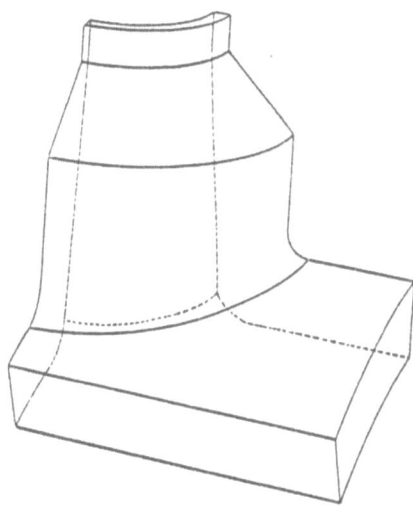

Figure 7 - Geometry of the BWR nozzle

Figure 7 shows the drawing by surfaces of one
quarter of the structure being analyzed. In Figure 8 two finite element models are presented.
The first is the mesh originally used to solve the
problem with a conventional general purpose finite
element program. The second is the mesh used with
FIESTA. Because of the capability of the program
to use high order interpolation functions, only 15
curved elements have been used to properly model
the structure. As previously mentioned, the first
step of the execution procedure is for the computation of the initial solution corresponding to a
uniform polynomial distribution of p-level=1 (constant strain elements).

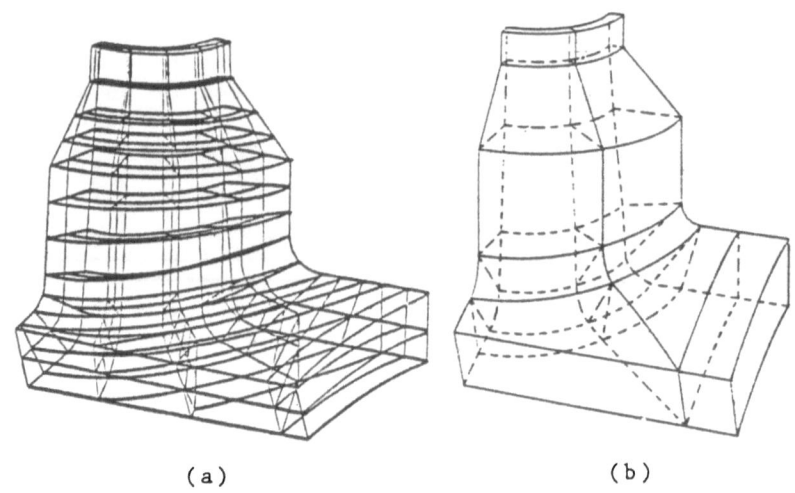

Figure 8 - F.E. meshes of the BWR nozzle
a) conventional grid
b) FIESTA grid

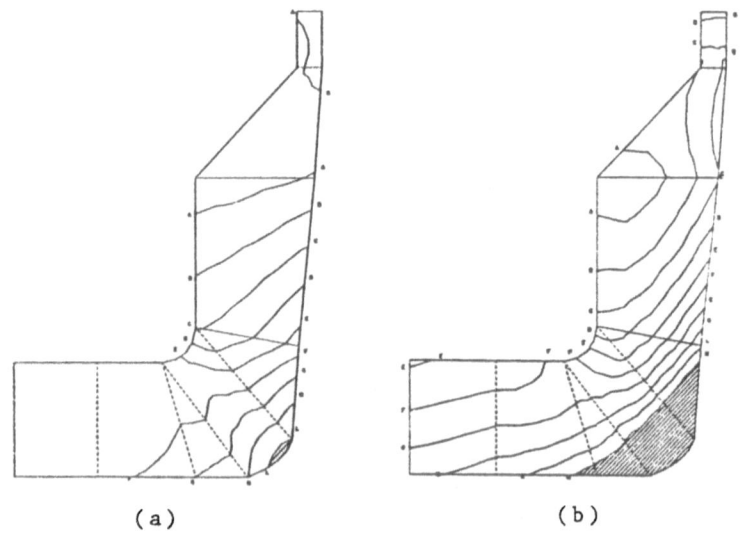

(a)   (b)

Figure 9 - Results of the initial solution (a)
and the uniformly upgraded solution
(b)

Results of the initial solution are presented in
Figure 9a where contour lines of the octahedral
shear stress are drawn on the later cross section
of the nozzle. The picture, that was required
with the automatic hatching of areas exceeding the
stress value of 100. units, indicates that the
peak stress is reached on the internal surface,
near the connection with the cylindrical shell.
The results of an uniform element upgrading up to
a cubic order of polynomial interpolation are
shown in Figure 9b. As expected, the new degrees
of freedom added to the ones used for the initial
solution produce a considerable increase of the
hatched area. The solution of Figure 9b is computed by using 912 degrees of freedom. Results
agree with those obtained by a conventional code
and the mesh of Figure 8 (1798 DOF) . Figure 10
presents the polynomial distribution and the results obtained after an execution of processor
*AUTOPL in which the maximum number of degrees of
freedom was limited to 600. The new results computed by using only 60 percent of the previously
used degrees of freedom fully agree with results
of the solution based on a uniform polynomial increase.

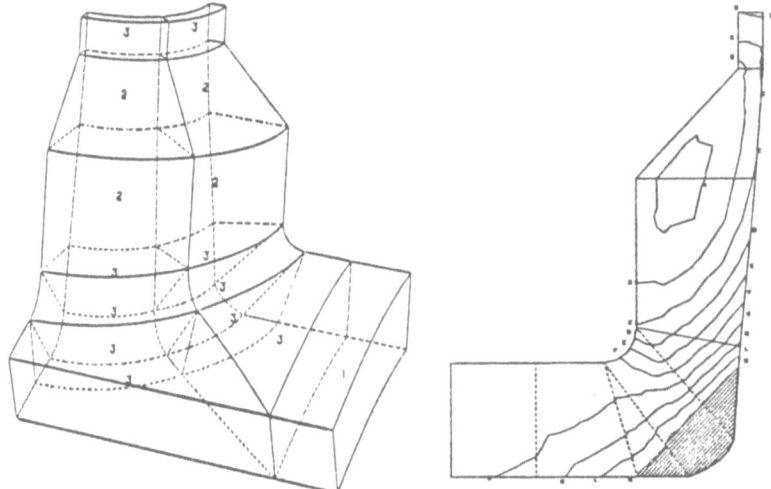

Figure 10 - Polynomial distribution and stress
output after the execution of
processor *AUTOPL with NDOF = 600

6.0 CONCLUSIONS

The p-version approach and the error indicators

for automatic (or semiautomatic) convergence are
now available in a commercial environment and are
demonstrated to provide validated results by simultaneously reducing both the need for user's interaction and the total computer and project cost.

7.0 REFERENCES

- Ismes (1984) FIESTA User's Manual

- Newman J.C., Raju I.S. (1979) Analysis of
  Surface Cracks in Finite Plates under
  Tension and Bending Loads
  NASA Tech. Paper 1578, 1-43

- Peano A. (1976) Hierarchies of Conforming
  Finite Elements for Plane Elasticity and
  Plate Bending
  Comp. and Maths. with Appls., Vol.2, 211-224

- Peano A. (1978) Energy Gradient Technique
  for Adaptive Finite Element Analysis
  Proc. of the 15th Meeting of the Soc. of
  Engineering Science

- Peano A., Pasini A., Riccioni R., Sardella L.
  (1979) Adaptive Approximations in Finite
  Element Analysis
  Computers and Structures, Vol.10, 333-342

- Peano A., Szabo B., Metha A. (1978) Self-Ada-
  ptive Finite Elements in Fracture Mechanics
  Comp. Methods in Appl. Mech. and Engineering
  Vol.16, N.1

- Peano A., Riccioni R. (1978) Automated
  Discretization Error Control in Finite
  Element Analysis
  Proc. 2nd World Cong. on Finite Element
  Methods, Dorset, England

- Peano A., Fanelli M., Riccioni R., Sardella
  L. (1978) Self-Adaptive Convergence at the
  Crack Tip of a Dam Buttress
  Proc. Int. Conf. on Num. Meth. in Fracture
  Mechanics, Swansea, U.K.

- Szabo B., Metha A. (1978) P-Convergent
  Finite Element Approximations in Fracture
  Mechanics
  Int. J. Num. Meth. Engig., Vol.12, 551-560

A POWERFUL FINITE ELEMENT METHOD FOR LOW CAPACITY COMPUTERS

Prof. J. Jirousek

IREM, Swiss Federal Institute of Technology, Lausanne, Switzerland

1. INTRODUCTION

The progress made in computer hardware over the last decades
has lead to the development of fairly powerful mini-computers
and for some time now there has been a strong demand to develop special versions of the existing FE programs for their
use. For obvious reasons, the practical application of conventional finite elements on micro-computers should, however,
be limited to comparatively simple problems which may be
solved with adequate accuracy by using relatively few elements
and a low total number of degrees of freedom. The method of
Large Finite Elements (LFE) removes this difficulty and enables to extend the range of practical applications to involved problems including various stress concentrations and/or
stress singularities.

Although both the theoretical basis of the LFE method
and some studies related to its practical efficiency have
already been published some time ago [1-4], the method is not
largely known and for the reader's convenience we start by
briefly summarizing the theoretical formulation.

2. PRINCIPLE OF THE METHOD

The principle of the method is to base the finite element
formulation on trial functions that satisfy, a priori , all
the field equations of the problem (Treffz's method) rather
than the essential boundary conditions and interelement continuity (Ritz).

Consider a boundary problem stated as follows : Let the
internal equilibrium of an elastic continuum occupying the
region Ω bounded by $\Gamma=\partial\Omega$ be expressed in terms of displacement

by a system of governing differential equations

$$\mathbf{Lu} = \bar{\mathbf{f}} \text{ on } \Omega ,\tag{1}$$

where $\mathbf{L}$ is a differential operator matrix, $\mathbf{u}$ is a vector of generalized displacement and $\bar{\mathbf{f}}$ is a conjugated vector of generalized body forces. The boundary conditions are defined in terms of the generalized boundary displacements $\mathbf{v}=\mathbf{v}(\mathbf{u})$ and/or the conjugated generalized boundary tractions $\mathbf{T} = \mathbf{T}(\mathbf{u})$ by

$$\mathbf{v} = \bar{\mathbf{v}} \text{ on } \Gamma_V \quad \text{and} \quad \mathbf{T} = \bar{\mathbf{T}} \text{ on } \Gamma_T ,\tag{1a,b}$$

where $\bar{\mathbf{v}}$ and $\bar{\mathbf{T}}$ are the prescribed quantities and $\Gamma_V + \Gamma_T = \partial\Omega$.

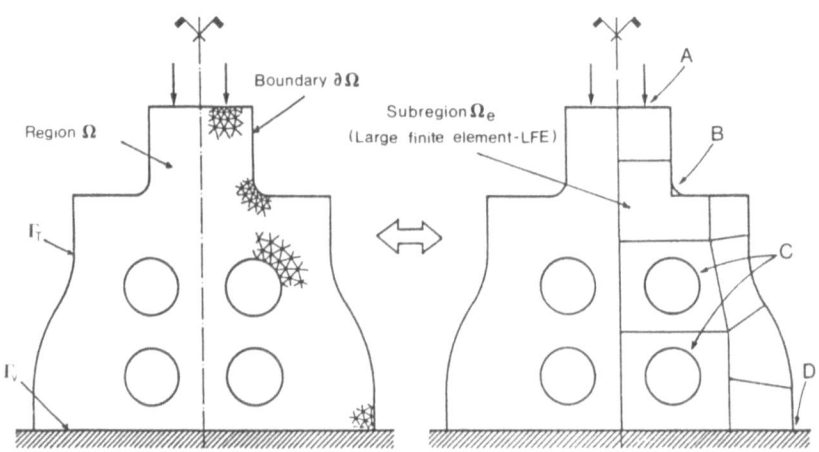

Figure 1. Typical conventional finite element (CFE) and large finite element (LFE) meshes: A. load dependent singularity; B. rounded off re-entrant corner stress concentration or sharp re-entrant corner stress singularity; C. stress concentration due to circular holes; D. clamped-free corner stress singularity.

The LFE method subdivides (Fig.1) the region Ω into a small number of subregions Ω_e (large finite elements - LFE) and assumes on each Ω_e an independent field of generalized displacements $\mathbf{u}_e$ expressed in terms of a particular integral $\overset{o}{\mathbf{u}}_e$ and a set of appropriate homogeneous solutions $\boldsymbol{\phi}_1 ,\ldots\boldsymbol{\phi}_2$ to (1):

$$\mathbf{u}_e = \overset{o}{\mathbf{u}}_e + \sum_{i=1}^{m} \boldsymbol{\phi}_i \mathbf{a}_i = \overset{o}{\mathbf{u}}_e + \boldsymbol{\phi}_e \mathbf{a}_e ,\tag{2}$$

where $\mathbf{a}_e$ is a vector of undetermined coefficients and the known coordinate functions $\overset{o}{\mathbf{u}}_e$ and $\boldsymbol{\phi}_i$ are such that

$$\overset{o}{L\mathbf{u}_e} = \overline{\mathbf{f}} \quad \text{and} \quad L\boldsymbol{\phi}_i = \mathbf{0} \quad \text{on} \quad \Omega_e \ . \tag{2a,b}$$

It is important to point out that $\overset{o}{\mathbf{u}}$ may be a "global" function, i.e. function extended over the whole Ω and, consequently, not confined to a single LFE.

For expository purposes, figures 1 and 2 represent simple
plane elasticity situations. However, it should be pointed out
that all considerations apply to any linearly elastic continuum (plate in bending, shell, etc.).

The goal is now to calculate the undetermined coefficients
$\mathbf{a}_e$ of all LFE so that the boundary conditions (1a,b) are
matched and the interelement continuity restored, as best as
possible, in some conveniently defined sense. Clearly, this
may be achieved in many different ways (collocation, least
square, etc., see e.g. [5-7] and others). Unfortunately, such
LFE cannot be implemented in the finite element (FE) library
of existing FE programs. Indeed, the way of obtaining the
system of simultaneous equations for the undetermined coefficients of all subregions fails to comply with the standard
finite element assembly rules of the direct stiffness method.

This important drawback may be overcome by the application of a technique similar to that used to enforce interelement continuity in various hybrid models, pioneered by
Pian, Tong, Lin and others. This involves assuming either an
independent interelement boundary displacement or boundary
traction fields, $\tilde{\mathbf{v}}$ or $\tilde{\mathbf{T}}$, defined in terms of conveniently
chosen nodal parameters, and constructing a convenient vari-
ational functional [2]. As opposed to common hybrid models,
this field may however be confined to the interelement portion
of the boundary.

From a practical point of view, the solution based on assumed interelement displacement appears as the most convenient:
If

$$\tilde{\mathbf{v}}_e = \mathbf{N}_e \mathbf{d}_e \quad \text{on} \quad \Gamma_e \ , \tag{3}$$

where $\mathbf{d}_e$ are nodal displacement parameters and where the interpolation functions of the matrix $\mathbf{N}_e$ are such that if the
corresponding nodal parameters of the adjacent LFE are matched,
$\tilde{\mathbf{v}}$ is the same for the two adjacent elements over their common
boundary, such formulation leads in the most straightforward
fashion to the conventional force displacement relationship

$$\mathbf{s}_e = \overline{\mathbf{s}}_e + \mathbf{K}_e \mathbf{d}_e \ . \tag{4}$$

In (4) $\mathbf{K}_e$ is the stiffness matrix of the element and $\bar{\mathbf{s}}_e$ stands for that part of the vector of equivalent nodal forces $\mathbf{s}$ that is independent of $\mathbf{d}_e$ (effect of distributed loads, variation of temperature...).

3. VARIATIONAL FORMULATION

From the method of matching the governing problem equations (1), it is obvious that any local solution representing a stress singularity of stress concentration may be straight-forwardly included in the expansion bases (2) of the particular LFE which contains this singularity or stress concentration. This enables the formulation of, for example, elements including angular corners, sharp V-shaped notches, arcs, circular holes, etc.

In general, the boundary, $\partial\Omega_e$ of a particular LFE (Fig.2)

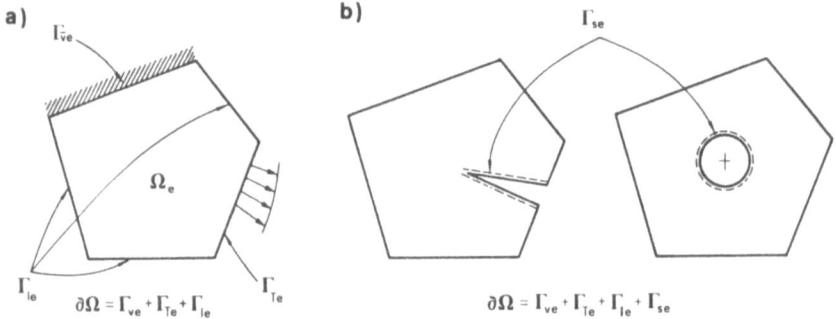

Figure 2. Regular a) and stress singularity or stress concentration b) types of the LFE. Note that the boundary conditions along the Γ_{Se} portion of the LFE boundary are satisfied by the approximating functions *a priori* and that all element types are produced by a single LFE routine associated with a library of optional approximating functions.

may be represented as a sum of four distinct portions

$$\partial\Omega_e = \Gamma_{Se} + \Gamma_{ve} + \Gamma_{Te} + \Gamma_{Ie} = \Gamma_e + \Gamma_{Se} , \qquad (5)$$

defined as follows:

- Γ_{Se} , portion of $\partial\Omega_e$ on which the prescribed boundary conditions are satisfied *a priori* (this is the case when the approximating functions are derived from a known local solution in the vicinity of a stress singularity or stress concentration);

- Γ_{ve} and Γ_{Te} , portion of the remaining part, $\partial\Omega_e - \Gamma_{Se}$, of the LFE boundary on which either displacements ($\mathbf{v} = \bar{\mathbf{v}}$) or boundary tractions ($\mathbf{T} = \bar{\mathbf{T}}$) are prescribed;
- Γ_{Ie} interelement portion of $\partial\Omega_e$;
- $\Gamma_e = \partial\Omega_e - \Gamma_{Se}$ (to simplify notation).

In the sequence we present shortly two variational functionals, $J(\mathbf{u},\tilde{\mathbf{v}})$ and $I(\mathbf{u},\tilde{\mathbf{v}})$. The main difference is the use of the independent $\tilde{\mathbf{v}}$ field. While in the first case, $\tilde{\mathbf{v}}$ is defined all over Γ_e , in the second one $\tilde{\mathbf{v}}$ is limited to the interelement portion Γ_{Ie} of the LFE. The LFE based on either the first or the second functional may be combined with each other in a LFE mesh.

3.1. Variational functional $J(\mathbf{u},\tilde{\mathbf{v}})$

The simpler of the two LFE formulations is based on the following proposition:

$$J(\mathbf{u},\tilde{\mathbf{v}}) = \sum_e \left[-\frac{1}{2}\int_{\Omega_e} \bar{\mathbf{f}}^t \mathbf{u}_e \, d\Omega - \frac{1}{2}\int_{\partial\Omega_e} \mathbf{T}_e^t \mathbf{v}_e \, d\Gamma + \int_{\Gamma_e} \tilde{\mathbf{v}}_e^t \mathbf{T}_e \, d\Gamma - \right.$$

$$\left. - \int_{\Gamma_{Te}} \tilde{\mathbf{v}}_e^t \bar{\mathbf{T}}_e \, d\Gamma \right] = \text{stationary}, \tag{6}$$

where the sum $\sum_e$ extends over all subregions (LFE) of Ω. The independent fields $\mathbf{u}_e$ and $\tilde{\mathbf{v}}_e$ of any particular LFE are subjected to the following subsidiary conditions:

a) the field $\mathbf{u}_e$ verifies the governing differential equations of the problem

$$\mathbf{Lu}_e = \bar{\mathbf{f}} \quad \text{on} \quad \Omega_e \tag{6a}$$

and, if relevant (Fig.2b), some boundary conditions

$$\mathbf{v}(\mathbf{u}_e) = 0 \quad \text{and/or} \quad \mathbf{T}(\mathbf{u}_e) = \bar{\mathbf{T}}$$

on appropriate portions of Γ_{Se} ; \tag{6b}

b) the boundary field $\tilde{\mathbf{v}}_e$ satisfies the kinematic boundary conditions

$$\tilde{\mathbf{v}} = \bar{\mathbf{v}} \quad \text{on} \quad \Gamma_{ve} \tag{6c}$$

and the kinematic interelement continuity requirements.

Now, if the variation of $J(\mathbf{u},\tilde{\mathbf{v}})$ with respect to $\mathbf{u}$ and $\tilde{\mathbf{v}}$ is performed and one observes that, by virtue of Betti's reciprocity,

$$\int_{\Omega_e} \delta\mathbf{u}_e^T \bar{\mathbf{f}} \, d\Omega = \int_{\partial\Omega_e} (\delta\mathbf{T}_e^T \mathbf{v}_e - \delta\mathbf{v}_e^t \mathbf{T}_e) \, d\Gamma$$

one obtains

$$\delta J(\mathbf{u},\tilde{\mathbf{v}}) = \sum_e \left[-\int_{\Gamma_e} \delta \mathbf{T}_e^t (\mathbf{v}_e - \tilde{\mathbf{v}}_e) d\Gamma + \int_{\Gamma_{Te}} \delta \tilde{\mathbf{v}}_e^t (\mathbf{T}_e - \overline{\mathbf{T}}_e) d\Gamma + \int_{\Gamma_{Ie}} \delta \tilde{\mathbf{v}}_e^t \mathbf{T}_e d\Gamma \right] \cdot (7)$$

Clearly, the stationary conditions associated with $\delta J(\mathbf{u},\tilde{\mathbf{v}})=0$ are the statical boundary conditions, $\mathbf{T}_e=\overline{\mathbf{T}}$ at Γ_{Te} , and the interelement continuity of displacements $\mathbf{v}$ and tractions $\mathbf{T}$.

Consequently, the stationary principle (6) implicitly imposes all the missing conditions and, as such, constitutes a suitable basis for a LFE formulation.

Examination of expression (7) of $\delta J(\mathbf{u},\tilde{\mathbf{v}})$ also shows that the solution becomes singular if any of the functions ϕ_i describes a rigid-body-motion mode of displacement and thus leads to a vanishing boundary traction term. Therefore, special care should be taken to form the matrix $\Phi = [\Phi_1, \ldots \Phi_m]$ as a complete set of linearly independent solutions associated with non-vanishing strains. Note that once the solution of the LFE assembly has been performed, the lacking rigid body modes may again be reintroduced in the internal fields $\mathbf{u}_e$ and their undetermined coefficients calculated by requiring e.g. the least squares adjustment of $\mathbf{v}_e$ and $\tilde{\mathbf{v}}_e$ at all nodes of any particular LFE.

3.2. Variational functional $I(\mathbf{u},\tilde{\mathbf{v}})$

If one wishes to restrict the use of the independent boundary displacement field $\tilde{\mathbf{v}}$ to the interelement portion of the boundary,

$$\tilde{\mathbf{v}}_e = \mathbf{N}_e \mathbf{d}_e \quad \text{on} \quad \Gamma_{Ie} , \tag{8}$$

the corresponding variational functional is simply

$$I(\mathbf{u},\tilde{\mathbf{v}}) = \sum_e \left[-\frac{1}{2} \int_{\Omega_e} \overline{\mathbf{f}}^t \mathbf{u}_e d\Omega - \frac{1}{2} \int_{\partial\Omega_e} \mathbf{T}_e^t \mathbf{v}_e d\Gamma + \int_{\Gamma_{ve}} \mathbf{T}_e^t \overline{\mathbf{v}} d\Gamma + \right.$$

$$+ \int_{\Gamma_{Te}} (\mathbf{T}_e - \overline{\mathbf{T}})^t \mathbf{v} d\Gamma + \int_{\Gamma_{Ie}} \tilde{\mathbf{v}}_e^t \mathbf{T}_e d\Gamma - \frac{1}{2} C \int_{\Gamma_{ve}} (\mathbf{v}_e - \overline{\mathbf{v}})^t (\mathbf{v}_e - \overline{\mathbf{v}}) d\Gamma -$$

$$\left. - \frac{1}{2} C \int_{\Gamma_{Ie}} (\mathbf{v}_e - \tilde{\mathbf{v}}_e)^t (\mathbf{v}_e - \tilde{\mathbf{v}}_e) d\Gamma \right] = \text{stationary} \tag{9}$$

where C is an arbitrary, but preferably small penalty coefficient. In addition to the statical boundary conditions, $\mathbf{T}_e=\overline{\mathbf{T}}$ at Γ_{Te} , and the interelement continuity of $\mathbf{v}_e$ and $\mathbf{T}_e$, rendering the functional $I(\mathbf{u},\tilde{\mathbf{v}})$ stationary imposes also implicitly the geometrical boundary conditions, $\mathbf{v}_e=\overline{\mathbf{v}}$ at Γ_{ve} .

Note that this formulation does not impose any condition on the rigid-body-motion terms among the approximating functions ϕ_i.

4. DEVELOPMENT OF THE LFE MATRICES $\overline{s}$ AND K AND PROGRAM IMPLEMENTATION

The evaluation of the LFE matrices only calls for integration along the element boundaries (section 3) which makes it possible, in two dimensional problems, to generate arbitrary polygonal or even curved sided elements. Choosing of convenient expansion set satisfying the governing problem equations and, if relevant, boundary condition on the portion Γ_{Se} of the element boundary, generally does not present a difficult problem. Whilst the standard regular displacement field for the classical Kirchhoff plate, for example, may be easily generated using biharmonic polynomials

$$\phi_{i+1} = r_0^2 \text{Rez}_0^k \; , \qquad \phi_{i+2} = r_0^2 \text{Imz}_0^k \; , \qquad \phi_{i+3} = \text{Rez}_0^{k+2} \; ,$$

$$\phi_{i+4} = \text{Imz}_0^{k+2} \; , \qquad (k=0,1,2,\ldots)$$

where x_0 and y_0 are local coordinates with origin at the center of gravity of the LFE, $r_0^2 = x_0^2 + y_0^2$ and the complex variable $z_0 = x_0 + iy_0$, the displacement field for a perforated or a singular corner LFE may be conveniently represented using the known local solutions in the vicinity of a circular hole and the Williams eigensolutions for the given apex angle respectively. The load term $\overset{o}{u} = \overset{o}{w}$ for uniform load may be e.g. set to

$$\overset{o}{w} = \frac{\overline{p}}{8D} x^2 y^2 \text{ where } D = \frac{Et^3}{12(1-\nu^2)} \text{ stands for plate stiffness.}$$

For a concentrated load $\overline{P}$ at x_p, y_p, the corresponding $\overset{o}{w}$ is equal to

$$\overset{o}{w} = \frac{\overline{P}}{16\pi D} r^2 \ell nr^2 \text{ with } r^2 = (x-x_p)^2 + (y-y_p)^2 \; .$$

It is essential to point out that this load term should be used as a global function, extending over the whole element assembly. This not only helps in improving the accuracy (the term $\overset{o}{w}$ being already continuous over the interelement boundary) but also simplifies the input data since the program user need not trouble to identify the elements concerned by the applied concentrated loads.

For the classical plate bending again, the conjugated vectors T_e and v_e may be readily defined e.g. as

$$T_e^T = [Q_n, \; M_n, \; M_{nt}] \text{ and } v_e = [w, \; -\frac{\partial w}{\partial n}, \; -\frac{\partial w}{\partial t}] \; ,$$

where Q_n, M_n and M_{nt} stand respectively for shear force, bending moment and twisting moment referenced in local cartesian frame n,t (n = external normal to the LFE boundary).

The approach to other problems of the linear theory of elasticity is quite analogical. The nature of the formulation makes it possible to write for each class of problems (plane elasticity, plate bending, etc.) a single element subroutine for LFE of a very general, say polygonal, form (variable number of sides) and provide it with a library of optional expansion sets. The program user selects the appropriate one according to the situation encountered by specifying a single control parameter. This possibility was successfully experimented within the general purpose program SAFE [8] and Table 1 presents the available LFE. Using 6 nodal degrees of freedom (DOF) for the plate bending LFE implies that the independent boundary displacement $\tilde{w}$ and normal slope $\partial\tilde{w}/\partial n$ are interpolated over each element side by Hermitian polynomials of order 2 and 1. Obtaining a 3 DOF/node version, if preferred, is a simple matter of appropriately changing the interpolation functions $\mathbf{N}_e$ at the LFE boundary.

Table 1. Families of Large Finite Elements available
in general purpose program SAFE.

FAMILY NAME	ELEMENT TYPE	NODAL DOF	Standard elements	Circular hole stress concentration elements	Angular singularity elements	REMARKS
LFEPLB	Plane elasticity elements	u, v $\varepsilon_x, \varepsilon_y$ γ_{xy}, ω_{xy} or u, v σ_x, σ_y τ_{xy}, ω_{xy}	1)		2) $\vartheta \geq 0$	Takes into account local effects of concentrated loads
LFEPEL	Plate bending elements	w, w_x, w_y w_{xx}, w_{xy}, w_{yy}	1)		2) $\vartheta \geq 0$	Various boundary conditions along the two sides adjacent to singular corner
LFESEC	Elements for cross-sectional properties and stress analysis of beams	Ψ, Ψ_x, Ψ_y (Ψ= warping function)			$\vartheta \geq 0$	

5. ASSESSMENT OF THE EFFICIENCY OF THE LFE METHOD

From the nature of the LFE concept it is obvious that the advocated approach is particularly well suited to solution of complicated problems involving various stress concentrations and/or singularities. Therefore for an unbiased assessment of the LFE efficiency, it is of interest to start with some quite

Table 2. Study of a clamped square plate ($\nu=0$) using
quadrilateral LFE : P...concentrated
central load; p...uniform load.

MESH	$Dw_A : Pa^2$	$M_{xA} : P$	$M_{xB} : P$	$Q_{xB} : P/a$
1 x 1	0.005611 (0%)	∞	-0.1244 (-1.1%)	-0.695 (-12.5%)
2 x 2	0.005600 (0.2)	∞	-0.1268 (0.8)	-0.780 (-1.8)
THEORY	0.005612	∞	-0.1258	-0.794
MESH	$Dw_A : pa^4$	$M_{xA} : pa^2$	$M_{xB} : pa^2$	$Q_{xB} : pa$
1 x 1	0.001275 (0.8)	0.0163 (-7.7)	-0.0549 (7.0)	-0.469 (6.0)
2 x 2	0.001266 (0.1)	0.0173 (-1.9)	-0.0516 (0.6)	-0.448 (1.6)
THEORY	0.0012653	0.01762	-0.0513	-0.441

A ... Plate center;   B ... Mid-edge

Table 3. Study of a simply supported square plate ($\nu=0$)
using quadrilateral LFE : P...concentrated
central load; p...uniform load.

MESH	$Dw_A : Pa^2$	$M_{xA} : P$	$Q_{xB} : P/a$	$V_C : P$
1 x 1	0.01156 (-0.3%)	∞	-0.365 (-12.5%)	0.173 (-0.6%)
2 x 2	0.01158 (-0.2)	∞	-0.430 (3.1)	0.175 (0.6)
THEORY	0.011601	∞	-0.417	0.174
MESH	$Dw_A : pa^4$	$M_{xA} : pa^2$	$Q_{xB} : pa$	$V_C : pa$
1 x 1	0.004075 (0.3)	0.0355 (-3.5)	-0.371 (9.8)	0.097 (3.9)
2 x 2	0.004063 (0)	0.0367 (-0.3)	-0.341 (0.9)	0.093 (0)
THEORY	0.0040624	0.03684	-0.338	0.093

A ... Plate center;   B ... Mid-edge;   C ... Plate corner

common standard test problem, such as e.g. the bending of a
thin square plate (side a), where the possible superiority of
the LFE solution is not *a priori* obvious. Tables 2 and 3 show
that already the crudest meshes (1x1 or 2x2 LFE over a symmetric quadrant) yield not only surprisingly accurate displacements w but also excellent moments (M), shear forces (Q) and
corner reactions (V).

The next examples demonstrate excellent capacity of the LFE to deal with various singular situations. The results shown in Table 4 represent the LFE reply to the skew plate challenge recently proposed by Finite Element News [9]. Again,

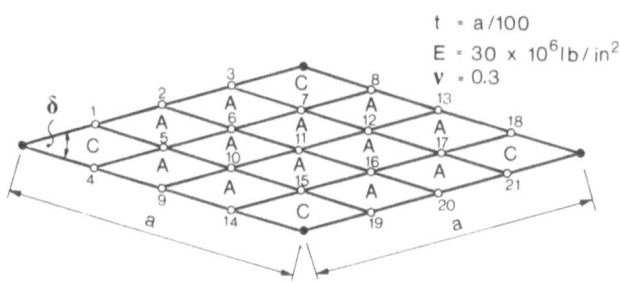

Figure 3. Typical LFE mesh for a skew plate study: A. Regular LFE; C. LFE with singular corner singularity.
o assembly nodes with 6 DOF; ● auxiliary nodes without DOF.

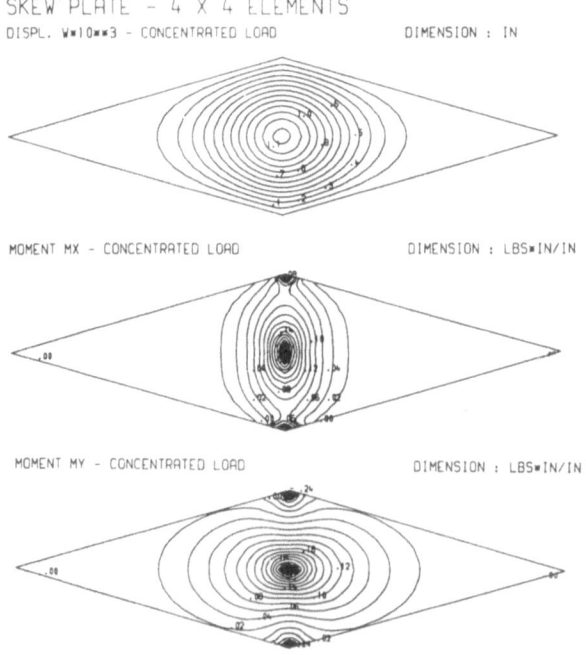

Figure 4. Simply supported skew plate from Fig.3 (a = 1 in) under concentrated central load (P = 1 lb).

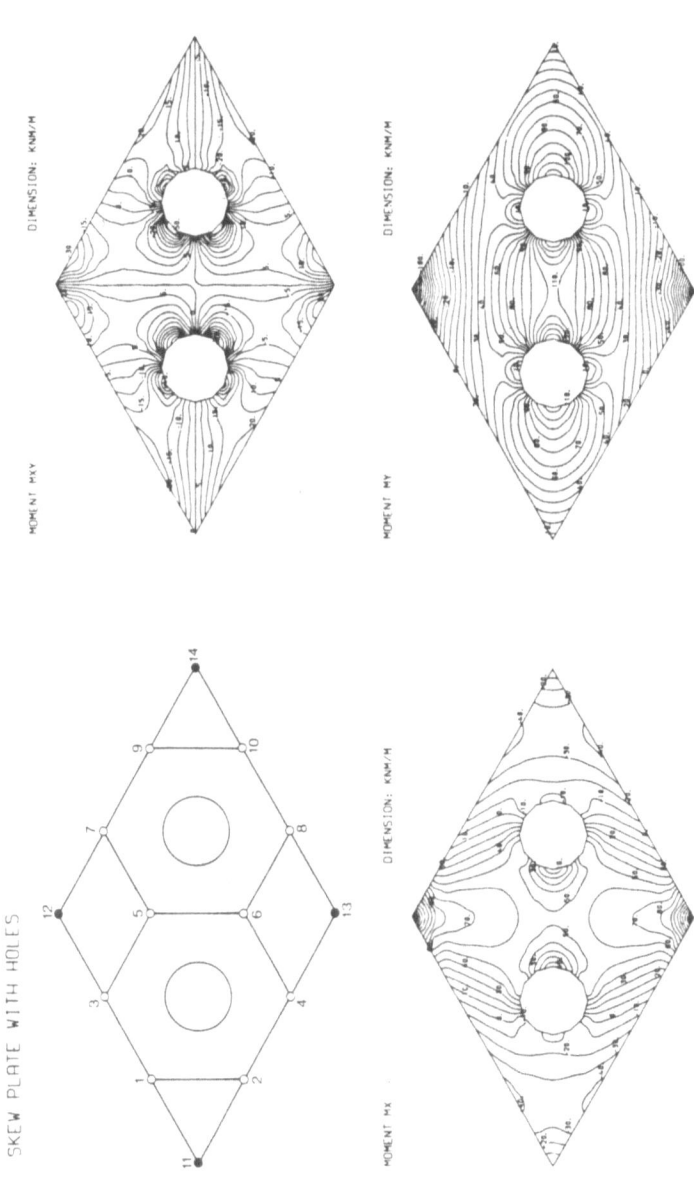

Figure 5. LFE mesh and corresponding results for simply supported uniformly loaded skew plate: ○ assembly nodes with 6 DOF; ● auxiliary nodes without DOF. 10 nodes x 6 DOF = 60 simultaneous equations.

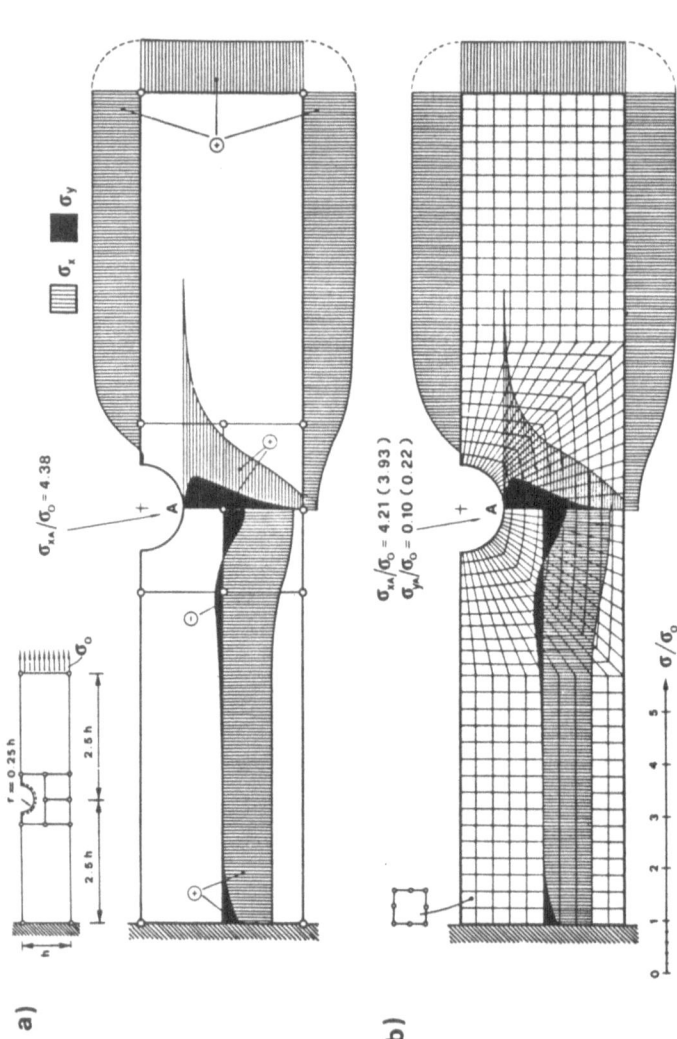

Figure 6. Tension of a notched steel plate ($\nu = 0.3$). a) solution using 5 plane elasticity LFE from Table 1 - 60 simultaneous equations; b) conventional FE solution using a fine mesh of 700 quadratic isoparametric elements - 4522 simultaneous equations. The results between brackets correspond to a two times coarser mesh of 175 elements.

Table 4. Convergence study for uniformly loaded simply supported skew plate. Uniform meshes of LFE (e.g. Fig.3).

Mesh over complete plate	2 x 2			4 x 4			8 x 8		
Percentage error in displacement and principal moments at centre (C)	w_C	M_{maxC}	M_{minC}	w_C	M_{maxC}	M_{minC}	w_C	M_{maxC}	M_{minC}
$\delta = 90°$	0	1.0	1.0	0	-0.4	-0.4	0	0	0
$\delta = 80°$	0	0.8	1.8	0	-0.4	-0.2	0	0	0.2
$\delta = 60°$	0	0.7	3.6	0	-0.2	-0.3	0	0	0
$\delta = 40°$	0	-3.9	-6.7	0	-0.7	-0.6	0	-0.4	0
$\delta = 30°$	-0.1	0	0.9	0.1	-1.0	0.9	0	-0.5	0

already the simplest 2 x 2 LFE mesh produces for any skew angle δ reliable results whereas the majority of the 33 plate bending elements reported in the Final Report [10] appear as unduly sensitive to the skew angle δ and present large errors despite very fine meshes (14 x 14) used. Figures 4 to 6 are other examples of efficiency of the LFE concept. It is of interest to point out that the overall cost of the analysis is frequently reduced in such cases by a factor of more than 20 as compared to analysis using more conventional elements in which case the user is faced with an additional difficulty of finding a FE mesh fine enough to yield reliable results yet not too fine or too difficult to generate to keep the computer and data preparation cost within reasonable limits.

The contour plots (Figs. 4 and 5) show that the interelement continuity upon which bears the only one approximation used is remarkably well satisfied not only be kinematic quantities but also by moments. Note that to allow obtaining good graphics with only few elements, the LFE subroutine evaluates results for optional auxiliary grids of up to 25 internal output points for a regular LFE (and more for a singular or stress concentration LFE) - see [11] for more information.

6. CONCLUSIONS

The LFE method combines the flexibility of the conventional FE for irregularly shaped regions with the accuracy and high convergence rate associated with the Trefftz's method.

The accuracy and practical efficiency of the LFE solution as compared with more conventional FE calculation, is already surprising e.g. in "regular" plate bending problems, and in cases presenting various stress concentrations or stress singularities, its superiority over conventional FE is obvious. The LFE also considerably reduces the input and output costs, since the mesh definition is much simpler and, when using graphic output facilities, full advantage may be taken of the knowledge of the analytic solution within each LFE and of the practically negligible discontinuities (Fig.4) at the interfaces.

The LFE method is related to both the so-called boundary solution procedures and the hybrid methods. For the time being, the numerical experimentation remains the basic method of analysing the LFE adequacy and a large amount of research is still necessary to exploit its vast possibilities.

REFERENCES

1. Jirousek, J. and Leon, N., (1977) A powerful finite element for plate bending. Comp. Meth. Appl. Mech. Engng., 12, 1:77-96.

2. Jirousek, J., (1978) Basis for development of large finite elements locally satisfying all field equations. Comp. Meth. Appl. Mech. Engng., 14, 1:65-92.

3. Jirousek, J. and Teodorescu, P., (1982) Large finite element method for the solution of problems in the theory of elasticity. Computers and Structures, 15, 5:575-587.

4. Jirousek, J. and Teodorescu, P., (1983) Large finite elements for the solution of problems of the theory of elasticity. Proceedings of the Third International Symposium on Numerical Methods in Engineering, March 1983, Paris, 2, 695-704. PLURALIS-PARIS.

5. Tolley, M.D., (1977) Grands éléments finis singuliers. Ph.D thesis, Université Libre de Bruxelles.

6. Tolley, M.D., (1978) Application de la méthode des grands éléments finis singuliers à la résolution de l'équation bi-harmonique avec conditions aux limites discontinues. C.R. Acad. Sci. Paris, t. 287, série A, 875-878.

7. Descloux, J. and Tolley, M.D., (1980) Approximation of the Poisson problem and of the eigenvalue problem for the Laplace operator by the method of large singular finite elements. Dept. of Math., Swiss Federal Institute of Technology, Lausanne and Fac. Appl. Sci., Université Libre de Bruxelles.

8. Jirousek, J., (to appear) Structural analysis program SAFE - Special features and advanced finite element models. Advances in Engineering Software.

9. Robinson, J., (1984) Skew effects - FEM User Project No. 2 - Morley's simply supported skew plate problem. Finite Element News, 3.

10. Robinson, J., (1985) An evaluation of skew sensitivity of thirty three plate bending elements in nineteen FEM systems. Robinson and Associates, Wimborne, Dorset BH21 6NB, England.

11. Jirousek, J., Bouberguig, A., Frey, F., (1985) An efficient post-processing approach based on probabilistic concept. NUMETA 85 - Numerical Methods in Engineering: Theory and Applications, January 1985, Swansea, 2, 723-732, A.A. Balkema, Rotterdam/Boston.

AN EXPLICIT FINITE DIFFERENCE SOLVER BY PARAMETER ESTIMATION

S.K. Dey   C. Dey

Mathematics Department   7th Grader
Eastern Illinois University   Charleston Jr. High School
Charleston, IL 61920   Charleston, IL 61920

1.   INTRODUCTION

Explicit finite difference schemes are possibly the simplest
methods to solve initial value problems. The main drawback,
however, is that in order to maintain stability of computation,
step size must be small. This often increases computer run
time, especailly when steady-state solutions are needed.   In
ref. [1], the second author combined a predictor which is an
explicit finite difference scheme with a one-step corrector and
solved Burgers' equation given in section 7. The step sizes
were chosen such that the stability criterion for the predictor
alone was violated. The corrector required estimation of a
filtering parameter such that stability properties may be
maintained.

For linear models, computation of this parameter is rather
simple. Nonlinear models are, however, first linearized at
the initial condition to compute the value of the filtering
parameter. The method is an explicit finite difference scheme
and as such it requires no matrix computation. Lomax [3] applied linearized stability analysis and found that the present
scheme may generate stability contours which could be significantly much larger than those for most explicit finite difference methods. In this article we will study some properties
of this method with some applications.

2.   THE ALGORITHM

For the initial value problem

$$du/dt = f(u,t), \ u(t_0) = U_0 \tag{1}$$

the present algorithm consists of the following two steps:

$$\hat{U} = U_n + hf(U_n,t_n) \quad \text{predictor} \tag{2}$$

$$U_{n+1} = (1-\gamma)\hat{U} + \gamma\{U_n + h \ f(\hat{U}, t_{n+1})\} \qquad \text{corrector} \qquad (3)$$

where h = step size, γ = filtering parameter. The predictor is Euler's forward extrapolation formula. If $\gamma = 0$, $U_{n+1} = \hat{U}$ (corrector is not used).

A simple application may be du/dt = -2u+1, t = 0   u = 1.5. (Analytical solution is u = 0.5 + exp(-2t)). Here f(u,t) = -2u + 1. With h = 0.1, $\gamma = 0.4667$, we got:

t	U(computed)	u(exact)
1.0	0.635232	0.635335
2.0	0.518288	0.518316
3.0	0.502473	0.502479
4.0	0.500334	0.500335
5.0	0.500043	0.500045

3.   CONSISTENCY, STABILITY AND CONVERGENCE

We consider a linear model:

$$du/dt = \lambda u \ , \ u(t_0) = u_0 \qquad (4)$$

Now using equations (3) and (4) and combining them we get,

$$U_{n+1} = \beta \ U_n \ , \ \beta = 1+z+\gamma z^2 \ , \ z = \lambda h. \qquad (5)$$

Assuming Taylor series expansion for $u_{n+1} = u(t_n + h)$ and using equation (4) we get:

$$u_{n+1} = (1+z+z^2/2! + z^3/3! + \ldots) \ u_n . \qquad (6)$$

Subtracting equation (5) from equation (6) we get:

$$\varepsilon_{n+1} = \beta \ \varepsilon_n + \delta_n \qquad (7)$$

where $\varepsilon_n = u_n - U_n$, $\delta_n = (-\gamma z^2 + z^2/2! + z^3/3! + \ldots) \ u_n$. δ_n is the truncation error and as h$\to$0, $\delta_n \to 0$ $\forall$n. This gives consistency. Now, iff δ_n is neglected, a necessary and sufficient condition for convergence to steady state is:

$$|\beta| < 1 \qquad (8)$$

This is our stability criterion. Since λ may be complex, so is z. If we set $|\beta| = 1$, z = x+iy and plot $|\beta| = |1+z+\gamma z^2| = 1$ in a complex plane (for a given γ) we obtain a stability contour. Three such contours for $\gamma = 0.095$, 0.13 and 0.25 are given by figures 1,2 and 3 respectively. The code was developed by Lomax [3]. In figure 1, the method is stable if (i) $-3 \le Re(z) < 0$ or (ii) $-10.6 \le Re(z) \le -8$.

Let $\lambda = -\mu$, $\mu > 0$. Then for stability $0 < h \leq 3/\mu$ or $8/\mu \leq h \leq 10.6/\mu$. Thus there are two distinct regions of stability, for one h is small, for the other h is considerably larger. This has been verified in Example 1 in section 6.

4. STEADY-STATE SOLUTION

This algorithm is primarily meant to compute steady-state solution. Here we prefer to use large time steps. It is important to analyze how such problems are affected by the present algorithm. For this we consider:

$$du/dt = \lambda u + a \ , \ u(0) = u_0 \tag{9}$$

The steady-state solution is $u = -a/\lambda$. The present scheme is:

$$\hat{u} = u_n + z \, u_n + h \, a + (TE)_n \tag{10}$$

$$u_{n+1} = (1-\gamma) \, \hat{u} + \gamma \{ u_n + z \, u_n + h \, a + (TE)_n \} \tag{11}$$

where $z = \lambda h$, $(TE)_n$ = truncation error at $t_n = 0(h^2)$.

As $h \to 0$, $(TE)_n \to 0$. Combining (10) and (11) we get:

$$u_{n+1} = (1 + z + \gamma z^2) \, u_n + (1 + \gamma z) \, ah + (1 + \gamma z) \, (TE)_n \tag{12}$$

Thus for the present scheme, truncation error is $(1 + \gamma z) \, (TE)_n \to 0$ as $h \to 0$. Neglecting this we get:

$$U_{n+1} = (1 + z + \gamma z^2) \, U_n + (1 + \gamma z) \, a \, h \tag{13}$$

where U_n = net function corresponding to $u(t_n)$. At the steady-state $U_{n+1} = U_n$. Thus from equation (13) $U_n = -a/\lambda$, the steady-state solution, which is the same as the analytical solution.

5. LINEAR SYSTEMS

The extension of the present scheme to the linear system is interesting. We consider:

$$dX/dt = AX \tag{14}$$

where $X = (x_1 \, x_2 \ldots x_J)^T \subset R^J$ and $A = a$ (JxJ) square matrix.

$(R^J$ = real J-dimensional space). The predictor is, in element form:

$$\hat{X}_j^n = X_j^n + h \sum_{k=1}^{J} a_{jk} X_k^n \ , \ j = 1,2,\ldots J \tag{15}$$

The corrector is:

$$X_j^{n+1} = (1 - \gamma_j) \, \hat{X}_j + \gamma_j \, \{X_j^n + h \, (\sum_{k=1}^{j-1} a_{jk} \, X_k^{n+1} + \sum_{k=j}^{J} a_{jk} \, \hat{X}_k)\}$$

(16)

Where X_j^n = the net function corresponding to $X_j(t_n)$.

If we combine equations (15) and (16) we will get:

$$P \, X^{n+1} = Q \, X^n$$

(17)

where P is a lower triangular and Q, an upper triangular matrix.

For stability $P^{-1}Q$ must be a convergent matrix, which is guaranteed if $\rho(P^{-1}Q) < 1$, where $\rho(P^{-1}Q)$ is the spectral radius of $P^{-1}Q$.

6. APPLICATIONS TO LINEAR MODELS

We now consider applications to linear models.

Example 1: $du/dt = -100(u - \sin t) + \cos t$, $u(0) = 0$.

The analytical solution is $u = \sin t$. Here $\lambda = -100$. If we choose $h = 0.1$, $z = \lambda h = -10$. From figure 1, $z = -10$ is within the stability contour for $\gamma = 0.095$. Computationally we found:

t	U(computed)	U(exact)
2.5	0.594931	0.598472
5.0	-0.957553	-0.958924
7.5	0.936782	0.938000
10.0	-0.543441	-0.544021
12.5	-0.066034	-0.066322
15.0	0.649246	0.650288

With $\gamma = 0$, the algorithm failed. If we reduce h and set $h = 0.06$, $z = \lambda h = -6.0$ which is exterior to the stability countour and thus the method failed.

Example 2: $dx/dt = y$, $dy/dt = -x$, $x(0) = 0$, $y(0) = 1$.

the exact solutions are $x = \sin t$, $y = \cos t$. The present algorithm uses $\hat{X} = X_n + hY_n$, $\hat{Y} = Y_n - h X_n$ and $X_{n+1} = (1-\gamma)\hat{X} + \gamma(X_n + h\hat{Y})$, $Y_{n+1} = (1-\gamma)\hat{Y} + \gamma(Y_n - h X_{n+1})$. Here,

$$P = \begin{bmatrix} 1 & 0 \\ \gamma h & 0 \end{bmatrix} \, , \quad Q = \begin{bmatrix} 1-\gamma h^2 & h \\ \gamma h - h & 1 \end{bmatrix} \, .$$

Eigenvalues of $P^{-1}Q$ are $\lambda = (1-\gamma h^2) \pm ih\sqrt{(1-\gamma^2 h^2)}$
If we choose $\gamma = 1/h$, $\lambda = 1-h < 1$. Here stability will be maintained although results could be grossly inacurate.

We chose $\gamma = 0.5125$, $h = 0.1$ and obtained:

t	x(computed)	x(exact)	y(computed)	y(exact)
1.1	0.891393	0.891207	0.452504	0.453596
2.1	0.861572	0.863209	-0.504388	-0.504846
3.1	0.039995	0.041581	-0.995331	-0.999135
4.1	-0.816288	-0.818277	-0.570285	-0.574824
5.1	-0.920220	-0.925815	0.377821	0.377978
6.1	-0.178161	-0.182163	0.976323	0.983268

Example 3 $u_t = -c\,u_x$, $u(0,t) = e^{-ct}$, $u(x,0) = e^x$. Analytically,

$u = e^{x-ct}$. Approximating u_t and u_x by two point forward and

backward differences respectively we get:

$$\hat{U}_j = U_j^n - a\,(U_j^n - U_{j-1}^n) \;,\; a = c\,\Delta t/\,\Delta x, U_j^n = U(x_j^\cdot, t_n) =$$

net function corresponding to $u(x_j, t_n)$. The corrector is

$U_j^{n+1} = (1-\gamma)\,\hat{U}_j + \gamma\{U_j^n - a\,(\hat{U}_j - U_{j-1}^{n+1})\}$ $j = 1,2,\ldots J$. Combining

these two formulas we get $P\,U^{n+1} = Q\,U^n$, where $P = \text{Tridiag}(-\gamma a, 1, 0)$

$Q = (1-\gamma-\gamma a)$. $\text{Tridiag}(a, 1-a, 0) - \gamma\,I$. The eigenvalues of $P^{-1}Q$

are given by $\lambda_j - (1-a)(1-\gamma-\gamma a) - \gamma = 0$ giving $\lambda_j = 1 - a + \gamma a^2$.

For stability $|\lambda_j| < 1$, giving $(a-2)/a^2 < \gamma < 1/a$, ($a = c\,\Delta t/\,\Delta x$).

Choosing $\Delta t = \Delta x$, $c = 2$, stability is obtained if $0 < \gamma < 0.5$.

Even if we choose $c = 2$, $\Delta t = 2\Delta x$, stability is guaranteed if

$0.125 < \gamma < 0.25$. The predictor itself ($\gamma = 0$) is stable only

when $|1 - a| < 1$ or $\Delta t < \Delta x$. In figure 4, $\gamma = 0$, $\Delta t = \Delta x$

$= 0.2$, $c = 2$. After 12 time steps, the predictor failed. In

figure 5, $\gamma = 0.13$, $\Delta x = 0.2$, $\Delta t = 0.4$, $c = 2$, the predictor-

corrector scheme showed no instabilities.

Example 4: $u_t = \nu u_{xx}$, $u(x,0) = \sin \pi x$, $u(0,t) = u(1,t) = 0$

Approximating u_t by a two-point forward difference formula and

u_{xx} by central differences we get $U_j^{n+1} = U_j^n + a(U_{j-1}^n - 2U_j^n + U_{j+1}^n)$

$a = \nu \Delta t/\,\Delta x^2$. For stability $0 < a \leq \frac{1}{2}$. This means, if $\nu = 1$

and $\Delta x = 0.05$, $\Delta t < 0.005$. The predictor-corrector algorithm

is: $\hat{U}_j = U_j^n + a\,(U_{j-1}^n - 2U_j^n + U_{j+1}^n)$, $U_j^{n+1} = (1-\gamma)\,\hat{U}_j +$

$\gamma\,\{U_j^n + a\,(U_{j-1}^{n+1} - 2\hat{U}_j + \hat{U}_{j+1})\}$.

Stability criterion is rather simple to compute. For $\nu = 1$, $\Delta x = 0.05$, $\Delta t = 0.005$ whereas the predictor itself ($\gamma = 0$) failed (figure 6), using $\gamma = 0.19$ perfect results were found (fig. 7)

7. APPLICATIONS TO NONLINEAR MODELS

Example 1: $du/dt = -25(u - 1/u)$, $u(0) = \sqrt{2}$. The analytical solution is $u = (1 + \exp(-50t))^{\frac{1}{2}}$. To obtain λ we linearize the model at $t = 0$. $f(u, t) = -25(u - 1/u)$, $\lambda = df/du\big|_{u_0} = -25(1 + 0.5) = -37.5$. If we choose $h = 0.1$, $z = \lambda h = -3.75$. This is inside the stability contour of $\gamma = 0.175$ [1]. Computationally steady-state solution $u = 1$ was found after 1.0 sec.

Example 2: $u_t + u u_x = \nu u_{xx}$, $u(x,0) = \sin \pi x$, $u(0,t) = u(1,t) = 0$.

This is Burgers' equation. The predictor-corrector algorithm is: $\hat{U}_j = U_j^n + a U_j^n (U_{j-1}^n - U_j^n) + b(U_{j-1}^n - 2U_j^n + U_{j+1}^n)$, $U_j^{n+1} = (1 - \gamma)\,\hat{U}_j + \gamma \{ U_j^n + a \hat{U}_j\,(U_{j-1}^{n+1} - \hat{U}_j) + b(U_{j-1}^{n+1} - 2\hat{U}_j + \hat{U}_{j+1}) \}$,

$j = 1, 2, \ldots J$ where $a = \Delta t / \Delta x$, $b = \nu \Delta t / \Delta x^2$. We used $\nu = 10^{-5}$, $\Delta x = 0.05$, $\Delta t = 4 \Delta x$. If $\gamma = 0$, the predictor failed. With $\gamma = 0.25$, results are plotted in figure 8. Results are not time accurate. Strong instabilities were present. But as time increases these instabilities are damped out, and computational results show correct values towards steady-state. More on these may be found in [1].

Example 3: The model for the one dimensional transonic flow is described by:

$$\phi_{xt} + 0.5\,(\phi_x^2)_x = 0; \quad \phi(x,0) = x, \ x \leq 0.25, \ \phi(x,0) = -x/3 + x/3, \ x > 0.25; \quad \phi(0,t) = \phi(1,t) = 0, \ \phi_x(0,t) = 1.$$

The steady-state solution is $\phi(x) = x$, $x \leq 0.5$, $\phi(x) = 1 - x$, $x > 0.5$ and at this stage residual is: $(\phi_x^2)_x = 0$. The model is approximated by:

$$\phi_j^{n+1} = \phi_{j-1}^{n+1} + \phi_j^n - \phi_{j-1}^n - a\,F_j(\phi^n) \tag{19}$$

where, $a = \Delta t / \Delta x^2$, $F_j(\phi^n) = (1 - \varepsilon_j)(\phi_{j+1}^n - \phi_{j-1}^n)(\phi_{j+1}^n - 2\phi_j^n + \phi_{j-1}^n)$

$+ \varepsilon_{j-1}(\phi_j^n - \phi_{j-2}^n)(\phi_j^n - 2\phi_{j-1}^n + \phi_{j-2}^n)$. $\varepsilon_j = 0$ if $\phi_{j+1}^n < \phi_{j-1}^n$, otherwise $\epsilon_j = 1$. The corrector is: $\phi_j^{n+1} = (1-\gamma)\hat{\phi}_j + \gamma$

$\{(\phi_{j-1}^{n+1} + \phi_j^n - \phi_{j-1}^n) - a\, F_j(\phi^{n+1}, \hat{\phi})\}$, where $F_j(\phi^{n+1}, \hat{\phi}) =$

$(1 - \epsilon_j)(\hat{\phi}_{j+1} - \phi_{j-1}^{n+1})(\hat{\phi}_{j+1} - 2\hat{\phi}_j + \phi_{j-1}^{n+1}) + \epsilon_{j-1}(\hat{\phi}_j - \phi_{j-2}^{n+1})$

$(\hat{\phi}_j - 2\phi_{j-1}^{n+1} + \phi_{j-2}^{n+1})$, where $\hat{\phi}_j$ is given by the right side of

equation (19). $j = 1, 2, \ldots J$. We used $\gamma = 0.095$, $\Delta x = 0.05$
and a variable Δt, $\Delta t = 0.02$ to 0.08 in 8 cycles. Results are
plotted in figure 9. They are not time accurate. Instabilities
appeared and were eliminated as steady-state was found in 45
iterations when $|\text{residual}|_t < 10^{-4}\, |\text{initial residual}|$. More
interesting results are given in [3].

8. DISCUSSIONS

Most explicit finite difference methods are severely restricted
by step sizes to maintain stability. Using appropriate values
of γ this may be avoided here. For linear models, the values
may be obtained from stability contours by Lomax [3]. For non-
linear models, linearization may be done using initial conditions. This produced some fruitful results [1]. However, this
can not be generalized for all nonlinear models. This is possibly the primary drawback of this method. Another drawback is
to find appropriate γ to obtain time accurate solutions. This
could be difficult even for linear models. However, if steady-
state solution is the main goal, a γ can be found from stability
contours and correct solution may be obtained. And for such
cases, the present scheme which requires no matrix computations
seems to be more attractive than many unconditionally stable
implicit schemes which either require computation of Jacobians
or have slow rates of convergence. If $\gamma \neq \frac{1}{2}$, the present
scheme has only first order accuracy. However, one may choose
a second or higher=order accurate finite difference approximation of a given model as predictor. Computational experiments
revealed that such predictors did not improve the results significantly. At present a search is being made to see whether
results may be bettered if instead of equation (3) we use:

$$U_{n+1} = (1 - \gamma_1)\,\hat{U} + \gamma_2\,\{U_n + hf(\hat{U}, t_{n+1})\}.$$

Suggestions on this topic are very welcome.

9. ACKNOWLEDGEMENT

This work has been partially supported by National Research
Council, Washington D.C. and by the Center of Advanced Studies
in Applied Mathematics of Calcutta University. Special thanks
to Dr. P.P. Chatterjee, Appd. Math Dept. Calcutta University.,
Dr. H. Lomax NASA-Ames Research Center for their support of
this project, and to Mrs. D. Burch for typing the article.

REFERENCES

1. Dey, S.K. and Dey, C. (1983) An Explicit Predictor-
    Corrector Solver with Applications to Burgers' Equation.
    NADA Tech. Memo. 84402.

2. Dey, C. and Dey, S.K. (1983) Explicit Finite Difference
    Predictor and Convex Corrector with Applications to Hyper-
    bolic Partial Differential Equations. Computer and Math.
    with Applications. 9, 1 : 549-557.

3. Lomax, H. (1983) Private Communication. NASA-Ames
    Research Center.

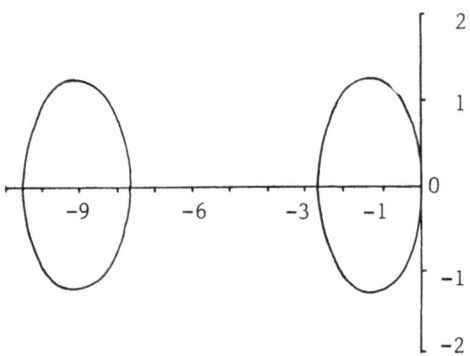

Figure 1. Stability contour for $\gamma = 0.095$

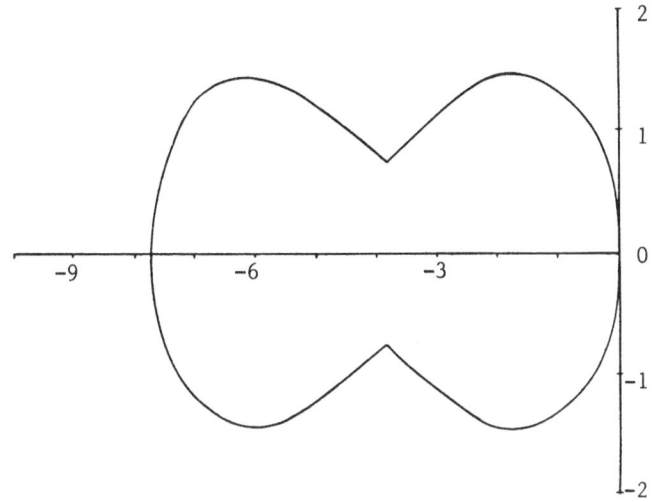

Figure 2. Stability contour for $\gamma = 0.13$

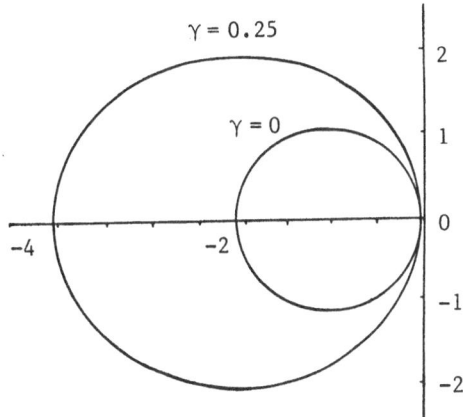

Figure 3. Stability contours for $\gamma=0.25$ and $\gamma=0$

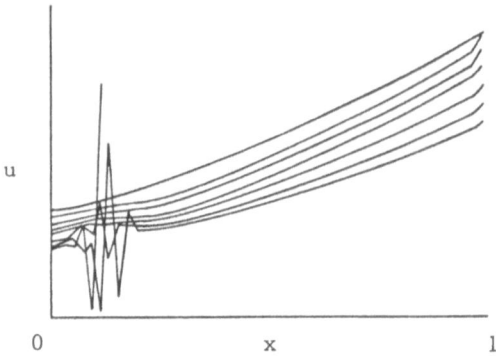

Figure 4.   In section 6, Ex. 3, Predictor
            failed after 12 time steps.
            ($\gamma=0$, $c=2$, $\Delta t = \Delta x = 0.2$)

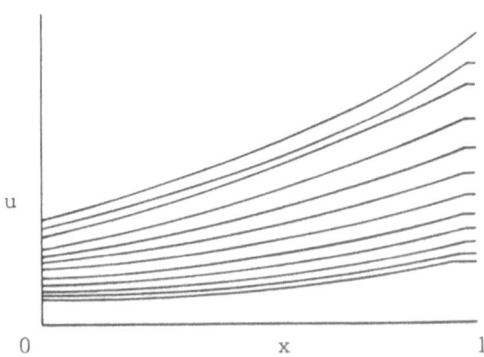

Figure 5.   In section 6, Ex. 3, predictor-
            corrector showed no instabilities.
            ($\gamma=0.13$, $\Delta x=0.2$, $\Delta t=0.4$, $c=2$)

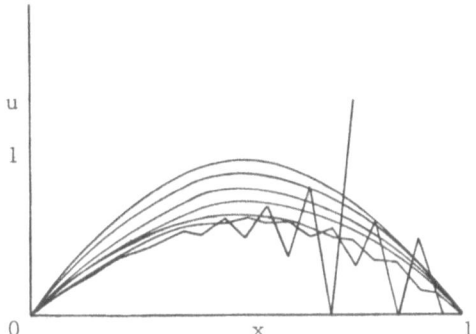

Figure 6. In section 6, Ex. 4, predictor failed.
($\gamma=0$, $\nu=1$, $\Delta x=0.05$, $\Delta t=0.005$)

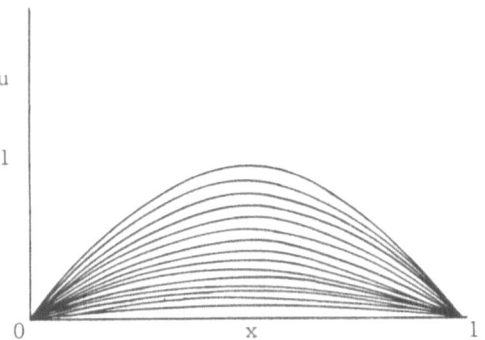

Figure 7. In section 6, Ex. 4, Predictor-
corrector showed no instabilities.
($\gamma=0.19$, $\nu=1$, $\Delta x=0.05$, $\Delta t=0.005$)

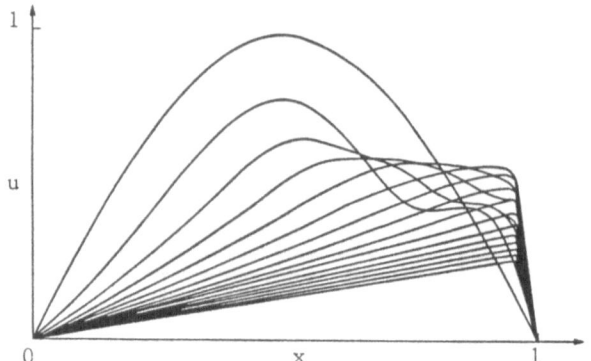

Figure 8. In section 7, Ex. 2, instabilities
are being damped out as steady-state
is reached.
($\gamma=0.25$, $\nu=10^{-5}$, $\Delta x=0.05$, $\Delta t=0.2$)

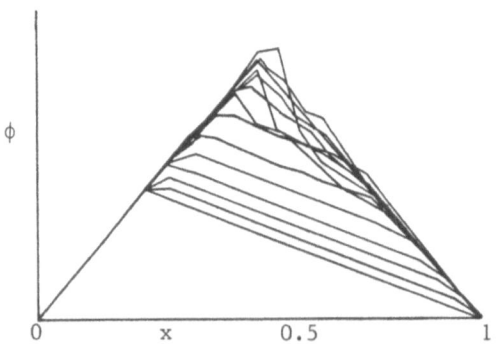

Figure 9. In section 7, Ex. 3 instabilities are
being eliminated and steady-state is
reached in 45 iterations.
($\gamma=0.0095$, $\Delta x=0.05$, $\Delta t_{min}=0.02$, $\Delta t_{max}=0.08$)

COMPARATIVE ANALYSIS BEM-FEM FOR STRESS CONCENTRATI-
ON AROUND THE HOLES IN ELASTO-PLASTIC INFINITE MEDI-
UM
V.F. Poterasu,N. Mihalache

Polytechnic Institute of Jassy,Romania

1. INTRODUCTION

The paper analyses the stress state around two
elliptical holes that form a head water channel.The
stress state is analized in the reinforced concrete
consolidation layer around the elliptical holes(the
gallerys)as well as in the adjacent rock to the ga-
llerys.The methods used in this analysis are the Finite Element Method and the Boundary Element Method.
A statical and a dynamical analysis of the holes is
done by means of the two methods.The analysis in the
elasto-plastic domain of the material is done by means of the Finite Element Method.The elastic and the
mechanical characteristics of the material will be
taken into account as well as Von Mises criterion
for the elasto-plastic domain.there will be done a
comparative analysis in the conditions when for the
reinforced concrete layer we admit Von Mises criterion and for rock Drucker-Prager criterion.In the pa
per will be presented the diagrams with stress concentrations near the holes too.

2. ANALYSIS OF THE HOLES IN ELASTO-PLASTIC INFINITE DOMAIN BY F.E.M.

The study of stress and displacement state for complex shape holes is done with great difficulties by
means of exact methods.The Finite Element Method can
be applied with good results to the study of the categories of problems that work in plane strain sta-
te.For these elements the expression of the virtual
mechanical work is given in the form,Owen,Hinton80

$$\int_{\Omega}[\delta\dot{\varepsilon}]^{T}\gamma\,d\Omega - \int_{\Omega}[\delta\dot{u}]^{T}b\,d\Omega - \int_{\Gamma_{t}}[\delta\dot{u}]^{T}t\,d\Gamma = 0 \qquad (1)$$

in which $\vec{\sigma}$ is the stress tensor, b - mass forces, t-
the stress vector on the contour, $\delta \varepsilon$ -specific strains vector, u-displacements vector, Ω -domain,
Γ_t -the boundary part where the exteriour stresses
work on.
In the problem of plane strains, the deformations vec
tor is known under the form:

$$\varepsilon = [\varepsilon_{xx}, \varepsilon_{yy}, \varepsilon_{xy}]^T \qquad (2)$$

The vector of the virtual displacements has the form

$$\delta u = [u_x, u_y]^T \qquad (3)$$

With the notations (3) the strains vector can be wri
tten

$$\delta \varepsilon = \left[\frac{\partial(\delta u_x)}{\partial x}, \frac{\partial(\delta u_y)}{\partial y}, \frac{\partial(\delta u_x)}{\partial y} + \frac{\partial(\delta u_y)}{\partial x} \right] \qquad (4)$$

From physical equations, taking into consideration
the relations(4) the stresses can be determined

$$\vec{\sigma} = D \varepsilon \qquad (5)$$

For linear elastic materials, the stress-strain or
constitutive matrix D is given as

$$D = \frac{E}{(1+\nu)(1-2\nu)} \begin{bmatrix} (1-\nu) & \nu & 0 \\ \nu & (1-\nu) & 0 \\ 0 & 0 & \frac{(1-2\nu)}{2} \end{bmatrix} \qquad (6)$$

Note that the stress normal to the xOy plane is non
zero and can be evaluated as

$$\sigma_{zz} = \nu[\sigma_{xx} + \sigma_{yy}] \qquad (7)$$

If a study is done, taking into account the material
elasto-plastic behaviour, the strains vector must have the elastic and plastic deformations components.

$$d\varepsilon_{ij} = (d\varepsilon_{ij})_e + (d\varepsilon_{ij})_p \qquad (8)$$

The elastic deformation component is given by

$$(d\varepsilon_{ij})_e = \frac{d\sigma_{ij}}{2\mu} + \frac{1-2\nu}{E} \delta_{ij} d\sigma_{kk} \qquad (9)$$

The plastic deformation component is given by:

$$(d\varepsilon_{ij})_p = d\lambda \frac{\partial Q}{\partial \sigma_{ij}} \qquad (10)$$

Equation(10)is termed the flow rule since it governs
the plastic flow after yielding. The potential Q must
be a function of J_2' and J_3' that represent:

$$J_2' = \frac{1}{2} \sigma_{ij}' \sigma_{ij}' \qquad J_3' = \frac{1}{3} \sigma_{ij}' \sigma_{jk}' \sigma_{ki}' \qquad (11)$$

In the problem of cavitys in the infinite domain
from two materials(for the example to be presented)
is necessary to be taken into consideration flowing

/running criteria corresponding to each material.

Thus for reinforced concrete running/flowing criterion Von Mises is taken into consideration and for rock Drucker-Prager flowing/running criterion is taken into account. Von Mises flowing/running criteri on takes into consideration the second rank invaria nt of the stresses tensor of the form:

$$J_2' = \frac{1}{2}[\gamma_{xx}'^2 + \gamma_{yy}'^2 + \gamma_{zz}'^2] + \gamma_{xy}^2 + \gamma_{yz}^2 + \gamma_{zx}^2 \qquad (12)$$

and is known under the form

$$(J'^2)^{1/2} = K_{(k)} \qquad (13)$$

in which k is a parameter of the material that can be determined experimentally.

Drucker-Prager criterion is a modified form of Von Mises criterion and is applied with good resu lts in rock mechanics

$$\alpha J_1 + (J_2')^{1/2} = k' \qquad (14)$$

in which

$$\alpha = \frac{2\sin\varnothing}{\sqrt{3}(3-\sin\varnothing)} \qquad k = \frac{6c\cos\varnothing}{\sqrt{3}(3-\sin\varnothing)} \qquad (15)$$

In the deviatoric plane $\gamma_1 + \gamma_2 + \gamma_3 = $const., Drucker-Prager running/flowing criterion corresponds to a circle inscribed in the hexagon that is associa ted to Coulomb criterion. In the elasto-plastic analy sis, the strain-stress relation is given

$$d\gamma = D_{ep}d\epsilon \qquad (16)$$

with

$$D_{ep} = D - \frac{d_D d_D^T}{A + d^T a} \qquad (d_D = Da) \qquad (17)$$

in which the scalar A is

$$A = -\frac{1}{d\lambda}\frac{\partial F}{\partial k}dk \qquad (18)$$

where A is obtained to be the local slope of the uni axial stress/plastic strain curve and can be determi ned experimentally from

$$H' = \frac{E_T}{1 - E_T/E} \qquad (19)$$

In the dynamic analysis, Bathe 74, Poterasu 85 of the elements we start from the virtual mechanical work principle written at the time t :

$$\int_\Omega [\delta\epsilon_n]^T \gamma_n d\Omega - \int_\Omega [\delta u_n]^T [b_n - \rho_n \ddot{u}_n - c_n \dot{u}] d\Omega - \int_{\Gamma_t} [\delta u_n]^T t_n d\Gamma = 0 \qquad (20)$$

The model of elasto-viscoplastic material is adopted where the constitutive relations are given in the form:

$$\dot{\varepsilon}_n = [\dot{\varepsilon}_e]_n + [\dot{\varepsilon}_{vp}]_n = [D]^{-1}\dot{\gamma}_n + \gamma'[\emptyset_n(F)]\frac{\partial F}{\partial \gamma_n} \qquad (21)$$

with the known semnifications.

3. ANALYSIS OF THE HOLES BY B.E.M.

Recent developments of the Boundary Element Method,Onishi 84 ,allow a numerical study of the elements made up of two or more materials.

If we consider each material of the element as being a subdomain having specific elastic and mechanic characteristics we can write the following notations(Fig.1)

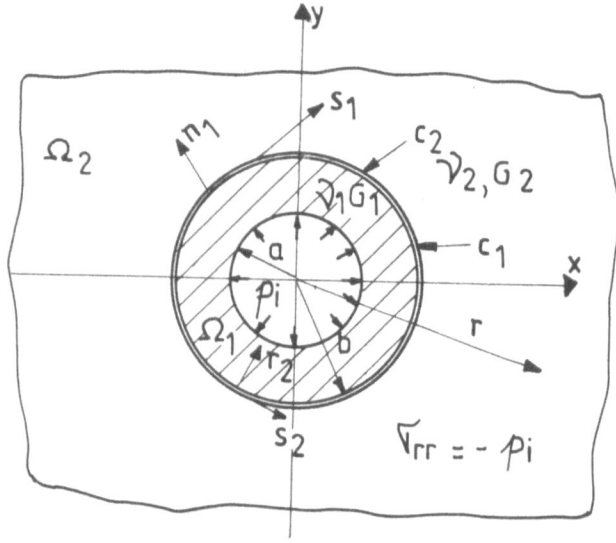

Fig.1 Inhomogeneous body with two subregions

Inhomogeneous body consisting of two subdomains.In this case the continuity conditions put in a point M are the form:

$$n_{12} = \frac{E_1}{E_2}$$

$$\gamma_s^{(1)}(M) = n_{12}\gamma_s^{(2)}(M)$$
$$\gamma_n^{(1)}(M) = n_{12}\gamma_n^{(2)}(M) \qquad (22)$$

for the stresses and

$$u_s^{(1)}(M) = -u_s^{(2)}(M) \qquad u_n^{(1)}(M) = -u_n^{(2)}(M) \qquad (23)$$

for the displacements.

According to Crouch 83 we can write the equations of a point on the boundary for each subdomain for stresses in the form:

$$\overset{i}{Y_s}{}^{(k)} = \sum_{j=1}^{N_k} \overset{ij}{A_{ss}}{}^{(k)} \overset{j}{P_s}{}^{(k)} + \sum_{j=1}^{N_k} \overset{ij}{A_{sn}}{}^{(k)} \overset{j}{P_n}{}^{(k)} \left.\begin{array}{c} \\ \\ \end{array}\right\} \quad k = 1 \div n$$

$$\overset{i}{Y_n}{}^{(k)} = \sum_{j=1}^{N_k} \overset{ij}{A_{ns}}{}^{(k)} \overset{j}{P_s}{}^{(k)} + \sum_{j=1}^{N_k} \overset{ij}{A_{nn}}{}^{(k)} \overset{j}{P_n}{}^{(k)} \left.\begin{array}{c} \\ \\ \end{array}\right\} \quad i = 1 \div N_k \qquad (24)$$

where i-segments on subdomain and k-the number of subdomains. Similarly, for displacements we can write

$$\overset{i}{U_s}{}^{(k)} = \sum_{j=1}^{N_k} \overset{ij}{B_{ss}}{}^{(k)} \overset{j}{P_s}{}^{(k)} + \sum_{j=1}^{N_k} \overset{ij}{B_{sn}}{}^{(k)} \overset{j}{P_n}{}^{(k)} \left.\begin{array}{c} \\ \\ \end{array}\right\} \quad k = 1 \div n$$

$$\overset{i}{U_n}{}^{(k)} = \sum_{j=1}^{N_k} \overset{ij}{B_{ns}}{}^{(k)} \overset{j}{P_s}{}^{(k)} + \sum_{j=1}^{N_k} \overset{ij}{B_{nn}}{}^{(k)} \overset{j}{P_n}{}^{(k)} \left.\begin{array}{c} \\ \\ \end{array}\right\} \quad i = 1 \div N_k \qquad (25)$$

where A_{em}, B_{em} are influence coefficients.

Calculating the displacements and the tractions on the contour and on interfaces by means of the known relations we can make the calculus for the dis placements and stresses in the specific interiour po ints. The Boundary Element Method allows us to determin the normal and shear stresses for elements worked by dynamic stresses. For elastic body worked by exteriour stresses that are varying in comparison with the time t, the displacements U_i and the stresses Y_{ij} can be given in the form, Takakuda 84:

$$Re(U_i e^{-iW t}) \quad \text{and} \quad Re(Y_{ij} e^{-iW t}) \qquad (26)$$

where W is the circular frequency. The displacement equation which must be satisfied in the elastic body takes the form:

$$C_{ijkl} U_{k,lj} + b_i = -\rho W^2 U_i \qquad (27)$$

where b_i denotes the body force, ρ the density and C_{ijkl} the elastic constant tensor in the form:

$$C_{ijkl} = [2 \nu \mu / (1-2\nu)] \delta_{ij} \delta_{kl} + \mu \delta_{ik} \delta_{jl} + \mu \delta_{il} \delta_{jk} \qquad (28)$$

In plane strain cases, the fundamental soluti on $U_i^{(j)}(P,Q)$ the x_i component of the displacement at a point P satisfies the displacement equation

$$C_{imkl} U_{k,lm}^{(j)}(P,Q) + \delta_{ij} \delta(P-Q) = -\rho W^2 U_i^{(j)}(P,Q) \qquad (29)$$

After the applications of Fourier transform we obtain

$$U_i^{(j)}(P,Q) = \frac{1}{4} \left[\frac{1}{\mu} H_0^{(1)}(k_T R) \delta_{ij} + \frac{1}{\rho W^2} (H_0^{(1)}(k_T R) - H_0^{(1)}(k_L R)]_{,ij} \right] \qquad (30)$$

in which R is the distance between points P and Q and $H_n^{(1)}(x)$ is a Hankel function of the first kind of order n. From the equation (27) and (29) together with the theorem of divergency of Gauss we obtain the identities of Somigliana that must be satisfied for the displacement U_i:

$$\int_S [t_j^{(n)}(Q)]^- U_i^{(j)}(P,Q)ds(Q) - \int_S [U_j(Q)]^- T_i^{(j)}(P,Q)\,ds(Q) +$$
$$+ \int_\Omega b_j(Q)U_i^{(j)}(P,Q)\,d\Omega(Q) = \begin{cases} U_i(P) & (P \in \Omega) \\ 1/2[U_i(P)] & (P \in \Omega) \\ 0 & (P \in \Omega_e) \end{cases} \quad (31)$$

Similar, equations for the displacement gradient $U_{i,j}$ are shown to be of the form:

$$\int_S [t_j^{(n)}(Q)]^- U_{i,j}^{(k)}(P,Q)\,ds(Q) - \int_S \int w^2 n_j(Q)[U_k(Q)]^- U_i^{(k)}(P,Q)ds(Q) +$$
$$+\int_S \left[\varepsilon_{pqr}n_p(Q)U_{i,q}(Q)\right]\varepsilon_{rkj}C_{klmn}U_{l,n}^{(m)}(P,Q)ds(Q) + \quad (32)$$
$$\int_\Omega b_k(Q)U_{i,j}^{(k)}(P,Q)\,d\Omega(Q) = \begin{cases} U_{i,j} & [P \in \Omega] \\ 1/2[U_{i,j}(P)]^- & [P \in S] \\ 0 & [P \in \Omega_e] \end{cases}$$

With the corresponding developments we can calculate the displacements and the stresses in elements actioned by dynamic stresses.

4. NUMERICAL RESULTS

An analysis of the stress and displacement state was done in two elliptical holes worked in a first variant by an internal pression p_i= 20 daN/cm^2 The numerical analysis was done by Finite Element Method (Fig.2)and by Boundary Element Method(Fig.3). In the study two materials were considered.The concrete with 25 cm. thickness has the following elastic characteristics:E_1 = 280 000 daN/cm^2 and $\mathcal{V}_1$ =0.16. The rock in which the elliptic gallery was made has E_2 = 190 000 daN/cm^2 and $\mathcal{V}_2$ =0.22.By Finite Element Method the domain was divided in 234 quadrilateral elements with 275 knots(Fig.2).
The computer program calculates the displacements in two directions of the nodal points and the stresses in the established points.To solve the problem by Boundary Element Method,the hole was divided in 60 segments and to compare the results there were calculated the stresses and the displacements in the points on the four segments(Fig.3).
It was possible to use the symmetry in comparison with the axis y.The values of the normal stresses $\mathcal{V}_{xx}$ and $\mathcal{V}_{yy}$ calculated by the two methods can be compared in the variation diagrams in the two materials on the segment 3 (Fig.3).We can notice close values obtained by the two methods(Fig.4). The stresses and displacements values on the segment 3 calculated by the two methods are given in

the Table 1.

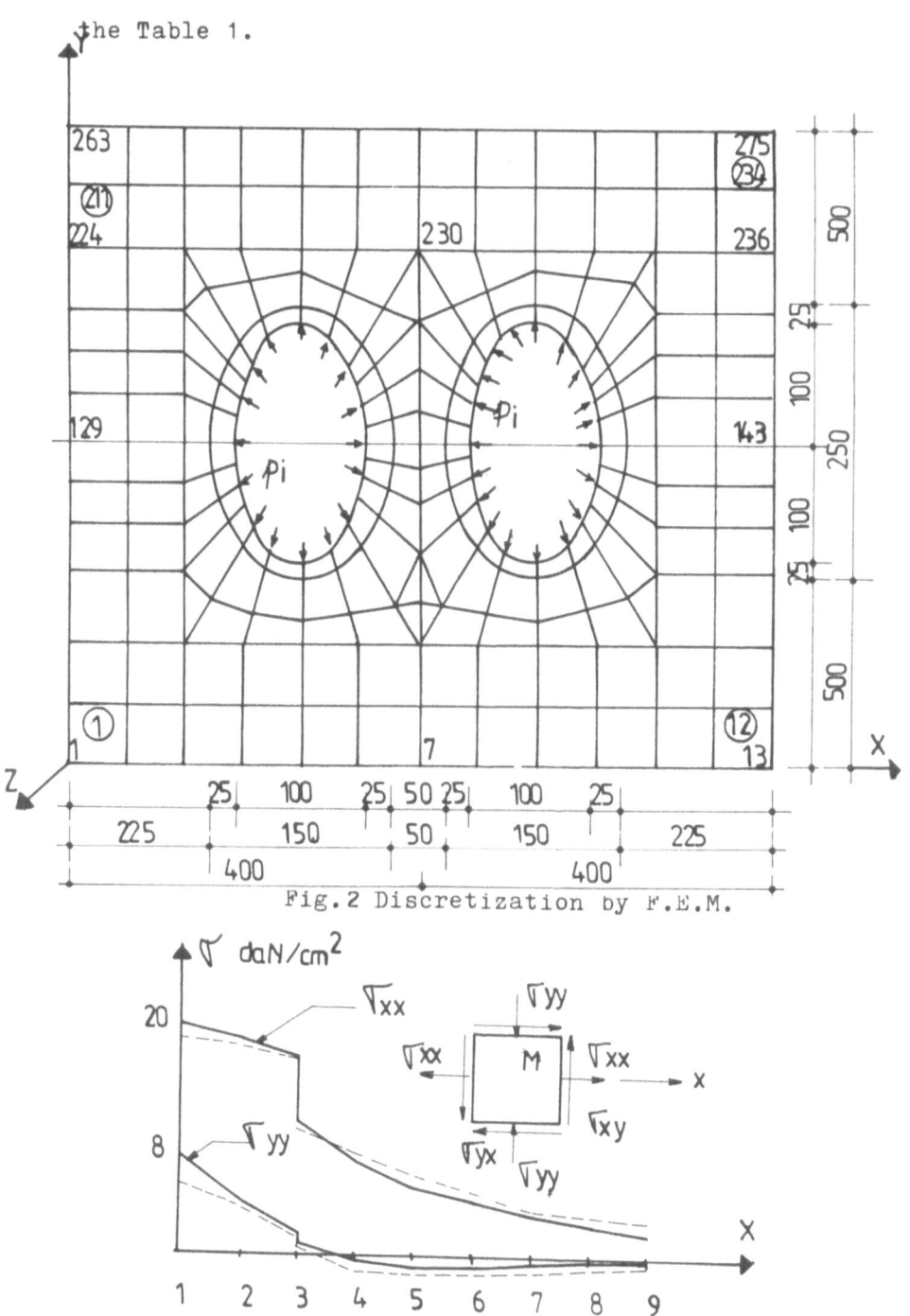

Fig.2 Discretization by F.E.M.

Fig.4 The normal stresses by —— BEM and ---- FEM

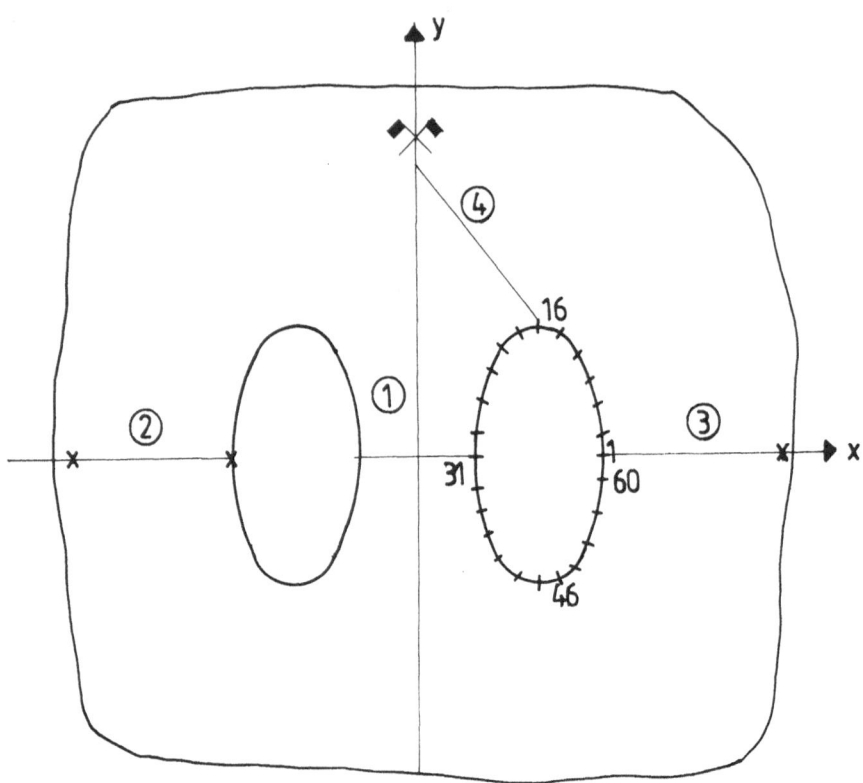

Fig.3 Discretization by B.E.M.

Table 1

KNOT	U_y mm.	U_x mm.	σ_{xx} daN/cm^2	σ_{yy} daN/cm^2	σ_{xy} daN/cm^2
B. E. M.					
1	0.109	0.393	19.39	7.606	1.388
2	0.103	0.397	18.18	4.03	1.76
3	0.095	0.401	16.6	1.73	1.75
3	0.095	0.401	11.26	1.17	1.19
4	0.078	0.413	7.98	-0.56	0.95
5	0.065	0.423	5.65	-0.77	0.87
6	0.059	0.429	4.58	-0.73	0.83
7	0.052	0.438	3.46	-0.62	0.79
8	0.045	0.445	2.70	-0.52	0.68
9	0.039	0.454	2.02	-0.41	0.69
F.E.M.					
1	0.110	0.416	18.01	6.92	1.32
2	0.101	0.420	17.32	3.85	1.71
3	0.089	0.428	16.21	1.58	1.69
3	0.078	0.436	10.72	1.02	1.14
4	0.065	0.446	8.45	-1.15	0.82

5	0.059	0.453	6.98	-1.38	0.80
6	0.051	0.460	6.05	-1.24	0.81
7	0.046	0.468	4.28	-0.96	0.72
8	0.042	0.472	3.70	-0.92	0.60
9	0.041	0.474	3.15	-0.81	0.48

The nodal points displacements values of the hole
were compared and it was traced the deformed shape
by the two methods.The stresses values and the ratio
between the two methods are shown in the Table 2.

Table 2

$\frac{t}{z}$	B.E.M.		F.E.M.		BEM/FEM	
	U_x mm.	U_y mm.	U_x mm.	U_y mm.	U_x %	U_y %
1	0.406	0.039	0.428	0.035	94.8	111
2	0.396	0.103	0.419	0.103	94.5	100
3	0.402	0.088	0.425	0.086	94.5	102
4	0.398	0.106	0.422	0.105	94.3	100.9
5	0.404	0.086	0.427	0.085	94.6	101.1
6	0.411	0.049	0.434	0.046	94.7	106.5
7	0.410	0.046	0.434	0.043	94.4	106.9
8	0.412	0.008	0.435	0.004	94.7	-
9	0.401	-0.083	0.423	-0.091	94.8	91.2
10	0.392	-0.111	0.416	-0.120	94.2	92.5
11	0.389	-0.116	0.415	-0.125	93.7	92.8
12	0.354	-0.195	0.395	-0.213	89.6	91.5
13	0.332	-0.139	0.371	-0.167	89.5	83.2
14	0.183	-0.348	0.183	-0.365	100	95.3
15	0.116	-0.369	0.117	-0.387	99.1	95.3
16	-0.079	-0.373	-0.085	-0.391	92.9	95.4

The hole deformation given by the interiour pression
is shown in Fig.5.

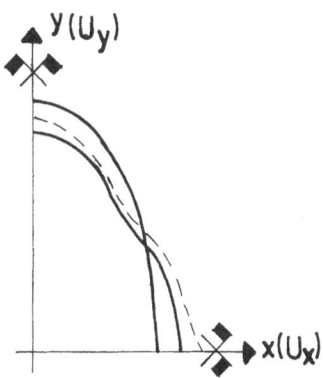

Fig.5 Deformation of the hole. ──B.E.M.
--- F.E.M.
For the study of the problem by the two methods,stu-
dying the data from Table 3 we notice some advanta-

ges that are shown by the Boundary Element Method in comparison with the Finite Element Method.

Table 3

Method	Date	2/1 %	Time sec	2/1 %	Memory Kb	2/1 %
1 B.E.M.	69		248		64	
2 F.E.M.	515	7.46	620	2.5	98	1.53

5. CONCLUSIONS

To solve such problems the Boundary Element Method has many advantages.In comparison with the Finite El ement Method the discretization is more simple,input data very few,computer time is reduced.By Finite El- ement Method the discretization was extended to the limit when the stresses and the displacements have values almost zero.By BEM it was necessary only the discretization of the hole boundary.The discontinui- ties between the stresses for the BEM and FEM and the displacements for FEM on the interface (Fig.4 and Table 1)in the knot 3 are in the ratio of the rock and concrete elastic modulus.

REFERENCES

Bathe, K.J., Ozdemir, H., Wilson, E. (1974) Static and Dynamic Geometric and Material Nonlinear Analy- sis,Report No.UCSESM 74-4,Univ.of California.

Brebbia, C.A., Futagami, T., Tanaka, M.(Eds.) (1983) Boundary Elements.Proceedings of the 5-th Internat. Conf.Hiroshima.Japan,Springer Verlag.

Crouch,   S.L., Starfield, A.M. (1983) Boundary Ele- ments in Solid Mechanics,George Allen& Unwin,USA.

Onishi, K., Kuroki, T., Ohura, Y. (1984) Kyokai-An Interactive BEM-FEM Solver,Report 814-01,Applied Mathematics Department,Fukuoka Univ.,Japan.

Owen, D.R.J., Hinton, E. (1980) Finite Elements in Plasticity:Theory and Practice,Pineridge Press.

Poterasu, V.F., Mihalache, N.(1985) Elasto-plastic Behaviour of the Seismic Reinforced Concrete Walls by FEM,NUMETA 85,Swansea.

Takakuda, K.(1984) On Integral Equation Methods for Elastic Boundary Value Problems,3-rd Report,Direct Methods for Dynamic Problems,Bull.JSME,27,226.

7. NUMERICAL METHODS POSTER SESSION

MICROCOMPUTER APPLICATION OF NON-RECTANGULAR
SPACE-TIME FINITE ELEMENTS IN VIBRATION ANALYSIS
- DIRECT JOINT-BY-JOINT PROCEDURE

C. I. Bajer

Department of Civil Engineering, Engineering College,
Podgórna 50, 65-246 Zielona Góra, Poland

1. INTRODUCTION

In dynamic analysis of structures carried on by dyscrete
methods we often apply the finite element method in
spatial partition and differential schemes in time integration. In this mean the partition of the structure,
rather artifical, is performed once in the begining and
concerns the whole time of computations. Such a limitation unables the solution of many problems.
    Considerable advantage can be obtained when the
space-time finite element method is introduced. In this
method the time dimension is treated in the same way as
spatial dimensions are. The time space must be defined
in which dynamic problems are to be described. The image
of the structure in time space can be underestood as
a prism in which the analyzed structure is a base (Fig. 1).
Now the space-time area can be divided into space-time
finite elements. It can be done with satisfaction of some
condition. For example in each moment of a time integration of the motion equation a changeable spatial partition
of the structure can be assumed. In that way the condition
of coincidence of the element edges with some characteristic curves during the whole time interval of the
investigation can be fulfilled. Such a possibility is
particularly useful in contact dynamic problems, in the
case of traveling support, in rolling problems when we
assign the mesh with a tool.

1.1. Outline of the method of space-time finite elements
The method was formulated by Kączkowski (1975). Short
introduction into simple non-rectangular space-time
elements was described by Bajer (1984). Below we shall
shortly sum up the main steps of space-time finite element

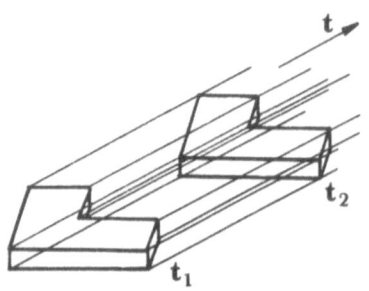

Figure 1. Space-time prism.

theory.

A space-time element has additional time dimension. For one-dimensional structures we obtain surface elements, for surface structure - 3 dimensional volumes and in the case of 3-D bodies - 4-D space-time F.E. should be described. Examples are depicted in Fig. 2. Displacements $\underline{\delta}_e(x_1, x_2, x_3, t)$ should be determined in all nodes of the S.T.F.E. Inside the displacement vector $\underline{f}(x_1, x_2, x_3, t)$ is obtained by interpolation

$$\underline{f} = \underline{N}\,\underline{\delta}_e \tag{1}$$

$\underline{N}$ is the matrix of shape functions. The stress $\underline{\varepsilon}(x_1, x_2,, t)$ and the strain $\underline{\sigma}(x_1, x_2, x_3, t)$ vector can be expressed by commonly known relations

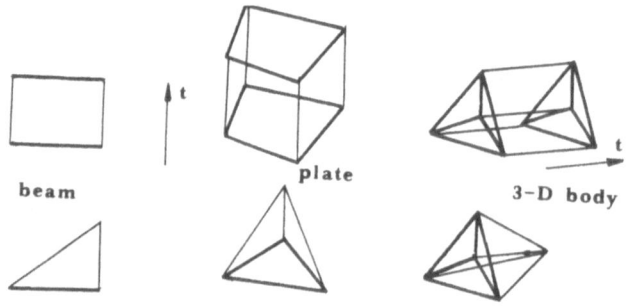

beam

plate

3-D body

Figure 2. Examples of S.T.F.E.

$$\underline{\varepsilon} = \underline{\partial}_x\,\underline{f} = \underline{\partial}_x\,\underline{N}\,\underline{\delta}_e \tag{2}$$

$$\underline{\sigma} = (\,\underline{E} + \eta_w \frac{\partial}{\partial t}\,)\,\underline{\varepsilon} \tag{3}$$

$\underline{\partial}_x(x_1, x_2, x_3)$ is the matrix of differential operators, $\underline{E}$ is the elasticity constant matrix and η_w is a coefficient of internal damping (the Kelvin-Voigt model was assumed). Deformation of the element in time $\underline{\varepsilon}_t(x_1, ..., t)$ can be determined as a velocity

$$\underline{\varepsilon}_t = \frac{\partial f}{\partial t} = \frac{\partial N}{\partial t}\,\underline{\delta}_e \tag{4}$$

The momentum of the material point $\underline{\sigma}_t(x_1,x_2,x_3,t)$ equals

$$\underline{\sigma}_t = -\underline{R}\,\underline{\varepsilon}_t \qquad (5)$$

$\underline{R}$ denotes the matrix of the elementary inertia coefficients. Denoting η_z as an external damping coefficient the virtual four-work (time-work, work integrated in time) of internal forces in the volume of element can be equated to the four-work of external forces

$$\int_V (\delta\underline{\varepsilon}^T\underline{\sigma} + \delta\underline{\varepsilon}_t^T\underline{\sigma}_t)\,dV = \delta\underline{d}_e^T\underline{F}_e - \int_V \delta\underline{f}^T\eta_z\frac{\partial f}{\partial t}\,dV \qquad (6)$$

The integration extends over the volume of the S.T.F.E. $\underline{F}_e$ is a vector of nodal impulses. Considering (2), (3),(4) and (5) it can be written

$$\int_V \left[(\underline{\partial}_x\underline{N})^T\underline{E}\underline{\partial}_x\underline{N} + (\underline{\partial}_x\underline{N})^T\eta_w\frac{\partial}{\partial t}\underline{\partial}_x\underline{N} - (\frac{\partial}{\partial t}\underline{N})^T\underline{R}\frac{\partial}{\partial t}\underline{N} + \underline{N}^T\eta_z\frac{\partial}{\partial t}\underline{N}\right]dV\,\underline{d}_e = \underline{F}_e \qquad (7)$$

We can notice that the same result can be obtained with the use of Galerkin method applied to the differential equation with time variable. Equation (7) can be extended over the whole structure. Then it can be written (sums are omited)

$$(\underline{K} + \underline{M} + \underline{W} + \underline{Z})\,\underline{d} = \underline{F} \qquad \text{or} \qquad \underline{K}^*\underline{d} = \underline{F} \qquad (8)$$

where:

$$\underline{K}= \int_V (\underline{\partial}_x\underline{N})^T\underline{E}\underline{\partial}_x\underline{N}\,dV \qquad \underline{M}=-\int_V (\frac{\partial}{\partial t})^T\underline{R}\frac{\partial}{\partial t}\underline{N}\,dV$$

$$\underline{W}= \int_V (\underline{\partial}_x\underline{N})^T\eta_w\frac{\partial}{\partial t}\underline{\partial}_x\underline{N}\,dV \qquad \underline{Z}= \int_V \underline{N}^T\eta_z\frac{\partial}{\partial t}\underline{N}\,dV \qquad (9)$$

Matrices $\underline{K}$, $\underline{M}$, $\underline{W}$ and $\underline{Z}$ are called stiffness, mass, internal damping and external damping matrices, respectively. The analysis of the joint connection in several succesive time layers leads to the infinite global matrix $\underline{K}^*$ in a form

$$\begin{bmatrix} \underline{A}_1 & \underline{B}_1 & & \\ \underline{C}_1 & \underline{D}_1+\underline{A}_2 & \underline{B}_2 & \\ & \underline{C}_2 & \underline{D}_2+\underline{A}_3 & \underline{B}_3 \\ & & & \ddots \end{bmatrix} \begin{Bmatrix} \underline{d}_1 \\ \underline{d}_2 \\ \underline{d}_3 \\ \vdots \end{Bmatrix} = \begin{Bmatrix} \underline{F}_1 \\ \underline{F}_2 \\ \underline{F}_3 \\ \vdots \end{Bmatrix} \qquad (10)$$

$\underline{A}_i$, $\underline{B}_i$, $\underline{C}_i$, $\underline{D}_i$ are square submatrices of the matrix $\underline{K}^*$ evaluated for one space-time layer. An order of each of them is equal to the total number of degrees of freedom in the structure. Such a formulation enables step-by-step formulation:

$$\underline{\delta}_2 = \underline{B}_1^{-1}(\underline{F}_1 - \underline{A}_1\underline{\delta}_1) \tag{11}$$

for the first step ($\underline{\delta}_1$ - vector of initial displacements)

$$\underline{\delta}_{i+1} = \underline{B}_i^{-1}\underline{F}_i - \underline{B}_i^{-1}(\underline{D}_{i-1} + \underline{A}_i)\underline{\delta}_i - \underline{B}_i^{-1}\underline{C}_{i-1}\underline{\delta}_{i-1} \tag{12}$$

In this meaning the space-time finite element method can be regarded as a time integration method. However, the time space can be divided into space-time elements of almost any shape. Although some restrictions are imposed on the shape, in practical use the arbitrary time space division can be applied.

2. APPLICATION OF TRIANGULAR AND TETRAHEDRAL SPACE--TIME ELEMENTS

Let us divide the space-time layer of uni-dimensional structureinto triangular elements. Global matrices of a system of Eq. (8) related to one space-time layer are filled by non-zero elements in normal way resulting from the topology of the division mesh. We can notice that the matrix of coefficients $\underline{B}_i$ in the equation of the layer equilibrium (12) has a triangular form. Lower left-hand part can be obtained under the condition of special numeration of nodes. Fig. 3 presents a space-time layer in the case of one-dimensional structure and the global stiffness matrix $\underline{K}^*$.

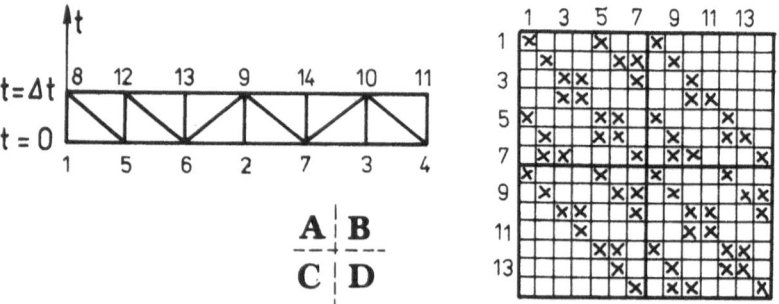

Figure 3. Space-time net for one space-time layer and its global matrix.

Now both the inversion of $\underline{B}_i$ in Eq. (12) and the alternative solution of the system of equations is unnecessary. Computations of nodal displacements for all joints in succesive steps can be performed joint-by-joint.

$$\delta_{i+1}^k = B_i^{k,k}{}^{-1}\left[F_i^k - \sum_{l=k}^{n} C_{i-1}^l \delta_{i-1}^l - \sum_{l=1}^{n} (D_{i-1}^{k,l} + A_i^{k,l})\delta_i^l - \sum_{l=1}^{k-1} B_i^{k,l}\delta_{i+1}^l\right]$$

$$k = 1, 2, \ldots, n \qquad (13)$$

Also upper right-hand triangular matrix B_i can be obtained
when we apply the mesh in which parallelograms are divided
into triangles in the opposite direction. Detailed consideration of the mesh topology leads to the more general
conclusion. Any numeration and any division into triangles
and tetrahedrons is possible to apply joint-by-joint
procedure. The only difference is the success of joints in
computations.

    To simplify the computational algorithm the special
way of space-time element generation was assumed. All
nodes of real division net are succesively considered.
Numbers of joints neighbour the actually considered one
in the spatial division of the structure, with itself, are
determined. Then finite elements are considered in which
actually considered joint is one of their joints. The
space-time finite elements are constructed. If the joint
number in sorrounding is greater than the number of actually considered joint its time coordinate t=0. In the case
when it is less than the number of the considered joint
its time coordinate t=Δt. When these numbers are equal we
obtain two nodes for which t=0 and t=Δt. The described way
of the space-time mesh generation in the case of surface
structure is shown in Fig. 4.

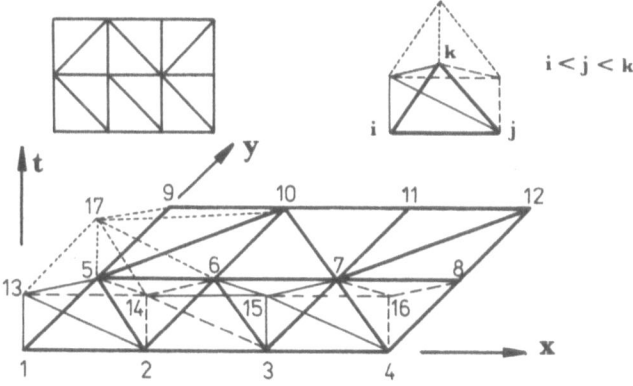

Figure 4. Construction of the space-time finite
elements for surface structures.

The global matrix related to the one space-time layer is
held in core in a form of a sequence of submatrices of
dimensions equal to the nodal number of degrees of freedom.
The constructed sequence of joints serves as a pattern of

submatrix location in the sequence of submatrices to be
stored. The method of global matrix recording is presented
in Fig. 5. The case of constent coefficient problem was
considered. Submatrices $\underline{A}_i$, $\underline{B}_i$, $\underline{C}_i$, $\underline{D}_i$ are constant for each
step. Also the problem with variable coefficients was
considered. For the sake of variable coefficients additional informations have to be stored. Respective sequence
of submatrices is also depicted in Fig. 5. Some observations about the matrix coefficients distribution make the

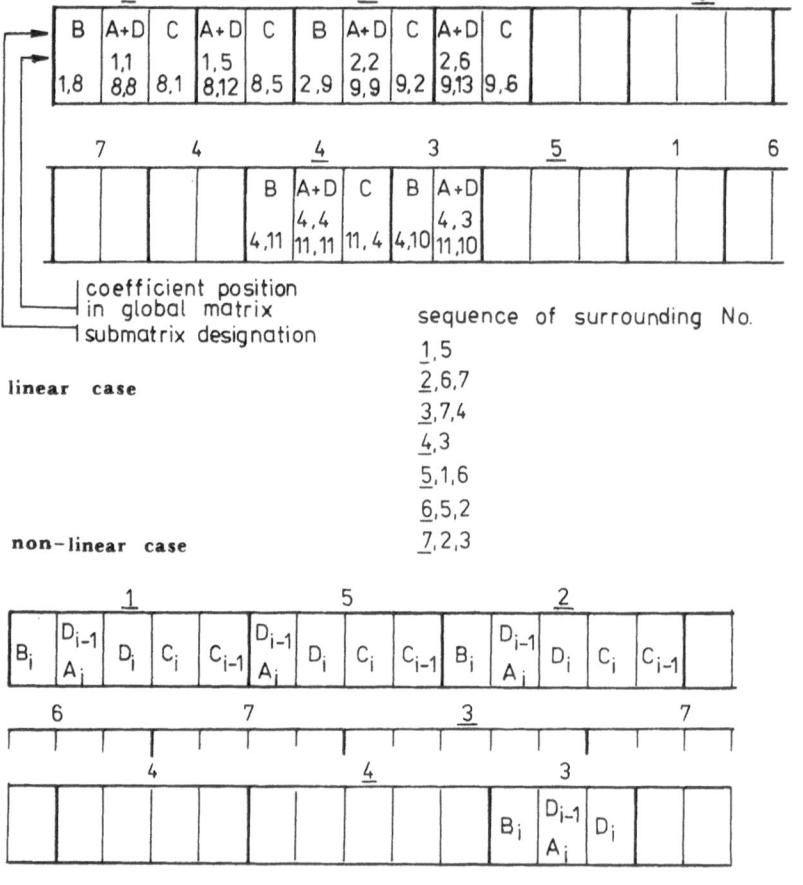

Figure 5. Location of matrix coefficients in core.

programing simple. The algorithm enables the storage of
only non-zero coefficients. High efficiency of the
memory use is caused by the mentioned feature and also
is resulted by the triangular form of matrices, with less
number of coefficients. It must be emphasized that in

this way <u>zero arithmetical operations are eliminated</u>. It considerably raises up the speed of computations. The storage efficiency can be estimated. The comparision with the amount of coefficients consumed in F.E. static problems has been done (it should be underestood as an approximation):

- static problems   = 1.
- S.T.F.E. linear problems   = 1.5
- S.T.F.E. non-linear problems = 2.7

Since the stored coefficients in each time step are used sequentially the whole sequence shown in Fig. 5 can be divided into parts and separately stored in external memory. Then they can be efficiently read-in and used. It is important that the computer program written in Fortran IV has less than 200 instructions and can be used in computations of <u>all types of problems without changings</u>. Two, three, four and so on dimensional problems can be computed under the condition of preparing appropriate element matrices.

3. TRIANGULAR SPACE-TIME FINITE ELEMENTS

3.1. Beam element.
Below triangular space-time finite element matrix will be derived. Linear distribution of displacements in the element area was assumed. It enables simple determination of the stiffness, mass and damping matrices.

Let a deflection w and an angle of rotation Θ be given by linear relations

$$w = a_1 x + a_2 t + a_3$$
$$\Theta = b_1 x + b_2 t + b_3$$

(14)

The shape matrix can be obtained by the inversion of nodal values of polynomial factors (14). In the case of beam the differential operator $\underline{\partial}_x$ has a form

$$\underline{\partial}_x = \begin{bmatrix} \dfrac{\partial}{\partial x} & 1 \\ 0 & \dfrac{\partial}{\partial x} \end{bmatrix}$$

(15)

The elasticity matrix $\underline{E} = \text{diag}\left[\dfrac{GA}{K}, EI\right]$.   (16)
G is the shear modulus, A - the cross sectional area, K - the shape factor for the cross section, EI - the flexural stiffness. The elementary inertia matrix $\underline{R} = \text{diag}\left[\rho A, \rho I\right]$ (17)
ρ is the mass density. Considering Eq. (7) or (9) we can compute a part of the space-time stiffness matrix $\underline{K}^*$. The integration over the element volume V could be accomplished analytically if the origin of the coordinates is placed at the center of the element gravity. x_i, t_i are the nodal coordinates and the indexes are changed periodically in a sequence i,k,l and j,m,n.

$$
\underline{K}^{*}_{i,j} = \begin{bmatrix}
\dfrac{GA}{4KV}(t_k-t_1)(t_m-t_n)- & \dfrac{GA}{4KV}(t_k-t_1)(x_m t_n - x_n t_m)+ \\[4pt]
-\dfrac{\varrho A}{4V}(x_k-x_1)(x_m-x_n)+ & +\dfrac{\varrho w}{4V}(t_k-t_1)(x_n-x_m) \\[4pt]
+\dfrac{\varrho z}{4V}(x_k t_1 - x_1 t_k)(x_n-x_m) & \\[4pt]
\hline
& \dfrac{EI}{4V}(t_k-t_1)(t_m-t_n)- \\[4pt]
\dfrac{GA}{4KV}(t_m-t_n)(x_k t_1 - x_1 t_k) & -\dfrac{\varrho I}{4V}(x_k-x_1)(x_m-x_n)+ \\[4pt]
& +\dfrac{\varrho z}{4V}(x_k t_1 - x_1 t_k)(x_n-x_m)
\end{bmatrix}
$$

$$
i,j=1,2,3 \tag{18}
$$

Although the element described has linear shape functions the obtained results are accurate enough. More precise derivation of triangular beam element was described by Bajer (1984).

3.2. Plane stress/strain element.
In the case of surface finite elements the S.T.F.E. has a tetrahedral shape. In plane stress/strain element a linear distribution of displacements $\underline{f}$ in the element is possible

$$
\underline{f}(x,y,t)=\begin{Bmatrix} u \\ v \end{Bmatrix}=\begin{Bmatrix} \underline{g}\,\underline{a} \\ \underline{g}\,\underline{b} \end{Bmatrix} \tag{19}
$$

where $\qquad \underline{g}=[x,y,t,1] \tag{20}$

and $\underline{a}$, $\underline{b}$ are vectors of four constants $a_i, b_i=1,..,4$. Then if the columns of matrix

$$
\underline{G}^{-1}=\begin{bmatrix} \underline{g}(x_1,y_1,t_1) \\ \vdots \\ \underline{g}(x_4,y_4,t_4) \end{bmatrix}^{-1} \tag{21}
$$

are denoted by $\underline{r}_i=[p_{1i},p_{2i},p_{3i},p_{4i}]^T$ the matrix of shape functions $\underline{N}$ have a form

$$
\underline{N}_i=\underline{g}\,r_i\begin{bmatrix} 1 & 0 \\ 0 & 1 \end{bmatrix} \tag{22}
$$

Now the matrices $\underline{\partial}_x$ (2) and $\underline{E}$ (plane stress) (3) have a known form

$$
\underline{\partial}_x=\begin{bmatrix} \dfrac{\partial}{\partial x} & 0 \\[4pt] 0 & \dfrac{\partial}{\partial y} \\[4pt] \dfrac{\partial}{\partial y} & \dfrac{\partial}{\partial x} \end{bmatrix} \qquad \underline{E}=\dfrac{E}{1-v^2}\begin{bmatrix} 1 & v & 0 \\ v & 1 & 0 \\ 0 & 0 & \dfrac{1-v}{2} \end{bmatrix} \tag{23}
$$

E - Young modulus, γ - Poissone coefficient. The inertia matrix $\underline{R}$ is diagonal and contains elements equal to mass density ρ. To simplify the integration over the volume V we can assume the nodal coordinates which satisfy the relations:

$$x_i = 0 \qquad y_i = 0 \qquad t_i = 0 \tag{24}$$

Submatrices $\underline{K}_{ij}$, $\underline{M}_{ij}$, $\underline{W}_{ij}$, $\underline{Z}_{ij}$ can be evaluated

$$\underline{K}_{ij} = V \left[\begin{array}{cc|cc} \dfrac{E}{1-\gamma^2}p_{1i}p_{1j} + \dfrac{E}{2(1+\gamma)}p_{2i}p_{2j} & \dfrac{E}{1-\gamma^2}p_{1i}p_{2j} + \dfrac{E}{2(1+\gamma)}p_{2i}p_{1j} \\ \hline \dfrac{E}{1-\gamma^2}p_{2i}p_{1j} + \dfrac{E}{2(1+\gamma)}p_{1i}p_{2j} & \dfrac{E}{1-\gamma^2}p_{2i}p_{2j} + \dfrac{E}{2(1+\gamma)}p_{1i}p_{1j} \end{array} \right] \tag{25}$$

$$\underline{M}_{ij} = -\rho V p_{3i} p_{3j} \underline{I}_2 \qquad \underline{W}_{ij} = \underline{O}_2 \qquad \underline{Z}_{ij} = \eta_z V p_{4i} p_{3j} \underline{I}_2$$

($\underline{I}_2$ - 2x2 unity matrix, $\underline{O}_2$ - 2x2 zero matrix).
It is seen that in the case of low polynomial order the internal damping can not be taken into considerations. Elements of higher order with mid-side nodes eliminate this imperfection.

3.3. Plate element.
Tetrahedral element of plate is considered in the same way as elements described above. Also linear functions describe the deflection w and angles of rotation Θ_x, Θ_y.

$$\begin{Bmatrix} w \\ \Theta_x \\ \Theta_y \end{Bmatrix} = \begin{Bmatrix} a_1 x + a_2 y + a_3 t + a_4 \\ b_1 x + b_2 y + b_3 t + b_4 \\ c_1 x + c_2 y + c_3 t + c_4 \end{Bmatrix} = \begin{bmatrix} \underline{a} \\ \underline{b} \\ \underline{c} \end{bmatrix} [x,y,t,1] = \begin{bmatrix} \underline{a} \\ \underline{b} \\ \underline{c} \end{bmatrix} \underline{g} \tag{26}$$

$$\underline{G}^{-1} = \begin{bmatrix} \underline{g}(x_1, y_1, t_1) \\ \vdots \\ \underline{g}(x_4, y_4, t_4) \end{bmatrix}^{-1} = \left[\underline{r}_1 ; \underline{r}_2 ; \underline{r}_3 ; \underline{r}_4 \right] = \begin{bmatrix} p_{11} & p_{12} \cdots & p_{14} \\ p_{21} & p_{22} \\ \vdots \\ p_{41} & \cdots \end{bmatrix} \tag{27}$$

The matrix of shape functions has a form

$$\underline{N}_i = \underline{g}(x,y,t) \, \underline{r}_i \, \underline{I}_3 \tag{28}$$

Now expresions (9) can be computed. Assuming the condition analogical to (24) we can integrate the expresions. To complete the description of derivation matrices $\underline{\partial}_x$ and $\underline{E}$ are listed below

$$
\underline{\partial}_x = \begin{bmatrix} 0 & \dfrac{\partial}{\partial x} & 0 \\ 0 & 0 & \dfrac{\partial}{\partial y} \\ 0 & \dfrac{\partial}{\partial y} & \dfrac{\partial}{\partial x} \\ \dfrac{\partial}{\partial x} & 1 & 0 \\ \dfrac{\partial}{\partial y} & 0 & 1 \end{bmatrix}
\qquad
\underline{E} = \begin{bmatrix} D & D & 0 & 0 & 0 \\ & D & 0 & 0 & 0 \\ & & \dfrac{1-\gamma}{2}D & 0 & 0 \\ & (\text{sym}) & & H & 0 \\ & & & & H \end{bmatrix}
\qquad (29)
$$

$$
D = \frac{h^3}{12}\frac{E}{1-\gamma^2} \quad ; \qquad H = \frac{5}{6}G\,h
$$

Finaly elements of submatrices $\underline{K}_{ij}$, $\underline{M}_{ij}$, $\underline{W}_{ij}$, $\underline{Z}_{ij}$ can be computed. Since the further derivation is simple the full form of matrices will not be given.

3.4. 3-dimensional body.
Let us define the components of strain vector $\underline{\varepsilon}$

$$
\underline{\varepsilon} = \begin{Bmatrix} \varepsilon_x \\ \varepsilon_y \\ \varepsilon_z \\ \gamma_{xy} \\ \gamma_{yz} \\ \gamma_{zx} \end{Bmatrix} = \begin{Bmatrix} \dfrac{\partial u}{\partial x} \\[4pt] \dfrac{\partial v}{\partial y} \\[4pt] \dfrac{\partial w}{\partial z} \\[4pt] \dfrac{\partial u}{\partial y}+\dfrac{\partial v}{\partial x} \\[4pt] \dfrac{\partial v}{\partial z}+\dfrac{\partial w}{\partial y} \\[4pt] \dfrac{\partial w}{\partial x}+\dfrac{\partial u}{\partial z} \end{Bmatrix}
\qquad (30)
$$

Assuming the displacement vector $\underline{f} = \mathrm{col}\begin{bmatrix} u,v,w \end{bmatrix}$ we can determine the differential matrix $\underline{\partial}_x$ that satisfies the relation (2). The distribution of displacements is given by linear functions

$$
\begin{aligned}
u &= a_1 x + a_2 y + a_3 z + a_4 t + a_5 \\
v &= b_1 x + b_2 y + b_3 z + b_4 t + b_5 \\
w &= c_1 x + c_2 y + c_3 z + c_4 t + c_5
\end{aligned}
\qquad (31)
$$

If we denote

$$
\underline{g} = \begin{bmatrix} x,y,z,t,1 \end{bmatrix} \qquad , \qquad \underline{G} = \begin{bmatrix} g(x_1,y_1,z_1,t_1) \\ \vdots \\ g(x_5,y_5,z_5,t_5) \end{bmatrix}
\qquad (32)
$$

and $\underline{r}_i$, p_{ij} are columns and elements of matrix $\underline{G}^{-1}$, respectively, the shape functions for joint "i" can be written:

$$
\underline{N}_i = \underline{g}\,\underline{r}_i\,\underline{I}_3 \quad , \qquad i=1,2,\dots,5
\qquad (33)
$$

Assuming the center of coordinates in the center of gravity of hyper-tetrahedron the integrals (9) can be computed

$$
\underline{K}_{ij} = V
\begin{bmatrix}
\begin{array}{l}
\alpha_{11}p_{1i}p_{1j}+ \\
+\alpha_{44}p_{2i}p_{2j}+ \\
+\alpha_{66}p_{3i}p_{3j}
\end{array}
&
\begin{array}{l}
\alpha_{12}p_{1i}p_{2j}+ \\
+\alpha_{44}p_{2i}p_{1j}
\end{array}
&
\begin{array}{l}
\alpha_{13}p_{1i}p_{3j}+ \\
+\alpha_{66}p_{3i}p_{1j}
\end{array}
\\[2mm]
\begin{array}{l}
\alpha_{21}p_{2i}p_{1j}+ \\
+\alpha_{44}p_{1i}p_{2j}
\end{array}
&
\begin{array}{l}
\alpha_{22}p_{2i}p_{2j}+ \\
+\alpha_{44}p_{1i}p_{1j}+ \\
+\alpha_{55}p_{3i}p_{3j}
\end{array}
&
\begin{array}{l}
\alpha_{23}p_{2i}p_{3j}+ \\
+\alpha_{55}p_{3i}p_{2j}
\end{array}
\\[2mm]
\begin{array}{l}
\alpha_{31}p_{3i}p_{1j}+ \\
+\alpha_{66}p_{1i}p_{3j}
\end{array}
&
\begin{array}{l}
\alpha_{32}p_{3i}p_{2j}+ \\
+\alpha_{55}p_{2i}p_{3j}
\end{array}
&
\begin{array}{l}
\alpha_{33}p_{3i}p_{3j}+ \\
+\alpha_{55}p_{2i}p_{2j}+ \\
+\alpha_{66}p_{1i}p_{1j}
\end{array}
\end{bmatrix}
\tag{34}
$$

α_{ij} -constant elements of the elasticity matrix $\underline{E}$.

$$
\underline{M}_{ij} = -\rho V p_{4i}p_{4j}\underline{I}_3 , \qquad
\underline{Z}_{ij} = \eta_z V p_{5i}p_{4j}\underline{I}_3 , \qquad
\underline{W}_{ij} = \underline{0}_3 \tag{35}
$$

Analogically as in the case of plane stress/strain the
internal damping matrix $\underline{W}$ is zero. It means that the order
of expresions (31) is too low. Other possibilities of
higher order elements will be shortly discussed below.

4. ELEMENTS OF HIGHER ORDER

It was shown that low order of interpolation unables the
accurate modelling of a phenomenon. Internal damping
matrix $\underline{W}$ /(25),(35)/ can not be appointed when Kelvin-
-Voigt model of visco-elasticity is assumed. In the case
of more complicated models low order elements are all the
more insufficient. Moreover, the plane stress/strain
element described in Par. 3.2 is parallel of the constant
strain F.E. and it in-
volves errors. Higher
order elements were pro-
posed. Some of them were
tested (Fig. 6a) but in
the method described in
the paper elements
depicted in Fig. 6b are
essential. Coefficients
of interpolation poly-
nomials for 2-D S.T.F.E.
can be chosen using
Pascal triangle. Poly-
nomial coefficients for
surface structures (3-D

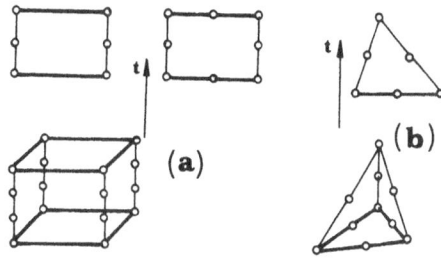

Figure 6. Higher order space-
-time elements.

S.T.F.E.) can be completed as shown in Fig. 7. The problem
is to work out a universal computer program for arbitrary
number of spatial dimensions and for any order of S.T.F.E.
An example of step-by-step and joint-by-joint procedure is
depicted in Fig. 8. Now in one time step the system of
equations is separated into several subsystems with small
number of unknowns in each of them. Figure 8 shows that
the solution of one layer of the problem with higher
order elements consists of 4 steps denoted A,B,C,D. Each
of the subsystem has not more than 4 joints.

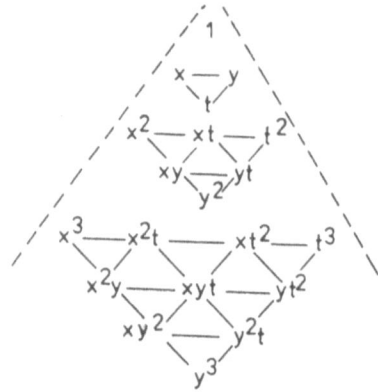

Figure 7. Coefficient triangle for polynomial of 3 variables.

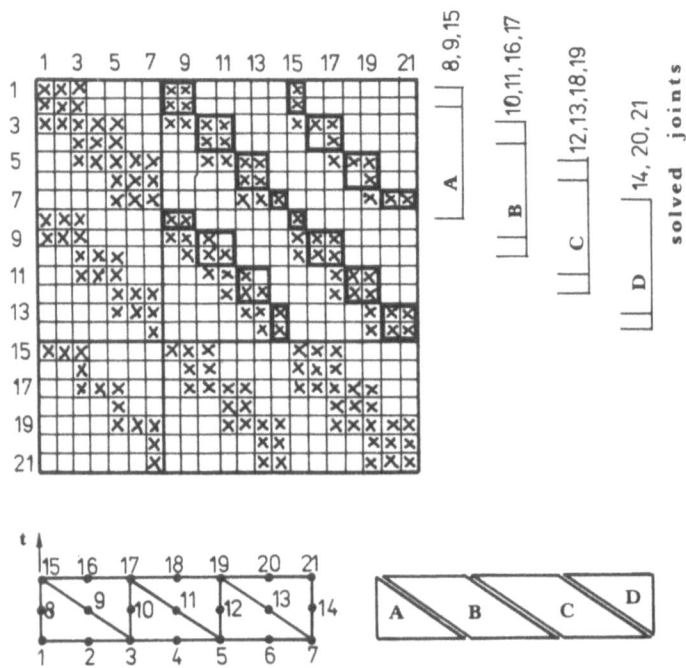

Figure 8. Stiffness matrix for higher order S.T.F.E. of a beam.

5. NUMERICAL EXAMPLES

Many problems were solved with the use of triangular, tetrahedral and hyper-tetrahedral space-time finite elements. Below some of them will be described.

Triangular beam elements described in 3.1 were
applied in vibration analysis of a cantilever beam divided
into 3 finite elements. A free end was subjected by a point
force 1·H(t) (H-Heaviside function). Results were compared
with analytical calculations and with the results obtained
with the use of conformed triangular elements described by
Bajer (1984). Results are collected in Table 1.

Table 1. Comparision of beam elements testing.

linear model		conformed model		exact value	
amplitude	period	amplitude	period	amplitude	period
0.894	1.353	1.	1.461	1.	1.

Plane stress elements were taken to a second test.
A deflection of a cantilever shaped structure composed of
6 triangular F.E. was measured. The load was the same as
in previous example. The value of amplitude 0.414 was
obtained while the static deflection obtained by F.E.M.
is equal 0.207. It prooves the exact results in a class
of constant strain elements.

Hyper-tetrahedral elements of 3-D strain were applied
in analysis of an elastic half-space. A sector of the
bounded half-space was considered and was composed of 384
finite tetrahedrons that gave 1536 S.T.F.E. The total
number of degrees of freedom was 375. Because of the great
number of unknowns the matrix coefficients were stored in
the external memory in several subvectors. Small time step
and great number of computational steps with 8 disc trans-
misions per one step (32K words for a task) considerably
lengthen the computations. Amplitudes of choosen points
can be compared with double displacements in static solution although during longer investigation a numerical
damping effect was observed. It was caused by poor representation of real number (32 bits).

In the last problem described in the paper vibrations
of a beam based on a unilateral and bilateral foundation
were investigated. In the first case quadrangular together
with triangular S.T.F.E. were used. The approach was
described by Bajer (1984). In the second case only triangular elements were applied. Displacements of choosen
points in time are shown in Fig. 9.

6. CONCLUSIONS

Triangular mesh involves the anisotropy of the space-time
area. The orientation of triangles in one direction causes
the transmision of the load influences with different
speed in different directions. Such an effect can be
removed by special numbering of nodes in two succesive
moments, but the influence on the result vanishes after
several steps.

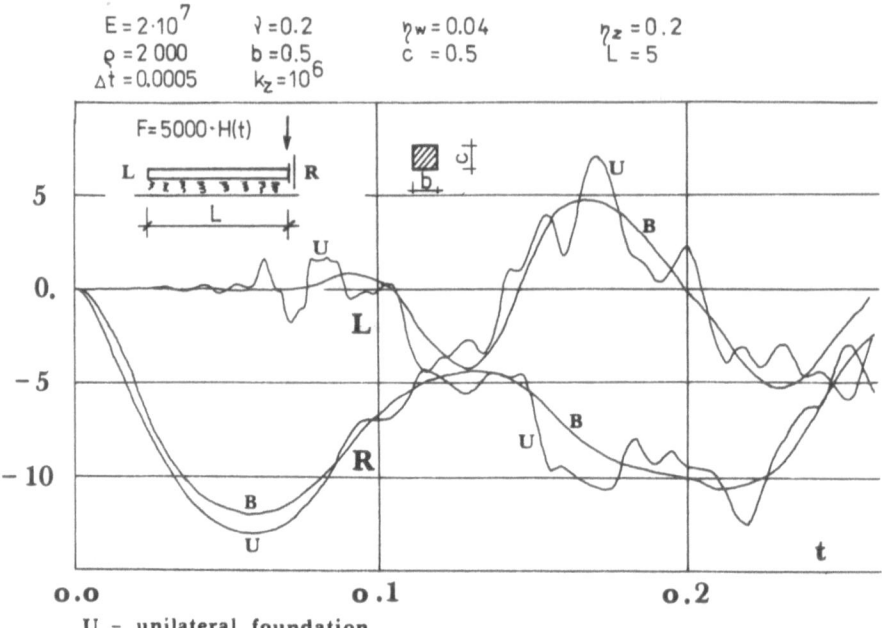

Figure 9. A beam based a unilateral and bilateral
foundation - numerical example.

High efficiency of presented method must be emphasized both in storage and in speed of calculations. The
stability problem is not described here but of high
importance. Many test problems were solved that proved the
efficiency of the method in engineering problems. Some
large structures were solved but in spite of it further
developing of the space-time finite element method is
necessary.

REFERENCES

Bajer, C. (1984) Non-stationary division by the space-time
finite element method in vibration analysis, Proc. Sec.
Int. Conf. Recent Advances in Structural Dynamics,
Southampton, ed. M. Petyt, H.F. Wolfe.

Kączkowski, Z. (1975) The method of finite space-time
elements in dynamics of structures, J. Techn. Physics,
16, 1.

MULTIEXTREMAL (GLOBAL) OPTIMIZATION ALGORITHMS
FOR ENGINEERING APPLICATIONS

J. Pintér and J. Szabó

Research Centre for Water Resources Development
(VITUKI), H-1453 Budapest, P.O. Box 27., Hungary

1. INTRODUCTION: THE GLOBAL OPTIMIZATION PROBLEM

In the course of engineering design and modelling, one often has to determine a number of parameters "optimally". As examples from the field of water
resources management, planning and operation of reservoir systems or water quality treatment plants,ca-
libration of different descriptive models, hydrologic
time-series analysis etc. can be mentioned.As a rule,
the optimality criterion is expressed by some specified objective function, while the feasible domain for
parameter selection is usually given by a number of
constraints. Hence, the mentioned type of tasks frequently leads to a mathematical programming problem
of the following form:

$$\min f(x) \tag{1}$$
$$g_i(x) \leqq 0 \quad i = 1,\ldots, m .$$

Here the vector x represents the decision variables (we suppose that x belongs to the n-dimensional
Euclidean real space R^n), while f and g_i i = 1,..., m
are the objective function and the constraints, respectively.

As it is well-known, under fairly elementary
(compactness and continuity) assumptions one can prove
that problem (1) has a (not necessarily unique) global
optimizer, i.e. such vector x^* for which there holds

$$g_i(x^*) \leqq 0 \quad i = 1,\ldots, m,$$

and for all other feasible vectors x we have the inequality

$$f(x^*) \leq f(x) \ .$$

However, most numerical methods in mathematical programming are capable to find only local optimizers. In other words, unless problem (1) is specified in advance to have only a single local (thus global) optimum, there may be significant differences between local optima, depending on the starting point of the applied locally convergent algorithm. This fact has motivated studies in the field of global optimization, where we seek for a globally optimal solution of (1), while it may have several local optima. Due to some inherent theoretical and numerical difficulties, this research became more intensive only recently, cf.e.g. the works of Archetti (1980), Boender et al. (1982), Dixon and Szegő (1975, 1978),Pintér (1984a), Strongin (1978) or Zilinskas (1982).

In the sequel, instead of the general problem (1), its specified form will be investigated:

$$\min_{a \leq x \leq b} f(x) \qquad f \in F[a,b] \ , \qquad\qquad (2)$$

where a,b are n-dimensional (finite) vectors, and f belongs to the class F [a,b] of those Lipschitz-continuous real functions which have only a finite number of global optima on the n-dimensional interval [a,b]. Thus,

$$F[a,b] = \bigcup_{L>0} F_L[a,b] \ ,$$

where $F_L[a,b]$ denotes the set of those functions, for which there holds

$$|f(x_1) - f(x_2)| \leq L \, ||x_1 - x_2|| ; \ a \leq x_1, \ x_2 \leq b \ .$$

Note that the special constraint-set in (2) is typical in many applications.

In Pintér (1983) we defined a general class of globally convergent one-dimensional (directional) optimization procedures for solving problem (2). The examples, presented there (different known and some new algorithms), indicated the applicability of the described axiomatic approach. The multi-dimensional extension of these results is given in Pintér (1984b).

In Section 2 we summarize the theory introduced in the mentioned papers. A global optimization algorithm and several numerical examples are briefly presented in Section 3.

2. BASIC CONCEPTS AND THEORETICAL RESULTS

Assume that the sequence of points $\{x^k\}$, generated for finding optimal solution(s) $x^* \in X^*$ to (2), is defined by the following type of algorithm, which gradually divides $[a,b]$ into adaptively generated n-dimensional subintervals.

GA(n) conceptual n-dimensional global optimization algorithm

Denote by ℓ, $m = m(\ell)$ and $k = k(\ell)$ the current number of iteration cycle, generated subintervals and points (vertices of subintervals), respectively. The lexicographically ordered vertices of the i-th subinterval will be denoted by

$$x(i) = \{x(i,1), x(i,2),..., x(i,2^n)\}$$

($x(i)$ is a matrix of $(n \times 2^n)$ size). The vector of respective function values is denoted by

$$z(i) = \{z(i,1), z(i,2),..., z(i,2^n)\}$$

$$z(i,j) = f(x(i,j)) \quad j = 1,..., 2^n$$

(hence, $z(i)$ is a vector, consisting of 2^n components). Further on, we assume that a real-valued interval selection function

$$R(i) = R(x(i), z(i))$$

and a real-valued point selection function

$$S(i) = S(x(i), z(i))$$

can be evaluated for each generated interval, respectively. Applying the above notations, the algorithm-scheme GA(n) consists of the following steps:

Step 0: Initialization. Set $\ell:=1$, $m:=1$, $k:=2^n$. The initial search information is represented by the vertices and corresponding function values of $[a,b]$.

Step 1: Interval sequencing. This can be accomplished e.g. by lexicographically ordering the "lower left corners" of the subintervals, generated up to the ℓ-th cycle.

Step 2: <u>Interval selection.</u> Choose an index t, for which there holds

$$R(t) = \max_{1 \leq i \leq m} R(i) \quad m = m(\ell) \; .$$

Step 3: <u>Point selection.</u> Choose a new point from interval x(t):

$$x^{k+1} = S(t)$$

(unless some termination criteria are fulfilled).Then x^{k+1} defines the partitioning of x(t) to 2^n new subintervals: these are generated by the intersections of the boundary of x(t) and the hyperplanes which contain x^{k+1} and are parallel to the boundary hypersurfaces of x(t). The algorithm continues at Step 1 of the next iteration cycle ($\ell := \ell+1$).

In Pintér (1983, 1984b) it has been proved that if the decision (selection) functions R and S satisfy certain analytical requirements (shortly referred later, cf. Section 3), then the set of points generated by an <u>arbitrary</u> GA(n)-type algorithm converges to X^*, the solution set of (2). Further on, an efficiency estimate has been given which shows that GA(n)-type methods may generate only a bounded number of points on subintervals of [a,b], which do not contain global optimizer(s). Both mentioned results are specified also for the case, when $f \in F_L[a,b]$ and L>0 is known. The details of these theoretical investigations are described in the mentioned papers.

3. NUMERICAL ASPECTS AND EXAMPLES

The general character of the mentioned theoretical results shows that, in principle, one has a great freedom when constructing GA(n)-type algorithms. On the other hand, there are some obvious difficulties, connected with the numerical realization of such algorithms: namely, the usual dimensionality and computational complexity problems. Therefore it is necessary to define such GA(n)-type algorithms, which have relatively small storage requirements and easily computable decision functions R and S. In our computations we used the following functions:

$$R(i) = K \cdot Vol(x(i)) +$$

$$+ \frac{[z(i,1)-z(i,2^n)]^2}{K \cdot Vol(x(i))} - 2(z(i,1)+z(i,2^n)), \quad (3)$$

$$Vol(x(i)) = \prod_{j=1}^{n} [x(i,2^n)_j - x(i,1)_j],$$

$$K=K(\ell) = (2 + \frac{C}{\ell}) \cdot \max_{1 \le i \le m} \frac{|z(i,1)-z(i,2^n)|}{||x(i,2^n)-x(i,1)||} \qquad (C>0);$$

$$S(t)=Q\{x(t,1), \frac{1}{2}[x(t,1)+x(t,2^n)], x(t,2^n)\}. \quad (4)$$

We remark that (3) is a generalization of Strongin's univariate interval-selection function (Strongin (1978)). It is easy to verify that R is continuous with respect to $x(i)$ and $z(i)$; translation-invariant with respect to $x(i)$, strictly monotonously decreasing in $z(i,1)$ and $z(i,2^n)$ and - with proper parameterization - its value is greater than the respective value of it for arbitrary degenerated (one-point) subinterval of $x(i)$. Basically, these properties are sufficient to assure global convergence of the algorithm, defined by (3) and (4). Formula (4) symbolizes a quadratic interpolation on the main diagonal of $x(t)$. We note that the function S has only a minor role in terms of theoretical convergence; however,it may significantly influence the local efficiency of the global search algorithm. (Actually, S is supplemented with some technical modification, in order to maintain convergence.)

The computational algorithm combines global and local optimization phases: the global search procedure outlined above is changed for local search, when some stopping criterion is fulfilled (e.g.the n-dimensional volume of the subinterval chosen at Step 2 is less, than a fixed threshold value). If one determines a priori the number M of (global or local) minima to be found, then M global and local search phases will be iteratively accomplished before the optimization procedure is ended. The applied local search consists of several cycles of a conjugated gradients type method, in which gradients are estimated by finite differences. (It is felt important that the algorithm should be applicable also in absence of higher order analytical information, as in many technical problems the analytical form of the figuring functions is unknown or can not be used to calculate derivatives: see e.g. the examples below.)

The combined global-local algorithm has been applied first to some known univariate (n=1) and multidimensional (n = 2,3,4) test problems, given in the works of Shubert (1972), Strongin (1978), Dixon and Szegő, eds. (1978), respectively. These results will appear elsewhere in details: here we only note that according to them, the above described algorithm is fairly stable (in all test problems each global optimizer was found) and efficient (the number of function evaluations, necessary to locate the optimal solution with prescribed accuracy, was in general significantly less, than in the results collected by the mentioned references).

For illustrating the real-world applicability of this methodology, we mention two examples from water resources engineering practice.

A model for streamflow forecasting

The National Water Monitoring Service (Budapest, Hungary) computes daily streamflow forecasts. These forecasts are based on an adequate discretized version of the so-called Kalinin-Milyukov-Nash cascade model (for details cf. e.g. Szöllősi-Nagy (1982)).The respective discrete state-space model is described by the vector equations

$$\underline{x}_t = F_t(\Delta t)\, \underline{x}_{t-\Delta t} + G_t(\Delta t)\, \underline{u}_t \quad (\underline{x}_o \text{ is known}), \quad (5)$$

$$\underline{y}_t = H\, \underline{x}_t \ ,$$

where the input process $u(t)$ and the output process $y(t)$ are sampled in equidistant time-moments $t = \Delta t$, $2\Delta t, \ldots, N\Delta t$; further on, $\underline{x}_t$ is the system state, $F_t(\Delta t)$ is the state-transition matrix, $G_t(\Delta t)$ is the input-transition matrix, H is the output generator matrix. (In the considered special application H is a vector, thus $\underline{y}_t = y_t$ is a scalar.) Our aim is to find an optimal parameterization of the model (5), which minimizes the deviation between observed $\{\hat{y}_t\}$ and computed $\{y_t\}$ output time-series in the sense

$$\min_{\substack{n_{min}\le n\le n_{max} \\ k_{min}\le k\le k_{max}}} f(n,k) \quad , \quad f(n,k) = \sum_{t=\Delta t}^{N\Delta t} (y_t - \hat{y}_t)^2 \ , \quad (6)$$

$$y_t = y_t(n,k) \ .$$

The parameters n,k explicitly figure in the definition of the matrices $F_t(\Delta t)$ and $G_t(\Delta t)$: hence,(6) can be regarded as a special case of the global optimization problem (2). From the physical interpretation of the model (5) we know that n is a positive integer, while k is a continuous variable. Consequently, for each feasible n, we have to solve a univariate optimization problem for finding k. In our case the bounds $1 \le n \le 10$, $0 \le k \le 10$ were applied; the solution of each univariate problem needed only some tens of steps (evaluations of $f(n,k)$).

Descriptive modelling of mixing processes

The second example is connected with a mixing process in a shallow lake (see Somlyódy(1982)). Assume that the temporal changes in the concentration of suspended solids can be described by the differential equation

$$\frac{dy}{dt} = - b_1 y + b_2 w^n, \tag{7}$$

where t is time, $y = y(t)$ denotes the concentration value, $w = w(t)$ is the velocity of wind; finally, b_1, b_2 and n are model parameters to be optimized.

Assume that the value of w is constant between t and $t + \Delta t$. Then, for $t = 0, \Delta t, 2\Delta t, \ldots N\Delta t$, the discretized solution of (7) can be explicitly calculated from

$$y_{t+\Delta t} = y_t\, e^{-b_1 \Delta t} + \frac{b_2}{b_1}\, w_t^n (1 - e^{-b_1 \Delta t}) \tag{8}$$

(the initial value y_o is supposed to be known). Using the expression (8) for computing the values x_t - given b_1, b_2 and n - and knowing the observed values $\hat{y}_t$, the parameter estimation problem can be formulated e.g. in the form

$$\min f(b_1, b_2, n), \quad f(b_1, b_2, n) = \sum_{t=\Delta t}^{N\Delta t} (y_t - \hat{y}_t)^2, \tag{9}$$

$$b_{1min} \le b_1 \le b_{1max}$$
$$b_{2min} \le b_2 \le b_{2max} \qquad y_t = y_t(b_1, b_2, n).$$
$$n_{min} \le n \le n_{max}$$

Again, this problem can be regarded as a special case of (2): now we have three continuous parameters, which are to be optimized. The feasible domain was given by the constraints $10^{-20} \le b_i \le 2$ $i = 1, 2$, $0 \le n \le 6$; the obtained three-dimensional global optimization problem was solved computing some 450 function values of $f(b_1, b_2, n)$.

At the end of this Section we note that, of course, one could apply other measures (norms), than least squares to calculate the distance between observed and calculated values. The essential need to apply global (rather than local) optimization stems from the fact that the respective objective functions (cf. (5)-(6) and (8)-(9)) are often non-convex (thus having, in general, several local optima). We think that this case is fairly typical in many problems, originated from engineering design and/or model calibration.

4. CONCLUDING REMARKS

In this paper the global optimization problem (2) and a class GA(n) of methods for its solution were briefly described. Following this, an algorithm was defined (as a specific element of GA(n)) and some numerical results were summarized.

Of course, a small number of experimental runs does not make possible to verify statements about the general efficiency and robustness of any particular method. We think, however, that - according to the obtained numerical results - the proposed method is reliable and rather fast. What is probably more important that the general theory of GA(n)-type methods makes possible to construct flexible procedures to solve global optimization problems, which are inherently far more difficult than local optimization problems of the same size.

All experimental runs, mentioned or described in this paper, were accomplished on an MO8X personal computer (with 32 kbyte memory) of the Research Centre for Water Resources Development (Budapest, Hungary).

REFERENCES

Archetti, F. (1980) "Analysis of stochastic strategies for the numerical solution of the global optimization problem", in: Archetti, F. and Cugiani, M. (eds.), Numerical Techniques for Stochastic Systems (North Holland, Amsterdam), pp. 275-295.

Boender, C.G., Rinnooy Kan, A.H.G., Stougie, L. and Timmer, G.T. (1982) "A stochastic method for global optimization", Mathematical Programming 22, 125-140.

Dixon, L.C.W. and Szegő, G.P. (eds.) (1975, 1978) Towards Global Optimization, Vol. 1-2. (North-Holland, Amsterdam).

Pintér, J. (1983) "A unified approach to globally convergent one-dimensional optimization algorithms", Technical report IAMI-83.5, Istituto per le Applicazioni della Matematica e dell'Informatica CNR, Milan.

Pintér, J. (1984a) "Convergence properties of stochastic optimization procedures", Mathematische Operationsforschung und Statistik, Series Optimization 14, 405 - 427.

Pintér, J. (1984b) "Globally convergent methods for n-dimensional multiextremal optimization", Mathematische Operationsforschung und Statistik, Series Optimization (to appear).

Shubert, B.O. (1972) "A sequential method seeking the global maximum of a function", SIAM Journal on Numerical Analysis 9, 379-388.

Somlyódy, L. (1982) "Water quality modelling: a comparison of transport-oriented and ecology-oriented approaches", Ecological Modelling 17, 183-207.

Strongin, R.G. (1978) Numerical methods for multiextremal problems (in Russian),(Nauka, Moscow).

Szöllősi-Nagy,A. (1982) "On the discretization of the continuous linear cascade by means of the state-space analysis", Journal of Hydrology 58, 223-236.

Zilinskas, A. (1982) "Axiomatic approach to statistical models and their use in multimodal optimization theory", Mathematical Programming 22, 104-116.

EIGENVALUE CALCULATIONS USING THE COLLOCATION FINITE ELEMENT
METHOD

N. G. Zamani and S. N. Sarwal

Department of Applied Mathematics, Technical University of
Nova Scotia, Halifax, Nova Scotia B3J 2X4, Canada

INTRODUCTION

The collocation finite element method has proven to be successful in treating many complex engineering problems [2,8,10,11,
13, 16].

The application of collocation in numerical solution of
differential equations dates back to the pioneering work of
Kantorovich in the early thirties [9]. However, the finite
element-based collocation is fairly recent and was mainly
initiated by the work of the following mathematicians [3, 4, 12,
15] during the early seventies. A comprehensive review of
collocation can be found in [5,6].

In conducting the present research, a literature search was
undertaken which turned up only one article dealing with
eigenvalue calculations using the collocation method [11].
However, its approach was not finite element-based.

In the present work a finite element-based collocation procedure
is proposed which is then used for calculation of eigenvalues of
several differential operators in one and two dimensions.

In order to assess the performance and the accuracy of the
method, three problems with engineering significance are
treated. To be more specific, second and fourth order boundary
value problems in one dimension and Helmholtz equation in two
dimensions were considered. Such problems arise in vibration
of strings, beams and membranes, respectively. The results are
promising and compare quite well with other numerical methods
such as the standard finite elements and finite differences.
The simplicity of collocation and its efficiency on digital
computers make it competitive with other techniques. All
calculations were made on a CDC Cyber 825 computer.

THEORETICAL FORMULATION

Consider the abstract boundary value problem in a bounded domain

$$Lu = 0 \text{ in } \Omega \quad , \quad \tilde{B}u = 0 \text{ on } \partial\Omega$$

where L and $\tilde{B}$ are linear differential operators and $\partial\Omega$ is the boundary of Ω. Clearly, due to the linearity of L and $\tilde{B}$, u = 0 is a solution to the above equations. The corresponding eigenvalue problem is described by

$$Lu = \lambda u \text{ in } \Omega \quad , \quad \tilde{B}u = 0 \text{ on } \partial\Omega$$

The objective is to determine λ such that the problem has a nontrivial solution u in Ω. Such a λ is called the eigenvalue and the corresponding u is called the eigenfunction for the operators L and $\tilde{B}$. The theorem below (which is stated for the sake of completeness) guarantees the existence of λ and u in a general framework.

Theorem
Let L be a compact self-adjoint positive operator in a separable Hilbert space H of infinite dimensions. Then the operator L has a countable set of eigenvalues, all of them being positive, and to each of them there corresponds a finite number of linearly independent eigenfunctions. The orthonormal system u_1, u_2, u_3, ... of eigenfunctions corresponding to the system $0 < \lambda_1 \leq \lambda_2 \leq \lambda_3, \leq ... < \infty$ is complete in H, furthermore $\lim_{n\to\infty} \lambda_n = \infty$. For a proof and terminology, see [14].

CONCRETE EXAMPLES

a)   The Rotating String Problem
Consider the problem of the deflection assumed by a stretched string of length ℓ and linear density ω, rotating with angular velocity ω about the x-axis. The two ends are assumed fixed. Denoting the displacement from the rotation axis by u(x) and the uniform tensile force by T, the governing equation is of the form

$$-\frac{d^2u}{dx^2} = \frac{\rho\omega^2}{T} u \quad 0 \leq x \leq \ell \quad , \quad u(0) = u(\ell) = 0$$

In this problem $L \equiv -\frac{d^2}{dx^2}$ and $\lambda \equiv \frac{\rho\omega^2}{T}$.

b)   The Rotating Shaft Problem
The deflection of an originally straight shaft of length ℓ, hinged at both ends, rotating with an angular velocity ω, is governed by the equation

$$\frac{d^4u}{dx^4} = \frac{\rho\omega^2}{EI} u \quad 0 \le x \le \ell \quad , \quad u(0) = u''(0) = u(\ell) = u''(\ell) = 0$$

here E is Young's modulus, I is the cross-sectional moment of inertia and ρ is the linear density of the shaft.

For this problem, L and λ are given by

$$L \equiv \frac{d^4u}{dx^4} \quad \text{and} \quad \lambda \equiv \frac{\rho\omega^2}{EI}$$

c) <u>Vibration Modes of a Cantilever Beam</u>
The free vibrations of a uniform cantilever beam of length ℓ is governed by

$$EI \frac{\partial^4 v}{\partial x^4} + \rho A \frac{\partial^2 v}{\partial t^2} = 0 \quad 0 \le x \le \ell$$

$$v(0,t) = v'(0,t) = v''(\ell,t) = v'''(\ell,t) = 0$$

As in the previous example, E is Young's modulus, I is the cross-sectional moment of inertia, A is the cross-sectional area and ρ is the linear density of the beam.

A solution of the form $v(x,t) = u(x) e^{i\omega t}$, leads to the following 4-th order eigenvalue problem

$$\frac{d^4u}{dx^4} = \frac{\rho A\omega^2}{EI} u \quad 0 \le x \le \ell , \quad u(0) = u'(0) = u''(\ell) = u'''(\ell) = 0$$

Here, $L \equiv \frac{d^4}{dx^4} \quad \text{and} \quad \lambda \equiv \frac{\rho A\omega^2}{EI} \quad .$

Remark: The above two problems are special cases of the following Sturm-Liouville equations to which our procedure is equally applicable.

$$- \frac{d}{dx} [p(x) \frac{du}{dx}] + q(x) u = \lambda r(x) u$$

$$- \frac{d^2}{dx^2} [p(x) \frac{d^2u}{dx^2}] + q(x) u = \lambda r(x) u$$

d) <u>Helmholtz Equation in Two Dimensions</u>
Consider the free vibration of a thin elastic membrane stretched over a bounded region Ω with tension T and density ρ. The equations governing the deflection w of the membrane are described by

$$\frac{T}{\rho} \nabla^2 w - w_{tt} = 0 \text{ in } \Omega \quad , \quad w = 0 \text{ on } \partial\Omega$$

Assuming

$$w(x,y,t) = u(x,y) \ e^{i\omega t}$$

leads to the two dimensional Helmholtz equation

$$- \nabla^2 u = \frac{\rho\omega^2}{T} \ u \text{ in } \Omega \quad , \quad u = 0 \text{ on } \partial\Omega$$

Here $L \equiv - \nabla^2$ and $\lambda \equiv \dfrac{\rho\omega^2}{T}$.

COLLOCATION FINITE ELEMENT FORMULATION

Given the eigenvalue problem

(1) $Lu = \lambda u \text{ in } \Omega \quad , \quad \tilde{B}u = 0 \text{ on } \partial\Omega$

An approximate solution $\hat{u}(x)$ of the form

(2) $\hat{u}(x) = \sum\limits_{j=1}^{M} a_j \ N_j(x)$

is sought for where $N_j(x)$ is assumed to satisfy the boundary condition in (1). In the finite element terminology, $N_j(x)$ is known to be a shape function. To obtain the coefficients a_j, M collocation points x_i are selected and equation (1) is forced to be satisfied at these points, namely:

$$L\hat{u}(x_i) = \lambda \ \hat{u}(x_i) \qquad i = 1, 2, \ldots, M$$

replacing $\hat{u}$ by its form in (2), one obtains

(3) $\sum\limits_{j=1}^{M} a_j \ L(N_j(x_i)) = \lambda \sum\limits_{j=1}^{M} N_j(x_i) \qquad i = 1, 2, \ldots, M.$

Expression (3) can be rewritten in the standard matrix eigenvalue form

$$Aa = \lambda \ Ba$$

where $A_{ij} = L(N_i(x_j)) \quad , \quad B_{ij} = N_i(x_j)$

It is easy to see that (3) is the outcome of a weighted residual procedure of the type

$$\int_{\Omega} [L(\hat{u}) - \lambda \hat{u}] \, W_i(x) \, d\Omega = 0 \qquad i = 1, 2, \ldots, M$$

where $W_i(x) = \delta(x-x_i)$ with δ being the Dirac delta function centered at x_i.

The choice of the shape functions N_i and the location of the collocation points have been the subject of intensive investigation in the past two decades. The "optimal" choice of these parameters are discussed in the next section.

CHOICE OF THE SHAPE FUNCTIONS AND COLLOCATION POINTS

First the construction of the shape functions and the collocation points for the 2nd order differential operator in one dimension is described. Let $0 = x_1 < x_2 < \ldots < x_p = 1$ be a one dimensional grid on the interval $[0,1]$. It is well known [1] that associated with each element $[x_i, x_{i+1}]$, there exists cubic polynomials ϕ_i, ϕ_{i+1}, ψ_i and ψ_{i+1} with the following properties:

$$D^k \psi_\ell(x_m) = \delta_{ko} \, \delta_{\ell m} \qquad \ell, m = 1, 2$$

$$D^k \psi_\ell(x_m) = \delta_{k1} \, \delta_{\ell m} \qquad \ell, m = 1, 2$$

where δ_{ij} is the Kronecker delta. In solid mechanics terminology, these functions are known as the beam bending shape functions.

Using the natural coordinate system $s = (x-x_i)/(x_{i+1}-x_i)$, the above shape functions are analytically described by

$$\phi_i = 1 - 3s^2 + 2s^3 \quad , \quad \phi_{i+1} = 3s^2 - 2s^3$$

$$\psi_i = s - 2s^2 + s^3 \quad , \quad \psi_{i+1} = s^3 - s^2$$

In the weighted residual formulation, the Hermite cubic polynomials just constructed play the role of the trial functions. The test functions are taken to be the Dirac delta distributions associated with the images of the Gaussian points in the element $[x_i, x_{i+1}]$. In this case the Gaussian points are roots of the 2nd degree Legendre polynomial in $[0,1]$.

In treating the 4-th order differential operator, the Hermite polynomials of degree seven are constructed in a similar fashion. For the collocation points, the images of the roots of a 4th order Legendre polynomial are used. This produces an approximate solution which is C^3, namely, three times continuously differentiable in the domain of interest. Regarding the Helmholtz equation in two dimensions, the tensor product of the Hermite cubic polynomials constructed initially in this section are employed. For the collocation points, the two Gaussian points in the x and y directions are used.

ALGEBRAIC EIGENVALUE COMPUTATION

In a previous section it was shown that collocation finite elements lead to the generalized matrix eigenvalue problem $Aa = \lambda Ba$. The two matrices A and B are the stiffness and the mass matrices, respectively. It can be shown that they are both invertiable and therefore one can rewrite the above equation as $B^{-1}A\, a = \lambda a$.

The inverse power method was applied to obtain the dominating modes of $B^{-1}A$. The existence of distinct eigenvalues are guaranteed by the theorem stated earlier.

Example 1
The parameters in the rotating string problem are chosen so that the governing equations are simplified to

$$- u'' = \lambda u \qquad 0 \leq x \leq 1 \quad , \qquad u(0) = u(1) = 0$$

The exact eigenvalues and eigenfunctions are $\lambda_n = n^2\pi^2$ and $u_n = \sin\sqrt{\lambda_n}\,x$, respectively.

The first four eigenvalues $\hat{\lambda}_n$ are computed with different uniform mesh spacings h = 1/2, 1/4, 1/8, 1/16 and 1/32. The relative errors $|\lambda_n - \hat{\lambda}_n|/\lambda_n$ are listed in Table 1.

A plot of $\log |\lambda_n - \hat{\lambda}_n|$ versus log h is depicted in Figure 1. This verifies the asymptotic error estimate $|\lambda_n - \hat{\lambda}_n| = O(h^4)$. The condition number of $B^{-1}A$ is plotted in Figure 2 and it can be seen that as $h \rightarrow 0$, $B^{-1}A$ becomes ill-conditioned.

Figures 3, 4, 5 and 6 show the approximate and the exact eigenfunctions u_n and $\hat{u}_n$, respectively for the case h = 1/32.

Figures 7, 8, 9 and 10 exhibit the relative pointwise errors in the calculated eigenfunctions where h = 1/32. Finally, Figures 11, 12, 13 and 14 give the asymptotic rates of convergence for

$$\|u_n - \hat{u}_n\|_\infty \qquad \text{and} \qquad \|Du_n - D\hat{u}_n\|_\infty \quad .$$

It can be seen that the following asymptotic rates are true,

$$\|u_n - u_n\|_\infty = \max_{0 \leq x \leq 1} |u_n(x) - \hat{u}_n(x)| = O(h^4)$$

$$\|Du_n - \hat{Du}_n\|_\infty = \max_{0 \leq x \leq 1} |Du_n(x) - \hat{Du}_n(x)| = O(h^3) \quad .$$

The errors are small even for the coarse grid corresponding to h = 1/2. The rates of convergence are consistent with the theoretical analysis of [3]. Although these theoretical results were not based on eigenvalue problems, they seem to hold for such cases.

Table 1

h	$|\lambda_1 - \hat{\lambda}_1|/\lambda_1$	$|\lambda_2 - \hat{\lambda}_2|/\lambda_2$	$|\lambda_3 - \hat{\lambda}_3|/\lambda_3$	$|\lambda_4 - \hat{\lambda}_4|/\lambda_4$
1/2	.22E-2	.21E0	.11E0	.82E-1
1/4	.16E-3	.22E-2	.66E-2	.21E+0
1/8	.10E-4	.16E-3	.78E-3	.22E-2
1/16	.67E-6	.10E-4	.54E-4	.16E-3
1/32	.43E-7	.68E-6	.33E-5	.10E-4

RATE OF CONVERGENCE. FIRST 4 MODES CONDITION NUMBER OF $B^{-1}A$

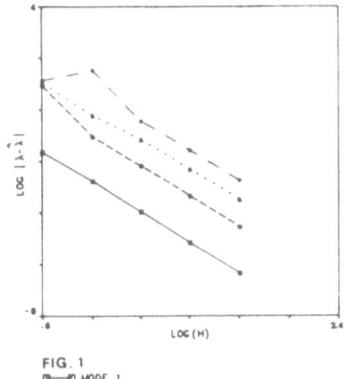

FIG. 1
□—□ MODE 1
⊖ — ⊖ MODE 2
△ · ·△ MODE 3
●— ● MODE 4

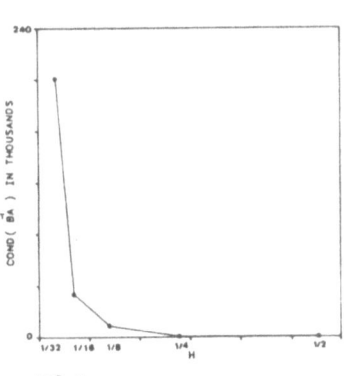

FIG. 2

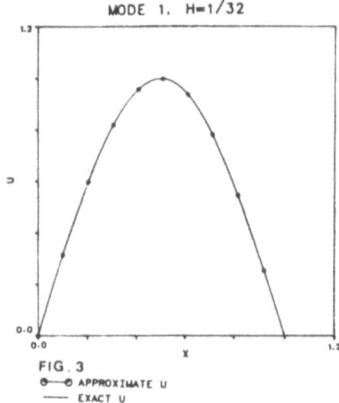

FIG. 3
o——o APPROXIMATE U
—— EXACT U

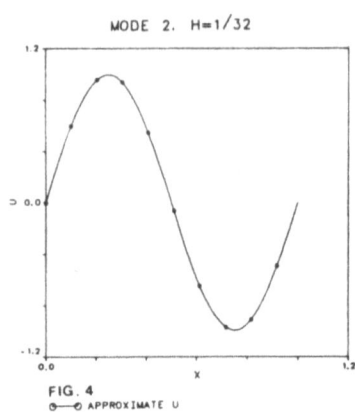

FIG. 4
o——o APPROXIMATE U
—— EXACT U

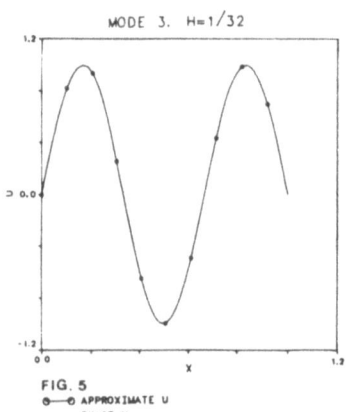

FIG. 5
o——o APPROXIMATE U
—— EXACT U

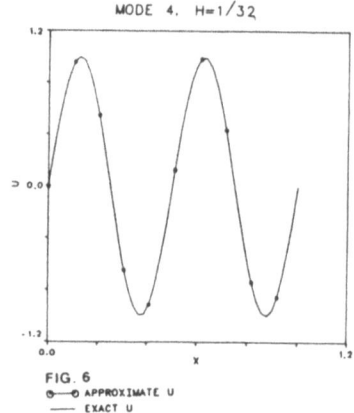

FIG. 6
o——o APPROXIMATE U
—— EXACT U

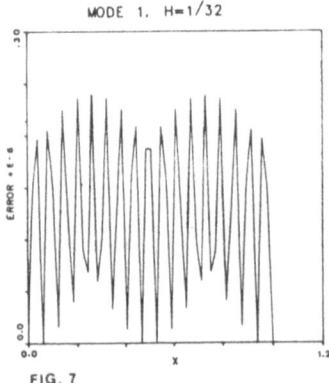

FIG. 7
—— POINTWISE RELATIVE ERROR IN U

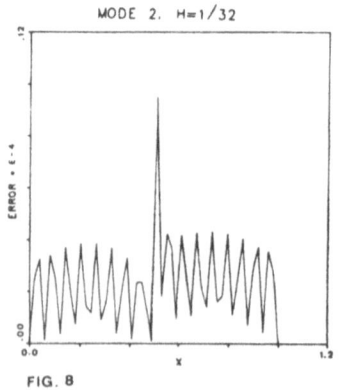

FIG. 8
—— POINTWISE RELATIVE ERROR IN U

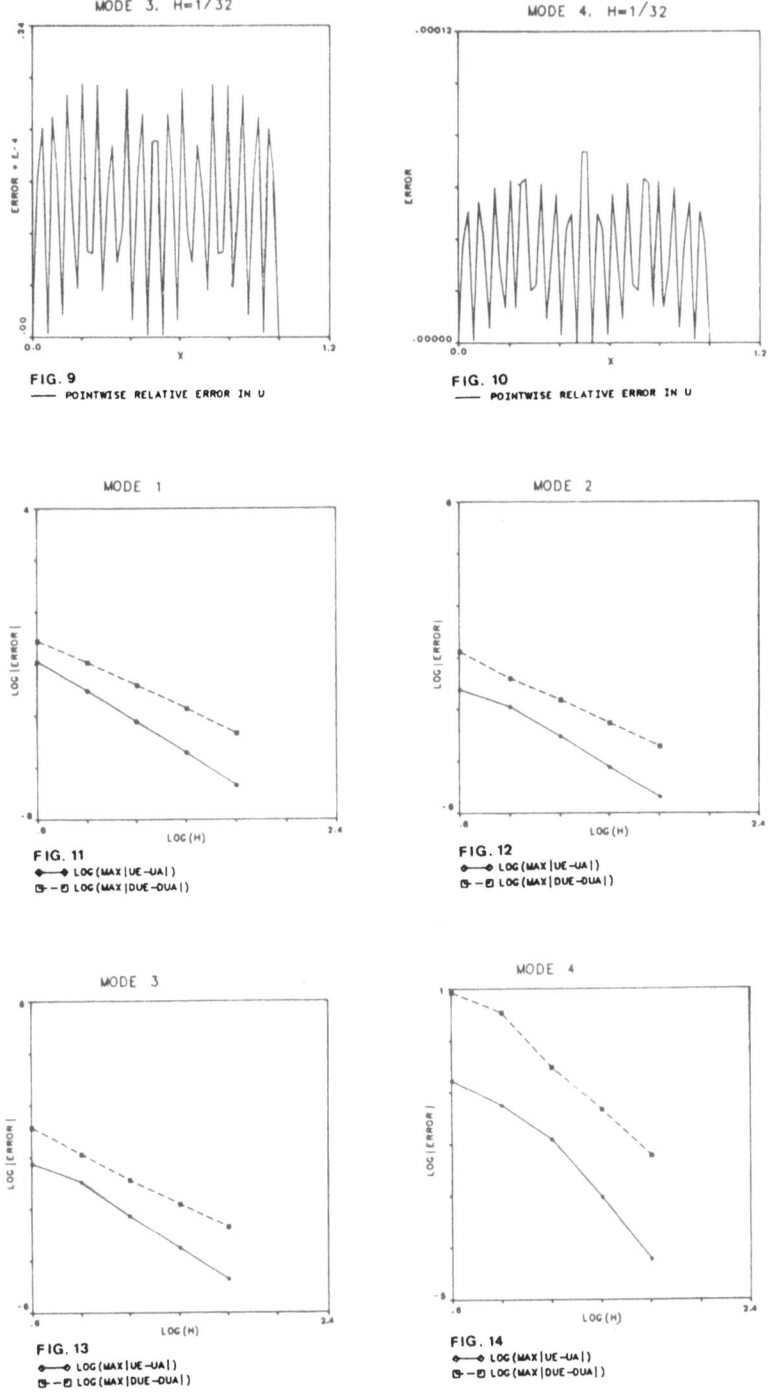

MODE 3. H=1/32

FIG. 9
— POINTWISE RELATIVE ERROR IN U

MODE 4. H=1/32

FIG. 10
— POINTWISE RELATIVE ERROR IN U

MODE 1

FIG. 11
◆—◆ LOG(MAX|UE−UA|)
◻−◻ LOG(MAX|DUE−DUA|)

MODE 2

FIG. 12
◆—◆ LOG(MAX|UE−UA|)
◻−◻ LOG(MAX|DUE−DUA|)

MODE 3

FIG. 13
◆—◆ LOG(MAX|UE−UA|)
◻−◻ LOG(MAX|DUE−DUA|)

MODE 4

FIG. 14
◆—◆ LOG(MAX|UE−UA|)
◻−◻ LOG(MAX|DUE−DUA|)

Example 2

The parameters in the rotating shaft problem are selected so
that the governing equations are reduced to

$$\frac{d^4 u}{dx^4} = \lambda u \qquad 0 \leq x \leq 1 \; , \quad u(0) = u''(0) = u(1) = u''(1) = 0$$

The analytical expressions for the eigenvalues and the eigenfunctions are

$$\lambda_n = n^4 \pi^4 \quad \text{and} \quad u_n = \sin \sqrt[4]{\lambda_n} \, x \quad .$$

Therefore, the deflection modes are as in Example 1 and to
avoid repetition, the problem was solved numerically with only
h = 1/2 and h = 1/4.

Table 2 shows the exact and the approximations to the first
four eigenvalues.

In Figures 15, 16, 17 and 18 the pointwise relative errors for
the corresponding eigenfunctions (with h = 1/4) are plotted.
In spite of the coarse grid used, the results are satisfactory.

Table 2

	λ_1	λ_2	λ_3	λ_4
h = 1/2	97.409153	1559.7934	7909.7351	23700.272
h = 1/4	97.409091	1558.5464	7890.2605	24956.695
exact	97.409091	1558.5455	7890.1364	24936.727

MODE 1. H=1/4

MODE 2. H=1/4

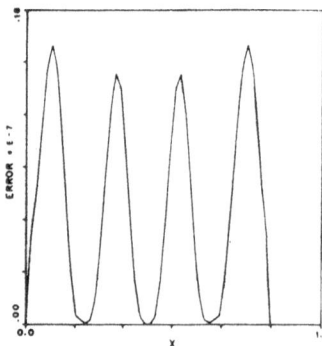

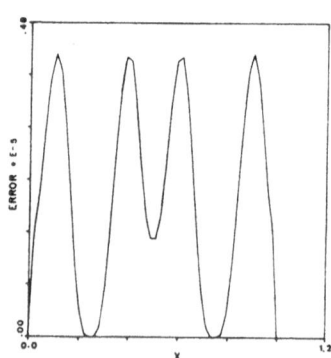

FIG. 15
——— POINTWISE RELATIVE ERROR IN U

FIG. 16
——— POINTWISE RELATIVE ERROR IN U

MODE 3. H=1/4

MODE 4. H=1/4

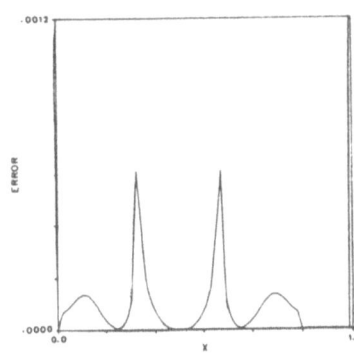

FIG. 17
—— POINTWISE RELATIVE ERROR IN U

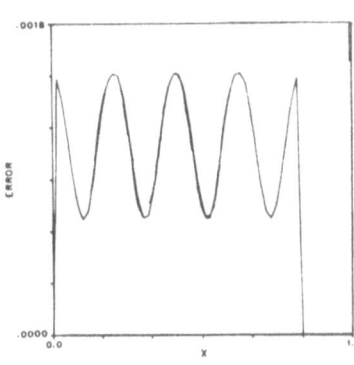

FIG. 18
—— POINTWISE RELATIVE ERROR IN U

Example 3
In this example, the problem of the free vibration of a
cantilever beam is reconsidered. The parameters are selected
so that the eigenvalue problem to be solved is of the form

$$\frac{d^4 u}{dx^4} = \lambda u \quad 0 \leq x \leq 1 , \; u(0) = u'(0) = u''(1) = u'''(1) = 0$$

There is no explicit formula for λ, however it satisfies the
transcendental equation $\cos \sqrt[4]{\lambda} \cosh \sqrt[4]{\lambda} = -1$. Collocation
finite element is used with h = 1/2, 1/4 and 1/8 and the first
four modes are calculated and listed in Table 3. Figures 19,
20, 21 and 22 show the exact and the approximate deflection
modes with h = 1/8.

Table 3

h	$\sqrt[4]{\lambda_1}$	$\sqrt[4]{\lambda_2}$	$\sqrt[4]{\lambda_3}$	$\sqrt[4]{\lambda_4}$
1/2	1.875104	4.694059	7.861149	11.081088
1/4	1.875104	4.694091	7.854758	10.995280
1/8	1.875104	4.694091	7.854757	10.995541
exact	1.875104	4.694091	7.854757	10.995541

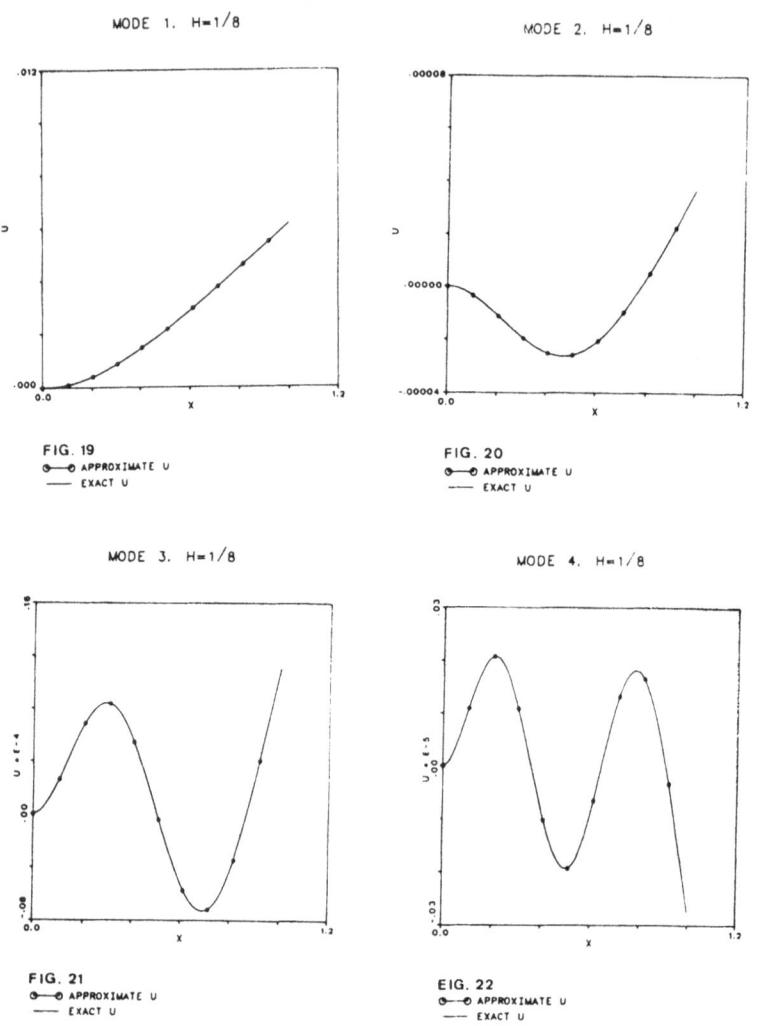

MODE 1. H=1/8

FIG. 19
○—○ APPROXIMATE U
—— EXACT U

MODE 2. H=1/8

FIG. 20
○—○ APPROXIMATE U
—— EXACT U

MODE 3. H=1/8

FIG. 21
○—○ APPROXIMATE U
—— EXACT U

MODE 4. H=1/8

FIG. 22
○—○ APPROXIMATE U
—— EXACT U

Example 4

Consider the Helmholtz equation in a unit square, given by

$$-\nabla^2 u = \lambda u \text{ in } \Omega \quad , \quad u = 0 \text{ on } \partial\Omega$$

The exact eigenvalues are known and described by $\lambda_{nm} = \pi^2(n^2+m^2)$. In order to apply collocation, a uniform grid of size h in the x and y directions is constructed. The dominating eigenvalues $\hat{\lambda}_{11}$, $\hat{\lambda}_{12}$, $\hat{\lambda}_{22}$ and $\hat{\lambda}_{13}$ are approximated, the results of which appear in Table 4. It can be seen that the approximations are quite accurate even for a coarse grid. The behaviour of cond $(B^{-1}A)$ is as given in Example 1.

Table 4

h	$\hat{\lambda}_{11}$	$\hat{\lambda}_{12}$	$\hat{\lambda}_{22}$	$\hat{\lambda}_{13}$
1	24.00000	48.00060	48.00060	71.99960
1/2	19.78351	57.89182	109.71395	153.89177
1/3	19.74914	49.56945	79.38967	117.87437
1/4	19.74275	49.43865	79.14363	99.29108
1/5	19.74065	49.38852	79.03640	99.06073
1/6	19.73994	49.36841	78.99695	98.89658
exact	19.73920	49.34802	78.95683	98.69604

Example 5
The Helmholtz equation in an L-shaped region is solved (see
Fig. 23). The boundary condition is u = 0 on the boundary.
There are no exact formulas for the eigenvalues of this problem.
However, a very accurate approximation to the smallest eigen-
value is obtained in [7] and taken to be λ_{min} = 9.639723846.

Initially, collocation is applied with a uniform mesh spacing
h, in the x and y directions. See Fig. 24 for the definition
of h. The approximations to the smallest eigenvalue are listed
in Table 5. One can conclude that a uniform mesh spacing does
not yield good results. This is not surprising as an L-shaped
region has a reentrant corner and there is a singularity in the
solution.

The condition number of $B^{-1}A$ is also listed in Table 5 and
extreme ill-conditionality is encountered for small values of
h. In order to improve the accuracy of the results, a non-
uniform grid is chosen which is made finer at the corner. The
calculated eigenvalue now appears in Table 6 (see Figure 25 for
the definition of h). With this nonuniform grid, a monotonic
convergence to the value in [7] is obtained. Again, the matrix
$B^{-1}A$ becomes extremely ill-conditioned for small values of h.

Table 5

h	$\hat{\lambda}_{min}$	Cond $(B^{-1}A)$
1/2	10.3154	1.9E+3
1/4	10.0357	3.0E+4
1/8	10.1151	8.7E+5
1/16	10.2263	5.4E+7
1/32	10.2825	2.3E+9
1/64	10.2968	5.0E+11
1/128	10.2953	2.1E+12

Table 6

h	$\hat{\lambda}_{min}$
1/4	9.8979
1/8	9.7467
1/16	9.6834
1/32	9.6470
1/64	9.6198
1/128	9.6066

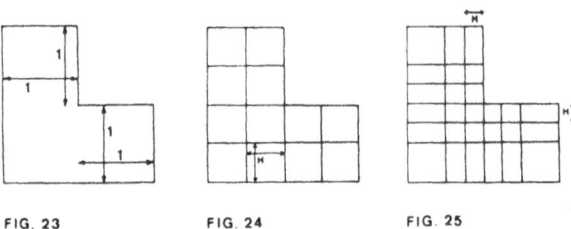

FIG. 23   FIG. 24   FIG. 25

CONCLUSIONS

In this article, the authors have investigated the application
of the collocation finite element method to the solution of
eigenvalue problems. Example 1 was studied in detail. This
included a verification of the rate of convergence of the eigenvalues and the corresponding eigenfunctions. The findings are
consistent with the theoretical results obtained in [3] for
other purposes.

In Example 5, the Helmholtz equation in an L-shaped region, which
is a hard problem to solve numerically, was treated. Even in
this case, collocation on a nonuniform grid reproduced the
smallest eigenvalue accurately.

The major advantage of collocation over the standard finite
elements is that no quadrature formulas are required. This is
one reason that collocation codes are very efficient. The
difficulty with collocation is associated with the fact that
it can only be applied to differential equations described in
an orthogonal coordinate system.

The authors are presently trying to extend their use of
collocation to nonlinear eigenvalue problems, non-orthogonal
coordinate systems and higher dimensions.

ACKNOWLEDGEMENT

This research was supported in part by the Nova Scotia
Department of Development under P.S.E.P. project number 7035650.

REFERENCES

[1]   Brebbia, C. A. and Connor, J. J. (1977) Finite Elements
        for Fluid Flow. Butterworths, second edition.
[2]   Chawla, T. C., Leaf, G. and Minkowycz, W. J. (1984) A
        Collocation Method for Convection Dominated Flows. Int.
        J. of Num. Methods in Fluids, 4, 271-281.
[3]   DeBoor, C. and Swartz, B. (1973) Collocation at Gaussian
        Points. SIAM J. of Num. Analysis, 10, 4:582-606.
[4]   Douglas, J. J. and Dupont, T. (1973) A Finite Element
        Collocation Method for Quasilinear Parabolic Equations.

Math. of Computation, 27, 17-28.

[5]   Dyksen, W. R., Houstis, E. N., Lynch, R. E. and
        Rice, J. R. (1984) The Performance of the Collocation
        and Galerkin Methods with Hermite Bi-Cubics. SIAM J. of
        Num. Analysis, 21, 4:695-715.

[6]   Finlayson, B. A. (1972) The Method of Weighted Residuals
        and Variational Principles. Academic Press, New York.

[7]   Fox, L., Hencrici, P. and Moler, C. (1967) Approximations
        and Bounds for Eigenvalues of Elliptic Operators. SIAM
        J. of Num. Analysis, 4, 1:89-102.

[8]   Frind, E. O. and Pinder, G. E. (1979) A Collocation
        Finite Element Method for Potential Problems in
        Irregular Domains. Int. J. of Num. Methods in Eng.,
        14, 681-701.

[9]   Kantorovich, L. V. (1934) A Method for Approximate
        Solution of Partial Differential Equations. Doklady
        Akademii Nauk SSSR. II, 532-536.

[10]   Kwok, W. L., Cheung, Y. K. and Delcourt, C. (1977)
        Application of Least Square Collocation Technique in
        Finite Element and Finite Strip Formulation. Int. J. of
        Num. Methods in Eng., 11, 1391-1404.

[11]   Lund, J. R. and Riley, B. V. (1984) A Sinc-Collocation
        Method for the Computation of the Eigenvalues of the
        Radial Schrodinger Equation. IMA J. of Num. Analysis,
        4, 1:63-98.

[12]   Phillips, J. L. (1972) The Use of Collocation as a
        Projection Method for Solving Linear Operator Equations.
        SIAM J. of Num. Analysis, 9, 1:14-28.

[13]   Pinder, G. E. and Shapiro, A. (1979) A New Collocation
        Method for the Solution of the Convection-Dominated
        Transport Equation. Water Resources Research, 15, 5:
        1177-1182.

[14]   Rektorys, K. (1977) Variational Methods in Mathematics.
        Science and Engineering, D. Reidel Publishing Company,
        Boston, U.S.A.

[15]   Russell, R. D. and Shampine, L. F. (1972) A Collocation
        Method for Boundary Value Problems. Numerische
        Mathematik, 19, 1-28.

[16]   Zamani, N. G. (1984) A Collocation Finite Element Method
        for Integration of the Boundary Layer Equations. J. of
        Applied Mathematics and Computation, 15, 2:109-122.

8. MICRO-APPLICATIONS

STRUCTURAL ANALYSIS USING MICROCOMPUTERS
STATE OF THE ART

D. NARDINI

UNIVERSITY OF ZAGREB, YUGOSLAVIA

1. INTRODUCTION

The development of microcomputers in recent years is making a
tremendous, and sometimes frightening impact on the whole human
society. It has brought about developments that were inconceivable to most people at the time of its appearance, and it is
still difficult to appreciate the full potential this phenomenon
for mankind.

Even after the birth of microcomputers in the early 1970's who
could have visualised todays personal computer explosion? Until
a few years ago, the microcomputer was viewed more as a toy for
amateurs and hobbysts. Most computer professionals looked down
on it, hoping that if it was ignored, it would go away. But it
didn't, and what a drastic change in views came about in a few
short years. A special thing about microcomputers, is that
they have made computers affordable, available and easy to use,
literally taking the latest technology to the masses.

While the psychologysts and futurologysts are busy in trying to
predict the overall consequences of this second industrial revolution on the human society, this article has a more modest
goal, namely to review the current state of the art in the use
of microcomputers in the field of structural analysis, with a
glimpse into the near future.

2. COMPUTERISED STRUCTURAL ANALYSIS - HISTORICAL BACKGROUND

Use of computers in the structural analysis is only some 30
years old, which is an extremely short period of time when compared to the multi century history of structural design. Methods of calculation of stresses that emerged in the pre-computer
era always needed a skilled engineer, in order to simplify the
structure in such a way, that it would be possible to solve it
by the aid of a pen, paper and a slide rule. Most of these procedures, some of them developed for very specialised cases, were

far from being general enough, to be programmed for an electronic computer. Also, at certain points of the analysis, the subjective engineering judgement was necessary, which made such methods quite impossible to implement. Realisation that an engineer (with his intelligence and experience) and the computer (with its enormous processing speed) are complementary, and that what is trivial for one is an unsurmountable task for the other, led to the complete reformulation of the structural analysis procedures. Retaining only the basic principles (virtual work, potential energy), it was possible to formulate conditions of equilibrium and continuity in terms of linear algebra, from where there was only a short step to utilising a computer.

Approximate methods, although derived from principles known for decades, were treated in a completely different manner. The computer was used not only to solve the resulting systems of linear algebraic equations, but also to formulate these equations from basic geometric and material data.

Parallelly with the emergence of more powerful computers, more sophisticated models of structural behaviour have been introduced. With still better numerical algorithms, the accuracy of the results was greatly improved. This on the other hand gave the impulse for new investigations of constitutive laws of different materials, because uncertainty in basic parameters became a limiting factor of the validity of the final results.

3. EMERGENCE OF MICROCOMPUTERS

When general purpose microcomputers became available in mid-seventies, their processing power was so inadequate, and the limitations so severe, that it was very hard to see their future role in any serious engineering analysis. Let us not forget that in those years a 16K core was considered enormous, and the flexible disk capacity was not more that 80K. If one did not know how to write in the machine language, his chances of getting anything programmed were quite slim.

However, an improved breed of microcomputers soon followed, with the emergence of a simple (later to become standard) operating system - CP/M, together with some high level languages (BASIC and FORTRAN). It then became possible to program, still modest, engineering applications on a microcomputer. A number of specialised structural analysis programs emerged at that time, dealing with portal frames, roof trusses, continuous beams and the like. It became quite obvious that in repetitive, everyday, calculations done in the engineering bureaus, a microcomputer could be extremely useful. Nevertheless, most software houses dealing with structural analysis still looked down upon microcomputers, considering them not worth their while.

In the last couple of years we are witnessing the birth of a new breed of microcomputers, built around powerful 16/32 bit processors with extremely large addressing spaces. What is specially interesting for engineering applications, is the emergence of the arithmetic (or floating point) co-processors, which speed up the arithmetic considerably. On todays sophisticated microcomputers, also referred to as work stations, the processing power per user is the same as that on a mainframe computer only some ten years ago. In that environment, it has become possible to develop serious engineering applications.

4. PERFORMANCE OF A MICRO IN ENGINEERING APPLICATIONS

Using a microcomputer in engineering has some advantages and some drawbacks when compared to a mini or a mainframe.

The main drawbacks are the low-processing speed, limited RAM and mass storage and slow disk I/O. We could also talk about lack of, some other things like standardisation, support and reliable documentation, but these problems worry the more the software developers than the end-users.

All the drawbacks mentioned are actually being remedied by the appearance of a new faster microprocessors, floating point co-processors, faster and cheaper memory chips and compact high capacity hard disks. In the wide spectrum of machines available it is not easy to draw a line where a micro ends, and a mini-computer begins.

The advantage of using a microcomputer for engineering computations are many. The main one is the microcomputer availability. While larger computers are still something that an average engineer feels uneasy with, microcomputers are increasingly starting to be considered as personal devices, similarly as calculators have been just a few years ago, and slide rules before that. The second important advantage is the qualitative improvement in the user-computer interaction that the micro has brought about. A well written data entry program will today use the bit-mapped graphics that a microcomputer offers, to full advantage, making data preparationm viewing and modification a trivial task. Also, in the post-processor phases, graphical presentation plays a crucial role.

5. PROBLEMS OF ENGINEERING SOFTWARE DEVELOPMENT FOR A MICRO-COMPUTER

There seems to be a reciprocial trend in the costs of hardware and software. Namely, it is a common knowledge that with the advance of micro-chip technology, prices of computer components are drastically reduced each year. On the other hand, newer software products tend to be more complex, more user-friendly, solving wider range of problems, and since they are mainly the result of human work, the price of producing them is on the rise. The days when the buyer of a computer was getting a bundle of free software with his machine are gone, and a good piece of software has to be paid for.

One way to bring down the price of software is to mass market it, and this is successfully done for microcomputers in the areas of business applications. The problem with engineering fields are that they are diverse, and there are not nearly as many potential users for each of them, as for general business packages.

It is not uncommon for a thorough structural analysis package to have more that a hundred man-years of investment. Also, it has to be supported and maintained by a highly qualified team of professionals. These factors make the end product very costly, and due to a limited user's community the unit price is high.

Most of todays best known structural analysis codes have been initialised some 15 years ago, and have since then been growing in size together with the growing capabilities of mainframe and mini computers. Suddenly, the world is swarming with microcomputers, and these huge programs start resembling dinosaurs. There is not an easy way to scale them down, and the software houses are confronted with a dilemma: whether to wait for micros to grow to the size of their programs, or to try and squeeze the programs into the existing machines.

It seems that neither of the approaches is an adequate one. The existing codes, as developed on mainframes, are best suited for machines of such size and power, and for the time being should be left there. What an engineer with a microcomputer needs is a number of specialised modules which will take off the burden of his everyday computations. And although none of the modules may have the capability to do earthquake analysis of a nuclear power plant, they may be adequate in 90% of other situations.

In spite of setting a lower goal than a general purpose structural analysis package on a micro, a software designer still has a hard task. Problems which he must overcome when developing engineering software for a micro are:

- Lack of overall standardisation
- Partial standardisation of programming languages
- Diverse operating systems
- Hardware dependency (monitors, disks)
- Core and mass storage limitations
- Slow processing and I/O
- Lack of support and documentation

As already mentioned, most of these problems are being overcome with the progress of microcomputers, but they are much more obvious from a software developer's than from the user's point. That is the reason why many applications are initially developed on a powerful computer (host) and then transferred to the micro (target).

An ideal micro for engineering software development and use would have to have the following properties:

- Floating point processor for fast number crunching
- Large core storage for code and arrays
- Large and fast mass storage for source programs and random access files
- Reliable FORTRAN
- File structure with subdirectories
- Multitasking
- Network communication

All of this is available today on some microcomputers at the higher end of the price scale, but most of the time one still has to compromise between the cost and the performance.

6. STRUCTURAL ANALYSIS PROGRAMS - AVAILABILITY

Today, there is a broad spectrum of structural analysis programs available for a microcomputer. Their capabilities vary widely, and one could categorise them into the following groups:

a. Specialised modules. These are the programs usually written in BASIC, with a very limited computer in mind. They tackle very specific tasks like design of a continuous beam, a portal frame or a plate. The input is usually organised as a series of questions to which the engineer types the answers. Due to a poor standardisation of BASIC, portability of such programs is not straight forward.

b. Integrated special purpose modules. These programs also solve specific tasks (structures with pre-defined geometry), but also include the designing and detailing. Different modules are available for reinforced concrete design or for steel design. They may also include tables of standard profiles and other useful parameters. The input to such programs tends to be of higher standard, usually "menu-driven".

c. General purpose structural analysis codes. These programs are quite big in size and they include possibilities to analyse framed structures with arbitrary geometry in plane or space. As a consequence, more input data are needed to describe the properties of the structure and the loads acting on it, and these are input using usually separate pre-processor modules. Some of these programs also have the post-processing capabilities with graphical display of geometry, internal forces, stresses and displacements.

d. Finite Element programs. In the last couple of years these have appeared as the scaled down versions of some well known FEM codes. Due to the complexity of such programs, their micro

versions are usually capable of solving only modest tasks. One
catch is that, though frame analysis programs can in majority
of applications get away with single precission arithmetic, FEM
codes must definitely resort to double precission, which drastically slows down the processing. Nevertheless, their performance on more powerful microcomputers has proven satisfactory.

7. FUTURE PROSPECTS

Quite a widespread misconception is that the computer, equipped
with the right software, will eventually be able to replace the
knowledge of an engineer, and that less qualified personnel
will be able to perform complicated tasks of structural design.
The author's experience is the contrary: although the computer
is capable of greatly increasing the analytical powers of the
engineer using it, the crucial decisions still rest with the
human. Therefore, the future designers will have to have not
less, but more knowledge than those of today, but their efficiency will be enhanced. Luckily enough, for at least some time,
we cannot expect a computer to replace a structural designer,
but rather to support him. The main role of the computer should
be to shield the engineer from the laborious and error prone
calculations, but still leave him enough freedom for his creative judgement.

Forecasting in the field of microcomputers tends to be optimistic for the near future, but often proves conservative for the
long term. With the technology of VLSI (Very Large Scale Integration) silicon chips, it is becoming possible for such tiny
devices to absorb more and more portions of the computer system.
It is rather obvious that with the advance of microprocessors
and memory modules, even today's largest structural analysis
programs will probably in just a few years be available on desk
top computers. But the crucial question is will we want them
the way they are? In contrast to the quantum leaps of hardware
progress, the software has long ago become the limiting factor
of any application. Is there other alternatives to the existing
painstaking software development?

REFERENCES

1. J. Dvornik, N. Bicanic and D. Nardini, "Computers in Structural Analysis", Proc. 7th Conference of Yugoslav Society of Structural Engineering, pp 181-201, 1983

2. N. Ivancic and D. Nardini, "Implementation of a Large Structural Analysis Program on a Microcomputer", Proc. 3rd International Conference "Computer at University", Cavtat, Yugoslavia, 1981

3. W.T. Bell and R.J. Plank, "Civil Engineering Design Programs for Microcomputers", Proc. ENGSOFT III Conference, pp 337-346, Ed. R.A. Adey, London, 1983

4. J. Seifert and N.J. Wyatt, "Practical Application of Micro Computers in a Structural Consultancy", Proc. ENGSOFT III Conference, pp 429-442, Ed. R.A. Adey, London, 1983

5. S.J. Fenves, "Computers in the Future of Structural Engineering", Proc. 8th Conference on Electronic Computation, pp 1-8, ASCE, 1983

6. J.M. Symms and H.R. Lundgren, "Interactive Structural Software for Micros", Proc. 8th Conference on Electronic Computation, pp 242-252, ASCE, 1983

7. P.L. Hazan, "Micro Computing in the 80's", Computing Journal, pp 137-143, October 1984

8. D.J. States, "Scientific Computing: The Shortcomings of the Micro", Byte Guide to IBM-PC, pp 220, September 1984

APPLICATION OF MICROCOMPUTERS IN THERMAL
INSULATION DESIGN AND ANALYSIS

S. N. Tay

School of Mechanical & Production Engineering
Nanyang Technological Institute
Upper Jurong Road, Singapore 2263

INTRODUCTION

Thermal insulation for hot surfaces of industrial installations
such as storage tanks, steam and hot water pipes and other
thermal process equipment is essential for their safe and
economical operation. With the present high cost of fuel,
inadequate insulation leads to excessive heat losses and higher
operating cost. On the other hand, the effectiveness of the
additional insulation diminishes as thickness increases. The
concept of determining an optimum insulation has long been
appreciated. However, the lengthy and laborious calculations
required to establish heat losses through varying insulation
thickness with a wide range of temperatures always discourage
such optimisation exercise. Many resort to estimation based on
previous experiences or guessed work. These will inevitably
lead to under estimation as the increase of fuel price is not
well incorporated.

Turner et al (1980) developed a procedure to determine the
economic thickness of insulation for flat surfaces and cylindrical pipes which required the use of three graphs and two tables
before arriving at the result. Computer programs written in
FORTRAN on mainframe computer ICL1904S for heat transfer
analysis and the calculation of optimal pipeline insulation have
been developed by Diamant (1977). However, it requires other
library stored subprograms to completely solve an insulation
problem. Moreover, mainframe computers are not readily
available to many industrial users.

With the widespread use of microcomputers, design and analysis
of thermal insulation can be effectively carried out on these
computers with better estimation of economic thickness or
payback period. This paper describes two computer programs
written in BASIC for microcomputers based on (i) economic
thickness method and (ii) payback period method for thermal

insulation design and analysis. Application of the programs to the designing of new insulation and the upgrading of existing pipeline insulation are presented. The abilities, merits and deficiencies of both methods are discussed.

BASIC EQUATIONS

Heat Losses Through Insulated Pipes

The heat transfer per unit pipe length q from the hot pipe surface through the insulation to the surrounding fluid (Figure 1) is given by

$$q = \frac{2\pi(T_1 - T_\infty)}{\frac{\ln(\frac{r_2}{r_1})}{k} + \frac{1}{r_2 h}}$$

(1)

where T_1 is the pipe surface temperature (K), T_∞ is the ambient temperature (K), r_1 is the outer radius of uninsulated pipe, r_2 is the radius of insulated pipe, k is the thermal conductivity of insulation and h is the heat transfer coefficient of the external surface.

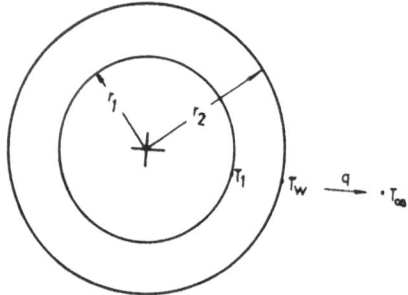

Figure 1   Pipe insulation

The overall heat transfer coefficient is the sum of heat transfer coefficients due to convection h_c and to radiation h_r.

$$h = h_c + h_r$$

(2)

Convective Heat Transfer Coefficient

For horizontal pipes, the value of h_c is given by the empirical equation for natural convection,

$$\frac{h_c d}{k_a} = 0.525 \, (\mathrm{GrPr})^{0.25}$$

(3)

for $10^4 < \mathrm{GrPr} < 10^9$

and

$$\frac{h_c d}{k_a} = 0.129 \ (GrPr)^{0.33} \tag{4}$$

for $10^9 < GrPr < 10^{12}$

where d is the diameter of insulated pipe, k_a is the thermal conductivity of air, Gr is the Grashof number and Pr is the Prandtl number.

Gr is defined by

$$Gr = \frac{\rho^2 g \beta (T_w - T_\infty) d^3}{\mu^2} \tag{5}$$

where T_w is the insulated pipe wall temperature (K), ρ is the density of air, β is the coefficient of volumetric expansion and μ is the dynamic viscosity of air.

Pr is defined by

$$Pr = \frac{\mu c_p}{k_a} \tag{6}$$

where C_p is the specific heat of air at constant pressure.

All the properties of air k_a, ρ, β, μ and C_p in Equations (3) to (6) must be evaluated at the mean film temperature T_f where

$$T_f = \frac{1}{2} \ (T_w + T_\infty) \tag{7}$$

Radiation Heat Transfer Coefficient
The heat transfer coefficient due to radiation h_r is given by

$$h_r = \varepsilon \sigma \ (T_w^2 + T_\infty^2) \ (T_w + T_\infty) \tag{8}$$

where ε is the emissivity of the outer pipe surface and σ is the Stefen-Boltzmann constant with a value of 5.67×10^{-8} W/m K.

Curve Fitting of Properties
Thermal conductivity of insulation   The thermal conductivity of insulation increases as the mean temperature rises.   In the normal range of working temperature, thermal conductivity can be approximated by a quadratic equation

$$k = at^2 + bt + c \tag{9}$$

where a, b and c are constants for an insulation material and t is the mean temperature in °C.

Properties of air   The thermal conductivity of still air within the normal working temperature (0 to 300°C) can be approximated by a linear equation

$$k_a = 7.919 \times 10^{-8} \, t + 2.404 \times 10^{-5} \text{ (kW/mK)} \qquad (10)$$

The product GrPr for air within the same temperature range can be approximated by

$$GrPr = (-1.0456 \times 10^6 \, t + 1.3173 \times 10^8)(T_w - T_\infty)d^3 \qquad (11)$$

Equations (9) to (11) are used directly to obtain the air and insulation properties at the working temperature.

ECONOMIC ANALYSIS

The return on investment of thermal insulation depends mainly on the prevailing fuel cost, installed cost of insulation, the working and ambient temperatures as well as the interest rate. A well known method is the so-called "economic thickness method" which determines the optimum insulation thickness based on the geometrical and operating conditions.   Another approach is to determine the payback period of a number of proposed insulation thicknesses and decide on the most suitable thickness according to the investment policy of the company.

Economic Thickness Method
The method is mainly used for the design of thermal insulation. Figure 2 illustrates the fundamental elements of the method. Curve A represents the cost of insulation (material, installation, interest and maintenance costs) which increases with the thickness.   Curve B represents the total cost of heat loss (fuel cost, maintenance, operation and depreciation costs of heat-producing and distributing systems) which decreases with additional insulation.   The sum of these two curves represents the total cost of a thermal installation.   The thickness corresponding to the minimum total cost is called the economic thickness of insulation.

The equation for economic thickness was first derived by B.L. McMillan in 1926.   For cylindrical pipe, the equation is given as

$$\left(r_2 \ln \frac{r_2}{r_1} + \frac{k}{h}\right) \sqrt{\frac{2r_2 - r_1}{r_2 - k/h}} = \sqrt{\frac{y(T_1 - T_\infty)C_e k}{C_{in}}} \qquad (12)$$

where y is the number of hours per year the pipe is running, C_e is the heat cost per kWh and C_{in} is the annual cost of insulation per m^3.

The value of C_{in} can be derived from

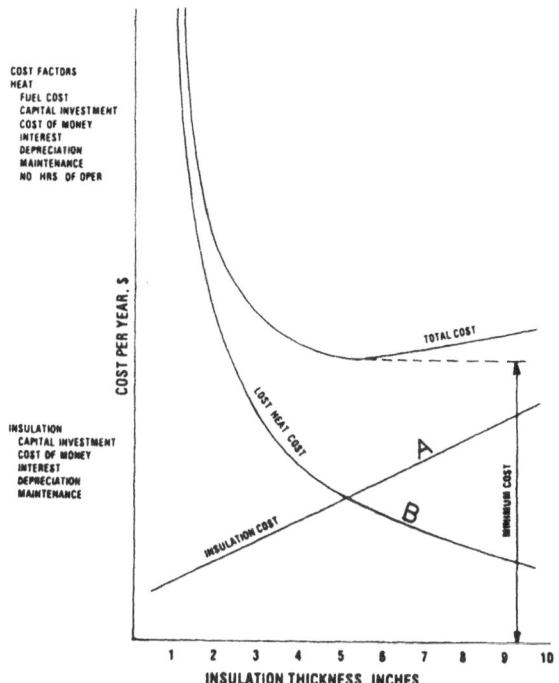

COST FACTORS
HEAT
  FUEL COST
  CAPITAL INVESTMENT
  COST OF MONEY
  INTEREST
  DEPRECIATION
  MAINTENANCE
  NO HRS OF OPER

INSULATION
  CAPITAL INVESTMENT
  COST OF MONEY
  INTEREST
  DEPRECIATION
  MAINTENANCE

COST PER YEAR, $

TOTAL COST

LOST HEAT COST

INSULATION COST

MINIMUM COST

A

B

INSULATION THICKNESS, INCHES

1 2 3 4 5 6 7 8 9 10

Figure 2 Economic thickness of insulation.

$$C_{in} = \frac{Ai\,(1 + i)^n}{(1 + i)^n - 1} \qquad (13)$$

where A is the total cost of insulation per m^3, i is the annual interest rate and n is the depreciation period in years.

Payback Period Method
The method uses a modified discounted cash flow technique based on "future worth value" developed by James (1978). The method calculates the payback period of insulation cost and is capable of determining the return on investment for new installation as well as upgrading of existing installation. A summary of the method is given as follows:

The first year saving S_1 with additional insulation is given by

$$S_1 = (Q_1 - Q_2)C_e y \qquad (14)$$

where Q_1 and Q_2 are the original and upgraded heat losses per metre of pipe length. For new pipe without existing insulation, Q_1 is the heat loss from the bare pipe.

The payback period is then given as

$$P = \frac{\ln \left[\dfrac{C}{S_1} (e^r - e^i) + 1 \right]}{(r - i)} \tag{15a}$$

for $i \neq r$

and

$$P = \frac{C}{S_1} e^r \tag{15b}$$

for $i = r$

where C is the total cost of additional insulation per metre pipe and r is the annual rate of increase of fuel. e is a constant having a value of 2.7183.

COMPUTER PROGRAMS

The two computer programs are both written in CP/M BASIC suitable for most microcomputers. They are written in a simple interactive "question and answer" manner which the users need only to key in the information requested. Iterative technique is used in both programs in order to obtain a coverged solution and the accuracy of the solution depends upon the convergent criteria being specified.

Economic Thickness Program
The economic thickness program (ECONTH) at first requests for and accepts the information and data as shown in Figure 3. It then solve the McMillan equation to obtain r_2 with some initial guessed values of insulation conductivity k and outer surface heat transfer coefficient h. With r_2, it then proceeds to calculate heat loss and obtain the new values of k and h. This process is repeated until h is converged. The latest values of k and h are then used to recalculate r_2. The entire iterative loop is repeated until r_2 is converged and a solution is thus obtained. The economic thickness is then equal to $(r_2 - r_1)$. The flow chart of the program is shown in Figure 4.

Payback Period Program
As for the ECONTH program, the payback period program (PAYBACK) at first requests for and accepts the required data as shown in Figure 5. It then calculates the heat loss Q_1 through the original insulation by setting initial values of k and h and work through the iterative loop until h is converged. The same procedure is repeated with an additional insulation to obtain

the upgraded heat loss Q_2. The program then proceeds to calculate the first year saving and payback period according to Equation (14) and (15). The flow chart of the program is shown in Figure 6.

APPLICATIONS AND RESULTS

To illustrate its application, the ECONTH program was used to calculate the economic thickness of a 50-mm nominal bore steam pipe running at 200°C. The output of the program is shown in Figure 7 which listed all the relevant input data and the heat loss and economic thickness calculated.

The ECONTH program can also be used to study the effect of changing any variable. The influences of pipe size and working temperature on the economic thickness are shown in Figure 8 and the effect of escalating fuel cost is shown in Figure 9.

The PAYBACK program has been used to study an existing insulated steam line of 40-mm nominal bore at 200°C. The output of the program is given in Figure 10 where all relevant input data together with the heat losses of original and upgraded insulations are printed. The payback period of additional insulation as well as that for re-insulation are also listed.

To demonstrate the capability of the PAYBACK program as an useful analytical tool, the program was used to study the variation of heat losses with thickness for two different insulation materials. The results which are plotted in Figure 11 reveal the relationship between heat loss, insulation thickness and cost for calcium silicate and ceramic fibre.

CONCLUSIONS

The two computer programs for the design and analysis of thermal insulation for hot pipes, ECONTH and PAYBACK, have been success-fully developed. The ECONTH program is more suitable for insulation design of new installation as it calculates the economic thickness and cost of insulation. However, the program does not take into account the effect of escalating fuel price. The PAYBACK program is more flexible to cope with varying conditions. The annual increase in fuel cost is being taken care of within the program. It is capable of analysing new design as well as upgrading or retrofitting existing insulation.

REFERENCES

Diamant, R.M.E. (1977) "Insulation Deskbook", Heating and Ventilating Publication, London.

James, B. (1978) "Evaluating Energy Saving Schemes", Chartered Mechanical Engineer, June, 1978.

Turner, W.C. and J.F. Malloy (1980) "Handbook of Thermal Insulation Design Economics for Pipes and Equipment", R.E. Krieger Pub. Co. and McGraw Hill., New York.

```
REFERENCE ?
ABC
UNINSULATED PIPE SURFACE TEMPERATURE (DEG.C.) ?
200
AMBIENT TEMPERATURE (DEG.C.) ?
30
UNINSULATED OUTER PIPE DIAMETER (M.) ?
0.0603
INSULATION MATERIAL ?
CALCIUM SILICATE
COEFFS A,B,C OF K-CURVE FOR INSULATION MATL--AT^2+BT+C(KW/MK) ?
1.15E-10,5.15E-8,5.12E-5
EMISSIVITY OF OUTER SURFACE ?
0.09
NUMBER OF OPERATING HOURS PER YEAR FOR PIPE ?
6000
HEAT COST PER KWH ($) ?
0.08
TOTAL COST OF INSULATION PER CUBIC METRE ($) ?
1600
INTEREST RATE ?
0.1
DEPRECIATION PERIOD OF INSULATION (YEAR) ?
10
CONVERGENT CRITERIA FOR H ?
0.01
ACCURACY REQUIRED FOR INSULATION THICKNESS (M.) ?
0.001
```

Figure 3 Input data for ECONTH program

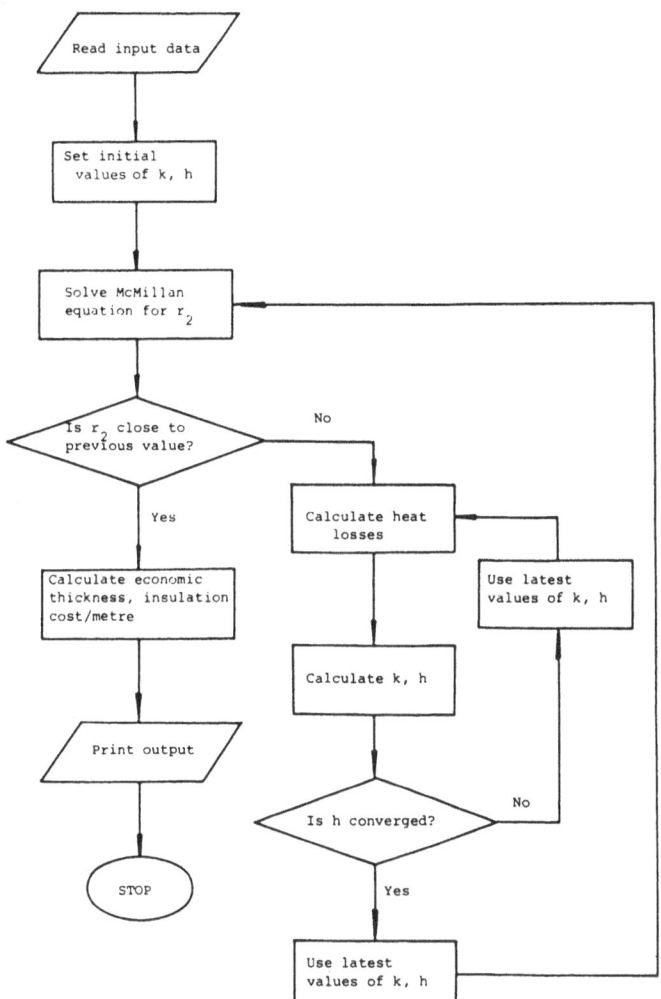

Figure 4 Flow chart of ECONTH program

```
REFERENCE?
F & N CO
PIPE SURFACE TEMPERATURE(DEG C)?
80
AMBIENT TEMPERATURE(DEG C)?
28
RADIUS OF UNINSULATED PIPE(M)?
0.0241
THICKNESS OF ORIGINAL INSULATION(M)?
0
COEFFICIENTS OF THERMAL CONDUCTIVITY CURVE--K=AT^2+BT+C?
1.15E-10,5.15E-8,5.12E-5
EMISSIVITY OF JACKET SURFACE?
0.09
CONVERGENCE CRITERIA FOR H?
0.01
PROPOSED TOTAL INSULATION THICKNESS(M)?
0.025
FRACTION OF TIME IN A YEAR THE PIPE IS RUNNING?
0.6
TOTAL INSULATION COST PER CUBIC METRE($)?
2000
HEAT COST PER KWH($)?
0.08
INTEREST RATE?
0.12
RATE OF ANNUAL INCREASE IN FUEL COST?
0.05
```

Figure 5   Input data for PAYBACK program

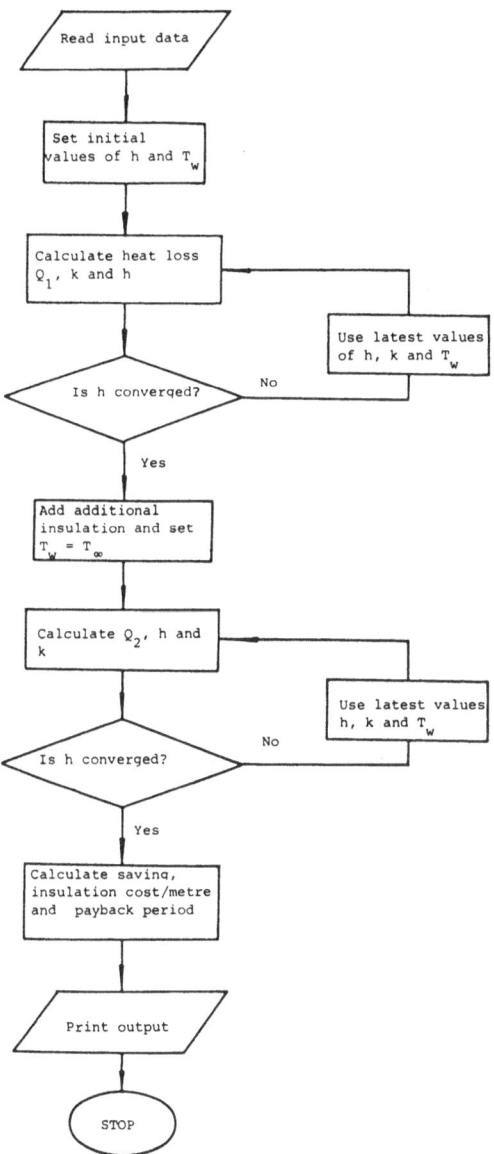

Figure 6 Flow chart of PAYBACK program

```
*** ECONOMIC  THICKNESS OF INSULATION FOR HORIZONTAL PIPE ***

REFERENCE : AAA

RADIUS OF UNINSULATED PIPE =  .03015 M.
UNINSULATED PIPE SURFACE TEMPERATURE =  200 DEG.C.
AMBIENT TEMPERATURE =  30 DEG.C.

INSULATION MATERIAL : CALCIUM SILICATE
NUMBER OF OPERATING HOURS PER YEAR =  6000 HOURS/YEAR
INSTALLED COST OF INSULATION PER CUBIC METRE =  1800 $/M3
EMISSIVITY OF OUTER SURFACE =  .09
HEAT COST PER KWH =  .06 $/KWH
INTEREST RATE =  .1
DEPRECIATION PERIOD OF INSULATION = 10 YEAR

  *** RESULT OF CALCULATION ***

OUTER RADIUS WITH INSULATION =  .0763414 M.
OUTER WALL TEMPERATURE =  52.5365 DEG.C.
THERMAL CONDUCTIVITY OF INSULATION =  5.95375E-05 KW/MK
OUTER SURFACE HEAT TRANSFER COEFFICIENT =  5.47663E-03 KW/M2K

INSTALLED COST OF INSULATION PER METRE PIPE =  27.8163 $/M

HEAT LOSSES =  .0592965 KW/M

ECONOMIC THICKNESS = 46.1914 MM.
```

Figure 7   Output from ECONTH program

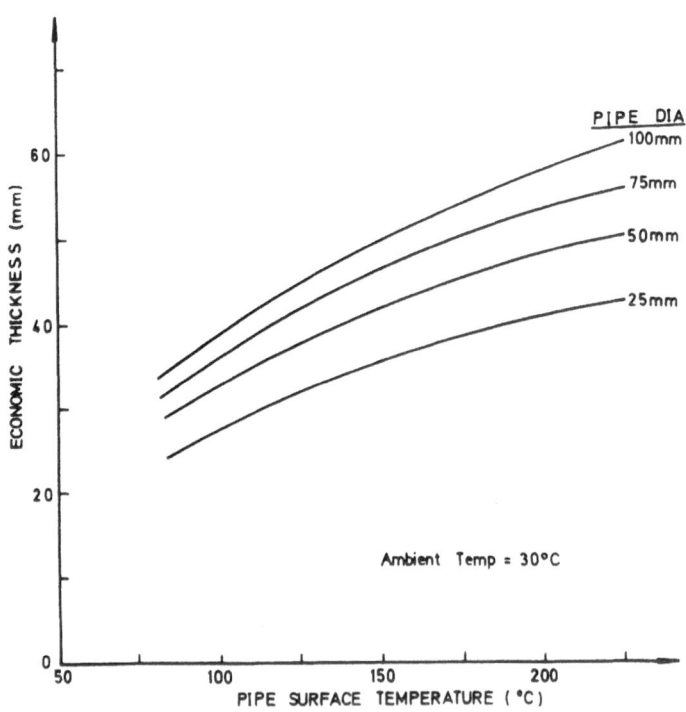

Figure 8   Influence of pipe size and
temperature on economic thickness

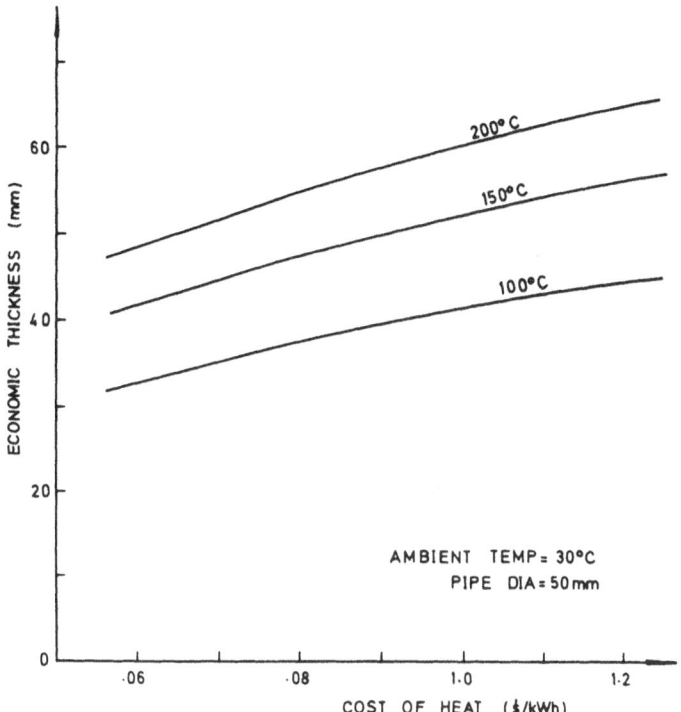

Figure 9 Effect of escalating fuel cost
on economic thickness.

```
*** PIPE INSULATION UPDATE ANALYSIS ***

REFERENCE : T.S. CO.

PIPE SURFACE TEMPERATURE =  200 DEG.C.
AMBIENT TEMPERATURE =  28 DEG.C.
RADIUS OF UNINSULATED PIPE =  .0245 M.
INSULATION COST PER CUBIC METRE =  $ 1600
ENERGY COST PER KWH =  $ .08
INTEREST RATE =  .12
RATE OF ANNUAL INCREASE IN FUEL COST =  .1
FRACTION OF PIPE RUNNING TIME =  .9

ORIGINAL INSULATION :
THICKNESS =  .025 M.
WALL TEMPERATURE =  63.1764945 DEG.C.
CONDUCTIVITY =  5.99757058E-05 KW/MC
OUTER SURFACE HEAT TRANSFER COEFFICIENT =  6.70097477E-03 KW/M2C
HEAT LOSS =  .0733122628 KW/M

PROPOSED INSULATION :
THICKNESS =  .05 M.
WALL TEMPERATURE =  48.307126 DEG.C.
CONDUCTIVITY =  5.93588588E-05 KW/MC
OUTER SURFACE HEAT TRANSFER COEFFICIENT =  5.35170284E-03 KW/M2C
HEAT LOSS =  .0508718591 KW/M

FIRST YEAR SAVING PER METRE PIPE LENGTH = $ 14.1536114

ADDITIONAL INSULATION COST PER METRE PIPE LENGTH = $ 15.582336
PAYBACK PERIOD OF ADDITIONAL INSULATION =  1.24433638 YEARS

RE-INSULATION COST PER METRE PIPE LENGTH = $ 24.8814721
PAYBACK PERIOD OF RE-INSULATION =  2.0019526 YEARS
```

Figure 10 Typical output from PAYBACK program

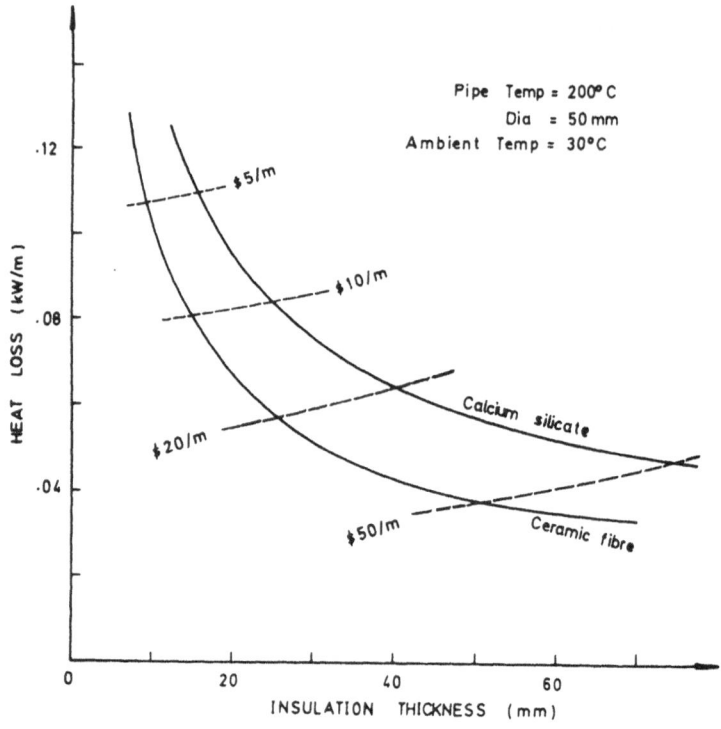

Figure 11   Relationship of heat loss,
            insulation thickness and cost

ON THE DEVELOPMENT OF A FINITE ELEMENT PROGRAM FOR MICROCOMPUTERS

Steven A. Brown
Department of Engineering Science and Mechanics
Virginia Polytechnic Institute and State University
Blacksburg, Virginia 24061, U.S.A.

and

Manohar P. Kamat
Professor of Engineering and Associate Director
Center of Excellence for Computer Applications
University of Tennessee at Chattanooga
Chattanooga, TN 37402

ABSTRACT

In this paper the authors highlight the hardware and software considerations which are crucial in the development of a reliable and meaningful finite element program operational on a microcomputer. The three guiding considerations in this development are user-friendliness, efficient memory utilization and speed of computation and perhaps in that order. These considerations formed the basis of a microcomputer based finite element program developed by the authors and known by the acronym SNAP/FE.

INTRODUCTION

Most modern day engineering problems do not lend themselves to exact solutions because of their complexity. However, approximate solutions of a large class of such engineering problems can be obtained most effectively utilizing the well-known finite element method (FEM). The finite element method is a domain discretization technique that uses piecewise approximations of continuous functions in going from the "part to the whole" utilizing the principle of virtual work or the methods of weighted residuals. The growth of this numerical technique has been accompanied by the development of some very versatile mainframe based computer codes such as ADINA, ANSYS, EAL, MARC, NASTRAN etc. Likewise, this growth has paralleled the advances made in computers. In particular, with a rapidly growing personal computer technology an acute need exists for adapting mainframe based finite element programs to microcomputers and making such programs user-friendly, efficient and capable of solving reasonably large scale problems.

As with the mainframe computers, several issues and software as well as hardware considerations dictate the nature and form of the final end product - the microcomputer based finite element (FE) program. The primary software consideration is the limit on the capabilities of the program in the form of the types of analyses the program is capable of performing, i.e., static, transient, vibration, buckling, thermal, etc. This in turn decides the finite element library, the type of material sets, the types of constraints (single point and/or multi-point) and the type of the solution processors (linear equations and eigenvalue solvers) that are required. Hardware considerations concern the type of the machine to be used for the development of the computer program which can, with perhaps minor modifications, run on as many other machines as possible. The IBM PC may be considered to be an ideal machine for such a development since it has a 16 bit microprocessor, a reasonably good memory capacity, good color graphics capability and most of all the feature that many other manufacturers make microcomputers that are IBM PC compatible, e.g., COMPAQ, TANDY 2000 etc. The use of Intel's 8087 math co-processor is imperative for any meaningful and decent microcomputer based FE program. The availability of a hard disk in addition to 360KB disk drives permits the use of out-of-core equation and eigenvalue solvers. Finally, miscellaneous I/O hardware like the mouse, light pen and digitizer are nice for input assistance in interactive sections of the program. For instance, the digitizer could be used to enter nodal data directly from pictures or drawings.

SNAP/FE - DETAILS AND DEVELOPMENT PHILOSOPHY

The single feature which distinguishes SNAP/FE from most other microcomputer based FE programs like those described in detail in reference [1] for instance, is its user-friendliness. The authors felt that it was more important to sacrifice some of its capabilities at the expense of its user-friendliness. Accordingly it was decided that, initially the program SNAP/FE will perform only three linear analyses namely the static, transient and vibration analysis with an element library consisting of a 3-D TRUSS, 3-D FRAME, 3-D MEMBRANE SHELL, 3-D THICKSHELL isoparametric elements and rotational and translational springs. It is expected however, that notwithstanding its good user-friendly features, SNAP/FE will eventually blossom into a mini-SAP type program of the kinds alluded to in reference [1].

SNAP/FE offers three processors, which are linked together by a DOS batch file. The BASIC preprocessor allows user friendly data input, and writes the data to program data files. The FORTRAN based solution processor reads the data from the program data files, performs the solution, and writes the results to program results files. The BASIC postprocessor allows for the output of the data through various print and plot options. For reasons already cited, it was decided to develop SNAP/FE using the popular IBM PC with 256K RAM. The program can run on the basic version of PC with DOS 2.0 or higher and requires no special device drivers or other hardware. Intel's 8087 math coprocessor is optional and highly recommended for the solution of large scale problems requiring a great deal of 'number crunching'. Again one 360 KB disk drive is acceptable although two such disk drives would be preferable. A hard disk may be necessary for large problems which generate large amounts of data. A graphics printer would allow hardcopy of graphics screens.

The backbone of the SNAP/FE's development was the use of split BASIC-FORTRAN operation. The superior data manipulation ability and the graphics capability of BASIC were ideal for input and output whereas the faster calculation speed made FORTRAN ideal for the core of the program.

Basic Pre and Post Processors

IBM BASIC is limited to a maximum of 64K bytes for program and variable space. Hence SNAP/FE stores only mesh input data in core while storing load information directly on disk. The entire data could have been stored directly on disk, but this would have made element and node generation very difficult and it would have been difficult to implement the built-in graphics.

User-friendliness means different things to different people, involving various degrees of computer and user participations. Regarding SNAP/FE, it was decided that the computer should let the user know exactly what input is required for the program. Accordingly, SNAP/FE provides a number of input screens with menus. Menu driven programs are user-friendly. Moving a cursor to the line of the desired activity is easy and quick. It is easier to point to an activity rather than type a command to perform the same function. The idea is similar to the use of a mouse on Apple's MACINTOSH. The menus used in SNAP are however not as sophisticated as those used in MACINTOSH, but accomplish the same task and are almost as easy to use.

Another user-friendly feature implemented in SNAP is the line type input which has a variable number of input entries. The program recognizes, by the number of values on a line, what particular input is being entered. This input type is ideal for use with sections containing generation capabilities.

Finally, a user-friendly program should 'expose or identify' as many input errors by the user as possible. For instance, improper cursor position, inappropriate input or incorrect number of values in a line input are a few typical errors which SNAP/FE identifies. When an error message is printed, the computer beeps its speaker and thereby brings user's attention to the error from another sense besides vision. This dual sense error indicator gives another level of user-friendliness, almost like having the computer identify errors through speech.

Many of the user-friendly features alluded to above could not have been used in any other programming language except BASIC. BASIC provides excellent character string handling capabilities along with good input/output features. IBM BASIC also provides some very nice graphics commands which facilitate the implementation of certain menu capabilities and some simple plot routines.

Solution Processor

FORTRAN was chosen for this portion of the code firstly because of familiarity with the language and secondly because it provides a reasonably fast code for the 'number crunching' operations requried in solving large scale finite element problems. Compared to minicomputers and mainframes, for most microcomputers, with the exception of perhaps the extended version of an IBM PC/AT, the Random Access Memory (RAM) is at a premium. Hence efficient memory utilization is imperative and solution schemes which optimize in-core stiffness matrix storage requirements are highly desirable. One such scheme is the one that exploits the sparsity and symmetry of the assembled stiffness matrix and stores the relatively few nonzero entries within what is known as the matrix skyline in a small vector in a compacted form. Of course, the skyline structure is dictated by the connectivity matrix of the finite element model. The symmetric stiffness matrix is factored into an LDL^T factorization with the L and D factors stored in exactly the same skyline structure of the upper triangular portion of the original matrix. This scheme [2] greatly reduces the memory requirements and provides a very efficient in-core solver that permits the solution of reasonably large scale problems. Currently SNAP/FE has such an in-core solver which permits the solution of certain types of problems with as large as about 2000 degrees of freedom or more.

For being able to solve even large problems SNAP/FE is currently in the process of implementing an element-by-element preconditioned conjugate gradient technique (EBE-PCG) proposed by Hughes et al. [3]. Like all conjugate gradient techniques, the EBE-PCG technique does not generate the entire assembled stiffness matrix. On the other hand it requires only the upper triangular portions of each element for the preconditioner and accordingly its storage requirements are much smaller than that of an in-core solver utilizing the total stiffness matrix in a compacted skyline form. This is especially true for those 3-D problems wherein the stiffness matrix has a lot of 'fill-in' within the skyline.

Another option for reducing storage requirements, that SNAP/FE is also currently considering is the use of overlays. Certain versions of FORTRAN for the PC support the use of overlays and can reduce the memory requirements of the code substantially if properly implemented.

The use of out-of-core solvers of the type proposed by Mondkar and Powell [5] was considered but not used because disk access in PC's is relatively slow even with a hard disk. The total solution time will thus involve a great deal of time for read/write operations during the factorization, forward and back substitution phases of the solution process. In view of these considerations and in view of the numerical experiments of reference [3] which indicate that EBE-PCG is a much more cost-effective alternative, the idea of implementing Mondkar and Powell's out-of-core solver was abandoned. Even with an EBE-PCG a hard disk is still desirable if not mandatory especially because of the limited amount of data that can be stored at one time on a floppy disk.

SNAP/FE currently employs a nonsingular lumped mass matrix for each of the four main elements as regards both the transient and vibration analyses. The linear transient analysis uses variants of the generalized Newmark-Beta method [7]. The transient analysis provides a restart capability that permits the analysis to be carried out for virtually any number of time steps. For vibration analysis SNAP/FE uses the subspace iteration method of reference [2] that exploits the skyline structure of both the stiffness and mass matrices. Future plans for SNAP/FE as regards the element library include the implementation of a shear-deformable plate bending and shell element. Likewise its analysis capabilities will be extended to permit thermal and buckling analysis.

DEMONSTRATION OF A TYPICAL CAPABILITY OF SNAP/FE

The basic architecture of the program SNAP/FE consists of a batch file that accesses the preprocessor followed by the solution and postprocessor sections. The preprocessor section which is written in BASIC contains two menus: one for input of data and the other for verification of these input data. The input data are written on several temporary disk files and then the program control is transferred to the FORTRAN solution section. This section, which can also operate as a stand alone section, reads the input data from the temporary files, calculates the element matrices, assembles them, performes the solution, and write the results (displacements, velocities, accelerations, stresses, nodal forces, etc.) to different temporary files. The program then begins the postprocessing section where various plots and printouts of the data can be obtained. Because of limitations of space, details on the menu structures, the architecture of the two programs and other details cannot all be elaborated upon sufficiently. Reference [5] however provides all such details in sufficient depth.

To enable the solution of large scale problems in a reasonable amount of time the IBM PC was equipped with an 8087 math coprocessor. The version of Microsoft FORTRAN currently implemented allows a labelled common block with a maximum length of 64K which translates into a single precision array of 16,000 variables. This in turn implies that with an in-core LDL^T solver that exploits the skyline structure of the matrix the total number of elements within the skyline may not exceed 16,000. Without any formal bandwidth optimization schemes and without EBE-PCG or out-of-core solvers [6] we are thus limited to problems with as many as 500 degrees of freedom assuming a mean semi-bandwidth of about 35. It is clear that using the EBE-PCG and out-of-core solvers we have in principle the capability of solving problems of unlimited size. Also, newer versions of FORTRAN allow for common blocks of unlimited size, which means the program would only be limited by the memory of the machine. However, all that is of interest to us in this paper is a demonstration of the menu driven features and the efficiency of the in-core solution process for the transient analysis of an uniformly loaded cantilever beam modeled using five, eight-noded isoparametric membrane elements [8]. Figure 1 is the actual graphic display of the beam discretization and Figure 2 is a graphic display of the beam response. Figures 3 through 18 show the actual screens during the input phase of the SNAP/FE for the transient analysis of the beam highlighting the menu structure and the user-friendly prompts.

CONCLUSIONS

The implementation philosophy of SNAP/FE outlined in the prceeding pages is by no means the only way to approach microcomputer-based finite element program. However, SNAP/FE is a successful, totally integrated program which provides a high degree of user-friendliness and ease of operation. Since many of today's engineers have little or no experience in operating a microcomputer, a truly user-friendly program is a tremendous advantage. SNAP/FE also gives the experienced user the option to develop custom tailored pre and post processing software using its open file structure.

By way of upgrading SNAP/FE's user-friendly features a number of plans exists. There are a few commercially available graphics packages which could be used in conjuction with file conversion utility programs so as the permit contour plots and hidden line plotting of data. Also the pre and post processors could be written in compiled BASIC thereby allowing faster execution. Finally, there is a plan to adapt SNAP/FE to operate on a MACINTOSH.

REFERENCES

[1] Falk, H., (1985), Finite Element Analysis Packages for Personal Computers, Mechanical Engineering, 54-71.

[2] Bathe, K. J. and Wilson, E. L., (1976), Numerical Methods in Finite Elements, Prentice Hall, Englewood Cliffs, N.J.

[3] Hughes, T. J. R., Winget, J., Levit, I., and Tezduyar, T. E., (1983), New Alternating Direction Procedures in Finite Element Analysis Based on EBE Approximate Factorizations. Computer Methods for Nonlinear Solid and Structural Mechanics, Eds. S. Atluri and N. PErrone, AMD-Vol. 54, 75-109.

[4] Brown, S. A., and Kamat, M. P., (1984), Structural Analysis of Large Scale Problems Using a Microcomputer. Proceedings of the Second National Conference on Microcomputers in Civil Engineering, Orlando, Florida, 302-306.

[5] Brown, S. A., and Kamat, M. P., (1984), SNAP: Structural Numerical Analysis Programs for Microcomputers, VPI&SU report, VPI-E-84-31.

[6] Mondkar, D. P., and Powell, G. H., (1984), Large Capacity Equation Solver for Structural Analysis. Computers and Structures, 4, 699-728.

[7] Zienkiewicz, O. C., (1977), The Finite Element Method, Third Edition, McGraw-Hill Book Company, London.

[8] Bathe, K. J., Ramm, E., and Wilson, E. L., (1975), Finite Element Formulations for Large Deformation Dynamic Analysis, International J. Num. Meths. Engrg., 9, 353-386.

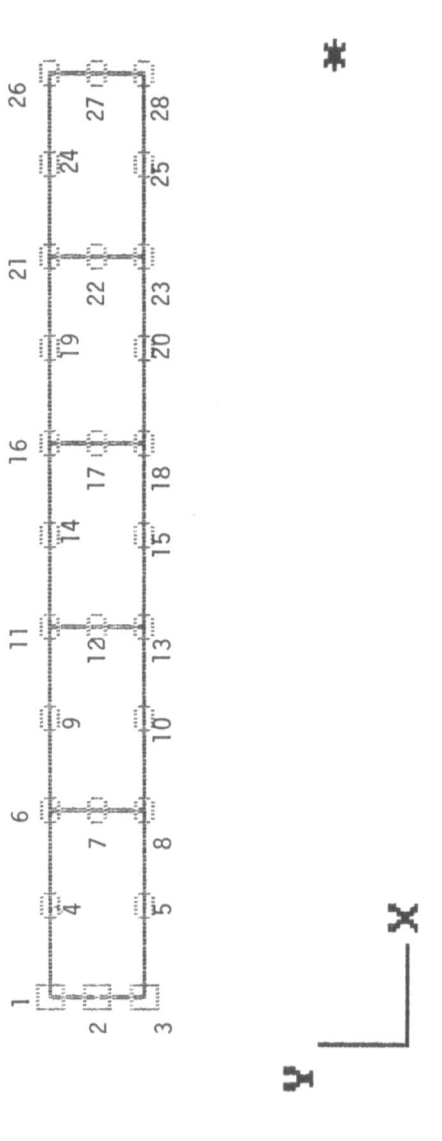

To move, 8=up 4=right 2=down 6=left
Enter Z to zoom C to compress E to exit

sample beam problem

FIGURE 1.

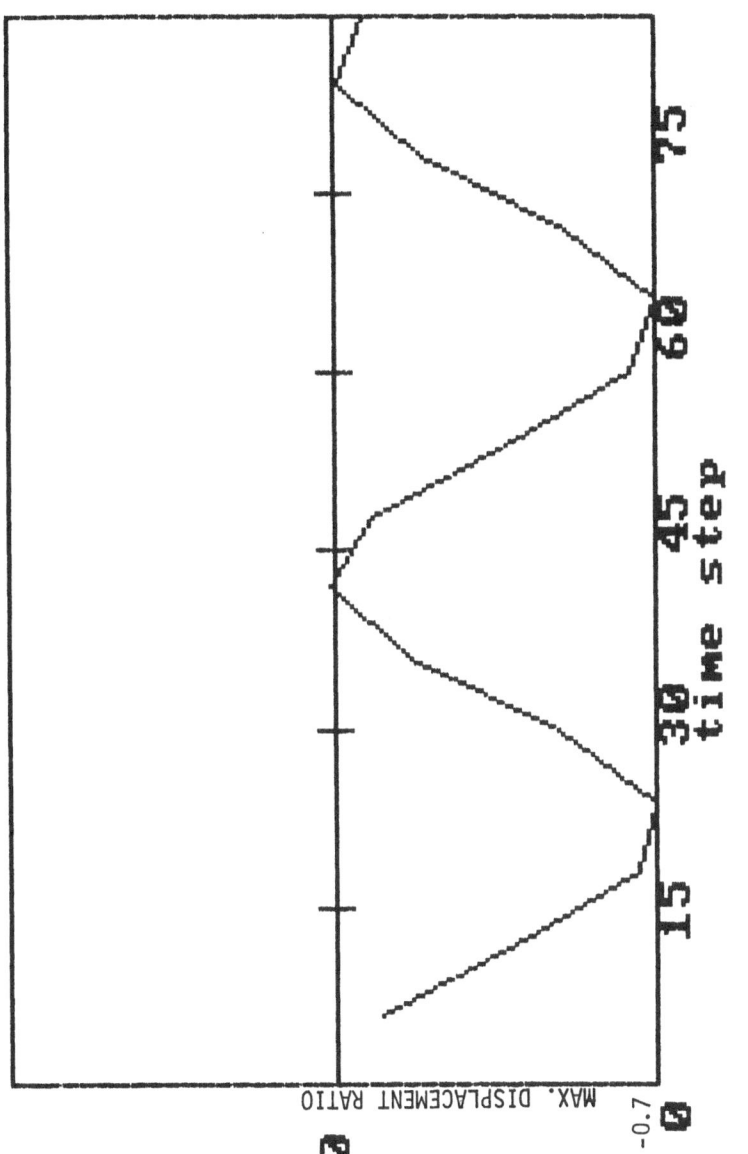

FIGURE 2.   transient response of beam

Figure 3. Screen 1

```
┌─────────────────────────────────────────────────────────┐
│ DATA INPUT MENU                                          │
│                                                          │
│ Recall existing data from files                          │
│ Enter new data                                           │
│ Exit program                                             │
│                                                          │
│ Move cursor to desired activity using the arrow keys.    │
└─────────────────────────────────────────────────────────┘
```

Figure 4. Screen 2

```
┌─────────────────────────────────────────────────────────┐
│ GENERAL INPUT                                            │
│                                                          │
│ Enter title for this problem    sample beam problem      │
│                                                          │
│ Enter number of nodes    28                              │
│                                                          │
│ Enter number of material sets    1                       │
│                                                          │
│ Enter number of multipoint constraints    0             │
│                                                          │
│ Enter the spatial dimensions of the problem              │
│                                                          │
│ Enter a 2 for 2-d or a 3 for 3-d    2                    │
└─────────────────────────────────────────────────────────┘
```

Figure 5. Screen 3

```
┌─────────────────────────────────────────────────────────┐
│ ELEMENT TYPE INPUT                                       │
│                                                          │
│ Move cursor to the desired element type, then enter the total│
│ number of elements of that type.  When finished, move the│
│ cursor to "Continue".                                    │
│                                                          │
│ Spring elements                                          │
│ Truss elements                                           │
│ Beam elements                                            │
│ 8 node plate elements-2d    # 2-d 8 node plate elements=5│
│ 8 node plate elements-3d                                 │
│ 16 node shell elements-3d                                │
│ Continue                                                 │
└─────────────────────────────────────────────────────────┘
```

Figure 6. Screen 4

```
NODAL COORDINATE INPUT

Separate input entries with commas

1  0,1,0,10,1,0,26,5
2  0,.5,0,10,.5,0,27,5
3  0,0,0,10,0,0,28,5
4  1,1,0,9,1,0,24,5
5  1,0,0,9,0,0,25,5

Node # - X1,Y1,Z1(,X2,Y2,Z2, Node 2, Increment)
```

Figure 7. Screen 5

```
NODAL CONSTRAINT INPUT

Enter node numbers constrained against motion, enter a null
line to continue.

X direction displacement constraint
1,3,1

Y direction displacement constraint
1,3,1

Node 1 (, Node 2, Increment)
```

Figure 8. Screen 6

```
2-D 8 NODE ELEMENT CONNECTIVITY INPUT

Number element nodes counterclockwise, corner nodes, then side
nodes.

Separate entries by commas

1  1,3,8,6,2,5,7,4,5,5

Element # - Node1, Node2,..., Node8, (# Elements, Increment)
```

Figure 9. Screen 7

```
MATERIAL SET #1 INPUT

Material Properties  ( )=Transverse Isotropy
      E (1)  Young's modulus          Enter E (1)  1.2e4
      G (12) Shear modulus            Enter G (12)   5e4
      (E(2))
      (Nu 12) Poisson's ratio
      (Nu 23)
      Rho - Mass/Volume              1.e-6
      Plane strain(=0)/stress(=1)  Enter 0 or 1   1
      User supplied material matrix
      Continue input - Material properties correct

Geometric Properties - ( )=Frame element, [ ]=8 Node element
      Area [Thickness]               Enter value   1
      (Iy)
      (Iz) Moments of inertia
      (J)
      (Py) Shear correction factors
      (Pz) Default=1
      Contiue - Material set values correct
```

Figure 10. Screen 8

```
ANALYSIS MENU

Move cursor to desired activity

Dynamic analysis only - No force data
Static or Dynamic analysis - Static force data
Transient or Dynamic analysis - Transient force data
Any analysis - Static and Transient force data

This menu determines which force data files to create.  For
example, choosing to create only a static force file allows
the program to perform either a staic analysis or a dynamic
analysis with existing data.
```

Figure 11. Screen 9

```
TRANSIENT ANALYSIS - GENERAL DATA

Enter number of time steps    90

Enter time interval of each step    1.35e-4

Enter number of intervals for output    5

Enter number of nodes at which are output (max=15)    3
```

Figure 12. Screen 10

```
INITIAL VALUE AND FORCING FUNCTION DATA

Move cursor to appropriate line and enter corresponding value

Number initial displacements
Number initial velocities
Number initial accelerations
Number periodic library forces
Number nonperiodic library forces
Number user supplied forces         Enter # forces    20
Continue input
```

Figure 13. Screen 11

```
Enter the first time interval for which results are to be
printed                                                    1

 :   :  :   :     :   :   :    :    :  :

Enter the fifth time interval for which results are to be
printed                                                   90
```

Figure 14. Screen 12

```
Enter the first node number for which results are to be
printed for each interval                                  5

 :   :  :   :   :   :   :    :    :  :

Enter the third node number for which results are to be
printed for each interval                                 18
```

Figure 15. Screen 13

```
USER SUPPLIED FORCE NUMBER 1

Enter node where force is applied    4

Enter direction (1-6) in which force acts    2

Enter time, force
0,25                          Another (y or n) ?    y
1,25

  .       .       .
  .       .       .
  .       .       .

User supplied force number 15

Enter direction (1-6) in which force acts    2

Enter time, force
0,50                          Another (y or n) ?
1,50                          Another (y or n) ?    n
```

Figure 16. Screen 14

```
ELEMENT MESH DATA OUTPUT MENU

Move the cursor to the desired activity using the arrow keys,
then press <RETURN>

Print out input data - to screen or printer
Plot finite element mesh to screen
Continue program execution
```

Figure 17. Screen 15

```
EXECUTION MENU

Move cursor to desired activity

Continue execution - perform dynamic analysis
Continue execution - perform static analysis
Continue execution - perform transient analysis
Halt execution - save data
Halt execution - no data saved
```

Figure 18. Screen 16

```
ELEMENT MESH RESULTS MENU

Move the cursor to the desired activity using the arrow keys,
then press <RETURN>

Print out input data - to screen or printer
Print out results only - to screen or printer
Print out specific node and element information
Plot proportional finite element mesh to screen
Exit program
```

DEVELOPMENT OF HARDWARE AND SOFTWARE FOR MACHINE CONTROL APPLICATIONS

G D Alford   BSc(Hons) DLC(Hons) CEng MRAcS

Teesside Polytechnic, Cleveland, U.K.

SUMMARY

The author has produced a number of programs for controlling a variety of machines, including machine-tools and industrial robots. Several different microcomputers have been employed in writing this software, usually in machine code in order to:-

        make the program transportable
        allow faster speed of operation
    and reduce the cost of the final system.

However in using machine code, particularly in the teaching situation, systems have to be developed to:-

        make the process more user-friendly
        give good testing and editing facilities
        reduce the development time
    and prepare appropriate firmware.

This paper describes the methods employed in developing such systems and summarizes:-

        actual machine control applications
        software aids employed
        ancillary hardware produced
    and necessary interfaces involved.

INTRODUCTION

For several years now the Department of Mechanical Engineeering at Teesside Polytechnic has been involved in the use of microprocessor systems for general computing, control of machines and data acquisition. Such systems form the basis of many of our new courses from Technician to Masters level, students are introduced to these systems through formal teaching, laboratory involvement and project work.

In several applications it is necessary to employ machine code
to get a suitable speed of response from the program. Thus the
use of assembler and hexadecimal code is included in the
teaching process. Simple programs are introduced for employing
input and ouput ports, to read sensors and transducers,
simulate systems with LEDs, and control cylinders and motors.
In project work students are given opportunity to develop these
ideas in real systems such as:-

> conversion of machine tools to computer control
> design and development of industrial robots
> data acquisition and processing.

Over this period many developments have taken place in software
and hardware. The history of these changes is now considered,
relating particularly to machine-control applications.

Acorn System I

Figure 1 shows the layout of this simple microcomputer which
has been used for machine control applications. As originally
purchased it consists of two Eurocards, one containing the 6502
microprocessor, ROM, RAM, I/O unit and EPROM socket and the
other hexadecimal keyboard, simple 7-segment display and
cassette interface. The unit has been built into a case to
include digital input switches, LED displays and standard
sockets for three 8-bit ports, together withspace for
additional RAM or EPROM boards.

The unit forms a very simple system for teaching machine code
at a basic level; the operating program provides
straightforward debugging facilities for display of the
registers at any stage. For longer programs, typically
required for machine control, the use of this system becomes
rather laborious and time-consuming. A disassembler was
developed to aid the debugging process. Programs fed into this
process can be sorted out automatically into single, double and
triple byte instructions and printed out in the form:-

address	instruction	
0020	8D 20 09	
0023	A9 00	
0025	8D 21 09	
0028	AD 21 09	
BR28	30 FB	branch to location 28
ER??	A3	error encountered here -
002E	40	A3 not in instruction
002F	8D 20 09	set.
0032	60	

Though not infallable, this utility can be used to pick out obvious errors and check branches which are a common source of problems. In addition it can be employed simply to list the program or data bytes on a printer in a choice of formats, such as:-

16 bytes per line: ADRS B0B1B2B3 B4B5B6B7 B8B9BABB BCBDBEBF

5 bytes of data per line: ADRS B0 B1 B2 B3 B4

The computer is extremely portable and can be powered from mains or battery sources. This makes it ideally suited to

on-board mobile robot applications, such as McK-9, developed as a student project. This robot is equipped with two model servo motors, one controlling speed and the other direction. In the learn mode the microcomputer reads the control over a given route. These signals are then reproduced by the computer in the repeat mode. Figure 2 shows the layout of this system.

Other robots such as Robertson (1) and Hydair are controlled by this unit. Robertson uses stepping motors for its main joints with model servos for the hand and gripper. It has no positional feedback but, is programmed for joint interpolation. Hydair is powered by air cylinders each of which has a crude speed contorl system in the form of a closed-circuit hydraulic cylinder as shown in Figure 3. Three 8-bit parts are needed to control this machine, one for speed control, one for directional power and the other for reading 8-bit absolute encoders.

Figure 4 shows the interface required for this unit. The logic components are used to economise on ports, using the speed and direction inputs efficiently. The decoder is used to select which encoder is to be read: only one is powered at any time. Solid state relays are used to switch the 240 V AC supplies for 15 solenoid valves controlling air and hydraulic cylinders.

The computer is equipped with cassette tape storage facilities, but for larger programs, such as required for these robots, some means of developing firmware is desirable. A particularly useful aid in this is an EPROM emulator or Instant ROM device. This is a CMOS memory with battery back-up (literally, as the battery is built on to the back of the chip). When connected externally to the read/write line of the computer and plugged into an EPROM socket the unit behaves as a RAM chip, but removing this line, as a ROM. The battery retains the program over a 3-4 month period. The program is easily debugged in this device and when correct is simply transferred to an EPROM in an appropriate machine.

The development of a long complex program on this computer does demand considerable concentration and dedication. The unit is obviously limited by its single memory display. Other simple computers such as Softy overcome this with a multiple memory display on a monitor screen. However, in practical terms, this computer has not been reliable in cassette storage and its instruction set is rather slow for machine control.

Acorn Atom

To develop longer programs, a more complex computer was required possessing an assembler to convert from mnemonic to machine code automatically. The Atom was one of the first low-cost computers to appear with this facility. The Department purchased a number of these machines primarily

for machine-code teaching but the BASIC and graphics was considered a useful bonus. The early machines had a poor keyboard and non-standard BASIC but generally fulfilled their purpose Special input/output units were produced (Figure 5.1) having similar D-socket connections to the System I. Thus demonstration hardware was interchangeable between the two systems.

Various additional interfaces have been developed to plug directly into the input/output units. Figure 6.1 shows a set of reed relays for switching up to 240 V AC/DC, typically used with solenoid valves. Figure 6.2 shows a decoder that can be used to drive up to 4 stepping motors from one 8-bit port: The motors can be driven in single or double-coil mode with this device. Two separate sets of 8 power amplifiers (Figure 6.3) can be plugged into this decoder; alternatively, one or both sets could be plugged directly into the input/output unit. Figure 6.4 shows a large-scale analogue-to-digital or digital-to-analogue converter. This can be used alone for teaching the principles involved or in conjunction with a separate proportional control unit (Figure 6.5) to demonstrate motor speed control and servo applications.

Addition Atoms were purchased later with Smart Arms 6E robots. These robots have six model servo motors whose position is determined by pulse-width control, see Figure 7. Theoretically it is possible to control up to 10 such motors directly with a single microcomputer by servicing the pulse output to one port bit every 20 ms. In practice a separate interface, Figure 8, is used to look after the regular supply of pulses to the six motors. The computer simply updates the positions in RAM from time to time and is otherwise used to perform other functions to read learning instructions or, in the repeat mode, set input and output controls from the robot to a model machining cell, for instance.

The original Smart Arms had an operating system loaded from cassette tape taking about 5 minutes. Again the value of firmware was noted leading to this and subsequent models being equipped with the program on EPROM.

BBC Microcomputer

Developed from the Atom, the BBC units offer many advances over its elder brother including:-

> Standard and greatly extended BASIC
> wide variety of input/output channels
> better graphics facilities with colour
> long variable names
> analogue inputs
> better assembler and machine code system
> sideways ROMs

All of these facilities make it more user-friendly and versatile in machine-control applications. Machine code can either be taught directly with screen memory facilities using the EXMON sideways ROM or through the assembler system.

The Department has again made modest investments in purchasing several of these computers. To maintain compatibility of input/output facilities, special units have been produced with our standard D-sockets. The first of these, see Figure 5.2 is similar to that for the Atom, running straight from the 6522 VIA. However, in this case only one 8-bit user port is available for input and output. The other port is normally connected to the printer and buffered before the socket outlet: this port must always be used as an output. Another input/output device, see Figure 5.3 has now been developed to run off the 1 MHz bus. This has its own 8154 I/O chip on board and provides two additional 8-bit I/O ports. Other interfaces have also been produced to utilize the analogue channels, see reference 2.

Simple Controllers

Because of its versatility it is both impractical and wasteful to permanently tie up one BBC machine to a small robot. Ideally each robot or machine should be driven by its own simple, but dedicated, microprocessor controller, similar to the original Acorn System I described above, with the operating program in machine code held in firmware; see reference 3 for instance. The problem is now to produce a system to allow development of the machine-code program on a computer like the BBC machine and then transfer this into firmware for use on the simple controller.

Another Cambridge firm, Control Universal, produce special systems to perform this function. A number of their CUBIT low cost controllers have been purchased by our Department. These are on a single Eurocard containing a 6502 microprocessor, 4K byte RAM, 4K byte ROM, and 20 I/O channels. To produce an EPROM to control this card a special device working from the 1MHz bus of the BBC has been invented. Being Mechanical Engineers, this has been christened a "RAM-ROM pump" and allows transfer of memory to and from an EPROM Emulator a page at a time. Page selection is, of course, performed mechanically using a rotary switch (see Figure 9).

At this point, it should be noted that the CUBIT card does not contain any operating system and thus the machine controlling program must provide all the essential features for reset, break, interrupts and start-up. To assist in these functions and also provide additional simpler computers for teaching, a special keyboard and display unit (Figure 10) has been developed. This provides the following useful functions on a menu basis, the selection being made on the edge switch. (See

Table 1.

CUBIT KEYBOARD/DISPLAY

Menu no.	Byte display left · right	Hex.key action	Go key action
0	AC PS	not active	
1	XR YR	for	
2	PP PL	registers	
3	SP SL		go from APAL
4	DY DZ	both bytes altered	
5	DA DX	from left to right	
6	DP DL	as four keys are	
7	DP DL	pressed, delay	insert temporary
8	NP NL	controls repeat	break at DPDL
9	IP IL		
A	AP AL		
B	BP BL		go from BPBL
C	CP CL		go from CRCL
D	DL DP	DA only changed	decrement DPDL
E	DL DP	by two entries	increment DPDL
F	DL DP		move memory length DA from IPIL to NPNL

A simple operating system to perform these functions has been produced which has a flow chart as shown in Figure 11.

Future Developments
Another method, now available, of transferring programs and data from the BBC computer is to use a device such as the BEEBEX unit which also works from the 1 MHz bus and provides a possible expansion of up to 2 Mbytes on the memory map. Sockets are provided to take standard Eurocards such as the CUBIT controller, ADC and DAC units, CMOS memory cards, digital

input/output units, etc.. These devices are currently being developed to form a separate rack unit to be used as a general purpose facility (Figure 12).

References
1.   Alford, G. D.   Machine-Code Program for Controlling and Industrial Robot.   Engineering Software for Microcomputers P.711 Ed. BA Schrefler, R. W. Lewis, S. A. Odorizzi, Pineridge Press 1984.

2.   Alford, G. D.   CAD/CAM Programs for a Microcomputer.   4th Polytechnics Symposium on Manufacturing Engineering P.350 City of Birmingham Polytechnic 1984.

3.   Alford, G. D. and Chadwick J. G.   A High-Level Robot Language.   Conference on Computer-Aided Engineering P.79 IEE Conference Pub. No. 243   1984.

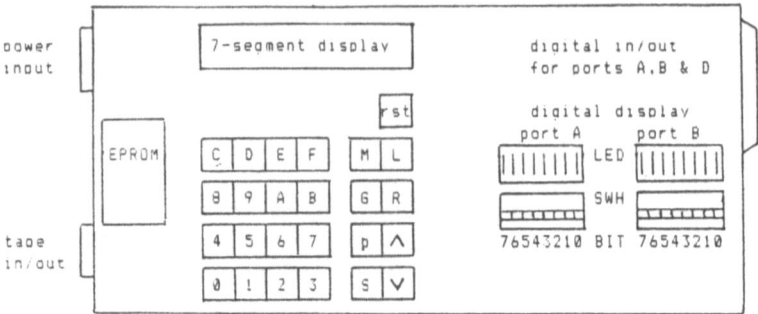

Fig.1 Acorn System 1 Computer

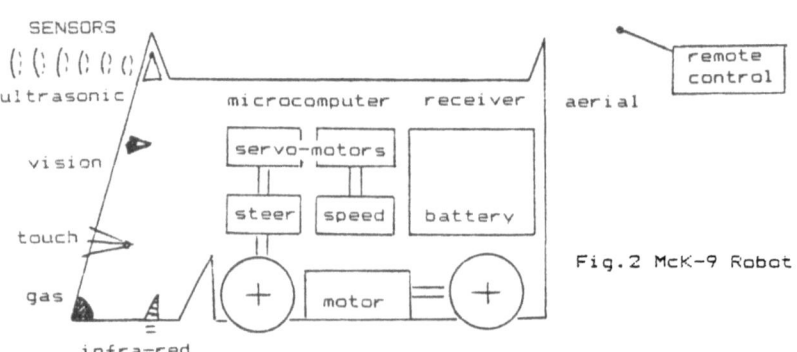

Fig.2 McK-9 Robot

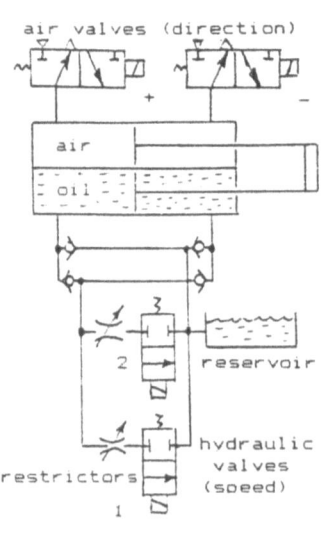

Fig.3 HYDAIR Robot Drive

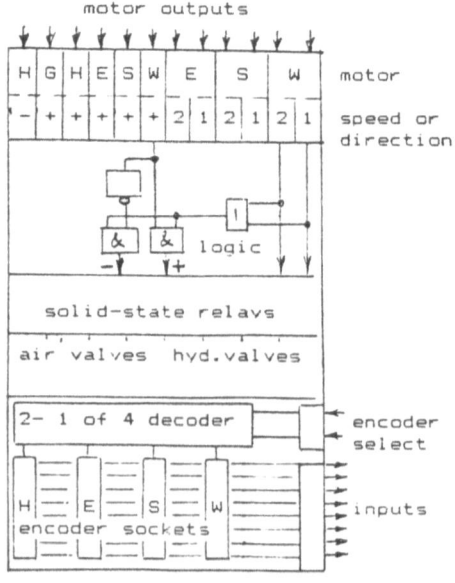

Fig.4 HYDAIR Robot Interface

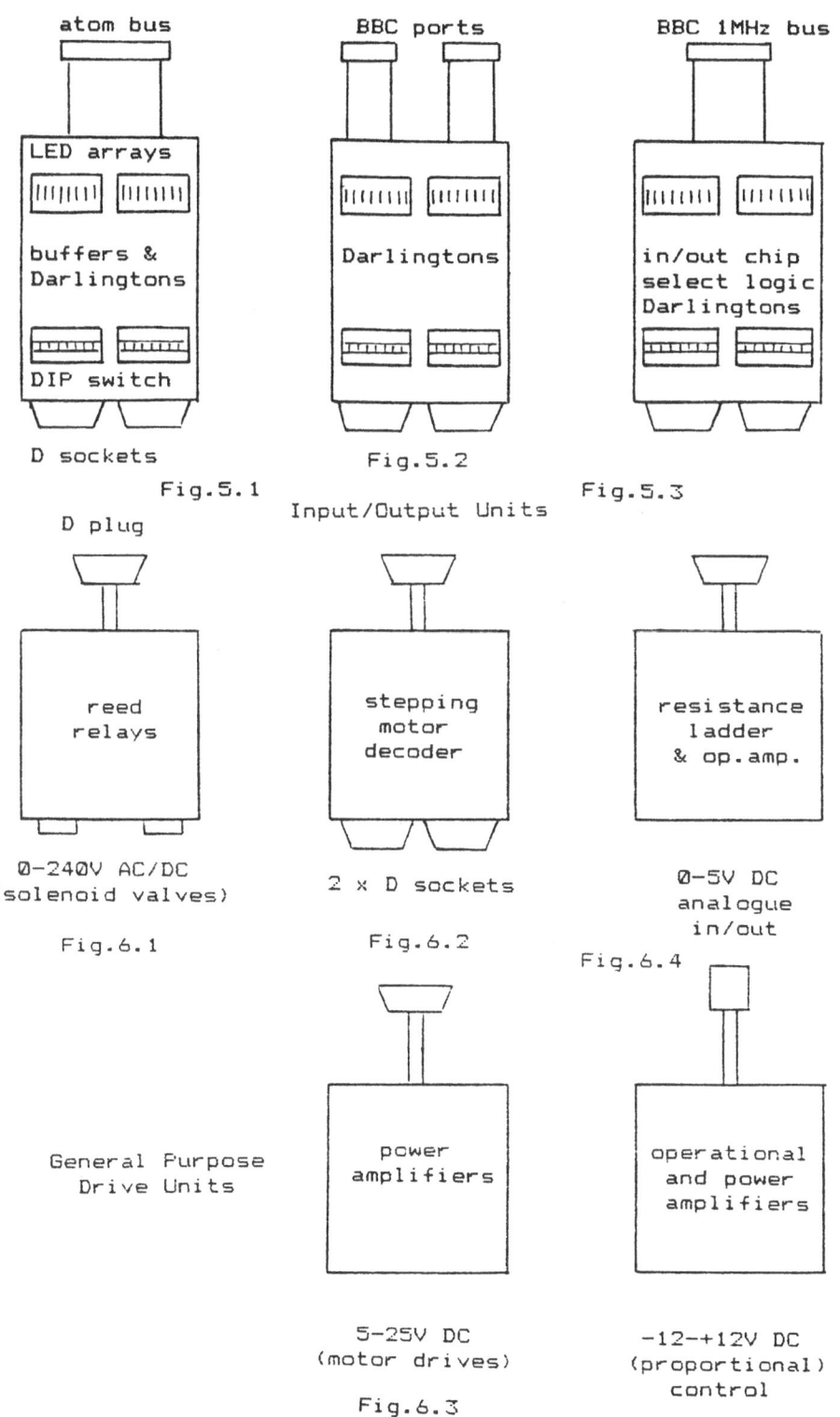

atom bus

LED arrays

buffers &
Darlingtons

DIP switch

D sockets

Fig.5.1

D plug

BBC ports

Darlingtons

Fig.5.2

Input/Output Units

BBC 1MHz bus

in/out chip
select logic
Darlingtons

Fig.5.3

reed
relays

0-240V AC/DC
(solenoid valves)

Fig.6.1

stepping
motor
decoder

2 x D sockets

Fig.6.2

resistance
ladder
& op.amp.

0-5V DC
analogue
in/out

Fig.6.4

General Purpose
Drive Units

power
amplifiers

5-25V DC
(motor drives)

Fig.6.3

operational
and power
amplifiers

-12-+12V DC
(proportional)
control

Fig.6.6

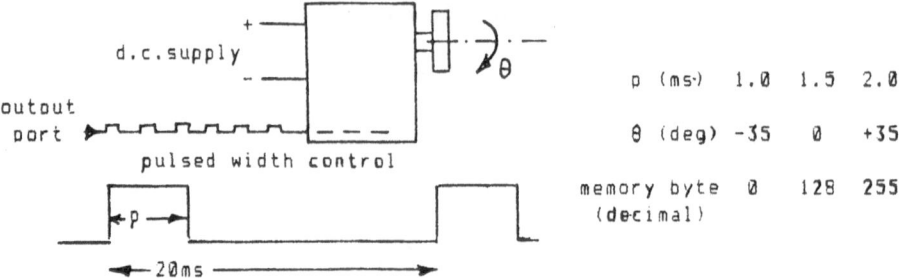

Fig.7 Pulse-Width Control

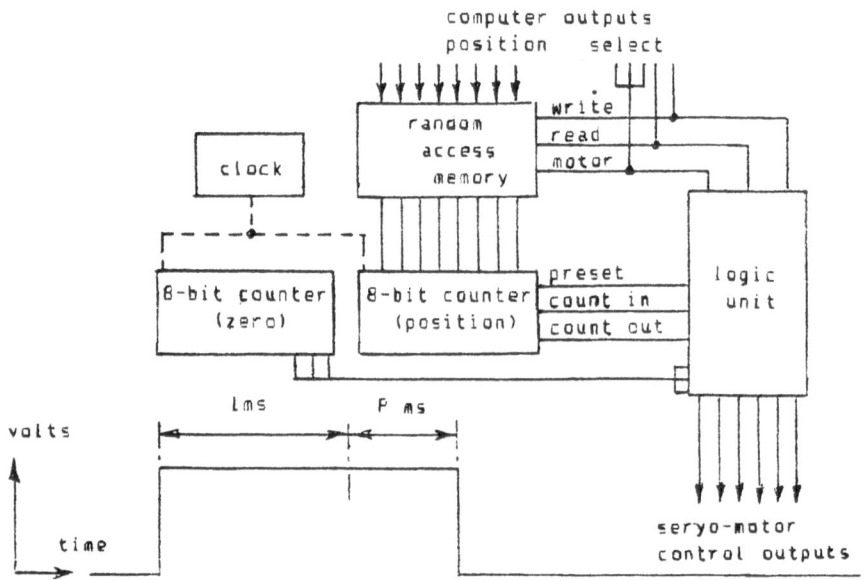

Fig.8 Smart-Arms Interface

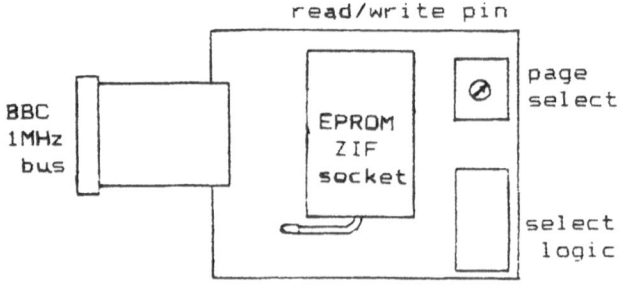

Fig.9 RAM-ROM Pump

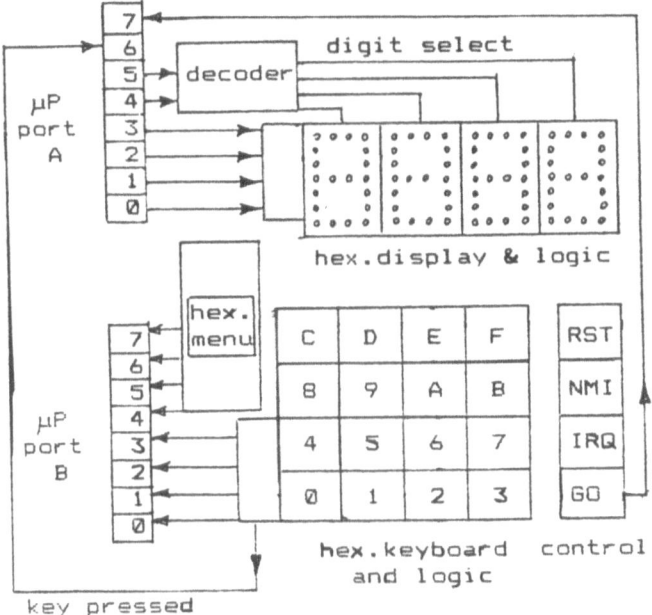

Fig.10 **Keyboard Unit**

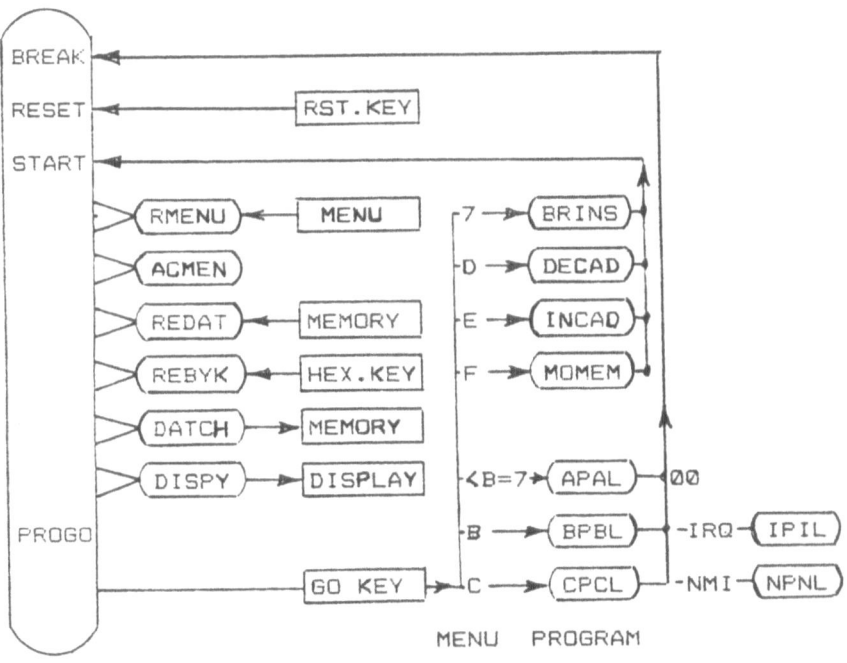

Fig.11 Operating System

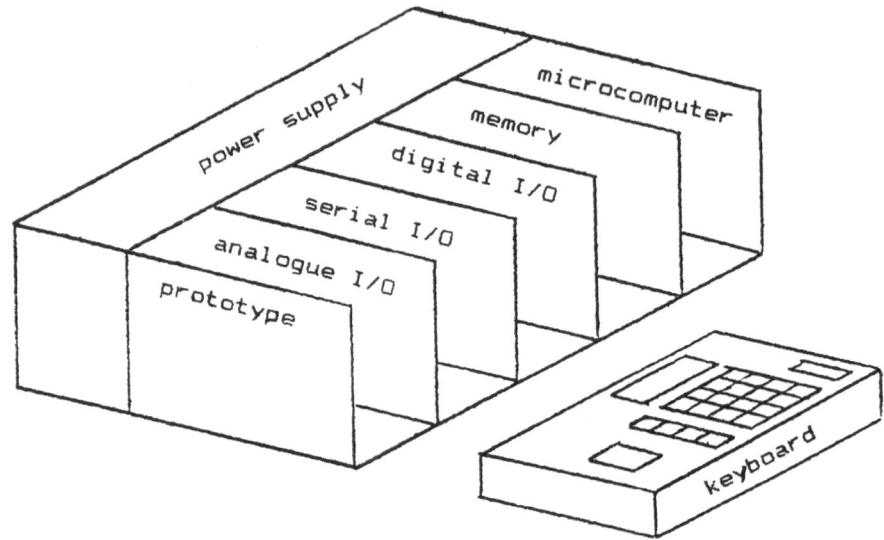

Fig.12 Rack System

FLAPS - A FATIGUE LABORATORY APPLICATIONS PACKAGE

S Dharmavasan, D R Broome, M Lugg, W D Dover

University College London

INTRODUCTION

FLAPS is a software tool for performing automated crack growth tests on components, inspection and crack monitoring in addition to the normal signal generation and data acquisition tasks. It is designed to provide engineers and materials scientists with a friendly and easy to use environment within which they can concentrate on their specific problems without having to write specialised computer programs.

Using FLAPS the sequence and operations to be carried out in a computer controlled test, namely signal generation, data acquisition, logical decisions etc may be set up, without the user having to write a single line of computer code. The set up is done by drawing a block diagram of the process on the computer screen. Once the sequence of operations is defined, the test design may be stored and executed repeatedly.

During the course of an experiment it will be necessary to record several pieces of information. The program uses up to four datafiles for storage of results. The structure of the datafiles and the information to be stored is designed by specifying the storage locations for the data on the video terminal. As the storage format defines the screen layout, printed results also take the same form. This allows the user to customise the output.

Once the experiment has been concluded, it is normally necessary to reduce the recorded results, re-calculate other parameters etc. The post-processing phase of FLAPS allows the user to selectively edit, reduce and re-calculate the results obtained. As the data is stored in the datafiles in cell locations, it is possible to specify relationships between different cells. This feature is similar to an electronic spreadsheet. Graphics are an integral part of the whole program.

Post-process of Test Results

The logged data from the run phase may now be displayed in either tabular or graphical form. Several fatigue specific calculations may also be performed automatically and the results plotted.

USER INTERFACE

One of the requirements for this program was that users of several levels of sophistication should be able to work with it. Therefore, it was essential for the program to be easy to use and that the learning period associated with the program should be short. One final requirement was that there should be substantial error checking to prevent data input errors causing expensive damage during the course of a test.

The user interface is implemented through a cursor addressable terminal. The screen is normally split into five windows, as shown in Figure 2, each of which has specific information related to it. This structure was chosen to aid the rapid assimilation and presentation of data.

The functions of the boxes or windows are as follows:

(a) Program location: Displays the current position of the user within FLAPS.

(b) Menu/Forms: This box displays the menu of choices for selection or a form for information required by a process to be entered. For instance, in the case of a constant amplitude sine loading the values for maximum load, minimum load, frequency, number of cycles etc can be filled by using a form type structure. Examples of menus and forms are shown in Figure 3.

(c) Help/Graphics/Calculator: This window is used to display information about the program at a specific point, provide some limited graphics and also draw the numeric keypad and display of results in the calculator mode.

(d) Instruction: Instructions at a particular point in the program are displayed in this box.

(e) Messages: Both error and system status messages are displayed in this window.

SYSTEM ARCHITECTURE

FLAPS is divided into three main phases which correspond to the main logical phases in an experimental test programme. They are:

(a) Design of test
(b) Running of test
(c) Post process of test results and reporting

This structure is shown schematically in Figure 1.

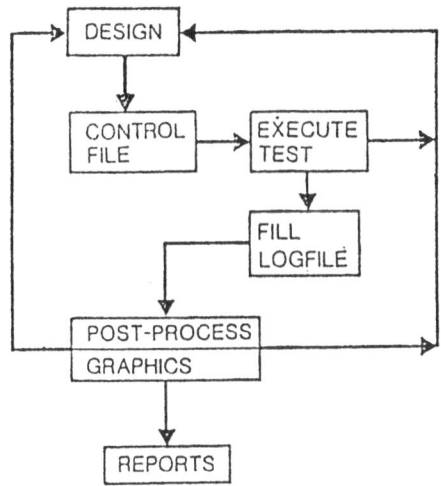

Figure 1: FLAPS System architecture

Design of Test
This phase allows the user to select from the different types of signal generation, data acquisition and other control functions. The data required for these functions is also input at this stage. This data is then stored in a CONTROL file which may be used several times or edited to modify some of the parameters for a slightly different test.

Running of Test
This phase interprets the data in the control file and executes the relevant real-time routines. Once the test is completed or halted for some reason the program reverts back to the design program.

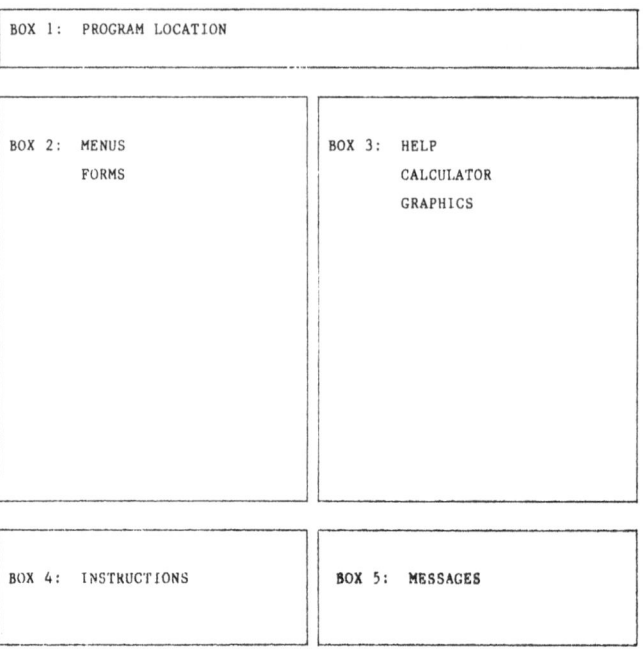

Figure 2: Functions of the multiple screen windows

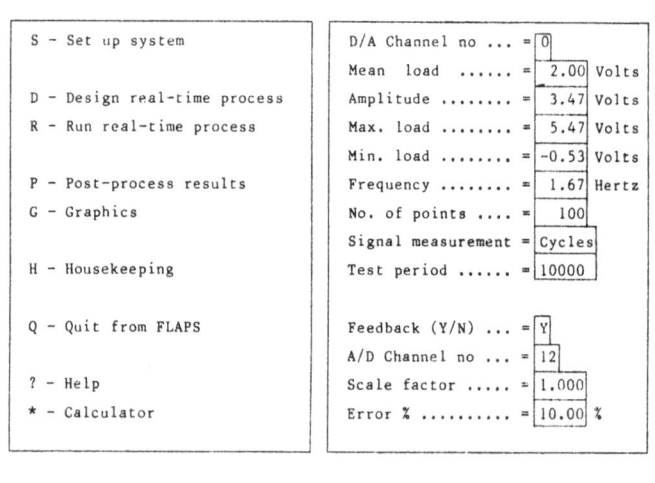

(a) Example of a menu   (b) Example of a form

Figure 3: Example of a Menu and Form

Extensive use is also made of the numeric keypad and the cursor control keys to input and edit data.

An online help program which is context sensitive is also available to provide information about the particular part of the program currently in operation.

In addition to these windows, under certain circumstances some other types of screens are also used. These are shown in Figure 4. The LOGFILE model specification provides a spreadsheet type screen for defining the locations for the logged data to be stored. The real-time process design is used to build up the sequences of the test and will be described later. In addition to these, graphics may also be displayed.

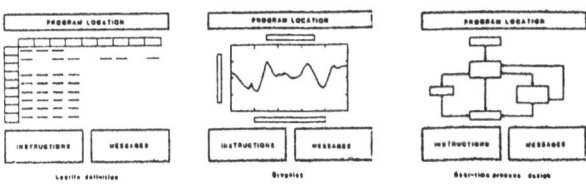

Figure 4: Some differernt types of screens used in FLAPS

DESIGN OF A REAL-TIME PROCESS

The Idea of Tasks

A fatigue test normally consists of several separate operations. In a relatively simple case it could be composed of the following tasks:
(a) signal generation for a specified length of time
(b) data acquisition
(c) termination of test if required number of cycles is reached or a critical value of crack depth is measured.

FLAPS allows different types of tasks to be constructed and the definition of the sequence in which these tasks are to be performed. Each task is an individual type of operation.

The major types of tasks supported by FLAPS are as follows:
(a) Signal generation
(b) Data acquisition
(c) Manual data input
(d) Timing operations
(e) Counting operations

(f) Calculations
It is also necessary to specify some logical decision making to allow re-direction of control during the course of a test.

The task is specified by typing a task name, around which a task box is drawn. The information associated with the task is entered by invoking the relevant form. An example of a task which is a constant amplitude sine wave generation and the associated information and controls are shown in Figure 5.

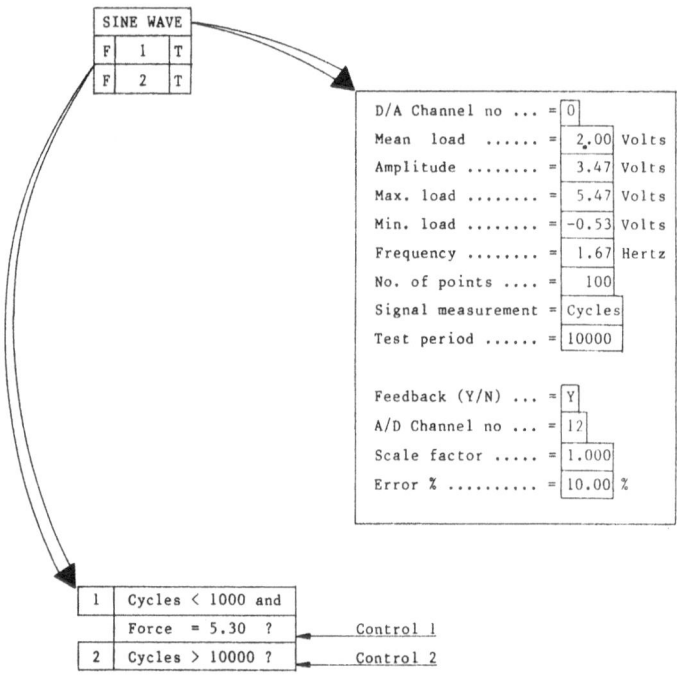

Figure 5: Details of information stored by a task

Specification of Controls

The requirement for a general-purpose control program is that it is possible to specify certain logical controls, which allows the sequence to be re-directed along several paths depending on the answer to a question which could be either TRUE or FALSE. The format and the relationships possible are shown below:

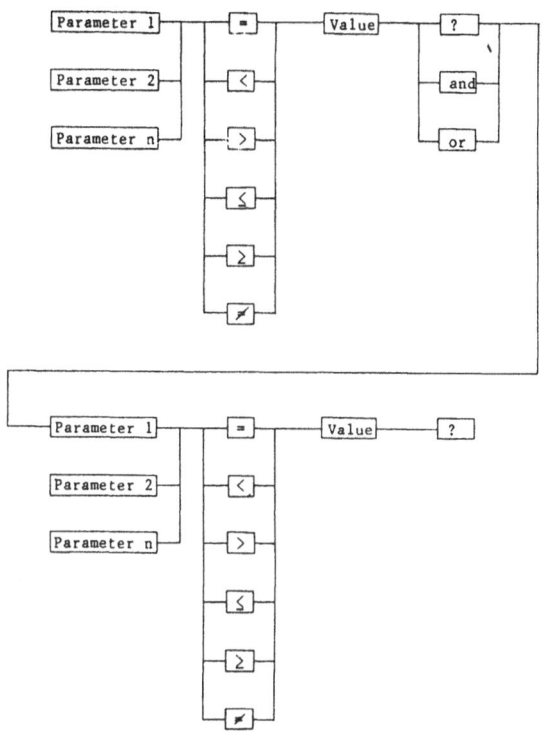

An example for a sine wave generation would be as follows:

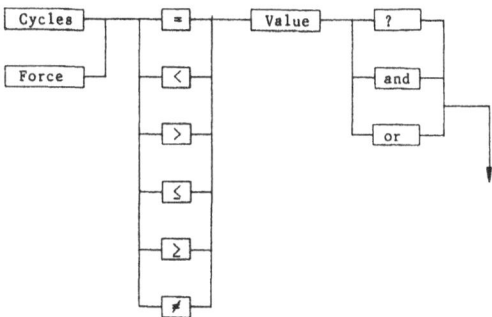

and the final choice might look as follows:

Cycles > 1000 ?

The appearance of the task box for the sine wave generation in the process block diagram before and after the control specification would be as follows:

```
┌─────────┐                    ┌─────────┐
│SINE WAVE│                    │SINE WAVE│
└─────────┘                    ├─┬────┬──┤
                               │F│  1 │ T│
                               └─┴────┴──┘
```

The F in the task box is connected to the task to which the control is to be transferred if the logical control is False. Similarly T is connected to the task to which control is to be passed if the logical question is True.

Specification of the Control Sequence
The sequence of the tasks is defined by connecting the task boxes. This is shown schematically in Figure 6. The steps required to specify the control sequence would be as follows:

(a) Name and specify task information (eg. SINE WAVE and parameters such as amplitude, mean value, frequency etc.)

```
┌─────────┐
│SINE WAVE│
└─────────┘

                    ┌─────────────┐
                    │MEASURE CRACK│
                    └─────────────┘

┌─────────┐
│STOP TEST│
└─────────┘
```

(b) Specify controls (eg. Cycles > 1000 ?)

```
┌─────────────┐
│ SINE WAVE   │
├──┬──────┬───┤
│F │  1   │ T │
├──┼──────┼───┤
│F │  2   │ T │
└──┴──────┴───┘
```

```
┌──────────────────┐
│ MEASURE CRACK     │
├──┬──────────┬─────┤
│F │    1     │  T  │
└──┴──────────┴─────┘
```

```
┌───────────┐
│ STOP TEST │
└───────────┘
```

(c) Move cursor to the task and select as the first task to be
     performed.

(d) Now move cursor to the task to which the control is to be
     transferred   and   select.   This   will   now   define   the
     connection   between   the   tasks   and   the   sequence.   The
     appearance of the block diagram after these operations are
     carried out is shown below.

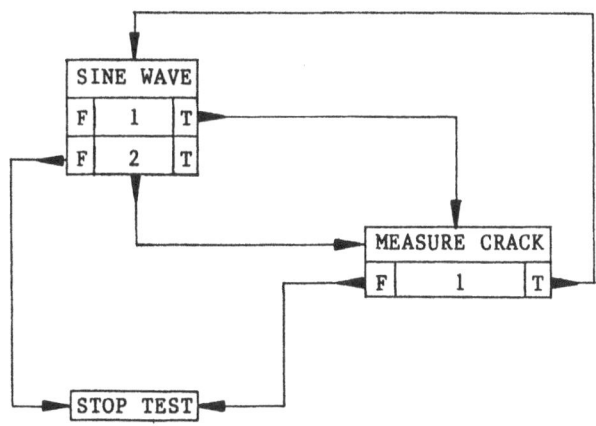

Details of Signal Generation Tasks
The following signal generation tasks are supported in the current version of FLAPS. They are:
(a) Square wave generation
(b) Sine wave generation
(c) Ramp generation
(d) Random signal generation: This is specified by defining a power spectral density.
(e) Block loading: This a combination of several different types of signal generation.
(f) Logic pulse output for driving stepping motors etc.
(g) Digital to analogue conversion on single channel
(h) Digital to analogue conversion on multiple channels
(i) Logic channel output for control of relays etc.

Details of Data Acquisition Tasks
The following data acquisition tasks are available:
(a) Single channel analogue to digital conversion
(b) Sequential channel analogue to digital conversion
(c) Analogue to digital conversion on pre-selected channels
(d) Peak monitor
(e) Crack measurement using an alternating current or direct current potential drop technique and switching unit
(f) Stepping motor controlled probe for potential drop crack measurement techniques.
(g) Strain gauge measurements.

Other Features
Several other processes also exist to control the test in a pre-determined way. For example using a timer and a counter it is possible to run tests for specified lengths of time and repeat certain processes.

DATABASE STRUCTURE

The number and nature of results stored is totally dependent on the actual test. In order to overcome the problem of having to write customised data storage routines, a methodology for the user to specify the structure of his database is implemented.

The datafile, called LOGFILE, is composed of two areas, one for static information such as probe sites, frequency of signal, test specimen number which do not change during the course of the test and the other for quantities that vary during the course of the experiment such as crack depths, temperature, number of cycles completed etc. The structure is shown schematically in Figure 7.

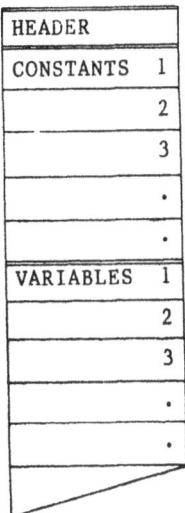

Figure 7: Structure of database

The specification of the structure is carried out by assigning
a cell location in a 52 x 2500 matrix, with each item of data
to be stored. Obviously, the blank cells are not stored.   The
user   accesses   the cells, as in the case of commercial
spreadsheet packages, by specifying a cell address, which is a
combination of a column identifier and a row number (eg., A23,
BB2345).

The screen that is displayed as a matrix of cells   is
essentially a record of data at a particular time. During the
course of a test there will be several records of data. It is
important to realise that the amount of data stored is limited
by the mass storage available.

There are several advantages in specifying the data to be
stored   in this manner.   Two of the main ones are   as
follows:
(a) The same format used for storing the data may also be used
     to print out results.
(b) As the results are pre-sorted into cells general purpose
     post-processing routines similar   to spreadsheet type
     calculations may be performed.

EXECUTION OF TEST

The execution of a test is carried out by a program written in assembly language. This program interprets the task information at the beginning of the test and then loads the relevant routines to carry out the real-time processes.

POST-PROCESSING

The use of a well defined data structure means that several general-purpose routines for analysis of the data may be constructed using simple spreadsheet type functions. In addition data may be selectively written to datafiles to be operated on by other analysis programs. This part of the program is currently under development.

CONCLUSIONS

A general purpose real-time control program has been developed. This program is extremely powerful in letting users set up a real-time process including the sequence of operations but is very easy to use. In conclusion two of the main features of the program described here are:

(a) Ease of use: This is achieved by an intelligent screen handler.

(b) Powerful and flexible: The use of the screen to draw block diagrams to define the sequence makes the program totally user configurable.

9. MICRO-APPLICATIONS POSTER SESSION

THE DESIGN OF NON-LINEAR STRUCTURES USING A SMALL
MICRO-COMPUTER.

Ian Davidson.

9 Dale Lane Appleton Warrington.

1.   INTRODUCTION.

Some structures combine a requirement for a high degree of
safety with the drawback of difficult structural analysis.
An example which may be cited is a prestressed concrete
pressure vessel for a nuclear reactor, and it is now some
23 years since this problem was first tackled.   Methods of
linear analysis were developed, principally by Dynamic
Relaxation and Finite Elements.   It was also essential to
establish an adequate safety factor for the vessel, and that
was met by making and testing a scale model.   This proved
to be expensive and time consuming; it could also lead to
uncertainty if the full size vessel had to differ from the
model.

In these circumstances the author attempted to develop a
method of structural analysis which could reproduce the
behaviour of the vessel up to and beyond the elastic limit,
almost to failure.   This requirement has now become more
common with the advent of limit state specifications.   In
general, the analysis will have to be applied several times
to a series of improving designs: it must therefore be
cheap.   It must also be capable of computing stresses
throughout the structure, to an acceptable accuracy,
whatever its shape, and with non-linearity.

Above all, the method of analysis and the programs which
apply it ought to be so simple that the design engineer can
readily visualise what is going on.   Non-linear analysis by
its nature is rather complex, and the results not always
easy to forecast.   Unless the engineer remains closely associated with the analysis and completely in control there is
a risk of unperceived errors.   The author's experience in
this field has left a strong preference for an improved
version of Dynamic Relaxation, which he calls PV, applied on

a small micro-computer.   The purpose of this paper is to
provide an outline of PV, with its application to non-
linearity.   The small personal computer, which is much
preferred to a main frame machine for its immediacy and
amazing economy, may be exemplified by the author's four-
year old PET, with 8-bit bytes, an available storage of 32k,
and no disks.

A major reversal from work with main frame computers is
that the micro costs nothing to run, and can do so unattended.
Hence speed of computation is not significant.   But storage
is now vitally important and the economy of storage with PV
is one of its principal advantages.   With the PET, using
in-core storage only, it is possible to compute 2,500
unknowns, which is sufficient for most two-dimensional and
axi-symmetric structures, but is on the small side for fully
three-dimensional structures.   However a modern 16-bit
personal computer, such as the IBM PC, would be able to
handle all requirements in core.

2.   P.V.

In PV the structure is treated as a continuum, divided into
an array of blocks;   consequently the method is not generally
suitable for frameworks.   The dimensions of the blocks may
vary, subject to the geometrical requirement that the dimension must be constant across any row or column.   All relevant stresses and deflections will be computed in each block,
to which loads may also be applied.   Hence the PV analysis
can be as comprehensive as required, depending on how many
blocks are chosen.   Figure 1 shows a typical block in a
plane stress (two dimensional) structure, with its parameters:
axi-symmetrical $(2\frac{1}{2}$ dimensional) and full three dimensional
blocks are very similar to this, and may be in whatever coordinate system is most convenient.

It is well known that the basis of Dynamic Relaxation
depends upon setting up finite difference equations relating
stresses to loads and deflections to stresses, in each block.
The equations are then solved by a process of successive
approximation, which has a simple physical analogy.   Consider
each block in turn and using Newton's Law calculate its
acceleration due to the external loads.   Choose a short
time interval and hence determine the deflection of each
block.   Now re-examine each block and compute the stresses
produced by the just determined deflections.   Then re-
calculate the deflections due to the just calculated stresses
plus the loads, and so on.   At every iteration when the
accelerations are calculated include a viscous damping, so
that the vibrations die down.

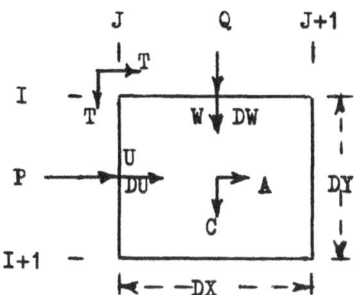

FIGURE 1.   Typical Two-dimensional Block.

(I,J)   co-ordinates of top left corner of block.
DX, DY   dimensions of block.
A, C   direct stresses in block.
T   shear stress in block.
U, W   velocities.
DU, DW   deflections.
P, Q   loads (pressures) applied to block.

The deflections are always compatible with the stresses, and at every stage the external loads are in equilibrium with the internal stresses plus the inertia loads.   When the vibrations die down the inertia loads have vanished, and the computed stresses and deflections are the correct solution.   In practise it is more convenient to work with stresses and velocities.   Note the great economy of storage because the parameters computed at each iteration ultimately become the required answers.

The physical analogy of the vibrating structure has two benefits: firstly it enables the appropriate equations to be determined in a very simple manner: secondly it shows that PV can also be used to compute the behaviour of the structure during and after the application of loads.   This dynamic response is particularly important when yielding occurs.   A fuller explanation of the fore-going will be found in chapters 2 to 6 of reference 1.

3.   CONVERGENCE.

For simplicity in what follows a two-dimensional structure will be assumed, but more complex blocks are to be treated in an analogous manner.   In order to ensure efficient convergence the time interval and the damping must be well chosen, and this can be done very easily with PV.

The time interval, TD, must not exceed :-

$$SQR (RO / EL / (1/XD/XD + 1/YD/YD)) \qquad (1)$$

where RO = mass density;   EL = elastic modulus;   and XD , YD are block dimensions.

The viscous damping, $KD = 1.6 * TD * CF$ \qquad (2)

where CF = fundamental circular frequency of the structure, which can generally be estimated from the usual formulae. The number of iterations required for convergence will be about 10 / KD.

In cases where the block dimensions are not constant, TD must be chosen to suit the smallest block, and fictitious densities can be applied very simply to maintain efficiency of convergence in the larger blocks.   In order to do this G2 (see figure 2) should be multiplied in each block by

$$SD / (1/XD/XD + 1/YD/YD) \qquad (3)$$

where SD is the value of $(1/XD/XD + 1/YD/YD)$ for the small-est block.   A similar expedient permits a different value of EL in every block, without loss of efficiency.

In the past, using Dynamic Relaxation at arms length on a main frame computer, it was found difficult to ensure

sufficient and efficient convergence.   Considerable wasted
expense occurred with overlong or abortive runs.   The
situation is entirely changed with the micro computer, which
can display key parameters during the convergence, and which
offers the possibility of interactive working.   It is
convenient to choose a velocity where the largest deflection
is expected to occur: this velocity is displayed, together
with the iteration number, current value of KD, and an error
signal.   A graph of the selected velocity is also presented
on the screen.

The error signal is simply the sign of the selected
velocity multiplied by the sign of its last increment.
The last 12 iterations will be displayed simultaneously,
and experience has shown that there should be between 3 and
7 negative error signals on the screen: if there are fewer
than this then the damping should be reduced, and vice
versa.   To do this it is only necessary to press the D key,
and the iterations will stop the next time round for the
new value of KD to be input.   Adequacy of convergence can
be determined from the velocities.   If none of these are
greater than 0.2 ins/sec then the stress errors are less
than 30psi in steel, or 6 psi in concrete.   These figures
are considered acceptable.   When the displayed velocity
is oscillating about zero, with excursions of less than
0.2 ins/sec, or 5 mm/sec, then the V key should be pressed.
All the velocities will then be displayed, and if these are
satisfactory the results may be output to screen or printer.

4.   BOUNDARY BLOCKS.

PV incorporates a fuller treatment of boundaries than was
included in the first Dynamic Relaxation programs, and this
results from determining the appropriate equations for every
possible shape of boundary block.   Each such block is given
a code number, and this array of code numbers, together with
the block dimensions, is input to the computer and provides
a complete description of the structure.   During the iterations the correct equations will be applied in each block,
hence structures of complex shape can be analysed accurately.

As seen in figure 1, deflections are calculated at top and
left edges of the block.   In order to calculate deflections
at bottom and right edges of the structure, it is convenient
to introduce dummy blocks, in which all the parameters are
zero except for one velocity.   A zero block in which all
the parameters are zero can be used to fill out the array
where there is no structure, and to provide a fixed point
of support.
It will also be seen from figure 1 that the shear is computed at the corner of a stress block, so that the shear
block occupies four quadrants of adjacent stress blocks.

At open boundaries and at salient corners it is assumed that there is no shear in the half blocks adjacent to the boundaries; this gives accurate results and simple equations. But there is a significant shear immediately adjacent to a re-entrant corner, which cannot be neglected. It has been found best to compute this shear in the normal way as in any solid block, and to apply this in full when computing velocities in adjacent blocks, even though on a boundary.

A two-dimensional boundary block is shown in figure 2, with its optional boundaries, which include a re-entrant corner. The derivation of a velocity equation is given, to illustrate the previous paragraph. The equation for calculating the shear stress is also shown. It will be noted that the derivation of the equations is elementary, and all the equations are very simple. Note also that velocities, deflections and stresses are computed incrementally, which is necessary when yielding occurs.

The validity of the above treatment of boundaries has been shown, inter alia, by computing the stresses in an infinite plate with a circular hole in the centre, loaded uniaxially. The boundary of the hole was simulated by an array of rectangular boundary blocks, and the results showed 95% accuracy over 95% of the plate, although this procrustean representation of polar co-ordinates by Cartesian ones was probably more extreme than in any structure likely to be met in practice.

The boundary block concept is very versatile; it permits for instance the accurate representation of folds in a folded plate, so that such a structure as a box girder can be analysed throughout webs, flanges and cross-diaphragm. Also special blocks have been derived with curvilinear co-ordinates, so that dome-like structures can be analysed efficiently and accurately. A full treatment of boundary blocks in two-dimensional, axi-symmetric and three-dimensional structures will be found in volumes 1, 3 and 4 of reference 1.

5.   CRACKS.

A continuum structure exhibiting non-linearity will often consist of concrete or masonry, sometimes reinforced or prestressed with steel. There are three possible causes of non-linearity, and in order of importance they are cracking, yielding of steel and multi-axial yielding of concrete or masonry in compression. In real life the loads will be applied in stages, and cracks will extend during this process: the compatible crack system must be established for each load stage.

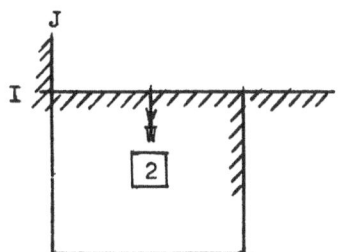

FIGURE 2.

Two-dimensional Boundary
Block. Code Number 2.

It can readily be shown that Ua = G1 * Ub + G2 * X   (4)

where   Ua = velocity after iteration
        Ub = velocity before iteration
        $G1 = (1-KD/2) / (1+KD/2)$
        $G2 = TD / (1+KD/2) / RO$
        X = force in direction of velocity /
            volume of block

Velocity Block for W.

Force in direction of W = (Q - C(I+1)) * XD + T * YD

Volume of block = XD * YD/2

Hence X = 2 * ((Q - C(I+1)) / YD + T / XD)   (5)

The value of Ta = Tb + G4 * X   (6)

where G4 = EL * TD / (1 + PO) / 2

and X = (U(I-1) - U) / YD + (W(J-1) - W) / XD   (7)

Note that G1, G2 and G4 are constants which can be calculated
once before the iterations commence.   But the X's must be
recalculated in each block at each iteration.

In order to represent each major crack correctly, PV introduces two new boundaries, forming the sides of the crack.
It is assumed that the cracks follow block boundaries, and
hence the width of the crack can be computed: it is also
assumed that the tip of the crack is at a block corner.   It
is difficult to determine crack extension by looking at the
stresses ahead of the tip, because it is a singularity and
the material has started to yield. But the author has found
that a reliable and simple criterion is the crack opening
displacement.   For the case of high-strength small-aggregate
concrete, as used in models, the author has found that the
crack should be extended when the nominal angle subtended at
the tip exceeds 1.2 milliradians.

A great advantage of PV is that insertion and extension
of each crack can be implemented merely by altering the
array of code numbers, thus maintaining accuracy without any
trouble: there is no stiffness matrix to be changed.   The
path of the crack is assumed by the engineer and applied to
the computation by the choice of code numbers: the computed
widths of the crack provide evidence as to whether the assumed path was correct.   If not, the crack may be extended
in the indicated direction for the next load increment.

Aggregate interlock is an important phenomenon that
should be included in the analysis.   If it is assumed that
the interlocking facets lie at an average angle of 45" to
the general line of the crack, then it will be seenthat a
free motion equal to + or - the width of the crack can occur
before any shear stress is mobilised.   When computing the
shear strain it is only necessaryto deduct the crack width
from the displacements along the crack (see figure 2 for
shear equation).   This produces remarkably accurate results.

Because the crack widths are computed at each iteration
it is also a simple matter to deal with cracks which close.
A comprehensive treatment of all these matters will be found
in chapters 21 to 26 of reference 1.   A comparison between
experimental results from a cracked, prestressed and reinforced concrete beam and a PV analysis is shown in figure 3.
These results were computed by program PV1B7, which is
supplied with reference 1.

6.   DYNAMIC LOADS.

As already explained in section 2 the iterative procedure of
PV in fact represents the time-response of the structure to
loads: if we are only interested in the final static result
we apply optimum damping to bring the structure to rest as
soon as possible.   But if the response is to be examined in
detail it is merely necessary to change the damping to what
is actually present in the structure, and to print out the
required parameters at appropriate intervals.

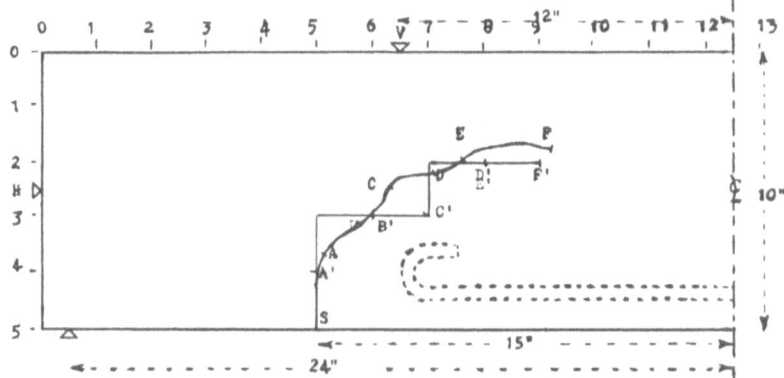

Figure 3

The figure shows half an experimental beam, 4 ins wide, which was supported as shown and prestressed by a constant force of 36500 lbs at H. Known deflections were applied at V, causing a crack to grow from the crack starter at S, in six stages from the onset of cracking (stage A) almost to failure (stage F).

The course of the experimental crack is shown on the figure with the position of the tip at each stage (A etc). The course of the crack as computed by PV is also shown with the position of the tip (A' etc).

At each stage the reaction at the support was measured, and also the central deflection. The measured values are compared with the computed values in the following table. Note that aggregate interlock and reinforced concrete played significant parts in the experiment and were computed satisfactorily by PV.

Load Stage	Load Defln.	Reaction		Error %	Cent. Defln.		Error %
		Exp.	PV		Exp.	PV	
A	.023	2290	2320	1	.030	.0293	-2
B	.0262	2320	2410	4	.0337	.0332	-1
C	.0294	2360	2490	5	.0375	.0369	-2
D	.0326	2390	2570	7	.0412	.0410	-1
E	.043	2480	2730	10	.0533	.0537	1
F	.0686	2600	3050	17	.0831	.0865	4

Generally speaking the application of dynamic loads will occur over some extended period as with earthquakes or wind loading. As the iterations of PV succeed each other at a constant time interval, TD, it is easy to insert suitable algorithms in the main iteration loop which will add increments of load to the structure, in accordance with the load time-history. The application of these techniques to elastic and to elastic-plastic structures forms the subject of chapters 36 to 39 of reference 1.

7. YIELDING.

Yielding in steel reinforcement bars or tendons can be computed very easily indeed, as it is uni-axial, but multi-axial yielding of concrete or masonry in compression provides a problem in computation, which PV is well suited to solve. The techniques explained in section 6 can be used to provide a picture of the stresses in each block during and after the application of the load. Then it is only necessary to choose some suitable constitutive law for concrete or masonry yielding, and to apply it to each block, in order to determine modified elastic parameters, appropriate to the stresses in that block at that instant.

A suitable constitutive law for concrete has been published by Professor Zienkewicz in reference 2. This states that the bulk modulus K remains at its normal elastic value, while the shear modulus G is a function of the second stress invariant J2. As PV calculates incrementally, it naturally accepts these tangent moduli. In order to calculate J2 at the centre of each block, values of T are averaged from the four corners. G is then calculated as a function of J2, and this is shown in figure 4, where SQR(J2)/cylinder strength is plotted against G/GO; GO is the normal elastic value of G. Hence modified values of the elastic modulus and Poisson's ratio can be calculated and applied in each block.

In pracice this routine works perfectly, and it is only necessary to compute modified elastic constants at intervals of ten iterations or so. But a slight problem also emerges because it is obviously necessary to apply the load gradually i.e. by an increment in each of the first 100 iterations, say. The problem is that the correct time interval for stability - see section 3 - will be something of the order of 2E-6 secs; in this case 100 iterations will be equal to 2E-4 secs. Evidently the application of the load in such a short time is tantamount to a violent impact, and the resulting stresses and deflections will be very great.

In this situation we can make use of fictitious densities
- see section 3.   If we arbitrarily increase RO by a factor
of 1E8, we can see from equation 1 that the time interval
will be increased to 2E-2 secs, and now if we apply the load
over 100 iterations, the loading period is 2 secs, which is
reasonable.   The use of these techniques is covered in
chapters 58 to 60 of reference 1, and program PV2B4 which is
supplied with reference 1 will produce the results shown in
figure 4.   They compare the strains computed by PV with
experimental measurements of axial strains in concrete
cylinders, which were subjected to uniform inward pressure,
while the axial load was increased in stages.   Four
typical experimental curves are shown in figure 4, derived
from two sets of experiments which used different concretes.

8.   CONCLUSIONS.

It has been shown that the basic idea of PV is extremely
simple, and figure 2 demonstrated that the derivation of the
necessary equations is elementary.   Because the solution is
by repeated iterations, in each of which velocities and then
stresses are computed in each block, it follows that the
programs to apply PV are short and simple;   they consist in
the main of loops.   Moreover, because each program contains
all the relevant boundary equations, they are sufficiently
versatile to analyse structures of any shape.

This versatility of PV has also been shown to extend over
structures in 2, $2\frac{1}{2}$ or 3 dimensions, in cartesian or polar
co-ordinates, cracked or not, yielding or not, with steady
or dynamic loads;   also including folded plate structures
and curvilinear co-ordinates.   In addition PV can also be
used to establish temperature distribution, either steady or
not, with the resulting thermal stresses.

The economy of storage with PV has been shown to allow
serious analysis of full-scale complex structures (2,500
unknowns) with a personal micro-computer.   Typically, a
prestressed concrete pressure vessel for a nuclear reactor,
with an electrical output of 600 MW, has been analysed in
sufficient detail to meet the requirements of the British
Standard Specification.   This analysis was performed on the
author's PET and is included in reference 1.   Analysing an
axi-symmetric structure with 250 nodes, it was found that
PV requires only 1/5 as much storage as Finite Elements.

9.   REFERENCES.

1.   Structural Analysis with a Micro-computer using
     Dynamic Relaxation.   Ian Davidson.   9, Dale Lane,
     Appleton.   Warrington.   WA4 3DX.
2.   Phillips and Zienkiewicz.   F. E. non-linear analysis
     of concrete structures. Proc. Inst. C. E. 2/61 Mar.
     1976. p. 72.

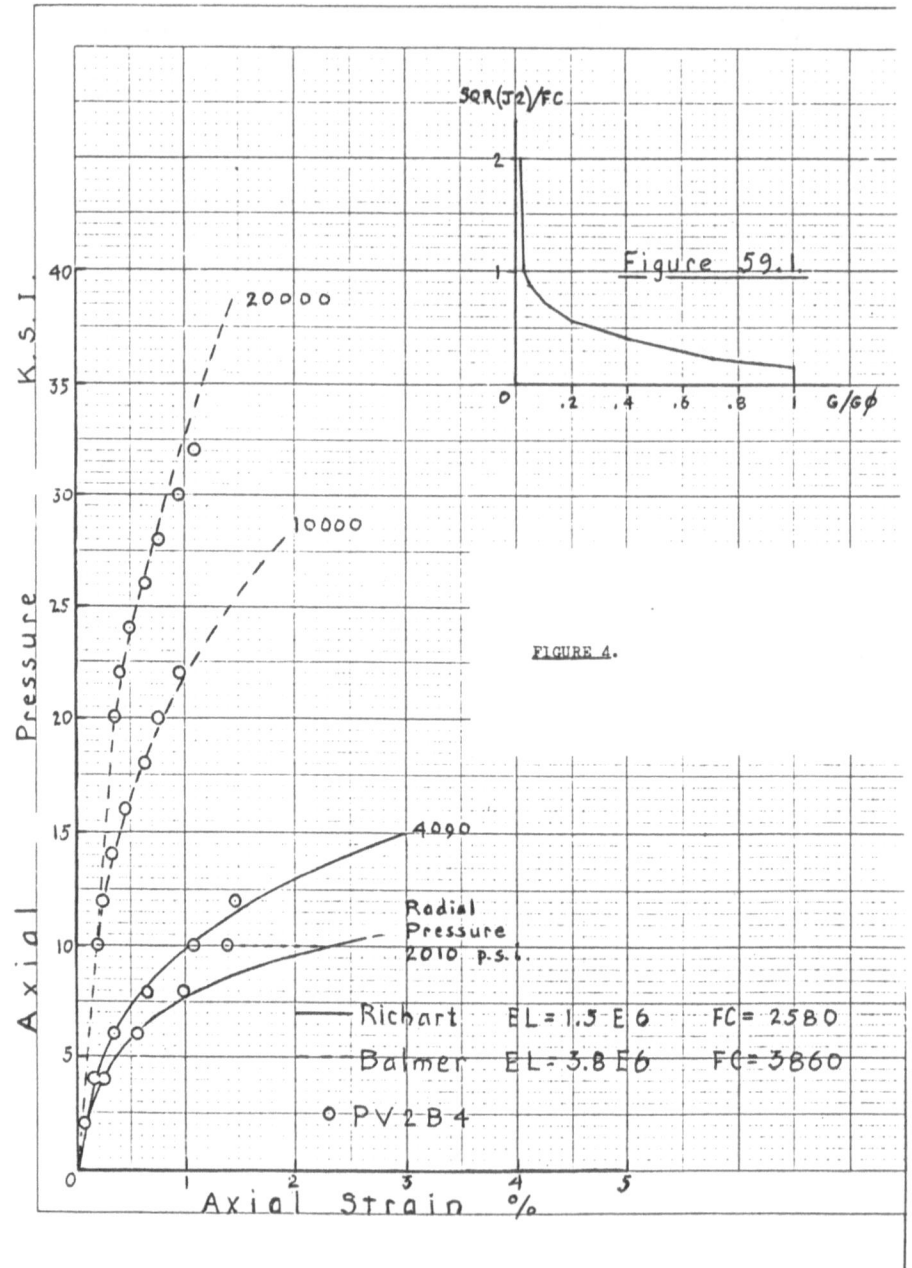

Figure 59.1

FIGURE 4.

PECI - CALCULATION OF BUILDING STRUCTURES

S.S. Kalmus

Centro de Informatica - FAAP - S. Paulo - Brasil

I.   INTRODUCTION

The PECI07 System is a programme which purpose is the complete
calculation of building structures of any geometry, and   any
number of floors, regardless of floor repetitions   or
transitions.

The current version PECI07, solves the   beam-column
structural system, starting from direct loads on   the beams,
the slab reactions, the mansonry load and other loads that may
have an influence on the beams.

The results are moments, shear forces,   reciprocal
reactions of beams, with values on the project scale, that is
1:50 (on the printer, too) and the reciprocal reactions on the
columns.

The data input is done in a simple interactive way.   At
this opportunity a series of predictable errors are   detected.
The final data-assurance and necessary corrections are done by
other modules of the system.

After designing the structure (beams, columns)   and
preparing the loads on the beams, the joints   (not columns)
must be numbered for definition of the reciprocal   reactions;
this numbering must follow the highest numbered column.

After putting the beams in correct order (see fig 1)   of
calculation, we prepare a worksheet (see fig 2) or, with
little practice, put it directly in the microcomputer.

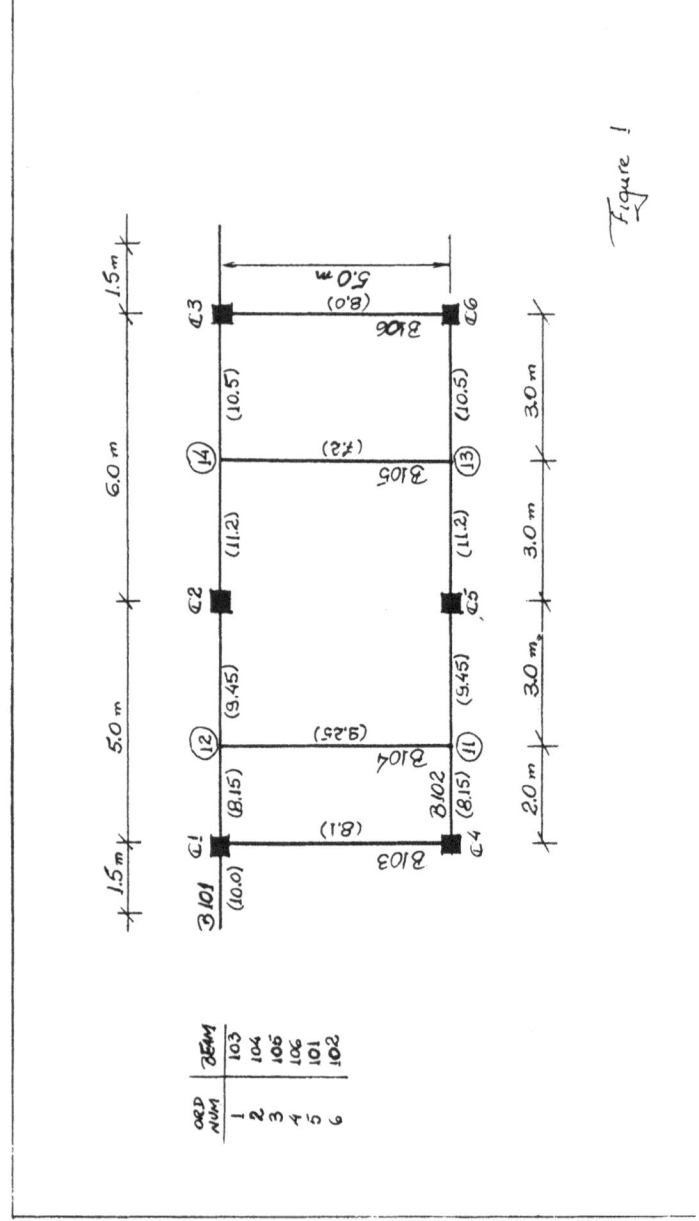

Figure 1

														DIST.	
Sign. Data	1st em. | 1st rev. | 2nd rev.	ORD. NUM.	BEAM	SPAN	SUPPORT LEFT	SUPPORT RIGHT	SPAN VAL.	REL. IN.	UNIF. LOAD	CONC. LOAD	DIST.	GENER. LOAD	DIST.	PARC. LOAD	START	END

FUNDAÇÃO A. A. PENTEADO — PROJETO PECI
DATA-INPUT WORKSHEET — SLAB # 5
PAG 1 / 1 — PR/007
CUSTOMER: SAMPLE
WORK: IV INTERNACIONAL CONFERENCE AND EXHIBITION

ORD. NUM.	BEAM	SPAN	SUPPORT LEFT	SUPPORT RIGHT	SPAN VAL.	REL. IN.	UNIF. LOAD	GENER. LOAD	DIST.	PARC. LOAD	DIST. START	DIST. END
1	103	1	4	1	5.	1.	8.1					
2	104	1	11	12	5.	1.	9.25	12	2.	8.15	0.0	2.0
3	105	1	13	14	5.	1.	7.2			9.45	2.0	5.0
4	106	1	6	3	5.	1.	8.0	14	3.	11.2	0.0	3.0
5	101	2	0	1	1.5	1.	10.0			10.5	3.0	6.0
		3	2	3	6.0	1.	0.0					
		4	3	0	1.5	1.	9.3	11	2.	8.15	0.0	2.0
		1	4	6	5.0	1.	0.0			9.45	2.0	5.0
6	102	2	5	6	6.0	1.	0.0	13	3.	11.2	0.0	3.0
										10.5	3.0	6.0

figure 2

II. WHAT IS NEEDED

1) A computer with 48kb of RAM memory with one or two   disk
drives.
2) A programme disk.
3) At least two empty disks-5 1/4".   This will be enough   to
start with, but probably a greater number will be   needed,
your planning will determine.
4) A 132 character printer (in direct or condensed form).

III. THE SYSTEM MENU

1) Data Input
2) Data Assurance
3) Data Correction
4) Execution/Output
5) End (finalization)

Data Input
If you select in MASTER MENU -1) Data Input, a   succession
of questions that define the structure will be shown in   a
clear interactive form; questions as: number of slabs,   the
highest column-number, number of beams for each storey, spans,
loads, inertias, etc.

    Finally this module asks if we want to accumulate   other
loads on the columns, if for any reason we have "run"   upper
slabs already, and now we want to continue with another "run"
or we want to define loads for future construction.

    In the event of the existence of symmetry in   the
structure a command called 'GENERATION' can accumulate loads
on the columns or the loads defined by the column-number   may
be accumulated on another column.

Data Assurance

This module will print, rapidly, on the printer, the image of
the data-input showing the errors or sending warning messages
such as: Column 8 transitioned, Beam 2 is an over-hang,
verify joints in "diagonals" (the end joint of any span in a
continuous beam is not the same as the inicial joint of the
following span), the distance value from the concentrated load
is greather than the span, load 135 is 'generated' but not
used or the reverse, etc.

Finally data-assurance will print the columns-number and
the number of beams or loads crossing on that column. This is
utmost importance in order to check that no beam has been
forgotten nor any error was made in any column.

Date Correction

In a very simple interactive way this module permits
correction according to the data-assurance list, of the errors
registered in data-assurance or in data-input.

Execution-Output

This is the main module of the system and through it will have
on the printer:

Figure 3, shows all details of the output for one beam.

Figure 4, shows the output of the plotting on the printer
(scale 1:50).
Figure 5, shows the output columns loads. The line "VAR."
represents the accidentability of the effect to be considered
or not by the designer (see below).

End(finalization)

This command finalize session and returns to the DOS.

IV. HYPOTHESIS

Beams: a) statically determinate with or without over-hang.

b) mono or bi-clamped with constant section with fixed
supports.

c) continuous with constant section between supports
but variation being possible on the sections between spans.

*** PROCESSAMENTO DE DADOS - ENG. SIMPSON S. KALMUS - *** MICPO ELSI ***   PAG. 6

* * * FOURTH INTERNATIONAL CONFERENCE AND EXHIBITION - LONDON -   * * *

* * * PECI - CALCULATION OF BUILDING STRUCTURES - SAMPLE -   * * *

--- BEAM 102   N# 6   SLAB 5   STOREY 6   TYPICAL   ---

------LOADING - FORCES------

SPAN	L	RI	UNIF.F	PARC.F	ST	END	C#	CONC.F	D	C#	CONC.F	D	C#	CONC.F	D	C#	CONC.F	D
1	5.00	5.00	0.000	8.150	0.00	2.00	11	23.125	2.00									
				9.450	2.00	5.00												
2	6.00	6.00	0.000	11.200	0.00	3.00	13	18.000	3.00									
				10.500	3.00	6.00												

------BENDING MOMENT------

BEAM

SPAN	L	RI	BXH										
1	5.00	0.00	10.74	19.45	26.11	21.61	30.74	10.11	-3.74	-19.97	-38.55	-59.49	102
2	6.00	-59.49	-30.32	-5.17	15.95	46.08	33.03	44.42	38.99	29.77	16.78	-0.00	102

------ SHEAR FORCES ------

N#

SPAN	L	RI	BXH										
1	5.00	23.52	19.45	15.37	11.30	7.22	-20.63	-25.35	-30.08	-34.80	-39.53	-44.25	6
2	6.00	51.99	45.27	38.55	31.83	25.11	0.39	-5.91	-12.21	-18.51	-24.81	-31.11	6

figure 3

```
*** FOURTH INTERNATIONAL CONFERENCE AND EXHIBITION - LONDON -    ***
*** PECI - CALCULATION OF BUILDING STRUCTURES  - SAMPLE -        ***
                        * PLOTTING-MOMENTS AND SHEAR FORCES *
BEAM SPAN  MOMENT                MOMENT-DIAG.           X-ORD  SHEAR              SHEAR - DIAG
 102  1   0.000                            *           0.000  23.521       I   *
                                           I                                I
 102  1   10.742                        *  I           0.500  19.446        I   *
                                           I                                I
                                           I                                I
 102  1   19.446                     *     I           1.000  15.371        I   *
                                           I                                I
 102  1   26.113                  *        I           1.500  11.296        I *
                                           I                                I
 102  1   30.742                *          I           2.000   7.221        I *
                                           I                                I
                                           I                                I
 102  1   21.609                   *       I           2.500 -20.629      *  I
                                           I                                I
 102  1   10.114                        *  I           3.000 -25.354     *   I
                                           I                                I
                                           I                                I
 102  1   -3.744                          I*           3.500 -30.079     *   I
                                           I                                I
 102  1   -19.965                  I------*            4.000 -34.804    *    I
                                           I                                I
 102  1   -38.548                  I------------*       4.500 -39.529   *     I
                                           I                                I
                                           I                                I
 102  1   -59.494                  I----------------------* 5.000 -44.254  *   I          * 51.991
                                           I                                I
 102  2   -30.315                  I----------*          0.600  45.271        I          *
                                           I                                I
                                           I                                I
 102  2   -5.169                   I-*                  1.200  38.551        I       *
                                           I                                I
                                           I                                I
 102  2   15.945                        *  I           1.800  31.831        I     *
                                           I                                I
                                           I                                I
 102  2   33.028                   *       I           2.400  25.111        I   *
                                           I                                I
 102  2   46.078                *          I           3.000   0.391        *
                                           I                                I
                                           I                                I
 102  2   44.422                *          I           3.600  -5.909      *I
                                           I                                I
 102  2   38.987                 *         I           4.200 -12.209      *  I
                                           I                                I
                                           I                                I
 102  2   29.771                  *        I           4.800 -18.509      *  I
                                           I                                I
                                           I                                I
 102  2   16.776                        *  I           5.400 -24.809      *  I
                                           I                                I
                                           I                                I
 102  2   -0.000                          *            .6.000 -31.109     *   I
```

figure 4
- out of scale -

*** PROCESSAMENTO DE DADOS - ENG. SIMPSON S. KALMUS - *** MICRO ELSI ***   PAG. 12

* * * FOURTH INTERNATIONAL CONFERENCE AND EXHIBITION - LONDON -   * * *

* * * PECI - CALCULATION OF BUILDING STRUCTURES - SAMPLE -   * * *

-------------------------------- SLAB NUM. 1 --------------------------------

--- COLUMNS - LOADS   SLAB 1 STOREY 2 TYPICAL   ---

COLUM	P 1	BXH	P 2	BXH	P 3	BXH	P 4	BXH	P 5	BXH	P 6	BXH
LOAD	326.042		473.901		355.450		229.799		505.283		268.324	
VAR.	22.488		0.000		19.084		31.234		0.000		26.029	
TOTAL	348.530		473.901		374.535		261.033		505.283		294.353	

SUM...... L= 2158.800
SUM...... V=   98.835
SUM...... T= 2257.636

figure 5

P.S.   You can have over-hangs in the mono-clamped   or
continuous-beams, fixed supports, at the ends of the
continuous-beams.
Loads: a) Uniformly distributed loads.
       b) Partially distributed loads.
       c) Concentrated loads.

V.   LIMITATION OF THIS VERSION OF THE PROGRAMME

Columns: 200
Spans   : 12
Loads partially or totally distributed, generated   or
concentrated (per span) : 10
Generated loads (reciprocal reactions): 500
Number of slabs: no limits
Number of Transitions: no limits.

VI. RESULTS:

* Titles, slabs enumeration, beam enumeration.
* Beam loads (initial data and generated reciprocal   loads),
lengths and spans/relative inertia in metres.
* Moments in kN.m for each tenth of the span and/or plotting
on the printer (scale 1:50).
* Shear forces in kN for each tenth of the span,   and/or
plotting on the printer (scale 1:50 - the same   lineordenates
as the moment).
* Load on the columns, in kN for each floor and the total load
on the floor.

   The loads on the columns are formad by beam reactions by
the accumulated loads per column, and by the accumulated load
of the building.

   The effect of the accidentability of the load on   the
reaction of the supports is indirectly taken   into
consideration, only half the statically   indeterminate
correction being subtracted from the statically   determinate
reaction (if the result is subtractive).

VII. FINAL CONSIDERATION

1) The system was develloped originaly in FORTRAN IV   in   a 370/145 IBM and has now addapted to the micros with   at   least 48 kb RAM.
2) The system equations solution of the continuous beam   was made by Cholesky-algorithm.
3) The average running time in interpreted-BASIC for one beam (3 span/beam - average) is 30 sec in a micro with a   2 Mhz clock and with an 80 c.p.s printer, reduced to half,   this time after compiling.

    This system is in use in S. Paulo Brasil for the last   2 years (on microcomputers).

MY ADDRESS:

Fundação Armando Alvares Penteado
Centro de Informatica
a/c Prof. Simpson S. Kalmus
R. Alagoas, 903
CEP 01242 S.Paulo   Brasil

COMPLEX WAVEFORMS from a PET and BASIC

D.A. Pirie

Aeronautics & Fluid Mechanics Department,
Glasgow University, Glasgow, U.K.

1. INTRODUCTION.

Most fatigue test machines can be driven by an
external voltage source. This report describes a
method of synthesizing suitable voltage waveforms
(of any desired shape) using a Commodore PET/CBM
microcomputer and a (12 bit) D/A converter connected
to the PET's IEEE port. The software is written in
BASIC.

The method could be adapted for use on any
microcomputer offering the following: a user port to
which a D/A converter could be connected, BASIC with
PEEK and POKE commands and a user-settable pointer
to the end of the BASIC program storage area in
memory.

The waveform is synthesized from a discrete set
of ordinates. BASIC is simple but slow, so if the
processes of calculating each ordinate, converting
to the 2-byte form appropriate to the D/A converter
and transmitting to the converter are carried out
"on-line" the desired frequency (~3Hz) can only be
obtained if the number of points used to synthesize
each cycle of the waveform is unacceptably small.

Much faster results are possible if these
processes are separated. Thus, all the required 2-
byte values are calculated and stored in memory;
these values are subsequently retrieved from memory
and transmitted to the D/A converter. The later
sections of this report discuss the implementation
of these processes in detail.

A WORD OF WARNING - the separation of these
processes of calculation and D/A conversion opens

wide the door to programming errors which could
result in the output from the D/A converter being
catastrophically greater than intended. It is
therefore strongly recommended that any test run on
a fatigue test machine should be immediately
preceded by a "proving run" in which the output from
the D/A converter is sent to a chart recorder.

2.   THE HARDWARE.

Any PET/CBM microcomputer of the 2000, 3000 or
4000 series may be used. The D/A conversion was
carried out by a PCI6300 interface (a multi-channel
D/A and A/D device by CIL Microsystems) connected to
the IEEE 488 instrument bus port on the PET. The D/A
sections of this interface provide a voltage output
in the range ±10V, to 12 bit (1 in 4096) resolution.
In principle, any 12-bit D/A converter would be
equally suitable (but might encode values in 2-byte
form in a different way!).

3.   THE SOFTWARE.

Four distinct processes must be programmed:

(1) calculation of the waveform ordinate
(2) conversion of this value from
    floating-point form to 2-byte form,
(3) storage of these 2 bytes in memory
(4) retrieval of these 2 bytes from
    memory and transmission to the D/A
    converter.

(Step (4) is really two steps but it is simple, and
faster, to combine the two into one).

Usually, steps (1),(2),(3) would be repeated a
number of times until all the required waveform
ordinates had been calculated and stored in memory.
The BASIC program containing these steps can be
quite distinct from the program containing step (4),
or one program can be written to contain all 4
steps.

Step (1) simply calls for an appropriate BASIC
assignment statement e.g.

linenumber Y = 1997
    or
linenumber Y = FS + AM * SIN(A)
   etc.

Step (2) is specific to the D/A converter

used. For the CIL PCI6300 interface the two bytes
(high and low) corresponding to the real value Y
(-2047<=Y<=2047) are obtained from

```
linenumber HI% = ABS(Y)/256
linenumber LO% = ABS(Y) - HI% * 256
linenumber IF Y>=0 THEN HI% = HI% + 8
```

(The % symbol indicates an integer variable.)

Step (3) utilises the BASIC POKE command, e.g.

```
linenumber POKE P,HI% : POKE P+1,LO%
```

which "pokes" (i.e. stores) the value HI% into
memory location P and the value LO% into memory
location P+1.

Step (4) uses the BASIC PEEK command to
retrieve the contents of the memory locations.
Transmission to the D/A converter is specific to the
converter - for the CIL interface the PET's IEEE bus
command PRINT# is used in the following format

```
linenumber PRINT#1,"O1"CHR$(PEEK(P))CHR$(PEEK(P+1))
                     ↑
                     !____letter O
```

Here the PRINT#1 command sends data over the IEEE bus
to logical file number 1. The data begins with the
string constant "O1" which selects Output channel 1
of the CIL interface as recipient of the two
following bytes CHR$(PEEK(P)) and CHR$(PEEK(P+1)).

The following example shows how steps (1) to (4)
may be combined to store NA points of a sinewave in
memory (starting at location P) and then transmit NC
cycles of this sinewave to the D/A converter. The
example also shows the OPEN and CLOSE statements
which must accompany the PRINT# statement (in PET
BASIC).

EXAMPLE PROGRAM:

```
(initial block of program assigns values to NA, NC,
FS, AM, P; resets "END of BASIC" pointer.)
40 REM             *******************
50 DA = 2 * π/NA
51 REM    DA = ANGLE BETWEEN SUCCESSIVE ORDINATES
60 REM             *******************
100 FOR A = 0 TO (NA - 1)
110 Y = FS + AM * SIN(DA * A)          : REM STEP 1
120 REM            *******************
```

```
200 HI% = ABS(Y)/256                          : REM
210 LO% = ABS(Y) - HI% * 256                  : REM STEP 2
220 IF Y>=0 THEN HI% = HI% + 8                : REM
230 REM                        *******************
300 POKE P+2*A,HI% : POKE P+2*A+1,LO%  : REM STEP 3
310 NEXT A
320 REM                        *******************
400 OPEN 1,10
410 FOR I = 1 TO NC
420 FOR A = 0 TO (NA - 1)
425 P0 = P+2*A : P1 = P+2*A+1
429 REM *                      NEXT LINE IS STEP 4
430 PRINT#1,"01"CHR$(PEEK(P0))CHR$(PEEK(P1))
440 NEXT A
450 NEXT I
460 CLOSE 1
470 REM                        *******************
```

The sinewave will repeat indefinitely if line 410 is
deleted and line 450 replaced by

450 GOTO 420

Line 400 shows the form of the OPEN statement, which
must precede the PRINT# statement. Here, logical
file number 1 (as used in the PRINT# statement) is
OPENed to device number 10 (the CIL interface). Line
460 shows the CLOSE statement which should be used
when transmission of data is complete.

5. PRACTICAL CONSIDERATIONS

The above example program illustrates all the
essential features of the BASIC programs used in
this method of synthesizing waveforms. Clearly,
that program falls a long way short of what would be
called for in a practical fatigue test program.
Among the additional features required would be the
conversion of specified load offsets and amplitudes
into voltages and hence into input values to the D/A
converter, counting of load cycles, provision of
"user-friendly" instructions for the test operator
(who might not be particularly "computerate"). It
might also be desirable to be able to store all the
calculated ordinates on disk or to have a simple
means of adjusting the time-scale of the synthesized
waveform.

5.1 Memory requirements
With a 12-bit D/A converter two memory locations are
needed to store each waveform ordinate. A 32K PET
should therefore offer the prospect of storing at

least 15000 ordinates. Waveforms of very considerable complexity can therefore be easily synthesized, particularly as a substantial amount of repetition (e.g. of sinewave cycles) will often feature in the desired waveform.

5.2 Speed
The execution speed of a BASIC program can be improved in various ways e.g. by careful attention to the order in which variables make their first appearances in the program, to the ordering of subroutines, to the use of multiple statements to each line number, omission of REMs (comments) etc. West [1] gives a comprehensive account.

The top speed (of retrieving 2-byte values from memory and passing them through the D/A conversion process) attainable from PET BASIC appears to be about 50 per second. This speed can be used to generate sinewaves of frequency 2Hz which look quite respectable on an oscilloscope. When used to drive a 20 tonne fatigue test machine it appeared that the dynamics of the machine itself filtered out the 50Hz "ripple". This frequency (2Hz) was close to the upper limit of this machine but the impression was gained that the method could probably be pushed to about 3Hz.

Figure 1 is a chart recorder trace of part of a "Gust Load" test program applied to an aircraft component.

6. CONCLUSIONS

Reasonably accurate waveforms (with sinewave components of frequency 2-3Hz) can be synthesized in a relatively simple fashion. The same method can synthesize square waves at about 25Hz (since the D/A converter holds each value till it receives a new one). The method should be accessible to anyone who can write simple programs in BASIC.

REFERENCES

1. West, R. (1982) Programming the PET/CBM. Level Ltd.

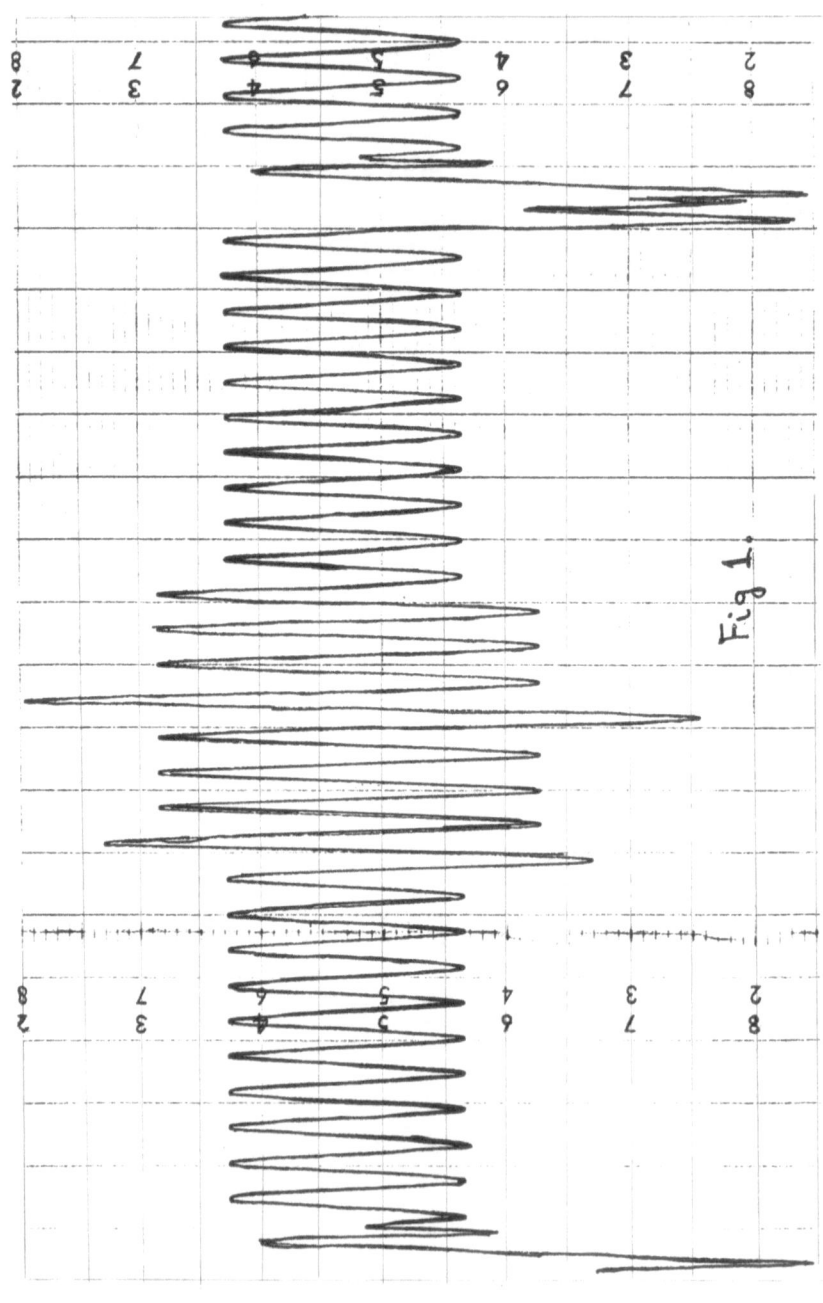

Fig 1.

THE ANALYSIS OF SPECKLE INTERFEROMETRIC RECORDS OF STRAIN BY
MICRO COMPUTER

I Grant and G H Smith

Heriot-Watt University, Edinburgh, UK

1. INTRODUCTION

Strain measurements are commonly made by attaching sensitive
gauges or films to suspect or critical regions of a particular
region of a structure or specimen under load test.

This paper describes a non-contacting optical method which
allows the rapid scanning of large areas to identify regions of
high strain. Measurements are made using photographic
techniques which allow assessment of strain levels in
identifiable "hot-spots". The naturally occurring granular
appearance of a surface exposed to laser light provides an
irregular optical grid for the measurement of surface
displacements.

A complementary apparatus is described which allows rapid,
computer controlled scanning and analysis of the speckle
photograph. The interference pattern produced on illuminating
the photograph with a coherent light source was swept by a
photosensitive detector driven by a micro-computer. The same
computer with suitable disk peripherals acted as a storage and
analysis device. A recent advance in the analysis hardware
using a vidicon system is described and illustrated.

The interference pattern produced was of a highly periodic
nature. Consequently a spatial analysis technique was adopted
as the main method of analysis. There were however occasions
when a knowledge of the full spatial frequency characteristics
of the signal were required and a two dimensional Fourier
Transform technique was also developed. The choice of method
was dependent on the nature of the problem.

The nature of speckle
Any optically rough surface will show an apparently granular
structure when coherent (laser) light is reflected from it.

This was due to the interference of the coherent light waves reflected from different points on the surface. The nature of the speckle pattern was due to the random variation in the roughness of the surface which produced a random variation in the phase of the interfering beams (1,2). The speckle can be thought of as 'fixed' to the surface and can be used as a method of measuring local displacements or strains on a loaded structure.

2. PROCEDURE FOR RECORDING SURFACE DISPLACEMENTS

i   The optical arrangement required to measure in-plane displacements is shown in Figure 1. A beam, from a Helium-Neon laser, was passed through a spatial filter to reduce optical noise. The beam was then expanded and directed onto the area under examination with a mirror. A 35mm camera was positioned normal to, and focussed on the surface (defocussing the camera allows the system to become sensitive to surface tilt and can be used in such a mode if required). The camera lens was covered with a narrow band diffraction filter which allowed only light of a wavelength corresponding to the He-Ne laser to expose the film. This enabled the measurements to be taken in ambient or subdued light. The intensity variation was recorded on holographic film. This being of very high spatial resolution was eminantly suitable for recording the speckle pattern. Normal low resolution negative film, for example Ilford FP4, could be used but did not provide data of as high quality.

ii   An exposure was made of the unstressed structure and related speckle pattern.

iii The structure was then loaded. This could be done by a variety of methods including mechanical or thermal stressing.

iv   A second exposure of equal time length was then taken on the same frame of film thus producing a double exposure.

Since the speckle moves with the surface any movement of the structure was mirrored by the second speckle pattern recorded on the film. Thus the double exposed negative contained all the information regarding the in-plane displacement of the surface. Strain data was derived from numerical differentiation of the displacement data taken over a region of the structures surface. The technique described here allows a point by point analysis of the negative.

3. ANALYSING THE DOUBLE EXPOSURE

A collimated laser beam was directed through the negative at the point of interest. If the displacement at that point was within the sensitivity governed by the optical arrangement fringes will be observed within a so-called diffraction halo, Figure 2 (see Appendix 1). If the fringes were viewed at a distance, L, from the negative plane the displacement, δx, on the structure was related to the fringe spacing, δF, by

$$\delta x = \lambda L / M \delta F \qquad (1)$$

where M was the magnification of the recording cameras optical system. Furthermore the inclination of the fringes was perpendicular to the direction of displacement.

Thus one arrives at a displacement vector for one point on the negative. To derive strain data it was necessary to analyse a grid of points on the negative making an automatic method of scanning and analysis desirable. The system described was built around the BBC model B microcomputer. This had the advantage of low cost but with a facility for upgrading.

Recording of intensity variations in the diffraction halo
The speckle negative was placed in its holder on an x-y traversing system. The two stepper motors were controlled from the BBC printer port. The direction of each stepper was set by two bits of the printer port register. The active motor was driven by setting another bit alternately to 0 and 1. In this way the negative was moved such that the laser beam could pass over its whole surface.

The resulting fringe pattern was analysed by a photo-transistor again mounted on an x-y traversing system in this case controlled via the BBC user port. A pinhole was placed over the photo-transistor to restrict the angle over which the light was gathered. Its size was such that it was larger than an individual speckle seen in the fringe pattern but smaller than the fringe spacing.

The output of the photo-transistor was amplified and input to the BBC analogue to digital convertor. The amplifier as well as providing a means of utilising the full range of the ADC ensured that the voltage input to the convertor did not become negative as this could result in its damage. It was found that calibration instability of the BBC ADC, caused by temperature drift, was insignificant in the present application.

The following procedure was then carried out, Figure 3. The slide was placed at the first point of interest and the photo-transistor moved across the pattern to record the intensity distribution and calculate the fringe spacing and inclination. This data was then stored on disk. The negative was then traversed to the next position and the procedure repeated. The resulting displacement data was then utilised to provide strain measurements over the surface.

Analysis of the intensity distributions

In the current application local fringe spacing and inclination provided sufficient detail of distortion under load. This indicated that a spatial technique would be an efficient approach. In other applications a full Fourier analysis of the two dimensional intensity distribution may be necessary and this technique was also investigated.

Spatial methods. Two spatial techniques were employed:

(a) A high density scan, in one direction, of 150 points at a spacing of 0.5mm in 5 low density scans spaced 2.5mm apart in the orthogonal direction. The points were smoothed over 3 adjacent values and the peak spacing in the high density direction was found (see below). The peak spacing in the other, low density direction, was found by calculating the difference in phase between peaks corresponding to the same fringe. The fringe inclination could then be calculated knowing the distance between low density scans.

(b) Two high density scans in each of the orthogonal directions. The peak spacings were then found explicitly for each direction.

Finding a peak

(a) An approximate value of peak separation was found by calculating the distance between the two maxima at each end of the scan and dividing by the total number of peaks found. A maximum was defined to be a point which was the largest of a group of ± 3 points.

(b) The intensity values were then reanalysed to find points which were within 95% of each intensity maximum and within a distance of ± 40% of the approximate value calculated for the fringe spacing. The second condition ensured that final averaging did not take place over more than one peak.

(c) An average position was then calculated from these points by weighting their positions according to their intensity values.

Fourier transform methods. This technique required a large amount of storage for the data points and thus a transforming technique which made efficient use of the computer RAM was necessary. The two dimensional FFT was carried out by transforming each row of the N*N array of points and then transforming the resultant columns (see Appendix 2). The algorithm used was a radix 2 method. The computationally efficient mixed radix methods were not possible due to the lack of storage space. The resulting peak values in the spectral density then provided an estimate of the dominant spatial frequencies in each direction and hence the fringe spacing in the x and y directions.

Three features of the FFT method employed worth noting are:

i   The BASIC language does not support complex arithmetic and variables and so two arrays were required to calculate the real and imaginary parts of the Discrete Fourier Transform separately.

ii   To calculate the full Fourier Transform on an array of N*N points required N + N/2 Discrete Fourier Transforms. This was because the transformed rows were symmetric about their middle due to the original data being real.

iii The maximum array size was restricted to 64*64 points to minimise the amount of data transfer to and from disc. Thus each row was input and transformed in turn but the resulting "half-row" remained in the computer ram. If larger array sizes were required it would be necessary to restore the transformed rows back to disc to free enough memory. The data for each column would then have to be brought back from the disc. This would greatly increase the computational time required.

4.   RESULTS

Tests were first conducted to determine the reproducibility of the fringe intensity pattern. It was found that any change in the absolute position of the fringes was unimportant as only relative spacings were required. Nevertheless the intensity distributions derived from two separate scans of the fringe pattern were comparable to within less than 0.5mm.

Each numerical method was tested by producing a doubly exposed negative of a structure displaced laterally by a known amount. Fringes of a specific inclination to the traversing direction could be produced by rotating the slide through a known angle.

Actual displacements of 100, 150 and 200 microns were used. The camera aperture was set at F5.6 and the optical magnification at 0.4. The negative to detector distance, L, was 0.665m. Since one was measuring the ability of the analysis system to determine the resultant fringe spacing these displacements were converted to corresponding fringe spacings, at the photodetector plane, by inserting the appropriate values into Equation (1). These were then used to calculate the fringe spacings in the x and y directions for a particular fringe orientation.

Figures 4-7 show the intensity distributions as measured by the traversing photo transistor. The results for each method are tabulated in Table 1. The figures illustrate the general features of the interference pattern.

i   The fringes were embedded in a   envelope, the 'halo',

Structural Displacement (microns)	Angle to x Traversing Direction	Expected Fringe Spacing(mm)		Spatial Method 1			Spatial Method 2			FFT Method		
		dx	dy	dx	dy	time	dx	dy	time	dx	dy	time
100	90°	10.52	∞	9.75	>80	46 secs	9.7	>70	50 secs	9.14	no peak	10.5 mins
150	90°	7.01	∞	6.60	>80		6.5	20		6.0	"	
200	90°	5.26	∞	5.00	>80		5.1	32		4.8	"	
100	60°	12.14	21.00	11.5	27.2		11.6	23.1		10.8	19.1	
150	60°	8.09	16.18	7.4	19.7		7.4	17.0		7.6	15.1	
200	60°	6.07	12.14	5.5	13.5		5.4	11.4		5.4	11.1	

TABLE 1

and thus increased in intensity as one approached the zero order, central region. It should be noted that these scans were carried out at some distance from the central area, the lowest scan in each figure being the closest.

ii The vertical bars show the positions that were derived from the data by the peak finding algorithm. As can be seen not all peaks can be expected to be detected. This was overcome by subtracting the initial approximation for the peak separation from the distance between two recorded peaks until that distance was of the same order as the initial approximation. The final peak spacing was calculated as an average of all the values from the scans.

iii Figures 4 and 6 show peak distributions for fringes which were set at 90° to the traversing direction. As can be seen there was no systematic trend in the fringe positions between each scan. Figures 5 and 7 however show the effect of fringes which were tilted with respect to the scan direction. In this case a clear change in phase from a peak in one scan to a corresponding peak in another scan was seen.

Figure 8 illustrates the result of applying an FFT to the data. As can be seen the spatial frequency was easily discernable. The peak spacing was calculated as the inverse of the spatial frequency.

Table 1 details the fringe spacings calculated by each method and compares them with the values derived knowing the structure displacement and experimental parameters. Also given are the times required to perform the analysis of the recorded intensity data.

5. DISCUSSION

i   Spatial method 1: The error in measuring the fringe separation in the x direction was typically 5%. This method had the advantage that the variation in light intensity over the traversing region was a minimum and permitted a relatively unsophisticated scanning procedure. The one difficult with this technique was the accuracy to which the fringe inclination could be found.

ii Spatial method 2: This method gave an average error of about 4%. Two problems were encountered with this technique. Firstly it was found that spurious peaks were liable to be found when the density of fringes in the particular scan direction was low. This led to an apparent displacement (strain) component in this direction. A more sophisticated algorithm for choosing the peak will remedy this. Secondly the positioning of the photo-transistor was more difficult. This was due to the need to scan a large distance in both the x and y directions.

iii Fourier Transform method: This was found to typically underestimate the displacements by 8% but changes in the algorithm are expected to improve this result markedly. It was however found to give more consistent measurements in both the x and y directions for the whole range of fringe inclinations. This method was better suited for coping with the random parts of the signal and less likely to pick up extra peaks. As can be readily seen the full two dimensional transform was inappropriate for the data required for the analysis of fringe spacing. The length of time required to calculate the N+N/2 transforms, exceeding 10 minutes, was much greater than the other two methods (under 50 seconds). The results do however show that the transform method does work and for the purposes of calculating fringe spacing two independent one dimensional transforms in the x and y direction would suffice. These would require only 48 seconds to transform two scans of 128 points.

It should be noted that the errors stated were typical for the analysis of data for displacements ranging from ~10-500 microns. Thus there was no loss of accuracy in the measurement of very small displacements. The accuracy was mainly dependent on the optical arrangement used to record the speckle pattern. Although the present results were presented in terms of the accuracy with which the fringe spacing could be determined the implications for strain recording are clear. The measurement of strains greater than 10-15 microstrain would be possible with an error of less than 10%.

The main problems to be overcome with this method are connected with the large variations in intensity over the diffraction halo. This factor necessitates careful positioning of the phototransistor before scanning. A new system is currently under development. In this the traversing photodetector is replaced with a Vidicon camera which is connected to the microcomputer via a digital interface. The image can consist of an array of 256*256 points with each pixel having any one of 256 grey levels. The advantage of this system is that it provides the ability to image the whole interference pattern in less than 10 seconds. The best region for analysis can then be readily found and the appropriate image processing applied where required. Figure 9 illustrates the new apparatus.

CONCLUSIONS

The point by point method used to analyse the double exposure speckle photograph meant that marked diffraction effects were evident in the viewing plane. Consequently an "operator-learning time" was required before rapid implementation of the analysis procedures was possible. After this period rapid and efficient computer aided processing of strain patterns was possible using the scanning technique. The major benefit of this method is the relatively low capital cost of the apparatus.

Preliminary studies with the Vidicon camera system indicate that this method of whole field analysis will greatly simplify the procedure allowing an even greater degree of automation.

APPENDIX 1 : THE NATURE OF THE VIEWED FRINGES

The intensity distribution resulting from the point by point analysis of the negative resulted from three properties of the double exposed negative:

i   The individual speckles, recorded as deposits of silver on the negative, diffract the light into a cone, the so-called Diffraction Halo. The angular extent of this cone is dependent on the size of the speckle recorded on the film. An approximate relationship between speckle size and halo angle is given by

$$\beta_H = \frac{\lambda}{D} \tag{2}$$

where D = speckle size on negative = $(1+M)*\lambda*F$.

ii   The pairs of speckle corresponding to the surface displacement diffract the light which interferes to produce cos**2 fringes. The angular extent of these fringes is given by

$$\beta_F = \frac{\lambda}{M*\delta x} \tag{3}$$

iii   A third feature of the intensity distribution is the appearance of Objective Speckle due to the negative being illuminated over a small region by the laser beam. The angular extent of an objective speckle is approximately

$$\beta_o = \frac{\lambda}{a} \tag{4}$$

where a is the diameter of the laser beam.

The three equations above can be used to calculate approximate values for the upper and lower bounds of measurement.

i   For a displacement to be measured at least two fringes must be within the diffraction halo. Thus the angular extent of the fringes must be less than the angular extent of the halo

$$\beta_F < \beta_H$$

$$\Rightarrow \delta X(min) = (1+M)*\lambda*F/M \tag{5}$$

ii The maximum displacement measurable is determined by the size of the objective speckle. The fringe spacing must be at least as great as the size of the speckle to be visible

$$\beta_F > \beta_o$$

$$\Rightarrow \quad \delta X(max) = a/M \tag{6}$$

Typical values
M = 0.5
F = 5.6
a = 1mm
λ = 632.8nm (for Helium-Neon light)

Thus $\delta X(min)$ = $5.3*10**-6m$
      $\delta X(max)$ = $2.0*10**-3m$

It should be noted that displacements less than that given by Equation (5) can be measured by introducing a deliberate shift, of a known amount, to the negative between exposures. This shift can then be subtracted out at the end.

APPENDIX 2 : THE TWO DIMENSIONAL FOURIER TRANSFORM

A two dimensional periodic signal is defined by

$$x_p(n_1,n_2) = x_p(n_1+m_1N_1, \; n_2+m_2N_2)$$

where N_1 and N_2 are the periods along each dimension.

This periodic signal can be represented by a linear combination of exponentials whose periods are sub-periods of N_1 and N_2 thus

$$x_p(n_1,n_2) = \frac{1}{N_1N_2} \sum_{k_1=0}^{N_1-1} \sum_{k_2=0}^{N_2-1} X_p(k_1,k_2) e^{i(\frac{2\pi}{N_1})k_1n_1} e^{i(\frac{2\pi}{N_2})k_2n_2} \tag{7}$$

The Fourier coefficients $X_p(k_1,k_2)$ represent the amplitude of $x_p(n_1,n_2)$ at the two-dimensional frequency

$$w = (\frac{2\pi}{N_1})k_1 \quad \text{and} \quad w = (\frac{2\pi}{N_2})k_2$$

The Fourier coefficients can be obtained by inverting Equation (7) above.

$$X_p(k_1,k_2) = \sum_{n_1=0}^{N_1-1} \sum_{n_2=0}^{N_2-1} x_p(n_1,n_2) e^{-i(\frac{2\pi}{N_1})n_1k_1} e^{-i(\frac{2\pi}{N_2})n_2k_2}$$

which can be written as

$$X(k_1,k_2) = \sum_{n_1=0}^{N_1-1} e^{-i(\frac{2\pi}{N_1})n_1k_1} [\sum_{n_2=0}^{N_2-1} x_p(n_1,n_2)e^{-i(\frac{2\pi}{N_2})n_2k_2}]$$

The bracketed term represents a series of N_1 one-dimensional discrete Fourier Transforms. Let $[\ ] = D_p(n_1,k_1)$ then

$$X(k_1,k_2) = \sum_{n_1=0}^{N_1-1} e^{-i(\frac{2\pi}{N_1})k_1n_1} D_p(n_1,k_2)$$

which is a series of N_2 one dimensional Discrete Fourier Transforms as k_2 varies from 0 to N_2-1.

Furthermore since the input series is real the Fourier coefficients are symmetric and thus only half the $D_p(n_1,k_2)$ points are required to evaluate the transforms in the second direction.

REFERENCES

1. Goodman J W (1975) 'Laser Speckle and Related Phenomena', ed Dainty J C, Springer Verlag.
2. Erf Robert K ed (1978) 'Speckle Metrology', Academic Press.

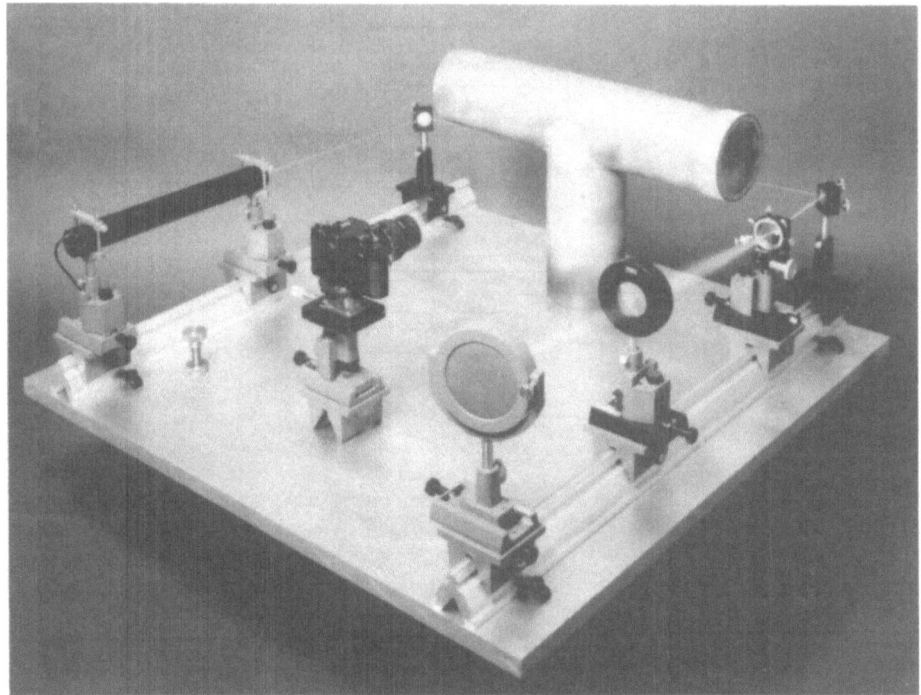

Figure 1: Apparatus Used to Record Speckle Pattern

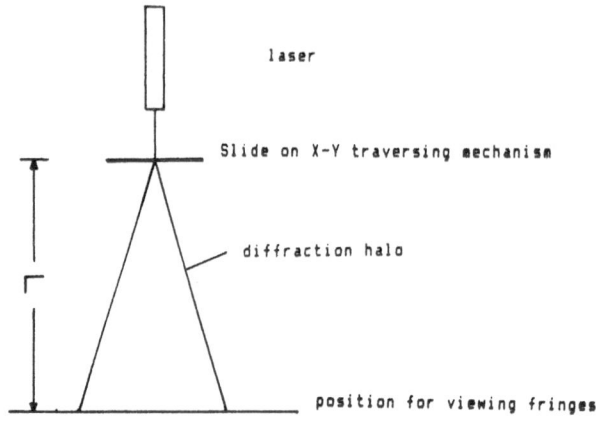

Figure 2: Arrangement for Reading Double Exposed Negatives

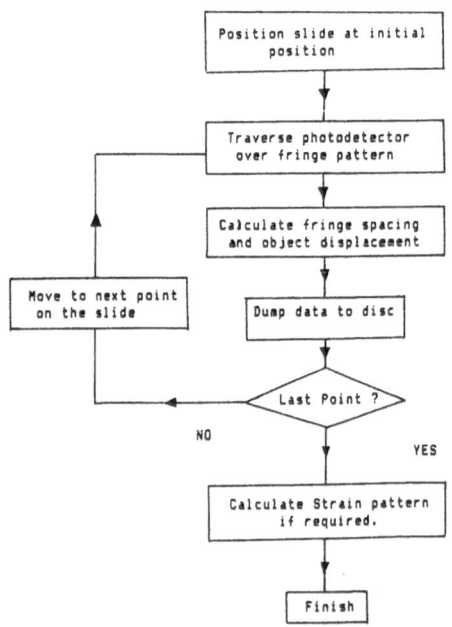

Figure 3: Flow Diagram of Data Acquisition

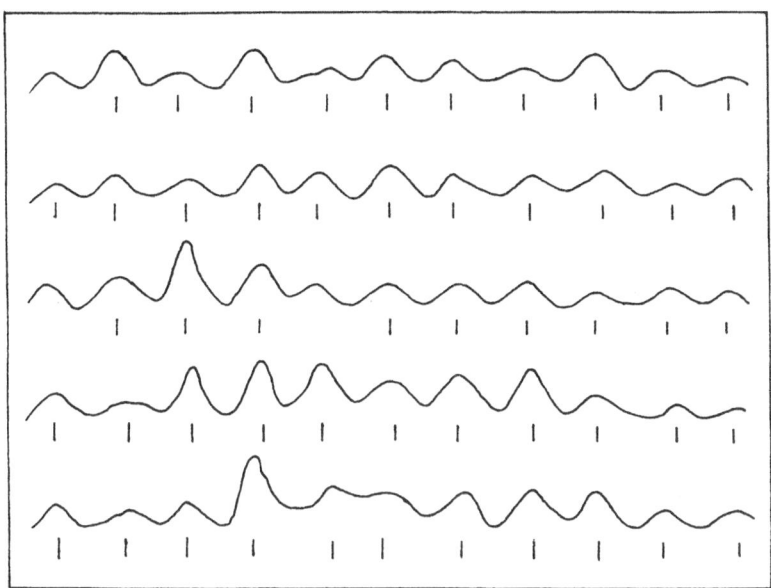

Figure 4: Intensity Distribution in Halo
Structure Displacement = 150 microns, angle = 90°

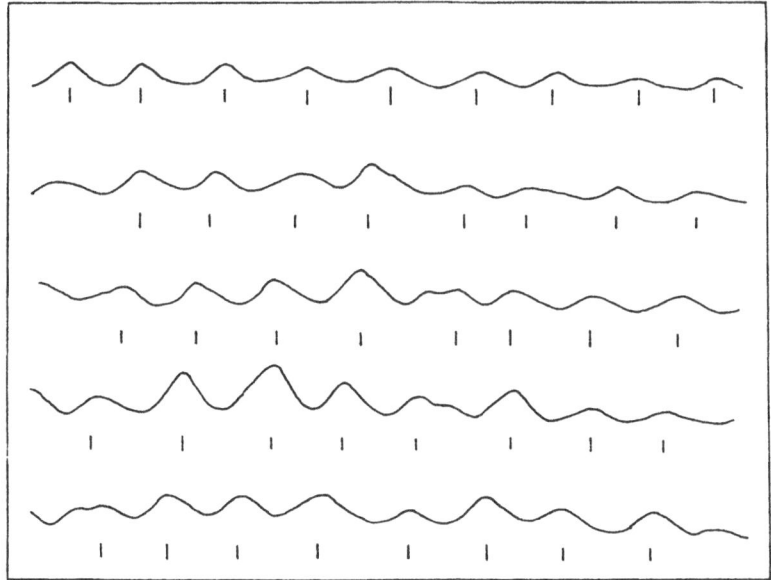

Figure 5: Intensity Distribution in Halo
Structure Displacement = 150 microns, angle = 60°

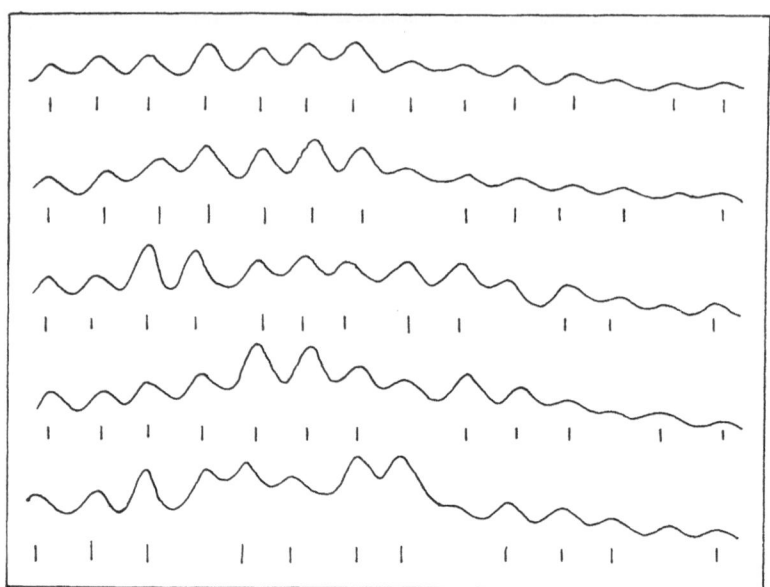

Figure 6:   Intensity Distribution in Halo
            Structure Displacement = 200 microns, angle = 90°

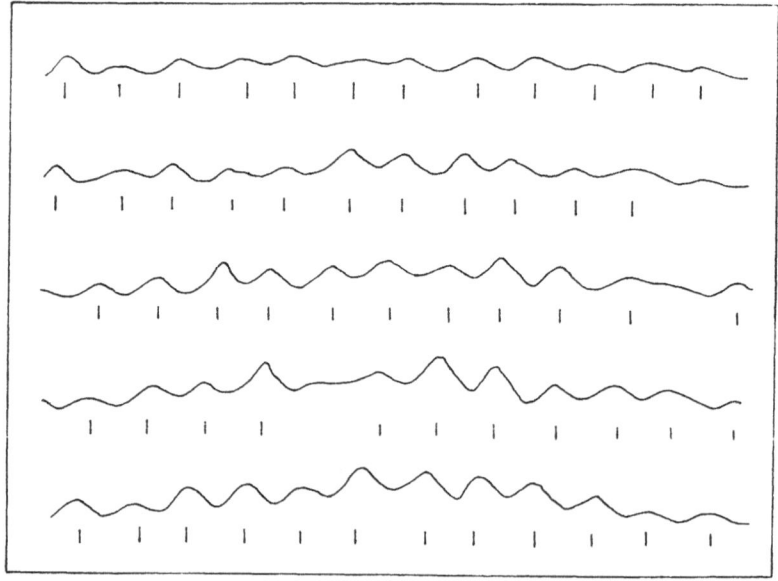

Figure 7:   Intensity Distribution in Halo
            Structure Displacement = 200 microns, angle = 60°

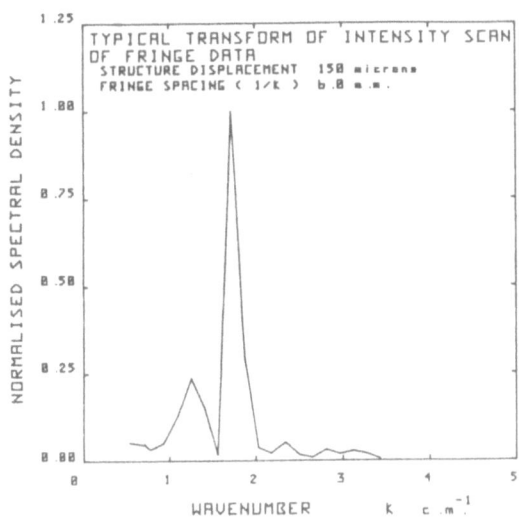

Figure 8: Spectral Density of Intensity Distribution

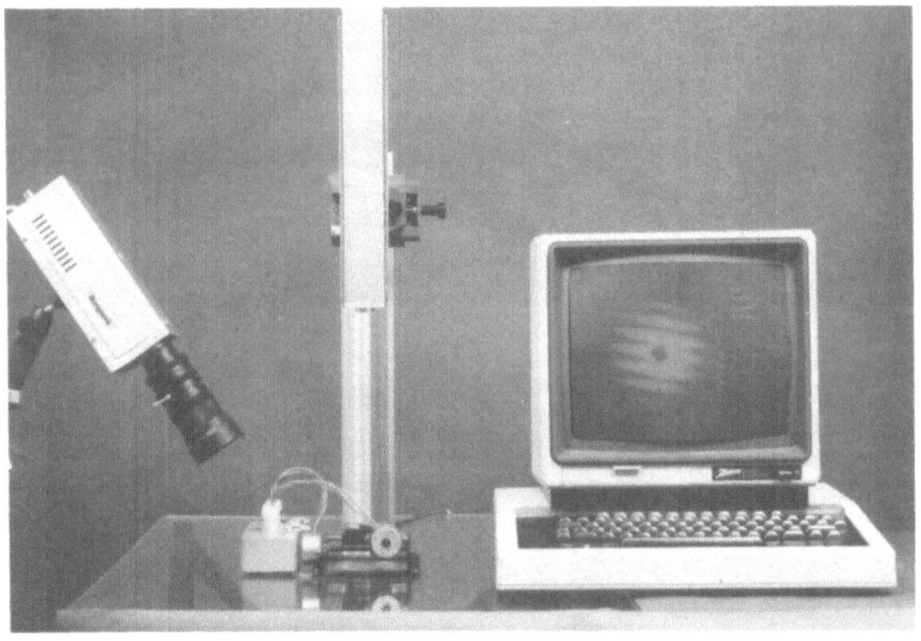

Figure 9: Modified System Employing a Video Camera and
Digital Interface for Data Acquisition

10. STRUCTURAL ENGINEERING

VAPSIMM: A COMPUTER PROGRAM FOR VIBRATION ANALYSIS OF PIPING
SYSTEMS INCORPORATING A MOVING MEDIUM

C.W.S. To

Department of Mechanical Engineering, The University of
Calgary, Calgary, Alberta, Canada T2N 1N4

ABSTRACT

The theory and structure of a computer program, VAPSIMM, for the
vibration and stability analysis of piping systems containing a
moving medium, are outlined in this paper. VAPSIMM is based on
the transfer matrix method in which the order of the transfer
matrix is kept unchanged with increasing number of piping
elements and complexity of the system, and every piping element
is considered as a distributed-parameter model. Computed and
experimental results of resonant frequencies for various flow
velocities of the contained medium from a laboratory model of
an elbow pipe are presented to show the accuracy. Selected
computed results for three-dimensional and helical piping elements are also presented to demonstrate the capability of the
program.

1. INTRODUCTION

The demand for dynamic analysis of piping systems containing a
moving medium has been increasing in the last two decades. This
is particularly so in the aerospace, nuclear power, oil and gas
industries. For instance, in the aerospace industry, the problem basically involves the necessity of delivering large quantities of fuel in a relatively short period of time [1]. Fuel
deliveries of 10^6 lb. in approximately 3 minutes are common and
these can result in Reynolds number as high as 3×10^7. This
creates fluctuating fluid forces on the containing pipe resulting from the passage of pressure transients, as, for example,
in starting and stopping the flow, and those forces associated
with periods during which the flow is quasi-steady.

To comply with the material strength requirements in the
design process a vibration and stability analysis is necessary.
A rigorous analysis can only be obtained by using shell or
three-dimensional theory. When the piping system contains
elements of various shapes and cross-sections one frequently

resorts to the application of the versatile numerical technique, the finite element method [2]. However, with the present state of the art a vibration and stability analysis employing the finite element method based on the shell or three-dimensional theory is very costly. Even when the finite element method is based on the simple beam theory the analysis of an entire piping installation can often prove to be expensive.

This paper describes a digital computer program, VAPSIMM, for the vibration and stability analysis of piping systems incorporating a moving medium. It is based on the transfer matrix approach in which the order of the transfer matrix is kept unchanged with increasing number of piping elements and complexity of the system [3]. It can also be implemented in a microcomputer. Consequently, it is much less expensive to use than those employing the finite element method. Moreover, compared with the finite element method based on simple beam theory, the adopted analysis, in which every piping element is considered as a distributed-parameter model, provides more accurate results.

In the next section a condensed version of the adopted analysis, on which the computer program is based, is presented. Section 3 deals with the computer program structure. Section 4 includes applications of VAPSIMM. Some computed results for an elbow piping element are compared with those obtained experimentally from a laboratory model. Computed results for three-dimensional and helical piping elements are also presented. Discussion on the results and concluding remarks on VAPSIMM are included in sections 5 and 6, respectively.

2. THEORY

The transfer matrix method employed in the development of VAPSIMM is the one presented by To and Kaladi [3]. In this method the piping system is modelled as one containing a series of piping elements. Every piping element may be described by the following equation of motion [4,5].

$$EI \frac{\partial^4 y}{\partial x^4} + \rho v^2 \frac{\partial^2 y}{\partial x^2} + 2\rho v \frac{\partial^2 y}{\partial t \partial x} + m \frac{\partial^2 y}{\partial t^2} = 0 \qquad (1)$$

where E = modulus of elasticity, I = moment of inertia of the pipe, m = mass of the pipe with fluid per unit length, ρ = mass of the fluid per unit length, v = steady mean flow velocity of fluid with respect to pipe, x = co-ordinate measured along the pipe length, y = lateral displacement of the pipe.

In Equation (1), the first term represents the elastic force; the second, the inertia force associated with the fluid following a curved path; the third, the inertia force associated with the Coriolis acceleration due to the relative motion of the fluid inside the pipe which has an angular velocity and the last term represents the inertia force due to lateral acceleration of the pipe including the moving medium.

By following the procedure described in [3] the transfer matrix with respect to the local co-ordinates for combined vibration, $[T_L]$, can be obtained as [3]

$$[T_L] = \begin{bmatrix} [F_A] & & & 0 \\ & [F_T] & & \\ & & [F_{xy}] & \\ 0 & & & [F_{xz}] \end{bmatrix} \quad (2)$$

where $[F_A]$, $[F_T]$ and $[F]$ denote transfer matrices for axial, torsional and flexural vibrations respectively. The subscripts xy and xz designate the planes in the rectangular co-ordinate system considered.

The above transfer matrix relates the state vector at the ends of the pipe element, with respect to the local co-ordinate axes, the x-axis being along the length of the pipe. This shall be referred to as the local transfer matrix.

If $[T_g]$ denotes the transformation matrix that relates the state vector with respect to the local co-ordinates, xyz, for any pipe element, to the state vector with respect to the global co-ordinates XYZ, then the transfer matrix for relating the state vectors in global co-ordinates, at the two ends of the pipe element, becomes

$$[T_G] = [T_g][T_L][T_g]^T \quad (3)$$

where the subscript G denotes global co-ordinates and the superscript T denotes the "transpose of".

In a given three-dimensional piping system, the overall transfer matrix connecting state vectors at the end points of the system is obtained by multiplying together the transfer matrices in global co-ordinates for all pipe elements. That is,

$$[T] = \prod_{i=1}^{n} [T_{Gi}] \quad (4)$$

where the first subscript, G, denotes the global co-ordinates and the second subscript, i, denotes the piping element number.

The transfer matrix [T] whose order is twelve for the three-dimensional system is used to determine the resonant frequencies with due consideration of the boundary conditions.

For example, for a piping system simply supported at

either end, all the moments and deflections at either end must
vanish. This gives

$$\{z\}_L = [T]\{z\}_0 \tag{5}$$

where

$$\{z\} = \{0 \quad \theta_X \quad 0 \quad V_X \quad 0 \quad \theta_Z \quad 0 \quad V_Y \quad 0 \quad \theta_Y \quad 0 \quad V_Z\}^T$$

Equation (5) can be rewritten as

$$\begin{Bmatrix} z_1 \\ z_2 \end{Bmatrix}_L = \begin{bmatrix} T_{11} & T_{12} \\ T_{21} & T_{22} \end{bmatrix} \begin{Bmatrix} z_1 \\ z_2 \end{Bmatrix}_0 \tag{6}$$

where

$$\{z_1\} = \{ -W_X \quad M_X \quad -W_Y \quad M_Z \quad -W_Z \quad M_Y \}^T$$

$$\{z_2\} = \{ \theta_X \quad V_X \quad \theta_Z \quad V_Y \quad \theta_Y \quad V_Z \}^T$$

$$[T_{11}] = \begin{bmatrix}
t_{1\ 1} & t_{1\ 3} & t_{1\ 5} & t_{1\ 7} & t_{1\ 9} & t_{1\ 11} \\
t_{3\ 1} & t_{3\ 3} & t_{3\ 5} & t_{3\ 7} & t_{3\ 9} & t_{3\ 11} \\
t_{5\ 1} & t_{5\ 3} & t_{5\ 5} & t_{5\ 7} & t_{5\ 9} & t_{5\ 11} \\
t_{7\ 1} & t_{7\ 3} & t_{7\ 5} & t_{7\ 7} & t_{7\ 9} & t_{7\ 11} \\
t_{9\ 1} & t_{9\ 3} & t_{9\ 5} & t_{9\ 7} & t_{9\ 9} & t_{9\ 11} \\
t_{11\ 1} & t_{11\ 3} & t_{11\ 5} & t_{11\ 7} & t_{11\ 9} & t_{11\ 11}
\end{bmatrix}$$

$$[T_{12}] = \begin{bmatrix}
t_{1\ 2} & t_{1\ 4} & t_{1\ 6} & t_{1\ 8} & t_{1\ 10} & t_{1\ 12} \\
t_{3\ 2} & t_{3\ 4} & t_{3\ 6} & t_{3\ 8} & t_{3\ 10} & t_{3\ 12} \\
t_{5\ 2} & t_{5\ 4} & t_{5\ 6} & t_{5\ 8} & t_{5\ 10} & t_{5\ 12} \\
t_{7\ 2} & t_{7\ 4} & t_{7\ 6} & t_{7\ 8} & t_{7\ 10} & t_{7\ 12} \\
t_{9\ 2} & t_{9\ 4} & t_{9\ 6} & t_{9\ 8} & t_{9\ 10} & t_{9\ 12} \\
t_{11\ 2} & t_{11\ 4} & t_{11\ 6} & t_{11\ 8} & t_{11\ 10} & t_{11\ 12}
\end{bmatrix}$$

and so on.

Assuming an excitation at the end 0 in the X-direction, the
boundary conditions given

$$\{z_1\}_L = \{\ 0 \quad 0 \quad 0 \quad 0 \quad 0 \quad 0 \ \}^T \tag{7a}$$

$$\{z_2\}_0 = \{\ -W_X \quad 0 \quad 0 \quad 0 \quad 0 \quad 0 \ \}^T \tag{7b}$$

so that

$$\{z_1\}_L = [T_{11}]\{z_1\}_0 + [T_{12}]\{z_2\}_0 \tag{8a}$$

$$\{z_2\}_L = [T_{21}]\{z_1\}_0 + [T_{22}]\{z_2\}_0 \tag{8b}$$

$$\{z_2\}_0 = - [T_{12}]^{-1} [T_{11}]\{z_1\}_0 \tag{8c}$$

At any point P along the pipe, if [F] is the transfer matrix relating the state vectors at the point to the state vectors at the end 0, then

$$\begin{Bmatrix} z_1 \\ z_2 \end{Bmatrix}_P = [F] \begin{Bmatrix} z_1 \\ z_2 \end{Bmatrix}_0 = \begin{bmatrix} F_{11} & F_{12} \\ F_{21} & F_{22} \end{bmatrix} \begin{Bmatrix} z_1 \\ z_2 \end{Bmatrix}_0 \tag{9}$$

Consider the first equation in Equation (9).

$$\{z_1\}_P = [F_{11}]\{z_1\}_0 + [F_{12}]\{z_2\}_0 \tag{10}$$

Using Equation (8c) gives

$$\{z_1\}_P = \left\langle \frac{|T_{12}|[F_{11}] - [F_{12}](\ adj[T_{12}]\ )}{|T_{12}|} \right\rangle \{z_1\}_0 \tag{11}$$

where adj[ ] designates the "adjoint of". Equation (11) gives the appropriate receptance expression. At resonance, the expression inside the angled brackets of (11) becomes a maximum. This means that the denominator, $|T_{12}|$ becomes a minimum. The expression $|T_{12}|$ shall be referred to as the frequency function.

Once the resonant frequencies are determined, the mode shapes are computed using the transfer matrix in Equation (4). The overall transfer matrix is first evaluated at the resonant frequency. Then, a value for one of the elements, not governed by the boundary conditions, of the state vector at one end is assumed, such that the remaining elements of the state vector at that end are solved after application of the boundary conditions.

The whole piping system is subsequently divided into a number of sections and the state vectors at the end points of each of the sections are obtained by multiplying the state vector at previous point with the transfer matrix for the section, computed at the resonant frequency. The state vectors found at different points along the length of the piping system give the mode shape.

3. STRUCTURE OF VAPSIMM

The computer program, VAPSIMM, written in Fortran 77 by making use of the theory briefly outlined in the last section, has been developed. The operational guidelines for the digital computer program were that the program was to (a) be aimed at vibration and stability analysis of pipe networks including a single-phase flowing medium, (b) be applicable to piping systems in aerospace, nuclear power, oil and gas industries, (c) be user-friendly in terms of input data preparation and output data manipulation, and (d) be applicable to microcomputer environment.

In view of the operational guidelines, VAPSIMM is organized into four modules, namely, the input module, the analysis module, the mode shape determination module, and output module. For brevity, the mode shape determination module is not included in this paper.

The input module consists of a series of subroutines that read in data such as the geometrical properties for every piping element and associated mechanical properties, fluid properties, and operational codes.

The analysis module also consists of a series of subroutines performing four tasks as described in the following:

Firstly, the input geometrical properties for every piping element are used to calculate the direction cosines, length, cross-sectional area, second moment of area and polar moment of area of cross-section of the pipe, for example. The computed results are stored in arrays.

Secondly, the transfer matrices for axial, torsional and flexural vibrations are obtained using the associated subroutines. The transfer matrices are used to form the local transfer matrix for every piping element by another subroutine. The stored direction cosines are applied to form the transformation matrix $[T_g]$ for every piping element using the transformation matrix formation subroutine. Another subroutine performs the formation of the global transfer matrix, $[T_{Gi}]$. The overall transfer matrix, $[T]$, for the piping system is formed by the overall matrix subroutine.

Thirdly, the application of boundary conditions for a simply-supported system, for example, is achieved by one of the boundary condition subroutines identified by one of the already input operational codes.

The final task of operations in the analysis module comprises two stages. During the first stage the frequency function is determined by a subroutine. In the second stage, use of the frequency function is made and the resonant frequency is located by applying the frequency sweep subroutine.

The last three tasks are repeated until the resonant frequencies of various fluid speeds for the specified number of vibration modes are determined.

The results are subsequently stored in various files for output or further data organization and presentation in the output module.

4. APPLICATIONS OF VAPSIMM

The computer program written in Fortran 77 has been applied to the determination of resonant frequencies for an elbow pipe with various boundary conditions. For the clamped-clamped case, experimental results are also provided to show the accuracy achieved. It has also been applied to obtain resonant frequencies of a simply-supported three-dimensional piping system, and a simply-supported helical piping system. The obtained resonant frequencies for all cases are plotted against fluid velocity so that the nature of stability for every system may be identified. The following sub-sections include some details of the examples considered.

4.1 Elbow Pipes

The elbow pipe is shown in Figure 1(a). The physical properties of the pipe are given in Table 1. The results for the cantilever elbow pipe are presented in Figure 2. Figure 3 provides results of the clamped-clamped elbow pipe. Figure 4 shows the influence of choosing a different number of straight piping elements in approximating the elbow. Here, the results for the simply-supported two-element representation of the elbow pipe are compared with those of the five-element representation which is shown schematically in Figure 1(b).

Experimental verification of a clamped-clamped elbow pipe was carried out. The properties of the laboratory model are those included in Table 1. The model was arranged in a vertical plane, with the supply of water from the mains connected to the top end. Water was drained off at the lower end through a flexible hose. The flow rate was determined using an orifice flowmeter. With the water flowing the pipe was excited by tapping lightly by hand. The resulting transient signals were picked up by the Bruel and Kjaer accelerometer type 4384. The signals were sent to an amplifier and monitored by an oscilloscope before reaching the analog to digital converter of the Hewlett Packard Structural Dynamics Analyser for processing. The resonant frequencies corresponding to the peaks of the

auto-spectral density plots were observed. These observed
values are presented in Figure 5 in which the theoretical results are also included for comparison. The theoretical results
predicted are based on a four-element representation.

4.2 Simply-supported three-dimensional pipe
The simply-supported three-dimensional pipe studied is shown in
Figure 6(a). The physical properties are those in Table 1. The
system is divided into four straight pipe elements as shown in
Figre 6(b) in which the local and global co-ordinates are rep-
resented by lower case and upper case alphabets, respectively.
Results are plotted in Figure 7.

4.3 Simply-supported helical pipe
The vibration and stability analysis of helical pipe conveying
fluid is of special interest due to its wide application in
industry. For instance, helical fuel rods containing flowing
fluid are extensively used in heat exchangers in nuclear power
plants [6]. The use of helical pipes in such heat exchangers
is reported to reduce the stress level, hence prolonging the
service life of the heat exchanger. Other examples where helical pipes find application include expansion loops in an oil
pipeline.

The simply-supported helical pipe considered here is divided into twelve straight pipe elements as shown in Figure 8.
The helix angle is 5°. The physical properties of every element are those of Table 1. The results are plotted in Figure 9.

5. DISCUSSION

Referring to Figure 2, it is interesting to note that divergent
type as well as oscillatory type instabilities occur. When the
Coriolis effect is included in the analysis the first mode exhibits divergent instability at a critical velocity of about 8.4
m/s. When the Coriolis effect is disregarded the curves corresponding to the first two modes loop to form a single curve
exhibiting an oscillatory or flutter type of instability.

For the clamped-clamped elbow pipe, however, the inclusion
of the Coriolis term in the analysis has no significant bearing
(see Figure 3). The experimental results tally excellently
with predicted values for the first two modes within the range
tested. The experimental results for the third mode are relatively higher than their corresponding theoretical values. This
reveals that the simple beam theory in Equation (1) cannot be
employed to accurately predict higher mode resonant frequencies.

Figure 4 indicates that results for the two-element representation are as accurate as those for the five-element approximation except for the second mode results.

6. CONCLUDING REMARKS

As was pointed out in the introduction, a rigorous vibration
and stability analysis for a piping system including a moving
fluid can only be obtained using shell or three-dimensional
theory. However, such analysis is very costly. The computer
program, VAPSIMM, described in the foregoing sections, provides
an economic alternative. It is user-friendly and applicable to
microcomputer environment.

Current and future efforts of VAPSIMM are concentrated in two
directions. The first direction is developing graph-plotting
capability while the second is further verification of the
program.

ACKNOWLEDGEMENTS

The experimental data and numerical results in this paper were
obtained by the author's former graduate research assistant,
V. M. Kaladi.

REFERENCES

1.   Clinch, J.M. (1970). *Prediction and Measurement of the
      Vibration Induced in Thin-Walled Pipes by the Passage of
      Internal Turbulent Water Flow*. Journal of Sound and Vibra-
      tion, 12 (4), pp. 429-451.

2.   Zienkiewicz, O.C. (1971). *The Finite Element Method in
      Engineering Science*. McGraw-Hill, London.

3.   To, C.W.S. and Kaladi, V. (submitted 1984). *Dynamic Char-
      acteristics of Piping Systems Containing a Steady Mean
      Flow*. Journal of Pressure Vessel Technology, Trans. ASME.

4.   Housner, G.W. (1952). *Bending Vibrations of a Pipeline
      Containing Flowing Fluid*. Journal of Applied Mechanics,
      Trans. ASME, 19, pp. 205-208.

5.   Benjamin, T.B. (1961). *Dynamics of System of Articulated
      Pipes Conveying Fluid: I, Theory*. Proc. Royal Society,
      Series A, 261, pp. 457-486.

6.   Liu, Y.C. (1981). *Flow-Induced Vibration of Curved Pipe
      Structures*. Ph.D. Thesis, Purdue University, U.S.A.

Table 1. Physical parameters

Young's Modulus	$= 5.519 \times 10^9$ N/m^2
Pipe inside diameter	$= 17.7$ mm
Pipe wall thickness	$= 1.9588$ mm
Pipe length	$= 2.52984$ m
Mass of pipe per unit length	$= 1.6668$ kg/m
Mass of fluid per unit length of pipe	$= 0.23811$ kg/m

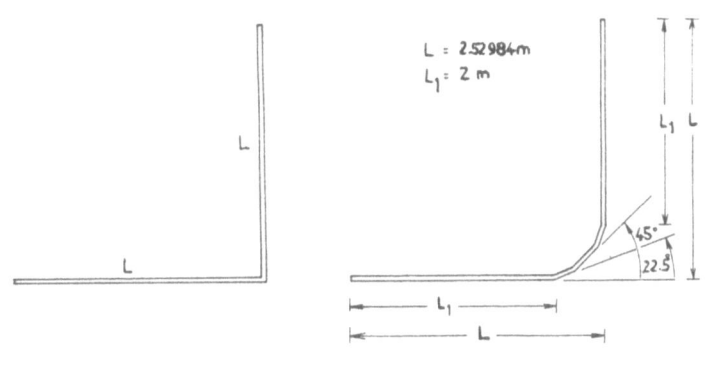

(a) 2 ELEMENT MODEL   (b) 5 ELEMENT MODEL

Figure 1.   Elbow

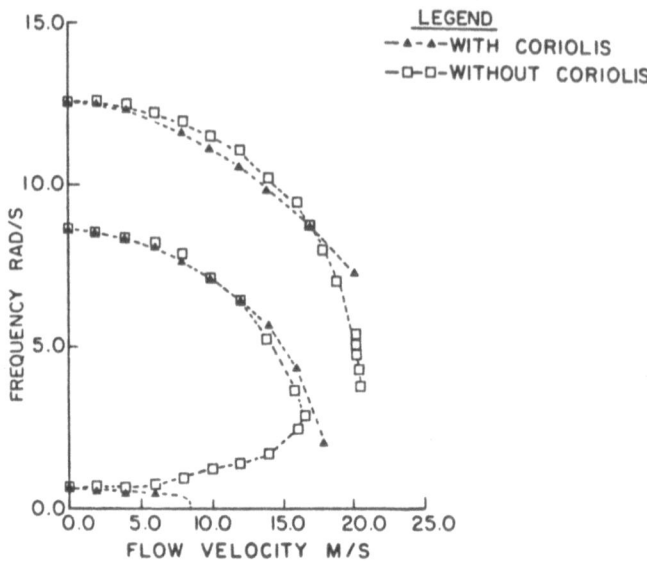

Figure 2.   Variation of resonant frequency
of cantilever elbow pipe with
flow velocity

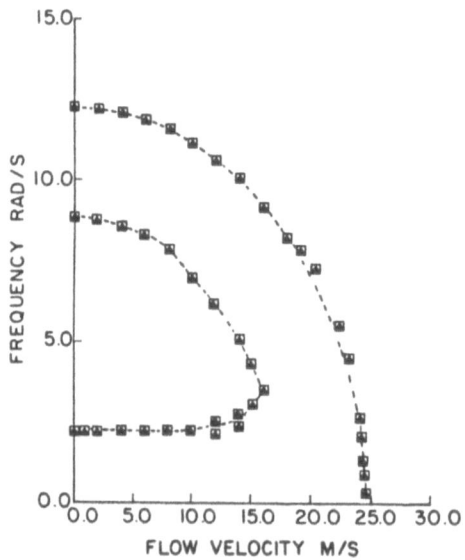

Figure 3.   Variation of resonant frequency of
clamped-clamped elbow pipe with
flow velocity

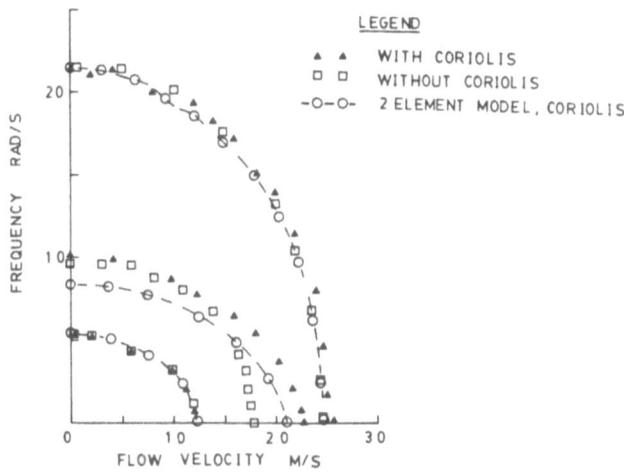

Figure 4.   Variation of resonant frequency
of the 5 element model with flow
velocity

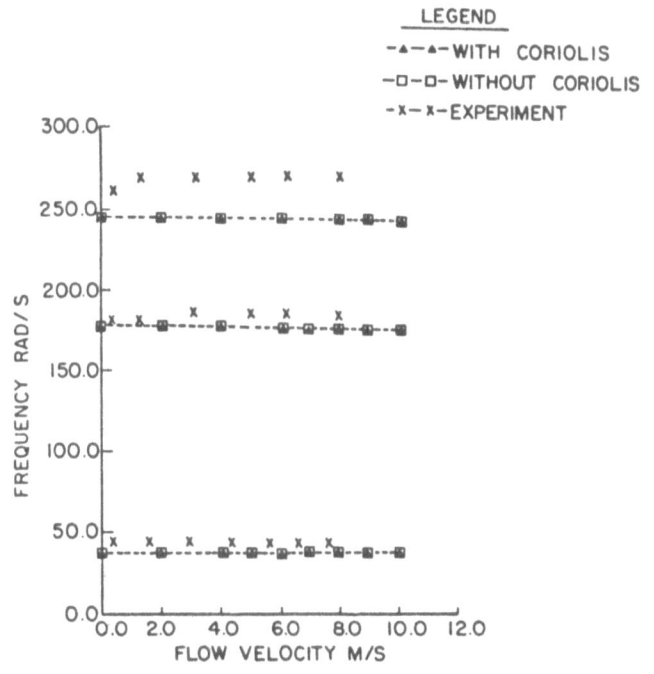

Figure 5.   Variation of resonant frequency
of clamped-clamped elbow pipe
with flow velocity

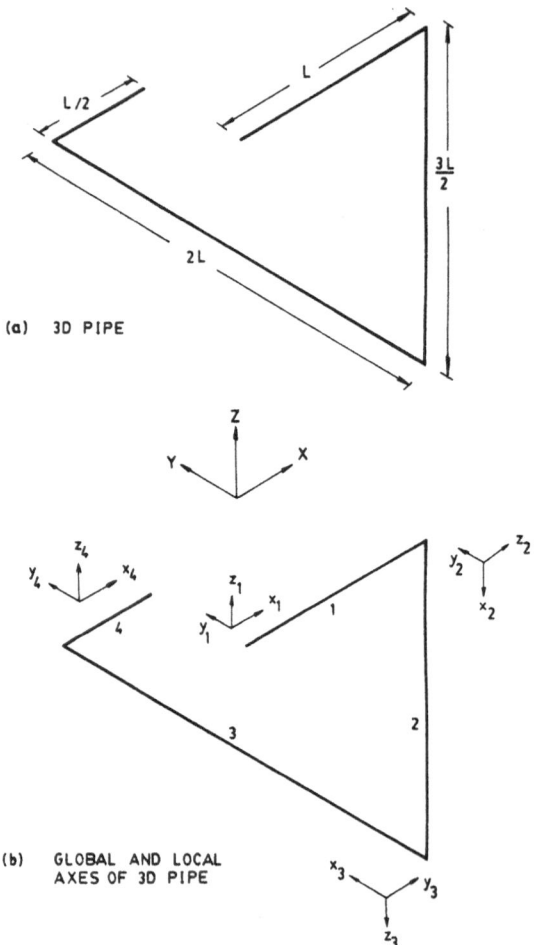

(a) 3D PIPE

(b) GLOBAL AND LOCAL
AXES OF 3D PIPE

Figure 6.   3 dimensional pipe

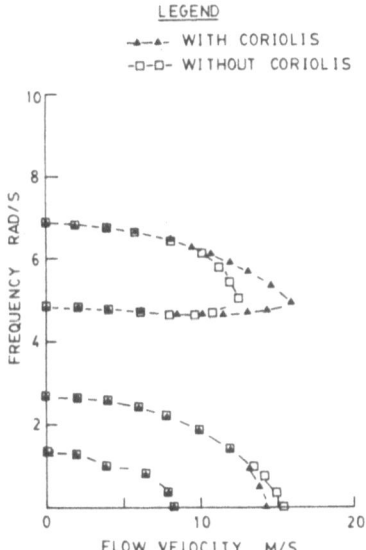

Figure 7.　Variation of resonant frequency
of 3D pipe with flow velocity

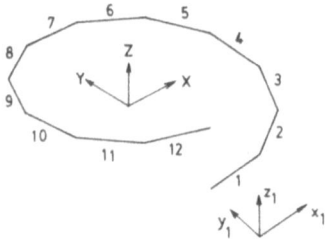

Figure 8.   12 element model of the
             helical pipe

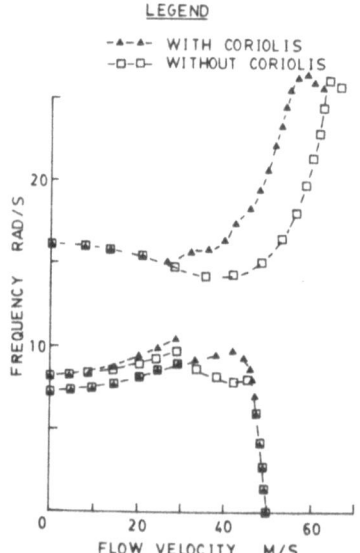

Figure 9.   Variation of resonant frequency
             of helical pipe with flow
             velocity

COMPUTER AIDED DESIGN OF SHELL STRUCTURES

H.Sardar Amin Saleh   Katholic University Leuven

Computational Mechanics Centre, Southampton, U.K.

INTRODUCTION

How far can we go in the automatic design of shell
structures and its numerical processing by computer?
It is our intention to design any kind of shell structure in civil engineering with the graphical and
numerical processing power of the computer. By design we mean to allow the user to define any shape of
shell, like cilinders, one-leaf hyperboloid, ellipsoids, hyperbolic paraboloids, conoids, cones, torus-
ses and helicoids, and to optimise it with regard to
its strength and dimensions(Fig. 1-8). In other words to analyse it with a finite element method and
to evaluate the result(graphically) in order to adjust the original design in a convergent way.

DESCRIPTION

It is our aim to make a proto-type package, and for
that purpose we developed already a great part. The
preprocessing part has been developed for hyperbolic
paraboloid shell structures. This choice is arbitrary, we could as well have chosen a cilindrical or a
spherical or any other type of shell.

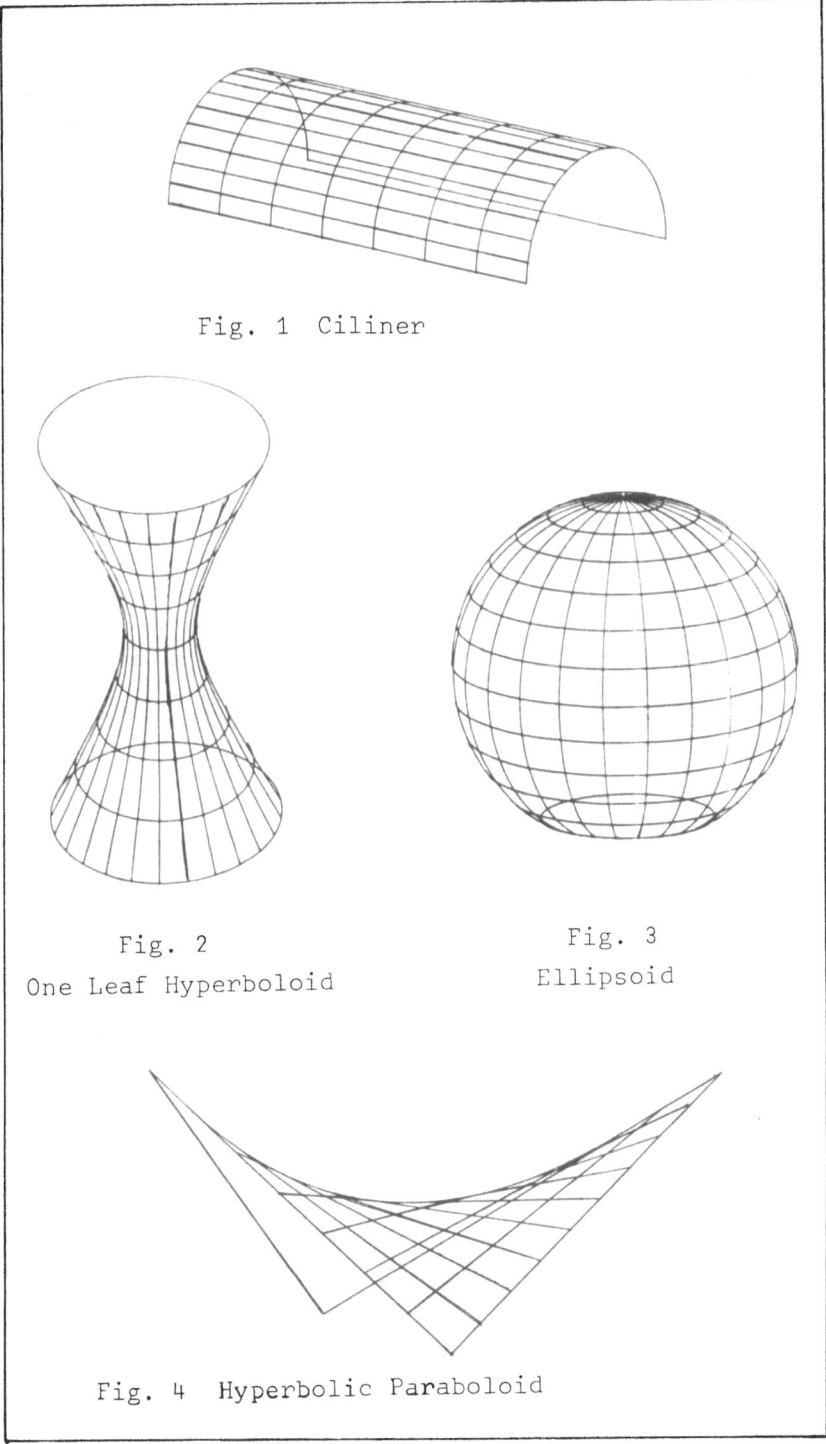

Fig. 1 Ciliner

Fig. 2
One Leaf Hyperboloid

Fig. 3
Ellipsoid

Fig. 4 Hyperbolic Paraboloid

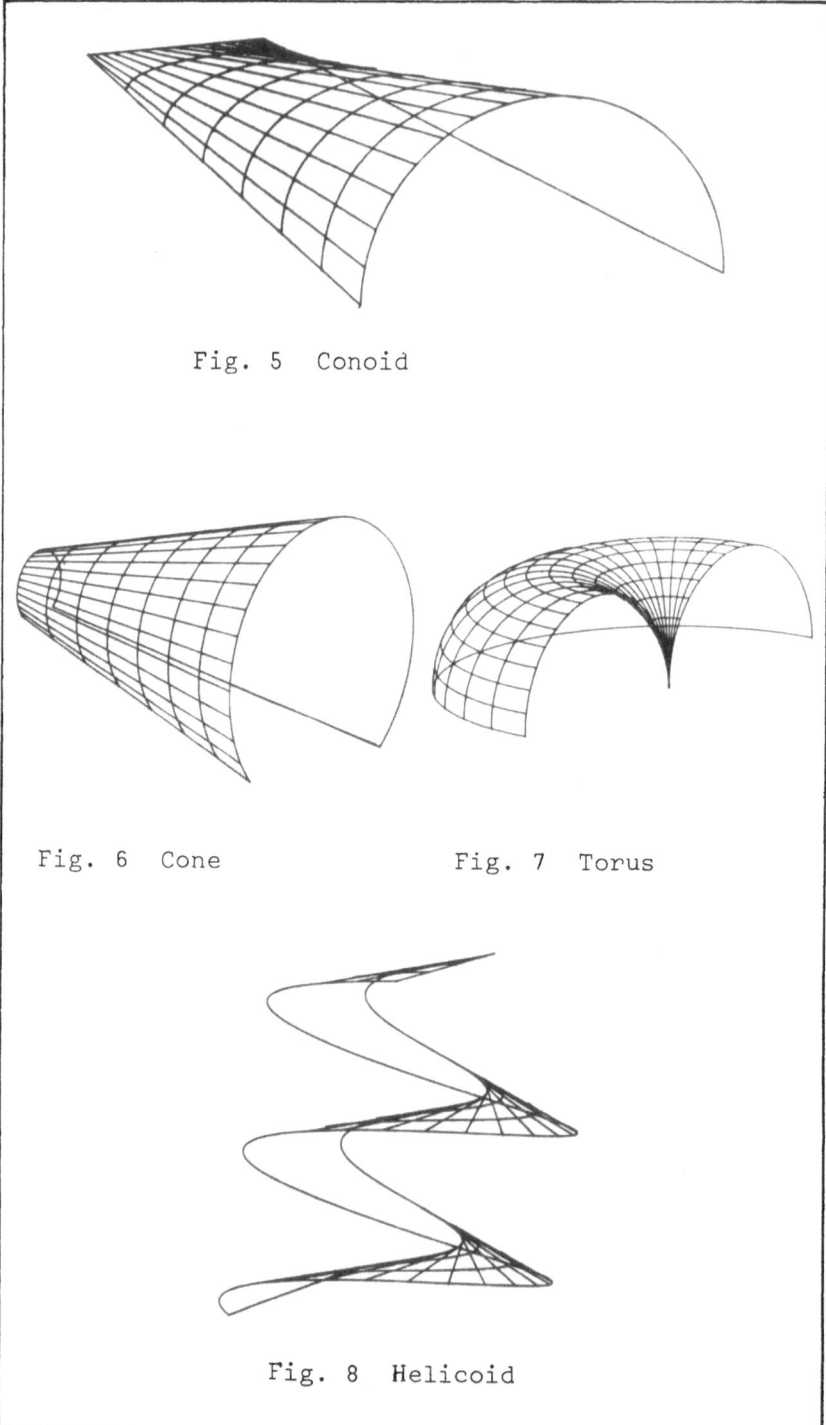

Fig. 5 Conoid

Fig. 6 Cone   Fig. 7 Torus

Fig. 8 Helicoid

DATA FLOW DIAGRAM

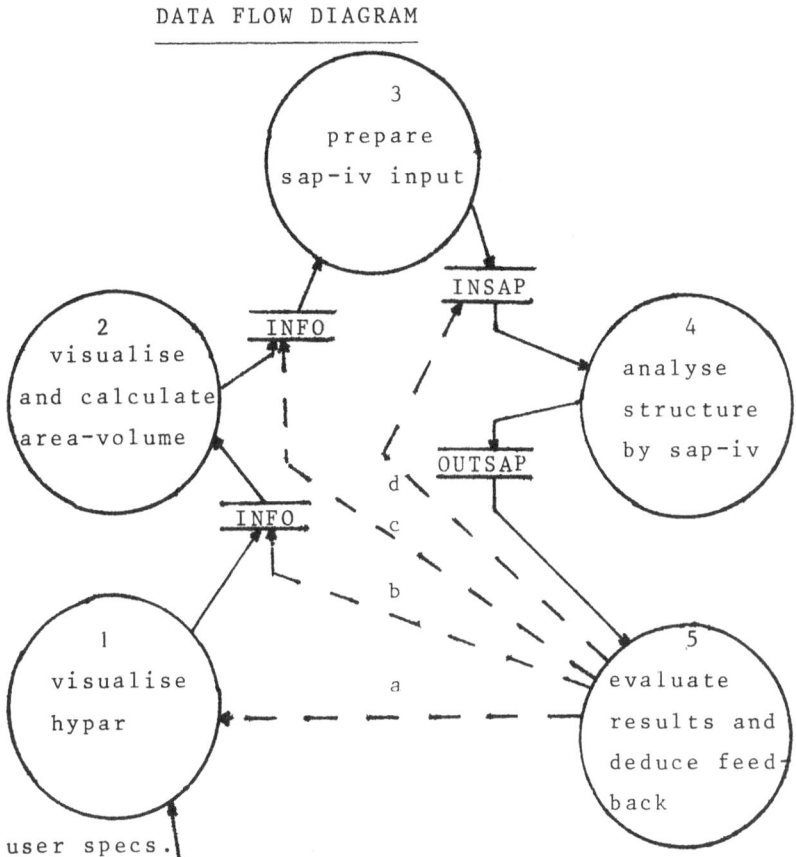

Fig. A

a........normal feed back way
b,c,d....internal ways

Starting our program, we can choose any type of shell(Fig. 1-8) and for a choosen type(e.g. Hypar) we get a menu on the display to have an idea of the possibility of the program,consisting of five parts as are illustrated in the data flow diagram(Fig. A).

1. Definition and visualisation of Hyperbolic Paraboloid.
2. Finite element model + Area / Volume calculation.
3. Preparation of SAP-IV input file.
4. SAP-IV analyse.
5. Postprocessing + Extraction of feed-back information.

At any time we have the possibility to :
-start a new project.
-repeat any part.
-continue with next part or stop.

The preprocesssing is divided in(the first) three parts, mainly for economic use of computer-resources and also for reason of splitting up the problem in to subproblems, afterwords we can always melt it back to one single part. Let's take a closer look to the different parts :

1. Definition and visualisation of Hyperbolic paraboloid.

Any Hypar can be defined by four corner points,in three dimensions,(a,b,c), (k,1,m), (p,q,r), (u,v,w) see Fig. C. The user has to specify also the number of lines(=number of intervals + 1) in each of the two directions " s " and " t " to describe the Hypar(Fig. B).
To give an idea about the used method we have to

refere to LAGRANGE Interpolation Formula. The formula
is the transfinite bilinear interpolation of which an
exact description can be found in 'surface for computer aided design of space forms ' from S.A.Coons. we
picted out some qualitative considerations in order
to have some feeling about the geometric model.
Consider a surface F in three-dimensional space. F
represents a relation from the (s,t)-surface in to
3-dimensional space F(s,t). Restricting the(s,t)-sur-
face to a square(0,0),(1,0),(0,1) and (1,1), relating
with a finite element that can serve to assemble a
shell surface(Fig. B).
We are looking for a formula that allows us to calculate points of the surface based on the knowledge of
sides of the square, i.e. F(s,0), F(s,1), F(0,t) and
F(1,t). Required formula is :

$R(s,t)=R1(s,t)+R2(s,t)-R3(s,t)$ ....(1)

With

$R1(s,t)=(1-s)xF(0,t)+sxF(1,t)$ ....(2)

$R2(s,t)=(1-t)xF(s,0)+txF(s,1)$ ....(3)

$R3(s,t)=(1-s)x(1-t)xF(0,0)+(1-s)xtxF(0,1)$

$\qquad +s(1-t)xF(1,0)+sxtxF(1,1)$ ....(4)

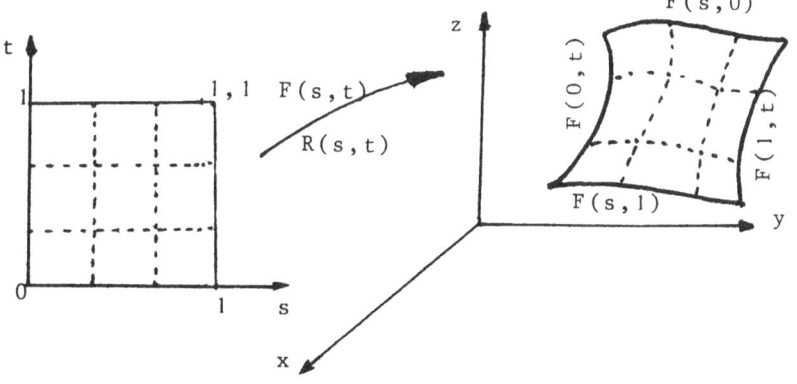

Fig. B

So,formula 2 is merely a linear interpolation formula between the points $F(0,t)$ and $F(1,t)$. These points are infact two arbitrary points of the sides $F(0,t)$ and $F(1,t)$ so that the formula is a transfinite interpolation formula between the given sides $F(0,t)$ and $F(1,t)$.

It is obvious in analogy, that formula 3 is an interpolation formula between the given sides $F(s,0)$ and $F(s,1)$.

Formula 4 is a bilinear interpolation formula between the four corner points of our surface, i.e. $F(0,0)$, $F(1,0)$, $F(0,1)$ and $F(1,1)$. required interploation formula is given by combination of R1,R2 and R3 as stated in formula 1. One can easily check that F and R are identical at the sides. However this method of describing surfaces is only exact for ruled surfaces, so far instance for the ellipsoids we use parameter equations instead.

As the visualisation is meant to provide the user a good view, the shell is displayed in axonometric projection (4/5/6) with hidden line removal(boundary lines always drawn) based on a simple algorithm using the scalar product of the view vector and the normal vector in a point on the surface. Infact the same algorithm is used to draw the the edge-lines where the visible part becomes invisible to avoid dead-end lines. The result is shown in Fig.C.

In order to have a better view on the shell, rotation can be performed by any angle around any axis in 3-D. Of cource at this stage it is quite easy to adjust any coordinate of the four corner points or to change the mesh-density. Finally we store all the information necessary for the following parts in a file named INFO.

In preprocessing part we have some checks based on ideas of Artificial Intelligence(AI) to perform a good predesign. For instance when we give unnormal value of thickness for a certain span for our shell, the AI program(depending on knowledge data base)does not accept the value and it propose a reasonable value for the thickness. At the end of this paper we give more detail about AI in our program.

2. Finite elment model + Area/Volume calculation.

Once a reasonable concept is set up, we can prepare for finite element modeling. Shells with ruled surfaces can be mathematically modelled with a trnsfinite linear interpolation method which is also used to devide the structure in finite shell elements.
We redraw our designed shell with complete(no hidden line removal) element modeling. This model can be completely visualised together with the projection of the hypar on a two dimensional plane to indicate the finite element generation with or without the automatic node and/or element numbering(Fig.D).
To have an idea about the area covered by the shell, the surface area of the shell and the volume between zero-level and the surface area we have the options to calculate any combination of these items. As all nodal data and shell-element data is generated in this part and visually checked, the data check option of the F.E. program (SAP-IV) is made superfluous regarding geometry.

3. Preparation of SAP-IV input.

Once part two has been finished we can write the data input file, named INSAP, for SAP-IV analysis interactiveley, with almost all the possibilities of SAP-IV

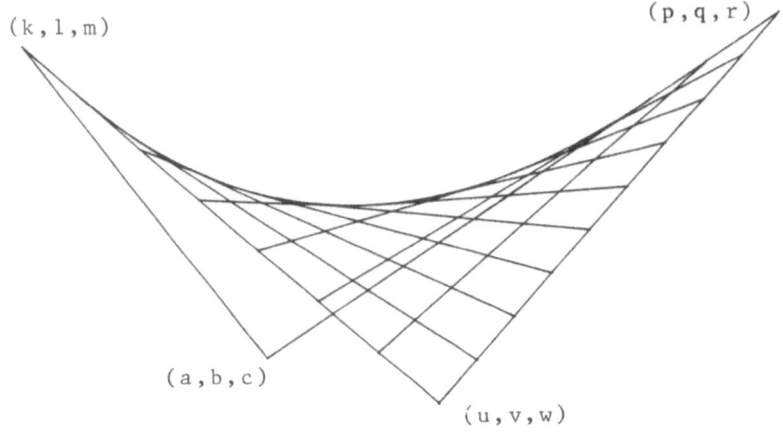

Fig. C

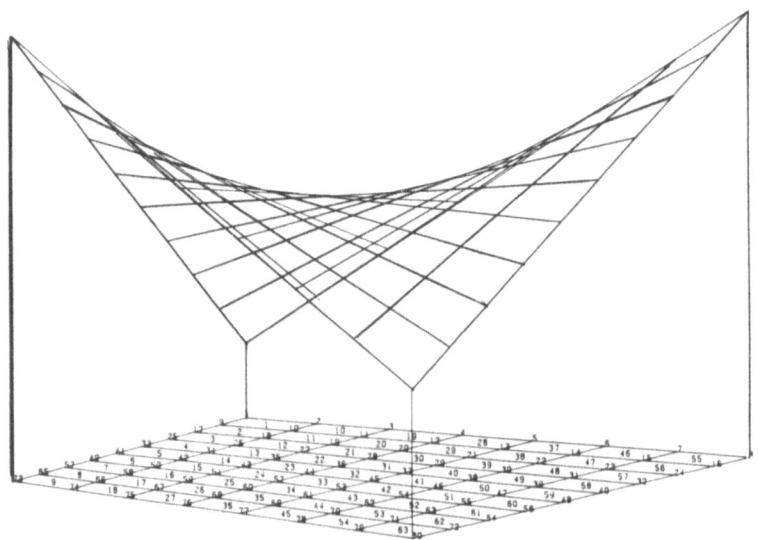

Fig. D

features. Geometrical information is retrieved automatically while the user still has to provide material
characteristics and load cases, although most of them
are default for stell and concrete.
As a special case we have the possibility of calculating any plate by this program.

4. SAP-IV analysis.

When we have excuted all three parts we are able to
submit the structure to SAP-IV, feedin it with the
data file INSAP, for analysing and evaluating the
results.

5. Postprocessing + Extraction of feed-back
information.

The output of SAP-IV analysis is directed to different files holding information about displacements of
the nodes,the moments and stresses in the elements.
As far as displacements are concerned we are able to
produce graphs of the deformed structure, eventually
superposed to the original structure. Such a case is
shown in Fig.E where a hypar has been submitted to a
vertical distributed load. Of cource the deformations
can be 'overscaled' to allow for a better view.
Now we are working to visualise bending moment components and membrane stresses in contour-maps.

        Our intention is to integrate design rules and
norms in a knowledge data base in order to deduce
some information regarding adjustment of our design.
The program will indicate to the user the dangerous
parts in the design but is left to the user to adjust
his consept in a sensible way.The implementation of
Artificial Intelligence to deduce some possible

actions, based on a knowledge data base, is our aim
to help the user in selecting the possibile changes,
by compairing values with build-in standards or experience rules, and to have an easy feed-back for a
good design. For this part a test program using the
very high level programming language PROLOG is in
progress. The AI-interface works on a VAX 750/Unix -
4.2BSD. The benefit from AI in this project is the
modularity of the knowledge and the separation from
its use so that it can be adapted very easily. The
design rules and norms are given to the expert system
controller Cricket. It asks the user for specific data, it searches for possible solutions by using the
knowledge base, and it can give explanations in a
kind of natural language(for a good solution).
A problem is caused by the fact the presently prepared preprocessor is written in FORTRAN and the postprocessing in PROLOG. So we have some difficulties to
make a link between both parts in two directions. The
PROLOG should use numerical results from the FORTRAN
program and the results of the PROLOG program have to
be used by the FORTRAN program to improve the design.
There is a possible solution by using the FORTRAN language for AI if we can make a good knowledge data
base. Else we can create a file in FORTRAN containing
the necessary information for the expert system and
use this file for interactive processing of AI with
PROLOG language. However this second method is not so
elegant because the user has to change from one computer to the other.

ACKNOWLEDGEMENT

The outher is most greatful to Prof.Dr.Eng.VAN LAETHEM
and Prof.Dr.Eng.Architecture NEUCKERMANS for their -

valuable guidance in my work. My special great thanks
are also due to Dr.Eng. E.BACKX for his continuous
help in computer. I express my thanks to Dr.Eng.
BRUYNOOGHE and Eng. KREKELS Bruno for their help in
Artificial Intelligence.

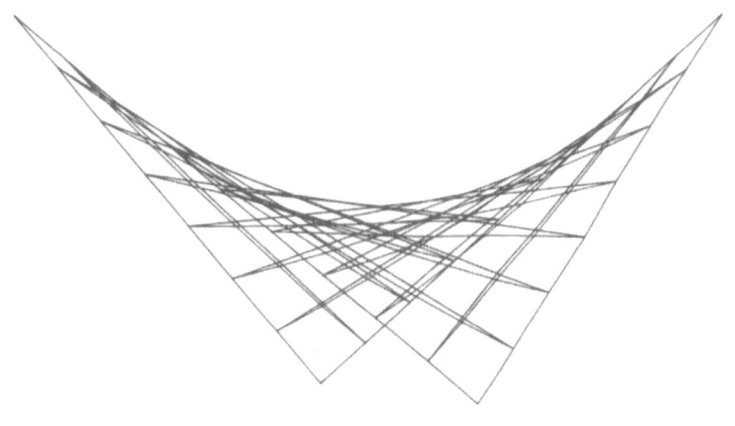

Fig. E

REFERENCES

1. E.BACKX, 'Projectie van 3-D naar 2-D' studiedag :
Grafische werken met computer v.v.T.J. 15/1/1982
2. Hugo callens, Piet handekijn, Hein lecluse, Lode
missiaen, Begeleider:E.BACKX maart 1983 K.U.L.
'Projectwerk Schaalkonstrukties'.
3. J.Encarnacao, E.G.Schlechtendahl: 'Computer Aided
Design', Fundamentals and system Architectures 1983.
4. Robert Kowalski, c.1979,'Logic For Problem Solving'

5. W.F.Clocksin, C.S.Mellish:'Programming in Prolog'
c. 1981.
6. William M.Newman, Robert F.Sproull :'Principles Of
Interactive Computer Graphics' second edition c.1979.

SSCAD ON A DESKTOP COMPUTER

J.J. Jonatowski and D. Koutsoubis

Space Structures Int. Corp., Plainview, NY, USA

INTRODUCTION

Space Structures International Corp. is a designer, fabricator
and installer of three-dimensional double-layer grid type structures. Up to two years ago, the analyses of such structures
were performed through a computer service bureau, utilizing a
well known, commercially available, general purpose finite element program. The program could only perform analyses; it had
no design capabilities. Although the program ran on the Cray-1
computer which had extremely fast processing power, the actual
turnaround time was not. Turnaround time was affected by the
following factors: user priority which affected cost, system
usage load on the service bureau and delivery of output results. Because Space Structures did not have an on-site high
speed printer and its location was somewhat distant from the
service bureau, computer output results were delivered by messenger service and were normally available for review 24 hours
after submittal of the computer run. In addition to the slow
turnaround, the costs of messenger service and service bureau,
which included batch processing, connect and storage charges,
became too high.

In order to reduce costs, shorten the analysis turnaround time,
and improve the quality of the design process, an extensive
study was conducted to determine alternatives to the service
bureau. Both present and future needs and requirements of the
company were defined in detail, and initial costs and schedules
of return on investment for different systems were investigated.
Finally, it was determined that a microcomputer with custom
software would be more effective and economical.

Thus, approximately two years ago, a Hewlett-Packard Series
200 Model 9836 microcomputer with a 10 meg. hard disk, 180 cps
line printer, and an 8-pen plotter were acquired. A three-
dimensional truss analysis program, originally developed by

Multitech Computer Services (Little Neck, NY), was converted to the HP computer. Custom pre- and post-processing modules were developed jointly by Space Structures and Multitech, and added to form a totally integrated computer-aided engineering system. This system has been named "SSCAD".

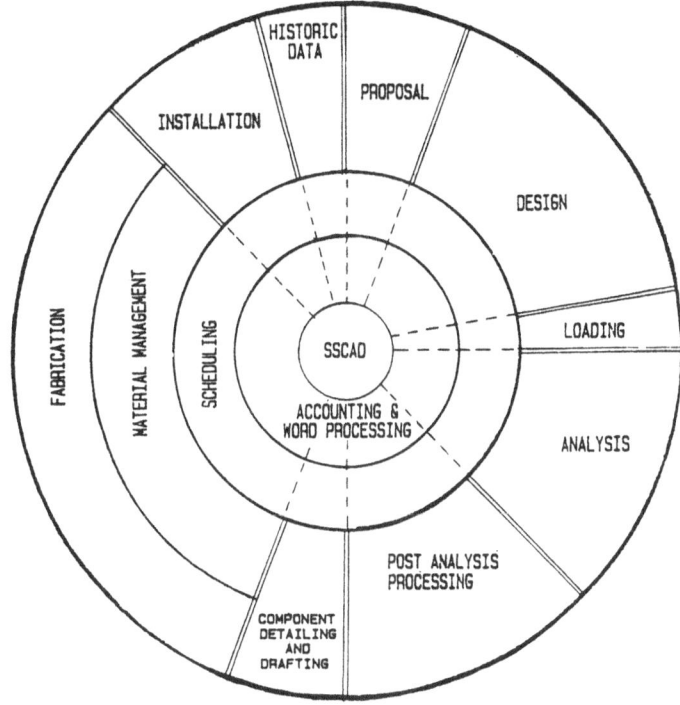

SSCAD KNOWLEDGE SYSTEM :

A TOTAL DESIGN, ENGINEERING, FABRICATING
AND MANAGEMENT SYSTEM.

Fig. 1 - SSCAD System

The SSCAD system has provided high productivity gains. For a project consisting of geometry and load generation, an average of three analyses, member and connection design, and final drawing generation, what used to take approximately 245 hours was reduced to 21 hours with the use of the above hardware (4).

A year later, after the above success, it was decided to augment the hardware configuration and install local access networking capabilities. The present configuration consists of:

- A shared resource manager;
- A 135 meg. storage unit;
- Two high speed spooler printers;

- One HP Series 500 Model 520 desktop computer (32 bit) with one disk drive, a 10 meg. hard disk, and local printer;
- The original HP 9836 (16 bit) with two disk drives, a 10 meg. hard disk, local printer and 8-pen plotter;
- Four HP 9816's (16 bit). Two of these units have two disk drives, local printer and 6-pen plotter.

It should be noted that the 32-bit Model 520 executes at least ten times faster that the HP 9836 used in the original productivity gains study (4).

SSCAD SYSTEM

The SSCAD system is not just a computer-aided engineering system, but includes non-engineering functions (Figs. 1,2). However, this paper will deal only with the engineering functions of the system.

```
 SSCAD 

SPACE STRUCTURES COMPUTER AIDED ENGINEERING AND DESIGN SYSTEM
STANDARD GEOMETRY VERSION: HP MODEL 500 SERIES 20 COMPUTER
SSCAD SESSION START:              Date: (25-Feb-85)   Time: (11:31)

 SSCAD MENU 

1. PRE-PROCESSING MENU
2. ANALYSIS AND DESIGN PROGRAM
3. POST-PROCESSING MENU

4. PROPOSAL DATA BASE PROGRAM
5. PROJECT MANAGEMENT MENU
6. WORD PROCESSING PROGRAM
7. TERMINAL MANAGER (DATACOMM) PROGRAM

8. EXIT SSCAD


CHOOSE DESIRED OPTION
```

Fig. 2 - SSCAD Main Menu

The menus and graphics used as illustrations herein were obtained as screen dumps from the Model 520. It should be noted that the HP 520 has color capabilities whose advantages in displaying data and graphics cannot be demonstrated in the black and white environment of this paper.

```
 SSCAD 

SPACE STRUCTURES COMPUTER AIDED ENGINEERING AND DESIGN SYSTEM
STANDARD GEOMETRY VERSION: HP MODEL 500 SERIES 20 COMPUTER
SSCAD SESSION START:              Date: (25-Feb-85)   Time: (11:31)

 PRE-PROCESSING MENU 

1. GEOMETRY PROGRAM
2. DOMEDESIGN PROGRAM
3. LOADING PROGRAM
4. SSCAD: EXPANDED GEOMETRY VERSION
5. MAIN SSCAD MENU
```

Fig. 3 - SSCAD Pre-Processing Menu

```
▐ SPACEFRAME GEOMETRY PROGRAM ▐

 GEOMETRY MENU

 1. SPACEFRAME GEOMETRY GENERATION
 2. SPACEFRAME GEOMETRY GRAPHICS
 3. SPACEFRAME GEOMETRY MODIFICATION
 4. SPACEFRAME GEOMETRY INFORMATION
 5. SPACEFRAME GEOMETRY RE-STRUCTURING
 6. SPACEFRAME GEOMETRY STORAGE TO/FROM DISC
 7. EXIT GEOMETRY PROGRAM




    INPUT DESIRED OPTION
```

Fig. 4 - SSCAD Geometry Menu

The engineering functions of SSCAD (Fig. 2) can be
divided in the following groups:

- Modeling, member properties, loading and support
  conditions generation (pre-processing);
- Ahalysis, and primary member and connection design;
- Review of analysis and design, secondary component design, and fabrication assembly and
  erection drawings (post-processing).

All three functions are integrated using a network of menus
and specially programmed keys. The engineer has no commands
to memorize because all the possible options available to him
at any given time are displayed and, if necessary explained
on the screen. The advantage of SSCAD over other systems is

CHOOSE SPACEFRAME SYSTEM:

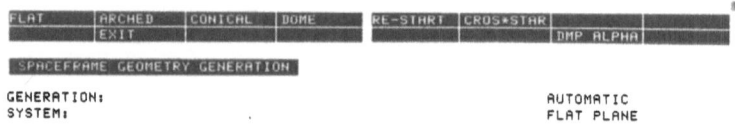

Fig. 5 - Type of Structures

CHOOSE GEOMETRY TYPE:

Fig. 6 - Type of Geometries

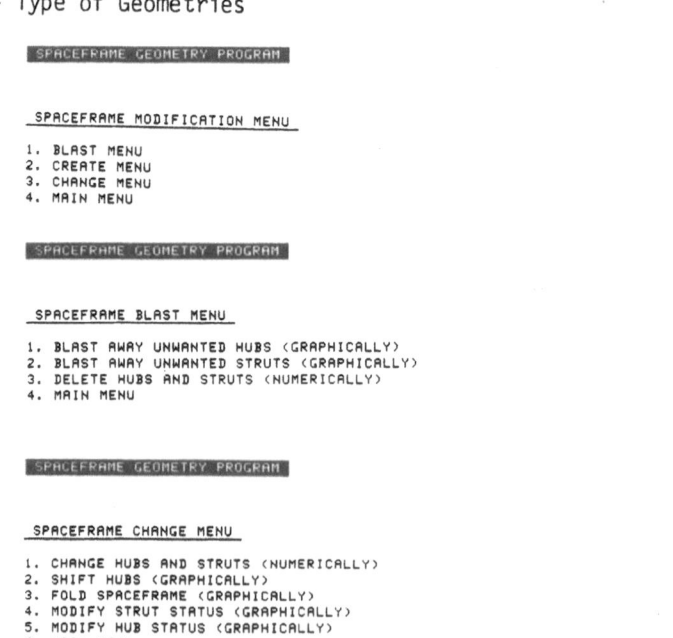

Fig. 7 - SSCAD Geometry Modification Options

extensive utilization of graphics to communicate information to the engineer. By graphically generating and
reviewing data, the engineer can input and interpret more
information, more efficiently, with less error.

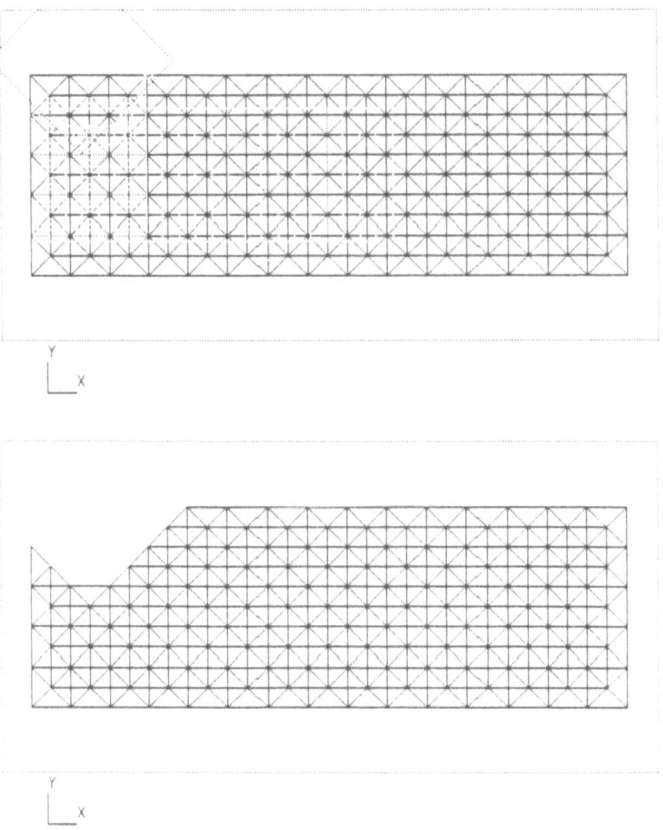

Fig. 8 - Use of Graphics Box to Delete Members and Joints

Modeling Function

The geometry generation subprogram may be accessed by
successively selecting option 1 from Figs. 2, 3 and 4.
The engineer then selects the type of structure (Fig. 5)
and geometry (Fig. 6) by pressing the appropriate function key. Overall dimensions of the structure are then
entered and within seconds, the system generates all the
coordinates and connectivities for that geometry. Using
a graphics subprogram, the engineer can graphically display the structure. Plan views, elevations, perspectives
with zooming capabilities are available. The engineer could
then use a variety of options to graphically modify his
original structure (Fig. 7). In order to quickly modify

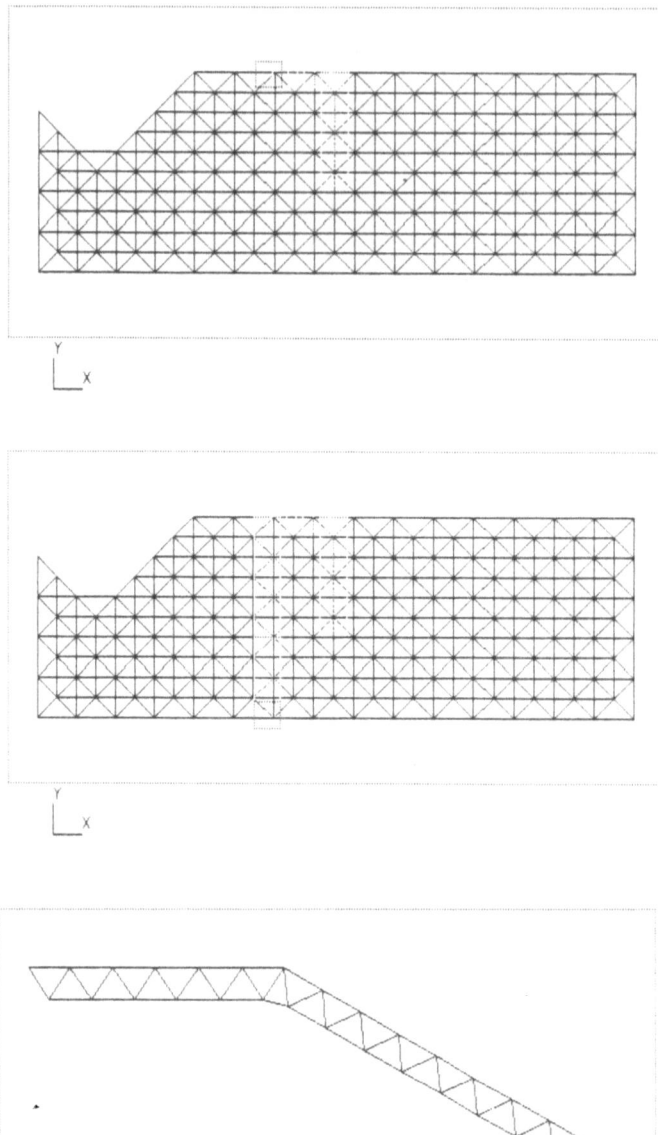

Fig. 9 - Example of Folding

the geometry, each subprogram includes a graphical box which
is superimposed over a chosen view. This box can be shifted,
scaled, or rotated using function keys to isolate any part
of the structure within the box. Once the desired part of the
structure is within the box, the engineer can specify any
of a number of different functions to be performed only on
those components. These functions, among others, include
deletion, creation, shifting and rotation of components
of the structure. Figure 8 illustrates the use of the
graphics box to delete a corner of the structure. The folding of the structure may be done by identifying graphically
two points (Fig. 9) about which the structure is to be bent
by a specified number of degrees.

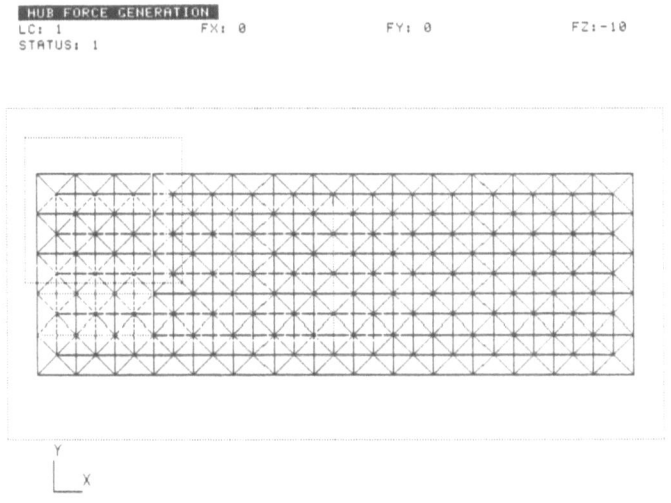

Fig. 10 - Loading of Structure Graphically

Once the geometry has been completed, the engineer can
graphically generate loads, supports and member properties.
For example, by specifying the type and magnitude of a particular force and isolating a section of the structure, the
engineer can generate the specified force on every component within the box (Fig. 10).

STRUT NO/TYP	HUB CONNECT.	P	GRO	DESCRIPTION	MAT	WEIGHT lbs/ft	NOMINAL	EFF.	NET TENS. AREA	BOLT/PIN	TYPE	SHEAR (1)	BEAR. IN EAR (2)	BEAR. STRUT (3)	ALLOW. COMP. BEARING (4)	ALLOW. (5)	ACTUAL	MOMENT	LC	ALLOW. (6)	ACTUAL	MOMENT	LC	BOLT TENS.	BOLT COMP.	STRUT TENS.	STRUT COMP.
45 I	18	19	15	2X5X.125	1	1.984	93.00	88.50	1.563	2-1/4	B	6.81	10.20	4.25	16.53	29.69	.77	0.00	8	21.76	3.34	0.00	1	18.2		2.6	20.2
46 D	11	20	6	2X2X.350	1	2.716	116.59	110.59	1.960	2-1/4	B	6.81	10.20	11.90		37.24	4.45	0.00	3	4.57	4.16	0.00	4	65.3		11.9	91.1
47 D	11	21	8	2X2 SPECIAL	1	5.123	116.59	110.59	1.960	4-1/4	B	13.63	20.40	23.80		37.24	3.22	0.00	8	11.38	9.22	0.00	8	67.6		8.6	81.0
48 D	12	21	3	2X2X.125	1	1.103	116.59	110.59	.812	4-1/4	B	13.63	20.40	8.50		15.43	6.95	0.00	5	2.30	1.84	0.00	8	81.8		45.1	79.8
49 D	12	22	8	2X2 SPECIAL	1	5.123	116.59	110.59	1.960	4-1/4	B	13.63	20.40	23.80		37.24	1.84	0.00	4	11.38	6.95	0.00	4	51.0		4.9	61.1
50 D	13	22	3	2X2X.125	1	1.103	116.59	110.59	.812	2-1/4	B	6.81	10.20	4.25		15.43	3.14	0.00	4	2.30	2.02	0.00	5	73.9		20.3	87.6
51 D	13	23	4	2X2X.188	1	1.603	116.59	110.59	1.175	2-1/4	B	6.81	10.20	6.39		22.32	2.02	0.00	1	3.14	3.14	0.00	1	49.2		9.0	99.9
52 D	14	23	6	2X2X.350	1	2.716	116.59	110.59	1.960	2-1/4	B	6.81	10.20	11.90		37.24	1.84	0.00	8	4.57	4.35	0.00	5	63.8		4.9	95.2
53 D	14	24	8	2X2 SPECIAL	1	5.123	116.59	110.59	1.960	2-1/4	B	6.81	10.20	11.90		37.24	.72	0.00	8	11.38	6.24	0.00	1	91.6		1.9	54.9
54 D	15	24	2	2X2X.095	1	.869	116.59	110.59	.629	2-1/4	B	6.81	10.20	3.23		11.95	.72	0.00	7	1.83	1.49	0.00	7	46.2		6.0	81.5
55 D	15	25	2	2X2X.062	1	.566	116.59	110.59	.419	4-1/4	B	13.63	20.40	4.22		7.96	3.47	0.00	8	1.31	.07	0.00	8	82.3		43.6	5.6
56 D	16	25	8	2X2 SPECIAL	1	5.123	116.59	110.59	1.960	2-1/4	B	6.81	10.20	11.90		37.24	.37	0.00	1	11.38	6.41	0.00	7	94.1		1.0	56.3
57 D	16	26	1	2X2X.062	1	.566	116.59	110.59	.404	4-5/16	B	21.29	25.50	5.27		7.67	4.30	0.00	8	1.31	.64	0.00	8	81.6		56.1	49.0
58 D	17	26	8	2X2 SPECIAL	1	5.123	116.59	110.59	1.960	4-1/4	B	13.63	20.40	23.80		37.24	.64	0.00	1	11.38	9.38	0.00	1	68.9		1.7	82.5
59 D	17	27	2	2X2X.095	1	.869	116.59	110.59	.582	4-3/8	B	30.66	30.64	9.69		11.05	9.38	0.00	1	1.83	.64	0.00	8	96.8		84.9	35.1
60 D	18	27	8	2X2 SPECIAL	1	5.123	116.59	110.59	1.877	4-5/16	B	21.29	25.50	29.75		35.50	.64	0.00	8	11.38	14.47	0.00	1	******			6
61 D	18	28	4	2X2X.188	1	1.603	116.59	110.59	1.081	4-3/8	B	30.66	30.64	19.18		20.54	14.46	0.00	1	3.14	5.01	0.00	1	75.4		70.4	28.4
62 D	19	28	7	2X2 SOLID	1	4.704	116.59	110.59	3.000	2-1/4	B	6.81	10.20	34.00		57.00	.78	0.00	8	5.55	5.01	0.00	1	73.5		1.2	90.3
63 D	19	29	5	2X2X.250	1	2.058	116.59	110.59	1.531	2-1/4	B	6.81	10.20	10.63		28.50	.97	0.00	7	3.80	3.17	0.00	6	46.6		3.4	83.5
64 I	11	30	15	2X5X.125	1	1.984	138.00	133.50	1.531	4-5/16	B	21.29	24.21	10.63	16.53	29.09	8.79	0.00	1	13.87	.81	0.00	8	82.7		30.2	.1
65 I	14	31	16	2X5X.1875	1	2.929	166.41	161.91	2.256	4-5/16	B	21.29	24.21	15.94	24.26	42.87	13.69	0.00	1	13.10	1.59	0.00	8	85.9		31.9	12.2
66 I	15	31	15	2X5X.125	1	1.984	138.00	133.50	1.563	4-1/4	B	13.63	20.40	8.50	16.53	29.69	.77	0.00	7	13.87	2.94	0.00	1	18.1		2.6	21.2
67 I	16	32	15	2X5X.125	1	1.984	166.41	161.91	1.563	4-1/4	B	13.63	20.40	8.50	16.53	29.69	3.81	0.00	1	9.43	.39	0.00	8	70.8		10.1	4.2
68 I	16	32	15	2X5X.125	1	1.984	138.00	133.50	1.563	4-5/16	B	21.29	24.21	10.63	16.53	29.09	9.40	0.00	1	13.87	.31	0.00	8	88.4		32.3	2.3
69 D	20	21	15	2X5X.125	1	1.984	93.00	88.50	1.563	2-1/4	B	6.81	10.20	4.25	16.53	29.69	1.02	0.00	7	21.76	.80	0.00	7	24.1		3.4	.0
70 D	22	21	15	2X5X.125	1	1.984	93.00	88.50	1.563	2-1/4	B	6.81	10.20	4.25	16.53	29.69	1.47	0.00	5	21.76	6.44	0.00	6	34.5		4.9	39.0
71 D	23	22	15	2X5X.125	1	1.984	93.00	88.50	1.563	2-1/4	B	6.81	10.20	4.25	16.53	29.69	2.93	0.00	4	21.76	10.32	0.00	5	49.0		9.9	42.4
72 D	23	23	15	2X5X.125	1	1.984	93.00	88.50	1.563	4-1/4	B	13.63	20.40	8.50	16.53	29.69	4.40	0.00	4	21.76	11.31	0.00	5	51.7		14.8	68.4
73 D	25	23	15	2X5X.125	1	1.984	93.00	88.50	1.563	4-1/4	B	13.63	20.40	8.50	16.53	29.69	4.97	0.00	5	21.76	14.10	0.00	6	98.5		16.7	85.3
74 D	25	24	15	2X5X.125	1	1.984	93.00	88.50	1.563	4-1/4	B	13.63	20.40	8.50	16.53	29.69	4.68	0.00	6	21.76	11.49	0.00	6	55.1		15.8	69.5
75 D	26	25	15	2X5X.125	1	1.984	93.00	88.50	1.563	4-1/4	B	13.63	20.40	8.50	16.53	29.69	4.65	0.00	7	21.76	6.89	0.00	7	54.7		15.7	41.7
76 D	27	28	15	2X5X.125	1	1.984	93.00	88.50	1.563	4-1/4	B	13.63	20.40	8.50	16.53	29.69	8.07	0.00	1	21.76	3.66	0.00	8	94.9		27.2	22.1
77 D	28	29	17	2X5X.250	1	3.822	93.00	88.50	2.938	4-5/16	B	21.29	40.36	21.25	31.63	55.81	19.45	0.00	1	42.93	3.14	0.00	8	91.5		34.8	9.9
78 D	30	21	8	2X2 SPECIAL	1	5.123	116.59	110.59	1.960	2-1/4	B	6.81	10.20	11.90		37.24	3.52	0.00	4	11.38	5.59	0.00	5	82.0		9.4	49.1
79 D	21	30	6	2X2X.350	1	2.716	116.59	110.59	1.960	2-1/4	B	6.81	10.20	11.90		37.24	1.38	0.00	6	4.57	4.31	0.00	5	63.2		3.7	94.3
80 D	24	31	1	2X2X.062	1	.566	116.59	110.59	.419	4-1/4	B	13.63	20.40	4.22		7.96	2.19	0.00	7	1.31	.00	0.00	6	51.9		22.5	.1
81 D	25	31	3	2X2X.125	1	1.103	116.59	110.59	.812	2-1/4	B	6.81	10.20	4.25		15.43	.44	0.00	7	2.30	2.14	0.00	2	50.4		2.9	93.1
82 D	28	32	8	2X2 SPECIAL	1	5.123	116.59	110.59	1.877	4-5/16	B	21.29	25.50	29.75		35.98	1.35	0.00	8	11.38	14.54	0.00	1	******			6
83 D	29	32	1	2X2X.062	1	.566	116.59	110.59	.419	2-1/4	B	6.81	10.20	2.11		7.96	1.38	0.00	3	1.31	1.20	0.00	4	65.7		17.4	91.7

Figure 11 - Sample Design Output Page

Analysis and Primary Design Functions

Analysis function The analysis function of SSCAD consists
of a three-dimensional truss analysis program utilizing
the conventional finite element displacement method (1,9).
The efficiency of a finite element program is influenced
by the method used for the solution of the equations. Some
of the available procedures that could have been used range
from iterative methods such as the Gauss-Seidel technique
(6) to direct Gaussian elimination schemes. SSCAD utilizes
the frontal equation solution technique originated by Irons
(3). This procedure minimizes computer core size requirements and the number of arithmetic operations (2).

Primary design function In the design phase, the structure
is designed to meet minimum strength requirements. Thus
every member is designed for the worst load cases according
to AAI (7) ro AISC (8) specifications depending upon whether
the material used is aluminum or steel. Also designed are
connections and joints. Figure 11 shows a typical page of
the printed design results. For each member, the worst
tension and compression load cases with the actual forces are
identified. The printout shows the size of sections selected
and its maximum allowable tensile and compressive carrying
capacity. Also shown is the computer designed connection
consisting of size and number of bolts required with its
allowable capacity in shear and bearing. The stress ratios
column on the right side of the table show, for each member
and its connection, the percentage of actual stress to allowable. This information gives an indication of reserve
strength for the structure.

Post-processing
After the analysis has been performed, the engineer may
choose to review the analysis results prior to performing
the primary design. The results may be presented numerically
or graphically. Displacements can be presented by graphically
displaying the deformed structure from any view using any degree of deflection magnification. Multi-colored plots of
member forces, where each color represents a different range
of magnitude of forces, can be used to identify different
stress levels and their location in the struture. This information combined with joint deflections are valuable for
comprehending the behavior of the structure. They may identify subtle problems or prompt changes resulting in a more
efficiently engineered structure which otherwise would not
have been noticed by reviewing numerical data.

Similar to analysis, primary design results may be presented
graphically and changed to account for material availability
and other factors. Graphical presentation of the design results consist of plots of the structure using different colors

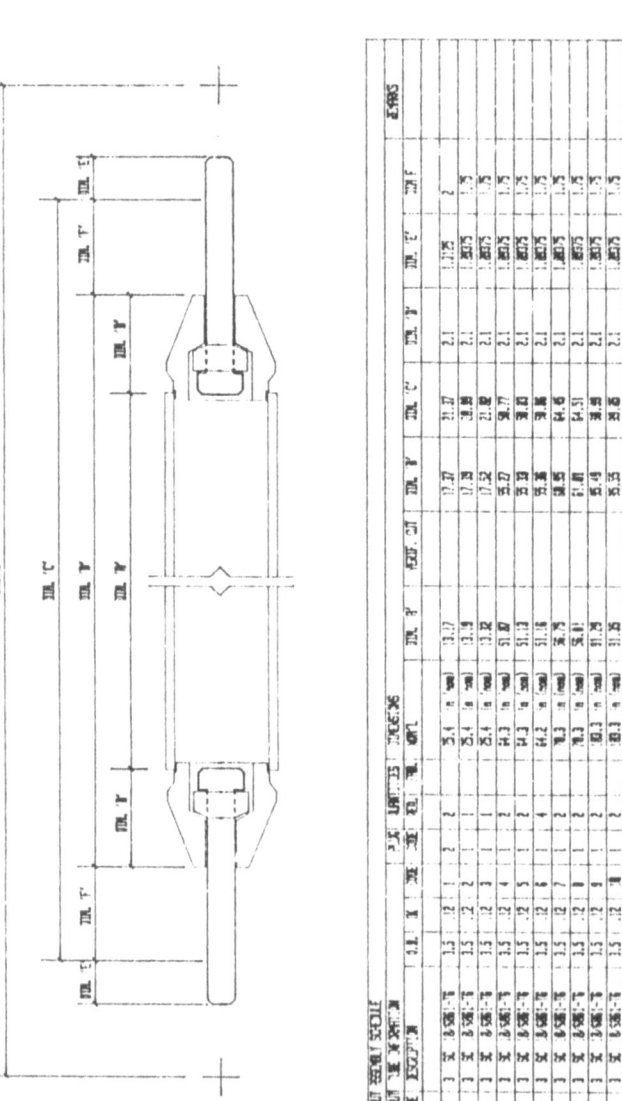

Figure 12 - Sample Fabrication Drawing

for the selected member sizes.

Secondary design is performed at this level of the SSCAD system. Secondary design consists of designing the different sub-components of what forms a joint and all the parts required to properly fasten the members to the appropriate joints. Summations of number of different components are generated for purchasing. Fabrication drawings (Fig. 12), bill of materials, assembly drawings and erection drawings are also generated.

CONCLUSION

SSCAD has proven to be an effective tool in the engineering and design of three-dimensional double-layer grid systems. This system is continually being expanded to perform as many functions as possible to automate Space Structures.

REFERENCES

1. Gere, J.M., and Weaver, W., Jr. (1956) Analysis of Framed Structures, D. Van Nostrand Company, Inc. Princeton, NJ

2. Hinton, E., and Owen, D.R.J. (1977) Finite Element Programming, Academic Press, London, England

3. Irons, B.M. (1970) "A Frontal Solution Program", Int. J. Num. Meth. Eng., Vol. 2, pp. 5-32

4. Jonatowski, J.J., and Koutsoubis, D. (1984) "Microcomputer Engineering of Large Space Strusses", Proc. of the Third Conf. in Computing in Civil Eng., San Diego, CA, pp 498-505

5. Jonatowski, J.J., and Koutsoubis, D. (1984) "A Microcomputer-Aided Design System for Large Space Structures", Third Int. Conf. on Space Structures, Univ. of Surrey, UK, pp. 301-305

6. Ralston, A (1965) A First Course in Numerical Analysis, McGraw-Hill, New York, NY

7. 1982 Specifications for Aluminum Structures, The Aluminum Association, Inc., Washington, D.C., 4th Ed.

8. (1978) Specifications for the Design, Fabrication and Erection of Structural Steel for Buildings, American Institute of Steel Construction, Chicago, IL, 8th Ed.

9. Zienkiewicz, O.C. (1971) The Finite Element Method in Engineering Science, McGraw-Hill, New York, N.Y.

COMPUTER PROGRAMME FOR ANALYSIS OF REINFORCED CONCRETE STRUCTURES

M.Sekulović, Ž.Praščević

Civil Engineering Faculty, University of Belgrade, Yugoslavia

1. INTRODUCTION

The computer programme for the analysis of plane reinforced concrete structures, which is developed at our Engineering Computing Centre, is described in this paper. The programme concerns reinforced concrete structures with straight and curvilinear elements and enables us to find the displacements and internal member forces according to the first and second order theory, taking into account creep and shrinkage effects of concrete. The theoretical basis for this programme is very detailly described in the authors previous works ((5),(6)(7)), and here will be presented the basic formulae only. The finite element method in isoparametric formulation is used in this analysis and this computer programme is written out in FORTRAN programming language and adopted for the DEC 20 computing system.

2. CONSTITUTIVE EQUATIONS FOR STEEL AND CONCRETE

A reinforced concrete member is composed from two materials steel and concrete with different rheological properties. Steel is assumed as an elastic material, while concrete under permanent loads exibits creep and shrinkage during time. For the periodic loads and effects (wind, snow, temperature changes etc.) concrete is also treated as an elastic material.
  According to this assumptions the constitutive equations are:
  - for steel

$$E_s \, \varepsilon_s(t) = \sigma_s(t) + E_s \, \varepsilon_\theta(t), \tag{1}$$

  - for concrete

$$E_{c\phi}(t)\varepsilon_c(t) = \sigma_c(t) + \varrho(t)\sigma_c(t_1) + E_{c\phi}(t)(\varepsilon_{sh}(t) + \varepsilon_\theta(t)) \tag{2}$$

t_1 is the age of concrete at the first loading, while t is the current time,

$\varepsilon_s(t)$, $\sigma_s(t)$ and $\varepsilon_c(t)$, $\sigma_c(t)$ are strains and stresses in steel and concrete respectively.

$\varepsilon_{sh}(t)$ and $\varepsilon_\theta(t)$ are strains due to shrinkage and temperature changes respectively.

$\varepsilon_{sh}(t)$ and $\varepsilon_\theta(t)$ are strains due to shrinkage and temperature changes respectively

E_s is the modulus of elasticity for steel.

Eq.(2) is the simplified algebraic stress-strain relationship for concrete proposed by several authors ((1),(2),(9),(10)). The coefficient $q(t)$ and the effective modulus $E_{c\phi}(t)$ for concrete depend upon a shosen rheological model for concrete and can be calculated by the next expressions:

$$E_{c\phi}(t) = \frac{E_c(t)}{1+\chi(t,t_1)}$$

$$q(t) = \frac{\phi(t,t_1)(1-\chi(t,t_1))}{1+\chi(t,t_1)\phi(t,t_1)}$$

(3)

$E_c(t)$ is the modulus of elasticity for concrete,

$\phi(t,t_1)$ is the creep coefficient.

$\chi(t,t_1)$ is the ageing coefficient which depends upon the age of concrete at loading, time under load and member dimensions. The values of this coefficient for different creep functions (ACI, CEB-FIP) can be found in literature ((1),(4)).

3. ISOPARAMETRIC CURVILINEAR ELEMENT AND STIFFNESS MATRICES

This computer programme is prepared for a structural system containing straight and curvilinear elements, which may be prismatic or nonprismatic. To formulate the stiffness matrices for the elements, the isoparametric formulation have been used, as it described in previous authors´ work (6).

Local coordinates x, y and displacements u, v of any point are expressed by the coordinates (x_i, y_i, x_j, y_j) and displacements (u_i, v_i, u_j, v_j) of the nodal points by the same shape functions

$$x = x_i(1-\xi)/2 + x_j(1+\xi)/2$$

$$u = u_i(1-\xi)/2 + u_j(1+\xi)/2$$

(4)

$$y = N_i(\xi)y_i + \overline{N}_i(\xi)\left(\frac{dy}{d\xi}\right)_i + N_j(\xi)y_j + \overline{N}_j(\xi)\left(\frac{dy}{d\xi}\right)_j$$

$$v = N_i(\xi)v_i + \overline{N}_i(\xi)\left(\frac{dv}{d\xi}\right)_i + N_j(\xi)v_j + \overline{N}_j(\xi)\left(\frac{dv}{d\xi}\right)_j$$

(5)

where

$$N_i(\xi) = (2-3\xi+\xi^3)/4, \qquad \overline{N}_i(\xi) = (1-\xi-\xi^2+\xi^3)/4,$$

$$N_j(\xi) = (2+3\xi-\xi^3)/4, \qquad \overline{N}_j(\xi) = (-1-\xi+\xi^2+\xi^3)/4. \qquad (6)$$

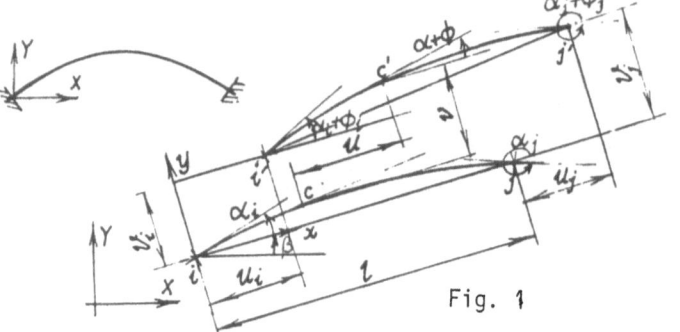

Fig. 1

To obtain the values of the stiffness matrices and vectors of the equivalent nodal forces for some element, the values of the areas of concrete and steel cross-section (A_c, A_s) at any point are expressed by the areas of cross-sections at nodal points $(A_{ci}, A_{cj}, A_{si}, A_{sj})$. The moments of the inertia of the concrete and steel cross-section (I_c, I_s) are expressed by the moments of inertia $(I_{ci}, I_{cj}, I_{si}, I_{sj})$ using shape function (6).

Further procedure for the member stiffness matrix formulation is similar as for the elastic element composed from two different materials: steel and concrete. Numerical values of the stiffness matrix elements and equivalent nodal forces, are obtained by the Gauss-Legendre quadrature for the local coordinate system xy, and then after transformation from the local to the global coordinate system XY are assembled the global stiffness matrices and vectors for the whole structure.

4. PHASES OF LOADING

The sustained load in the first two phases is only taken into account. The first phase corresponds to the moment of time $t=t_1$ when coefficients $\phi=0$, $q=0$ and shrinkage $\varepsilon_s=0$, so that the structure behaves as an elastic system. The equilibrium equations in this phase are

$$\left((S)_c + (S)_s + (S)_g \right) \{\Delta\}_1 = \{Q\}_1 \qquad (7)$$

where
$(S)_c$ and $(S)_s$ are contributions of the concrete and steel parts to the stiffness matrix of the assembled structure,
$\{Q^r\}_1$ is the vector of generalized nodal forces due to

external load in this phase.

$\{\Delta\}_1$ is the vector of generalized nodal displacements in phase 1.

The second phase of the sustained load corresponds to the given coefficient $\phi(t, t_1)$ at curent time t with equilibrium equations:

$$\left((S)_{c\phi}+(S)_s+(S)_g\right)\{\Delta\}_2 = \{Q^P\}_2+\{Q^{sh}\}_2+q(t)(S_c)\{\Delta\}_1 \qquad (8)$$

$\{Q^P\}_2$ and $\{Q^{sh}\}_2$ are the vectors of generalized nodal forces due to external sustained load shrinkage respectively.

The elements of the matrix $(S_{c\phi})$ are calculated in the same way as the elements of the matrix (S_c) substituting $E_c(t)$ by $E_{c\phi}(t)$.

In the subsequent phases $(m=2,3,4...)$ the structure is assumed as an elastic system under periodic loads (wind, snow, temperature shanges, siesmic forces etc.), so that equilibrium equation are

$$\left((S_c)+(S_s)\right)(\{\Delta\}_k-\{\Delta\}_1) = \{Q^P\}_m+\{Q^\theta\}_m \qquad (9)$$

$$(m=2,3,4...)$$

$\{Q^P\}_k$ and $\{Q^\theta\}_k$ are the vectors of the generalized nodal forces due to external loads and the temperature change in the phase k.

Solving the equilibrium equations, taking into account prescribed joint displacements, we obtain the values of the generalized displacements $\{\Delta\}_k$ $(k=1,2,3...)$ in the global coordinate system in every phase. Transforming these displacements from the global coordinates $(0,X,Y)$ to the local coordinates of the member k we can determine the internal member forces by the following expressions.

$$\{r\}_1 = \left((s)_c+(s)_s+(s)_g\right)\{\delta\}_1 - \{q^P\}_1 \quad phase\ 1$$

$$\{r\}_2 = \left((s_{b\phi})+(s)_s+(s)_g\right)\{\delta\}_2-q(t)(s)_c\{\delta\}_1-\{q^P\}_2 \quad phase\ 2$$

$$\{r\}_3 = \left((s)_c+(s)_s\right)\{\delta\}_m - \{q^P\}_m - \{q^\theta\}_m - \{q^P\}_2$$

$(s)_c$ and $(s)_s$ are the stiffness matrices of the member k, (q_p) and (q_θ) are the vectors of the equivalent nodal forces due to external load and temperature changes.

$\{r\}$ is the vector of the member internal forces.

$\{\delta\}$ is the vector of the generalized nodal displacements in local coordinates.

These vector written in the transposed form are

$${q}^T = {q_{i1}, \; q_{i2}, \; q_{i3}, \; q_{j1}, \; q_{j2}, \; q_{j3}}$$

$${r}^T = {N_i, \; T_i, \; M_i, \; N_j, \; T_j, \; M_j}$$

$${\delta} \;\; = {u_i, \; v_i, \; \phi_i, \; u_j, \; v_j, \; \phi_j}$$

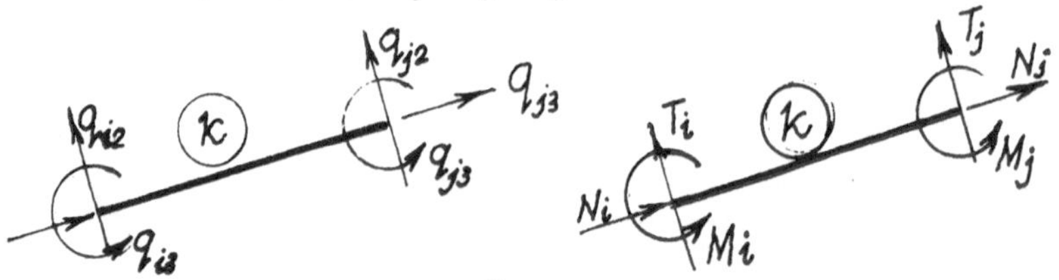

Fig. 2

5. FLOW CHART OF COMPUTER PROGRAM AND EXAMPLE

The flow chart of the computer program is shown in Fig. 4. For each phase of loading the problem is solved firstly according to the theory of first order and after that according to the theory of second order.

The program contains nine subroutines.

Input data are:

Card (1): Title of the problem.

Cards (2): Number of members, joints, supports and loads.

Card (3): JOINT COORDINATES - leading card.

Cards (4): Number of joint, global joint coordinates X,Y.

Card (5): MEMBER INCIDENCES AND MEMBER TYPE - leading card.

Cards (6): Number of member, number of left joint, number of right joint, member type (0 - straight element with fixed ends, 1 - straight element with the left hinged and the right fixed end, 2 - straight element with the left fixed and the righd hinged end, 3 - straight element with the both hinged ends, 4 - curvilinear element).

Card (7): GEOMETRICAL PROPERTIES - loading card.

Cards (8): Number of member, then at left and right end areas of concrete corss-section, areas of the steel cross-section moments of inertia of concrete cross-section, moments of inertia of steel cross-section, statical moments of concrete cross-section statical moments of steel cross-section, and for the curvilinear element: slopes of tangent on the element axis at left and right joint in the global coordinates (dX/dY).

Card (9): MATERIAL PROPERTIES - leading card.

Card (10): Moduli of elasticity for concrete and steel, creep coefficient, coefficient of agening and shrinkage of concrete.

Card (11): PRESCRIBED JOINT DISPLACEMENTS - leading card.

Cards (12): Number of joint, prescribed joint displacement in
X and Y directions and rotations. If some displacement is unlimited then put any number greater then
1.

Card (13): PHASE OF LOADING - loading card.

Card (14): NUMBER OF LOADED JOINTS AND MEMBERS - loading card.

Card (15): Number of loaded joints and members. If temperature changes are taken into account then third number
is 1, else 0.

Card (16): JOINT LOADS - leading card.

Cards (17): Number of joint, components of external forces in
global coordinates (F_x, F_y) and concentraded moment.

Card (18): MEMBER LOADS - leading card.

Cards (19): Number of member, components of the external distributed loads in local member coordinates and distributed external moments at left and right joint
respectively.

Cards (20): TEMPERATURE CHANGES - leading card for loading 3,4..

Card (21): Number of member, temperature rise of the member
axis, temperature diference Δt between the top and
botton edges of the cross-section divided by the
depth $h(\Delta t/h)$.

In the third and next phases of loading the periodic, loads
and temperature changes have to inputed only.

Input and output data are presented for one reinforced concrete arch (Fig. 3) in the next example.

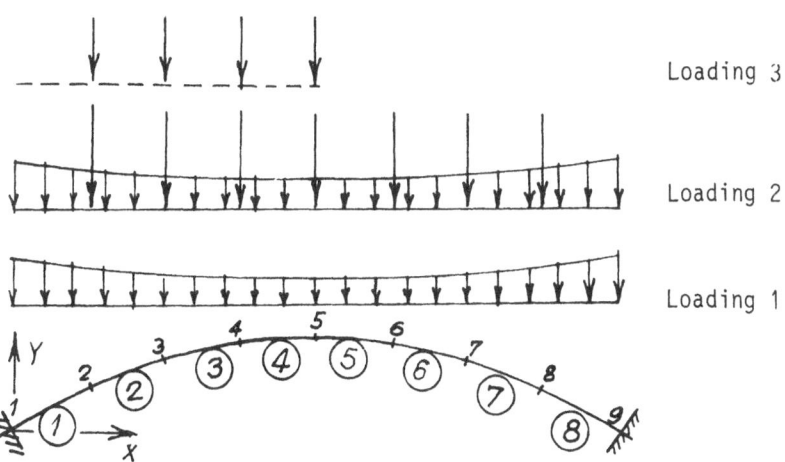

Loading 3

Loading 2

Loading 1

Fig. 3

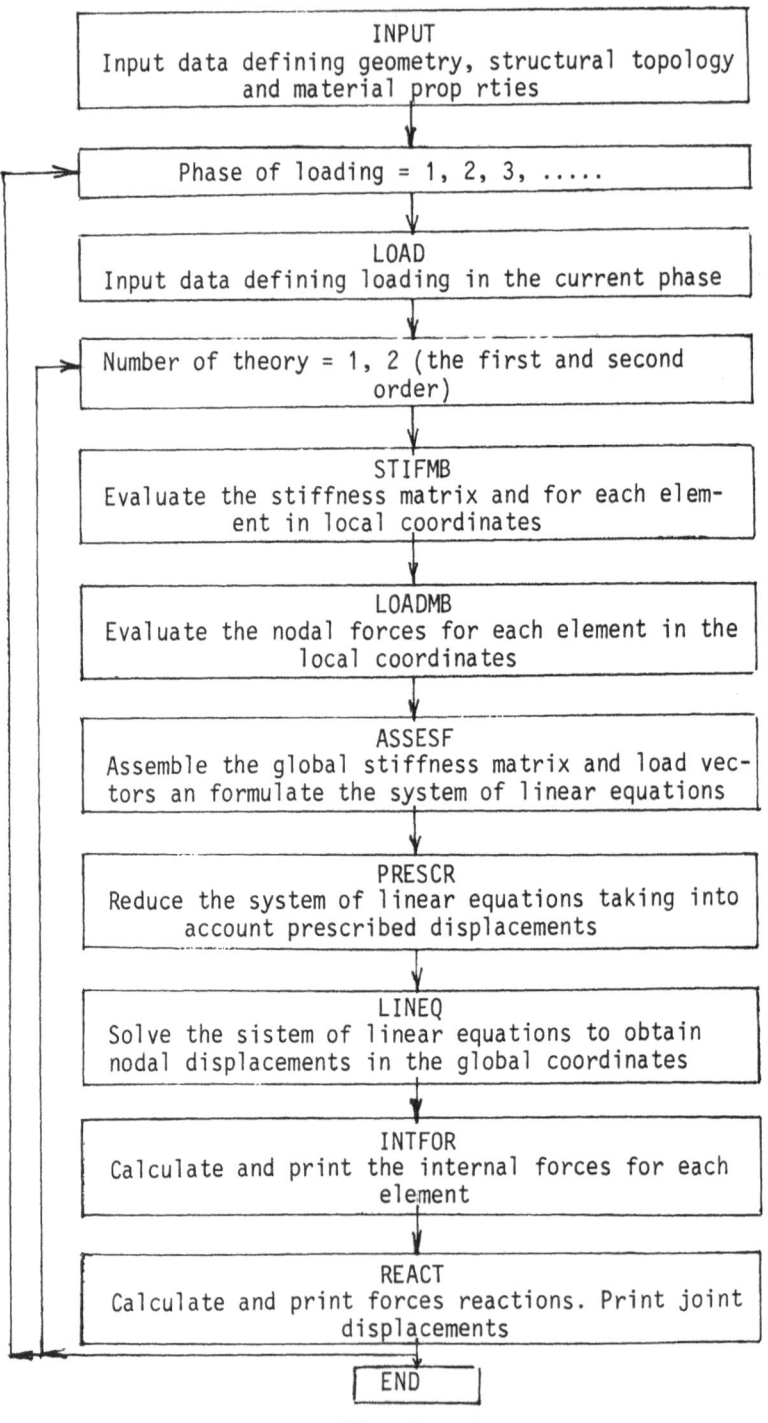

Fig. 4

```
REINFORCED CONCRETE ARCH - EXAMPLE
NUMBER OF JOINTS,MEMBERS,SUPPORTS AND LOADS
NUMBER OF JOINTS     9
NUMBER OF MEMBERS    8
NUMBER OF SUPPORTS   2
NUMBER OF LOADINGS   3
JOINT COORDINATES
1   0.    0.
2   24.15  12.075
3   48.3   20.700
4   72.45  25.875
5   96.60   27.600
6  120.75  25.875
7  144.90  20.700
8  169.05  12.075
9  193.20   0.000
MEMBER INCIDENCES AND MEMBER TIPE
1   1   2   4
2   2   3   4
3   3   4   4
4   4   5   4
5   5   6   4
6   6   7   4
7   7   8   4
8   8   9   4
GEOMETRICAL PROPERTIES
1  7.8  4.7  0.062  0.045  5.83  5.25   0.07502  0.0635
0.571429  0.42857
2  4.7  4.49 0.045  0.041  5.25  5.47   0.0635   0.0661
0.42857   0.28571
3  4.49 4.35 0.041  0.039  5.47  5.70   0.0661   0.0700
0.28571   0.14286
4  4.35 4.20 0.039  0.041  5.70  6.00   0.0700   0.0736
0.14286   0.
5  4.20  4.35  0.041  0.039  5.47  5.25  0.0661  0.0635
0.  -.14286
6  4.35  4.49  0.039  0.041  5.7  5.47  0.070 0.0661
-0.14286  -0.28571
7  4.49  4.7  0.041  0.045  5.47  5.25  0.0661  0.0635
-0.28571  -0.42857
8  4.7  7.8  0.045  0.062  5.25  5.83  0.0635  0.07502
-0.42857  -0.571429
MODULI OF ELASTICITY,CREEP COEFICIENT AND SHRINKAGE
 38000000.  200000000.  2.5  0.8  0.00015
PRESCRIBED JOINT DISPLACEMENTS
1  0.   0.   0.
9  0.   0.   0.
```

```
PHASE OF LOADING 1 - DEAD LOAD
NUMBER OF LOADED JOINTS AND MEMBERS
0  8  0
MEMBER LOAD
1 -96.74   -46.29   -169.31   -108.0  0.
2 -46.29   -30.84   -108.00   -107.93 0.
3 -30.84   -15.38   -107.93   -107.66 0.
4 -15.38    -0.     -107.66   -105.00 0.
5   0.      15.38   -105.00   -107.66  0.
6  15.38    30.84   -107.66   -107.93  0.
7  30.84    46.29   -107.93   -108.00  0.
8  46.29    96.74   -108.00   -169.31  0.
PHASE OF LOADING 2 - COMPLETE DEAD LOAD
NUMBER OF LOADED JOINTS AND MEMBERS
 7  8  0
JOINT LOAD
2  0.   -2476.   0.
3  0.   -2093.   0.
4  0.   -1868.   0.
5  0.   -1824.   0.
6  0.   -1868   0.
7  0.   -2093.   0.
8  0.   -2476.   0.
MEMBER LOAD
1  -96.74   -46.29   -169.31   -108.0   0.
2  -46.29   -30.84   -108.00   -107.93  0.
3  -30.84   -15.38   -107.93   -107.66  0.
4  -15.38   -0.      -107.66   -105.00  0.
5   0.       15.38   -105.00   -107.66  0.
6  15.38    30.84   -107.66   -107.93  0.
7  30.84    46.29   -107.93   -108.00  0.
8  46.29    96.74   -108.00   -169.31  0.
PHASE OF LOADING 3 - DEAD AND LIVE LOAD
NUMBER OF LOADED JOINTS AND MEMBERS
4  0  0
JOINT LOADS
2  0   -1087.   0.
3  0.   -1087.   0.
4  0.   -1087.   0.
5  0.    -543.   0.
```

THEORY OF 1 ORDER, PHASE OF LOADING 1

MODULI OF ELASTICITY

FOR STEEL: ES = 0.20E+09

FOR CONCRETE: EC = 0.38E+08 ECF = 0.38E+08 RO = 0.

INTERNAL FORCES

ELE-MENT	JO-INT	AXIAL FORCE	SHEAR FORCE	BENDING MOMENT
1	1	22325.18	1588.96	16649.52
1	2	-20290.70	-273.15	1015.14
2	2	20290.70	273.15	-1015.14
2	3	-19286.03	-5.98	2584.48
3	3	19286.03	5.98	-2584.48
3	4	-18720.76	67.13	1134.15
4	4	18720.76	-67.13	-1134.15
4	5	-18539.80	0.02	51.52
5	5	18539.80	-0.02	-51.52
5	6	-18720.77	-67.09	1126.00
6	6	18720.77	67.09	-1126.00
6	7	-19286.04	6.02	2581.01
7	7	19286.04	-6.02	-2581.01
7	8	-20290.71	273.18	1013.17
8	8	20290.71	-273.18	-1013.17
8	9	-22325.20	1588.99	-16651.60

FORCE REACTIONS

JOINT	FORCE X	FORCE Y	MOMENT
1	18595.34	12456.00	16649.52
9	-18595.34	12456.04	-16651.60

JOINT DISPLACEMENTS

JOINT	DISPL X	DISPL Y	ROTATION
1	0.00000	0.00000	0.00000
2	0.00328	-0.01193	-0.00053
3	0.00437	-0.02296	-0.00026
4	0.00258	-0.02697	-0.00005
5	-0.00000	-0.02753	-0.00000
6	-0.00258	-0.02700	0.00005
7	-0.00437	-0.02298	0.00026
8	-0.00328	-0.01194	0.00053
9	0.00000	0.00000	0.00000

THEORY OF 2 ORDER, PHASE OF LOADING 1

MODULI OF ELASTICITY

FOR STEEL: ES = 0.20E+09

FOR CONCRETE: EC = 0.38E+08 ECF = 0.38E+08 RD = 0.

INTERNAL FORCES

ELE-MENT	JO-INT	AXIAL FORCE	SHEAR FORCE	BENDING MOMENT
1	1	22355.90	1571.40	16639.74
1	2	-20323.22	-259.21	1009.22
2	2	20323.22	259.21	-1009.22
2	3	-19320.05	3.74	2640.15
3	3	19320.05	-3.74	-2640.15
3	4	-18755.79	72.13	1137.78
4	4	18755.79	-72.13	-1137.78
4	5	-18575.18	0.02	17.05
5	5	18575.18	-0.02	-17.05
5	6	-18755.79	-72.09	1130.59
6	6	18755.79	72.09	-1130.59
6	7	-19320.06	-3.70	2636.86
7	7	19320.06	3.70	-2636.86
7	8	-20323.23	259.24	1007.16
8	8	20323.23	-259.24	-1007.16
8	9	-22355.92	1571.44	-16641.99

FORCE REACTIONS

JOINT	FORCE X	FORCE Y	MOMENT
1	18630.72	12456.00	16639.74
9	-18630.72	12456.04	-16641.99

JOINT DISPLACEMENTS

JOINT	DISPL X	DISPL Y	ROTATION
1	0.00000	0.00000	0.00000
2	0.00329	-0.01196	-0.00053
3	0.00439	-0.02304	-0.00026
4	0.00258	-0.02700	-0.00005
5	-0.00000	-0.02753	-0.00000
6	-0.00259	-0.02704	0.00005
7	-0.00439	-0.02305	0.00026
8	-0.00329	-0.01196	0.00053
9	0.00000	0.00000	0.00000

THEORY OF 1 ORDER, PHASE OF LOADING 2

FOR CONCRETE: EC = 0.38E+08 ECF = 0.13E+08

CREEP COEFFICIENT = 2.50 SHRINKAGE 0.00015

COEFFICIENT OF AGENING = .800 RO = 0.167

INTERNAL FORCES

ELE- MENT	JO- INT	AXIAL FORCE	SHEAR FORCE	BENDING MOMENT
1	1	37594.71	1327.71	23264.26
1	2	-35490.45	-1754.44	7089.44
2	2	34515.11	-521.36	-7089.44
2	3	-33496.93	-1013.76	9273.21
3	3	32921.95	-998.71	-9273.21
3	4	-32366.64	-791.63	5435.84
4	4	32102.46	-1057.59	-5435.84
4	5	-31927.08	-911.97	3393.66
5	5	31927.08	-912.03	-3393.66
5	6	-32102.47	-1057.52	5426.57
6	6	32366.65	-791.70	-5426.57
6	7	-32921.96	-998.64	9269.12
7	7	33496.95	-1013.83	-9269.12
7	8	-34515.14	-521.30	7086.87
8	8	35490.48	-1754.50	-7086.87
8	9	-37594.74	1327.77	-23267.37

FORCE REACTIONS

JOINT	FORCE X	FORCE Y	MOMENT
1	31982.61	19804.98	23264.26
9	-31982.61	19805.05	-23267.37

JOINT DISPLACEMENTS

JOINT	DISPL X	DISPL Y	ROTATION
1	0.00000	0.00000	0.00000
2	0.01655	-0.05951	-0.00272
3	0.02206	-0.11367	-0.00119
4	0.01268	-0.13024	-0.00013
5	-0.00001	-0.13161	-0.00000
6	-0.01271	-0.13035	0.00013
7	-0.02207	-0.11372	0.00119
8	-0.01656	-0.05952	0.00272
9	0.00000	0.00000	0.00000

THEORY OF 1 ORDER, PHASE OF LOADING 3

MODULI OF ELASTICITY

FOR STEEL: ES = 0.20E+09

FOR CONCRETE: EC = 0.38E+08 ECF = 0.38E+08

INTERNAL FORCES

ELE- MENT	JO- INT	AXIAL FORCE	SHEAR FORCE	BENDING MOMENT
1	1	42464.66	1875.53	41186.95
1	2	-40266.20	-2853.90	10048.95
2	2	38862.67	-421.01	-10048.95
2	3	-37797.02	-1662.22	20498.30
3	3	36923.42	-1395.43	-20498.30
3	4	-36384.90	-942.69	15680.64
4	4	35966.99	-1982.61	-15680.64
4	5	-35883.58	-542.78	4662.88
5	5	35883.58	-1824.22	-4662.88
5	6	-36148.22	-714.06	-1562.75
6	6	36412.40	-1135.17	1562.75
6	7	-36976.84	-1208.48	1941.29
7	7	37551.82	-803.99	-1941.29
7	8	-38511.07	-1241.41	5543.25
8	8	39486.41	-1034.39	-5543.25
8	9	-41482.52	156.79	-5779.60

FORCE REACTIONS

JOINT	FORCE X	FORCE Y	MOMENT
1	35939.12	22696.80	41186.95
9	-35939.12	20717.23	-5779.60

JOINT DISPLACEMENTS

JOINT	DISPL X	DISPL Y	ROTATION
1	0.00000	0.00000	0.00000
2	0.02439	-0.07675	-0.00368
3	0.03584	-0.14876	-0.00151
4	0.02529	-0.16102	0.00053
5	0.01049	-0.13826	0.00114
6	-0.00103	-0.11149	0.00088
7	-0.00948	-0.08689	0.00105
8	-0.00924	-0.04541	0.00193
9	0.00000	0.00000	0.00000

THEORY OF 2 ORDER, PHASE OF LOADING 3

MODULI OF ELASTICITY

FOR STEEL: ES = 0.20E+09

FOR CONCRETE: EC = 0.38E+08 ECF = 0.38E+08

INTERNAL FORCES

ELE- MENT	JO- INT	AXIAL FORCE	SHEAR FORCE	BENDING MOMENT
1	1	42483.29	1874.98	42385.77
1	2	-40284.77	-2855.48	9939.43
2	2	38881.24	-419.43	-9939.43
2	3	-37815.25	-1666.13	21433.04
3	3	36941.64	-1391.52	-21433.04
3	4	-36402.42	-949.04	16548.16
4	4	35984.51	-1976.26	-16548.16
4	5	-35900.03	-551.55	4697.99
5	5	35900.03	-1815.45	-4697.99
5	6	-36163.26	-725.06	-2382.61
6	6	36427.44	-1124.17	2382.61
6	7	-36990.24	-1221.43	1009.44
7	7	37565.23	-791.05	-1009.44
7	8	-38522.73	-1255.95	5599.29
8	8	39498.07	-1019.86	-5599.29
8	9	-41492.45	141.02	-4480.19

FORCE REACTIONS

JOINT	FORCE X	FORCE Y	MOMENT
1	35955.57	22705.57	42385.77
9	-35955.57	20708.47	-4480.19

JOINT DISPLACEMENTS

JOINT	DISPL X	DISPL Y	ROTATION
1	0.00000	0.00000	0.00000
2	0.02528	-0.07879	-0.00387
3	0.03766	-0.15362	-0.00154
4	0.02664	-0.16398	0.00068
5	0.01147	-0.13761	0.00126
6	-0.00012	-0.10963	0.00087
7	-0.00889	-0.08597	0.00102
8	-0.00890	-0.04491	0.00195
9	0.00000	0.00000	0.00000

REFERENCES

1. Bažant, Z.P. (1972) "Prediction of Concrete Creep Effects Using Age-Adjusted Modulus Method", Journal of ACI, Vol.69 pp.211-217.

2. Djurić, M. (1963) "Theory of Composite and Prestressed Concrete Structures" (in Serbo-Croatian). Naučno delo, Beograd.

3. Hinton, E. and Owen, D.R.J. (1977) "Finite Element Programming". Academic Press, London.

4. Neville, A.M., Digler, W.H. and Brooks, J.J. (1983) "Creep of Plain and Structural Concrete". Construction Press. London, New York.

5. Praščević, Ž. (1983) "Analysis of Plane Reinforced Concrete Structures According to Second Order Theory". In Engineering Sofware III (Ed.R.A.Adey) Springer Verlag, Berlin, pp. 566-579.

6. Sekulović, M. and Praščević, Ž. (1984) "Analysis of Reinforced Concrete Arches by Finite Element Method". Proc. of Int. Conf. on Computer Aided Design. Split, pp. 317-330.

7. Sekulović, M. (1984) "The Finite Element Method" (in Serbo-Croatian). Gradjevinska knjiga, Beograd.

8. Trost, H. (1967) "Auswirkungen de Superpozitionprinzips auf Kriech und Relaxationsprobleme bei Beton und Spanbeton". Beton und Stahlbetonbau, Vol. 62, pp. 230-238.

9. Ulitcki, I.I. (1962) "A Method of Computing Creep and Shrinkage Deformation of Concrete for Practical Purposes" (in Russian). Beton i železobeton No. 4, Moskow.

10. Zienkiewicz, O. C.(1978)"The Finite Element Method". Mc Grow-Hill, London.

EXPLOITING PARALLEL COMPUTING WITH LIMITED PROGRAM CHANGES
USING A NETWORK OF MICROCOMPUTERS

J. L. Rogers, Jr. and J. Sobieszczanski-Sobieski

NASA Langley Research Center

INTRODUCTION

As the speed of a single processor computer approaches a
physical limit, computer technology is beginning to advance
toward parallel processing to provide even faster speeds.
Network computing and multiprocessor computers are two
discernible trends in this advancement. Given the two
extremes, a few powerful processors or many relatively simple
processors, it is not yet clear how engineering applications
can best take advantage of parallel architecture. Neither is
it clear at this time the extent to which engineering analysis
programs will have to be recoded to take advantage of this new
hardware. Initial investigations of these questions can begin
immediately by exploiting the physical parallelism of
selected problems and the modular organization of existing
programs to solve these problems.

To gain experience in exploiting parallel computer
architecture without making major changes to the code, an
existing program was adapted to perform finite element
analysis by distributing substructures over a network of four
Apple IIe microcomputers connected to a shared disk (Rogers
and Sobieszczanski-Sobieski, 1983). This network of
microcomputers is regarded merely as a simulator of a parallel
computer because it should be obvious that substructure
analysis of a practical problem of significant size should be
performed on a computer with much more power than this
particular microcomputer. In this network, one microcomputer
controls the entire process while the others perform the
analysis on each substructure in parallel.

After the substructure analysis was implemented in parallel, a
newexperiment was planned using this system. In this

experiment, the substructure analysis is used in an iterative, fully-stressed, structural resizing procedure to evaluate resizing in which the analyses of all substructures are not completed during a single iteration. Methods to handle the resulting mixture of old and new analysis data, referred to as asynchronous parallelism, need to be developed for parallel computing applications. Although the present work involves only structural analysis, this research gives some initial insight on how to configure multidisciplinary analysis and optimization procedures for decomposable engineering systems using either high-performance engineering workstations or a parallel processor supercomputer. In addition, the operational experience gained will facilitate the implementation of analysis programs on these new computers when they become available in an engineering environment.

BACKGROUND

In 1975 a feasibility study (Universal Analytics, 1975) was performed to determine the effort required to convert NASTRAN (NASTRAN User's Manual, 1983) to execute on the ILLIAC IV computer, and to assess the advantanges that would be gained from such a conversion. The projected advantages in speed improvement were significant. For example, the decomposition of a 10,000 degree-of-freedom matrix on the ILLIAC IV was estimated to be 40-100% faster than on a CDC 6600 computer when the matrix could not be contained in central memory. The problem that the study pointed out was that the code conversion effort would require 110-140 man months over a period of 36-50 months. If funds had been supplied and the project begun in 1976, it would probably not have been completed until 1980. About two years later, in 1982, the ILLIAC IV was taken off-line. This historical example illustrates the difficulties that could be encountered in future wholesale conversions of engineering analysis codes to parallel processor computers which appear to be the wave of the future (Siewiorek, 1982; and Noor, Storaasli, and Fulton, 1983)

While such wholesale conversion efforts should ultimately provide the greatest increases in efficiency, it is also important in the interim to benefit from the speed improvements offered by parallel computing without the cost, manpower, and time involved in the conversion of major analysis codes. This research demonstrates that for structural analysis, the current investment in sequential, modular structural analysis programs can be used to exploit parallelism of a network of computers.

APPROACH

The approach taken for this project was to establish reference results using the Engineering Analysis Language (Whetstone, 1980), called EAL, to analyze a finite element model that was not substructured. An existing small finite element analysis code was then modified to handle substructures and applied to the same model on a CYBER mainframe computer. Next, this program was implemented on a microcomputer to test the substructure method sequentially. The program was then distributed over a network of these microcomputers with little change to the analysis code to test the substructure method executed in parallel. A Fully Stressed Design (FSD) capability was added to test the behavior of substructure analysis in an iterative process in which some of the analyses were completed before others.

The Model
The finite element model used for testing is shown in figure 1. This model contains 16 joints, 21 beam elements, and 42 degrees-of-freedom (the size of the model was limited by the memory of the microcomputer). The framework has three substructures with each substructure composed of seven beams. The cross-sections and material properties are identical for all beams. A load is applied at one of the boundary nodes as shown in figure 1.

The Small Finite Element Program
Input for the model was written for a small, undocumented finite element program developed in the past for a CYBER computer without any intent to ever use it for parallel processing. It did not even have an explicit substructuring capability. In this study, this program represented an "investment in existing software" that was to be salvaged. The results from the unchanged program were verified against the reference run. New code for substructuring based on equations from (Przemieniecki, 1968) was then added to the program. The model was divided into three substructures with the new code used to compute the boundary stiffness matrix for each substructure using equation 1:

$$K_b = K_{bb} - K_{bi} K_{ii}^{-1} K_{bi}^T \qquad (1)$$

Each of the three 18x18 substructure stiffness matrices was reduced to 6x6 equivalent beam stiffness matrices (figure 2). These three stiffness matrices were input to the program, assembled to represent a stiffness-equivalent framework composed of three beams, each beam representing one substructure. The forces and displacements at the boundary

nodes were computed for each such beam. Modifications were
made to the program for reading these forces from a file and
applying them to the corresponding substructures. By applying
support conditions to the substructures, solutions were
obtained for the interior node displacements, internal forces,
and elemental stresses. These results were also verified
against the reference run. It should be noted that the
substructure analysis was simplified because the external
loads were applied only to the boundary nodes. Should any
loads be applied to the interior substructure nodes, it would
have been necessary to add code to transfer these loads to the
boundary nodes.

Conversion to the Microcomputer

At this point, the program for sequentially performing
substructure analysis existed on a CYBER mainframe computer.
The next step was to convert the program to the microcomputer.
Since the entire program was written in FORTRAN=77, the move
was quite simple and the program was contained in the
microcomputer's 64K byte memory without overlay. Although the
program itself was entirely core resident, the test case shown
in figure 1 was too large for analysis without substructuring.
Therefore, the first step on the microcomputer was to run the
substructuring sequentially. The problem took 57 minutes to
execute. The results were verified against the reference run
with little loss in precision (less than 1%).

Distributing the System

The approach selected for distributing the system was to use
one microcomputer to execute a controller program and three
microcomputers to analyze each of the substructures. All of
the microcomputers were connected to a 20MB Corvus hard disk
which was used for data communication among the computers.
The operations assigned to each computer are shown in
figure 3. The controller program started the system
(operation 0), assembled the substructure stiffness matrices
and solved for the forces on each substructure at the boundary
nodes (operation 2), and output the data (operation 4). The
substructure programs computed the substructure stiffness
matrices (operation 1), and used the forces from the
controller program to solve for internal forces, node
displacements, and elemental stresses for each substructure
(operation 3). Note that parallelism only exists in
operations 1 and 3. The output of the data (operation 4) also
could have been distributed, but it was found to be easier to
keep it centrally located.

When distributing the system to four microcomputers, the
purpose was to minimize changing the original analysis code.
Only the procedures involved in operations 0, 2, and 4 were
retained in the controller program while only those procedures

involved in operations 1 and 3 were retained in the substructure program.

A subroutine was added to both the controller program and the substructure program to schedule their execution. This scheduling was accomplished by using three files on the shared disk; one file for each substructure program. When it was time for the controller program to execute each of the three files contained a zero, and when it was time for a substructure program to execute, its respective file contained a nonzero number. Each program queried its file and if it was not its turn for execution it was put in a "holding pattern" by performing a simple multiplication loop before querying again. The system could have been implemented on only three processors with one processor doubling for executing the controller and substructure programs.

The ideal is to reduce the time required to solve the same problem sequentially on a single processor to (time/n) where n is the number of processors used to solve the problem. However, it is seen in figure 3 that not all of the calculations can be executed in parallel. In addition, some time was lost in an inevitable overhead such as checking and looping while waiting for a substructure or controller program to finish executing. Thus, the parallel system with substructures took about 27 minutes to complete execution, .47 of the time required for the reference run. This is short of the ideal, .25, but is still more than twice as fast as the sequential system.

The particular division of the structure from figure 1 into substructures is, of course, not the only one possible. If a larger number of smaller substructures was used, larger numbers of parallel computers could have been employed. However, the larger the number of substructures the larger the dimensionality of the assembled structure stiffness matrix (ultimately, if each substructure represents a single beam component, the assembled stiffness matrix would return to the size it would have had if no substructuring was used). Consequently, to minimize the overall computer time, an attempt should be made to balance the size of the assembled structure stiffness matrix against the size and number of the substructure stiffness matrices. The degree of the time reduction depends also on the number of substructuring levels (Sobieszczanski-Sobieski, James, and Dovi, 1983). Thus, tailoring the analysis process for a particular application to take advantage of multiprocessor efficiency is an important issue that faces an analyst using a multiprocessor system.

Resizing
An FSD algorithm was added to examine the behavior of parallel substructure analysis in an iterative process (figure 4). The FSD was performed by resizing all the beams in a given substructure according to the ratio of the maximum absolute normal stress occuring in the substructure to a specified allowable stress. The stress ratio was used as a scale factor to modify the beam cross-sectional moment of inertia. Consistently, the cross-sectional area was multiplied by the square root of the scale factor, and the cross-section linear dimensions were all multiplied by the scale factor to power 1/4. At this point, resizing was synchronized, which means the process would always wait until all data were updated before processing rather than mixing old and new data. An iteration history of the changes in the design variable (plotted as the factor on cross-section linear dimension) for each substructure is shown in figure 5.

Asynchronous Resizing
Since most of the engineering calculations performed in support of design
are iterative in nature, the computational behavior of an iterative distributed process in which some subtasks are completed later than others because of unequal computational requirements for various subtasks is of significant interest (Baudet, 1978; and Sobieszczanski-Sobieski, 1982). If such an imbalance of computational requirements occurs, a choice can be made to let the iterative process continue, temporarily using old data for those processes which are late. The process then becomes asynchronous as it mixes new and old data. The effect of this mixing on the convergence and efficiency can easily be tested in a parallel system such as described above. The tests are conducted by bypassing analysis of selected substructures in some iterations. Obviously during the first loop through the system, all of the substructures will be analyzed to provide a starting point.

There is a large number of different ways in which an asynchronous iterative process can proceed. Using the framework structure from figure 1 as an example, it is conceivable to have at least the following variants.

1. Referring to figure 3, consider being at the outset of iteration "i". Operations 1.1, 1.2, and 1.3 are expected to yield the boundary stiffness matrices for substructures 1,2 and 3, all of which having been resized as a result of an FSD operation at the end of the previous iteration "i-1". Assume that operation 1.1 is late but the process moves on anyway using the old boundary stiffness matrix from iteration "i-1", that does not reflect the "i-1" resizing. That means that operation 2 combines the updated matrices for substructures 2

and 3 with an outdated matrix for substructure 1. In operations 3.1, 3.2, and 3.3, consistently, an old stiffness matrix that does not reflect the "i-1" resizing is used, while the updated stiffness matrices are used in operations 3.2, and 3.3. After this analysis based on the partially incorrect data, all substructures, including substructure 1, are subject to the FSD resizing.

2. Proceed as above, but do not resize that particular substructure for which the old stiffness matrix was used in the analysis (substructure 1 in this example).

3. Complicate variants 1 and 2 by changing: the number of substructures that are assumed to be "late", the number of iterations over which the old data are being used for each substructure, etc. Obviously, a very large number of possibilities can be considered.

It was expected that, for this particular model, the asynchronous processing would have little effect on the convergence other than slowing it down by going through more iterations. In fact, the asynchronous operation in this case may be regarded as a continuation of the FSD process from an artificially injected new starting point. In addition, an analogy can be made between this process and other iterative methods such as the Gauss-Seidel iterative algorithm for solving linear algebraic equations. These methods are error insensitive, i.e. if an error is entered into the process, the process recovers after several iterations and proceeds as if no error had been introduced. One may speculate that a different behavior will be observed in case when nonlinear programing is used instead of FSD for nonconvex cases. Then, there will be a potential for such an asynchronous operation to trigger a switch to another path through the design space that could end up at a different local minimum.

To determine if the above expectations were correct, numerous test cases were executed. The existence of coupling among the substructures was demonstrated by holding substructure 1 constant. As seen in figure 6, this case converged to different results than is seen in the following figures thus showing that the substructures are coupled. Next, variants 1 and 2 were tested. The results shown in figures 7 and 8, respectively, indicate that the asynchronous operation, shown as connected lines, had only a slight influence on the convergence as manifested in small discrepancies that can be seen between the lines and symbols from the synchronous sizing. For instance, asynchronous results for substructures 2 and 3 (figure 7) are above the synchronous ones but both results converge after about eight iterations.

Variant 3, complicating variant 2, was tested using seven different cases. It was decided to complicate variant 2 rather than 1 because of the ease of implementing the variables. Table 1 lists the substructure(s) and iteration(s) that were delayed for each test case. In each of the test cases the results led to the expected behavior. Figures 9 and 10 (test cases 6 and 7 respectively) demonstrate typical results.

Table 1 Test Cases for Asynchronous Processing

CASE	SUBSTRUCTURE DELAYED	ITERATIONS DELAYED
1	2	2
2	1,2	2
3	1	Every other iteration
4	2	Every other iteration
5	1,2	Every other iteration
6	1	Even iterations beginning with 2
and	2	Odd iterations beginning with 3
7	Random combinations	Random

CONCLUDING REMARKS

An experiment was conducted to determine if advantage can be taken of parallel processing without making major changes to an analysis code. This experiment used a network of four microcomputers to simulate a parallel processing computer. A small finite element analysis computer program with a substructuring capability was applied to a framework of beams. One microcomputer controlled the system while the other three analyzed the substructures. The results verified that the computer time was indeed reduced relative to the time required for solution on a single computer. As expected, the reduction factor was proportional to the number of computers minus corrections for data communication and incomplete parallelism of the problem. The reduction was achieved with almost no change to the analysis portion of the code. The experiment also included resizing of the design variables using a Fully Stressed Design algorithm to simulate an iterative optimization to obtain an indication of the effect of asynchronous parallel computing on the convergence of an iterative process. Results from 10 test cases indicated that, for this model, asynchronous processing did not affect convergence other than possibly causing the process to go through more iterations.

These results are not completely general in that they only apply to an iterative process which is monotonically

These results are not completely general in that they only apply to an iterative process which is monotonically convergent. This process is typical of many processes in design. In general, if an iterative process is nonconvex (dependent on the starting point and the path taken) then this conlusion would not apply and the asynchronous process may lead to different results.

REFERENCES

Baudet, G. (1978) The Design and Analysis of Algorithms for Asynchronous Multiprocessors. Ph.D Thesis, Department of Computer Science, Carnegie Mellon University.

Noor A.; Storaasli, O.; and Fulton, R. (1983) Impact of New Computing Systems on Finite Element Computations. ASME Publication H00275, pp. 1-32.

Przemieniecki, J. (1968) Theory of Matrix Structural Analysis. Ch. 9, McGraw-Hill Book Co.

Rogers, J., Jr. and Sobieszczanski-Sobieski, J. (1984) Initial Experiences with Distributing Structural Calculations Among Computers Operating in Parallel. NASA CP 2335, pp. 45-54.

Siewiorek, D. (1982) State of the Art in Parallel Computing. Abstracted from Computer Structures: Principles and Examples, McGraw-Hill Book Co.

Sobieszczanski-Sobieski, J. (1982) A Linear Decompostition Method for Large Optimization Problems. Blueprint for Development. NASA TM-83248.

Sobieszczanski-Sobieski, J.; James, B.; and Dovi, A. (1983) Structural Optimization by Multilevel Decomposition. AIAA Paper No. 83-0832.

The NASTRAN User's Manual (1983) NASA SP-222.

Universal Analytics Inc. (1975) Feasibility Study for the Implementation of NASTRAN on the ILLIAC IV Parallel Processor. NASA CR-132702.

Whetstone, D. (1980) EISI-EAL: Engineering Analysis Language. Proceedings of the Second Conference on Computing in Engineering, ASCE, pp. 276-285.

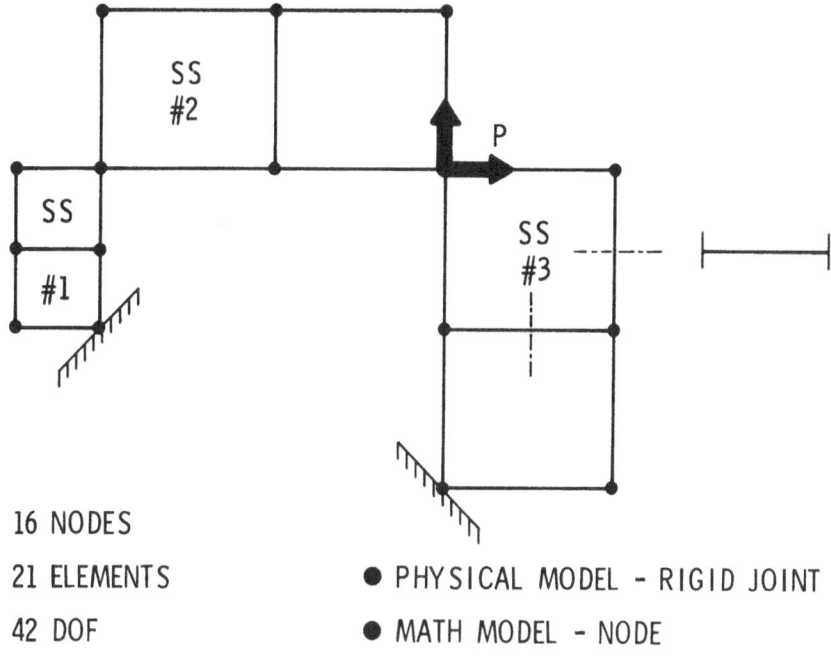

16 NODES

21 ELEMENTS   ● PHYSICAL MODEL - RIGID JOINT

42 DOF   ● MATH MODEL - NODE

Figure 1 - Framework used for testing

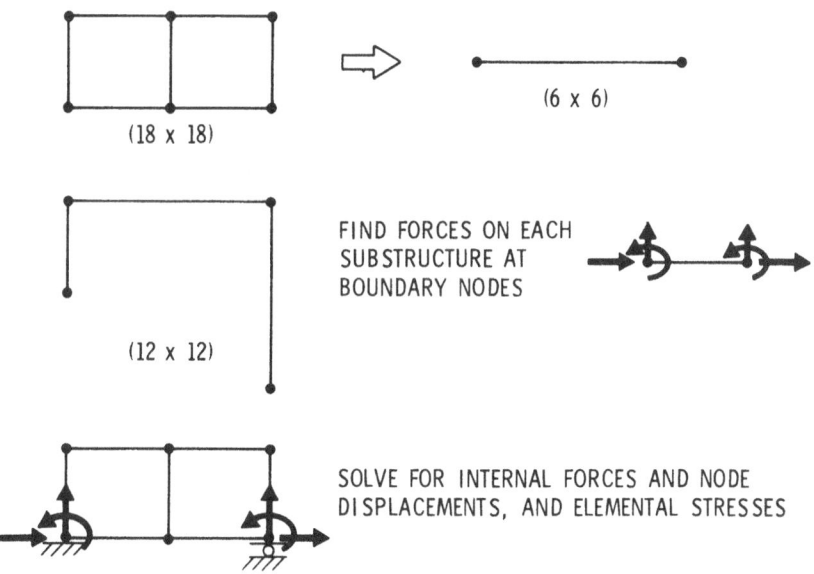

Figure 2 - Actions being taken at
each step

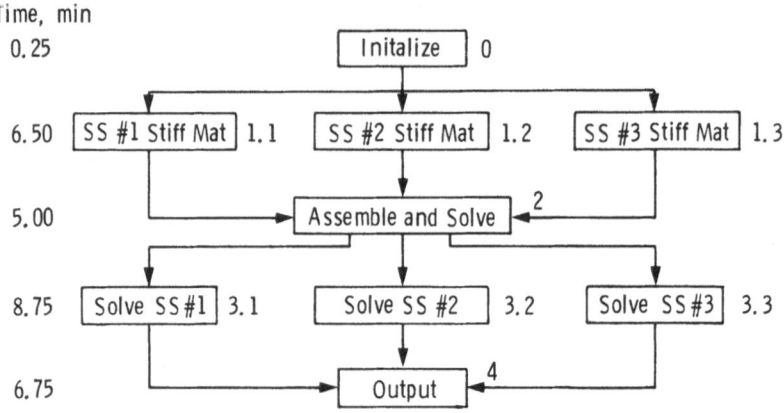

Figure 3 - Flowchart of substructure
analysis

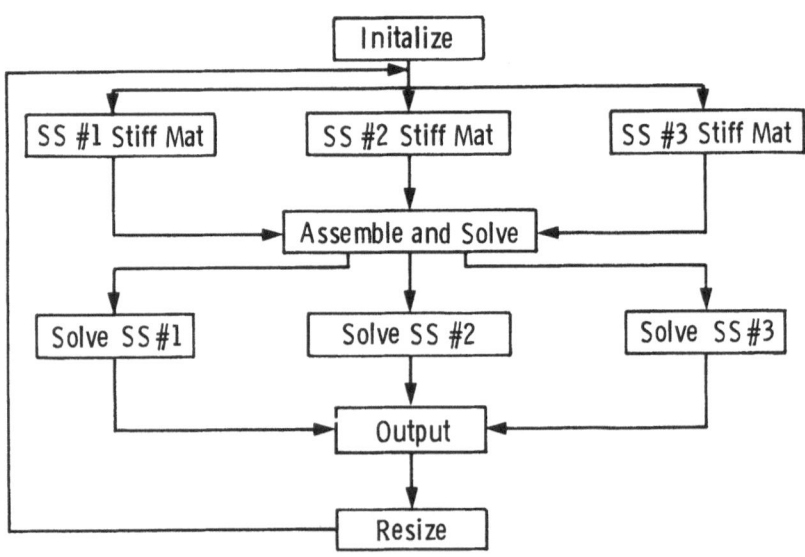

Figure 4 - Flowchart of substructure
analysis with resizing

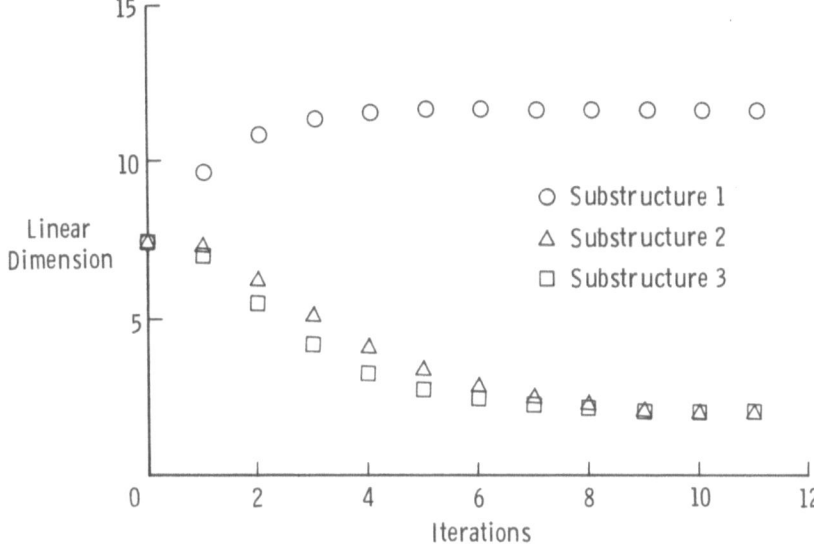

Figure 5 - Synchronous resizing

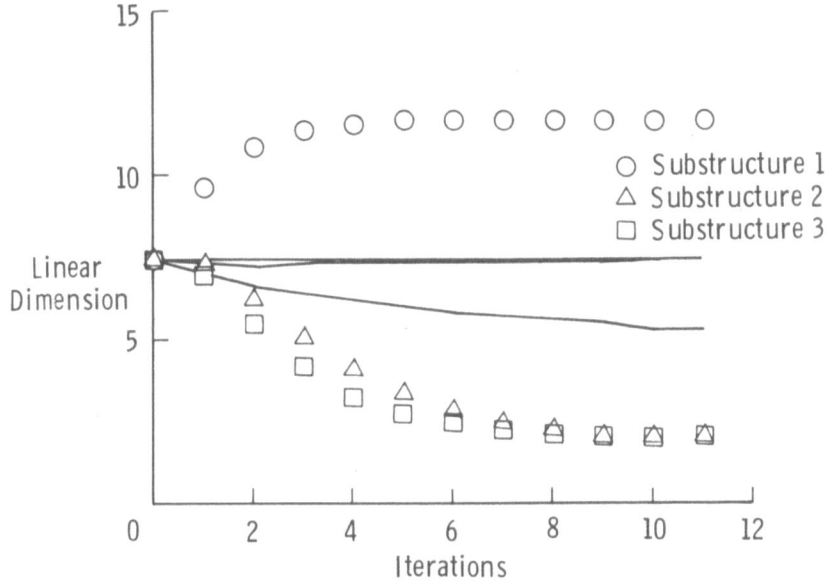

Figure 6 - Asynchronous resizing
showing coupling

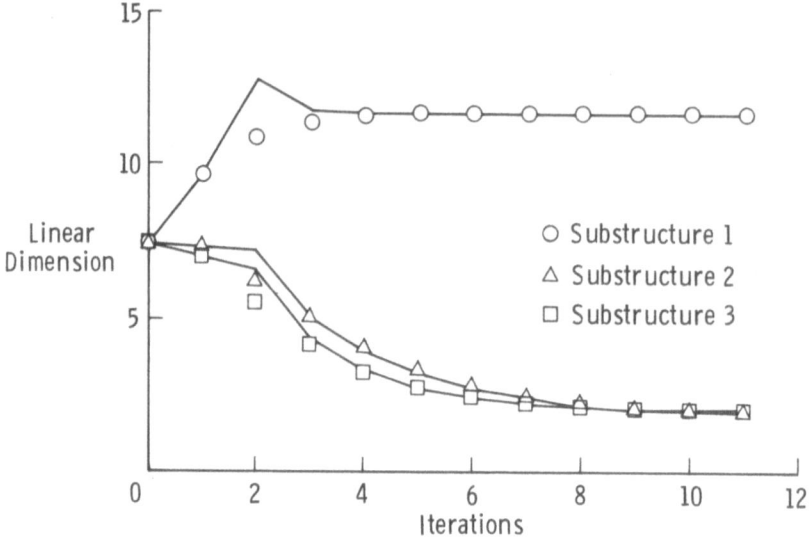

Figure 7 - Asynchronous resizing
(variant 1)

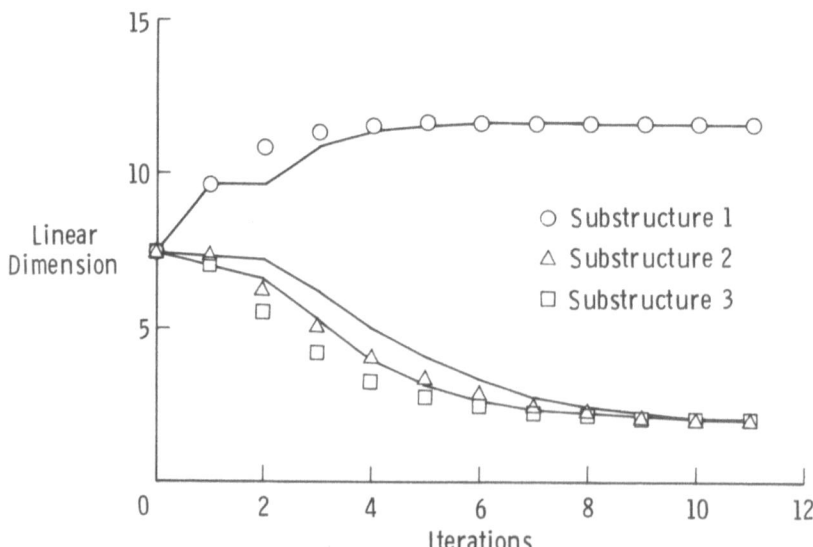

Figure 8 - Asynchronous resizing
(variant 2)

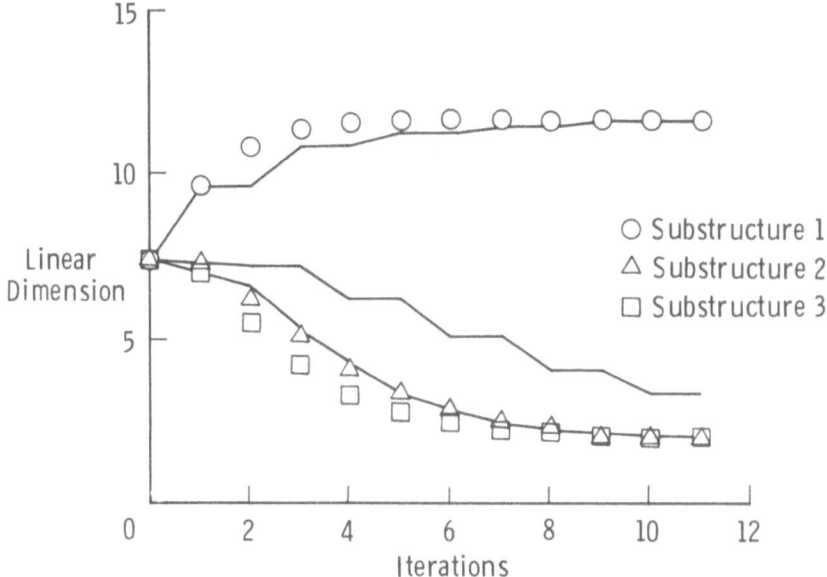

Figure 9 - Asynchronous resizing
(delay substructure 1
on even iterations and
substructure 2 on odd
iterations)

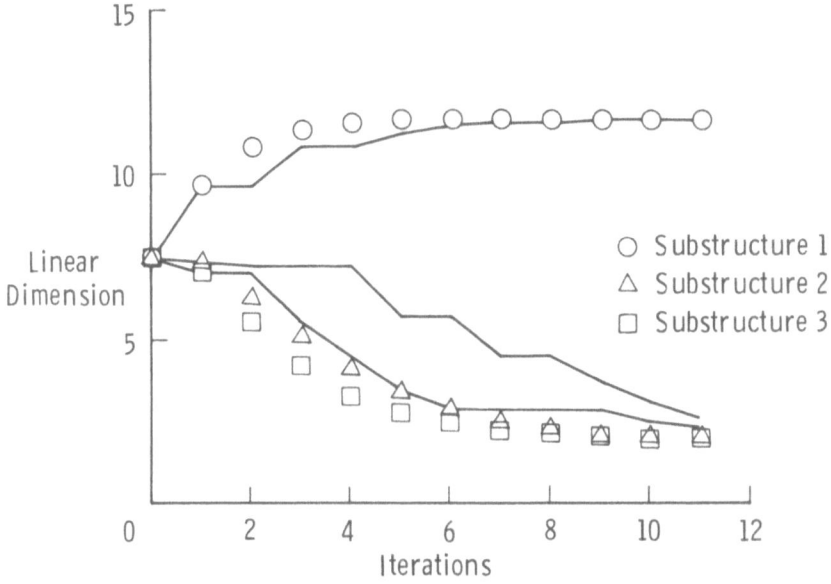

Figure 10 - Asynchronous resizing
(random)

THROUGH THICKNESS AND SURFACE STRESS DISTRIBUTION FOR WELDED
TUBULAR T-JOINT USING FINITE ELEMENT ANALYSIS

G. S. Bhuyan, K. Munaswamy and M. Arockiasamy

Faculty of Engineering & Applied Sciences,
Memorial University of Newfoundland, Canada

1. INTRODUCTION

The finite element method is considered to be the most
powerful and versatile discretization technique available for
numerical solution of complex tubular joints, for the
offshore structures. Existing literature on the analysis of
tubular joints is limited to the use of two dimensional
plate/shell elements, which neglect the displacement
variation across the wall thickness. Also, the plate/shell
element idealization does not permit consideration of the
weld profile and the actual chord/brace interaction. The
present investigation describes the analytical studies on the
variation of stresses along the chord/brace surfaces at
critical points and across the chord wall of the tubular
T-joints loaded through the brace, axially and in in-plane
bending.

2. MODELLING OF THE JOINT

The joint is discretized using an automatic mesh
generation technique. The joint parameters are given in
Table 1. The weld dimensions used for the modelling are
shown in Figure 1. The chord/brace surface and the weld
region are discretized using 8-node nad 6-node three
dimensional elements respectively. Since the loads acting on
the tubular T-joint could be either axial or in-plane
bending, the advantage of symmetry is taken into
consideration in discretization and only half of the T-joint
is considered for modelling. The discretized model is shown
in Figure 2.

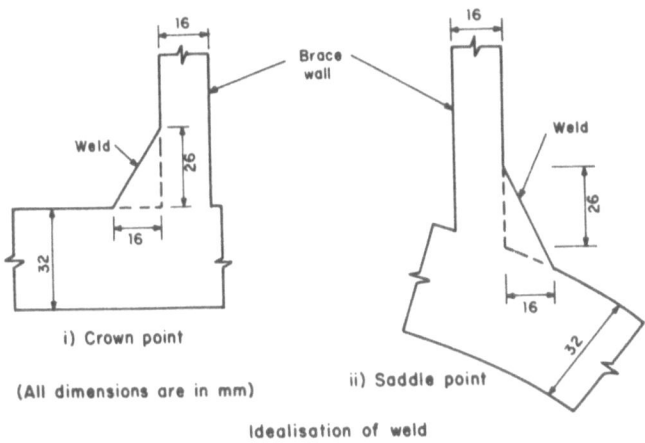

i) Crown point

(All dimensions are in mm)

ii) Saddle point

Idealisation of weld

Figure 1 Modelling of Weld at Chord/Brace Intersection

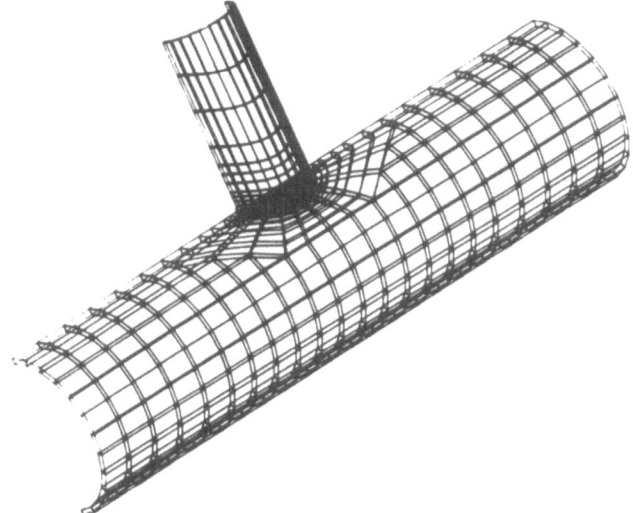

Figure 2 Three Dimensional Finite Element Mesh for the Joint
with Plug

Table 1
Joint Parameters

Type of Joint	D (mm)	T (mm)	$\beta = \frac{d}{D}$	$\tau = \frac{t}{T}$	$\gamma = \frac{D}{2T}$	$\alpha = \frac{2L}{D}$
T	850	32	0.5	0.5	13.2	8.4

D = Chord diameter
d = Brace diameter
T = Chord thickness
t = Brace thickenss
2L = Length of chord between supports

3. ELEMENT STIFFNESS FORMULATIONS

The displacements u_i, v_i and w_i along x, y and z axes at each node are taken as the nodal degrees of freedom. The faces of the 8-node incompatible brick element are defined by local coordinates ξ, η, ζ = ±1. The global coordinates are defined as

$$x = \sum_{i=1}^{8} N_i x_i$$

$$y = \sum_{i=1}^{8} N_i y_i \qquad (1)$$

$$z = \sum_{i=1}^{8} N_i z_i$$

where x_i, y_i and z_i are the element nodal point coordinates and N_i is the shape function which is given by

$$N_i = \frac{1}{8} (1 + \xi\xi_i)(1 + \eta\eta_i)(1 + \zeta\zeta_i) \qquad (2)$$

To improve the flexural characteristics of 8-node isoparametric brick elements, three incompatible modes are introduced in the displacement interpolation functions. The global displacements u, v and w at any point within the element can be expressed as

$$u = \sum_{i=1}^{8} N_i u_i + \sum_{j=1}^{3} g_j \alpha_j$$

$$v = \sum_{i=1}^{8} N_i v_i + \sum_{j=1}^{3} g_j \beta_j \qquad (3)$$

$$w = \sum_{i=1}^{8} N_i w_i + \sum_{j=1}^{3} g_j \gamma_j$$

where $g_j(j=1,2,3)$ are the additional shape functions for the incompatible element and α_j, β_j and γ_j are the generalized displacement coordinates. The first term of the right hand side of Equation 3 represents the polynomial corresponding to the basic element. The displacements due to the terms in the polynomial of the basic element are continuous across the interface of the element, but those due to the second terms in Equation 3 are not necessarily continuous across the element boundaries. The incompatible shape functions g_j are given as

$$g_1 = (1-\xi^2)$$

$$g_2 = (1-\eta^2) \tag{4}$$

$$g_3 = (1-\zeta^2)$$

The stress-strain relations are

$$\{\sigma\} = \{\sigma_x \ \sigma_y \ \sigma_z \ \tau_{xy} \ \tau_{yx} \ \tau_{zx}\}^T = [D] \ \{\epsilon\}$$

$$= [D] \ \{\epsilon_x \ \epsilon_y \ \epsilon_z \ \gamma_{xy} \ \gamma_{yz} \ \gamma_{zx}\}^T \tag{5}$$

where

$\quad [D]$ = the elasticity matrix

The strain-displacement relation of the element can be written as

$$\{\epsilon\} = [B_1 \vdots B_2] \begin{Bmatrix} a_1 \\ \overline{a_2} \end{Bmatrix} \tag{6}$$

where

$$\{a_{1i}\} = \{u_i \ v_i \ w_i\}^T, \text{ for } i=1 \text{ to } 8 \tag{7}$$

are the element nodal point displacements, and

$$\{a_2\} = \{\alpha_1 \ \beta_1 \ \gamma_1 \ \alpha_2 \ \beta_2 \ \gamma_2 \ \alpha_3 \ \beta_3 \ \gamma_3\}^T \tag{8}$$

are the element generalized coordinates due to the incompatible modes. The submatrices $[B_1]$ and $[B_2]$ contain the derivatives of functions N and g respectively.

The equilibrium equations for the element can be written as

$$\begin{bmatrix} [K_{11}] & \vdots & [K_{12}] \\ \text{---} & \text{+} & \text{---} \\ [K_{21}] & \vdots & [K_{22}] \end{bmatrix} \begin{Bmatrix} \{a_1\} \\ \text{---} \\ \{a_2\} \end{Bmatrix} = \begin{Bmatrix} \{F_1\} \\ \text{---} \\ \{F_2\} \end{Bmatrix} \tag{9}$$

where

$$[K_{ij}] = \int\limits_{-1}^{1} \int\limits_{-1}^{1} \int\limits_{-1}^{1} [B_i]^T [D] [B_j] |J| d\xi \, d\eta \, d\zeta \qquad (10)$$

are the submatrices of the element stiffness matrix. $\{F_1\}$ is the element load vector containing equivalent nodal forces. $\{F_2\}$ represents the forces in terms of the element generalized coordinates contributed by only thermal loading; in the present case $\{F_2\} = \{0\}$, $|J|$ is the determinant of Jacobian matrix.

Before assembling the global stiffness matrix, the incompatible degrees of freedom $\{a_2\}$ in Equation 9 are condensed out to ge the resulting equation

$$[\bar{K}_{11}] \{a_1\} = \{F_1\} \qquad (11)$$

where

$$[\bar{K}_{11}] = [K_{11}] - [K_{12}] [K_{22}]^{-1} [K_{21}] \qquad (12)$$

Equation 11 is used to assemble the global stiffness matrix and nodal displacements are then computed. The incompatible degrees of freedom $\{a_2\}$ are calculated as

$$\{a_2\} = [K_{22}]^{-1} [K_{21}] \{a_1\} \qquad (13)$$

Knowing the displacements, $\{a_1\}$ and $\{a_2\}$ the element strains and corresponding stresses are obtained using Equations 6 and 5 respectively.

The 6-node incompatible prism element stiffness matrix is derived in a similar manner from the 8-node incompatible brick element by coalescing the nodes.

4. RESULTS AND DISCUSSIONS

4.1 General

A computer program based on the formulation presented in Section 3 is used for the 3-dimensional analysis of the tubular T-joint. The joint is analysed for axial and in-plane bending loads of 10^6N and 10^5N respectively. The total number of elements, nodes and the degrees-of-freedom in the discretized model with plug are 516, 1102 and 3306 respectively. To obtain the solution for such a system in one run in the VAX system, the problem is solved in stages by i) storing the stiffness matrix on a disk, ii) decomposing the stiffness matrix in blocks of 500 equations or 1000 equations and iii) finally back substituting to arrive at the displacement vector in stages.

   The stresses can be accurately calculated at the Gauss
integration points than at the nodal points. Since the main
interest is to obtain the stress distribution at the outer
surface of the shell, the stresses are computed at the nodes
and averaged between the elements that are connected at the
nodes.

4.2 Stresses due to axial loading

The variation of maximum surface stresses at saddle point due
to axial load is shown in Figure 3. For the chord side the
analysis gives similar variation of stresses as that reported
by Parkhouse (1981). The hot spot stress at the weld toe at
saddle point in the chord side obtained from three
dimensional analysis without plug is 20 percent higher than
that obtained from the two dimensional analysis (Arockiasamy
et al., 1984). But the stress obtained from the
3-dimensional analysis with plug is 10 percent less than the
corresponding value from the two dimensional approach. Three
dimensional anlaysis without plug gives a hot spot stress on
the brace side which is only 3 percent higher than the two
dimensional analysis. The stresses near the weld toe from
the two dimensional analysis are higher than those obtained
from the three dimensional analysis with and without plug.
The hot spot stress concentration factors for joint are given
in Table 2.

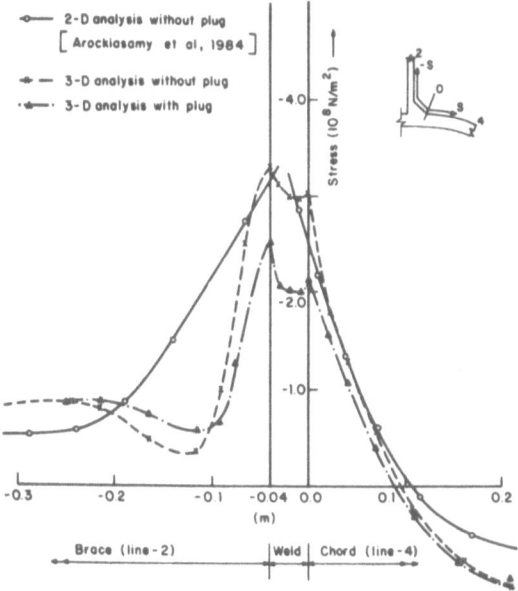

Figure 3 Maximum Surface Stress Variation at Saddle Point due
         to Axial Load

Table 2
Hot Spot Stress Concentration Factors

Methods	Axial				In-plane Bending			
	Chord		Brace		Chord		Brace	
	Crown	Saddle	Crown	Saddle	Crown	Saddle	Crown	Saddle
Present 3-D analysis	2.2 2.09*	6.448 4.57*	2.55 2.34*	7.08 5.41*	1.306 1.06*	- -	1.79 1.304*	- -
Clayton et al (1980)	-	5.8	-	7.1	-	-	1.7	-
Irvine (1981)	-	7.7	-	-	-	-	-	-

*In the analysis plug stiffness is included

The stress variation along the weld surface is shown in
Figure 3. From chord weld toe the stress decreases up to
certain point and then starts increasing and reaches a
maximum value at the brace weld toe, which compares well with
that reported by Morgan (1979).

Figure 4 shows the stress distribution across the weld
reinforcement and brace wall at the saddle point due to axial
load. A rapid decrease in stress across the weld leg is
observed in contrast to the nearly constant value through the
brace wall. The stress values at the crown point obtained
from the two dimensional analysis are lower on the chord side
but higher on the brace side when compared to the three
dimensional analyses with and without plug (Figure 5).
Gradual increase in stress is observed along the weld surface
from the chord to brace weld toe. The through thickness
variation of stress across weld and brace (Figure 6) is
similar to that observed at the saddle point (Figure 4).

The variation of strain concentration factors (SNCFs)
near the saddle point for axial load, obtained from the three
dimensional analysis compares well with the measured values
reported by Back et al (1981) for the tubular joints having
the same dimensions as that under study in this report
(Figures 7 and 8).

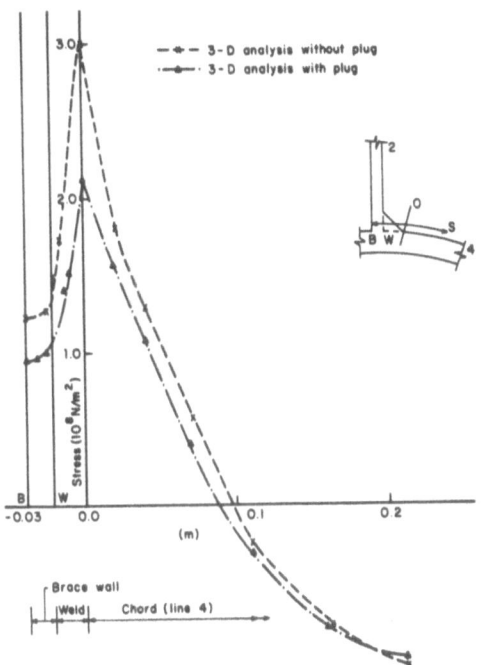

Figure 4 Stress Distribution Across the Weld and Brace Wall at
Saddle Point due to Axial Load

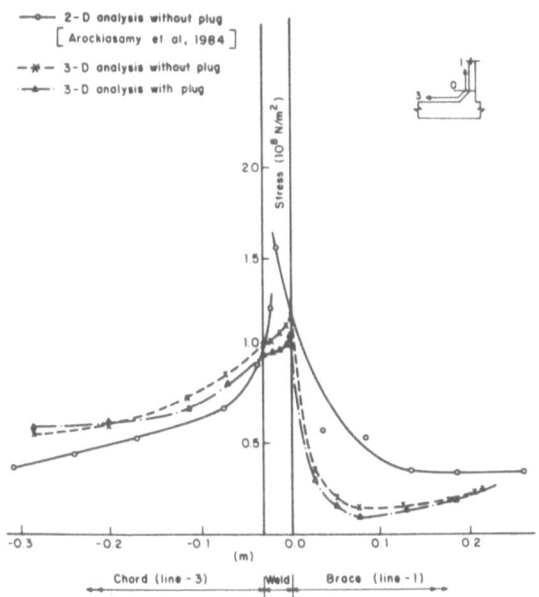

Figure 5 Maximum Surface Stress Variation at Crown Point due
to Axial Load

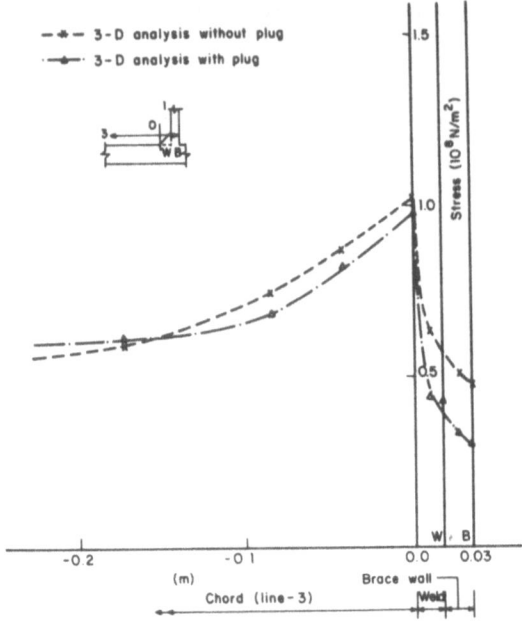

Figure 6 Stress Distribution Across the Weld and Brace Wall at

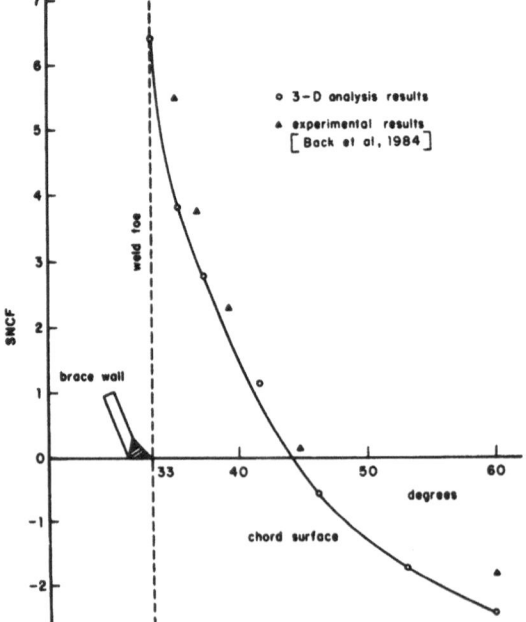

Figure 7 Strain Distribution Along Chord Surface at Saddle
Point due to Axial Load

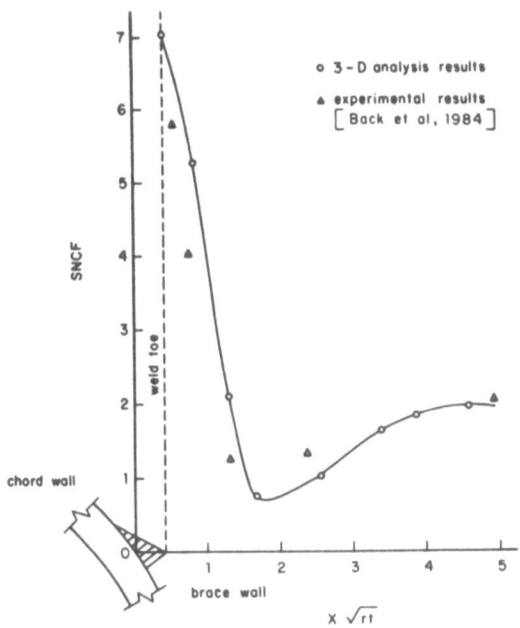

Figure 8 Strain Distribution Along Brace Wall at Saddle Point due to Axial Load

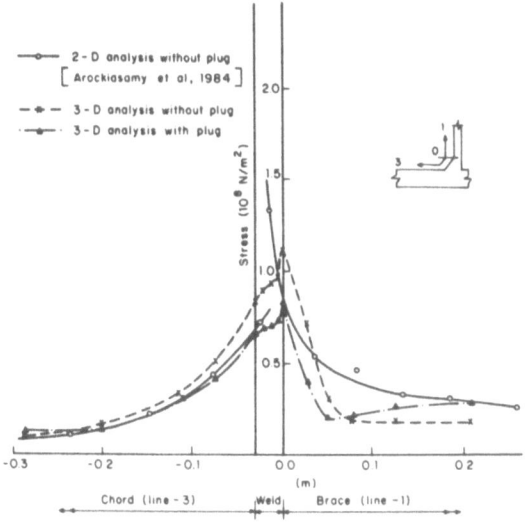

Figure 9 Maximum Surface Stress Variation at Crown Point due to In-Plane Bending Load

4.3 Stress due to in-plane bending

The hot spot stresses at the weld toe computed from the three
dimensional analyses are higher on both the chord and brace
sides at the crown point compared to those obtained from the
two dimensional analysis (Figure 9). However the stress
gradient obtained from the three dimensional analysis on the
brace side is steeper than that given by the two dimensional
analysis. Stress variations across the weld and brace
(Figure 10) are similar to those obtained for the case of
axial load (Figure 6).

5. ACKNOWLEDGEMENTS

The authors would like to thank Dr. G. R. Peters, Dean,
Faculty of Engineering and Applied Sciences, and Dr. I.
Rusted, Vice-President, Memorial University of Newfoundland
for their continued interest and encouragement. Thanks are
due to Dr. Vosikovsky for his helpful suggestions and
discussions. The financial support from Canada Centre for
Mineral and Energy Technology (CANMET) is gratefully
acknowledged. Appreciation is due to Mrs. Vatcher for the
careful typing of the manuscript.

REFERENCES

Arockiasamy, M., Vosikovsky, O., Bhuyan, G. S. and Munaswamy,
K. (1984) Stress Concentration Factors for Welded Tubular
T-Joints Using Fintie Element Analysis with Automatic Mesh
Generation, proc. of Second Int. Conf. on Welding of Tubular
Structures, Boston.

Back, de and Vaessen, G. H. G. (1981) Fatigue and Corrosion
Fatigue Behaviour of Offshore Steel Structures, ECSC
Convention, 7210-KB/6/602 J.7.1f/76, Final Report, Delft.

Bathe, K. J. (1982), Finite Element Procedures in Engineering
Analysis, Prentice-Hall, Inc., New Jersey.

Clayton, A. M. and Parkhouse, J. G. (1980), The Stress
Analysis of a Large Diameter Brace T-Joint, Interim Technical
Report, 2/01, UKOSRP.

Clayton, A. M., Martin, T. (1980), Comparison of Stress
Levels Obtained on Some Early Welded T-Joint Tests, Interim
Technical Report, 2/07, UKOSRP.

Irvine, N. M. (1981), Comparison of the Performance of Modern
Semi-emperical Parametric Equations for Tubular Joint Stress
Concentration Factors, Proc. of Int. Conf., Steel in Marine
Structures, Paris.

Marshall, P. W. (1984), Connections for Welded Tubular
Structures, proc. of Second Int. Conf. on Welding of Tubular
Structures, Boston.

Morgan, E. F., (1979), Use of High Order Isoparametric Solid
Finite Elements for the Stress Analysis of Welded Tubular
Joints, Southwest Research Institute, Project 03-9217-001,
Texas.

Parkhouse, J. C. (1981), Improved Modelling of Tube Wall
Intersections Using Brick Elements, Proc. of Int. Conf.,
Steel in Marine Structures, Paris.

Robert E. Hoffman, Earl and Wright (1980), The Accuracy of
Different Finite Element Types of the Analysis of Complex,
Welded Tubular Joints, Proc. of the Offshore Technology Conf.
Houston.

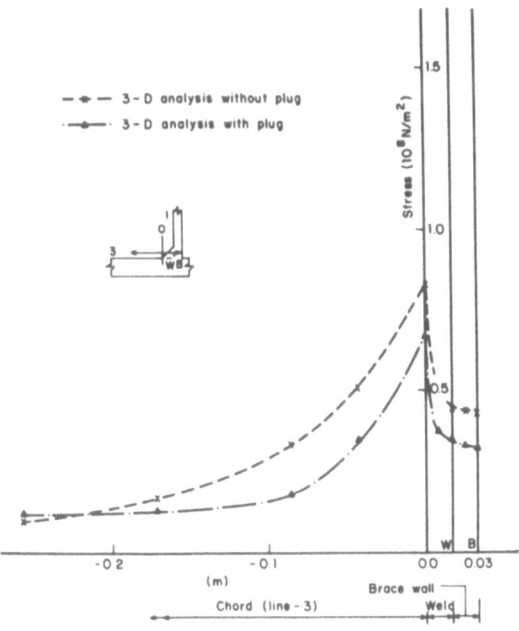

Figure 10 Stress Distribution Across the Weld and Brace Wall at
Crown Point due to In-Plane Bending Load

DOLMEN: A COMPLETE CAD SYSTEM FOR MICROCOMPUTERS

Francesco Biasioli
Dipartimento Ingegneria Strutturale – Politecnico di Torino

ABSTRACT :

This paper discusses in details the concepts employed in the development of the CAD DOLMEN system, and its implementation on microcomputers.

DOLMEN is a CAD system of 80's generation, especially conceived for computer-aided analysis and design of steel and r.c. structures in an integrated low cost environment.

The availability of high-performance processors – namely Motorola 68000 – and graphic terminals has lowered price/performance ratio for microcomputers.The implementation of refined CAD software on such hardware is now really possible.

General aims in development, actual level of implementation, future trends of work are subsequently outlined.

1 - INTRODUCTION

In a general engineering design problem, the following steps are usually performed:

1) definiction of actions and materials
2) definiction of the real structure to be designed
3) definiction of the ideal structure to be analyzed
4) structural analysis
5) design
6) drawings, bill of quantities, etc.

At every stage, computers can help man's work, not only in fastening repetitive calculations, but also suggesting and controlling alternative solutions.

It is possible to separate steps previously indicated into five main blocks, according to the results each step has to achieve. So we can call steps from 1 to 3 "General description stage", step 4 "Analysis stage", step 5 "Design stage" and step 6 "Document stage".

DOLMEN, conceived as an integrated CAD system, assists the designer at every stage, with different features which will be extensively examined later.

2 - GENERAL DESCRIPTION STAGE

Actions, materials and general dimensions of structural members are usually defined according to Codes and/or contractor's general requirements. In general, several types of actions must be dealt with - e.g. self weight and superimposed loads, wind, earthquakes, settlements, indirect actions, etc. A preliminary design choice is usually made about location of supports, floor framing, foundations, which is left to designer's judgement and experience. Initial dimensions are stated, usually according to simplified calculations.

DOLMEN has a wide possibility of defining in a summarized form tables of loads, actions, cross-section properties and materials, directly from Code data or manufacturer tables. This is performed on a separate basis from the problem under investigation, so it is possible to call the general tables and create a working table of data.

The geometrical definition of the structure is accomplished starting from plant positions of supports,
using macro-intructions where single items look as : 1>3 6 10,7,3*3 5 2.5, 1.2 3*2 4,1 4 3*2 , which generates nodes 1 2 3 6 10 in a local X-Y coordinate system with origin in previously defined node 7, at intervals of (3,3,3,5,2.5) mt. in X-direction, and (1.2,2,2,2,4) mt. in Y-direction,and assigns to every node a local coordinate system defined from conventional numbers 1,4 and 2.

The same pattern of commands is widely used also in member and load definitions: so it is possible to define in a single instruction all the beams pertaining to the same frame between different supports at different levels, having the shape and dimensions previously stated in the relative working table of cross-sections.

Every set of data is stored and accessed separately in a common data-base, with large possibilities of changing, modifying, adding, deleting data, and immediate graphic presentation of results; the designer so can follow the creation of the skeleton of the structure, with immediate visual control of data correctness. General functions are usually performed via user-definable-keys, and printed or plotted output is always possible.

Before starting analysis stage, a complete control of data is performed: data are tested for completeness, consistence and correctness.

As a major help to the designer, a "rational design" facility is provided according to several points:

a) Lateral stiffnes control: the rigidity of cross walls, if provided, must withstand lateral forces, otherways sway frame model is assumed.

b) Slenderness ratio calculations for buckling check.

c) Vertical loads bearing capacity of columns

d) Correctness of in-plane disposition of resisting elements for horizontal forces.

The principal aim is to avoid "trial and error " calculations, and to control initial dimensions and dispositions of resisting elements. In general, suggestions and schemas provided by Eurocodes or other major National Codes are followed.

3 - ANALYSIS STAGE

The structural model assumed for the whole structure is the well known "pseudo-tridimensional model", in which separate elements like frames, shear-walls, cores are interconnected through rigid floor diaphragms. Finite joint sizes, shear deformations of members are considered.

The choice of such a simplified model has been done for two definite reasons:

a) relatively large and complex structures can be solved without excessive micro-computer power and calculation time requirements: as a matter of fact, it is possible with a 240Kb available random access memory to solve structures formed with more than 80 columns, 20 frames, 15 floors under seismic loads in less than two hours elapsed time.

b) dealing with separate and relatively complex elements, more load cases can be analyzed: different load patterns on frames for accidental loads are generated in an automated way by the program.

Linear static and dynamic analysis can be performed: in the latter case, three-dimensional frequences and mode shapes are evaluated and a response spectrum approach is used.

The simplified model has however been modified for taking into account some major problems - e.g. congruence of vertical displacements in ortogonal shear walls - and has been tested in a large amount of cases, with satisfying results, with more complex F.E.M. models.

Loads can be added, multiplied, factorized at will, so giving appropriate stress resultants envelopes for subsequent design. All diagrams and envelopes, as well as a deformed shape of the elements of the structure under different load conditions can be traced on video and copied or plotted to paper.

4 - DESIGN STAGE

Design is at the moment developped for R.C. members.

The design stage is logically divided into two main steps, separating designer's "habitudes" from actual problem specifications.

A designer's habitude is considered a possible choice, among others, of:
- limit-states or allowable stresses design
- moment redistributions, assigned or calculated
- moment and shear reduction, from center to face of supports
- shear resisting model: stirrups only and/or bent bars

Actual problem specification is the definiction of:
- characteristic and working stresses for materials
- number of layers and diameters of preferred bars
- magnification and/or reduction factors for theoretical areas of steel
- distance between bars, covers etc.

At design stage a great effort has been made to "aid" the designer to develop a rational optimization of the structure: load bearing capacity of beams and columns, according to previously stated geometrical dimensions are controlled to avoid excessive or inadequate performance of elements.

"Optimized" dimensions are always indicated, according to different criteria, e.g. ductility requirements for beams on limit state design, complete material exploitation in allowable stress design, and so on.

Sections of different shapes - rectangular, T, circular and others - can be considered: for every shape appropriate interaction diagrams are generated, to speed up load bearing capacity control.

As an example, every column has theoretically to be tested under eight different load cases, related to all different combinations of maximum and minimum normal effort and biaxial bending moments. This can be accomplished in reasonable time looking at critical surfaces in a pre-defined interaction diagram.

After dimensional check and effective reinforcement arrangement in a definite number of sections - min. 20 - of a beam, serviceability limit state analysis is performed.

Deflections are calculated by a double-curvature integration, with stress-strain Sargin law for concrete; crack widths are calculated according to C.E.B. models, with due account of "tension-stiffening".

A large number of print report options are available: theoretical and effective reinforcement areas and distribution, resisting moment diagrams and general reports can be easily obtained.

5 - DOCUMENT STAGE

Three major steps aree performed at this stage, graphical editing, drawing and technical report composition.

General sketches of resisting elements given from previous stage can be modified in an interactive way, according to designer's judgement. Lenghts and numbers of bars, positions, geometrical dimensions of beams and columns are directly changed on CRT with mixed use of a mouse or a rotatory knob for selecting elements, and of keyboard input for numerical data.

Missing or undefined elements - e.g. secondary beams not previously considered- can be described at this stage. Each element is separately stored, with all related data, for computational purposes.

After graphical editing, interactive arrangement of stored elements is performed to give final drawing. Elements can be moved on a given grid, to avoid superposition; notes, headings and comments are added. Every drawing can be split in several parts, according to plotter drafting dimensions.

Bills of quantities, summary tables and technical reports are also generated. Every file can also be treated by a separate word-processor, for better result.

6 - FUTURE DEVELOPMENTS

Work is in progress to improve system capabilities at every of previously defined stages. Non-structural members like masonry walls, various floor framing systems, cores with non-uniform torsional behaviour, math foundations on elastic soil are going to be introduced at the analysis stage.

At the design stage, steel members design is goig on, and a new approach is going to be estabilished to take account of National Codes in a simple way via a particular problem-oriented language.

7 - BIBLIOGRAPHY

1. WILSON,E.L., DOVEY,H.H., Three dimensional analysis of building systems - TABS, Report n. EERC 72-8, December 1972, Berkeley.

2. HEIDEBRECH,A.C., STAFFORD-SMITH,B., Approximate analysis of open section shear walls subject to torsional loading, Journal of Structural Division, Proceedings of ASCE, Vol. 99, ST12, December 1973

3. HUMAR,J.L., KANDOKER,J.U., A computer program for three dimensional analysis of buildings, Computer and Structures Vol. 11 n. 5, 369-87, May 80

4. TESCHLER,L, Low-cost drafting on personal computers, Machine design, Vol. 55, n. 19, 65-69, August 1983.

11. STRUCTURAL ENGINEERING
    POSTER SESSION

DYNAMICS OF MULTI-STOREYED BUILDINGS BY MEANS OF SHEAR WAVES

A. Mioduchowski and M.G. Faulkner

Department of Mechanical Engineering, The University of
Alberta, Edmonton, Alberta, Canada.

1. INTRODUCTION

An exact description of the complex physical effects and processes which occur in buildings during dynamic loading is
virtually impossible. It is therefore important to choose a
physical model representing a simplified structure which on
one hand describes behaviour of a real building and on the
other can be treated by the known mathematical apparatus.

A method of investigation is proposed in this paper for the
analysis of displacements and strains of the column cross-
sections in a multi-storey building subject to dynamic loading.
It is based on the theory of propagation of one dimensional
elastic shear waves, together with the appropriate initial and
boundary conditions [1]. This reduces the problem to a system
of ordinary differential equations with shifted arguments which
must be solved in a certain sequence. A simple, effective and
stable numerical procedure is proposed for solving the system
of equations and some numerical results are presented in graphical form.

2. GOVERNING EQUATIONS

Consider a multi-storeyed building having steel columns rigidly
connected with the foundation and floors. It is assumed that
the foundation and floors are undeformable and are displaced
by plane motion during the time of dynamic loading. The masses
of the foundation and ceilings are denoted by m_0 and m_i,
$i = 1,2,...N$ respectively. It is assumed that for all vertical
columns the velocities, strains and displacements for the given
height of a building are the same [2]. It is further assumed
that the column cross-sections remain plane and parallel to
each other during dynamic loading. These columns are characterized by the following parameters: G-shear modulus, A-column
cross-section area, k-shear coefficient, ρ-density and ℓ_i-
height of the column.

For the continuous model of a building discussed here the investigation of velocities $\partial y_i(x,t)/\partial t$, strains $\partial y_i(x,t)/\partial x$ and displacements $y_i(x,t)$ of the columns, in the case of a three storey building is reduced to the solution of 3 wave equations

$$\frac{\partial^2 y_i(x,t)}{\partial t^2} - c^2 \frac{\partial^2 y_i(x,t)}{\partial x^2} = 0 \ , \ i = 1,2,3 \tag{1}$$

where $c^2 = kG/\rho$, with the following boundary conditions

$$kAG \ \frac{\partial y_1(x,t)}{\partial x} - m_o \ddot{y}_o - D_o \ \frac{\partial y_1(x,t)}{\partial t} = 0 \ , \ x = 0$$

$$y_1(x,t) = y_2(x,t) \ , \ x = \ell$$

$$y_2(x,t) = y_3(x,t) \ , \ x = \ell + a\ell \tag{2}$$

$$kAG \left[\frac{\partial y_2(x,t)}{\partial x} - \frac{\partial y_1(x,t)}{\partial x} \right] - m_1 \ \frac{\partial^2 y_2(x,t)}{\partial t^2} - D_1 \ \frac{\partial y_2(x,t)}{\partial t} = 0$$

$$x = \ell$$

$$kAG \left[\frac{\partial y_3(x,t)}{\partial x} - \frac{\partial y_2(x,t)}{\partial x} \right] - m_2 \ \frac{\partial^2 y_3(x,t)}{\partial t^2} - D_2 \ \frac{\partial y_3(x,t)}{\partial t} = 0$$

$$x = \ell + a\ell$$

$$kAG \ \frac{\partial y_3(x,t)}{\partial x} + m_3 \ \frac{\partial^2 y_3(x,t)}{\partial t^2} + D_3 \ \frac{\partial y_3(x,t)}{\partial t} = 0 \ , \ x = \ell + a\ell + b\ell$$

and initial conditions

$$y_i(x,t) = \frac{\partial y_i(x,t)}{\partial t} = 0 \ \text{ for } t = 0 \ , \ i = 1,2,3 \tag{3}$$

where a and b are coefficients and $\ddot{y}_o$ represents horizontal acceleration of the foundation. This horizontal acceleration which represents external loading of the structure may be either the following function

$$\ddot{y}_o(t) = \sum_{k=0}^{m} H(t-t_k) \ [a_k + b_k(t-t_k)] \ ,$$

where H(t) is the Heaviside function, or it might be of the Dirac type

$$\ddot{y}_o(t) = \sum_{k=0}^{m} a_k \delta(t-t_k)$$

The damping forces which are loading the rigid foundation and rigid ceilings are assumed in the form

$$F_{D_i} = - D_i \frac{\partial y_{i+1}(x,t)}{\partial t}$$

where D_i is a coefficient of equivalent viscous damping of the i-th column. These damping forces take into account internal as well as external damping.

In order to simplify presentation of the proposed method of analysis equations (1), (2) and (3) refer to a three storey building only, with storeys of various heights. It should be stressed however, that the analysis as outlined above is easily applicable to a N-storeyed building, and in an APPENDIX to this paper the final equations are given for such a case.

Upon the introduction of non-dimensional quantitites:

$$\bar{x} = x/\ell \ , \ \tau = ct/\ell \ , \ \bar{y} = y_i/y_o \ , \ F(\tau) = m_o \ddot{y}_o \ell^2/c^2 y_o \tag{4}$$

$$\kappa_i = A\rho\ell/m_i \ , \ \bar{D}_i = D_i c/kAG \ , \ i = 1,2,3$$

the relations (1), (2) and (3) become

$$\frac{\partial^2 y_i}{\partial \tau^2} - \frac{\partial^2 y_i}{\partial x^2} = 0 \ , \ i = 1,2,3 \tag{5}$$

$$\kappa_o \frac{\partial y_1}{\partial x} - F(\tau) - \kappa_o D_o \frac{\partial y_1}{\partial \tau} = 0 \ , \ x = 0$$

$$y_1 = y_2 \ , \ x = 1$$
$$\tag{6}$$
$$y_2 = y_3 \ , \ x = 1 + a$$

$$\kappa_1 \left[\frac{\partial y_2}{\partial x} - \frac{\partial y_1}{\partial x} \right] - \frac{\partial^2 y_2}{\partial \tau^2} - \kappa_1 D_1 \frac{\partial y_2}{\partial \tau} = 0 \ , \ x = 1$$

$$\kappa_2 \left[\frac{\partial y_3}{\partial x} - \frac{\partial y_2}{\partial x} \right] - \frac{\partial^2 y_3}{\partial \tau^2} - \kappa_2 D_2 \frac{\partial y_3}{\partial \tau} = 0 \ , \ x = 1 + a$$

$$\kappa_3 \frac{\partial y_3}{\partial x} + \frac{\partial^2 y_3}{\partial \tau^2} + \kappa_3 D_3 \frac{\partial y_3}{\partial \tau} = 0 \ , \ x = 1 + a + b$$

and
$$y_i = \frac{\partial y_i}{\partial t} = 0 \ , \ \text{for } t = 0 \ , \ i = 1,2,3 \tag{7}$$

where for convenience all $y_i(x,t)$ functions are written as y_i and all bars are omitted.

The solutions of equations (5) are sought in the nondimensional form

$$y_i = f_i \left[(t-t_{oi}) - (x-x_{oi}) \right] + g_i \left[(t-t_{oi}) + (x-x_{oi}) \right]$$

where the function f_i represents a wave propagating to the right and g_i represents a wave propagating to the left side of the i-th column as a result of application of the force $F(\tau)$. It is assumed that the functions f_i and g_i are continuous and for negative arguments equal zero. By taking into account that the first perturbation in the first column starts at the instant $t_{01} = 0$ at the bottom of that column, $x_{01} = 0$, and the first perturbation in the i-th column starts at its bottom as well, one gets:

$$y_1 = f_1(\tau-x) + g_1(\tau+x)$$

$$y_2 = f_2(\tau-x) + g_2(\tau+x-2) \tag{8}$$

$$y_3 = f_3(\tau-x) + g_3(\tau+x-2(1+a))$$

By substituting (8) into the boundary conditions (6) a system of equations is obtained for functions f_i and g_i. One easily observes that there occur simple relationship between arguments of the functions appearing in the same equation. Upon denoting the largest argument in each equation by the variable ξ the arguments of the remaining functions are then shifted by a constant. This procedure leads to the following final system for six linear, ordinary differential equations of the first and second order for the unknown functions $f_i(\xi)$ and $g_i(\xi)$, $i = 1,2,3$:

$$g_1'(\xi) = - f_1'(\xi-2) + f_2'(\xi-2) + g_2'(\xi-2)$$

$$g_2'(\xi) = - f_2'(\xi-2a) + f_3'(\xi-2a) + g_3'(\xi-2a)$$

$$g_3''(\xi) + r_3 g_3'(\xi) = - f_3''(\xi-2b) + r_4 f_3'(\xi-2b)$$

$$f_1'(\xi) = g_1'(\xi)(1-D_o)/(1+D_o) - F(\xi)/(\kappa_o(1+D_o))$$

$$f_2''(\xi) + r_1 f_2'(\xi) = - g_2''(\xi) - \kappa_1 D_1 g_2'(\xi) + 2\kappa_1 f_1'(\xi)$$

$$f_3''(\xi) + r_2 f_3'(\xi) = - g_3''(\xi) - \kappa_2 D_2 g_3'(\xi) + 2\kappa_2 f_2'(\xi)$$

(9)

where

$$r_1 = \kappa_1(2+D_1) \; , \; r_2 = \kappa_2(2+D_2)$$

$$r_3 = \kappa_3(1+D_3) \; , \; r_4 = \kappa_3(1-D_3)$$

The system of equations (9) should be solved in the given sequence in the successive intervals of the argument ξ. Because functions f_i and g_i equal zero for negative arguments, hence when solving equations (9) in this sequence the right hand sides of these equations are always known if force $F(\xi)$ is a known function of time.

3. NUMERICAL PROCEDURE AND EXAMPLE

One notes that all second order differential equations in the set (9) have the form

$$f''(\xi) + rf'(\xi) = - g''(\xi-\ell) + sg'(\xi-\ell) + h(\xi-k)$$

where $g'(\xi-\ell)$ and $h(\xi-k)$ are given functions, and whose solution for $\xi \geq \xi_o$ is as follows

$$f'(\xi) = e^{-r(\xi-\xi_0)} \left\{ \int_{\xi_0}^{\xi} \left[-g''(\xi-\ell) + sg'(\xi-\ell) + h(\xi-k) \right] \right.$$

$$\left. e^{r(x-\xi_0)} \, dx + f'(\xi_0) \right\}$$

After integrating by parts this finally gives

$$f'(\xi) = f'(\xi_0) \, e^{-r(\xi-\xi_0)} + g'(\xi_0-\ell) \, e^{-r(\xi-\xi_0)} - g'(\xi-\ell) +$$

$$e^{-r(\xi-\xi_0)} \left[(r+s) \int_{\xi_0}^{\xi} g'(x-\ell) \, e^{r(x-\xi_0)} \, dx + \right. \qquad (11)$$

$$\left. \int_{\xi_0}^{\xi} h(x-k) \, e^{r(x-\xi_0)} \, dx \right]$$

Since both integrals in equation (11) can be easily evaluated because $g'(x-\ell)$ and $h'(x-k)$ are known functions, one can now apply this equation to transform all second order differential equations in (9) and this gives the final system of equations suitable for numerical integration:

$$g_1'(\xi) = -f_1'(\xi-2) + f_2'(\xi-2) + g_2'(\xi-2)$$

$$g_2'(\xi) = -f_2'(\xi-2a) + f_3'(\xi-2a) + g_3'(\xi-2a)$$

$$g_3'(\xi) = g_3'(\xi-\Delta) \, e^{-r_3\Delta} + \left[(r_3+r_4) \frac{\Delta}{2} - 1 \right] f_3'(\xi-2b) +$$

$$+ \left[(r_3+r_4) \frac{\Delta}{2} + 1 \right] e^{-r_3\Delta} \, f_3'(\xi-2b-\Delta)$$

$$f_1'(\xi) = g_1'(\xi)(1-D_0)/(1+D_0) - F(\xi)/\left[\kappa_0(1+D_0) \right] \qquad (12)$$

$$f_2'(\xi) = f_2'(\xi-\Delta)e^{-r_1\Delta} + \left[(r_1-\kappa_1 D_1) \frac{\Delta}{2} - 1 \right] g_2'(\xi) +$$

$$+ \; [(r_1 - \xi_1 D_1) \frac{\Delta}{2} + 1] \; e^{-r_1 \Delta} \; g_2'(\xi - \Delta) \; +$$

$$+ \; \Delta \kappa_1 [f_1'(\xi) + f_1'(\xi - \Delta) e^{-r_1 \Delta}]$$

$$f_3'(\xi) = f_3'(\xi - \Delta) \; e^{-r_2 \Delta} + [(r_2 - \xi_2 D_2) \frac{\Delta}{2} - 1] \; g_3'(\xi) \; +$$

$$+ \; [(r_2 - \xi_2 D_2) \frac{\Delta}{2} + 1] \; e^{-r_2 \Delta} \; g_3'(\xi - \Delta) \; +$$

$$+ \; \Delta \xi_2 [f_2'(\xi) + f_2'(\xi - \Delta) \; e^{-r_2 \Delta}]$$

where $\Delta = \xi - \xi_o$. The system of equations (12) can now be solved in the successive intervals Δ. Any numerical integration procedure, e.g. Simpson Integration can be used, giving a simple, stable and effective numerical procedure for solving equations (9).

As an example a three-storey building is considered with the following parameters: $D_0 = 6$, $D_1 = D_2 = D_3 = 3$, $\kappa_0 = 0.15$ and $\kappa_2 = \kappa_3 = \kappa_4 = 0.5$. Both column length coefficients a and b were varied between 0.8 and 1.2, and some results are plotted in Fig. 1 and Fig. 2. In Fig. 1 the strain $\partial y_1 / \partial x$ is plotted versus time t for the column cross-section x = 1 and a = b = 1, for the Dirac type of loading. In Fig. 2 the displacement y_3 is plotted versus time t, for x = 1 + a + b, a = 1 and b = 0.8, 1.0, 1.2, for the loading shown.

REFERENCES

[1] Graff, K.F., Wave Motion in Elastic Solids, Ohio State University Press, Ohio (1975).

[2] Clough, R.W. and Penzien, J., Dynamics of Structures, McGraw-Hill (1975).

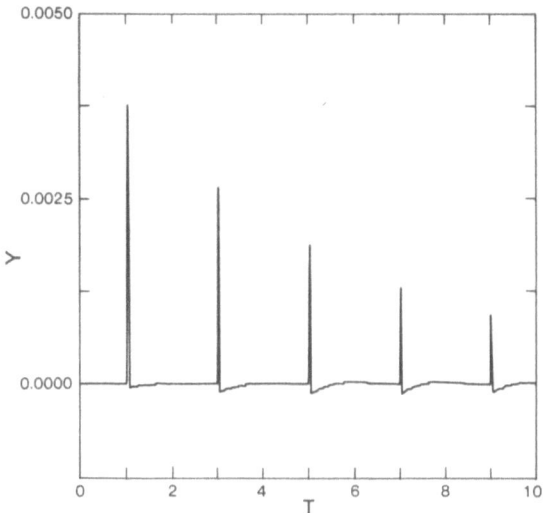

Fig. 1. Strain $\partial y_1 / \partial x$ for $x = 1$ and $a = b = 1$

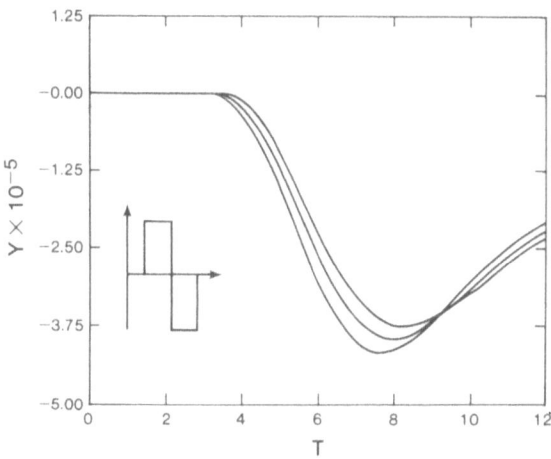

Fig. 2. Displacement y_3 for $x = 2 + b$, $b = 0.8$, 1.0, 1.2

APPENDIX

A similar procedure to the one outlined above but applied to the case of a N-storeyed building with storeys of various heights leads to the set of 2N differential equations for the functions $f_i(\xi)$ and $g_i(\xi)$, $i = 1,2,\ldots,N$:

$$g_i'(\xi) = - f_i'(\xi-2a_i) + f_{i+1}'(\xi-2a_i) + g_{i+1}'(\xi-2a_i)$$

$$i = 1,2,\ldots, N-1$$

$$g_N''(\xi) + r_N g_N'(\xi) = - f_N''(\xi-2a_N) + r_{N+1} f_N'(\xi-2a_N)$$

$$f_1'(\xi) = g_1'(\xi)(1-D_o)/(1+D_o) - F(\xi)/(\kappa_o(1+D_o)) \tag{10}$$

$$f_i''(\xi) + r_{i-1} f_i'(\xi) = - g_i''(\xi) - \kappa_{i-1} D_{i-1} g_i'(\xi) + 2\kappa_{i-1} f_{i-1}'(\xi)$$

$$i = 2,3,\ldots, N$$

where $\quad r_i = \kappa_i(2+D_i)$, $i = 1,2,\ldots, N-1$

$$r_N = \kappa_N(1+D_N) \text{ , } r_{N+1} = \kappa_N(1-D_N)$$

and a_i, $i = 1,2,\ldots,N$, are length coefficients for all storeys.

COMPUTER PROCEDURE FOR OPTIMAL DESIGN OF STRUCTURES
UNDER TRAFFIC LOADS

Lj. R. Savić

Civil Engineering Faculty, Belgrade, Yugoslavia

1. INTRODUCTION

The concern of a structural engineer is today extended to the
area of the optimum structural design. Between many different
approaches to structural design an important role plays the
optimality criteria approach based on the assumption that a
criterion related to the behaviour of the structure is satisfied at the optimum (Kirsch, 1981). Fully stressed design is
an example of such an approach, in which we start from the
assumption that in an optimal structure each structural element is subjected to its limiting stress under at least one of
the loading condition. The chosen procedures for solution are
usually iterative, where each iterative cycle consists of two
steps: An analysis of the structure at the current design followed by a redesign operation. Continuing the iterative process
until the predetermined optimality criteria are satisfied,
we usually in a simple way obtain a reasonable design, not far
from the optimum.

In the present paper we pay attention to some problems in
the organization of computer programs when fully stressed design is applied. By way of background, it may be recalled that
the fully stressed design procedures are relatively efficient
in comparison with many mathematical programming methods. However, engineering experience undoubtly indicates that such a
procedure usually yields a satisfactory good design. Our aim
is to give some suggestions for the organization of a file base
when fully stressed procedures are applied to structures assembled from line elements. In Isreb's papers (1977, 1984) is shown
how a file base of an optimum design problem can be generated
in the case when the member sectional properties, moments of
inertia and section moduli are simple power functions of section
area. Starting from Isreb's considerations which are related to
analysis of structures under loads with fixed locations, we give some additional remarks concernig the fully stressed design
of structures under traffic loads, what is more complex than

the analysis in the previous case.

2. DESIGN VARIABLES AND DESIGN INVARIANT QUANTITIES

To give a general background, it is necessary to recall that we operate with two sets of quantities during a design process:
    (i) design variables (quantities that can be changed in the design process)
    (ii) preassigned (or invariant) parameters.
    Design variables can represent the following physical properties of the structure:
    (1) the cross sectional dimensions or the member sizes,
    (2) the configuration of the structure,
    (3) the topology of the structure,
    (4) the properties of the material.
    The simplest design variables are the cross sectional dimensions. Since many structural problems with a practical nature have fixed configurations, topology and material properties, we restrict our considerations to the size type of design variables.
    In what follows we confine attention on a straight uniform member with only one design variable - the cross sectional area A. We further restrict the element moments of inertia and section moduli to the forms

$$I_j = i_j (A_j)^n \tag{1}$$

and

$$Z_j = z_j (A_j)^m \tag{2}$$

respectively, where n and m are positive constants called element inertia exponents. The indices j stand for the numbers of the members. The unit moment of inertia i_j and unit section modulus z_j are positive constants that are independent from design variables. As a consequence of the above given size-inertia relation (1), the stiffness matrix of a beam element can be expressed as follows:

$$(K_j) = \left(k_j^{(1)}\right)A_j + \left(k_j^{(2)}\right)A_j^n \tag{3}$$

what holds for the case when the influence of shear forces on the deformation of the beam is neglected. Matrices $\left(k_j^{(1)}\right)$ and $\left(k_j^{(2)}\right)$ are called unit element stiffness matrices. The first part of the expression on the right hand side of the equation (3) represents the bar action, whereas the second part is related to beam action.
    In regard to the above mentioned, the load vector for a beam element can we write in the following form:

$$\{P\} = \{p^{(1)}\}A + \{p^{(0)}\} \tag{4}$$

where $\{p^{(1)}\}$ is the unit load vector due to size dependent loading, such as gravity, and $\{p^{(0)}\}$ represents the dead loads.
    When the basic unknowns in the displacement finite element method - the nodal displacements are calculated, it is possible to find the force quantities related to a member cross section, according to the following equation:

$$\{s\} = (\ (s^{(1)})A + (s^{(2)})A^n\ )\{U\} \tag{5}$$

where $\{U\}$ are nodal displacements of the considered member and $\{s\}$ is the vector of cross section forces. Matrices $(s^{(1)})$ and $(s^{(2)})$ are independent of the design variable A and are called unit force recovery matrices.

Within the scope of a computer program for the iterative fully stressed design procedure, the design invariant quantities given in equations (1)-(5) can be calculated only once and stored on proper storage devices undisturbed throughout the program. That can be done for each structural element and obviously represents a departure from the traditional approach in finite element computer programs, where such file bases are not generated.

3. FULLY STRESSED DESIGN OF STRUCTURES UNDER TRAFFIC LOADS

In the case when the structure is subjected to traffic loads, which represent the motion of vehicles along the structure, the problem of finding a fully stressed design becomes apparently more complex and requires more arithmetyc operations. For each set of design variables, that means in each iteration step, we need to know the extreme values due to traffic load for certain force quantities. These quantities which are essential for the design correspond to the so called dangerous positions of loads. The efficiency of a computer program for solution of such problems becomes an important task. As it will be shown in present paper, for structures assembled from straight uniform members, in a relatively simple way we can extend the file base described in Section 2 with several quantities, that correspond to the analysis of the structure under traffic loads.

3.1. Influence functions for force quantities

To find the extreme values due to traffic loads for a section force quantity we can start from the concept of influence functions. On the basis of reciprocity theorems the influence function for a force quantity can be determinated as a displacement function of the structure. This displacement function is due to a given unit generalized displacement in the considered cross section, that corresponds to the force which we need to know. Let C and P be the considered cross section and the position in which we calculate the value of the displacement function, respectively. Denoting a value of the displacement function with v, we can write:

$$v_{PC} = v_{PC,i} + v_{PC,d} \tag{6}$$

where the displacement v_{PC} (what means the displacement v in P due to given unit displacement in C) is expressed according to two values, $v_{PC,i}$ and $v_{PC,d}$. The value $v_{PC,i}$ is the value of the displacement function in P in what we call displacement indeterminate structure and depends directly on the values of the basic unknowns in a displacement finite element procedure.

Let $U_{j,C}$ be a basic unknown quantity, i.e. displacement or rotation, due to the given displacement in C. For $v_{PC,i}$ we can use the following equation:

$$v_{PC,i} = \sum_{j=1}^{M} v_{P,j} U_{j,C} \qquad (7)$$

where M stands for the total number of basic unknowns in the displacement finite element method. The value $v_{P,j}$ is the displacement in P when $U_{j,C}=1$ and $U_{k,C}=0$, $k \neq j$.

In the situation when all basic unknowns are equal to zero, we in fact have to deal with a structure where still displacements are possible, but only along separate members. Such a structural system we call displacement-determinate structure. For a given displacement in position C of a considered member only the displacements along this member are nonzero in such a structure. In the equation (6) this is represented with $v_{PC,d}$. Thus, the value $v_{PC,d}$ is nonzero only when C and P belong to the same member.

Examples for the above mentioned displacement functions are given in an other paper (Nikolić, Savić, 1983) and it will here not be repeated. However, it is important to observe that in the case when we neglect the influence of shear forces on the deformation, the functions $v_{P,j}$ and $v_{PC,d}$ along a beam element are independent of the cross sectional properties of a straight uniform member. Thus, we can view these functions as design invariant functions.

The complete displacement function v_{PC} depends on the values of unknowns $U_{j,C}$, which will be found as a solution of the basic set of the finite element linear equations, where the external nodal forces are caused by the given unit displacement in section C. Examples of expressions for these forces are also given in the above mentioned paper by Nikolić and Savić, where can be seen that the nodal forces, similarly as quantities described in Section 2, can be expressed as simple functions of the basic design variable of the considered member.

3.2. Calculation of force quantities

We now proceed to describe a suitable procedure for calculation of the extreme values for sectional force quantities in the case when the structure is subjected to a set of external concentrated forces, which represent the motion of vehicles along the structure. We denote the given forces as follows:

$$q_j, \quad j=1, 2, \ldots, R$$

where R stands for the total number of forces. Let s_C be the force quantity (in the cross section C) that we will to calculate. Keeping in mind the definitions and notations introduced in the Section 3.1., for a given position of the external forces we can write:

$$s_C = \sum_{j=1}^{R} q_j v_{q_j C} \tag{8}$$

where according to the equation (6), $v_{q_j C}$ stands for the displacement related to the position where the force q_j acts. Using the expressions on the right hand sides of the equations (6) and (7) we write:

$$s_C = \sum_{l=1}^{M} (\sum_{j=1}^{R} q_j v_{q_j,l}) U_{l,C} + \sum_{j=1}^{R} q_j v_{q_j C,o} \tag{9}$$

For a given unit displacement related to section C, the nodal displacements $U_{l,C}$ depend on the design variables, but all other quantities in the expression on the right hand side of the equation (9) are independent on the design variables.

To calculate the extreme values of s_C , we must consider a sufficient number of different positions of the given set of concentrated forces. Under the assumption that all members of the structure are straight and uniform, this problem can easily be solved. First, we observe that the force quantity s_C can be considered as a function of the given loads q_j, $j=1,2,\ldots,R$ and a chosen coordinate x that defines the positions of the external forces:

$$s_C = s_C(q_j,x) \tag{10}$$

The form of this function is obviously not constant for various values of x. Only for x varying within the limits of an interval, we have a function with a fixed form. Which separate intervals must be considered depends on the nature of the influence function for s_C and on the distribution of the given loads. In such an interval we have to deal with a function that is cubic in x. Consequently, to determine this function it is sufficient to know the values of s_C for four different values of the argument x in this interval. When the function is determined it is easy to find the extreme values of the function in the considered interval. Repetitions of the above described operations for all intervals lead to the extreme values of s_C that we need. However, it is usually before any calculation obvious that only a few intervals must be considered to find extreme values. Consequently, these intervals can be precisely specified before computations.

When in this sence the positions of the given loads are defined, the corresponding values of the force quantity can we consider as

$$\{S_C\}^T = \{s_C^{(1)} \quad s_C^{(2)} \quad \ldots \quad s_C^{(t)}\} \tag{11}$$

where superscript stand for the numbers of the load cases and t is the total number of load cases that we must to consider. Keeping the form of the equation (9) in mind, we can now write a following relationship:

$$\{S_C\} = (Q_C)\{U_C\} + \{S_{C,0}\} \tag{12}$$

where (Q_C) denotes a matrix with t rows and M columns. This matrix is independent on the design variables and for the pre-assigned positions of external load the elements in the matrix will be calculated according to the first expression on the right hand side of the equation (9). In the column vector $\{S_{C,0}\}$ all elements will be calculated starting from the second expression on the right hand side of the equation (9), what represents design invariant quantities.

When for a considered design the nodal displacements $\{U_C\}$ due to given displacement in C are determinated, we can proceed to the computation of the values stored in $\{S_C\}$. The next step consists of the calculation of extreme values for s_C in the sence of the foregoing discussion.

Within the scope of one design iteration step the procedure that is described for a particular force quantity s_C can be be repeated for any other force quantity which is important for decisions in the design process. Design invariant matrices and column vectors related to the above mentioned force quantities will be computed only once before the iteration process starts and stored on proper storage devices together with design invariant quantities which are shown in the Section 2.

Similar analysis is possible also for some structures that are assembled not only from straight uniform members, but that have simple size-stiffness and size-inertia relations. Such problems are not the subject of the present paper.

On the basis of the considerations given in this paper an efficient computer program for a fully stressed design can be developed. The results that we obtain in such a computer analysis can be further used, if it is necessary, as a good starting point for optimum design procedures based on mathematical programming.

3.3. An additional comment on reanalysis methods

To obtain the basic unknowns $\{U_C\}$ for a given unit displacement related to a considered cross section C, the structural equilibrium equations

$$(K)\{U_C\} = \{R_C\} \tag{13}$$

must be solved. Here (K) stands for the global stiffness matrix of the structure. The column vector $\{R_C\}$ represents the nodal forces due to given unit displacement in C. But often in the iterative structural optimization procedures only approximate solutions of the equilibrium equations can be reasonably applied (Kirsch, 1981). In the case when we have to solve equation (13) in various iteration steps, such an approach can lead to an efficient computer program. Proceeding from one iteration step to the next the considerable modifications of the design variables can be restricted to a small part of the structure and we can, for instance, use the reduced basis method to solve the equation (13).

In the reduced basis technique we start from the assumption that the displacement vector of a new design can be approximated by a linear combination of a number of linearly independent vectors related to previously analyzed designs. The number of later vectors is much less than that of the degrees of freedom in the structure. The basic question in using reduced basis method lies in the choice of a set of these vectors. One possibility, that is suggested by Noor, A.K. and Lowder, H.E., 1974, is based on the first-order Taylor series approximation of the displacement vector.

Let $\{r_1^*\}$, $\{r_2^*\}$,..., $\{r_\alpha^*\}$ be a set of α linearly independent vectors. Next, denote we $\{U_C^*\}$ the displacement vector of an original analyzed design with variables

$$A_1^*, A_2^*, \ldots, A_\beta^*$$

where β stands for the total number of design variables. Then, we can assume the following:

$$\{r_1^*\} = \{U_C^*\}$$

(14)

$$\{r_{j+1}^*\} = \{\partial U_C^*/\partial A_j\}, \; j = 1, 2, \ldots, \beta$$

The vectors $\{\partial U_C^*/\partial A_j\}$ can easily be obtained by differentiation of the equilibrium equation (13) with respect to A_j:

$$\left(K^*\right)\{\partial U_C^*/\partial A_j\} = - \left(\partial K^*/\partial A_j\right)\{U_C^*\}$$

(15)

where the current design stiffness matrix $\left(K^*\right)$ is used for calculation of $\{\partial U_C^*/\partial A_j\}$. We further observe that under assumptions adopted in Section 2 the matrices $\left(\partial K^*/\partial A_j\right)$ have simple forms. Such a matrix can be assembled after differentiating the stiffness matrices of individual elements with respect to A_j. Equation (3) yields

$$\left(\partial K_j^*/\partial A_j\right) = n\left(k_j^{(2)}\right)A_j^{*(n-1)}$$

(16)

and

$$\left(\partial K_i^*/\partial A_j\right) = 0, \; i \neq j,$$

(17)

what indicates that $\left(\partial K^*/\partial A_j\right)$ is a sparce matrix with only one nonzero submatrix $\left(\partial K_j^*/\partial A_j\right)$, which can be calculated in a simple way multiplying the design invariant matrix $\left(k_j^{(2)}\right)$ by $nA_j^{*(n-1)}$.

The displacement vector $\{U_C\}$ of a new design can be expressed in terms of the chosen basis vectors (14) as

$$\{U_C\} = \left(r_B^*\right)\{y\}$$

(18)

where

$$\left(r_B^*\right) = \left(\ \{U_C^*\} \ \{\partial U_C^*/\partial A_1\} \ldots \ \{\partial U_C^*/\partial A_\beta\} \ \right)$$

$$\{y\}^T = \{y_1 \ y_2 \ \cdots \ y_{\beta+1}\}$$

(19)

The vector of undeterminated coefficients can be obtained solving a small system of equations, that can be easily derived substituting equation (18)$_1$ in (13) and premultiplying by $\left(r_B^*\right)$:

$$\left(r_B^*\right)^T (K) \left(r_B^*\right) \{y\} = \left(r_B^*\right)^T \{R_C\} \ .$$

(20)

REFERENCES

Isreb, M. (1977), Three-Dimensional Beam Elements Synthesis Applications with Stress Constraints, Non-linear Size-Stiffness Relationships and Various Size-Inertia Powers. Comput. Structures, Vol. 7, pp. 565-569.

Isreb, M. (1984), Software and Synthesis-Oriented-Structural Analysis Education. Comput. Structures, Vol. 18, pp. 641-646.

Kirsch, U. (1981), Structural Design, Concepts, Methods and Applications. McGraw Hill Book Company.

Nikolić, D., Savić, Lj. (1983), A Method for Solving Influence Functions of Structures Assembled from Line Elements (in serbo-croatian). Publication of the 7th Conference of the Yugoslav Association of Structural Engineers, Cavtat, pp. 199-206.

Noor, A. K., Lowder, H. E., (1975), Approximate Techniques of Structural Reanalysis. Comput. Structures, Vol. 5, pp. 9-12.

OPTIMIZATION OF STRUCTURES WITH RANDOM PARAMETERS

S.F. Jóźwiak

Politechnika Warszawska, Warszawa, Polska

1. INTRODUCTION

Structural design problems can be usually formulated in the form of corresponding optimization problems. Genarally recognized nondeterministic nature of structural design and the necessity of making decisions under conditions of uncertainity suggest that such problems should be formulated as stochastic optimization problems. Uncertainity may arise either from randomness in the structural properties and loading or from arbitrary decisions made in the process of idealization, or both. The main reason for neglecting the probabilistic /random/ character of parameters defining the structure and therefore limiting the analysis to the deterministic model, are difficulties in dealing with the stochastic programming problems, especially in the case of more complex structures. Formulation of the optimization problem presented in the paper is based on the concept of the expected value. Solution of the corresponding mathematical programming problem has been obtained by means of indirect method. The basic idea of the method is to convert the probabilistic problem into an equivalent deterministic one. In the paper the chance constrained programming technique has been applied. The technique was oryginally developed by Charnes and Cooper [1] and has been adopted and applied to solve engineering problems, among others, by Rao [6], Davidson, Felton and Hart [2],[3].

2. FORMULATION OF STOCHASTIC OPTIMIZATION PROBLEMS

In the formulation of optimization problems, involving systems described by random variables /stochastic optimization/, two main approaches can be mentioned. In the first the optimization problem consists in finding the decision variables for which the objective function reaches extrema /minimum/ on the assumption that random variables adopt such values for which the criteria reach the most unfavourable value. This approach leads to the minimum-maximum optimization problem. The second approach is connected with the concept of expected value. The objective function and /or/ the constraints are formulated as expected values of function of random variables. Expected value of function $f(x,y)$ is [4]:

$$Ef(x,y) = \int_{-\infty}^{+\infty} f(x,y) \; h(x) \; dx \; , \tag{1}$$

where $f(x,y)$ - function of random variable x and deterministic variable y, E - expected value, $h(x)$ - probabiliy density function.

Stochastic optimization problem can be stated in the form of the following stochastic programming problem:

$$\text{minimize} \quad Ef(\mathbf{X},\mathbf{Y}) \tag{2}$$

subjected to

$$P_j(X,Y) = P\left[g_j(X,Y) \geq 0\right] \geq p_j, \quad j=1,2,\ldots,k \tag{3}$$

where $f(X,Y)$ - objective function, X - vector of N random variables in which it is assumed that x_l l=1, 2,...,n, $n \leq N$, are decision variables, Y - vector of m deterministic decision variables, k - number of constraints.

Equation (3) denotes that probability of realizing $g_j(X,Y)$ greater or equal to zero must be greater than or equal to specified probability p . The stochastic programming problem stated in Equations (2) and (3) can be converted into an equivalent deterministic nonlinear programming problem by applying the chance constrained programming technique. For this purpose constraint functions $g_j(X,Y)$ are expandet about the mean value of $\overline{X}$

$$g_j(X,Y) = g_j(\overline{X},Y) + \sum_{i=1}^{N} \frac{\partial g_j}{\partial x_i}\bigg|_{X=\overline{X}} (x_i - \overline{x}_i) + \begin{array}{l}\text{higher} \\ \text{order} \\ \text{derivative} \\ \text{terms,}\end{array} \tag{4}$$

where $\overline{X}$ - vector of mean values of random variables $\overline{x}_i$ i=1,2,...,N.

If standard deviations σ_{x_i} of x_i are small, $g_j(X,Y)$ can be approximated by the first two terms of equation (4). The mean $\overline{g}_j$ and standard deviation σ_{g_j} are given by:

$$\overline{g}_j = g_j(\overline{X},Y), \qquad \sigma_{g_j} = \left[\sum_{i=1}^{N}\left(\frac{\partial g_j}{\partial x_i}\right)^2\bigg|_{X=\overline{X}} \sigma_{x_i}^2\right]^{1/2}. \tag{5}$$

Carrying out transformations, as shown in [6], the inequality constraints (3) can be finally written in the form

$$g_j - \phi^{-1}(p_j) \sqrt{\sigma_{gj}^2} \geq 0, \qquad (6)$$

where $\phi^{-1}(p)$ is the value of the standard normal variate corresponding to the probability p.
Similarly to constraint functions the objective function $f(X,Y)$ can be expandet about the mean value of $\overline{X}$. Neglecting the terms of order higher than two, objective function can be written as:

$$f(X,Y) \cong f(\overline{X},Y) + \sum_{i=1}^{N} \frac{\partial f}{\partial x_i}\Big|_{X=\overline{X}} (x_i - \overline{x}_i) = \Psi(X,Y). \qquad (7)$$

If all x_i, $i=1,2,\ldots,N$ follow normal distribution, the function $\Psi(X,Y)$ also follows normal distribution. The mean $\overline{\Psi}$ and standard deviation σ_Ψ are:

$$\overline{\Psi} = \Psi(\overline{X},Y), \qquad \sigma_\Psi = \left[\sum_{i=1}^{N} \left(\frac{\partial f}{\partial x_i}\right)^2\Big|_{X=\overline{X}} \sigma_{x_i}^2 \right]^{1/2}. \qquad (8)$$

For the purpose of optimization the new objective function F_d can be constructed as:

$$F_d = k_1\overline{\Psi} + k_2\sigma_\Psi, \qquad (9)$$

where $k_1 \geq 0$, $k_2 \geq 0$ indicate the relative importance of $\overline{\Psi}$ and σ_Ψ for minimization.
Thus the stochastic optimization problem stated in the form of stochastic programming problem has been converted into an equivalent deterministic nonlinear programming problem stated in equations (9) and (6). An inconsistence must be mentioned here between the definition of the objective function given by equation (2) and equivalent deterministic function (9). Equation (9) is consistent with the definition (2) only for $k_1=1$ and $k_2=0$. For $k_1=0$ and $k_2=1$ F_d is equal to σ_Ψ and the problem is formulated as optimization with the criteria concerning standard deviation of objective function.

3. OPTIMIZATION OF ENGINEERING STRUCTURES WITH RANDOM PARAMETERS

For the purpose of the design of engineering structures, the constraints are usually formulated as:

$$g_j(X,Y) = b_j(X,Y) - B_j(X,Y) \leqslant 0, \qquad (10)$$

where $b_j(X,Y)$ - j-th response of the system, $B_j(X,Y)$ - allowable limit for j-th response.
Constraint (6) takes the form

$$\overline{b}_j + \phi^{-1}(p_j)\left[\sigma_{b_j}^2 + \sigma_{B_j}^2\right]^{1/2} - \overline{B}_j \leqslant 0, \qquad (11)$$

where standard deviations σ_{b_j}, σ_{B_j} are given by

$$\sigma_{b_j} = \left[\sum_{i=1}^{N}\left(\frac{\partial b_j}{\partial x_i}\right)^2\Big|_{X=\overline{X}} \sigma_{x_i}^2\right]^{1/2}, \quad \sigma_{B_j} = \left[\sum_{i=1}^{N}\left(\frac{\partial B_j}{\partial x_i}\right)^2\Big|_{X=\overline{X}} \sigma_{x_i}^2\right]^{1/2}. \qquad (12)$$

For some typical cases, the formulae for the determination of the mean values of response and derivatives $\partial b/\partial x_i$ occuring in formulae for standard deviations are given in table 1.
Basing on the presented approach, programm modules have been prepared for the optimization of engineering structures. The following types of structures can be optimized by the modules available at present
1. structures under static loads,
    -plane and space trusses,
    -beams on elastic foundation,
    -plane strain, plane stress and axisymmetric structures,
2. free vibrating structures,
    -plane and space trusses,
    -plane strain, plane stress and axisymmetric structures,
3. structures under dynamic loads,
    -plane and space trusses,
    -frames,
    -plane strain, plane stress and axisymmetric structures.

4. EXAMPLES

Two examples are presented to illustrate the technique. Inequality-constrained problems in the form of Equations (9), (11) were solved by application of Mathematical Nonlinear Programming System [6] using Powell's method. The structures were assumed to be linearly elastic with Young modulus E=210000 MPa. In both cases it was also assumed that random variables are statistically independent and follow normal distribution.

<u>Truss example</u> The truss shown in Figure 1 is designed to transmit loads P_1, P_2. The mean volume $\overline{V}$ is taken as objective function and element cross-section areas a_i, $i=1,2,\ldots,7$ are taken as design variables. Mean values of random decision variables are restricted to lie between prescribed values and constraints are placed on probabilities p_1, p_2 of displacements $/u_7$, $u_9/$ exceeding allowable values d_1, d_2. The stochastic programming problem can be stated as follows:

minimize EV   (13)

with constraints

$$a_i^{min} \leq \overline{a}_i \leq a_i^{max},$$ (14)

$$P\left[u_7 - d_1 \leq 0\right] \geq p_1 ,$$

                                                (15)

$$P\left[u_9 - d_2 \leq 0\right] \geq p_2 ,$$

where u_7, u_9 displacements of joints 4 and 5 in x direction.

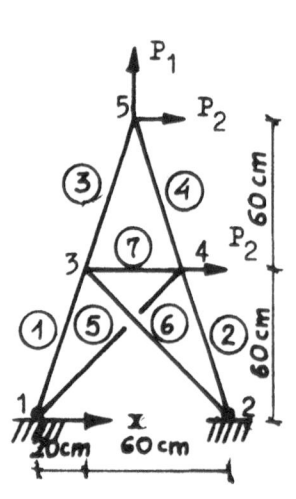

$2cm^2 \leq \overline{x}_1 \leq 10cm^2$
$1cm^2 \leq \overline{x}_2 \leq 5cm^2$
$1cm^2 \leq \overline{x}_3 \leq 5cm^2$
Figure 1. Plane truss

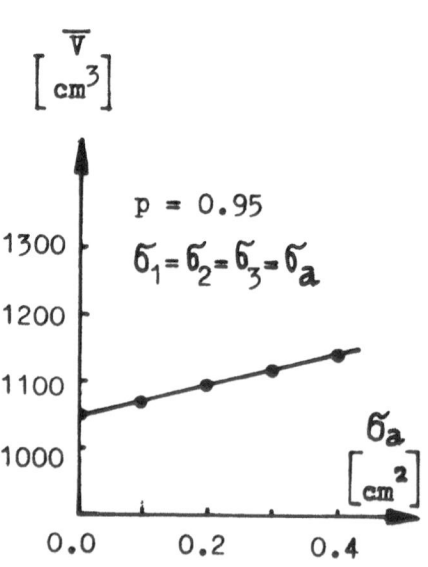

Figure 2. Influence of standard deviation σ_a on the optimal volume of the structure

Corresponding deterministic problem has the form:

$$\text{minimize} \quad \overline{V} = \sum_{i=1}^{7} a_i l_i \tag{16}$$

with constraints

$$a_i^{min} \leq \overline{a}_i \leq a_i^{max}, \tag{17}$$

$$\overline{u}_7 + \Phi^{-1}(p_1) \widetilde{6}_{u_7} - d_1 \leq 0,$$

$$\overline{u}_9 + \Phi^{-1}(p_2) \widetilde{6}_{u_9} - d_2 \leq 0, \tag{18}$$

where l_i - i-th element length.
Displacements $\overline{u}_7$, $\overline{u}_9$ and $\widetilde{6}_{u_7}$, $\widetilde{6}_{u_9}$ can be calculated according to formulae given in table 1.
Figures (2) and (3) show results obtained on the assumption that cross-section areas a_i are only random variables. To simplyfy calculations it was also assumed that $a_i = x_1$ for i=1,2,3,4, $a_i = x_2$ for i= 5,6, $a_7 = x_3$ and that $\widetilde{6}_{x_j} = \widetilde{6}_a$ for j=1,2,3. In numerical calculations it was assumed that: $p_1 = p_2 = p$, $d_1 = 0.05$ cm, $d_2 = 0.10$ cm, $P_1 = 30$ kN, $P_2 = 10$ kN.

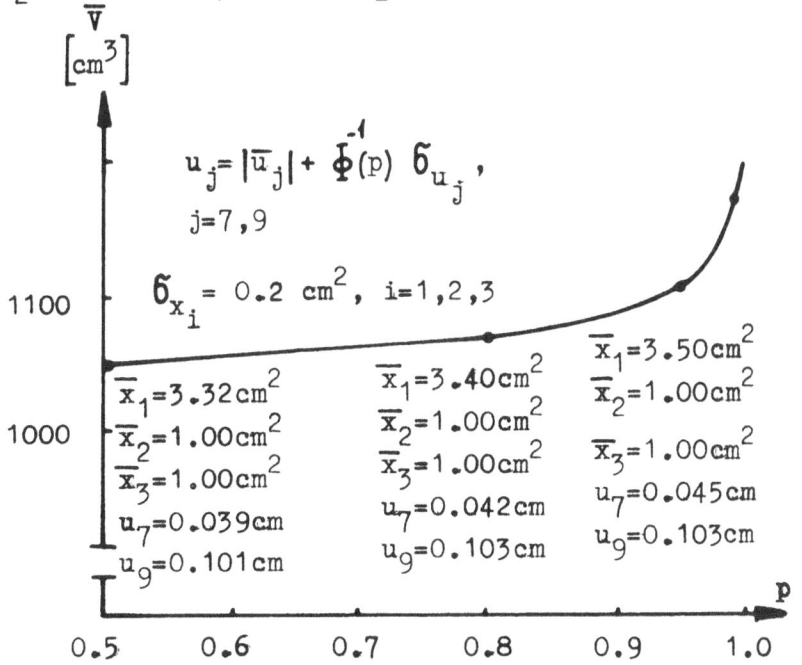

Figure 3. Influence of probability p on the optimal volume of the structure

The influence of standard deviation of cross section area 6_a on the optimum volume is shown in Figure 2. Optimum structure volume $\overline{V}$ v/s probability p is presented in Figure 3. The optimal values of decision variables and joint displacements u_7, u_9 are also given in the figure.

<u>Free vibrating truss</u>   The example concerns the design of the truss presented in Figure 4. To avoid resonance it is prescribed that the lowest fundamental frequency must be grearer than the excitation frequency Ω. Joints coordinates y_i, i=1,2,3 are decision variables $/y_1 = x_2$, $y_2 = z_2$, $y_3 = x_3/$. The mean volume $\overline{V}$ is taken as objective function with the following constraints

$$y_i^{min} \leq y_i \leq y_i^{max}, \quad i=1,2,3 \qquad\qquad 19$$

$$P\left[\omega_{min} - \Omega \geq 0 \right] \geq p , \qquad\qquad 20$$

where ω_{min} can be calculated according to formulae given in table 1 /problem no. 4/, y_i^{min} , y_i^{max}, Ω, p - are given values. In this problem it was assumed that the only random variables are element lengths.

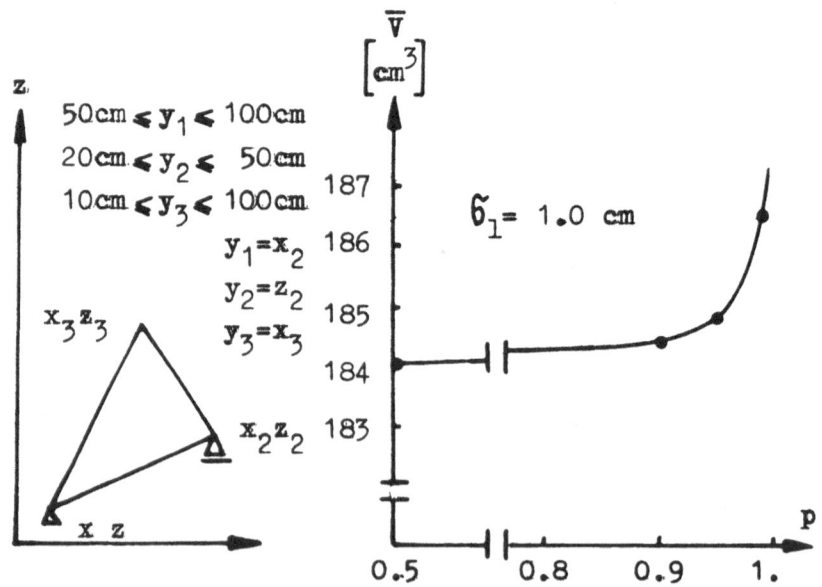

Figure 4. Free vibrating truss

Figure 5. Influence of probability p on the optimal volume $\overline{V}$ of the structure

The deterministic programming problem has the form:

$$\text{minimize} \quad \overline{V} = \sum_{i=1}^{3} a_i l_i (\overline{X}, Y) \qquad\qquad 21$$

with constraints

$$y_i^{min} \leq y_i \leq y_i^{max} \qquad\qquad 22$$

$$\omega_{min} - \Phi^{-1}(p)\, \sigma_{\omega_{min}} - \Omega \geq 0 , \qquad\qquad 23$$

where l_i - i-th element length, a_i - area of the cross-section of i-th element.
The numerical results presented in Figures 5 and 6 were obtained for $a_i = 1$ cm^2, i=1,2,3, material density $\varrho = 7.8 \; 10$ Nsek$^{}$/cm , $\Omega = 3125$ rad/sek. In Figure 5 the influence of probability p on optimal volume $\overline{V}$ is presented The optimal volume v/s standard deviation σ_l is shown in Figure 6. The optimal values of decision variables are also given in this figure.

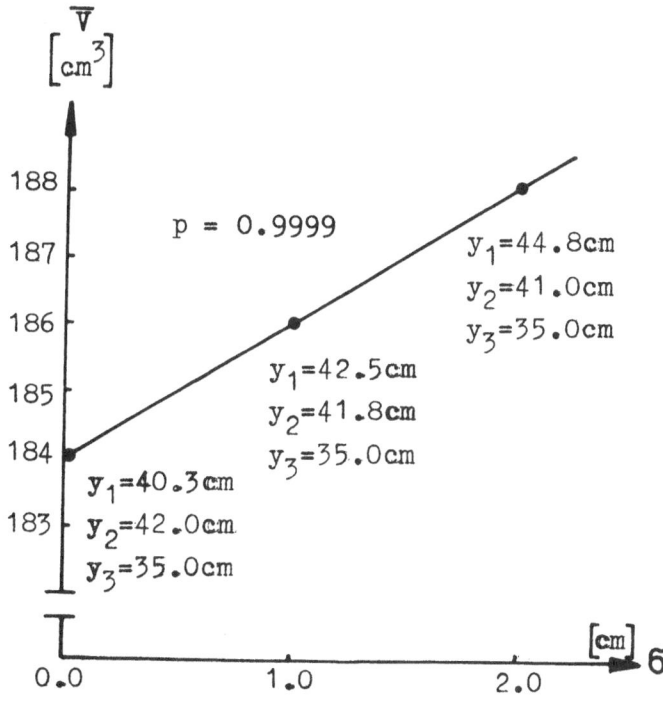

Figure 6. Influence of standard deviation σ_l on the optimal volume of the structure

Table 1. Estimation of mean values of $\bar{b}$
and derivatives $\partial b / \partial x_i$

Problem no.	System response	Mean value of response for $X=\bar{X}$	Response derivatives
1.	$b = u$ u-joint displacements	$\bar{u}=K^{-1}F$	$\dfrac{\partial b}{\partial x_i} =K^{-1}\left[\dfrac{\partial F}{\partial x_i} - \dfrac{\partial K}{\partial x_i}u\right]\Bigg\vert_{X=\bar{X}}$
2.	$b = \gamma$ γ-stresses in elements	$\bar{\gamma}=T\,\bar{u}$	$\dfrac{\partial \gamma}{\partial x_i} = \dfrac{\partial T}{\partial x_i}u + T\dfrac{\partial u}{\partial x_i}\Bigg\vert_{X=\bar{X}}$
3.	$b_j = \lambda_j$ λ_j^j - j-th eigenvalue	$(K-\lambda_j M)\Lambda_j=0$	$\dfrac{\partial \lambda_j^j}{\partial x_i} = \dfrac{\Lambda_j^T\left(\dfrac{\partial K}{\partial x_i}-\lambda_j\dfrac{\partial M}{\partial x_i}\right)\Lambda_j}{M_j^*}\,;$
4.	$b_j = \omega_j$ ω_j^j - j-th natural frequency	$\omega_j=\sqrt{\lambda_j}$	$M_j^*=\Lambda_j^T M \Lambda_j$

where $\partial b/\partial x_i$ - vector with elements $\partial b_j /\partial x_i$,
K- stifness matrix, M - mass matrix, F - force vec-
tor, T - matrix of element geometric and material
properties, Λ - j-th eigenvector, δ_{ij} -Kronecker delta.

References

1. Charnes, A. Cooper, W.W. (1959) Chance constrained programming, Managament Science,6, 73-79.
2. Davidson, J.W. Felton, L.P. Hart, G.C. (1977) Probability based optimization for dynamic loads, J.
   Struct. Div., Proc. ASCE, ST10, 2021-2033,
3. Davidson, J.W. Felton, L.P. Hart, G.C. (1977) Optimum design of structures with random parameters
   J. Comp. Struct.,7, 481-486.
4. Haugen, E.B. (1980) Probabilistic Mechanical Design, John Wiley, New York.
5. Kręglewski, T. Rogowski, T. (1984) Metody optyma-
   lizacji w języku Fprtran, PWN Warszawa,in polish.
6. Rao, S.S. (1978) Optimization- Theory and Application, Wiley Eastern Ltd, New Delhi.

ANALYSIS OF PLATES BY THE INITIAL VALUE METHOD

H.A. Al-Khaiat

Kuwait University, Kuwait

1. INTRODUCTION

The analysis of plates may be either "rigorous" or "approximate". The rigorous analysis consists of techniques for seeking
direct solutions to the partical differential equation of
plates [1]. A rigorous solution of rectangular plates can be
obtained only for a limited number of cases. For the majority
of practical problems, a rigorous solution either can not be
found or is of such a complicated nature that it can be applied
only with great difficulty in a practical computation. For many
cases, approximate methods are the only approaches that can be
employed. Of the many numerical methods used, finite-element
[2] and finite-difference [3, 4, 5] techniques are the most
frequently used methods.

The initial-value method is an approximate method that has
proven its efficiency in the analysis of beams [6] and cables
[7]. In this study, the method is applied to plate problems;
the results of the method and its efficiency can be shown to be
better than the other numerical methods for many plate problems.

2. INITIAL VALUE FORMULATION

The initial-value method is applied to the two-dimensional
plate problems. The method consists of solving the differential
equation of plates (1) by the step-by-step integration procedure.

$$\frac{\partial^4 w}{\partial x^4} + 2 \frac{\partial^4 w}{\partial x^2 \partial y^2} + \frac{\partial^4 w}{\partial y^4} = \frac{P}{D} \tag{1}$$

Since integration is possible for only one direction, the finite
difference equations are employed to replace the derivatives in

the other direction with deflections.

The integration procedure for the initial-value method can be summarized for the plate shown in Figure 1 as follows:

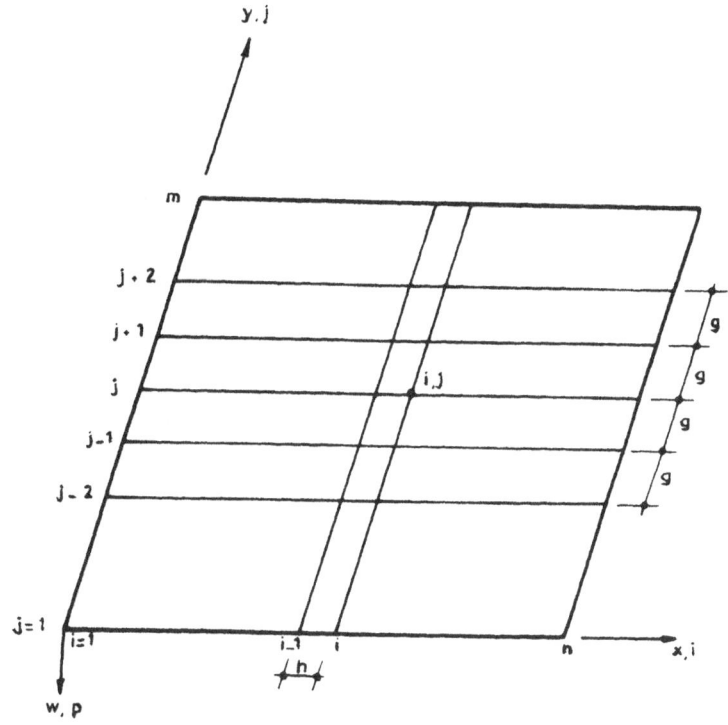

Figure 1. Mesh of a Rectangular Plate

1. The fourth derivative is written as follows:

$$(\frac{\partial^4 w}{\partial x^4})_{i,j} = \frac{P_{i,j}}{D} - \frac{2}{g^2} [(\frac{\partial^2 w}{\partial x^2})_{i,j-1} - 2(\frac{\partial^2 w}{\partial x^2})_{i,j}$$

$$+ (\frac{\partial^2 w}{\partial x^2})_{i,j+1}] - \frac{1}{g^4}[w_{i,j-2} - 4w_{i,j-1}$$

$$+ 6 w_{i,j} - 4 w_{i,j+1} + w_{i,j+2}] \tag{2}$$

2. Along edge i=1, certain quantities of the deflections and their derivatives are fixed by the boundary conditions and other values are assumed. Application of equation (2) at i=1 will yield the corresponding values of $(\frac{\partial^4 w}{\partial x^4})_{i,j}$.

3. At section i=2, it is first assumed that

$$(\frac{\partial^4 w}{\partial x^4})_{i,j} = (\frac{\partial^4 w}{\partial x^4})_{i,j-1} \tag{3}$$

4. All lower order partial derivatives can be determined from the trapezoidal-rule integration procedure. These value are then substituted into equation (2) to establish new value of $(\frac{\partial^4 w}{\partial x^4})_{i,j}$.

5. If any of the resulting values of $(\frac{\partial^4 w}{\partial x^4})_{i,j}$ determined in Step 4 does not agree as closely as desired with the values assumed in Step 3, then the new values are taken as the assumed values and Step 4 is repeated.

6. The integration is continued over the next interval (next i) and Steps 3, 4, and 5 are repeated. The same procedure continues until the terminal section (edge i=n) is reached.

In Step 6, deflection and their derivatives were found at i=n. Some of these values or some combination of them should be zero in order to satisfy the terminal boundary conditions.

Thus it is necessary to carry one particular solution and 2m independent homogeneous solutions across the structure. The true solution results from combining the particular solution and some linear combination of the homogeneous solutions that will satisfy the terminal boundary conditions.

3. DISCUSSION AND RESULTS

For a tupical plate problem with a mesh nxm, and with all edges
simply supported or fixed, the finite-difference method produces n-1 x m-1 simultaneous equations. The finite-element method, on. the other hand, has at least 3(n-1)x(m-1) degrees of
freedom. But, depending on the direction of integration only
2(m-1) or 2(n-1) simultaneous equations are needed to solve
the same problem by the initial-value method.

Example 1
The method has been applied to a square plate with uniformly
distributed load. The results for two different boundary conditions are given for the coefficient of the maximum deflection
(a_m) and the coefficient of the maximum moment (c_m) in Table 1.

Table 1. (a) $a_m (x10^{-6})$ with Simply Supported Edges

Method	4x4 Mesh		8x8 Mesh	
	a_m	% Error	a_m	% Error
Exact	4062	–	4062	–
Finite Element	3939	-3.03	4033	-0.71
Finite Difference	4030	-0.79	4055	-0.17
Initial Value	4044	-0.44	4059	-0.07

(b) $C_m (x10^{-4})$ With Simply Supported Edges

Method	4x4 Mesh		8x8 Mesh	
	c_m	% Error	c_m	% Error
Exact	479	--	479	--
FE	502	+4.80	--	--
FD	457	-4.59	473	-1.25
IV	475	+0.84	479	0.00

(c) $a_m (x10^{-5})$ With Fixed Edges

	4x4 Mesh		8x8 Mesh	
	a_m	% Error	a_m	% Error
Exact	126	--	126	--
FE	140	+11.1	130	+3.2
FD	180	+42.9	143	+13.5
IV	133	+5.6	128	+1.6

Example 2

In Table 2, the results for a square clamped plate with a con-centrated load at the centre are shown.

Table 2. Values of a_m

Method	$a_m (x10^{-5})$	% Error
Exact	560	--
FE	580	+3.57
IV	552	-1.43

Example 3

The method has been, also, applied to a rectangular plate $(\ell_y = 2\ell_x)$ which is subjected to a uniform load and a uniform tension (taken as $\dfrac{4\pi^2 D}{L_y^2}$). The coefficient of the maximum deflec-tion is:

Exact Method : a_m = 0.0670

Initial Method: a_m = 0.0676

% Error = +0.90

4. CONCLUSION

The method considered has given the lowest magnitudes of errors in all the examples and requires less time in the solution of the problem than the other numerical methods.

REFERENCES

Al-Khaiat, H (1979) Analysis of Beams on Elastic Foundation by the Initial-Value Method. M.Sc. Thesis. The Pennsylvania State University

Florin, G. (1979) Theory and Design of Surface Structures and Slabs and Plates. Trans. Tech. Publications.

Karlasam, S. (1980) An Automated Energy Based Finite Difference Procedure for the Elastic Collapse of Rectangular Plates and Panels. Journal of Computers and Structures, 11, 3: 239-249.

Reddy, J. (1979) Improved Finite Difference Analysis of Bending of the Rectangular Plates. Journal of Computers and Structures, 10, 3:431-438.

Szilard, R. (1974) Theory and Analysis of Plates: Classical and Numerical Methods. Prentice-Hall, Inc., New Jersey.

Timoshenko, S. and Woinowsky-Krieger, S. (1959) Theory of Plates and Shells. McGraw-Hill Book Co., New York.

Ugural, A. (1981) Stresses in Plates and Shells. McGraw-Hill Book Co., New York.

West, H., and Robinson, A. (1968) Continuous Method of Suspension Bridge Analysis. Journal of the Structural Division, ASCE, Proc. Paper 7280, 94, 12:2861-2883.

Zienkiewicz, O.C. (1977) The Finite Element Method. McGraw-Hill Book Co., New York.

COMPUTER-AIDED DESIGN OF STEEL FLOOR BEAM FRAMEWORK

M.C. Thakkar and S.J. Shah

Elecon Engineering Co. Ltd,
Vallabh Vidyanagar 388 120, Gujarat.

INTRODUCTION

In Industries like cement, petro-chemical, steel, fertilizer, power generation; structural floors in tall buildings in steel, like Crusher Houses, Junction Towers, Winch Towers, Take-up Tower and such other buildings, play prominent role in the overall structural designs of such buildings. Machineries like crushers, motors, gears, winch, conveyor: head-end & tail-end chutes, hoppers, belt feeders, maintenance appliances etc., are the important mechanical items to be supported on floors. Requirement of floor beam framework grid type and size vary from application to application and also to the size and type of equipment to be supported.

If rational approach for design of these floor beam framework is adopted, then it would save considerably in steelwork and also in design time.

Conventional design approach in floor design, adopted by the average design engineer is much more cumbersome, laborious and time consuming and would not normally permit the choice from various alternative feasible designs.

Structural design requirement for the steel-floor framework are in general as follows.

1.   Load requirements.

2.   Fabrication and erection methods.

3.   Design code constraints as applicable in the area.

4.   Material availability of steel section and utilisation.

Keeping above requirement in mind, the authors have made a major break-through and developed CAD for steel floor beam framework named as CAD-FLOR.

Preliminary work
The floor beam framework is not repetitive in steel buildings. The floor beam grid is decided by an experienced design engineer [Fig 1]. Loads on beam (say dead load, live load, material load, spillage, machinery loads etc, say uniformly distributed loads, point loads) are also decided by designer, [Fig 1] taking into account, machineries under various conditions of operation.

CAD-FLOR DESCRIPTION

The loads on the beam can be of any number of point loads and/or uniformly distributed load of different load intensities. The beams are assumed to be simply supported for finding the reactions, bending moments etc.

The beams can have the unknown point loads which are to be picked up from the previously solved beam reactions - may be left hand side or right hand side reactions.

Atomatic numbering can also be taken care for any no. of similar beams with respect to span, load positions etc.

Input data
Input data for a problem with a view to fix up span, loadings, unknown point loads, load positions; are kept to minimum. Input data consist of span, point loads, uniformly distributed loads, and their dimensional locations, unknown point loads in the form of serial no. and reaction notations.

Analysis
The floor beam grid is not repetitive in steel buildings. The span may repeat but not the loads to carry. But if one desires to go for rigorous structural analysis to check deflection, then it takes too much time or consumes more computer time both of which adds to cost very often it so happen that such rigorous analysis is not at all required.

Instead, the authors have taken different approach. From their experience they developed the formula to check deflection & calculations are carried out for minimum moment of inertia requirement.

Optimal design of sections
The available structural sections are grouped keeping in view the functional and structural (flexural members) requirement, the available structural sections are grouped in weight ascending order so that the analysis is not required to repeat but the

computer will go on checking the members arranged from the group. Computer will select the member when all the constraints such as Indian Standards design codes [1] [2] restrictions, functional requirements, minimum modulus of section required, minimum moment of inertia required - are met with. The result will be the optimal solution.

Design Output

Design output is presented in such an understandable format in which principal values such as reactions, bending moments, the maximum moment points, the modulus of section and second moment of inertia required and provided are printed in such a way that the non-conversant engineer can follow and the manual scrutiny of design is also possible. The total weight of steel required for the floor under consideration is also calculated.

The output from IBM360/44 for the floor Fig:1 is indicated by Fig:2.

CAD-FLOR APPLICATIONS

The CAD-FLOR system developed by the authors has been successfully used in many buildings for material handling projects in India, e.g. Satna Cements (Madhya Pradesh), Korba Thermal Power Station (Madhya Pradesh), Ramagundam Super Thermal Power Project (Andhra Pradesh), Wanakbori Thermal Power Station (Gujarat).

Further Development

The authors are planning to develop system further to use as preprocessor to get the optimal computerised designs for complete building.

CONCLUSIONS

1. With CAD-FLOR, structural designs for steel-floor beam framework for different projects will be consistant.

2. Chances of numerical errors are eliminated.

3. Approach for analysis and design in CAD-FLOR has helped in getting the optimal design solution.

UNITS

1 kg = 9.80665 N

ACKNOWLEDGEMENTS

The authors are thankful to their management for providing an opportunity to present this paper at ENGSOFT 85

LIST OF REFERENCES

1   IS-800-1962, Indian Standard Code of Practice for Use of Structural Steel in General Building Construction (Revised), Indian Standard Institution, New Delhi, April 1982.

2   IS-875-1964, Indian Standard Code of Practice for Structural Safety of Building : Loading Standard (Revised), Indian Standard Institution, New Delhi, April 1982.

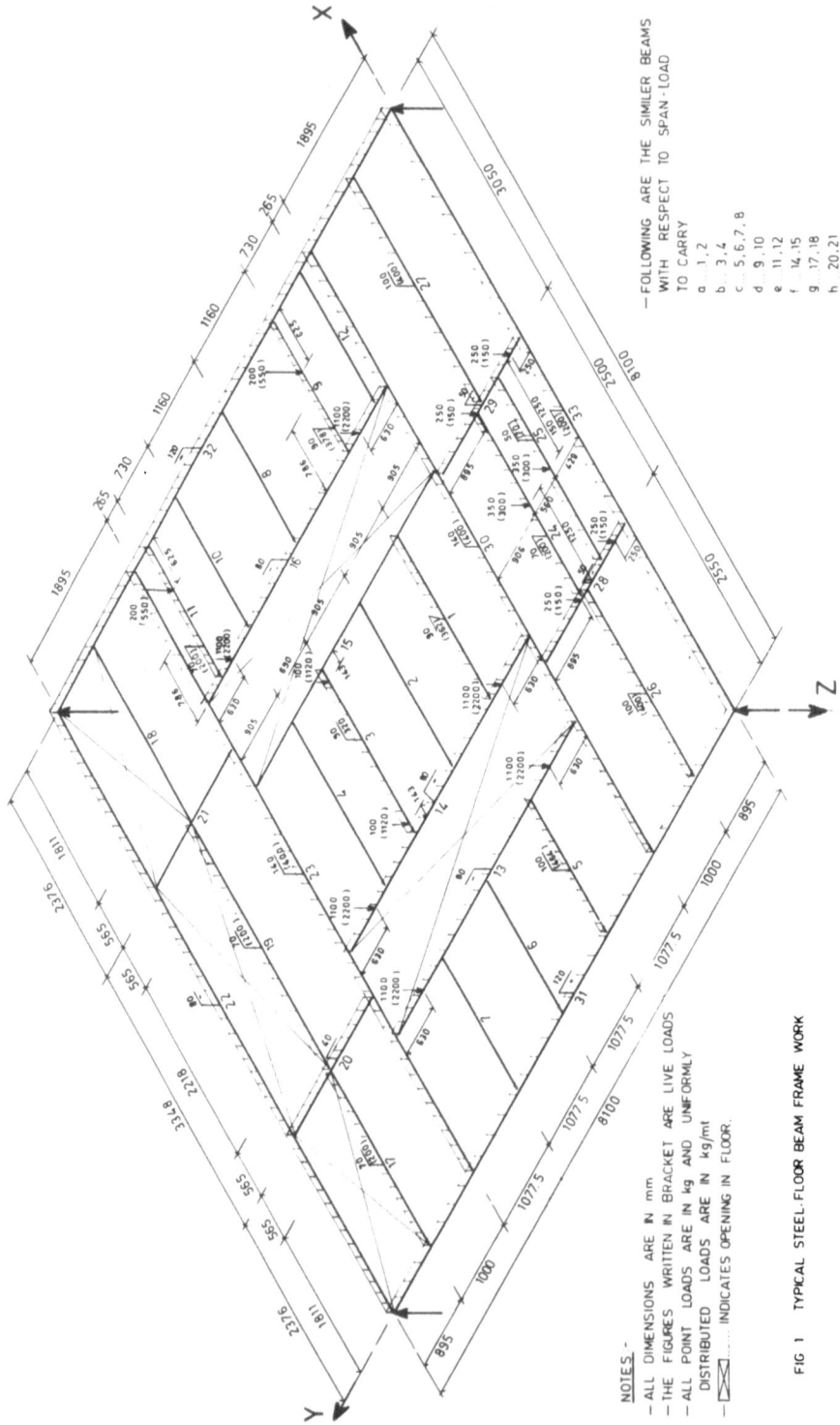

NOTES :-
- ALL DIMENSIONS ARE IN mm
- THE FIGURES WRITTEN IN BRACKET ARE LIVE LOADS
- ALL POINT LOADS ARE IN kg AND UNIFORMLY
  DISTRIBUTED LOADS ARE IN kg/mt
- ⬚ INDICATES OPENING IN FLOOR.

FIG 1 TYPICAL STEEL·FLOOR BEAM FRAME WORK

- FOLLOWING ARE THE SIMILER BEAMS
  WITH RESPECT TO SPAN·LOAD
  TO CARRY
  a 1,2
  b 3,4
  c 5,6,7,8
  d 9,10
  e 11,12
  f 14,15
  g 17,18
  h 20,21

CD/844 RAMAGUNDAM STPP C.M. FLOOR AT RL 150.60 M LVL   PAGE   1

BM SPAN NO	SPAN M	LEFT-HAND SIDE REACTION			RIGHT-HAND SIDE REACTION			POINT OF ZERO SHEAR M	MAXIMUM BENDING MOMENT KG-M	PERMISSIBLE BENDING STRESS IN COMP. KG/SQ.CM	Z REQD CM3	I REQD CM4	SECTION PROVIDED	Z PROVIDED CM3	I PROVIDED CM4
		DL KG	LL KG	TOTAL KG	DL KG	LL KG	TOTAL KG								
1	2.218	100	401	501	100	401	501	1.109	278	1650	16	99	ISMB 125	72	449	SO.OK
2	2.218	100	401	501	100	401	501	1.109	278	1650	16	99	ISMB 125	72	449	SO.OK
3	2.218	200	1475	1675	200	1475	1675	1.109	427	1650	25	152	ISMB 125	72	449	SO.OK
4	2.218	200	1475	1675	200	1475	1675	1.109	427	1650	25	152	ISMB 125	72	449	SO.OK
5	1.811	91	420	511	91	420	511	0.905	231	1650	14	67	ISMB 125	72	449	SO.OK
6	1.811	91	420	511	91	420	511	0.905	231	1650	14	67	ISMB 125	72	449	SO.OK
7	1.811	91	420	511	91	420	511	0.905	231	1650	14	67	ISMB 125	72	449	SO.OK
8	1.811	151	532	683	212	702	915	1.186	480	1650	29	140	ISMB 125	72	449	SO.OK
9	1.811	151	532	683	212	702	915	1.186	480	1650	29	140	ISMB 125	72	449	SO.OK
10	1.811	132	371	503	194	541	736	1.186	407	1650	24	118	ISMB 125	72	449	SO.OK
11	1.811	132	371	503	194	541	736	1.186	407	1650	24	118	ISMB 125	72	449	SO.OK
12	4.310	1408	2830	4238	1408	2830	4238	2.154	3366	1650	204	2336	ISMB 250	411	5131	SO.OK
13	4.310	1525	3606	5131	1619	4547	6166	2.715	5397	1650	327	3749	ISMB 250	411	5131	SO.OK
14	4.310	1525	3606	5131	1619	4547	6166	2.715	5397	1650	327	3749	ISMB 250	411	5131	SO.OK
15	4.310	1601	3313	4914	1601	3313	4914	2.155	3631	1650	220	2522	ISMB 250	411	5131	SO.OK
16	2.376	83	238	321	83	238	321	1.188	191	1650	11	72	ISMB 125	72	449	SO.OK
17	2.376	83	238	321	83	238	321	1.188	191	1650	11	72	ISMB 125	72	449	SO.OK
18	3.348	134	335	469	134	335	469	1.674	392	1650	23	211	ISMB 125	72	449	SO.OK
19	1.895	140	269	410	153	303	456	1.005	390	1650	23	119	ISMB 125	72	449	SO.OK
20	1.895	140	269	410	153	303	456	1.005	390	1650	23	119	ISMB 125	72	449	SO.OK
21	6.100	477	303	780	477	303	780	4.071	1742	1650	105	2274	ISMB 300	574	8603	SO.OK
22	6.100	3777	9375	13152	3884	9642	13525	4.333	31648	1650	1930	41585	ISMB 600	3060	91613	SO.OK
23	2.500	263	400	663	263	400	663	1.250	617	1650	37	248	ISMB 125	72	449	SO.OK
24	2.500	238	238	475	238	238	475	1.250	500	1650	30	201	ISMB 125	72	449	SO.OK
25	2.550	128	510	638	128	510	638	1.275	407	1650	24	167	ISMB 125	72	449	SO.OK
26	3.050	153	610	763	153	610	763	1.525	582	1650	35	286	ISMB 125	72	449	SO.OK
27	1.895	759	845	1604	463	602	1066	0.989	941	1650	57	287	ISMB 125	72	449	SO.OK
28	1.895	772	898	1670	475	650	1125	0.895	996	1650	60	304	ISMB 125	72	449	SO.OK
29	6.100	4040	8021	12061	4092	9025	13117	4.495	30936	1650	1874	40394	ISMB 600	3060	91613	SO.OK
30	6.100	4723	10061	14783	4549	10143	14692	4.050	28616	1650	1734	37364	ISMB 500	1809	45218	SO.OK
31	6.100	5126	11192	16318	4961	11230	16191	4.050	31986	1650	1938	41764	ISMB 600	3060	91613	SO.OK
32	6.100	1418	1727	3146	1328	1636	2964	4.407	7487	1650	453	9775	ISMB 350	779	13630	SO.OK

TOTAL WEIGHT OF BEAMS OF THIS FLOOR=   5753.90KG

FIG: 2   DESIGN OUTPUT

AN ESTIMATING PACKAGE FOR STEEL FABRICATED STRUCTURES

H C Ward

CAD Unit, Teesside Polytechnic, Middlesbrough.

1. INTRODUCTION

In order to make a profit, or for that matter even to stay in business, it is necessary for manufacturing concerns to be more competitive than their rivals are. This is usually attained by the installation of new equipment, reduction in manpower, bonus incentives etc. Basically all these methods are directed towards one end - increased productivity, or to put it more bluntly - 'more output for the same number of employees' or 'maintain the same output with fewer number of employees'.

However no matter how the productivity is achieved unless orders are obtained the firm will suffer dire consequences. Therefore in order to succeed it is essential to get out and win orders. This can only be achieved by the preparation of tenders accurately and quickly, which in turn depends upon the efficiency and size of the estimating department. Efficiency generally controls the accuracy of the estimate, size controlling the number of tenders which may be handled in any given time.

The larger firms, although having reduced manpower over the years, have generally retained reasonably sized estimating sections. However efficiency is generally lost by each estimator having to spend time searching through standard data in order to maintain uniformity. Alternatively each estimator may be allowed to do his/her 'own thing' in which case lack of accuracy and uniformity is likely to be predominant.

The smaller firms, in a bid to reduce manpower, are often reduced to one estimator, or even to share this job with some other function. Due to working under pressure the net result is the production of ill-prepared, inaccurate estimates. To make matters worse if the person concerned is involved in other functions then these too tend to suffer. It follows therefore that there is a need for fast, efficient and uniform tendering procedures.

2. CURRENT PRACTICE

Present methods of preparing estimates for fabricated structures
may vary slightly in the detail, but in general will involve
the following major features.

(i)   Identification of the sections involved, along
      with the length and number-off for each section size.
      This may be eased if a bill of materials is
      available, however in many instances at the
      tendering stage outline drawings only are available.
      This, therefore, involves careful study of what
      may turn out to be not very carefully prepared
      drawings.

(ii) Using this information it is now possible to
      calculate the total weight of material being
      used, and its total cost. To do this requires
      constant reference to standard section tables,
      and must also take into account the source of
      the material itself, eg. stockholder, mill or
      any other source.

(iii) At this point some firms may resort to the
      practice of taking the total weight of the
      material and multiply this by some set cost/tonne.
      This figure is then used as the estimated price.
      While this may be acceptable on small jobs it
      can be extremely dangerous for larger or more
      complex jobs. If the multiplying figure is too
      low then the firm may be given this contract
      and in due course find themselves in a job 'loss'
      situation. On the other hand if the figure is too
      high then they will probably lose the contract to
      some rival company.

(iv) In order to overcome the dangers of employing
      stage (iii) it becomes necessary to determine
      the various manufacturing operations which need
      to be carried out on each section, plus
      determining any other actions necessary.
      These may then be costed and totalled.

(v)   At this stage all costs, materials, labour,
      transport and any other additonal costs may be
      added together to form a total cost for the
      project. To this is added some margin for
      profit in order to make up the estimate price.

It can be seen that this can be a lengthy process, and one which is prone to human error. The author can remember a test case in which five estimators in the same firm were each given the same estimate to prepare. The result was five prices which were so vastly diverse as to be almost unbelievable. It was for this reason that the use of computers for estimating purposes was considered. The criteria to be met was speed, accuracy, uniformity and the ability to be used on microcomputers as well as larger mainframe or super-mini computers.

3. COMEST

Based on the recommendations of a short feasibility study Comest is a computer package which was developed to assist in the speeding up of the estimating process, with the additional advantages of less probability of error and more uniformity of prices, irrespective of which estimator prepares the estimate. The package is modular in concept, Figure 1 giving an overview in line diagram mode.

It can be seen that there are basically two main areas of concern in the package.

The first concerns the creation of standard data-bases, which in this instance concerns details of standard sections, the welding times, drilling times, and a shorter base holding details such as labour rates, steel prices, etc.

Under normal conditions the estimator does not require direct access to these data-bases and hence they are password protected. This ensures that the data cannot be modified or deleted by unauthorised personnel.

The second area is that of producing the actual estimate itself, and is so designed that in operation the estimator follows the same logical train of thought as in manual estimating, but without any of the mental hassle involved in that process.

Estimate
Input On entering the estimating module the screen is cleared, and the estimator is presented with a typical estimating sheet heading, as shown below in Figure 2.

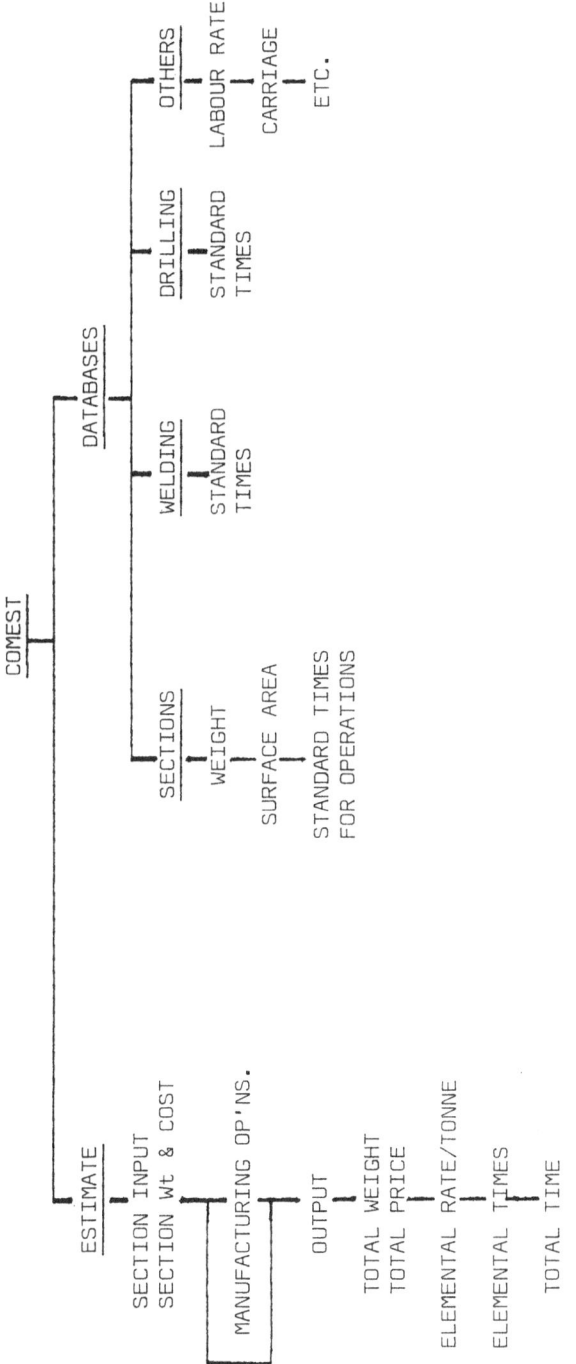

FIGURE 1 COMEST STRUCTURE

XYZ FABRICATING COMPANY

MATERIAL ESTIMATING SHEET

CUSTOMERS NAME	DATE	ESTIMATE NO
ANYBODY FABRICATORS	02-04-1985	AF-123

SECT NO	SECTION TYPE & SIZE	LENGTH	NUMBER OFF	PRICE CODE
1	UB686X254X152	3	6	S/H
2	UB686X254X125	4	4	S/H
3	Fini			

A list of entries will follow. Check them and note any Section Numbers which require modifying.

SECT NO	SECTION TYPE & SIZE	LENGTH	NUMBER OFF	PRICE CODE
1				
2				
↓				

FIGURE 2.

It will be seen that it is quite a straightforward task to fill in this sheet, simply filling in the details required, the cursor automatically moving to the next required piece of information.

On completion of the header the cursor moves down to the section input requisition. Again it can be seen that it is a simple matter to enter data.

The first column represents the number of entries made, and it is automatically printed on the screen by the program as the estimator progresses. One of the major points was the format to be adopted for entering a section type and size. The format eventually chosen is believed to be typical of the manner in which an engineer would describe the section in words. The example shown is the format for a Universal Beam 686 x 254 x 152. A similar format is used for plain plate and chequer plate. When asking for the price code it was felt safer to ask the estimator to actually enter 'MILL', 'S/H' or 'SPE' rather than a simple code such as 1, 2, or 3.

Having completed one section the program immediately asks for the next section, and will continue until the estimator signals that he has no further entries to make. This is done by entering 'Fini' instead of a section identity.

At this point the estimator is given the opportunity to examine, modify or delete any entered item.

When satisfied that all is correct the program will calculate each entry's weight and cost, the total weight and cost, and will produce a printout of all details up to that point.

XYZ FABRICATING COMPANY

MATERIAL WEIGHT & COST SHEET

CUSTOMERS NAME   DATE   ESTIMATE No.
ANYBODY FABRICATORS.   02-94-1985   AF 123

TOTAL MATERIAL WEIGHT ....= 12 TONNES.
TOTAL COST ....= £3000.00

FIGURE 3.

The estimator now has the option of ending the program, or of going on to fill in the manufacturing requirements.

Manufacturing Operations On entry into this part of the program the estimator is given the opportunity of entering the manufacturing requirements for each section in turn. Displayed on the screen is a work content sheet as shown in Figure 4.

WORK CONTENT SHEETS

WORK CONTENT FOR ITEM 1 - SECTION-UB686X254X152

1 SAW-No.CUTS & BEVEL ANGLE..(2)	14 DRILL-STD.CONN No.ENDS(1)
2 CROP-MULTIPLE FACTOR.......(2)	15 DRILL M/C-No.HOLES......(1)
3 BURN M/C-THICKNESS, LENGTH(2)	16 DRILL RAD.DIA, THICK, No(3)
4 BURN Q - THICKNESS, LENGTH(2)	17 PUNCH- STD.CONN, No.ENDS(1)
5 BURN HAND - LENGTH.........(1)	18 PUNCH - No. OF HOLES...(1)
6 BURN - END PLATE...No.ENDS.(1)	19 SCREW - DIA & LENGTH....(2)
7 PLATER - MARK TIME.........(1)	20 MILL - NO.ENDS..........(1)
8 PLATER - ASSEMBLE TIME.....(1)	21 LATHE - TIME............(1)
9 PLATER - S.E.PLATE.No.ENDS.(1)	22 GRIND - STD..No.ENDS....(1)
10 WELD - PROFILE STD, No.ENDS(1)	23 GRIND - LENGTH..........(1)
11 WELD - THICKNESS & LENGTH..(2)	24 D.O. - TIME.............(1)
12 WELD PREPn. - LENGTH.......(1)	25 GALV. - PRICE/TONNE.....(1)
13 STD. NOTCH - No.ENDS.......(1)	26 PAINT - PRICE/METRE.....(1)
	27 HANDLE - TIME...........(1)

FIGURE 4.

Again completion of the sheet is a straightforward operation and requires practically no training for the estimator. Consideration of the first item gives:

W1 SAW-No of RNDS-ANGLE(2)?

W1 is symbolic for the first working operation for that particular section. SAW is requesting whether or not the section is to be sawed to length

No CUTS - one end or both

Angle - is the cut straight across or at an angle? If at an angle enter the number of degrees.

(2) - tells the estimator he has two entries to make at this point.

Each operation is queried in turn. If the operation is necessary the estimator enters the detail required. If it is not then 'Return' only needs to be entered.

Provision is made for the entry of additional saw or crop time, and for any other time deemed necessary.

The estimator then has the option of modifying all these entries in a manner similar to the 'Input' program.

This procedure is followed for each section type entered until all sections have been dealt with.

Finally the possibility of other general items is dealt with eg. bought out parts, bolts, etc., followed by transport costs, commission and profit margins.

Output On completion of all the entries the output is generated automatically by the computer. Typically the output presentation is of the form shown in Figure 5. Initially this is presented on the VDU screen, but may be sent to the printer for a hard copy output if required. Provision is also made for storing this information on disk for future reference.

XYZ FABRICATING COMPANY

ESTIMATE OUTPUT LISTING

CUSTOMERS NAME   DATE   ESTIMATE No.
ANYBODY FABRICATORS 02-04-1985   AF 123

MATERIAL REQUIRED

SECT NO	SECTION TYPE & SIZE	LENGTH	NUMBER OFF	PRICE CODE	WEIGHT
1	UB686X254X152	3	6	S/H	
2	↓	↓	↓	↓	
3					
↓					

TOTAL WEIGHT & COST

  TOTAL WEIGHT 462 TONNE TOTAL PRICE £274,428.
                  F.O.B.

  RATES PER TONNE

MATERIAL	220
WORKMANSHIP	185
PAINTING	40
BOLTS	15
TRANSPORT F.O.B.	19
PROFIT	89
COMMISSION	26

        £594/TONNE F.O.B

  TOTAL WORKMANSHIP (HOURS)

W1	SAW	240	W14	DRILL M/C	420
W2	CROP	115	W15	DRILL RADIAL	
W3	BURN M/C	300.3	W16	PUNCH	80.8
W4	BURN QUICKLY	140.2	W17	PUNCH	0
W5	BURN HAND	54	W18	SCREW	0
W6	BURN STD.NOTCH	64	W19	MILL	36
W7	PLATER MARK	821.3	W20	LATHE	0
W8	PLATER ASS.	884	W21	GRIND STD	100
W9	PLATER STD. END	640	W22	GRIND - LENGTH	50
W10	WELD STD	420	W23	D.O.	620
W11	WELD LENGTH	94	W24	GALV	-
W12	WELD - PREP	0	W25	PAINT	-
W13	DRILL - STD		W26	HANDLE	50
			W27	PACK	-

                4690 HOURS

FIGURE 5

CONCLUSION

The program has been in use for some time now, and has certainly come up to all expectations in terms of time saved, and accuracy and uniformity of the results.

Furthermore the program does not require the use of a large computer installation for its implementation. Typically versions of the program are now running on systems using Prime, Apollo, Alpha and Apricot computers.

12. HYDRAULICS

OPTIMAL DESIGN FOR THE RUN-OF-THE-RIVER PLANT SYSTEM

M. Petričec[*], J. Margeta[**] and N. Mladineo[**]

[*] Institut za Elektroprivredu, Proleterskih Brigada 37, 41000 Zagreb, Yugoslavia
[**] Faculty of Civil Engineering, V.Masleše bb, 58000 Split, Yugoslavia

1. INTRODUCTION

Today, because of increasing energy needs and growing energy costs, it is necessary to exploit previously unexploited energy potentials with low effiency which exist in almost every country. A characteristic aspect of the water system in Yugoslavia is that there are numerous streamflows with gentle slope, which have been considered unsuitable from an economic standpoint for generating electric energy. But increasing costs of energy call for a new approach to such streamflows and arouse interest in their exploitation. Investigations show that the best solution for such streamflows is a series of run-off-the-river plants (RRP). The relatively high costs of the construction of such structures result in a low degree of efficiency, and this means, that special attention should be paid to the maximum exploitation of water potential, i.e., to the optimal dimensioning of the system. Usually there are many different technical solutions (construction variants) for the construction of these structures, particularly if they are multi-purpose (e.g., if they can be used for flood protection, agriculture, recreational facilities etc., apart from energy supply).

The great number of variant solutions possible for implementing the system makes it difficult to choose optimal values, especially if an evaluation of all the parameters resulting from the multi-purpose exploitation of the system is wanted.

2.   PROBLEM CHARACTERISTICS

The characteristics of the system, i.e., the fact that there
are several run-of-the-river plants (RRP) with their basic
parameteres having different values, make it possible to model
the whole system as a multistage process, where every RRP corresponds to one stage. By reducing the system to a spatial
multi-stage model, it is possible to apply dynamic programming
(DP) in order to define optimal project parameters. Such a
model was used in writing a tender for a project of run-of-the-
river plants on the Mura river where it was necessary to
determine two project parameters which could make it possible
to achieve maximum efficiency (B/C) for several RRP. In order
to define the task, it was necessary to begin with the fact that
the geometry of the system makes it possible to exploit the
slope of the streamflow by building several RRPs with a uniform
system of operation. The results of the analysis showed four
possible locations for RRPs for which the input data were taken
(geometric computation, hydrology, construction costs, etc.).
Since the economic efficiency of the system, i.e., the cost-
benefit relation, depends directly on the banked-up water level
(H_j) and the installed discharge (QI_i), it was necessary to find
the dimensions of these parameters for each RRP so that the
whole system would yield optimum results.

Theoretically, for each RRP the banked-up water level can
vary from the minimum value (i.e. RRP are not built) to a
maximum value (H_{iMAX}) which is either equal to or smaller than
the slope of the system (DH). Evidently, by using greater
unlevelling on the downstream RRP the upstream plants are either
eliminated or have a low banked-up water level, i.e., are more
gently sloped. The interaction between parameters H_i and QI_i
calls for a great number of possible variants of the system
dimensions, which can be approximately illustrated by an appropriate network (DP method) as in Figure 1. Using dynamic programming the network is "searched" by examining all the possible
paths through the network, where every "path" is one variant of
the system dimensions, and by choosing the optimum one. It
should be noted that the cost-benefit relation for every variant
(B/C) and for the whole duration of the exploitation of the
structure is computed.

3.   PROBLEM FORMULATION

Evidently, the problem consists of two design variables (QI_i,
H_i). However, one of the significant constraints imposed upon
the problem is that the installed discharge, QI, should be the
same for all RRP, i.e.:

$$QI_1 = QI_2 = QI_3 = QI_4 \qquad (1)$$

with regard to this constraint and a restricted selection of
constructive values, QI, the problem can be formulated as one

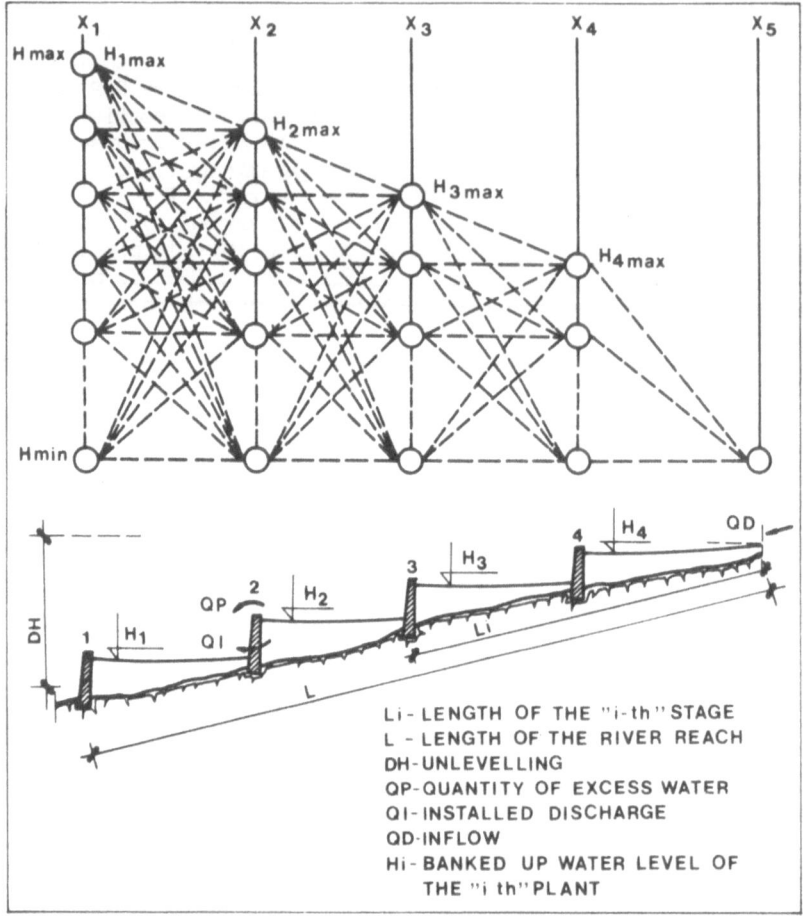

Figure 1. Schematic presentation of the problem

dimensional with an iterative method of choice of the optimal value QI.

In this case the possible locations of RRP were taken as a stage. The stage variable represents the banked-up water level (H_i), Figure 1. For each stage, the stage variable should be

$$H_i^{MIN} \leqslant H_i \leqslant H_i^{MAX} \tag{2}$$

where

$$H_i^{MAX} = \frac{DH \cdot L_i}{L} + \triangle h_i \tag{3}$$

$\triangle\,h_i$ - constructive safety banking

H_i^{MIN} - minimum structural level

The computation is accomplished by an iterative method starting from the first stage (Figure 2). The benefit-cost ratio (B/C) is analyzed for the given QI at each stage for each feasible combination of H_i according to the discrete value $\triangle\,H_i$ for the whole period of the plant exploitation.

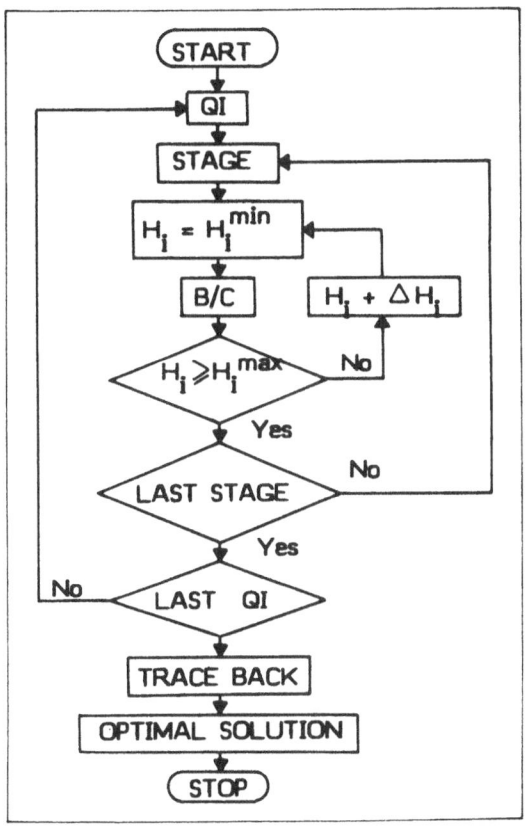

Figure 2. Model flow chart

Selection of the optimum construction values (GI and H_i) has been performed with a view of achieving:

$$\max \sum_{i=1}^{N} \sum_{j=1}^{T} \frac{B_{i,j}}{C_{i,j}} \tag{4}$$

where:

i - stage
j - time step
$B_{i,j}$ - benefit at stage i and time j
$C_{i,j}$ - cost at stage i and time j
T - project life
N - number of stages

Construction costs and benefits of all structures are the function of banked-up water level and the installed power.

$$C_i = f_i(QI, H_i)$$
$$B_i = f_i(QI, H_i) \tag{5}$$

The plan for construction and investments over time is also defined for each structure.

In order to make a more detailed analysis indirect costs and benefits resulting from the construction of energy structures were introduced into the computation. All the costs and benefits for each water-energy step occurring at time "t" were reduced to the given value by using discount computation with a discrete discount factor.
The "i-th" structure can be written as follows:

$$C_{SG} = \sum_{t=0}^{N_G+N_E} C_{tG} (1 + r)^{-t}$$

$$C_{SE} = \sum_{t=N_G}^{N_G+N_E} C_{tE} (1 + r)^{-t} \tag{6}$$

$$B_{SK} = \sum_{t=N_G}^{N_G+N_E} B_{tE} (1 + r)^{-t}$$

where:

N_G, N_E - number of years of construction and exploitation
C_{SG}, C_{SE}, B_{SK} - present value of construction and exploitation costs and benefits
C_{tG}, C_{tE}, B_{tE} - construction and exploitation costs and benefits for a time increment (Δt) of 1 year

Benefits in the exploitation period include produced energy.

Since the main objective was to select the optimum value of the basic construction parameters (banked-up water level and installed discharge) all costs occurring in the construction and exploitation of the structure were divided into two groups:

- costs depending on the banked-up water level RRP (embankment, spillway etc.)

$$C_{Hi} = f_i (H_i) \tag{7}$$

- costs depending on the installed discharge (water power plant, long distance pipeline)

$$C_{Qi} = f_i (QI) \tag{8}$$

The energy production was calculated using the hydrological input data obtained after 30 years of daily discharge measurements. The discharge duration curve was defined statistically according to these data.

The energy production in the period of system operation was calculated with the time increment DT over the discharge duration curve. The time step was defined when determining the flow frequency curve and the duration curve (Figure 3). The energy production was calculated according to the following formula:

$$E = \sum_i^N \sum_j^T E_{i,j} = 9,81 \cdot QW \cdot HN_i \cdot \gamma \cdot DT_j \tag{9}$$

where:

$\quad$ QW $\quad$ - operation discharge
$\quad$ HN_j $\quad$ - netto head
$\quad$ DT_j $\quad$ - time increment
$\quad$ γ $\quad$ - combined efficiency of the turbine and generator

The computation of the possible energy production is carried out with the following conditions:

$\quad$ $QD \geqslant Q_{MAX}$ $\qquad$ - plant not operating ($QW = 0$)

$\quad$ $Q_{MAX} > QD > QI$ - plant operating with the installed discharge; excess water spills ($QW = QI$)

$\quad$ $QD \leqslant Q_{MIN}$ $\qquad$ - plant not operating ($QW = 0$)

In the period when:

$$QI > QD > Q_{MIN}$$

the plant operation, i.e. production is based on two methods of (operation) production:

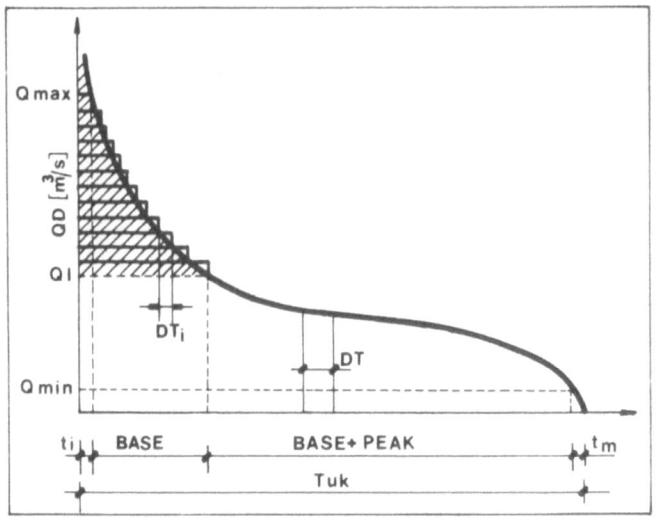

Figure 3. The discharge duration curve

 - peak energy
 - base energy

        The period of peak energy production covers the
maximum of 9 hours a day with two startings during the day. The
rest of the day the HE has base energy production if the
hydrological conditions make it possible.

        In order to define the energy benefits during each time
increment for the given discharge "QD", the water surface along
the system of RRP has to be computed, the total difference
defined and the possible energy production computed.

4.   EXAMPLE

According to the mathematical model employed the DINP
programme was developed and used to treat the problem numerically. Tables A, B and C present the input data.

        Optimum data as presented in Figure 4, show that the
optimum value of the installed discharged is:

$$QI = 1,35 \ Q_{sr} \tag{10}$$

where:

        Q_{sr} - average inflow for the calculation period

TABLE A   GENERAL DATA

	I	II	III	IV
INSTALLED DISCHARGE	$200 \leqslant QI \leqslant 400$ m³/sec			
CONSTRUCTION TIME	0-4	2-6	4-8	6-10
DISCOUNT RATE	10,5%			
PEAK OPERATION TIME	up to 9 hours (with two starting points)			
INTERCALAR INTEREST RATE	7,5%			
DIRECT ANNUAL OPERATION COSTS	5,6% of the invested sum			
INDIRECT ANNUAL COSTS	2,5% of the invested sum			

TABLE B   GEOMETRY AND HYDROLOGICAL DATA

	I	II	III	IV
MINIMUM AND MAXIMUM ANALYSIS LEVEL (m) (elevation)	137-140	147-151	155-158	164-167
DURATION CURVE	Figure 3			
MORPHOLOGICAL DATA	Measured river-bed cross sections in each 1 km			
MANNING'S RESISTANCE COEFFICIENT	0.020-0.033			
USEFUL RESERVOIR (STORAGE) CAPACITY	4.10^6	$3.2.10^6$	$4.0.10^6$	$5.5.10^6$

TABLE C   INVESTMENT COSTS

COSTS OF BANKED-UP WATER LEVEL OPERATION	Figure 5
COSTS OF POWER FUNCTION	Figure 6

Table D presents the optimum banked-up water levels for this QI:

       RRP-1   139,50 m o.s.
       RRP-2   148,00 m o.s.
       RRP-3   156,50 m o.s.
       RRP-4   166,00 m o.s.

System parameters, dimensioned in this way, together with the costs give the maximum value (B/C) ranging from 0,696 and, consequently, it can be concluded that the given energy costs do not ensure efficient operation of the system.

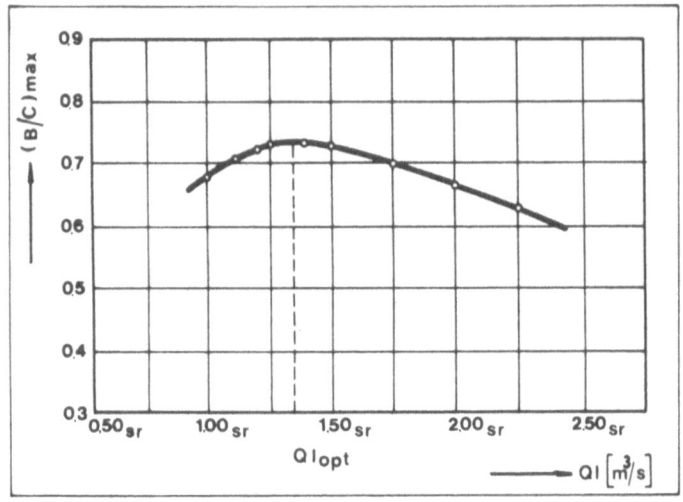

Figure 4.   Optimal value of QI

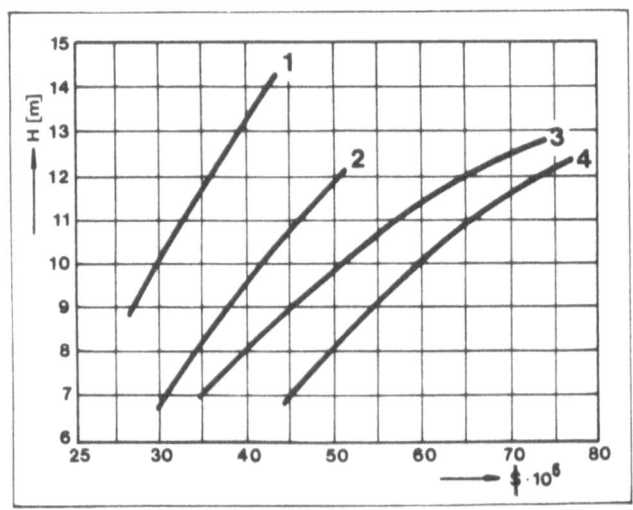

Figure 5. Cost of the banked-up water level for each plant

However, if the computation includes other benefits such as: flood control, recreation, irrigation, etc., then the benefits exceed the costs of the system construction and maintenance. This paper, however, does not include such a computation.

TABLE D   COMPUTATION OF OPTIMUM PROJECT PARAMETERS MAX (B/C)

Lowest Level	Highest Level	Production (GWH) Run-of	Base	Variable	Efficiency Local	Cumulative
R u n - o f - t h e - r i v e r p l a n t (RRP-1)						
132.54	138.00	78.48	35.36	36.42	0.596	0.596
132.54	138.50	85.16	38.38	39.76	0.616	0.616
132.54	139.00	91.84	41.39	43.09	0.628	0.628
132.54	139.50	98.52	44.41	46.42	0.634	0.634
132.54	140.00	105.20	47.43	49.76	0.634	0.634
R u n - o f - t h e - r i v e r p l a n t (RRP-2)						
139.50	147.00	76.41	34.86	35.80	0.690	0.781
139.50	147.50	83.09	37.88	39.13	0.699	0.800
139.50	148.00	89.77	40.89	42.47	0.706	0.809
139.50	148.50	96.45	43.91	45.80	0.709	0.811
139.50	149.00	103.14	46.93	49.13	0.710	0.807
R u n - o f - t h e - r i v e r p l a n t (RRP-3)						
148.50	155.00	78.22	35.07	38.08	0.682	0.624
148.50	155.50	84.90	38.09	41.41	0.691	0.653
148.00	156.00	94.07	42.31	45.53	0.697	0.680
148.00	156.50	100.75	45.33	48.86	0.700	0.689
148.00	157.00	107.43	48.35	52.19	0.700	0.690
R u n - o f - t h e - r i v e r p l a n t (RRP-4)						
156.50	164.50	78.32	35.44	36.24	0.693	0.660
156.50	165.00	85.00	38.46	39.58	0.695	0.672
156.50	165.50	91.68	41.48	42.91	0.696	0.680
156.50	166.00	98.37	44.50	46.24	0.696	0.683
Optimum Parameters for the Energetic System						
RRP-1		98.52	44.41	46.42	0.634	0.634
RRP-2		89.77	40.89	42.47	0.809	0.706
RRP-3		100.75	45.33	48.86	0.689	0.700
RRP-4		98.37	44.50	46.24	0.683	0.696

Banked-up Water Level

139.50 m (el.)
148.00 m (el.)
156.50 m (el.)
166.00 m (el.)

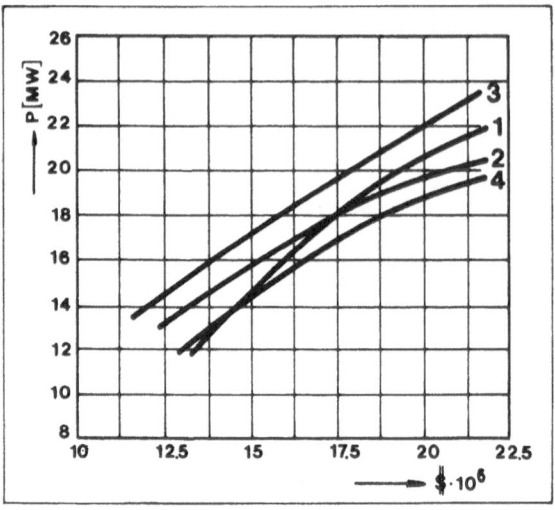

Figure 6. Cost of the power function for each plant

5. CONCLUSION

The presented model of dynamic programming has proved very
useful for the selection of the basic project variables for the
RRP system. This model was used to analyze a series of alternative solutions. The selection of the basic project variables
should be approached this way whenever the economic efficiency
of the plant is questionable. The presented example demonstrated that the economic efficiency of the plant is not satisfactory if only energy benefits are considered without taking into
account the other benefits. The multipurpose exploitation of
such plants justifies their construction. The objective of this
paper was not to discuss the characteristics of the aforementioned problems but to present a model to solve them. The advantage of this model is the possibility to analyze efficiently a
great number of possible alternative solutions using the actual
data obtained by measurements, statistical treatment and other
in situ investigations. Consequently, the obtained values are
real and reliable as they entirely represent the actual conditions as well as the possible technological solutions for the
construction of the analyzed structures.

REFERENCES

1. Haimer, Y.Y. and Nainis, W.S. (1974) Coordination of
   Regional Water Resource Supply and Demand Planning Models.
   Water Resources Research, Vol. 10, No. 6.

2. Jansen, P.PH. (1979) Principles of River Engineering.
   Pitman – London.

3. Murray, N. and Weissbeck, K. (1980) Computer Program for a
   River Hydro Development. Water Power & Dam Construction,
   London, March.

4. Quentin, W.M. (1980) Optimal Operation of Surface Water
   Resources Systems for Water Supply and Hydroelectrical
   Power Generation. Present at the ORSA/TIMS Joint National
   Meeting, Colorado Springs.

THE INFLUENCE OF ENGINEERING SOFTWARE FOR THE DESIGN AND
ANALYSIS OF URBAN STORM DRAINAGE

Roland K Price

Hydraulics Research, Wallingford, Oxon, UK

1. INTRODUCTION

The national capital investment in sewerage has been estimated
as being in excess of £42 billion [4]. With an average number
of sewer collapses of 16/100 km per year and a total length of
sewers the order of 234000 km a large cost is being born by
the nation in maintaining existing sewers. Nationally the
annual expenditure on sewers is of the order of £200m of which
£70m is spent on structural deficiencies and £120m on problems
due to lack of capacity. Because many of our sewers are more
than 50 years old the problem of maintaining existing sewers
is increasing. Consequently considerable research investment
has been made into sewer rehabilitation techniques leading to
the publication of a manual by the Water Research Centre [6].
Two key phases of the procedure recommended in the manual
involve the assessment and remedy of structural and hydraulic
deficiencies in a sewer system. Integral to the hydraulic
assessment is the ability of the engineer to study the
performance of a given system in fulfilling its designed
intention to convey sewage economically and efficiently to its
intended destination. Since most sewers in the UK are
combined, that is, they convey both foul and storm water
flows, the performance of a sewer system is assessed in terms
of its ability both to convey the normal foul flow to
treatment and to discharge excess storm water to receiving
streams. Such an assessment requires the frequencies of
surface flooding in the catchment and pollution incidents in
the receiving streams to be kept to an acceptable level.

The analysis of the performance of a sewer system requires an
appropriate model. For many years the only type of model
available to describe temporal changes in flow depended on the
assumption that the sewage flows with a free surface in the
sewer: when the sewer becomes surcharged any calculation of
the resulting effects has had to be grossly simplified. More

recently, the emphasis on improving the design of storm sewers and on analysing the behaviour of existing sewers has lead to the development of models which can reproduce with acceptable accuracy the effects of surcharging and surface flooding in sewers. The advent of such models has subsequently lead to radical changes in drainage design and, in particular, to the rehabilitation of existing sewers.

The purpose of this paper is to identify the changes that are taking place within the water industry because of the availability of the new models and accompanying software, and to explore the implications of the software for the future.

2. SEWER DESIGN

Most storm sewers throughout the world have been sized using what is known as the rational method or a variant of it (such as the Lloyd-Davies method in the UK). This method is in essence very simple in that the design discharge used to size a sewer is calculated from a simple explicit algebraic formula relating the discharge to a proportion of the contributing area for the runoff and the design rainfall appropriate for the sewer. The sewer is sized to convey the calculated discharge without surcharging. Because the design rainfall value used in the calculation depends on the sewers upstream (but not downstream) the calculations begin at the topmost sewer and progress downstream to the outfall of the system.

With the advent of computers more sophisticated calculations could be introduced into the design procedure, including the temporal flow in the sewers due to a rainfall event. In this case a design rainfall profile (or hyetograph) is used to generate the runoff as discharge hydrographs in the sewers. The sewers are then sized to convey the peak flow, again without surcharging. The most notable method using this approach was developed in the early 1960s by the Transport and Road Research Laboratory [5].

As computers became commercially available more engineering design offices began to use the RRL method as it was called, such that by 1973 75% of design offices were using the method [7].

The key deficiency with all the design methods above is that they assume the sewer is sized to convey the design discharge without surcharging. However a sewer system will generally surcharge at least one or more pipes for rainfall events rarer than the design event. Indeed the primary reason for installing storm sewers in the first place is to limit the frequency of surface flooding to acceptable levels. Storm sewers have therefore been sized for a design rainfall event of, say, one year return period with the implicit assumption that surface flooding will be limited to a frequency of, say, ten years or more. This assumption is generally born out in experience, but the designer does not know ab initio the true frequency of surface flooding from his system. For example, he may have grossly overdesigned his system such that smaller sewers would have given adequate protection to the community. Correspondingly he could have underdesigned the system and induced an unacceptable frequency of flooding at a particular location which can only be relieved by expensive alterations to the system.

Ideally the engineer would like to directly size his sewers to convey a design rainfall event without causing significant surface flooding. Such an approach would require a model which reproduces surcharging in the sewers. Unlike the traditional models above the calculation of surcharging in the sewers and therefore pressurised flow depends not only on flows in sewers upstream but also on flows in sewers downstream. This introduces an extra order difficulty in calculating the flows and therefore in calculating water levels in the manholes and the corresponding sizes of the sewers. When it is also realised that the specification of design runoff is much more complex in this case than with the simple rational method the concept of designing for pressurised flow becomes unmanageable even with powerful computer hardware. Instead an alternative procedure can be devised in which the traditional concept of sizing the sewers for pipe full flow using a design rainfall with small return period is retained, and then the designed system is analysed for rainfall events rarer than the design rainfall. In this way the engineer can assess the performance of his designed system and either make minor modifications to the design to limit significant surface flooding to an acceptable level, or redesign his system for a rainfall event with higher return period.

This approach forms the basis of the Wallingford Procedure for the design and analysis of urban storm sewers [2]. Besides leading to a more objective design of storm sewers to achieve the required service the Procedure also builds on the traditional approach followed by designers. Although the Procedure is therefore radical in that it attempts to design for a frequency of surface flooding rather than just pipe full flow, it nevertheless has continuity with the way design of sewers has been done in the past. This of course is of considerable importance if engineers are to be pursuaded to adopt and implement a new methodology.

A major innovation in the Procedure is the extensive use of computing required to carry out a design. Unlike the traditional approach the complete design of storm sewers using the Procedure has to involve a computer at some stage.

Drainage engineers have not in the past made much use of
computers except for the specialised few who have run the RRL
and similar computer-based methods. The implementation of the
Wallingford Procedure depends partly however on as many
engineers as possible being involved with the calculations.
This is because the phenomena of surcharging in sewers leads
to different effects on water levels and flows in complicated
sewer networks than would previously have been anticipated by
engineering intuition. It is important therefore that
engineers appreciate these effects and their physical basis so
that they can improve designs accordingly. The best way this
can be achieved is by giving each individual engineer access
to the models, not only to carry out a design or simulation
but also to extend their experience of the nature of
surcharging and thereby to improve their understanding of
flows in sewers. In this sense the Procedure, or more
particularly the accompanying software, has an educative as
well as a practical value.

3. DESIGN SOFTWARE

Any calculation of surcharging in sewers and time dependent
pressurised flow is made complicated by the rapid changes in
discharges and water levels in the manholes. Unlike the time
scale for changes under free surface flow in the pipes, which
is the order of a minute or more, the time scale for changes
under pressurised flow, particularly in the transition from
free surface to pressurised flow, is the order of a few
seconds. Consequently the calculation of pressurised flow
requires a time step of typically one second. Additionally
the calculation of levels in the manholes and discharges in
the interconnected group of surcharged pipes is best done by
inverting a matrix at each time step. Considering these
complications and the necessary algorithms to keep track of
the groups of surcharged pipes, the software to simulate flows
in a complex sewer system is itself a major product requiring
powerful computer hardware both to accommodate the size of the
program and to be capable of carrying out the calculations in
a reasonable time. Standard mainframe computers can of course
be used to run such software efficiently. However the
mainframe computers of many organisations such as the UK Water
Authorities are used more for billing and accounts rather than
for engineering design. Consequently some engineers have only
been able to have limited access to existing software and have
not therefore been able to exploit the new Procedure fully.
For this reason a major advance has been made in mounting the
software, called WASSP (Wallingford Storm Sewer Package), onto
16-bit micro-computers. The software, including seven
distinct programs for:-

> data capture and edit
> data checking
> modified rational method
> hydrograph method
> optimising method
> simulation method
> graph plotting

has been mounted on Sirius 1, ACT Apricot and IBM micros [10].
The major program is the simulation method which utilises
almost all of the available size allowance using a standard
linker, namely 384K bytes. The package therefore contains one
of the largest programs written in Fortran and mounted on a
micro. The simulation program can accommodate more than 300
separate sewers in a system and simulate an event up to 8
hours duration. With a maths co-processor one hour of event
simulation in a system of 100 pipes which is extensively
surcharged takes about one hour of processor time. Although
this is a sizeable time in practice, particularly for a large
system, it is entirely feasible within normal design practice
and lengthy simulation runs can be left overnight.

The advantages to the designer of operating with the software
on a micro to which he has immediate access are several. He
can interact with the machine and do a number of designs and
simulations rapidly. He can explore alternatives with
confidence, and build up his experience of how his system
performs. All this depends of course on the software
providing the facilities he needs, and, in particular, being
'user friendly' with extensive help facilities. These
requirements are satisfied by MicroWASSP.

5. SEWER SIMULATION

As with the introduction of any new procedure its implications
are often more far reaching than are conceived by its
originators. The emphasis of the Wallingford Procedure is on
design and the assessment of frequency of surface flooding for
the design. However the current situation in the UK, as in
many developed countries, is that fewer new sewer systems are
being constructed, instead there is mounting concern about the
ageing of extensive existing sewers with higher costs of
maintenance and emergency repair of sewer collapses. Because
the rehabilitation of existing sewers is cheaper than their
renewal considerable attention has been given over the last
decade to improving techniques for relining and strengthening
existing sewers. Relining sewers can reduce their capacity
and hence may make the performance of the system worse. This
may be unacceptable in systems with a history of flooding.
Additionally many old systems have incomplete records, with
unrecorded overflows, unknown connections, and reduced
capacity due to blockages or siltation. Over the years the

sewers may have been required to drain a larger area than they were originally designed for and the spare capacity between the design flow and significant surface flooding has been reduced. Any simulation model, such as incorporated in the Wallingford Procedure, therefore becomes a valuable tool to analyse the performance of an existing system and the effect of changes to that system designed to improve its performance.

The procedure for hydraulic analysis proposed in the Sewerage Rehabilitation Manual published by the Water Research Centre includes the complementary use of flow surveys and a computational hydraulic model such as in WASSP. The purpose of the flow surveys is to verify the model, such as identifying missing overflows and interconnections, and establishing contributing areas and sewer roughnesses. Once verified the model can be used to develop improvements to the system. At this stage a whole range of design changes to the system can be explored including reduction of flows entering the system, attenuation of flows in the system and the impact of flows on receiving streams. In the past many of these changes could only be estimated qualitatively, now the tools exist to predict the changes quantitatively. This can only serve to make alterations to sewer systems more economic and thereby to provide a better service to the community.

6. IMPROVED SOFTWARE

The advent of micro computers has already had a widespread effect on drainage design in the UK, giving engineers ready access to powerful software to explore economic alternatives for the improvement of inadequate sewer systems. However, such access has brought its own attendant difficulties. Despite the user-friendliness of micros and of suitable software most traditional software, including MicroWASSP, is prescriptive in nature. The simulation in MicroWASSPin particular incorporates some sophisticated hydraulic modelling which is generally beyond the training experience of most users. In itself this is not a problem, except that the results from the simulation can in rare circumstances exhibit oscillations or inaccuracies (such as in conservation of volume) which may invalidate the results. In these circumstances the user needs to appreciate why the results were obtained and how to reconstruct the model or change the input data to improve the results. It is because of these difficulties that some hydraulicians argue that such software should not be made generally available except to those who have been specially trained in computational hydraulics. To restrict the use of the software would however militate against the use of the Wallingford Procedure. This therefore raises a dilemma which requires resolution. In that the software has been made generally available some means need to be found of assisting inexperienced and inexpert users to implement the software correctly and responsibly. The obvious means available are manuals and training courses. But much of the expertise to be transferred, particularly in selecting data and interpreting results, is qualitative in nature. Such information may therefore be best built into an intelligent advisor using techniques of artificial intelligence. This advisor would provide an interface between the user and the original software.

Some preliminary work on such an advisor has been done in using PROLOG to implement the modified rational method [3]. This advisor employs the strategy of asking the user a series of questions to establish essential data to evaluate the peak discharge for a given sewer. The advisor can also accept queries such as 'why' and 'how'. The main part of advisor consists of a consultation program forming the interactive core of the system, the dynamic and static databases, and some components required for the second part of the advisor, namely the explanation program. Each consultation is driven with the requirement of evaluating a value for the context type 'qp'(the peak discharge) for a sewer by asking a series of questions which in turn require the evaluation of other context types by asking more questions, or at root taking a stored value of a parameter. The advisor then backtracks to evaluate the original context types in turn. The dynamic database is set up afresh for each consultation and includes

> labdata given by the user for the specific site;

> dynamic data supplied by the user in response to the questions;

> context tree which records the path and values for the context types encountered in the consultation

In contrast the static database is inherent to the system and includes the rules of inference, every context type, the parameters and the values used, information on which questions to ask the user, the type of answers to expect, and the structure of the context tree. These properties enable the advisor to make the correct inferences as each rule is invoked.

This intelligent advisor addresses a simple problem, but nevertheless highlights a possible approach to the development of an appropriate advisor for WASSP. Besides the features identified above the advisor needs to include:-

* advice on collecting relevant data

* treatment of uncertainty in input data

* selection of appropriate models/methods

* operation of selected method(s)

* interpretation of results (including data
  and model uncertainties)

* advice on application of results

The development of such an advisor should enhance still
further the implementation of the Wallingford Procedure and
improve the reliability and efficiency of designs and
improvements to sewerage systems.

ACKNOWLEDGEMENTS

This paper is published with the permission of the Managing
Director of Hydraulics Research Ltd. The preliminary
intelligent advisor was developed by D M Obene under Dr K
Ahmad at the University of Surrey as part fulfilment for the
degree of MSc.

REFERENCES

1. Hydraulics Research (1984) MicroWASSP User Guide, HR,
    Wallingford, Oxon, U.K.
2. National Water Council (1981) Design and Analysis of
    Urban Storm Drainage. The Wallingford Procedure - in five
    volumes. National Water Council, London (available from
    Hydraulics Research, Wallingford, Oxon, U.K.)
3. Obene D M (1984) Expert systems in civil engineering.
    Thesis in part fulfilment of requirement for degree of MSc
    in Structural Engineering of the University of Surrey.
4. Reed E C (1982). The assessment of the problem in the
    U.K. Proc. ICE Int. Conf. on Restoration of Sewerage
    Systems Paper 1, pp 3-8[2], Thomas Telford Ltd, London.
5. Transport and Road Research Laboratory (1963). A guide
    for engineers to the design of storm sewer systems, Road
    Note 35 (second edition 1976).
6. Water Research Centre (1984). Sewerage Rehabilitation
    Manual, WRC.
7. Working Party on the Hydraulic Design of Storm Sewers
    (1976). A review of progress March 1974 - June 1975,
    National Water Council, London.

IRRIGATION NETWORK MODELLING

S. Selvalingam, S.L. Ong and S.Y. Liong

Department of Civil Engineering, National University of Singapore, Singapore

INTRODUCTION

The modern agricultural practices such as the multiple cropping of high yielding varieties and the use of fertilizers depend very much on the irrigated water. Insufficient irrigation supply can drastically reduce the yields while excess irrigation and improper use may lead to waterlogging and excessive accumulation of salts. Therefore a properly designed water distribution system is essential for easy and efficient irrigation. An irrigation distribution system may consist of the canals, structures for distribution of water to various fields and other cross-drainage structures such as culverts, siphons etc. The canals may comprise of main, branch, secondary, tertiary canals and field channels. In such a complex system, the computation of water surface elevations and the corresponding discharges at various locations within the system can be very tedious and time consuming especially when the design calculations have to be repeated for all possible alternatives. But the exercise is an essential component in the design of the irrigation system. Efficient conveyances and structures save labour, land and water, which in turn increases the agricultural production.

A computer software has been developed to calculate progressively the water levels and discharges at locations starting from the field channels to the head works. Numbering of the branches and the different reaches of the canals follow the procedure adapted in the urban runoff model ILLUDAS (Watson, 1981). The program also allows for the calculation of the head losses across canals, regulators, weirs and cross-drainage structures. This mode of the program is useful in the design of the irrigation layout and a separate mode of the software does the computations for an existing irrigation layout. In the latter case, calculations start from the head works and proceed to the field channels. This mode of the program is also useful to check the water surface elevations at various locations when the actual discharge at the head works is less than the designed value.

The software is written in FORTRAN and it could be run on microcomputers as well. The results derived for the several distribution system alternatives would be useful in arriving at an efficient design of the system and good water management practices in irrigated agriculture. Possible applications and the scope for the expansion of the software are discussed along with the limitations.

DISTRIBUTION SYSTEM - ESSENTIAL FEATURES

An irrigation canal takes its supplies from a stream or a river. In order to divert the water into the canal, it is necessary to construct the canal head works and thereby raise the normal water level. This canal intake elevation is to be designed by running mode I of the software. In mode II, this elevation is an input to the program.

Canals and diversion box
Water flows into the main canal from the head works through a regulator which controls the discharge and/or the elevation of the water upstream. Main canal may branch into distributories or secondary canals and/or feed the plots through the field channels. In either case a diversion box structure is usually provided. The water in the upstream canal flows into the box structure and then through regulators or weirs into the downstream canals and field channels. The reach of the canal as used herein is defined between two box structures and this reach would have the same cross-sectional characteristics and longitudinal slope. It may include a regulator just after the upstream box and any number of cross drainage works (Figure 1).

Field channels
Discharge into the field channel is regulated by a weir at the diversion box structure. For example, the type used by the Department of Irrigation in Indonesia is known as the Romijn weir, which also acts both as a regulating and measuring device. The software application need not be limited to this type, any other type of regulators can easily be incorporated. Length of the field channel and its upstream water elevation depends on factors like the extent of plot, slope of the land, distance and elevation of the highest point in the plot and the distance and elevation of the furthest point in the plot.

Cross-drainage works
Whenever a canal crosses a road or other canals or a natural

drainage on its passage, it is necessary to construct masonary
works. Broadly speaking the cross-drainage works may be
classified into the following types; circular or box type of
culverts, troughs carried on beams and slabs over piers and
arches, siphons and aquaducts. Drop or fall structures in a
canal can also be included into this category. The head loss
across these structures would depend upon the typical
dimensional parameters of the structures as well as the
discharge through them.

DESIGN MODE (MODE I)

In this mode of calculation, the water surface elevations and
discharges are progressively computed from the fields to the
head works. A typical network of the irrigation system is
shown in Figure 1. The numbering scheme followed is similar
to that used in the ILLUDAS model (Watson 1981). The
longitudinal section between two box structures would have a
branch number and a reach number, with the most downstream
reach in a branch bearing the number zero for the reach.
Reach numbers are increased in the upstream direction. The
downstream diversion box in a reach would carry the same
identification tag as that reach. Whenever two branches meet
at a junction, one of the branch numbers would continue
upstream and the other branch number would terminate. This
information is given to the program as "continuing branch" and
"ending branch".

Field channels always take off from a diversion box and
hence the number of channels, regulator details, the necessary
information for each field are all read with respect to that
particular diversion box structure. The number of cross-
drainage works, their types, the regulator at the upstream end
of a reach are all given as inputs along with the reach
details such as cross-section, slope, etc.

Water surface elevations

For each field (F_i), the elevation of the highest point (say,
A), its distance from the diversion box (L_1), the elevation of
the furthest point (say, B), its distance from the diversion
box (L_2) are known. Hence the water surface elevation within
the box required to feed this particular field is given by,

$$
\text{El. for } F_i \text{ (in m)} = \text{Max of} \begin{cases} \text{El. of A} + L_1.I + 0.15 + \text{regulator/} \\ \quad \text{weir head loss} \\ \\ \text{El. of B} + L_2.I + 0.15 + \text{regulator/} \\ \quad \text{weir head loss} \end{cases} \quad (1)
$$

I is the given slope of the field channels and 0.15 m is the
additional depth of water necessary for paddy. This is not
required for other crops.

12-26

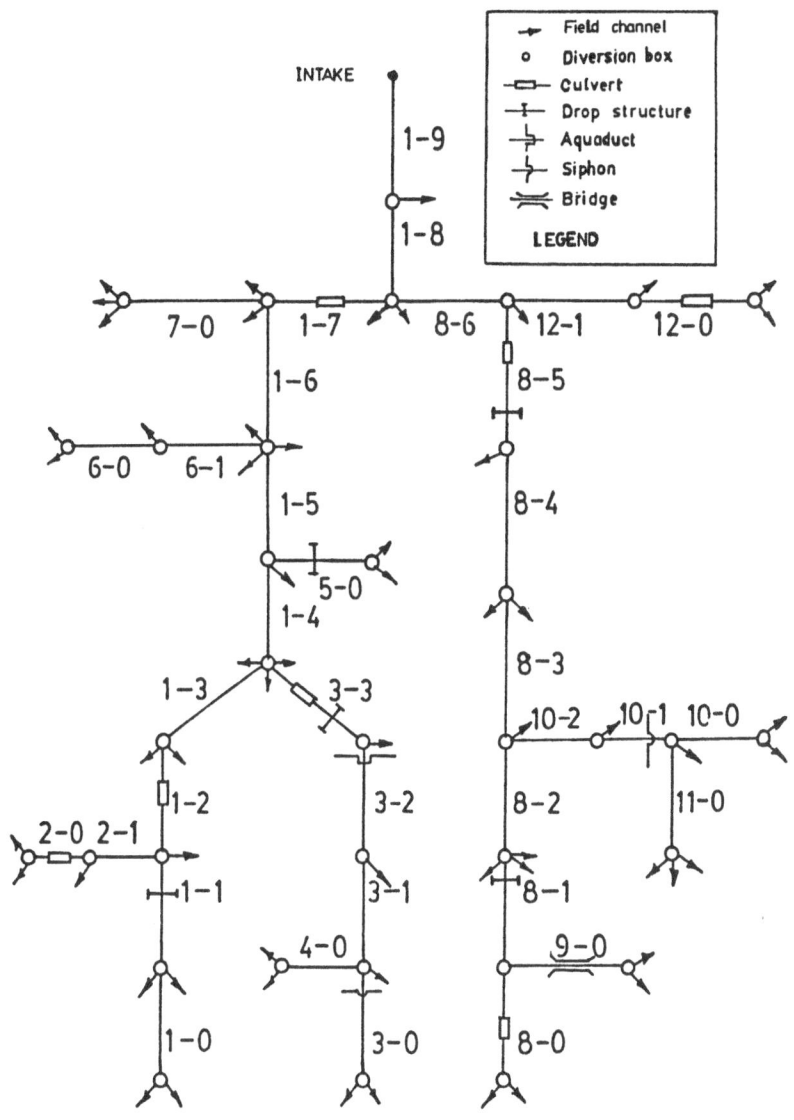

FIGURE 1   NETWORK—IRRIGATION SYSTEM

The discharge to each field in computed by

$$Q = \alpha . A . a \qquad (2)$$

where a is the water requirement of the crop in ℓ/s/ha., A is the area in ha., α is a coefficient which is equal to 1 for crops other than paddy and for paddy, the value comes either from a formula or a table depending on the cultivation season. The regulator or weir for this field channel is designed based on the computed discharge and with the given constraints on the maximum allowable number of openings, widths of weirs and head losses.

The above procedure is repeated for each of the n fields (F_i) taking off from the box. If other canal reaches also take water from this diversion box, then the design water surface elevation for this box structure is given by

$$\text{Eℓ. within box = Max of} \begin{cases} \text{Eℓ. of fields } F_i \ (i = 1, 2, \ldots n) \\ \text{Eℓ. of upstream ends of the reaches} \end{cases}$$

$$(3)$$

In order to determine the water surface elevation at the upstream end of the canal reaches, the design discharge in the canal has to be estimated first. Canal discharge is not calculated using the continuity equation but it is based on the extent of the area supplied. The equation (2) used earlier is employed here, but it should be noted that the coefficient α decreases with increasing areas when obtained from the given formula or table. This is based on the view that all the fields would not be irrigated at the same time of the day and crops in the various fields may be at different stages of growth.

The canal details can be calculated based on the design discharge using the computational options provided. In some instances there are limitations on the allowable velocities, side slopes and on the ratio of based width/depth, while in others there may be limits on the longitudinal slopes. The canal design provides also the head loss in that particular reach. If any cross-drainage works and regulators are required on the canal, they may be based upon the design discharge of the canal. Hence,

Eℓ. upstream end of a reach
= Eℓ. of downstream box + canal head loss + cross-drainage
   head loss + regulator/weir head loss   (4)

The above procedure would be repeated for all upstream reaches and diversion boxes until the water surface elevation for the head works is computed.

Flow chart

Flow chart of the computer program is shown in Figures 2 and 3. Input details for each set of reach and the corresponding downstream box are read one after the other, starting from the most downstream reach (Table 1). Canal and diversion box output details are printed before the program goes ahead with the next set of reach and box. In this manner the number of variable values to be stored are very much minimized.

Table 1 Extracts from branch number system inputs

Design Mode				Supply Mode			
BRANCH	REACH	ENDBR	CONBR	BRANCH	REACH	STABR	CONBR
1	0	0	0	1	9	0	0
1	1	0	0	1	8	8	1
2	0	0	0	8	6	12	8
2	1	2	1	12	1	0	0
1	2	0	0	12	0	0	0
1	3	0	0	8	5	0	0
3	0	0	0	8	4	0	0
4	0	4	3	8	3	10	8
3	1	0	0	10	2	0	0
3	2	0	0	10	1	11	10
3	3	3	1	11	0	0	0
1	4	0	0	10	0	0	0
5	0	5	1	8	2	0	0
1	5	0	0	8	1	9	8
	etc.				etc.		

Each field is fed from the box through the weir, the number of openings and head losses of which are calculated using the subroutines. Other type of outlets can easily be incorporated through additional subroutines. Similarly each type of the cross-drainage works and the regulator on the canal reach may require different input details and additional subroutines. Canal junctions are identified by the continuing and ending branch numbers, "CONBR" and "ENDBR" respectively. At each junction, comparison of elevations in the canals and the accumulation of areas supplied by the canals are carried out. The maximum number of canals set by the program at each junction can easily be increased, if necessary. Typical branch, reach number inputs are shown in Table 1 for some portion of the distribution system of Figure 1.

It is possible to suppress the detailed outputs of all the canals and reaches and print only the details for main canal intake. In the current version, the program does not permit a user to obtain output for a specified canal reach or a box. This is a compromise made so as to reduce the amount of data

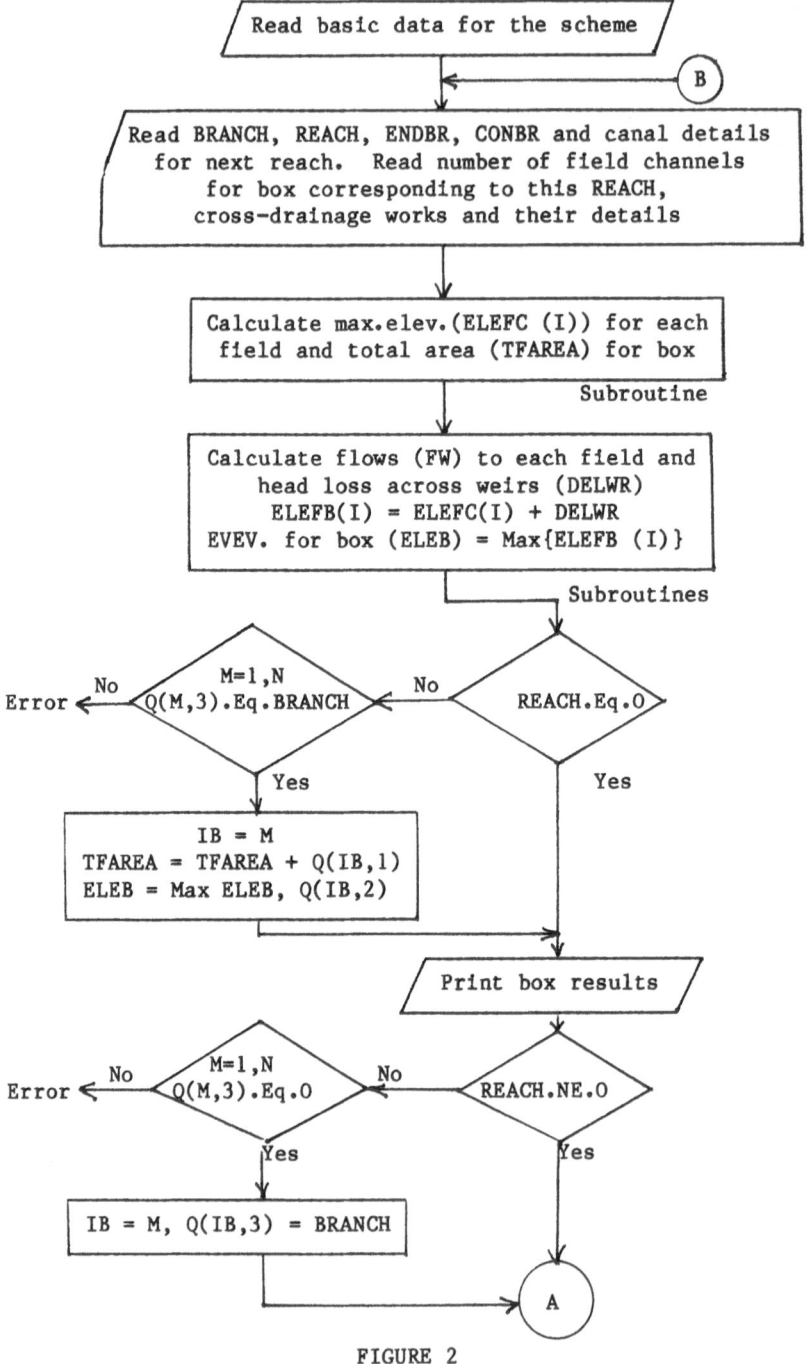

FIGURE 2

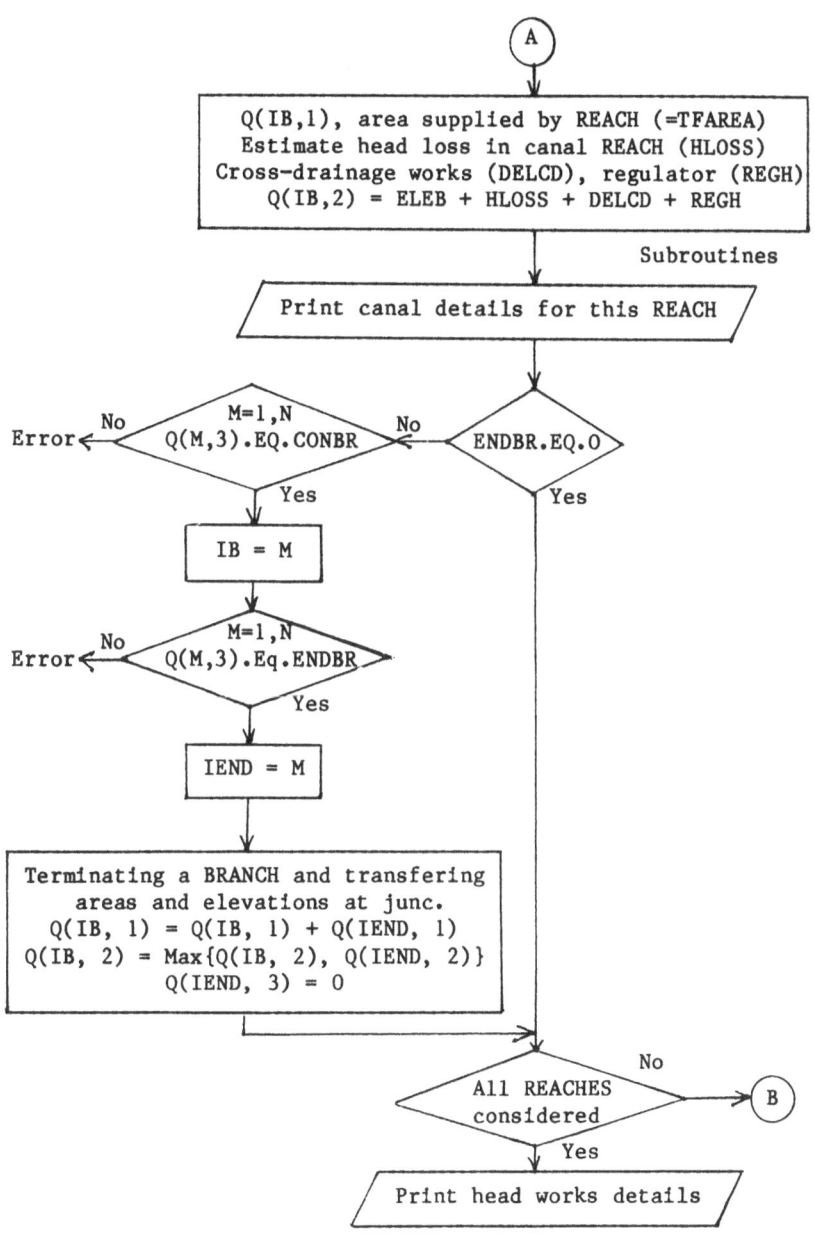

Note: Q(M,1) = Area supplied by reach
        Q(M,2) = El. for the reach
        Q(M,3) = Branch number
        N   = Max possible number of branches at a junc.

FIGURE 3

to be entered by the user. Moreover, it is judged that the two options of output provided should be adequate for most circumstances. However, this restriction can easily be rectified, if necessary.

SUPPLY MODE (MODE II)

When designing an irrigation layout, it may be possible that the necessary design elevations and discharges are not met at the head works at all times of the year. In this case it is desirable to know the impact of the reduced elevation and discharge on the discharge to each of the individual fields. Alternatively an existing old irrigation layout which has been either abandoned or functioning with low supply may have to be brought to full use. The question here again is to study the impact of various intake elevations and discharges on the system. The supply mode of the program is meant to execute this type of study.

Flow chart
Flow charts are shown in Figures 4 and 5, the program structure is similar to the design mode except that the computations have to start from the head works and proceed progressively downstream to the field channels. At the junction boxes, there are branches which originate rather than terminate. Extracts of the branch, reach number inputs for the system of Figure 1 are shown in Table 1.

It is to be noted that the flow continuity equations are not imposed as a constraint at the junction boxes in the design mode. Flows were calculated based on the given formula corresponding to the extent of the area supplied. Similarly in the supply mode, discharges for uniform flow conditions in the canal reaches are computed using the given upstream elevations and the bed slope. Therefore the continuity equations are not always satisfied at the junctions. This means in the actual operation, surface profiles may become non-uniform to satisfy the continuity equation. The run of the supply mode being a feasibility exercise, the impact of the above assumptions may not be serious. If necessary, appropriate subroutines can be modified to accomodate the most imposing constraint.

Head losses in the canals and regulators have to be computed iteratively based on the given uniform slope conditions. It may be necessary to check the given slope against mild, critical or steep slopes corresponding to the discharge. Drop structures are to be suggested if the slope is other than "mild". Cross-drainage losses are computed based on the calculated discharge.

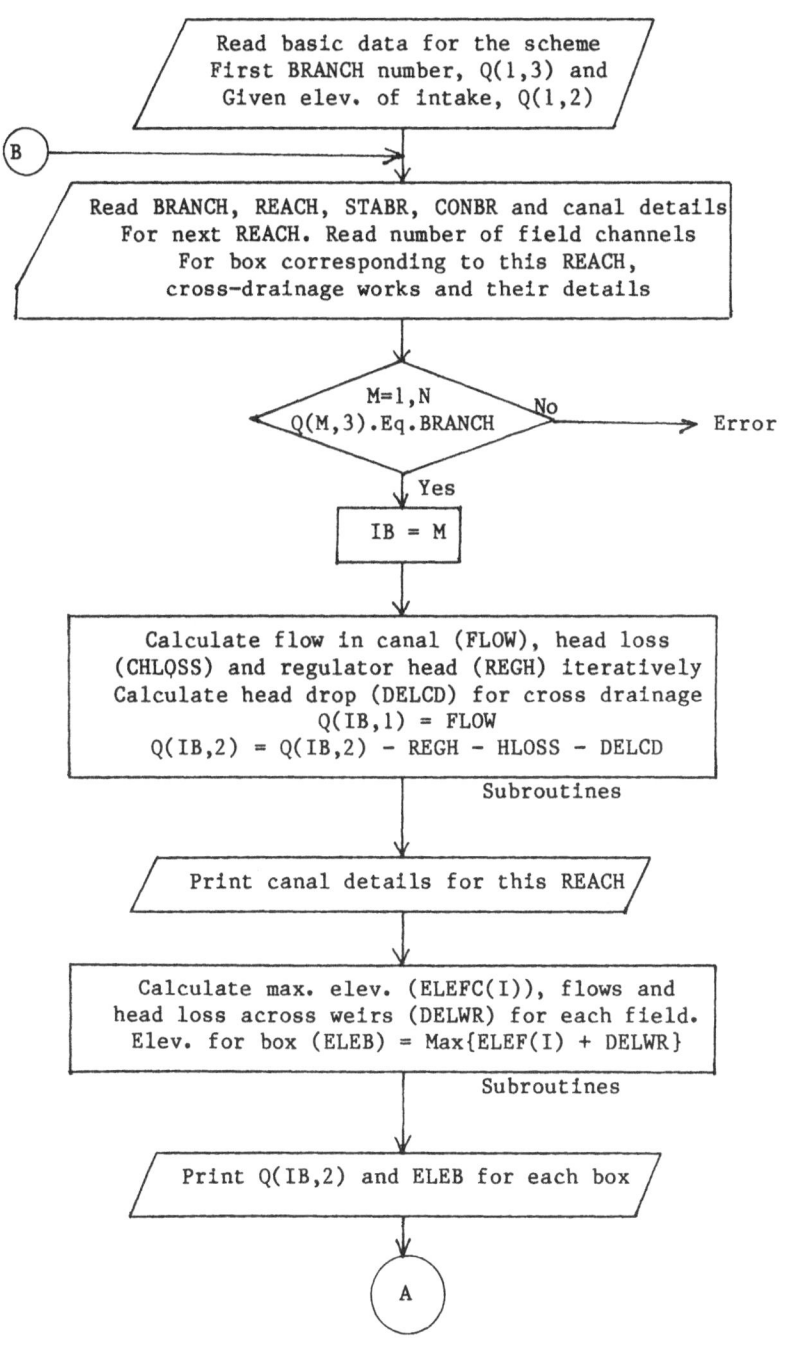

FIGURE 4

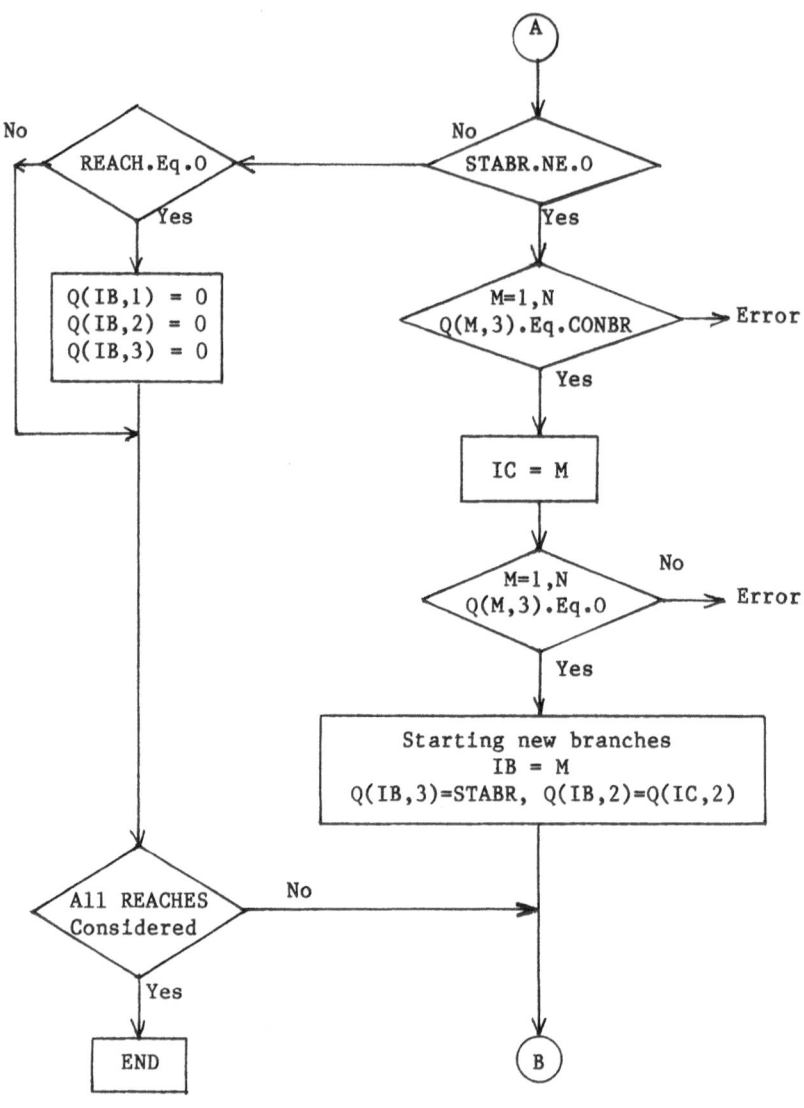

Note: $Q(M,1)$ = Flow in a reach
       $Q(M,2)$ = Eℓ. for the reach
       $Q(M,3)$ = Branch number
       N   = Max possible number of branches at a junc.

FIGURE 5

CONCLUSIONS

Irrigation distribution network modelling software with the
design and supply modes are explained through flow charts.
The former is useful in the design of the entire distribution
system and to estimate the necessary water surface elevation
at the head works to irrigate this system. The supply mode is
useful in exploring the impact of various intake elevations
and study the feasibility of supplying an existing system.
Different types of canal cross-sections, regulators, weirs and
cross-drainage works can be accomodated through additional
subroutines. The program could be run on microcomputers as
well.

ACKNOWLEDGEMENTS

The program described herein was developed for P.T. Dacrea,
Design & Engineering Consultants, Jakarta, Indonesia. Authors
wish to acknowledge Mr Ongko Sutjahjo and Mr Bernard Rasli of
P.T. Dacrea for their assistance in providing the basic
information of the irrigation system and the design procedures
followed in their country. The opportunity provided to the
authors by P.T. Dacrea to work on this project is greatly
appreciated.

REFERENCES

Watson, M.D. (1981) Application of ILLUDAS to Storm Water
Drainage Design in South Africa. Urban Hydrology Series,
Hydrological Research Unit, No. 1/81.

COMPUTERISED FORECAST SYSTEM FOR RIVER FLOWS

J.M. Dujardin, J.L. Rahuel, P. Sauvaget

Engineers, SOGREAH, Grenoble, FRANCE

INTRODUCTION

The aim of this paper is to present the computer forecasting system CFS which was developed by SOGREAH (5) in 1984 for application to hydrological forecasting. Since its conception and several of its algorithms are of a general nature, it could however be applied to other forecasting problems. Moreover, its modular design makes it easy to add new algorithms. The system manages in conversational mode the use of data and of forecasting methods. The advantage of CFS is that it can perform forecasts for a sequence of cells calibrated by different methods, according to a chosen scenario. Certain cells may be brought into play concurrently, according to criteria of quality and availability of information. These forecasts are performed for different times.

The CFS is at present used for real time forecasting of river discharges and levels, on the basis of water level data. Rainfall data from recording stations situated in the upstream catchment area could also be used.

This paper presents the forecasting methods and the architecture of CFS, as well as its application to the forecasting of discharges on a major African river, the Niger.

PRESENTATION OF THE SOFTWARE

The computer forecasting system creates an interface between data processing and various forecasting methods. The CFS processes the available discharge data for each station, automatically prepares the data necessary for model calibration by a chosen method and calculates simple statistics for the data bank in use (number of missing values, common periods of observation for gauging stations, etc). In the case of the Niger river basin, the CFS uses the following forecasting methods:

The CLS method

This is based on the constrained linear system optimisation algorithm (Todini (4)), and is a particular case of Wiener's multichannel input/one output transfer function estimation with physically-based constraints on each input series (fig 1).

Figure 1

The linear transfer function for each cell is represented by:

$$Q(t) = \sum_{j=1}^{P} \alpha j \; Q(t-j) + \sum_{j=1}^{N} hI_j(\tau) \; I_j(t-\tau) + \epsilon(t)$$

where the first sum represents the lagged output and the second is a sample convolution function of all inputs. The weights hIj are estimated with respect to a set of constraints imposed on the inputs (eg continuity and non-negativity conditions) by the least squares method.

The Muskingum-Cunge method (2)

This method simulates flood propagation in rivers when inertia forces can be neglected as compared with the resistance which the river opposes to the flow. It is essentially based on a diffusive non-linear equation:

$$\frac{\partial Q}{\partial T} + a \frac{\partial Q}{\partial X} - b \frac{\partial^2 Q}{\partial X^2} = 0$$

which is discretised according to the classical form of Muskingum's formula:

$$Q(t) = C_1 \; I(t) = C_2 I(t-\Delta t) + C_3 Q(t-\Delta t)$$

where

$Q(t)$ = discharge at the downstream station
$I(t)$ = discharge at the upstream station
  t = computational time step.

$$C1 = \frac{KX - 0.5\Delta t}{K(1-X) + 0.5\Delta t} \; ; \qquad C2 = \frac{KX + 0.5\Delta t}{K(1-X) + 0.5\Delta t}$$

$$C3 = \frac{K(1-X) - 0.5\Delta t}{K(1-X) + 0.5\Delta t}$$

K = propagation time between upstream and downstream station

X = attenuation coefficient expressed in terms of conveyance, celerity and slope of the river.

The originality of the method, as compared to classical Muskingum, is that the coefficients K and X are linked to the physical characteristics of the river.

Both the CLS and Muskingum-Cunge methods are propagation methods, typically applicable to hydrological studies. Both can be associated, in the context of the CFS, with one of the two general statistical approaches: ARIMA and SAPHARI.

The ARIMA
(p,d,q) (AutoRegressive Integrated Moving Average) model (Box and Jenkins (1)) is a univariate time series model given by:

$$\phi(B)\nabla^d Q(t) = \Theta(B)\, a(t)$$

where

$\phi(B)$ is an autoregressive operator of order p

$\Theta(B)$ is a moving average operator of order q

∇^d is a differentiation operator of order d

$Q(t)$ is an observed or transformed time series of discharges
$a(t)$ is a residual series, assumed to be white noise.

The ARIMA model can be considered as the output from a dynamic system, which transforms an input white noise series $a(t)$ into an observed series by two kinds of filters:

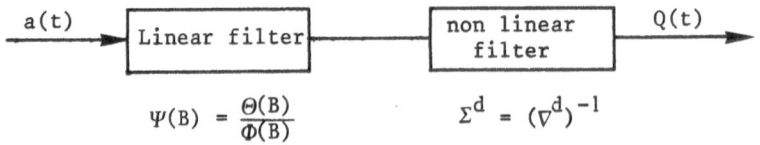

Figure 2

The SAPHARI method
This is a multivariate, multi-lead, seasonal forecasting model (6), which has been developed by SOGREAH. The model's parameters can be varied with time, and it can take into account non-

linearities of the total and partial links. Each forecast produced is associated with a probabilised confidence interval which enables automatic competition between different models in real time.

Each equation (1) is calibrated by range s (ranges of discharge-rainfall or seasonal intervals) and by lead time i.

For a given range s and lead time i:

$$Q(t+i) = a_s^i Q(t) + \sum_{j=1}^{K} b_{s,j}^i \times \nabla_{s,j} X_j (t - \tau_{s,j}^i)$$

$$+ d_s^i + \alpha_s^i T + z^i(t+i)$$

$Q(t+i)$ Discharge to be forecast at lead time i

a_s^i Coefficient depending on range s and lead time i

$Q(t)$ Known discharge

$\displaystyle\sum_{j=1}^{k}$ Index of the K explanatory stations

$b_{s,j}^i$ Coefficient of the variable j for range s and lead time i

$\nabla_{s,j} X_j$ Function of the explanatory variable X_j (first difference of linearisation or indentical function)

$(t - \tau_{s,j}^i)$ Dephasing of the explanation X_j relative to Q

d_s^i Coefficient depending on range s and lead time i

$\alpha_s^i T$ Slow, linear "tendential" process depending on the year T

$z^i(t+i)$ Error (noise) of the general model for the lead time i, which can be modelled on the complete model

One forecasting method may be concatenated with the others in a forecasting scenario. Moreover, the ARIMA method may be used in each of the other methods as a complementary model of their residuals.

The SAPHARI method is also used, in the context of CFS, as a procedure for filling of missing values in real time.

In the case of forecasting of river discharges, one of these methods is used to adapt a forecasting cell for each section on the basis of past data. The adjustment parameters are stored in a so-called "calibrated models bank". During the forecasting operation, in accordance with the scenario provided by the operator, the CFS will progressively select from memory these parameters for each of the cells used. Forecasting will thus be effected from upstream to downstream, the forecast produced for an upstream cell being used by a downstream cell.

The advantage of this system is that the best adapted forecasting method can be chosen for each section of the river.

Forecasting is effected in real time. This means that each time the data required for forecasting is updated, a new forecast can be produced by the CFS. The number of forecasts produced thus depends on the frequency of data revision. This frequency is independent of the CFS, but related to the frequency of data transmission.

The CFS has to use a specific interface enabling it to read, evaluate and store this data. The data is stored in a buffer file pending updating of the historical data bank. This updating will be periodic (for example, annual for daily dischar-ges). After this operation, it is possible to effect a further calibration of the forecasting cells, which will be improved on account of the more numerous data.

ARCHITECTURE OF THE COMPUTERISED FORECASTING SYSTEM (CFS)

Schematically, the main tasks and files of the CFS may be represented as in fig. 3.

In order to achieve its final objective (real-time forecasting), the CFS must be capable of executing several tasks, following instructions from an operator. The main tasks correspond to the four operational modes of the CFS:

The data interrogation mode provides access to a data bank of past observations (eg discharges) in order to obtain the series of observations necessary for the calibration (adjustment) of model parameters.

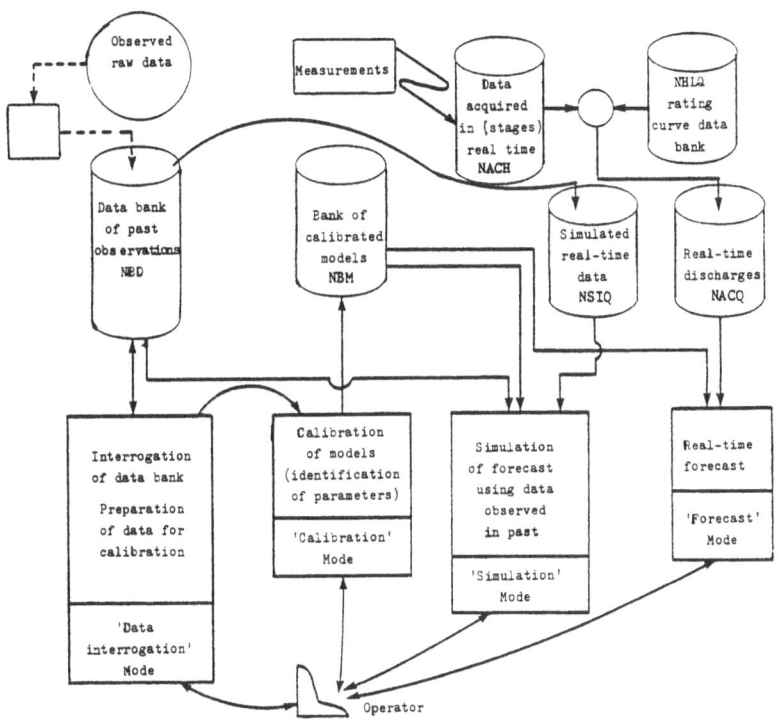

Figure 3

The **calibration** mode is used to calibrate model parameters by different selected methods (eg CLS, Muskingum, SAPHARI, ARIMA) and to file the calibrated models in the model bank. This mode also provides access to the model bank to carry out the following tasks: printing, destruction of outdated files, ordering of various items in the model bank. An item corresponds to a forecasting cell (or a model).

The **simulation** mode simulates forecasts by using past observations. For example, using the observations concerning a former flood (that is, using the data bank in order to create a file of real-time observations), the CFS activates the forecasting cells with the preset updating times and forecasting periods. With this mode it is possible to judge the quality of the models and forecast results on the basis of data which did not participate in the calibration process.

The _forecast_ mode provides for the forecast itself by the concatenation of different forecasting cells using data acquired in real time and stored in a special file. It provides forecasts in printed or graphic form.

Architectural principles of CFS Certain sub-routines constitute switches towards other sub-
routines which are the extremities of the arborescence. The switches are established by conversation with the program user. This structure is such that, in the example of figure 4, it is only possible to go from terminations Cl to C4 by passing through the switches Bl and Al and then B2.

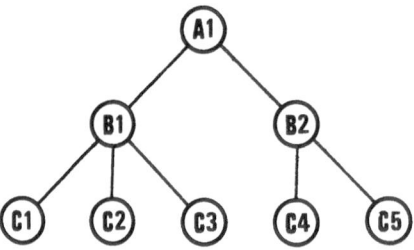

Figure 4

A certain number of checks and messages are provided for the user in interactive mode to help him make decisions.

The arborescent structure of the CFS facilitates the sectioning (overlay) of the program during its execution.

In addition, in preparation for a smaller memory on certain computers, the CFS has adopted the dynamic allocation method. As far as possible, the size of the subroutines is limited by using utility subroutines.

At the same time, with the prospect of installation of the software on small computers requiring overlay, the various branches were divided into sub-branches, so that the longest branch does not occupy too much memory capacity.

The various branches address common files which serve as interfaces between these branches. Thus the calibrated models bank,

built up in the calibration branch, is called up during operation of the CFS in the forecasting simulation and forecasting
branches.

The language used is Fortran 77, for which the norms were res-
pected, so as to be able to adapt the software for use on different types of computer (in particular, extensive use was made
of the CHARACTER statement and the terms IF - THEN - ELSE -
ENDIF).

The programme is conversational. At each important switch a menu is proposed to the user. For the operating files, in which
only a few variables are likely to be altered from one test to
another, a semi- conversational presentation was adopted. The
user has to fill in the zones of a grid formated in advance, in
which the meaning of each of these variables appears.

Particular care was taken in the presentation of the print-
outs, by adding lines of comments, by grouping all formats together at the end of the subroutines, and by systematically in-
denting all lines to the right within the loops. At the same
time, all numbers of logical units or table dimensions are expressed in the name of a variable, thus facilitating comprehension of the print-outs and modifications.

The names of the subroutines and variables have been chosen so
that the user can readily identify what they are referring to.

A user's manual has been produced, setting out the methods used
and the architecture of the system, detailing the structure of
the files and their contents, and presenting the general organisation of the software. This document is completed by detailed operating instructions.

In each of the program's subroutines, tests are performed to
check the quality of the data or the replies to the proposed
menus. If the reply to a menu is not correct, the menu is displayed again (returning to a higher hierarchical level). If an
error is detected at the level of a calculation method, a message is displayed to point out and explain this error.

The software runs to a total of 13 000 instructions.

Peripheral units - Files
- 1 terminal, 80 characters
- 1 printer, 132 characters
- 1 Benson tracing table (optional).

The files used are of two types: sequential disk files and
non-formatted direct access disk files.

The sequential files are used in semi-conversational mode, for entering the calibration parameters of a method, for development of a forecasting scenario (one line per forecasting cell), or for storing forecast standard deviations related to a scenario. The non-formated direct access files have been used for each of the software banks, on account of their flexibility as regards formating rules and the rapidity of access to the required information, whatever the size of the bank.

Each of the banks is preceded by a catalogue which groups together the essential parameters of the bank and each of its units, and indicates the connection number of each unit.

The CFS uses six non-formated direct access files:

- the historic data bank (discharges) NBD, which is used to generate the data series and to adjust statistically the model parameters,

- the bank of height-discharge laws NLHQ, which groups the parameters of the calibration curves of the water level gauging network,

- the calibrated models bank NBM used for storage of all the parameters describing a forecasting cell,

- the bank of water level data acquired in real time NACH, which is updated each time a forecasting operation is started, and which contains all the gauge heights trans- mitted in real time, with a sufficient time lag,

- the bank of discharge data acquired in real time, obtained by transforming the heights under NACH by the height- discharge laws NLHQ,

- the central forecasting file NPRE used to store the forecasting results for each of the stations and for each of the processed scenarios.

Depending on the user's choice, outputs are either displayed on the screen or printed out. The formats and presentation of tables have been designed with particular care.

The graphic outputs also appear according to the user's choice on the console (graphic output over 80 characters) or on the print-out (graphic output over 132 characters).

If a Benson tracing table is also available, it is possible, by means of an external drawing utility, to provide for graphic outputs on the plotter.

The CFS software is designed for maximum operating flexibility. Its conversational character and the ludic aspect of its presentation greatly facilitate operator training, the time required to learn how to use the system being reduced to the minimum. Furthermore, with the automation of certain tasks and the use of interfaces, much time can be saved during the forecasting cell calibration phase. It is of course advisable, as for all software, that the user be initiated in the architecture of the system and especially in the structure of the input-output files. To this effect, he must know the Fortran 77 language and have basic knowledge of data processing. However, it is above all important for the user to have attained a reasonable level of instruction in hydrology, so as to have a critical eye open on the results obtained at all stages of operation of the CFS.

USE OF THE CFS FOR FORECASTING DISCHARGES OF THE NIGER

The Niger is a major African river, more than 4000 km long. It rises in Guinea, in the Fouta Djalon mountains, situated in a wet, tropical zone, then describes a vast arc, passing through the arid zones of Mali and Niger, before falling into the Atlantic in Nigeria. The Niger basin extends over several African countries which have grouped together in a Niger Basin Authority (Autorité du Bassin du Niger, ABN), to ensure harmonious management of the common resource represented by the river. However, it is not possible to manage a system efficiently without having forecasts. Thus the ABN and the World Meteorological Organisation (WMO) set up a vast cooperative project aimed at producing hydrological forecasts in real time, for various gauging stations situated on the Niger and its tributaries. In view of the vast extent of the basin and the poor communications, the water level recording stations of the network were equipped with Argos beacons. These send daily water level readings by satellite to an international forecasting centre situated in the middle of the basin at Niamey, capital of Niger. This data is checked before processing by the CFS software in order to produce real time forecasts of water levels and discharges. The forecast results obtained by the CFS are sent by satellite to the national forecasting centres of each of the member countries of the ABN (fig. 5).

The historical data bank groups together the discharges for about 100 stations distributed throughout the basin. Of this total, 48 stations equipped with Argos beacons are at present used for forecasting by the CFS.

The calibrated models bank at present contains 123 forecasting cells.

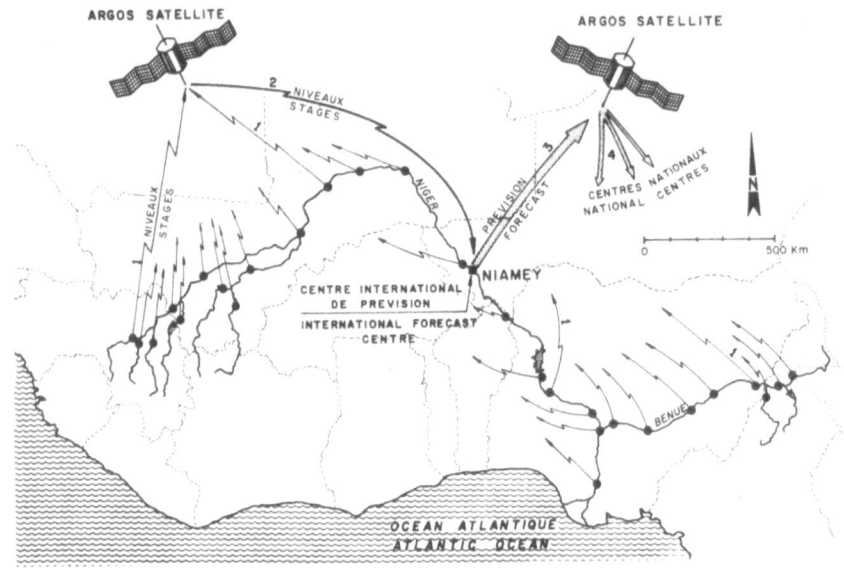

Figure 5

NIGER BASIN - GENERAL DIAGRAM OF
THE HYDROLOGICAL FORECASTING SYSTEM

1   Water levels transmitted from the beacons to the ARGOS
     satellite

2   The satellite stores then transmits this data to the in-
     ternational forecasting centre at Niamey

3   The forecasts are produced by the CFS and sent back to the
     satellite

4   The satellite transmits these forecasts to each of the na-
     tional forecasting centres

●   Water level recording stations equipped with ARGOS beacons

TRAINING OF PERSONNEL AND TRANSFER OF SOFTWARE TO NIAMEY

In order to train the African engineers who will operate the
system, several training courses were organised. The trainees
were selected for these courses by the various Hydrological
Departments of the ABN member states.

There were nine trainees in all, one from Benin, two from
Cameroon, two from Guinea, one from the Ivory Coast, one from
Mali and two from Nigeria. Since the Nigerians are English-
speaking, all the courses were given in French and English.

A first recruitment course was organised in Niamey by the WMO officer seconded to the project. The selected trainees then came to France for a six-week training session with SOGREAH. Their training was finally completed in Niamey during installation of the software, for a period of three weeks. During the first session, general training was given in the fields of data processing and hydrology. During the subsequent sessions, the trainees were initiated specifically in the hydrological methods used in the CFS, in the architecture of the software and in the application of the system to the concrete case of the Niger basin.

The facility of application of the CFS software, thanks to its conversational character, enabled it to be used as a teaching aid during the courses given for the African trainees. The hydrological methods and structure of the software, which were the subjects of the courses, were thus illustrated didactically on the console, making it much easier for the trainees to assimilate these methods.

The CFS was developed on an IBM 43-31 at SOGREAH's computer centre. This is a 32 bit machine with a main frame memory storage of one million octets. However, since the CFS is a real-time forecasting facility, it had to be installed on a computer in Niamey, that is the Digital PDP 11-60 at the Centre Aghrymet. This is a 16 bit small computers of 256 Koctets capacity, but with only 64 Koctets of addressable memory. The main difficulties which had to be overcome in transferring the software are related to this limited main memory storage.

It was essential to solve these problems before leaving for Niamey. Thus SOGREAH's engineers went several times to the Digital offices in Lyons to work in an environment comparable to that of the Centre Aghrymet in Niamey. The problems were solved as follows:

a)   by using a very ramified overlay, each branch of which is kept sufficiently short, so as to respect the limitation on use of memory storage, that is a maximum of 56 Koctets for each branch,

b)   by creating five main modules for the CFS, selected by the "SPAWN" control which is specific to the PDP system,

c)   by limiting to the strict minimum the simultaneous opening of buffers relating to external files,

d)   by using to the maximum the 256 Koctets of main memory. To this effect, a facility specific to the PDP 11-60 was used, that is the statement VIRTUAL instead of DIMENSION. Thus for tables it is possible to use memory storage outside the 64 Koctets of addressable memory. However, the use of this instruction implied a substantial limitation on the FORTRAN 77 norm. Each of the routines thus had to be adapted to this new situation relative to the norm.

All the files were transferred to Niamey on tape. Previously, the non- formated direct access files had to be transformed into formated sequential files, while retaining the maximum of significant numbers. All the files coded in EBCDIC (IBM) then had to be transformed into ASCII files (Digital). Once these problems had been solved in Grenoble and Lyons, it was possible to install the software in Niamey.

The Aghrymet computer centre is particularly well maintained and managed by competent personnel. The PDP 11-60 computer is backed up by a second machine in case of failure of the first. It is used on a shared time basis by the various operators. It took a week to store the files on disks, compile the FORTRAN files, reconstitute the non-formated direct access files, create the modules and carry out tests on each of the branches, with correction of a few errors of detail. The tests showed that the running times of the CFS are very short. Finally, additional training was given to the future users on site.

CONCLUSION

The CFS is a large, high-performance software product in the forecasting field. It has been successfully applied to real-time forecasting of the floods of the Niger. In this context, it is an essential link in a whole forecasting chain which makes use of the most modern techniques. Despite its large size, it was possible to install the CFS on a micro-computer in Niamey, where it has proved its efficiency and its rapidity of execution. The care taken in design of the software and its conversational character make this a readily transferrable facility, which is accompanied by a very detailed user's manual.

ACKNOWLEDGMENTS

The development of the CFS was made, for the specific purpose of Niger River, under contract with WMO (HYDRONIGER Project) within cooperation frame of World Meteorological Organisation - Niger Basin Authority. The authors acknowledge the role of WMO - ABN in elaboration and implementation of CFS.

REFERENCES

(1) Box, G.P., Jenkins, G.M (1976) Time Series Analysis, Forecasting and Control, Holden-Day.

(2) Cunge, J.A. (1969) On the subject of a flood propagation computation method (MUSKINGUM method), Journal of Hydr. Research, ASCE, n° 2.

(3) Salas, J.D. et al. (1980) Applied modeling of hydrologic time series', WRP.

(4) Todini, E., Wallis, J.R. (1977) Using CLS for daily or longer period rainfall-runoff modelling, in mathematical models for surface hydrology, J. Wiley & Sons, NY, 1977, pp. 149-168.

(5) Cunge, J.A., Chené, J.M. Dujardin, J.M., Erlich, M., Rahuel, J.L, Sauvaget, P. (1984) Computer forecasting system of discharges for Niger River basin - User's Manual - SOGREAH.

(6) Chéné, J.M. (1985) Modèle de prévision stochastique SAPHARI - Application au fleuve Niger et à la rivière l'Isle à Périgueux (SAPHARI stochastic forecasting model - Application to the river Niger and to the Isle at Périgueux), SOGREAH.

(7) Curry, K.D., Bras, R.L., (1985) Multivariate Seasonal time series forecast with application to adaptative control, MIT R.253.

MATHEMATICAL MODELLING SOFTWARE FOR RIVER MANAGEMENT : 'CARIMA'
AND 'CONDOR' SYSTEMS

Ph. Belleudy, J.A. Cunge, J.L. Rahuel

Engineers, Applied Mathematics Dept., SOGREAH, Grenoble, FRANCE

INTRODUCTION

Engineers have always needed predictive simulation tools to
assess the consequences of hydraulic projects before their
implementation. Industrialisation added a new dimension to
engineering problems: that of the transport and fate of pollutants in rivers and, more generally, water quality problems.
Physical reality is extremely complex and engineers face a
whole spectrum of river sizes: from small urban or rural
streams to water courses several thousands kilometres in
length.

Figs. 1a and 1b show two examples of the kind of differences in
scale of such phenomena. Local flow simulation and faithful representation of details is vital to assess the results of flood
protection of the inundated Adour River valley in France (Fig.
1a). A grasp of essential problems in the face of an enormous
geographical area, volume of handled data, and various hydrological regimes along the river are the main points for managers
of the water resources of the Niger River basin (Fig. 1b).

The way in which the consequences of our actions is assessed
has evolved over many generations from intuitive comprehension
of situations, through rough calculations, reduced scale models
to mathematical models. With the constant development of tools
(computers, i.e. hardware as well as the methods, that is to
say computational hydraulics) it has been possible recently to
lay out ambitions design objectives for river simulation systems, namely:

(i) The system must be capable of dealing with a model in
which both inertial channel flow and non-inertial flood plain
flow and storage are represented with no fundamental restrictions on the ways the two flow regimes may be linked. The
pollutant (or, more generally, dissolved matter) transported

must be properly simulated, including reactions between different pollutants transported, their decay, etc.

(ii) The user must be able to schematise the real situation
and build this model without having to worry about its topological make-up and calculation sequence. Multi-connected networks of streams or flow exchanges must be accepted. The user
must be able to modify the original layout (by introducing
proposed structures, new reaches, transforming an arborescent
network of streams and channels into a multiconnected looped
one etc.) without difficulty.

(iii) The system must provide a high level of automatic data
checks and convenient procedure for correcting and modifying
model parameters. The system must also equate needs and computing resources: one does not need the same core memory or even
the same computer for a model of small dimensions (say 200
computational points) and for a large model of 2000 to 3000
points.

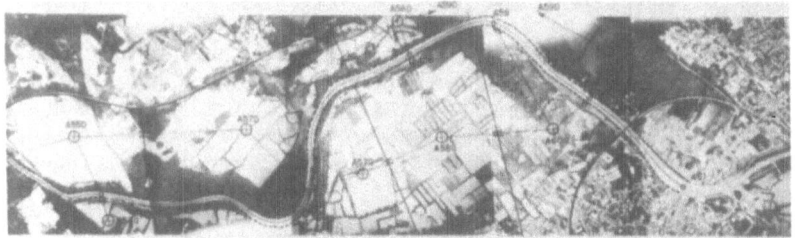

Fig. 1a   Part of the Adour River valley (inundated)

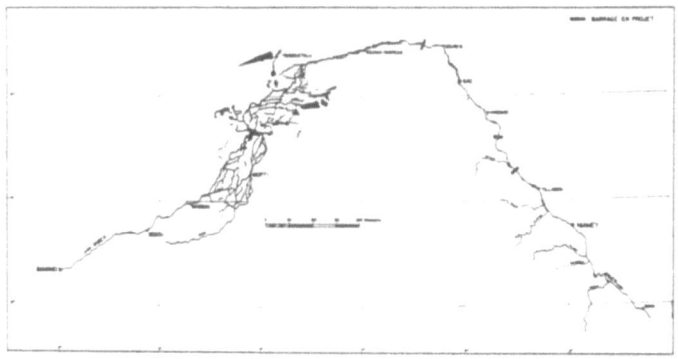

Fig. 1b   Niger River basin

The purpose of the paper is to present the software developed
for management of water flow and transport of matter in net-

works of rivers or man-made channels according to the above-mentioned requirements. The software, developed by SOGREAH, consists of two distinct but completely compatible modelling systems (or codes):

CARIMA (CAlcul RIvières MAillées = Computation for Multiconnected River Networks) is a modelling system which enables simulation of unsteady flow (e.g. flood wave propagation, water transportation for irrigation or town supply purposes, drainage) in rivers and open channels.

CONDOR (CONvection - Diffusion and/Or Reaction) is a system which simulates unsteady convective-diffusive transport of matter (e.g. salinity, pollutants, heat, suspended fine sediments) undergoing decay and reactions, in rivers and open channels.

ENGINEERING PROBLEMS AND FORMULATION

Both systems are simulation tools. CARIMA is used to study the effects of natural or man-made modifications in river/flood plain features, or to study the effectiveness of water conveyance canal networks. CONDOR simulates the behaviour, i.e. variations essentially in time and space of the concentration of different substances transported by water. The substances may have a natural origin (e.g. salt, suspended silt) or be the result of human activities, e.g. thermal and organic (sewerage) pollution, toxic pollutants, etc. Simulation of events observed in the past (in order to validate models) and of the consequences of projected modifications makes it possible to choose optimum solutions, discard dangerous or insignificant designs, and predict the changes in water quality, etc.

CARIMA formulation There are two general flow regimes treated by CARIMA: one-dimensional channel flow (1-D), in which the full flow equations are considered, and two-dimensional flood plain flow (2-D) for which non-inertial, resistance-dominated flow equations are used.

One-dimensional flow is modelled by a series of so-called 'computational points' along the river, each of which corresponds to a measured or assumed cross-section. These points are linked by one-dimensional computational reaches over which the average section properties at either end are assumed to govern the flow. These reaches may form a looped structure, for example in the case of flow division around a large island, a river cut-off, or interconnected delta channels; Fig. 2a is an example of a one-dimensional looped network.

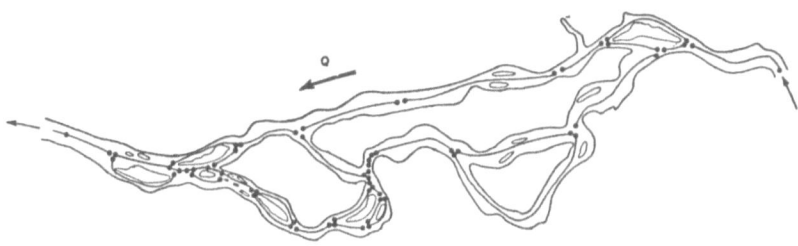

Fig. 2a Topological scheme of a model
for navigation studies

The equations of Barré de Saint-Venant for one-dimensional unsteady river flow are written as follows:

$$\frac{\partial y}{\partial t} + \frac{1}{B}\frac{\partial Q}{\partial x} = 0; \qquad \frac{\partial Q}{\partial t} + \frac{\partial}{\partial x}(u^2 A) + gA\frac{\partial y}{\partial x} + S_f = 0 \qquad (1)$$

where:

$B(y,t)$: channel width

$A(y,t)$: cross-sectional area

g : gravitational acceleration

$y(x,t)$: water surface elevation

$Q(x,t)$: discharge

x : longitudinal axis (independent variable)

t : time (independent variable)

S_f : 'friction slope', expressing the resistance opposed by
      the river bed to the flow

$u(x,t)$: Q/A = mean flow velocity in the section

Two-dimensional flow on the flood plain is modelled as a series
of interconnecting cells, for each of which the relation between storage volume and water surface elevation is known or
assumed. Cell emplacement and links between cells are chosen so
as to follow natural features as closely as possible (roads,
dikes, etc.). Fig. 2b is an example of a two-dimensional flood
plain zone adjacent to a river; the arrows indicate possible
flow paths between cells and between cells and the river.

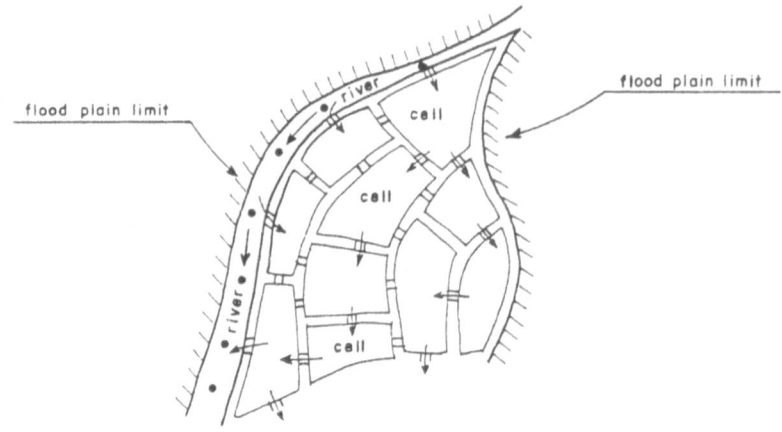

Fig. 2b   Topological scheme of a model having both riverine
and flood plain flow

Two-dimensional flow calculation is based on continuity of
volume for each cell and non-inertial flow laws between cells.

Continuity of volume j in a flood plain cell j is expressed as:

$$\frac{\partial V_j}{\partial t} = \sum_{i=1}^{n} Q_{ij} \tag{2}$$

where:

$V_j = V_j(y_j,t)$: volume of water in cell j
$Q_{i,j}(t)$: discharge from cell i to cell j
n : number of cells i communicating with cell j

Discharge $Q_{i,j}$ is function of water levels in two adjacent
cells: $Q_{i,j} = f(y_i, y_j)$. These functions represent in CARIMA
basic hydraulic features such as: weirs, sills, short channels,
ori- fices, flood gates, hydroelectric plants, etc.

The system of non-linear partial differential equations (1),
(2) is solved numerically using an efficient implicit finite
difference scheme. The numerical characteristics and solution
algorithm were published elsewhere (Cunge et al. 1980). The
results obtained, as a function of time, are: water stages y,
discharges Q and mean velocities u at all computational points,
cells and links.

CONDOR formulation
The CONDOR system is built around numerical solution of the
unsteady transport equations in one space dimension for N
pollutants:

$$\frac{\partial C_r}{\partial t} + u \frac{\partial C_r}{\partial x} - \frac{1}{A} \frac{\partial}{\partial x} \left(AD \frac{\partial C_r}{\partial x} \right) = \sum_{=1}^{n} (a_{rp} C_p) + b_r \tag{3}$$

where:

C_r (x,t): concentration of the pollutant r, considered uniform
in the river cross-section

D(u,y,x): longitudinal diffusivity, encompassing turbulent and
cross-sectional differential convection effects

a_{rp}, b_r : coefficients defining reactions between pollutants,
their decay and possible sources and sinks.

To solve Eq. 3 in terms of C_r(x,t) one needs the flow variables
u(x,t), y(x,t) which are furnished by the CARIMA system. These
variables are computed while satisfying the flow continuity
equation, hence Eq. 3 in concentrations is also conservative as
far as pollutant masses are concerned. The system of Eq. 3 is
solved for all computational points and cells by a numerical
method based on splitting operators: the advection part of Eq.
3 is solved by the method of characteristics with hermitian
interpolation almost completely eliminating spurious numerical
damping (cf. Holly and Preissmann, 1977); the diffusion part is
solved by an implicit finite difference method which leads to a
double sweep algorithm; the reactions and decay part is solved
by local implicit finite differences. The algorithm is described by Sauvaget, 1985 and Belleudy and Sauvaget, 1985.

ARCHITECTURE OF THE SYSTEMS

Link between CARIMA and CONDOR
Fig. 3 showns a schematised link between the two systems. For
example, CARIMA may be used alone if pollution transport is not
to be simulated. The use of the CONDOR system alone, however,
is difficult since it requires a coherent set of flow characteristics such as velocities, depths and discharges, which must
satisfy the flow continuity equation. These are automatically
furnished by CARIMA, but are difficult to obtain from, say,
field measurements.

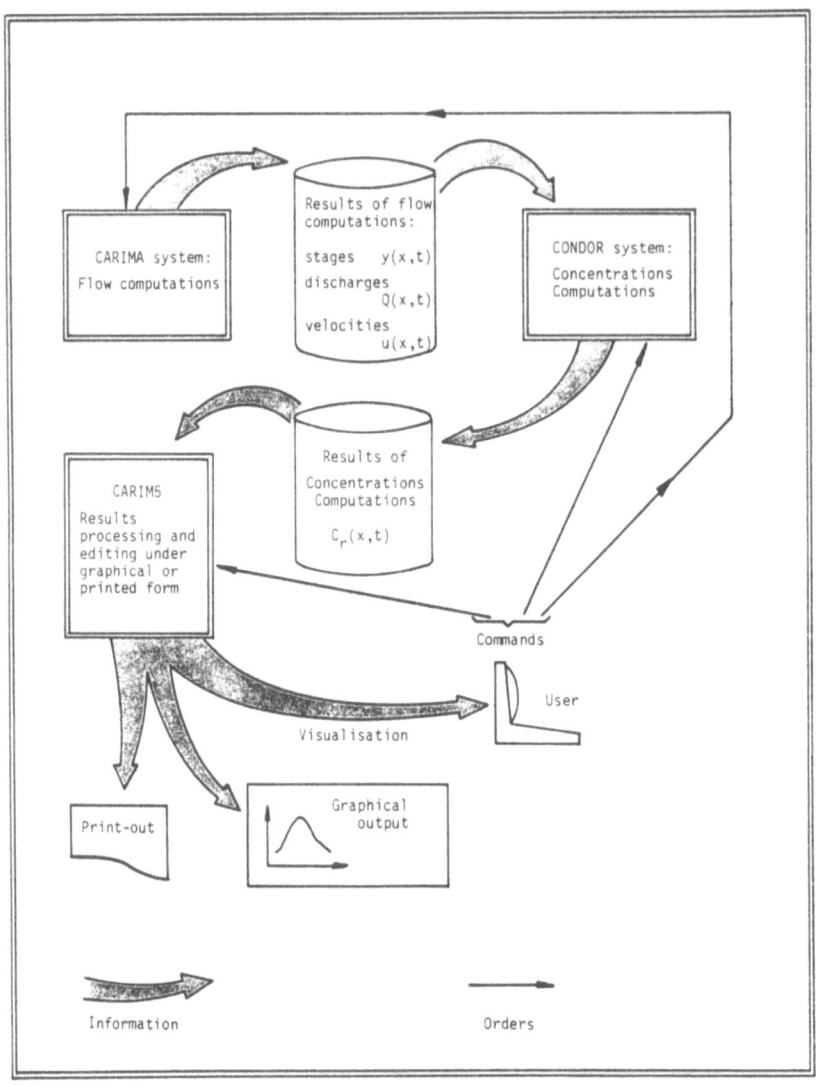

Fig. 3   CARIMA and CONDOR: schematised link
between the two systems

Structure of CARIMA system

The general programme flow chart is shown in Fig. 4. It is
adapted to the way in which a modeller proceeds in his work.
The user must first assemble the topographical data necessary
to define the river system, including maps, cross-sectional
areas, serial photographs, and data necessary to define any
structures in the model (weir crest elevations, flood gate

12-56

dimensions, etc.). To these data must be added information on inflow hydrographs, downstream water levels, rating curves, lateral inflows and outflows.

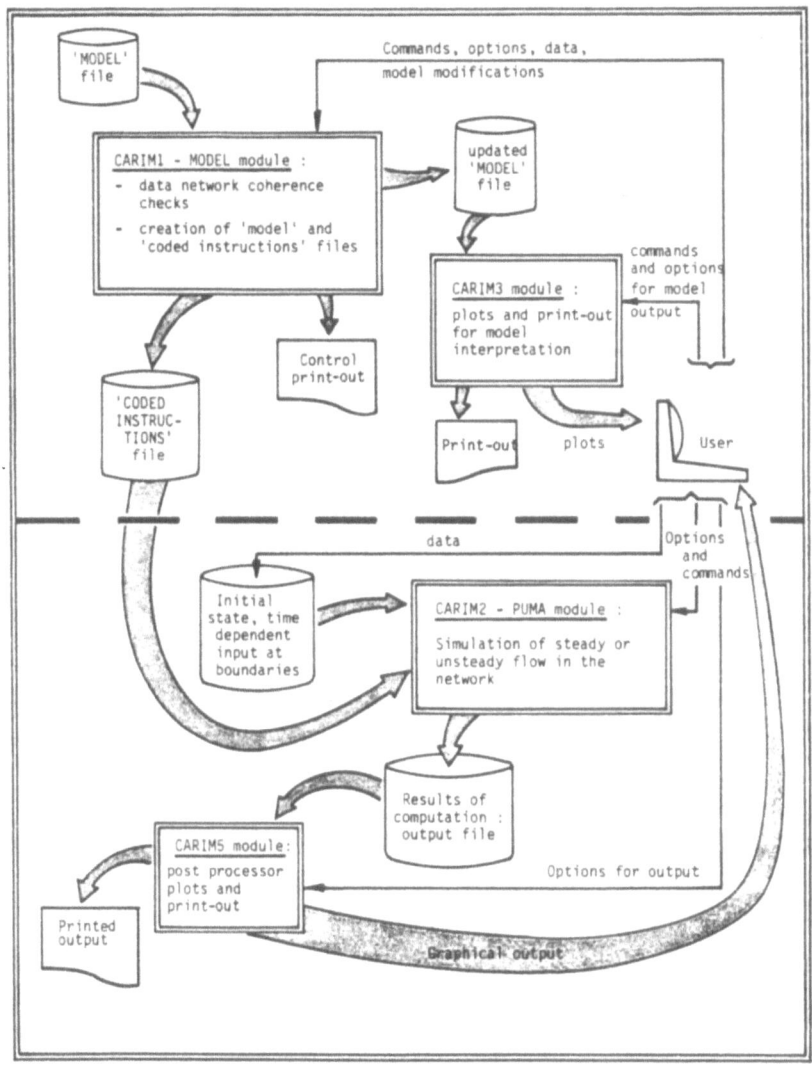

Fig. 4   CARIMA system: general flow chart

Having collected all the data the user <u>defines</u> his model by furnishing input information which establishes the topological links and physical data for each hydraulic element of the network. In Fig. 5 is shown an example of a possible real life situation and its schematisation.

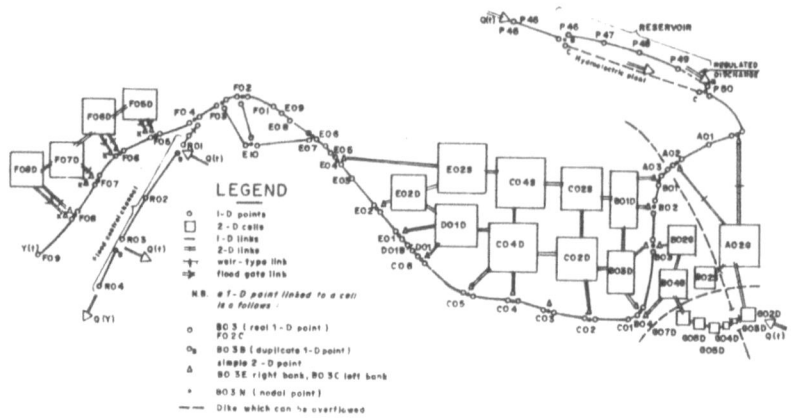

Fig. 5   Adour River: topological discretisation

The programme CARIM1-MODEL (cf. Fig. 4) processes all this information, checks for data and topology inconsistencies, establishes the calculation order and creates two output files:

- updated model file, which contains the detailed description
of the model, i.e. the schematised situation. This data is in
the form generally used by hydraulics specialists, who can
easily understand how to modify it in accordance with their
needs;

- a condensed data file containing the physical and logical
data necessary for the calculation, the so-called (cf. Fig. 4)
'coded instructions', which are not to be read by users but
only by the programme.

The function of the MODEL programme is to build and to update
the models. No flow computations are made, but the result (the
updated model file) may be edited at will by the user on the
screen or by printing or in graphic form. Editing is carried
out with the CARIM3 module (cf. Fig. 4). Thus all checks can be
made at this stage.

The second stage is based on the use of the CARIM2-PUMA programme, which carries out the unsteady or steady flow computations actually desired using the 'coded instruction' file and
initial and time-varying boundary conditions furnished by the
user.

MODEL and PUMA are two modules of the same system and their
consecutive or separated use is decided by options taken by the

user at the time the system is run. E.g., if the model is already built and only boundary condition input (flood hydrograph) varies, the user would indicate the PUMA option only, thus saving his own and the computer's time needed to recreate a 'coded instruction' file from engineer's data (such as cross-sections, etc.).

The 'Coded instruction' principle enables optimum use of computer resources without bothering the user. Each 'coded instruction' is composed with a number of items of information, or 'words', defining an element of the model: its hydraulic characteristics, all the data it needs, all operations which should be made by PUMA to process it. Coded instructions are stored sequentially in the optimum order for processing hydraulic elements. As the result, the highest efficiency of computational means is obtained. The coded instruction file may be stored in the core memory or on the outside, direct access or magnetic tape file. Thus, the user can trade-off computation speed against memory requirements in accordance with his model size or simulation time. The best order for processing different elements depends upon the numerical method used in PUMA and is very different depending on what is considered easy or handy by the user. The MODEL module frees the user completely from these considerations. The price, however, is the computer time needed to create an optimised coded instruction file. Hence the separation between MODEL and PUMA which makes it possible to avoid running MODEL except when modifications of the river model itself are needed.

Structure of CONDOR system

The CONDOR system is structured in an almost totally analogous way to CARIMA. It makes use of two CARIMA files: 'Coded instructions', in order to define compatible topology and topographic data and 'Results of computations', because it needs depth, velocities and discharges. For lack of space the CONDOR flow chart is omitted here, CARIMA being a sufficient example.

Characteristics of CARIMA and CONDOR systems

Both systems are programmed in FORTRAN IV. Several subroutines were also doubled in on IBM 370 series ASSEMBLER in order to increase processing speed. The CARIMA system (without CARIM5 postprocessor) represents more than 14,000 FORTRAN statements. The MODEL module occupies around 273 K bytes (100 K with an overlay structure) of the central core memory, the PUMA module about 352 K bytes (respectively 75 K); (space needed for working files, data tables, etc. is not included in these numbers). The CONDOR system (without CARIM5) consists of more than 8,000 FORTRAN statements and occupies about 325 K bytes of core memory. Both systems are being currently run on IBM 4331 and a larger IBM-equivalent computer (corresponding to IBM 3032/33).

INPUT-OUTPUT FACILITIES

Input is file-oriented using full screen facilities. All data are introduced under the form familiar to hydraulics specialists, e.g.: cross-sectional profiles by a sequence of (abscissa, ordinate) or (elevation, width) couples, roughness of natural cross-sections by Strickler-Manning coefficients, etc. All computational points can be named mnemotechnically (4 characters are available). Once the model file is built, it is edited under a double form: detailed print-out and graphic output, which enables the hydraulic soundness of the model to be checked immediately. The outputs are accompanied by warnings or control messages concerning the model, e.g., it is not possible to adopt absurd roughness coefficients or to have conveyance decreasing for an open channel, etc.

Output is taken care of by a special separated module CARIM5. All computational results are stored on the output file and CARIM5 is a real postprocessor furnishing, at will and according to the user's option: detailed printed output; $y(t)$, $Q(t)$, $v(t)$, $dy/dt.(t)$, $dv/dt.(t)$, $C_r(t)$ plots for any computational points; free surface lines $y(x)$ for any reach; concentration profiles $C_r(x)$; min-max water stage envelopes for any given reach. All these outputs come on the screen and are then plotted, if so desired, in print-ready form.

MAN-SOFTWARE INTERACTION

Both systems are designed to be used by hydraulics engineers who do not know the algorithmic details or the way the systems are programmed. Nevertheless, the users must have a good understanding of the hydraulic and hydrodynamic hypotheses on which all modular elements of the software are based, e.g., an engineer who builds a mathematical model of a river valley with CARIMA may have to choose, for a particular reach, between a rough canal, an orifice, a sill or an overspill weir. Or again, he has to choose between an accumulating polder-like element or a river reach of high roughness. It is his responsibility to choose. Hence he has to know with what hypotheses each feature was simulated in CARIMA or CONDOR, but it is not necessary to know how it was translated in FORTRAN. This concept is very different from what is sometimes called user-friendliness in engineering software, which what might result in leaving the user in total dependence upon the software. He has no choice, but follows the hydraulics defined for him by somebody else who might not be even a hydraulics specialist ! Such unfortunate situations arise most often when the conversational software designed for massive commercialisation on micro-computers is used. In the CARIMA-CONDOR system concept, the software must not replace the hydraulics engineer. This concept, however, implies that the user must have double liberty:

(i) to be able, for a given model, to modify the initial schematisation rapidly and have a choice of several hydraulic solutions for simulation of a given situation;

(ii) to be able to complete the system with elements which do
not exist yet in it, e.g. to add new types of gates, to introduce an inclined sill while originally only a horizontal one
was provided, to introduce a new type of decay of a pollutant
(e.g. function of the depth and the velocity).

We have shown above how possibility (i) was provided. Both systems were developed in such a way that possibility (ii) can be
provided with ease, but of course, not by the user. New elements are constantly being added to the system and average
period required for one such addition is a week. It must be
stressed, however, that the process of such addition involves
three consecutive phases and requires people who know how the
software is built, and who therefore have different qualifications from these of average user. The three phases are:

a) hydraulic analysis and formulation,

b) formulation in the algorithmic form required by the general
   CARIMA or CONDOR algorithm,

c) implementation in the programme.

SOFTWARE PORTABILITY, EXPLOITATION EXPERIENCE

Because of the modular concept and sequential coded instruction
file processing both systems are, theoretically at least, easily portable the computers which have a minimum memory capacity
of 352 K bytes and accept FORTRAN. Actually there are at least
two reasons why the portability may be seriously limited:

The first is linked to the fact that what is called 'standard'
FORTRAN is not standard at all. The CARIMA/CONDOR system have
been implemented on several IBM systems, but one can imagine
that the same difficulties as mentioned by Dujardin, Rahuel and
al. (this conference) would appear with any other system. The
second is of a broader nature and is related to the remarks
made in the MAN-SOFTWARE INTERACTION section. Quite obviously,
the CARIMA/CONDOR system should be implemented only where a
satisfactory hydraulic environment and computational hydraulic
back-up are guaranteed.

Excellent documentation and a propicious environment within
SOGREAH have made the system an everyday tool, used by more
than 30 engineers. Since its creation in 1977 several hundreds
of models have been built and run, 90% of them by hydraulics
engineers who know next to nothing about programming and data
processing. We have gone so far as to print a special abbrevia-
ted user's manual to enable rapid construction of simple models

while using only a fraction of the system's possibilities. However, all the users are hydraulics engineers, most often with experience of laboratory modelling as well. The back-up is ensured by a team of computational hydraulics specialists. Care is taken that two of these, who know all the details of the programme, be present at any given time in head office, so that new addition may be implemented.

ACKNOWLEDGEMENTS

Both CARIMA and CONDOR were developed by SOGREAH's Applied Mathematics Department. The development was guided by Alexandre Preissmann, whose ideas, basic and detailed, are acknowledged. The bulk of work concerning CARIMA was carried out by Forrest M. Holly, Jr. while most of the effort behind CONDOR was made by P. Sauvaget and Ph. Belleudy.

REFERENCES

Belleudy, Ph. and Sauvaget, P. (1985) A New System for Modelling Pollutant Transport in River and Canal Networks. IAHR, 21st Congress, Melbourne, Australia.

Cunge, J.A., Holly, F.M., Jr. and Verwey, A. (1980) Practical Aspects of Computational River Hydraulics. PITMAN, London.

Dujardin, J.M., Rahuel, J.L. and Sauvaget P. (1985) Computerised Forecast System for River Flows. ENGSOFT 85, London.

Holly, F.M., Jr. and Preissmann, A. (1977) Accurate Calculation of Transport in Two Dimensions. Journal of Hydr. Div., ASCE, vol. 103, no. HY3, Nov. 1977.

Sauvaget, P. (1985) Numerical Computation of Dispersion. Dispersion in Rivers and Coastal Waters. Chapter 3 of Developments in Hydraulic Engineering, edited by P. Novak.

OPERATIONAL CONTROL OF A WATER DISTRIBUTION SYSTEM
UTILIZING MICRO COMPUTERS

George W. Lackowitz

Consultant Engineer, Yorktown Hgts., N.Y. U.S.A.

ABSTRACT

Micro Computer technology provides a low cost
solution to the complex problems encountered in
efficiently operating a water distribution system.
This paper describes how micro computer technology
can be used to:

- develop a operating plan for system pumps and
  automatic control valves.

- control the operation of a system's pumps and
  control valves.

- monitor and evaluate the operation of system
  components.

This paper also discusses how micro computer
technology has been used, or is being considered for
use to perform these functions in operating the water
system of the City of Yonkers, New York. The Yonkers
water system provides an average of 30 million
gallons per day (mgd) to the City's 200,000 residents
utilizing four (4) high lift pump stations. At the
present time all the pumps in the system are operated
manually, thus system personnel must be sent to each
of the remote pump stations several times a day to
turn pumps on and off.

As part of a Water System Master Plan Study performed
for the City, a micro computer based extended period
water distribution model was developed. The model
was utilized to develop control set points for each
of the system's pumps. Implementing these set-points
will minimize energy costs while maximizing the

pumps's benefit to the system. The City is presently investigating the feasibility of utilizing a micro computer based control system to implement the pump control plan.

DESCRIPTION OF THE EXISTING WATER DISTRIBUTION SYSTEM

The Yonkers Water Distribution System supplies 200,000 residents with an average of 30 million gallons per day (mgd). Due to the significant elevation differences within the City, the water system, is divided into high and low pressure districts. The low pressure districts receive their flow by gravity from upland reservoirs owned by the City of New York. Water service to the high pressure zones is provided through four (4) pump stations scattered throughout the City. Three (3) of the pump stations are unmanned with the fourth, the treatment plant pump station, providing a manned control center for the water system.

Existing Water System Controls
Each of the pump stations have a local flow recorder with a 24 hour recording chart. At two of the pump stations, Hillview and Crisfield, station output is telemetered back to the control center where it is also recorded on 24 hour charts. In addition, the liquid level in the two water storage tanks is telemetered to the control center where they are recorded on 24 hour charts.

At the present time the operation of the system is completely manual. As a result, an operator must be sent to a pump station each time a pump is turned on or off. This mode of operation has the following inadequacies:

- Since it may take an operator as long as 30 minutes to get from the control center to a remote pump station, the system cannot respond quickly to emergency conditions such as the need for extra flow during a fire.

- Utilizing personnel simply to turn pumps on and off is an inefficient use of manpower.

- Existing operating procedures tend to increase energy costs.

- Existing equipment provides the system data in such a form that it is difficult to analyze for underlining trends.

Water System Master Plan Study
To identify existing problems within the system, the Yonkers City Wide Master Plan was performed. In addition to determining solutions to the identified structural problems within the system, the study also examined water system operating procedures. Among the analyses performed utilizing the micro computer based model were:

- Determining the effect of new water mains on the availability of fire flow.

- Determining the optimum elevation for a new water storage tank.

- Determining the effect of closing a water main upon the system without having to physically send a crew to close the valves. This provides a means of identifying the potential for low pressure and discoloration (due to reversal of flow) when the line is closed.

Both the structural reinforcement analysis and system operating analysis were performed utilizing a micro computer based extended period water distribution model.

UTILIZATION OF MICRO COMPUTER EQUIPMENT TO DEVELOP A LEAST ENERGY COST OPERATING SYSTEM

As has been previously stated, the Master Plan Study included the development of a micro computer based extended period water distribution model. The ability to utilize a micro computer to perform extended period water distribution modeling for a system of this size is a recent event made possible by the increased speed of micro computers and the increased efficiency of FORTRAN compilers. The computer program used for this project was originally developed at an American University in 1970 for use on an I.B.M 360 computer. With some minor modifications to the original FORTRAN code, this program now runs on a MS-DOS based micro computer utilizing an INTEL 8088 processor. The current version of the program requires 384K of memory and can be used on models with up to 550 pipes. The major practical constraint to even larger models is the amount of time that such simulations would take.

The use of micro computers for these numeric intensive "number-crunching" applications was not feasible until the development of the 8087 NUMERICAL COPROCESSOR and FORTRAN compilers which allowed a program to dimension greater than 64K of data. For a typical hydraulic calculation performed on a 8086/8088 based micro, the 8087 NUMERIC COPROCESSOR can decrease the run time by over one-half. Obviously the faster the time of computation, the larger the model which can be simulated practically on the micro computer. The ability to dimension more than 64K in a program, a capability compilers marketed before 1983 did not possess, makes it possible to download mainframe hydraulic programs such as a water model onto a micro computer.

Description of the Yonkers Extended Period Model
Hydraulic modeling for the City of Yonkers was performed utilizing a model with 525 pipes, 3 elevated water storage tanks (reinforced conditions) 11 pumps (reinforced conditions) and 2 pressure regulating valves (reinforced conditions). The extended period feature of the model allows it to simulate hydraulic conditions over the entire 24 hour diurnal pattern. An extended period model is equivalent to a motion picture, in that many "hydraulic snapshots" of the water system are put together to simulate the changes over time in water system pressure, storage tank liquid level, and pump station output.

Utilizing this model we were able to simulate for both existing pipeline conditions and future pipeline conditions:

- The effect of operating a particular pump upon water system pressure.

- The effect of alternative pump set points controlled by the liquid level in an adjacent water tank upon water system pressure.

The goal of this analysis is to develop an operating plan which minimizes electrical energy costs while maintaining adequate water pressure throughout the City. The City is billed on the basis of two measurements of electrical consumption:

(1) Energy usage in kilowatt hours.

(2) Peak power demand in kilowatts.

In addition the City is charged a delivery fee which is related to the kilowatt demand. Electrical usage charges are direct charges for the volume of electric power consumed. An operating procedure to reduce the usage charge maximizes the use of the system's highest efficiency pumps and minimizes the hydraulic lift needed. Demand charges are based upon the maximum amount of power consumed in a short period of time (e.g., thirty minutes) during the billing period. Therefore, an operation's procedure to reduce demand charges will attempt to maintain power consumption at a constant rate.

Utilizing a trial and error solution, model simulations were performed with various pump control set points to determine the set points which achieved the goal of least energy cost.

IMPLEMENTATION OF THE RECOMMENDED PLAN

Due to the complexity of the operating plan, implementation of the plan requires the installation of an automatic pump controller system. At this point in time two types of automatic control systems are being considered for the City of Yonkers. These types are:

- Non-Intelligent Mechanically Based Operating System.

- Micro computer based Operating System.

Both systems would make it possible to automatically operate the Yonkers water system. In addition under either system, the pre-programmed control plan could be overridden from the control center from where pumps could be turned on or off. But, the micro computer based operating system has several significant advantages over a non-intelligent mechanical system.

A micro computer based operating system allows the engineer to incorporate complex operating scenarios into the system. As an example a plan can be incorporated which automatically rotates a station's lead and lag pumps. Thus each pump can get equal run time unless it is decided to run one pump more than the others. In that case, the program can be modified to achieve the desired distribution of run time for each of the pumps.

Thus, over a week's time, the micro compute will record flow and pressure data for each of the station's pumps. This data can be automatically compared to previously observed readings to determine if the pump's performance is deteriorating. A deterioration in pump performance can be an indicator that the equipment is in need of maintenance. Thus each pump is continuously monitored without the need for operator intervention.

Finally, the micro can automatically produce daily, weekly, monthly and yearly system reports. In a complex water system the laborious task of record keeping and report generation can require a significant amount of manpower. Since most of the data necessary to generate these reports has been automatically input to the computer, the system requires little if any additional manual data input to generate the reports. Thus, a major source of error in computer generated reports, the manual input of data, is eliminated.

Description Of Mechanical Systems

The system which has been traditionally used to control a water system are of the mechanical type. These systems multiplex numerous tones signals over a single telephone line. Signals can convey both digital information such as a relay switch setting and analog data such as pump station flow and water tank liquid levels. Pumps will be turned on and off by mechanical level switches located in the water tank chart recorder enclosure box. The typical configuration would be to have the water tank level transmitted to the remote pump station where, the pump controllers would be located. In this configuration, the system would continue to function even if the communications link to the central control station malfunctions. The central control station in this configuration would receive over a single telephone line the status of pumps, pump station flow, pressure and water tank liquid level. An alarm for power failure and communications line malfunctions would also be enunciated at the central control station. The pre-set pump controls could also be overridden using manual controls located at the central control station.

The system would maintain pump station records on 24 hour paper charts. Analysis of data from these charts is typically a time consuming and labor intensive process.

Description Of A Micro Based System

A Micro computer based system typically has a masterstation micro computer with a micro computer installed at each of the remote pump stations. The remotely located micro computers maintain in their memory the operational programs for the pump station. Therefore, if the communications link to the master

station should be interrupted, the system would
continue to operate. The remote computer also has
the capability to redial the master computer when the
communications link is lost. The remote micro is
enclosed in an industrial grade cabinet. Within the
cabinet the remote micro typically has removable
memory boards which allow for the future expansion of
the system.

The master station can be an off the shelf micro
computer or a custom designed micro. In either case
the master station has the following functions:

- In emergency conditions to allow manual
  override of the pre-set operating plan.

- Display visual and audio alarms.

- Compile the data collected by the remote
  computers.

- Analyze the data to determine if a pump is
  operating at its optimum flow and pressure.

- Maintain preventative maintenance schedules
  for the water system. Also, analyze actual
  pump station flow and pressure to determine
  pumps with decreased performance that will
  require maintenance inspection.

- Maintain an inventory of consumable chemicals
  and spare parts.

- Produce daily, weekly, monthly and yearly
  reports summarizing the data collected by the
  remote computers. Analytical results can be
  provided in both graphical and tabular form.

Cost Comparisons
A comparison of cost estimates provided by vendors
for both systems reveals that they have nearly equal
capital costs. This can be attributed to the
following factors:

- The sharp decrease in cost of micro computer
  power which has occurred over the last few
  years.

- The development by each manufacturer of a
  control software which can be inexpensively
  customized to each application.

- The high cost of manufacturing precision mechanical equipment.

Since the micro computer based systems can perform all the functions of the mechanical systems plus provide system analysis and record keeping at the same cost as a mechanical system, the micro computer based system is the most cost effective.

RECOMMENDATIONS FOR CONTROL SYSTEM SPECIFICATION DOCUMENT

Both mechanical and micro based control systems are complex unions of electrical and mechanical components. Both systems rely on relays to open and close contacts which start and stop pumps. The micro based systems rely on the proper operation of computer chips and software for dependable operation. With mechanical systems delicate mechanical equipment must function properly for the system to operate correctly. In either case, initial system debugging will be necessary to achieve a properly operating system.

The specification document should be geared toward describing the performance which will be expected of the equipment with payment made upon achievement of that performance. In summary, a specification document for a control system should include the following components:

- Exact identification of what the system will be expected to monitor and control.

- Identification of the records the system will produce.

- A payment schedule linked to performance milestones in the installation and debugging of the equipment.

- Identification of the long term maintenance which the vendor will supply. Included in this item should be the maximum response time once a problem is identified, the responsibility of the Owner to service the equipment, the maximum number of emergency service calls which the vendor will supply (this should be unlimited), the routine maintenance which is included in the service contract.

In addition, if the document is for a micro computer based system it should include:

- Detailed description of the information to be included in the daily, weekly, monthly and yearly reports.

- Sample drawings of the screen displays which the control software will be capable of producing.

The micro computer spec should also identify how the master station functions. System complexity must be appropriate for the caliber of the operating staff. Therefore, included in the master station operating features should be:

- A password system which prevents unskilled operators from modifying either accidentally or intentionally operating software. The entry level control system should only allow the operator to turn on or off pumps. All other functions of the system should be password protected.

Finally, the document must include a description of system training which the vendor will provide. All personnel should be instructed in how to use the system to turn pumps on and off, what each of the alarms indicate and who to contact in case of a problem. Several operators should be selected for detailed instruction in what the software is controlling and how system controls can be modified. The vendor must also be required to provide ongoing response to system questions as they arise. This will probably take the form of periodic visits, answers to telephone calls, and responses to written questions.

SUMMARY

Micro computers can be cost effectively used by water systems for both analytical and control functions. These machines can now utilize software written for powerful mainframe computers to perform extended period analyses of complex water systems. From these analyses, engineers can develop operating plans for a variety of water demand scenarios. System operating plans can then be implemented utilizing micro computers with specially designed control software. In addition to actually controlling the system, micro computers can perform ongoing analyses and record keeping functions which mechanical based systems cannot perform. This is truly significant when it is considered that micro computer based control systems are approximately equal in cost to mechanical based systems. Regardless of the control system selected, initial problems can be expected. These problems should be rectified during the debugging period. But, incorporating system performance to the equipment payment schedule will assure the vendor's interest in turning over a completely debugged operating system.

ACKNOWLEDGEMENT

The author wishes to thank the Yonkers Water Bureau and its Superintendent, James A. Neary, for guidance and assistance on this project. A special thank you to my wife Debbie, for typing and editing this paper.

13.   HYDRAULICS POSTER SESSION

UNSTEADY FLOW COMPUTATION FOR PLANNING FLOOD
CONTROL PROJECTS

S. Bognár and L. Somlyódy

Research Centre for Water Resources Development
(VITUKI), H-1453 Budapest, P.O. Box 27., Hungary

1. INTRODUCTION

One of the most important information for planning flood
control projects is the design flood level which is generally
determined either on the basis of observations or by calculating the steady water surface profile. The flood levels are,
however, rarely associated with steady flow since at the time
of flood peak the water level is also affected by the variation
of discharge. When flood control measures - like the design
of flood dikes, diversions, reservoirs etc. - are under consideration their effect on flood level should be predicted,
otherwise proper sizes /e.g. elevation of dyke crests/ can
not be determined with satisfactory accuracy. The effect of
alluvial rivers may also affect the design flood levels, and
the prediction of these changes is a very important part of
the design procedure.

Unsteady flow computations are, therefore, necessary to
derive design flood level and changes in flood levels. There
are a number of methods to solve the one-dimensional unsteady flow equations /Mahmood and Yevjevich 1975, Kozak 1977
etc./. Thus, their application was, hindered especially for
small-scale projects.

By the introduction of microcomputers the application of
the unsteady flow calculation becomes more economical, but
software changes are necessary due to limited RAM capacity
of microcomputers.

In the present case a "small" HP-85 microcomputer of
32 kByte RAM was available, thus necessitating to employ an

effective solution procedure of the Saint-Venant equations.

The requirements of the calculation were as follows:

- the irregularity of the natural channel had to be taken into account,
- the boundary conditions of the model had to be consistent with the available geometric and hydrological data,
- the procedure had to lead to an acceptable execution time on the microcomputer employed.

Instead of the traditional method for solving the system of linear equations, resulted in from the use of the implicit finite difference method, the computationally effective "double - sweep" method was applied. Data required for the computations were stored on mini discs. The computer program was written in BASIC language in an interactive fashion. The results of calculations were plotted on the graphic display and external printer.

2. BASIC EQUATIONS OF 1D UNSTEADY FLOW

The flow in rivers is assumed to be one-dimensional and the dependent variables are the streamflow rate Q, and the local water level /or depth/ Z. The flood plain is assumed to contribute only to storage.

Thus the well known continuity and momentum equations are

$$\frac{\partial Q}{\partial x} + B \frac{\partial Z}{\partial t} = q \qquad (1)$$

$$\frac{1}{g} \frac{\partial v_x}{\partial t} + \frac{1}{2g} \frac{\partial (v_x)^2}{\partial x} + \frac{\partial Z}{\partial x} + S_f = 0 \qquad (2)$$

where B is the width of the entire channel /including flood plain/ v_x the average velocity of the stream in the main channel, S_f is the friction term.

After differentiating the momentum equation (2), one can write

$$\frac{1}{g\,F_s}\frac{\partial Q}{\partial t}+c\,\frac{Q}{g\,F_s^2}\frac{\partial Q}{\partial x}+\frac{\partial Z}{\partial x}+\frac{Q^2}{g\,F_s^3}\frac{\partial F_s}{\partial x}+\frac{Q\,a\,q}{g\,F_s^2}+S_f=0 \quad (3)$$

where F_s is the cross-sectional area of the main channel.

The interaction between the riverbed and the inundated valley is one of the most important factors affecting flood propagation. It was taken into account in the above equation with coefficient a, / a= B/B_s and c=1 +a /, which characterizes the relation of different widths / B_s is the width of the main channel /.

3. FINITE DIFFERENCE SCHEME

After linearizing the resistance term in Eq. (3), a four - point implicit procedure / Preissmann (SOGREAH) implicit scheme / was employed [3]. This scheme is unconditionally stable, when the weighting coefficient fulfils the relation $\theta \geq 0.5$ and its accuracy is satisfactory for the present type of problem.

4. SOLUTION OF THE LINEAR ALGEBRAIC EQUATIONS

The finite-difference form of the Saint-Venant equations can be written for any pair of computational points. If there are N grid points along the longitudinal axis x, one can write 2(N-1) + 2 = 2N algebraic equations for 2N unknowns. This system can be solved and unknown variables at the next time level / t +Δt / can be obtained.

For the solution of the system of algebraic equations, the " double-sweep " method [3] was employed.

Consider the system of Eqs. (1) and (3) which may be transscribed to

$$- A_{1i}\,u_i + A_{2i}\,v_i + A_{1i}\,u_{i+1} + A_{2i}\,v_{i+1} = A_{5i} ,$$

$$B_{1i}\,u_i + B_{2i}\,v_i + B_{3i}\,u_{i+1} + B_{4i}\,v_{i+1} = B_{5i},$$

(4)

where $u_i = Q_i^{j+1} - Q_i^{j}$ and $v_i = Z_i^{j+1} - Z_i^{j}$.

Assume that there is a linear relationship of the type

$$u_i = E_i v_i + F_i \qquad (5)$$

for a point i. If this approximation is valid, then one can prove that the analogous linear relationship also holds for the next point i + 1

$$u_{i+1} = E_{i+1} v_{i+1} + F_i \qquad (6)$$

As a demonstration, Eq. (5) is substituted into Eq. (4) to yield

$$A_{1i} u_{i+1} + A_{2i} v_{i+1} = A_{1i} u_i - A_{2i} v_i + A_{5i}$$

and $\qquad (7)$

$$B_{3i} u_{i+1} + B_{4i} v_{i+1} = - B_{1i} u_i - B_{2i} v_i + B_{5i}$$

From the first Eq. (7) one can find the relation between u_i and the increments of dependent variables u_{i+1} and v_{i+1}

$$v_i = (A_{1i} u_{i+1}) / d + (A_{2i} v_{i+1}) / d - (A_{5i} + A_{1i} E_i) / d \qquad (8)$$

where $d = A_{1i} E_i - A_{2i}$.
This equation can be written as

$$v_i = L_i u_{i+1} + M_i v_{i+1} - N_i \qquad (9)$$

Eleminating u_i from the second Eq. of (7), and then expres - sing u_{i+1} as a function of v_{i+1}, we obtain that

$$u_{i+1} = v_{i+1} \frac{-B_{4i} - M_i / e}{B_{3i} + L_i / e} + \frac{B_{5i} - B_{1i} F_i + N_i / e}{B_{3i} + L_i / e} \qquad (10)$$

where $e = B_{1i} E_i + B_{2i}$.

This is obviously a linear relationship of the form indicated by Eq. (5).

This way an explicit type, two-step procedure can be reali - zed for solving the system of linear equations. The major advan- tage of the algorithm that the number of elementary operations, and consequently the computer time, and the RAM capacity required is proportional to the number of grid points N rather than to N^3 which is the case when using standart matrix inversion techniques.

5. EXTERIOR AND INTERIOR BOUNDARY CONDITIONS

They can be given in the form of $Z = Z(t)$, $Q = Q(t)$ or $Q = Q(Z)$, for any of the two boundaries. The calculation procedure of coefficients E_1 and F_1 at the upper boundary and the water stage Z_N or the discharge Q_N increment at the lower boundary of the modelled river reach is given in [3].

Reservoirs, inflows, tributaries etc. belonging to this category can be easily handled as shown is [3].

6. INITIAL BOUNDARY CONDITIONS

$Z(o, x)$ and $Q(0, x)$ are derived from a soubrutine computing gradually varying steady-state flow.

7. APPLICATION OF THE MATHEMATICAL MODEL TO PRACTICAL PROBLEMS

A valuable agricultural area of about 12 km^2 in the triangle of Zagyva River (Hungary), and its tributary has been saved from medium floods by the transfer of the mouth of the tributary. Additionally, a flood protecting dyke along the lower stretch of the tributary and another dyke / with total length of 12 km / along the Zagyva River were constructed / Fig. 1. /.

Hydraulic calculations were performed to study and answer the following questions

- what is the possible effect of the dyke system / built in 1967, Fig. 1. b / and the transfer of the mouth of tributary / Fig 1. c/ on flood level,

- what further rise in design flood level can be expected due to to the exclusion of a part of the flood plain / which is then utilized for agricultural purposes/,

- how to determine the " optimum " crest level of the new dyke system / Fig. 1. d/ , letting the extreme floods to surpass it and inundate the previously protected area / this way, an emergency reservoir could be created /.

In order to analyze the above problem, a data base has been established using the past and altered geometry of the affected river reaches and flood plains, the vegetation conditions and the observed flood levels. The upstream boundary condition contained the input discharges time series, while the downstream boundary condition consisted water level time series of past flood waves / the above conditions have been selected, since the main problem was to determine the flood level changes at the upstream boundary /.

The mathematical model was first calibrated for various past periods by using the time series of the recording gauge " B " / Fig. 1. /, situated between the upstream and downstream boundaries of the investigated river reach. The objective of calibration was to find the Manning roughness coefficients resulting in the the best fitting between model simulation and measurements. These coefficients could not be directly cal - culated, because no steady water level observations were conducted in the field.

Following calibration, the model was validated by using independent observations as compared to the callibration phase. For design purposes a " critical " flood wave / actually monitored / characterizing the present situation was selected. This " scenario " was employed for both, the original "natural " and the existing conditions. Comparison of the results / Fig. 2. / gives an impression on the effect of human intervention of flood levels.

Subsequently the effect of enlarging the protected flood-plain area and higher dyke crest level was analyzed. Finally, the degree of gauge-height decrease was determined for the case when the flood -plain is used as an emergency reservoir / Fig. 2. /.

8. RESULTS OF CALCULATIONS

Simulation results applying the validated model agree fairly well with the observed hydrographs / where $\Delta t = 4$ h was employed /. Model results for various situations clearly demonstrate the influence of regulation on the flood level / Fig. 2. /.

- the effect of the tributary transfer and the flood dyke on flood levels / at the gauge " B" / is about 45 - 60 cm / increase /,

- the higher crest level and construction of protected flood-
plain can further rise the flood level by 15-20 cm, but
this increase can be entirely eliminated by using the protected flood-plain, as an emergency reservoir.

9. CONCLUSION

Based on improved numerical methods , a program was
developed for microcomputers to calculate the unsteady river
flow. The application of this software allows to determine
design flood levels for different water projects and to evaluate alternative projects for a given purpose.

The software outlined here and others similar in character are expected to have a growing interest in hydrological
practice in the future. The major reasons are the extremely
fast propagation of microcomputers outside the scienfitic
world, the friendly nature of such machines and economic
effiency.

REFERENCES

1. M. Kozak /1977/ Computation of unsteady flow / in
                    Hungarian /
                    Akadémia Kiadó, Budapest

2. J.A. Ligett and J.A. Cunge / 1975 / Numerical methods
                    of solution of unsteady flow equations
                    Water Resources Publication Vol. 1.

3. K. Mahmood and V. Yevjevich / 1975 / Unsteady flow
                    in open channels
                    Water Resources Publication Vol. 1.

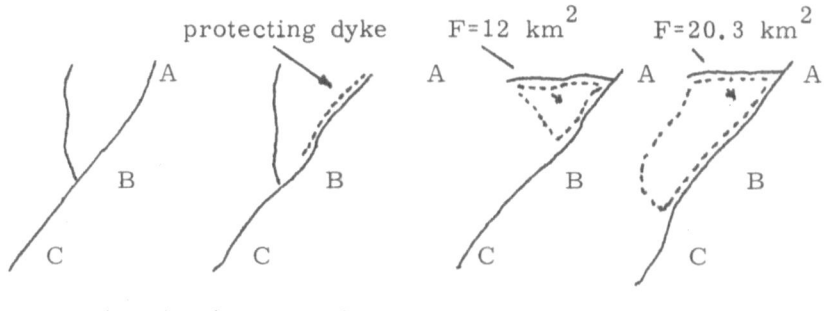

a. past situation   b. past situation   c. present   d. planned
                                            situation   situation
   (before 1964)   (before 1967)

Figure 1. Studied reach of Zagyva River

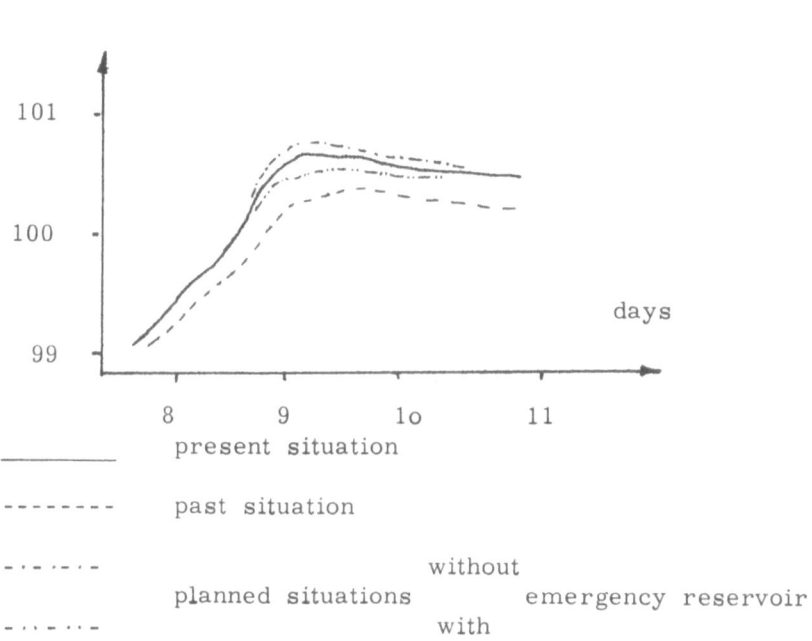

H (m a.As)

_____ present situation

- - - - - - - -   past situation

- · - · - · -   without
              planned situations   emergency reservoir
- · · - · · -   with

Figure 2. Calculated flood levels of different situations at
              gauge " A "

MODELLING OF AN UNSTEADY FLOW IN A COMPLEX WATER RESOURCES SYSTEM APPLIED TO MANAGEMENT

Miloš Baošić and Branislav Djordjević

"Jaroslav Černi" Institute for the Development of Water Resources, Belgrade, Yugoslavia and Faculty of Civil Engineering, University of Belgrade, Yugoslavia

1.   INTRODUCTION

The management of a complex water resources system (WRS) in real time in Yugoslavia resulted in the development and improvement of the mathematical models and numeric procedures presented in this paper.

The WRS that has here been considered is located in the north eastern part of Yugoslavia in the Panonic lowlands. It is characterized by a very complex hydrography as a large number of artificial canals intersect with many natural watercourses and now together form a new hydrographic unity. This system is managed by using weirs located at the most important inflow and outflow profiles. The weirs maintain the required water levels while the discharges can be directed towards the desired outflow profiles, depending on the hydrologic situation in the recipients. The system is a multi-purpose one but is mainly used for flood protection, drainage, irrigation and water supply purposes as well as for enabling navigation. The most delicate task of all is flood protection. This is so since the system is located in a lowlying area while the natural watercourses that are part of it flow in from the surrounding mountains and are of a typically torrential type with big flood waves of a short time concentration. This is why the management of water resources during the occurrence of flood waters actually consists of determining the optimum operation of weirs, in order to enable the most efficient discharge of the flood waters towards the outflow profiles that connect the system with the rivers Danube and Tisza, these being the main recipients of the system.

In order to efficiently manage a WRS in real time, a mathematical model of an unsteady flow was developed and the following were the main problems encountered in this case:

- the existence of a large number of junctions along the major canals and tributaries;

- in order to be able to manage the system in real time, both the operation of the model and the use of the numeric procedure should be easy, since it is only through very quick computations possible to, on time, determine the optimal strategy for the management of weirs;

- a simulation model should be such that it can easily be transformed into an optimization model by introducing an appropriate management and limitation criterion.

2. GENERAL STATEMENT OF THE MODEL

In order to solve the problem of wave routing or in other words solve the system of two partial

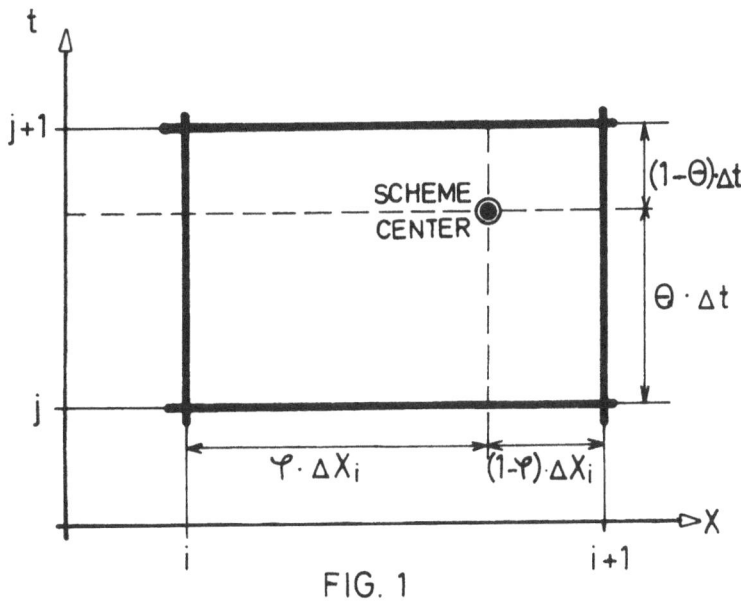

FIG. 1

differential equations, the implicit computational scheme of the finite difference method of the Preissman type was used. The most general type of scheme was used with the weighting coefficients Θ and φ given as a function of time (t) and space (x) [1].

The system of two partial differential St. Venant's equations (the continuity equation and dynamic equation) has been presented in matrix form as:

$$
\begin{bmatrix} 1 & 0 & 0 & B \\ 2Q/A^2 & 1/A & g & 0 \end{bmatrix} \cdot \begin{bmatrix} \partial Q / \partial x \\ \partial Q / \partial t \\ \partial z / \partial x \\ \partial z / \partial t \end{bmatrix} = \begin{bmatrix} -q \\ -WQ \end{bmatrix} \tag{1}
$$

$$
\left(W = -\frac{Q}{A^3} \cdot \frac{\partial A}{\partial x} + g \cdot \frac{Q \cdot n^2}{A^2 \cdot R^{4/3}} \right)
$$

where:

z - water level, Q - water discharge, B - width of cross section, A - cross-sectional flow area, q - continuous lateral inflow per unit length, R - hydraulic radius, n - Manning's coefficient, g - acceleration due to gravity.

These two partial differential equations can also be written in a simpler form as:

$$
\Delta (Q, CQ_1) + \Delta (Z, CZ_1) = 0
$$
$$
\Delta (Q, CQ_2) + \Delta (Z, CZ_2) = 0 \tag{2}
$$

where Δ is the differential operator given in the form of:

$$
\Delta (f, c) = C_1 \cdot \frac{\partial f}{\partial x} + C_2 \cdot \frac{\partial f}{\partial t} + C_3 f + C_4
$$
$$
C = \{ C_1, C_2, C_3, C_4 \} \tag{3}
$$

where

$$CQ_1 = \{ 1, 0, 0, 0 \}$$

$$CZ_1 = \{ 0, B, 0, q \}$$

$$CQ_2 = \{ 2Q/A^2, 1/A, W, 0 \} \qquad (4)$$

$$CZ_2 = \{ g, 0, 0, 0 \}$$

The already mentioned Preissman 4-point scheme with a discretization of the shape

$$f = \varphi \left[\Theta \cdot f_{i+1}^{j+1} + (1-\Theta) \cdot f_{i+1}^{j} \right] + (1-\varphi) \left[\Theta \cdot f_i^{j+1} + (1-\Theta) \cdot f_i^{j} \right]$$

$$\frac{\partial f}{\partial x} = \frac{\Theta}{\Delta x} \left(f_{i+1}^{j+1} - f_i^{j+1} \right) + \frac{1-\Theta}{\Delta x} \left(f_{i+1}^{j} - f_i^{j} \right) \qquad (5)$$

$$\frac{\partial f}{\partial t} = \frac{\varphi}{\Delta t} \left(f_{i+1}^{j+1} - f_{i+1}^{j} \right) + \frac{1-\varphi}{\Delta t} \cdot \left(f_i^{j+1} - f_i^{j} \right)$$

was used in the course of the further development of the operator $\Delta (f, C)$ which in its discretized form becomes:

$$\Delta(f, C) = D_i \cdot f_i^{j+1} + D_{i+1} \cdot f_{i+1}^{j+1} + E_i \qquad (6)$$

$$(i = 1, \ldots, n)$$

The expressions for D_i, D_{i+1} and E_i are quite complex and will not be presented here because of the limited amount of space.

By using expression (6) for the operator $\Delta (f, C)$,

the system of equations (2) can be written in a developed form as an algebraic system consisting of (2n- 2) equations with (2n) unknowns. Two of the functions are determined based on the boundary conditions while the values of the remaining unknowns and of the coefficients in all the points of the initial time moment are determined based on the initial conditions. The steady state is considered to be the initial condition.

The algebraic system of equations has the following form:

$$A \cdot X = \alpha \qquad (7)$$

where:

A - two-dimensional matrix of the non-linear coefficients

X - one-dimensional matrix of the unknown values Q and z

α - one-dimensional matrix of the free terms.

Special attention has been paid to the solving of this system of algebraic equations and more detailed information on this has been given in part 4 of this paper.

3. THE PROBLEM OF TRIBUTARIES AND IRRIGATION INFLOWS

As has already been said, the configuration of the WRS that has been considered is very complex. The watercourses within the system, intersected at many locations, are often times torrential streams with a marked torrential regime at the time of the occurrence of flood waves. Besides this, there are many inflow and water intakes within the irrigation system (the so-called internal waters). Since the internal flood waters usually coincide with the external flood waters from tributaries, the complex flow pattern within the WRS is then even more delicate.

The adopted scheme (Fig. 2) for solving this problem is similar to the one for the interior boundary conditions given in Ref. [4].

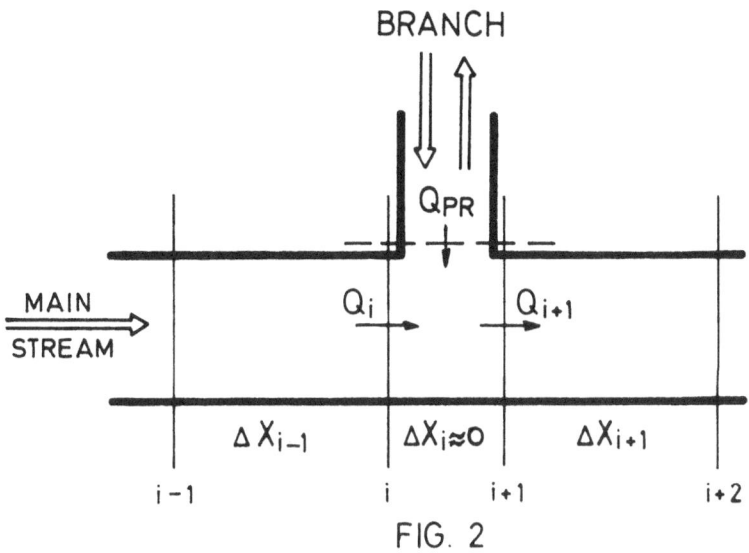

FIG. 2

Based on the assumption that the distance between profiles (i) and (i + 1) is small, it is possible to say that:

$$Q_{i+1}^{j+1} = Q_i^{j+1} + Q_{PR}^{j+1}$$
$$Z_{i+1}^{j+1} = Z_i^{j+1}$$

(8)

The values of Q_{pr}^{j+1} have been defined by the
following equations : $Q_{pr}(t) > 0$ for the case of
tributaries and drainage and $Q_{pr}(t) < 0$ for the case of
irrigation. These last two cases have been defined by
the operation of pumping stations. When the inflow is
due to gravity, drainage is then considered as a
tributary.

A suggestion as to how this problem should be
solved has been given in this paper and is based on
the expansion of the system of equations (7) for each
of the mentioned locations where the changes in the
discharges occur for $Q_{pr}(t)$.

In this way, the system of equations (7) given
in matrix form is:

$$\begin{bmatrix} x & x & x & & & & & \\ x & x & x & & & & & \\ & & x & x & x & x & & \\ & & x & x & x & x & & \\ & & \boxed{\begin{matrix} -1 & 0 & 1 & 0 \\ 0 & -1 & 0 & 1 \end{matrix}} & & \\ & & & x & x & x & x & \\ & & & x & x & x & x & \\ & & & - & - & - & - & \\ & & & - & - & - & - & \\ & & & - & - & - & - & \\ & & & & & x & x & x \\ & & & & & x & x & x \end{bmatrix} \cdot \begin{bmatrix} x \\ x \\ x \\ Q_i \\ Z_i \\ Q_{i+1} \\ Z_{i+1} \\ x \\ - \\ - \\ - \\ x \\ x \end{bmatrix} = \begin{bmatrix} x \\ x \\ x \\ x \\ Q_{PR} \\ 0 \\ x \\ x \\ - \\ - \\ - \\ x \\ x \end{bmatrix} \quad (9)$$

This inclusion of two fours of coefficients (-1,0,1,0) and (0,-1,0,1) into matrix A, as well as the introduction of an appropriate pair of values $(Q_{pr}, 0)$ into the matrix α of the system of equations (7) is, as has already been said, done at as many locations within the system as there are changes in the discharges $Q_{pr}(t)$.

This results in the increase of the system of equations which in turn makes the numerical aspects of this work especially important.

4.   NUMERICAL ASPECTS

The algebraic system of equations (7) is non-linear and the first problem that was encountered therefore consisted of determining how it should be solved. The method consisting of the linearization of the non-linear coefficients in matrix A is well known in literature [1] and [4]. The authors of this paper chose the iterative procedure using the known values from the previous time interval.

By introducing $f_{i+\varphi}$ and $f^{j+\theta}$ to denote the unknown values and the values of the coefficients at the $(i + \varphi)$ and $(j + \theta)$ points in the net , this paper suggests the implementation of a somewhat more generalized form of the iterative procedure similar to Verwey's variant of the Preissman scheme [2] in which $\varphi = \theta = 1/2$.

For the known time interval (j), the coefficients

in the points of the net $(i + \varphi)$ are computed:

$$f^j_{i+\varphi} \approx \varphi \cdot f^j_{i+1} + (1 - \varphi) \cdot f^j_i \qquad (10)$$

For the time interval $(j + 1)$ and for iteration (k) it is assumed that:

$$f^{(k)j+1}_{i+\varphi} \approx \varphi \cdot f_{i+1}\left(z^{(k-1)j+1}_{i+1}\right) + (1 - \varphi) \cdot f_i\left(z^{(k-1)j+1}_i\right) \qquad (10')$$

where $z_i^{(k-1)j+1}$ are the values obtained by solving the system of equations (9) in the previous iteration $(k-1)$. For the first iteration $(k=1)$ it was assumed that $f^{j+1}_{i+\varphi} \approx f^j_{i+\varphi}$. In order to solve the system of equations (9), it is the values of the coefficients in the points of the time intervals $(j + \theta)$ that are considered:

$$f^{(k)j+\theta}_{i+\varphi} \approx \theta \cdot f^{(k)j+1}_{i+\varphi} + (1 - \theta) \cdot f^j_{i+\varphi} \qquad (11)$$

By substituting (11) and solving the system of equations (9) the following new approximate values $z^{(k)j+1}_i$ and $Q^{(k)j+1}_i$ are obtained for all the numerical points (i).

The procedure is continued until the given accuracy (δ) is obtained:

$$\max \left| z_i^{(k)j+1} - z_i^{(k-1)j+1} \right| \leq \delta \qquad (12)$$

where $1 < i < N$ and δ is the predefined small number.

Another problem that the authors encountered is the problem of solving large systems of equations. Because of the iterative procedure used for determining the non-linear coefficients, it was essential to determine a simple and easy direct procedure that could be used for solving the system of equations.

The authors have in this paper used the Cholesky [3] scheme, well known and used for solving large sparse systems of equations when applying the finite element method. What was actually done was that only the basic idea of this scheme was taken. The complete algorithm is original and consists of compressing the matrices with the new indexing. The procedure is a very quick one and the operations have been reduced to the smallest possible number. Instead of the matrix

of the coefficient having the following dimensions
(2n-2) x (2n-2), the dimensions of the matrix of the
coefficients used in the procedure are (2n-2) x 4.
The top and bottom auxiliary triangular matrix in the
Cholesky scheme, the dimensions of which are (2n-2) x
(2n-2), have here also been compressed and newly indexed with the new dimensions (2n-2) x 3. The total
number of elements in these 3 matrices is in this way
significantly reduced and with (12.x (n-1) x (n-1))
elements it equals (20.x (n-1)) so that it is actually
reduced (0.6 x (n-1)) times. If the number of computational points (i) is 101 for example, the number of
elements in these three matrices is reduced 60 times.
This saves the computer memory from memorizing large
square matrices with a large number of zero-elements
and it also eliminates the computations with these
elements.

The essence of this procedure can be seen on the scheme of the matrix of coef- ficient A of the system of equations (7) as well as on the schemes of the aux- iliary triangular matrices B and C.

$$\begin{bmatrix}
a_{11}\ a_{12}\ a_{13} & & & & \\
a_{21}\ a_{22}\ a_{23} & & & 0 & \\
 & a_{32}\ a_{33}\ a_{34}\ a_{35} & & & \\
 & a_{42}\ a_{43}\ a_{44}\ a_{45} & & & \\
 & & a_{54}\ a_{55}\ a_{56}\ a_{57} & & \\
 & & a_{64}\ a_{65}\ a_{66}\ a_{67} & & \\
0 & & & a_{76}\ a_{77}\ a_{78} & \\
 & & & a_{86}\ a_{87}\ a_{88} &
\end{bmatrix}$$

Instead of matrix A, B and C are given as (example given of an 8 dimen- sional matrix):

$$\begin{bmatrix}
b_{11} & & & \\
b_{21}\ b_{22} & & 0 & \\
 & b_{32}\ b_{33} & & \\
 & b_{42}\ b_{43}\ b_{44} & & \\
 & & b_{54}\ b_{55} & \\
 & & b_{64}\ b_{65}\ b_{66} & \\
0 & & & b_{76}\ b_{77} \\
 & & & b_{86}\ b_{87}\ b_{88}
\end{bmatrix}
\qquad
\begin{bmatrix}
1\ C_{12}\ C_{13} & & & \\
1\ C_{23} & & 0 & \\
 & 1\ C_{34}\ C_{35} & & \\
 & 1\ C_{45} & & \\
 & & 1\ C_{56}\ C_{57} & \\
 & & 1\ C_{67} & \\
0 & & & 1\ C_{78} \\
 & & & 1
\end{bmatrix}$$

After compressing and newly indexing, the new matrices are obtained in the following form:

$$
\begin{bmatrix}
0 & a_{11}^{12} & a_{12}^{13} & a_{13}^{14} \\
0 & a_{21}^{22} & a_{22}^{23} & a_{23}^{24} \\
a_{32}^{31} & a_{33}^{32} & a_{34}^{33} & a_{35}^{34} \\
a_{42}^{41} & a_{43}^{42} & a_{44}^{43} & a_{45}^{44} \\
a_{54}^{51} & a_{55}^{52} & a_{56}^{53} & a_{57}^{54} \\
a_{64}^{61} & a_{65}^{62} & a_{66}^{63} & a_{67}^{64} \\
a_{76}^{71} & a_{77}^{72} & a_{78}^{73} & 0 \\
a_{86}^{81} & a_{87}^{82} & a_{88}^{83} & 0
\end{bmatrix}
\quad
\begin{bmatrix}
0 & 0 & b_{11}^{13} \\
0 & b_{21}^{22} & b_{22}^{23} \\
0 & b_{32}^{32} & b_{33}^{33} \\
b_{42}^{41} & b_{43}^{42} & b_{44}^{43} \\
0 & b_{54}^{52} & b_{55}^{53} \\
b_{64}^{61} & b_{65}^{62} & b_{66}^{63} \\
0 & b_{76}^{72} & b_{77}^{73} \\
b_{86}^{81} & b_{87}^{82} & b_{88}^{83}
\end{bmatrix}
\quad
\begin{bmatrix}
1 & C_{12}^{12} & C_{13}^{13} \\
1 & C_{23}^{22} & 0 \\
1 & C_{34}^{32} & C_{35}^{33} \\
1 & C_{45}^{42} & 0 \\
1 & C_{56}^{52} & C_{57}^{53} \\
1 & C_{67}^{62} & 0 \\
1 & C_{78}^{72} & 0 \\
1 & 0 & 0
\end{bmatrix}
$$

where the bottom numbers represent the old indexes while the upper ones represent the new indexes.

The solution of the system of equations (7) obtained in this way represents the direct solution (X_d) with a certain error Δx. The use of the proposed algorithm would, it is assumed, give an exact solution (x_t) that can in matrix form be given as:

$$X_t = X_d + \Delta X \tag{13}$$

ΔX is corrected by solving the system of equations:

$$A \cdot \Delta X = \Delta \alpha \tag{14}$$

by applying the same direct procedure where:

$$\Delta \alpha = \alpha - A \cdot X_d \tag{15}$$

The uniform programme procedure makes it possible to, in a simple manner, get the correct solution of the system of equations (7). This procedure for solving a system of equations has been separately tested and has given very good results.

5. THE USE OF MATHEMATICAL MODELS FOR MANAGEMENT OPTIMIZATION

The simulation models of this water resources system can be transformed into optimization models by introducing appropriate criteria for evaluating management decisions as well as the limitations of the conditions of management and of the system. In accordance with the numerous objectives of this WRS, different criteria can be defined so that the task can be considered as a multi-criteria (vector) optimization problem. Some of the possible criteria for management optimization can be defined as follows:

a) The main objectives of management during actual flood protection can be defined as the need to minimize water levels (Z_i) at the central reaches of the WRS by implementing adequate measures, or as the need to reduce the costs of exploitation of the WRS (T) as well as the possible flood damages (R). In accordance with this, the criteria can be defined as:

$$Z_i \longrightarrow min , \quad OR : (T+R) \longrightarrow min \qquad (16)$$

An efficient numeric procedure makes it possible to, right away and for each management activity, determine the conditions in all the characteristic locations as well as the discharges in all the reaches of the WRS. In some hydrologic situations, by varying the management within a certain permissible range, the computer memorizes the matrices showing the management conditions, damages, costs, etc., and it is possible to, by reviewing them and in accordance with the defined criteria for evaluating management, determine the best form of management.

b) At times of normal exploitation, when the management meets the interest of most of the water resources users, a certain amount of the profit of multi-purpose WRS (B), of the costs (T) and damages (R) correspond to each of the management actions so that the criterion for maximizing the net benefit can be given as:

$$(B - T - R) \longrightarrow max \qquad (17)$$

c) In some incidental situations in a WRS, the criterion can be formulated in accordance with requests for a quick intervening action in order to minimize the possible damages or realize some other requests that are then a priority (the protection of a settlement, etc.).

Bearing in mind the numerous criteria, it is possible to check how sensitive to the changes in the criteria a solution is, or, it is also possible to apply the vector multi-criteria optimization procedure.

*   *
*

The simulation model of the considered WRS has been numerically completed and tested and is now being introduced for actual use.

REFERENCES:

1. Abbott M.B. (1980). Elements of the Theory of Free Surface Flows - Computational Hydraulics, London

2. Cunge J.A., Holly Jr F.M., Verwey A. (1980). Practical Aspects of Computational River Hydraulics, London

3. George A., Liu J. (1981). Computer Solution of Large Sparse Positive Definite Systems, New Jersey

4. Liggett J.A., Cunge J.A. (1975). Numerical Methods of Solution of the Unsteady Flow Equation, in Unsteady Flow in Open Channels (ed. K. Mahmood and V. Yevdjevich), Chap. 4, Water Research Publications, Fort Collins

IRRIGATION WATER REQUIREMENT MODEL

Shie-Yui Liong
Dept. of Civil Engrg., National Univ. of Singapore,
Singapore 0511

Ongko Sutjahyo, Bernard Rasli
PT Dacrea, Bendungan Hilir Raya, KAV. 36A, Jakarta, Indonesia

INTRODUCTION

Reasonably good estimates of consumptive use, farm and irrigation water requirements are essential in planning, design and operation of irrigation adnd drainage systems, and for evaluation and management of hydrologic and water resources systems. The availability of climatic records for many planned irrigation areas, particularly in developing countries, is often a major problem. With the assigned climate data one then has to estimate the seemingly endless repetitive calculations of consumptive use for different types of crop, the farm and irrigation water requirements.

This paper describes a computer model which estimates the irrigation water requirement. An example of an output print-out follows the description of the model and the flow chart.

CONSUMPTIVE USE

Consumptive use of water by a crop is defined as the depth of water consumed by evaporation and transpiration during crop growth, including water consumed by accompanying weed growth except deep percolation [5]. In this section parameters involved in the determination of consumptive use are introduced.

REFERENCE EVAPOTRANSPIRATION (ETo)
The reference evapotranspiration (ETo) is "the rate of evapotranspiration from an extensive surface of 8 to 15 cm tall, green grass cover of uniform height, actively growing, completely shading the ground and not short of water" [2]. Depending upon the climate data available ETo in this model

can be determined by the modified Penman method, the Radiation method, or the Blaney - Criddle method [1,3]. The climate data required to compute ETo are the mean daily data for 10-day period.

CROP FACTORS

Crop factor is defined as the ratio between the consumptive use and ETo. Crop factors for many different types of crop are readily found in literatures and for each crop usually the crop factors of every 10% of growth period are given.

The consumptive use for every 10-day period can now be estimated by multiplying ETo with the corresponding crop factor.

FARM WATER REQUIREMENT

The farm water requirement is not only function of the consumptive use but also of the cropping intensity, percolation, and, for some types of crop, nursery and puddling water requirements.

CROPPING INTENSITY

From the view point of labor resources and water requirements a farm field is divided into several divisions. Crops are planted at and harvested from different divisions at different time levels. Table 1 shows the cropping intensity of a crop with 70-day growth period in a field divided into 6 divisions. In this example only half of the first division, or 1/12 of the entire field, is planted at the end of the first 10-day. At the end of the second 10-day the first division is completely planted whereas only half of the division is planted. At the end of the sixth 10-day planting on the entire field has been completed. Harvesting would naturally be started on the first half of the first division; next would be the second half of the division 1 and the first half of the division 2; etc.

With the above described cropping scheme labor forces can be utilized more efficiently and, more important perhaps, the demand for water is gradually increased, rather than suddenly, to the peak demand.

PERCOLATION

For some soil types percolation loss can be quite severe and should be taken into consideration. The computer model can accomodate a constant percolation loss or the time- and space-variant percolation loss for every 10-day period.

NURSEY AND PUDDLING WATER REQUIREMENTS

Some crops such as paddy require nursery and puddling. The water requirement for nursery depends on the amount of water

Table 1 Cropping Intensity in a Farm with 6 Divisions

CROPPING INTENSITY IN DIVISION NO.	APRIL			MAY			JUNE			JULY			AUGUST		
	E	M	L	E	M	L	E	M	L	E	M	L	E	M	L
1		$\frac{1}{12}$	$\frac{1}{6}$	$\frac{1}{6}$	$\frac{1}{6}$	$\frac{1}{6}$	$\frac{1}{6}$	$\frac{1}{6}$	$\frac{1}{12}$						
2			$\frac{1}{12}$	$\frac{1}{6}$	$\frac{1}{6}$	$\frac{1}{6}$	$\frac{1}{6}$	$\frac{1}{6}$	$\frac{1}{6}$	$\frac{1}{12}$					
3				$\frac{1}{12}$	$\frac{1}{6}$	$\frac{1}{6}$	$\frac{1}{6}$	$\frac{1}{6}$	$\frac{1}{6}$	$\frac{1}{6}$	$\frac{1}{12}$				
4					$\frac{1}{12}$	$\frac{1}{6}$	$\frac{1}{6}$	$\frac{1}{6}$	$\frac{1}{6}$	$\frac{1}{6}$	$\frac{1}{6}$	$\frac{1}{12}$			
5						$\frac{1}{12}$	$\frac{1}{6}$	$\frac{1}{6}$	$\frac{1}{6}$	$\frac{1}{6}$	$\frac{1}{6}$	$\frac{1}{6}$	$\frac{1}{12}$		
6							$\frac{1}{12}$	$\frac{1}{6}$	$\frac{1}{6}$	$\frac{1}{6}$	$\frac{1}{6}$	$\frac{1}{6}$	$\frac{1}{6}$	$\frac{1}{12}$	
TOTAL		$\frac{1}{12}$	$\frac{3}{12}$	$\frac{5}{12}$	$\frac{7}{12}$	$\frac{9}{12}$	$\frac{11}{12}$	$\frac{12}{12}$	$\frac{11}{12}$	$\frac{9}{12}$	$\frac{7}{12}$	$\frac{5}{12}$	$\frac{3}{12}$	$\frac{1}{12}$	

NOTE: E = the first 10- day of the month
       M = the second 10- day of the month
       L = the last 10- day of the month

required for nursery bed preparation, losses through percolation and evapotranspiration, and, of course, on the nursery intensity. Area assigned for nursery is usually 1/20 of the entire farm area. The length of the nursery period is 20 days. The nursery area is divided into the same number of divisions as that for cropping. Table 2 shows the nursery intensity with 6 divisions.

Puddling water requirement is a function of the amount of water required to saturate the soil profile, the water losses through percolation and evapotranspiration, the amount of standing water after puddling, and on the puddling intensity. The length of puddling is 10 days. Table 3 shows the puddling intensity for a field with six divisions.

The farm water requirement is then determined by summing the consumptive use, percolation loss, nursery and puddling water requirements.

IRRIGATION WATER REQUIREMENT

In determining the irrigation water requirement several factors such as effective rainfall, irrigation efficiency and other parameters discussed in earlier sections should be taken into consideration.

EFFECTIVE RAINFALL
Not all rainfall is effective and part may be lost by surface runoff, deep percolation or evaporation. Hence, effective rainfall is that part of rainfall which is effectively used by the crops to meet their consumptive need [2,4].

The effective rainfall is a function of several parameters such as (1) the characteristics of the rain; (2) the antecedent moisture contents; (3) crop type; (4) climatic conditions; (5) conditions of storage in the fields, etc. A guideline of how to estimate the effective rainfall is given in [2].

This irrigation water requirement model does not include the contribution from the groundwater table which, depending on its distance from the root zone, might further reduce the irrigation water requirement.

IRRIGATION EFFICIENTY
In each application of irrigation water from the headworks to the point where it is directly available to the crop losses of water incurred should be considered. This model incorporate the conveyance efficiency and the field application efficiency.

The conveyance efficiency is defined as the ratio of

Table 2 Nursery Intensity in a Farm Field with 6 Divisions

NURSERY INTENSITY IN DIVISION NO.	MARCH			APRIL			MAY			JUNE		
	E	M	L	E	M	L	E	M	L	E	M	L
1			$\frac{1}{12}$	$\frac{1}{6}$	$\frac{1}{12}$							
2				$\frac{1}{12}$	$\frac{1}{6}$	$\frac{1}{12}$						
3					$\frac{1}{12}$	$\frac{1}{6}$	$\frac{1}{12}$					
4						$\frac{1}{12}$	$\frac{1}{6}$	$\frac{1}{12}$				
5							$\frac{1}{12}$	$\frac{1}{6}$	$\frac{1}{12}$			
6								$\frac{1}{12}$	$\frac{1}{6}$	$\frac{1}{12}$		
TOTAL			$\frac{1}{12}$	$\frac{3}{12}$	$\frac{4}{12}$	$\frac{4}{12}$	$\frac{4}{12}$	$\frac{4}{12}$	$\frac{3}{12}$	$\frac{1}{12}$		
TOTAL x $\frac{1}{20}$			$\frac{1}{240}$	$\frac{3}{240}$	$\frac{4}{240}$	$\frac{4}{240}$	$\frac{4}{240}$	$\frac{4}{240}$	$\frac{3}{240}$	$\frac{1}{240}$		

NOTE : Total nursery area is $\frac{1}{20}$ of the total farm area; nursery period is 20 days

13-27

Table 3 Puddling Intensity in a Farm Field with 6 Divisions

PUDDLING INTENSITY IN DIVISION NO.	APRIL			MAY			JUNE		
	E	M	L	E	M	L	E	M	L
1	$\frac{1}{12}$	$\frac{1}{12}$							
2		$\frac{1}{12}$	$\frac{1}{12}$						
3			$\frac{1}{12}$	$\frac{1}{12}$					
4				$\frac{1}{12}$	$\frac{1}{12}$				
5					$\frac{1}{12}$	$\frac{1}{12}$			
6						$\frac{1}{12}$	$\frac{1}{12}$		
TOTAL	$\frac{1}{12}$	$\frac{2}{12}$	$\frac{2}{12}$	$\frac{2}{12}$	$\frac{2}{12}$	$\frac{2}{12}$	$\frac{1}{12}$		

NOTE : Puddling period is 10 days

water received at inlet to a block of fields to that released at the headworks. The field application efficiency is the ratio of water directly available to the crop and that received at the field inlet [2].

The irrigation water requirement, ND, is then estimated as follows:

$$ND = \frac{(CU + P - ER) \times C1 + (NR - ER) \times C2 + (PR - ER) \times C3}{(Ec \times Ea)} \tag{1}$$

where
CU = consumptive use (mm)
P = percolation (mm)
ER = effective rainfall (mm)
NR = nursery water requirement (mm)
PR = puddling water requirement (mm)
C1 = ratio of nursery area to the farm area
C2 = ratio of puddling area to the farm area
C3 = cropping intensity
Ec = conveyance efficiency
Ea = field application efficiency

FLOW CHART

Flow chart of the irrigation water requirement model is shown in Fig. 1

EXAMPLE

An example is given to elucidate the above mentioned computations. In this example the farm field is divided into six divisions; the crop type is paddy with 70-day growth period; percolation loss is assumed to be constant; the cropping coefficients at each 10% growth period are 1.00, 1.01, 1.02, 1.05, 1.07, 1.18, 1.30, 1.38, 1.38, and 1.28; a value of 0.64 is assigned to the irrigation efficiency.

Modified Penman method was used to calculate ETo. Table 4 shows the step-by-step computation of farm water requirement. Table 5 shows the computation of the net irrigation water requirement.

The computational time for this example was 0.07 second on IBM 3081.

REFERENCES

1. Blaney, H.F. and W.D. Criddle (1962) Determining Consumptive Use and Irrigation Water Requirements. USDA-ARS Technical Bulletin 1275.

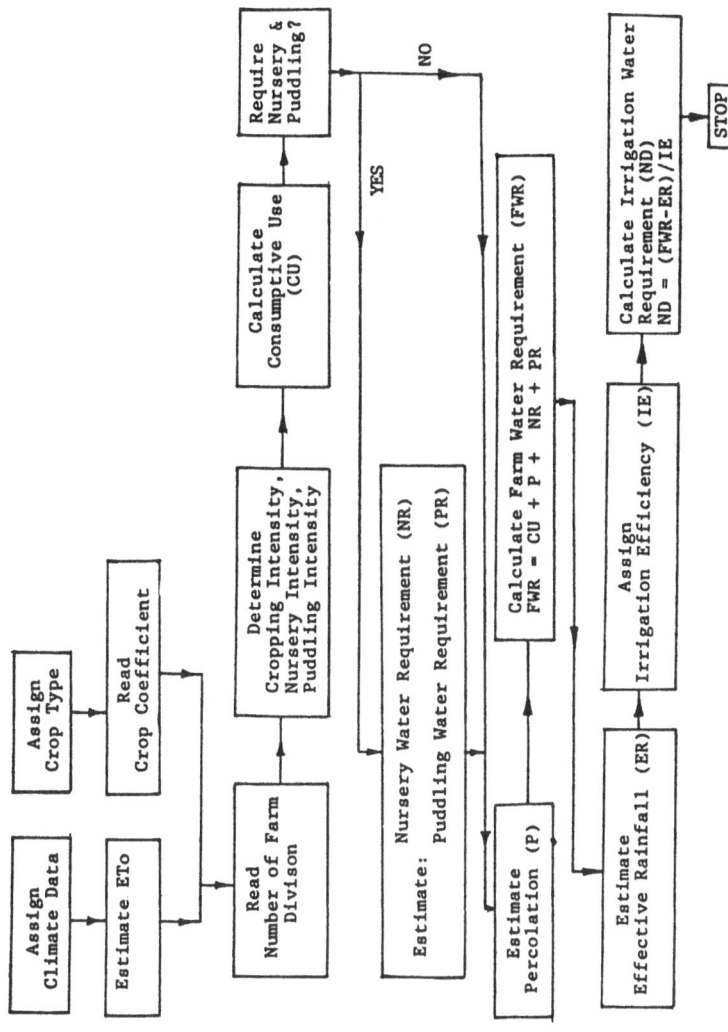

Fig 1   Flow Chart of the Irrigation Water Requirement Model

Table 4 Farm Water Requirement

(1) TIME	MARCH			APRIL			MAY			JUNE			JULY			AUGUST		
	B	M	L	B	M	L	B	M	L	B	M	L	B	M	L	B	M	L
(2) CROPPING INTENSITY OF THE PROJECT AREA					0.08	0.25	0.42	0.58	0.75	0.92	1.00	0.92	0.75	0.58	0.42	0.25	0.08	
(3) CROPPING COEFFICIENT (WEIGHTED AVERAGE)					1.00	1.01	1.02	1.03	1.06	1.11	1.17	1.22	1.26	1.30	1.34	1.34	1.30	
(4) POTENTIAL EVAPOTRANSPIRATION (MODIFIED PENMAN)					38.6	38.6	40.5	40.5	40.5	37.4	37.4	37.4	44.1	44.1	44.1	44.3	44.3	
(5) CONSUMPTIVE USE (3)*(4)					38.7	38.9	41.2	41.7	43.1	41.5	43.9	45.7	55.5	57.6	59.0	59.4	57.6	
(6) PERCOLATION					20.0	20.0	20.0	20.0	20.0	20.0	20.0	20.0	20.0	20.0	20.0	20.0	20.0	
(7) (5)+(6)					58.7	58.9	61.2	61.7	63.1	61.5	63.9	65.7	75.5	77.6	79.0	79.4	77.6	
(8) WATER REQUIREMENT (7)*(2)					4.9	14.7	25.5	36.0	47.3	56.4	63.9	60.2	56.6	45.2	32.9	19.8	6.5	
(9) WATER REQUIREMENT FOR NURSERY			0.5	1.5	2.0	2.0	2.0	2.0	1.5	0.5	0.0	0.0	0.0	0.0	0.0	0.0	0.0	
(10) WATER REQUIREMENT FOR PUDDLING				15.8	31.7	31.7	31.7	31.7	31.7	15.8	0.0	0.0	0.0	0.0	0.0	0.0	0.0	
(11) FARM WATER REQUIREMENT (8)+(9)+(10)			0.5	17.3	38.6	48.4	59.2	69.7	80.5	72.7	63.9	60.2	56.6	45.2	32.9	19.8	6.5	

UNIT FOR (4) TO (11) : MM

Table 5 Net Irrigation Water Requirement

TIME		NR	PR	CU	P	ER	C1	C2	C3	ND
MAR	E									
	M									
	L	120.00				10.80	0.00			0.46
APR	E	120.00	190.00			94.72	0.01	0.08		7.94
	M	120.00	190.00	38.70	20.00	72.12	0.02	0.17	0.08	20.99
	L	120.00	190.00	38.93	20.00	13.52	0.02	0.17	0.25	46.86
MAY	E	120.00	190.00	41.20	20.00	14.88	0.02	0.17	0.42	51.33
	M	120.00	190.00	41.69	20.00	41.00	0.02	0.17	0.58	38.91
	L	120.00	190.00	43.10	20.00	76.22	0.01	0.17	0.75	19.78
JUN	E	120.00	190.00	41.51	20.00	94.75	0.00	0.08	0.92	7.73
	M	0.00	0.00	43.87	20.00	66.00	0.00	0.00	1.00	0.00
	L	0.00	0.00	45.65	20.00	70.86	0.00	0.00	0.92	0.00
JUL	E	0.00	0.00	55.47	20.00	67.07	0.00	0.00	0.75	6.30
	M	0.00	0.00	57.57	20.00	5.84	0.00	0.00	0.58	41.60
	L	0.00	0.00	59.04	20.00	11.00	0.00	0.00	0.42	28.58
AUG	E	0.00	0.00	59.37	20.00	122.02	0.00	0.00	0.25	0.00
	M	0.00	0.00	57.60	20.00	64.00	0.00	0.00	0.08	1.09
	L	0.00	0.00	0.00	0.00	0.00	0.00	0.00	0.00	0.00

2. Doorenbos, J. and W.O. Pruitt (1977) Guidelines for Predicting Crop Water Requirements. FAO Irrigation and Drainage paper 24, Food and Agriculture Organization of the United Nations, Rome, Italy.

3. Linsley, R.K., M.A. Kohler and J.L.H. Paulhus (1949) Applied Hydrology. McGraw Hill, New York.

4. Mazumder, S.K. (1983) Irrigation Engineering. Tata McGraw Hill Publishing Company Limited, New Delhi, India.

5. Varschney, R.S., S.C. Gupta and R.L. Gupta (1972) Theory and Design of Irrigation Structures. Vol. I, Nem Chand & Bros, Roorkee, India.

14.   GEOMETRIC MODELLING

SOLID MODELLING : A TEENAGED ART - WHEN WILL IT MATURE?

M.J. Pratt

Cranfield Institute of Technology.

1. INTRODUCTION

The history of solid modelling now spans some fifteen years.
Although progress was initially rapid, more recently the pace
of development seems to have slowed except in the area of
graphical rendering, where advances have been driven by the
increasing availability of sophisticated graphics hardware.
In fact the last five years has been a time of consolidation,
as may be seen by a study of two state-of-the-art seminars
(Carter [1979], Faux [1983]) in which the capabilities of a
range of solid modelling systems were examined.   In the
interval between these events few major changes have occurred
in solid modelling.   There are certainly more systems in
evidence, but most of them are based on techniques which
have been known for some time.   Systems are, in general, more
robust.   A wider range of engineering applications is
available, but we have not yet come close to realising the
ultimate potential, which was recognised by the earliest
workers in this area, for the true integration of engineering
design and manufacture based on the use of a solid model as
the product definition.   This paper will examine some of the
problems which hamper the attainment of this ideal, and will
outline some of the issues which have so far prevented the
wholehearted industrial adoption of solid modelling techniques.

2. GEOMETRIC COVERAGE

From the point of view of geometric coverage, one of the
simplest solid modellers is PADL-1, developed by Rochester
University (Voelcker et al [1978]).   This system models
objects which can be represented in terms of rectangular
blocks and circular cylinders; both types of primitive
volume are restricted to having their axes aligned with the
coordinate axes of the modelling space.   In spite of these
severe restrictions PADL-1 is capable of modelling about 40%

of the parts manufactured by a range of mechanical engineering
companies. Its later development PADL-2 (Brown [1982]) extends
the range of primitive volumes to cover cones, spheres and
toruses; it additionally allows unrestricted orientation of
these primitives in space. Over 90% of parts in the same
range of companies may now be modelled. As might be expected
the remaining 10% of parts pose as many problems, if not more,
than the initial 90%. These are parts which exhibit 'free-
form' surfaces of various types from blends and fairings, which
are relatively highly constrained, to surfaces where aesthetics
are the primary consideration and geometric constraints are few.
The modelling of objects requiring free-form geometry is
currently a major development area. Two main classes of
techniques are emerging, corresponding to the two major types
of solid modelling systems.

The constructive solid geometry (CSG) modellers are those
in which the primary data structure is a tree whose 'leaves'
are primitive volumes and whose nodes are set operations (or
boolean operations). The modelled object is thought of as
being defined in terms of volumes of material, some of which
are unioned together and some subtracted to leave holes or
voids (Requicha and Voelcker [1982], Pratt [1985]). Various
groups of researchers have demonstrated that it is possible to
define a free-form primitive in the context of this type of
system as the volume swept out by the motion of some simple
object along a space curve. One example is RAYMO (van Wijk
[1984]), in which the motion of a sphere with varying radius
sweeps out a volume. The PADL-2 system previously mentioned
effectively uses volumes of this type to define fillets of the
'rolling ball' type, in which the fillet surface is the
envelope of the motion of a sphere rolling in contact with
the two surfaces to be filleted (Rossignac & Requicha [1984]).
The Japanese TIPS system also now allows the use of primitive
volumes defined by sweeping (Shiroma et al [1982]). All these
are systems of the CSG type, and it therefore appears that some
consensus has been reached that the approach described is
appropriate for these systems. The second major class of
solid modelling systems is based on boundary representation, i.e.
the explicit representation of the surface of the modelled
object in terms of its individual faces, edges and vertices,
together with 'topological' information concerning the inter-
connections of all these elements (Requicha & Voelcker [1982],
Pratt [1985]). For modellers of this type conventional surface
modelling techniques which have been in use in the aircraft and
other industries for nearly twenty years (Faux and Pratt [1979])
are proving suitable. Examples include BUILD, with double-
quadratic surfaces (Jared and Varady [1984]), EUCLID, with
Bezier surfaces (Brun [1984]) and GEOMOD, with rational B-spline
surfaces (Tiller [1983]). It should be noted that several
boundary representation systems with free-form surface capabilities represent such surfaces in terms of assemblies of plane

facets for visualisation purposes and for the calculation of surface/surface intersections. EUCLID and GEOMOD fall into this class, while BUILD always operates in terms of the exact surface representation.

This section is only intended to give some indication of trends in free-form surface geometry for solid modellers; a fuller discussion of this important research area can be found in Varady and Pratt [1985], where the capabilities of a wider range of systems are outlined. One major associated problem is that of the development of robust, accurate and economical algorithms for the computation of curves of intersection between free-form surfaces. This is essential in a solid modelling context since object edges are often defined by such intersections. Various approaches to the intersection problem are surveyed in Pratt and Geisow [1985].

3. DATABASE ISSUES

Most solid modelling systems are provided with their own specialised database which is specifically designed to support efficient interactive use of the system. Boundary representation systems usually use a highly redundant network structure with nodes representing individual faces, edges and vertices while pointers represent their topological interconnections. There are conflicting requirements in the design of such a database. Firstly it is desirable to store a large amount of redundant data explicitly, despite the fact that much of it can be derived by computation from other information in the system. This allows efficient responses to various types of interrogations; speed is, of course, at a premium in an interactive system. However, if the degree of redundancy is too high a very large number of changes in the datastructure will result from even the simplest change to the model, so that although interrogations may be answered efficiently the response to modelling commands is poor. There is a considerable art in balancing these two aspects of database design. Experiments in using general-purpose databases for solid modelling have shown that they are not very suitable.

CSG Modellers are based on a primary data structure of the tree type, as already mentioned. This provides compact storage for the model at a high conceptual level, but for most applications purposes low-level information is required such as occurs explicitly in a boundary representation data structure. This leads to the necessity for 'evaluation' of the CSG tree, and in many cases the setting up of a secondary data structure of the boundary representation type when required for any specific purpose. In this case, however, the network structure is evaluated from scratch and not built up incrementally as modelling proceeds. The secondary data structure therefore does not have to be optimised in the manner described in the last paragraph; since it is largely ephemeral it can be fairly crude in nature.

Most solid modellers are now fairly good at modelling individual parts, and many can also model assemblies of parts, in the sense of collections of objects positioned and oriented in space. However, for tolerancing purposes it is important to know which are the pairs of mating faces in an assembly, and this is an issue to which little attention has so far been given. What is required is the ability to set up logical relationships between mating faces of different components, and to attach associated information describing the nature of the fit between them. The German system COMPAC has some capability in this respect (Dassler et al [1982]), but the developers of most other systems have not yet fully addressed this problem.

It has been suggested that the solid model may eventually completely replace the drawing as the primary definition of a product to be manufactured. This may well be so, but attempts to move in this direction have revealed further database problems. It will be necessary to relate other entities to the model itself, including for example views, sections, bills of materials, process plans, machine tool and tool data files. If the model is of an assembly, then the assembly model must clearly be related to the component part models. There is a major problem here in maintaining consistency in all this related data; means must be devised which will ensure that a change in any one of the associated data files is accompanied by consistent changes in all the related files. Aspects of this difficulty are discussed in Eastman [1981] and Lee and Fu [1983].

4. LINKS TO AUTOMATED MANUFACTURE

Most existing commercial solid modellers provide an interface to at least one system for the generation of data for numerically controlled (NC) machine tools. The transfer of geometric data is usually in terms of parallel plane cross-sectional contours of the modelled object; the choice of sections is under the control of the user. Typically, the NC system accepts the contour information, calculates the appropriate offsets for the centre of the cutting tool and outputs data which will cause the machine tool to drive the cutter in contact with the part surface round the contour. This method is most suitable for parts which are $2\frac{1}{2}$D in nature, though doubly curved or sculptured surfaces can also be rough machined using the same terracing strategy.

The approach described involves some interaction on the part of the user and generally does not cause the part to be machined in an optimal way. Certain research teams have achieved a rather higher degree of automation by using a cellular decomposition of some volume containing the part to be made, in which cells are classified as internal to the part, external to it or containing some area of its boundary (Carey & de Pennington [1983], TIPS Working Group [1978], Yamaguchi & Kunii [1984]). Again, the machining process generated is not likely to be optimal.

Other approaches to the automatic generation of NC data from a solid model include CAPSY (Spur et al [1978]), AUTOMAC (Parkinson [1984]) and ROMAPT (Chan [1982]). In the first case the processes concerned are limited to turning and drilling, which are essentially 2D in nature. ROMAPT allows the user to select faces from a solid model, and then outputs the geometry of each chosen face as a line of source code in the NC programming language APT. When all the necessary geometry has been expressed in this way the user can add further statements to control the motion of the cutter with respect to the geometry already defined, the eventual outcome being that the desired shape is machined.

By contrast with most of the approaches described, true manufacturing automation must take into account a very large number of diverse factors and must generate a process plan which is optimal in some well-defined sense. The relevant considerations include all of the following:

- machine tools (and cutting tools) available in a particular organisation, together with their capabilities as regards power, precision etc;

- jigs, fixtures, clamps necessary to hold the part while it is being machined;

- feeds and speeds (linear and rotational velocities of the cutting tool) according to part material, desired finish etc;

- time (and hence cost) estimation for each machining process;

- number of parts to be made, since this has a bearing on what will be the most economical process.

The information supplied by solid modellers is at the moment limited mainly to part geometry, and this in itself is not sufficient to determine the machining processes for machining the part. A part has a function; to enable it to achieve that function a designer imposes tolerances on some of its dimensions and specifies surface finishes where appropriate. The manner in which individual features of a part are machined should depend upon this information. However, there is no commercial solid modeller today which can represent tolerance data in a way which is useful for process planning. How to do this is one of the major outstanding problems of solid modelling,

and until it is solved there can be little further worthwhile
progress towards true integration of CAD and CAM.

Apart from the tolerance problem there are other difficulties in using a solid modeller to represent part geometry for
manufacture.   The boundary representation systems seem to be
best suited for this purpose, but they represent the shape of
the part at a very low level, in terms of individual faces, edges
and vertices.   Process planning works in terms of higher level
constructs, however;   these are part features (or form
features) such as holes, pockets, slots, grooves and bosses.
Associated with any one such feature of a part is a small
number of possible manufacturing options.   The choice between
these options requires the presence of tolerance data in
particular, though size and surface finish also need to be
considered.   If tolerance information were available, then,
a knowledge of the features of a part would go a considerable
way towards determining the individual machining operations
needed to make it.   These operations would then have to be
sequenced in some optimal manner in order to generate the
overall plan.

If process planning is to be based upon this approach
then the solid modeller should have some means of representing
features.   Most (but not all) commercial systems lack this
capability.   The ROMULUS system (Shape Data Ltd. [1983]) allows
the user to associate groups of faces as features and subsequently manipulate them in various ways.   The automatic recognition
of various simple types of features has also been demonstrated
(Parkinson [1983], Staley et al [1983]).   However, it is the
present writer's opinion that the designer should be allowed
from the outset to design in terms of features.   If this is so,
then every feature he defines in the model can be represented
as a feature automatically on creation, and this information
later used in the determination of the process plan.   At
present the designer's intent is largely lost because the modeller's data structure is at too low a level;   it seems highly
wasteful to use automatic recognition techniques to reconstruct
his original intent at some later stage when it could easily
have been captured at the outset.

5.   USER INTERFACES TO SOLID MODELLERS

Most solid modellers are graphical interactive systems, and the
design process with such a system involves a two-way interchange
of information between the user and the computer.   The system
can help the user in various ways;   for example, it can prompt
him as to his possible choices of next action at any stage, and
can answer queries of various kinds regarding the model he has
constructed.   The most important feedback from the system is
visual, however.   The designer is provided with a visual
representation of what he is designing;   he can furthermore

choose the nature of this representation in most cases, and
this is a feature of inestimable value. Wireframe representations are quick to compute and are often useful. They can
be enhanced for better 3D visualisation by such devices as
local hidden line removal (a 'quick and dirty' approximation
to full hidden line removal) or depth cueing (intensity
modulation of lines as they recede from the viewer).   On
some refresh and raster terminals there is also the possibility
of real-time rotation of wire-frame representations which,
particularly when combined with true perspective projection,
gives a very strong sense of 3D perception.   Full hidden line
removal is a computationally expensive process, but again
greatly aids correct visualisation.   Shaded surface graphics
is becoming increasingly common in use, often on terminals
with colour capability.   The use of specialised hardware will
in the next few years allow real-time rotation of shaded
surface pictures with full hidden surface removal.

Various devices may be made available to the user to
enable him to communicate with the system.   Some modellers are
entirely keyboard-oriented, while others use light-pens,
tablets with styluses and all the other devices conventionally
associated with interactive CAD/CAM systems.   Menu-based
interfaces are popular, although some modellers are mainly
command driven.

In some respects the most interesting component of user
interface design is the procedural aspect.   This is concerned
with the user's 'conceptual model' of what he is designing and
the processes he is using to design it.   For example, most
early CSG modellers required the user to work entirely in
terms of 3D volumetric elements which he created, scaled and
positioned appropriately before invoking the boolean operations
of addition or subtraction.   Such a mode of operation requires
genuine 3D thinking, which does not come easily to many traditionally trained draughtsmen accustomed to working in 2D.
Furthermore, most systems (and this is also true of boundary
representation modellers) do not provide good facilities for
aiding the designer to position his volumetric elements
precisely where he wants them with respect to each other.
There is much scope for improvement in this respect.

Most boundary representation systems (and, increasingly,
CSG systems) provide the user with some facilities for initial
design in 2D.   Profiles can be created and then swept, linearly
or rotationally, to define volumes which can then be combined
using boolean operations.   Some systems, for example BUILD
(Anderson [1983]) go further, and allow the definition of
profiles on plane faces of existing objects, which may then be
swept inwards or outwards to give a depression or a projection
respectively.   This approach is not only powerful but is also
computationally economical since it does not require the use of

expensive boolean operations. Such techniques should be more widely provided. The BUILD system mentioned earlier also exhibits several other user interface features which are powerful yet economical. This includes a set of 'local' operations which affect only a limited part of the modeller data structure. Some of them do not even affect the topology of the object at all, but merely effect local modifications to its geometry. Such methods are not without disadvantage, however; it is the user's responsibility to ensure that the use of a local operation results in a meaningful object, since checks against undesired changes in object topology would be expensive to provide.

One suggestion which is re-emerging after a period in limbo is the provision of a user interface design in terms of the form features mentioned in the last section. The advantage is that once the designer has specified a feature the existence of that feature is known to the system and the information can subsequently be used for process planning. This apparently avoids the need for automatic feature recognition. The chief problem is that design features do not always correspond with manufacturing features; where they do matters are straight-forward, but where they do not the manufacturing features are usually in some sense complementary or inverse to the design features and some form of automatic recognition must be used to identify them. Another advantage of designing in terms of features is that it may be possible to create local datum frames automatically whenever features are called into existence. This will aid not only in the positioning of features in the model but also in the association of tolerance data with them. As already pointed out, such information is vital for process planning.

6. OTHER SOLID MODELLING INTERFACE ISSUES

There is widespread interest internationally at present in the problems of transmitting product data between dissimilar CAD/CAM systems. Communication of this type is necessary both within individual companies using more than one system for different purposes and in a contractor/subcontractor situation where dissimilar systems are involved. The most highly developed medium for inter-system data transmission is currently IGES (ANSI [1982]). This is capable of transmitting data represent-ing wireframe and surface models in its present form. The organisation CAM-I has developed a neutral format called the Experimental Boundary File (XBF) based on IGES, for the transmission of solid modelling data (CAM-I [1981]). The requirements include the representation of topological as well as geometrical and other associated data. The IGES Advanced Geometry Subcommittee in the USA has also now defined a trans-mission format to cope with solid modelling data. This is the Experimental Solids Proposal (ESP), which is at present under-going practical tests to verify its viability before being

incorporated in Version 3 of IGES some time late in 1985. The
design of the IGES ESP was heavily influenced by the earlier
work of CAM-I already mentioned. Another U.S. proposal in this
area is PDDI (Product Data Definition Interface), which is part
of the output of a major program in Integrated Computer Aided
Manufacture (ICAM) financed by the U.S. Air Force (U.S. Air
Force [1984]). There are further proposals stemming from
France and Germany which will influence efforts towards international standardisation of means for product data transmission.
Finally, the Commission of the EEC has recently agreed to fund
a major investigation of the problem, covering the entire
spectrum of systems from 2D draughting systems up to solid
modellers, under their ESPRIT scheme. The work will be carried
out by ten organisations in five countries, and will involve
a substantial element of wide-area networking of the transmitted
data. It is likely to be several years before all the interested parties reach agreement on the best transmission format, so
that international standardisation is likely to be a long drawn-
out process.

There is a growing realisation of the importance of yet
another type of interface. This involves communication between
a CAD system and an application program rather than between one
CAD system and another. The advantage of a standard interface
for application programs would be that a single application could
be written which would interface equally well with any one of a
range of CAD/CAM systems. The CAM-I organisation has been
active in this area also, and has commissioned the design of an
Applications Interface Specification (AIS), which consists of
specifications of a number of FORTRAN subroutines reflecting the
creation and interrogation facilities to be expected of a typical
solid modeller (CAM-I [1980]). If the AIS is implemented for a
selection of modellers then a FORTRAN application program (for
example a finite element mesh generator or an automatic process
planning system) should be able to interface with any of them
equally well, communicating by means of the standard set of
subroutine calls. Both the AIS and the XBF mentioned earlier
have been tested by the Lucas Group [Pratt and Wilson 1984].
Shape Data have just released a Kernel Interface for their
ROMULUS modeller; this is based heavily on the AIS, and contains
in its first version 101 subroutines (Shape Data [1984]). Other
vendors of solid modelling systems have been approached by CAM-I
and several are very enthusiastic about the AIS concept, which is
likely to be refined during 1985 before testing of a number of
implementations. It is intended eventually to seek international
standardisation for an interface of this type.

7. CONCLUSION

Solid modelling is currently in a state of transition. It is
not so long ago that the first such system emerged from the
universities, and many current modellers are already quite

accomplished as regards pure geometric definition. Geometry
is only part of the story, however. The purpose of this paper
has been to point out that there are several other areas where
much intensive research is needed before solid modellers can
begin to play their destined key role in truly integrated
design and manufacturing systems.

8. REFERENCES

Anderson C (1983) The New BUILD User Guide. CAE Group
Document No. 116, Computer Aided Engineering Group, Cambridge
University Engineering Dept., July 1983.

ANSI (1982) Digital Representation for Communication of
Product Definition Data. ANSI Y14.26M, American National
Standard, American Society of Mechanical Engineers, New York.

Brown CM (1982) PADL-2: A Technical Summary. IEEE Computer
Graphics & Applications, March 1982, 69 - 84.

Brun JM et al. (1983) The Use of EUCLID in the CAM-I Test.
In Faux |1983|, 203 - 238.

CAM-I (1981) CAM-I Geometric Modelling Project Boundary File
Design (XBF-2). Report No. R-81-GM-02.1, CAM-I Inc.,
Arlington, Texas.

CAM-I (1980) An Interface between Geometric Modellers and
Applications Programs (3 vols). Report No. R-80-GM-04,
CAM-I Inc., Arlington, Texas.

Carey CG, de Pennington A (1983) A Study of the Interface
between CAD and CAM using Geometric Modelling Techniques.
Proc. 3rd Anglo-Hungarian Seminar on Computer Aided Geometric
Design, Cambridge, September, 1983, published by Cambridge
University Engineering Dept.

Carter WA, ed. (1979) Proc. Geometric Modelling Seminar,
Bournemouth, November 1979. Document No. P-80-GM-01, CAM-I
Inc., Arlington, Texas.

Chan BTF (1982) ROMAPT: A New Link between CAD and CAM.
Computer Aided Design 14, 261 - 266.

Dassler R et al. (1982) Databases for Geometric Modelling and
their Applications. In Filestructures and Databases for CAD,
eds. J. Encarnacao & FL Krause, North-Holland Publ. Co.

Eastman CM (1981) Database Facilities for Engineering Design.
Proc. IEEE 69,10, 1249 - 1263.

Faux ID, ed. (1983) Proc. 2nd CAM-I Geometric Modelling Seminar,
Cambridge, December 1983. Document No. P-83-GM-01 (2 vols.),
CAM-I Inc., Arlington, Texas.

Faux ID, Pratt MJ (1979) Computational Geometry for Design and
Manufacture, Ellis Horwood, Chichester.

Jared GEM, Varady T (1984) Synthesis of Volume Modelling and Sculptured Surfaces in BUILD. Proc. CAD 84 Conf., Brighton, April 1984, IPC Science and Technology Press.

Lee YC & Fu KS (1983) A CSG-based DBMS for CAD/CAM and its supporting Query Language. Proc. Database Week, San Jose, May 23 - 26, 1983, Computer Society Press.

Parkinson A (1983) Feature Recognition and Parts Classification in BUILD. CAD Group Document No. 112, Cambridge University Engineering Laboratory.

Parkinson A (1984) An Automatic NC Data Generation Facility for the BUILD Solid Modelling System. Proc. 16th CIRP International Seminar on Manufacturing Systems, Tokyo, 1984.

Pratt MJ (1984) Eurographics 1984 Tutorial Notes, Springer Verlag.

Pratt MJ, Geisow AD (1985) Surface/Surface Intersection Problems. Proc. IMA Conf. on Mathematics of Surfaces, Manchester, Sept. 17 - 19, 1984, ed. JA Gregory, Oxford University Press.

Pratt MJ, Wilson PR (1984) IGES-based Transmission of Solid Modelling Data. Proc. MICAD 84 Conf., Paris Feb/March 1984, Hermes Publishing, Paris.

Requicha AAG, Voelcker HB (1982) Solid Modeling - A Historical Summary and Contemporary Assessment. IEEE Computer Graphics & Applications, March 1982, 9 - 24.

Rossignac JR, Requicha AAG (1984) Constant-Radius Blending in Solid Modeling. Computers in Mechanical Engineering, July 1984, 65 - 73.

Shape Data Ltd. (1983) Modeller Test - ROMULUS. In Faux |1983|, 324 - 367.

Shape Data Ltd. (1984) ROMULUS Kernel Interface Reference Manual, Release 1.0, November 1984.

Shiroma Y, Okino N & Kakazu Y (1982) Research on 3D Geometric Modelling by Sweep Primitives. Proc. CAD 82 Conf., Brighton, March 30 - April 1, 1982, Butterworths.

Spur G et al. (1978) CAPSY - A Dialogue for Computer Aided Manufacturing Process Planning. Proc. 19th MTDR Conference, Manchester.

Staley SM, Henderson MR & Anderson DC (1983) Using Syntactic Pattern Recognition to Extract Feature Information from a Solid Geometric Database, Computers in Mechanical Engineering, Sept. 1983, 61 - 66.

Tiller W (1983) Rational B-Splines for Curve and Surface Representation. IEEE Computer Graphics and Applications, Sept. 1983, 61 - 69.

TIPS Working Group (1978) TIPS-1 Technical Information Processing System. Institute of Precision Engineering, Hokkaido University, Japan.

US Air Force, Department of (1984) Product Definition Data Interface (5 vols.). Obtainable from Computer Aided Manufacturing International, Inc., Arlington, Texas, as documents DR-84-GM-01 to -05.

Varady T, Pratt MJ (1985) Design Techniques for the Definition of Solid Objects with free-form Surfaces. To appear, Computer Aided Geometric Design.

Voelcker HB et al (1978). The PADL-1.0/2 System for Defining and Displaying Solid Objects. Computer Graphics $\underline{12}$, 3 (Proc. SIGGRAPH 1978), 257-263.

Van Wijk JJ (1984) Ray Tracing Objects defined by Sweeping a Sphere. Proc. Eurographics 84.

Yamaguchi K et al (1984) Computer Integrated Manufacturing of Surfaces using Octree Encoding. IEEE Computer Graphics & Applications, January 1984, 60 - 65.

EUCLID(*): A POWERFUL TOOL FOR INTERACTIVE CAD AND SPECIALIST SOFTWARE DEVELOPMENT.

K.R.Colman.

European Organization for Nuclear Research, Geneva.

1. INTRODUCTION

The Euclid Computer Aided Design (CAD) system, supplied by Matra Datavision of France, was installed at the European Organization for Nuclear Research (CERN) in September 1982. Two of the principal reasons for its selection were that it is a full 3D solid modelling package, and that it has a large potential for user software development.

The 3D interactive system has been responsible for a tangible increase in the quality of drawings, but due to the response time of Euclid, an increase in production is more difficult to assess. The programming potential has been fully exploited, with real increases in quality and production being experienced.

To take advantage of Euclid's potential, some 28 existing members of the Installation and Mechanical engineering group of the Large Electron Positron collider division (LEP-IM), have been trained to use the interactive system. In addition, personnel with previous programming experience have been trained in the principles of Euclid programming.

2. THE LEP PROJECT

Euclid's principal role at CERN is to assist in the design and installation of a Large Electron Positron Collider (LEP) [1]. The LEP machine is to be housed in an underground tunnel of circumference 27km, cross-sectional diameter 3.8 metres at an inclination of 1.4%. It will lie partially beneath the Jura mountains and traverse the Franco-Swiss frontier four times.

(*) EUCLID is a trade mark of Matra Datavision.

Large underground experimentation halls will house complex
particle detectors that encompass the machine. LEP itself will
consist of a circular vacuum chamber surrounded by accelerating
cavities and magnets, all of which must be adequately supported,
cooled and powered.

Euclid therefore has to cope with the full range of civil,
mechanical and electrical engineering. The examples in this
paper, taken from the LEP project, reflect this requirement.

3. HARDWARE

CAD can not exist solely as software; the performance of any
system is greatly influenced by the hardware on which it is
designed to run. In this aspect Euclid is fairly flexible, it
can be made to run under many configurations.

Central computers
The majority of installations are based on a central computer.
Among the manufacturers supported are DEC, IBM, SEL, PRIME and
UNIVAC. The most popular being the DEC VAX.

Work stations
Four basic types of terminal can be used. Virtually any make of
graphics terminal with a storage or cathode ray tube can be used.

The monochrome work station, based on a Tektronix 25inch GMA125
storage tube, graphics pad and some local intelligence is the
most widespread. However it is to be discontinued since
Tektronix announced their decision to phase out their tube.

The colour work station, based on a 19inch Lexidata raster screen
is the colour alternative to the monochrome workstation.

All the above work stations must be connected to a central
computer. For small installations a stand-alone system will soon
be available. Essentially the same as the colour station, the
mono station also has a Micro VAX 2 and 50 megabytes of disk
storage.

Plotters
Almost any plotter can be used, both electrostatic and pen. If a particular model is not supported then Matra Datavision, as was the case with CERN, may be persuaded to release the appropiate source codes for user development.

CERN installation
CERN, which at present has the largest single Euclid installation, is based on two VAX 11/785's connected in a VAX cluster, each with 8 megabytes of memory. 20 monochrome workstations and 4 Tektronix 4016 graphic terminals grouped in remote clusters provide interactive access to the system. 8 Hewlett Packard pen and 4 Versatec electrostatic plotters provide the means of drawing output. In the near future one of the 785's will be replaced by the new VAX 8600.

4. SOFTWARE

The Euclid system first came into existance in the early 1970's in the form of a programmable package. Since then it has undergone considerable development so as to introduce interactive use, comprehensive data bases, 2D for drawings and numerous useful extensions. The power of the basic 3D package has also been augmented.

Today, Euclid is a full 3D solid modelling system that can be used not only for mechanics, but also supports electrical schematic design, kinematics and numerical control. It is however the 3D mechanics that have proven to be the most useful feature in constructing LEP.

Full 3D allows the creation of accurate projections of an object from any view point, not just the three principal views of a classical drawing. All Euclid designs at CERN therefore contain a fourth 3D view which gives an overall idea of the shape of the object. This has proved very useful for people, who while being experts in their own field do not possess the skill to read a 2D drawing.

Full 3D also aids in the design of objects, as the user can not use the system as an electronic drawing board. He is forced to think in terms of the 3D object. To enforce this, a range of creation and manipulation tools are available. At the top end are functions to fuse objects together or find their common volume. Use of these tools allows the user to create successively more complex 3D objects until he arrives at the desired result. He can then pass into 2D mode and add textual information to make his plan.

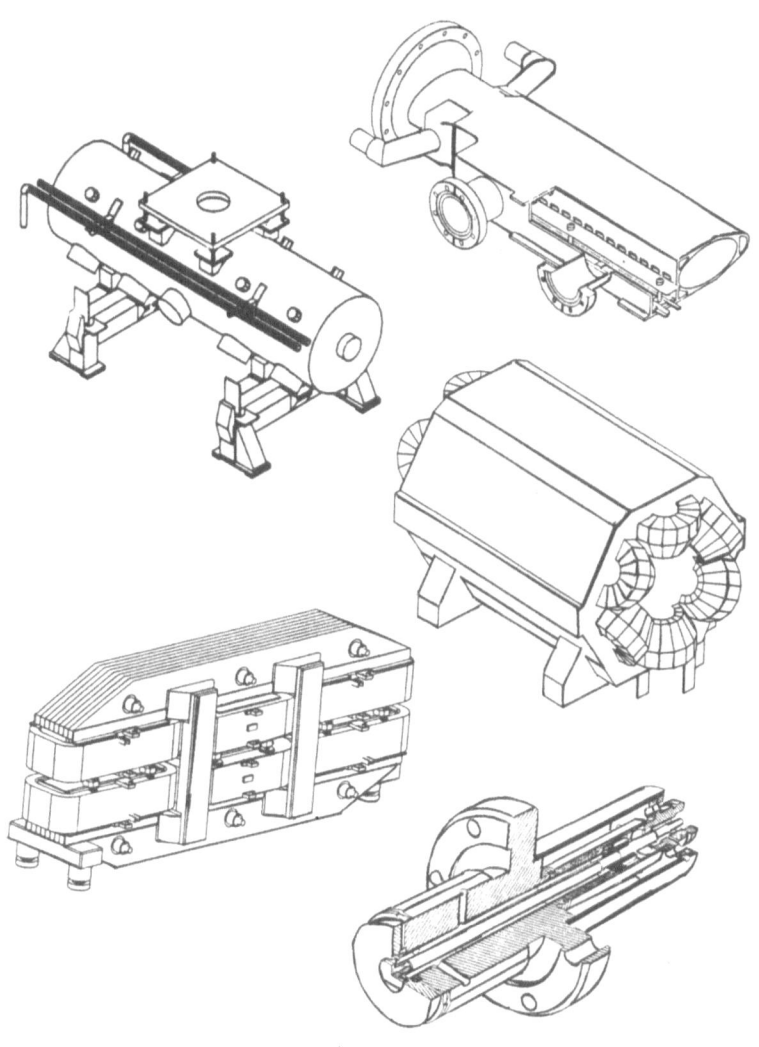

Figure 1. 3D solids. A vacuum chamber, accelerating cavity,
defocusing sextupole magnet, dipole wiggler magnet,
and beam position monitor.

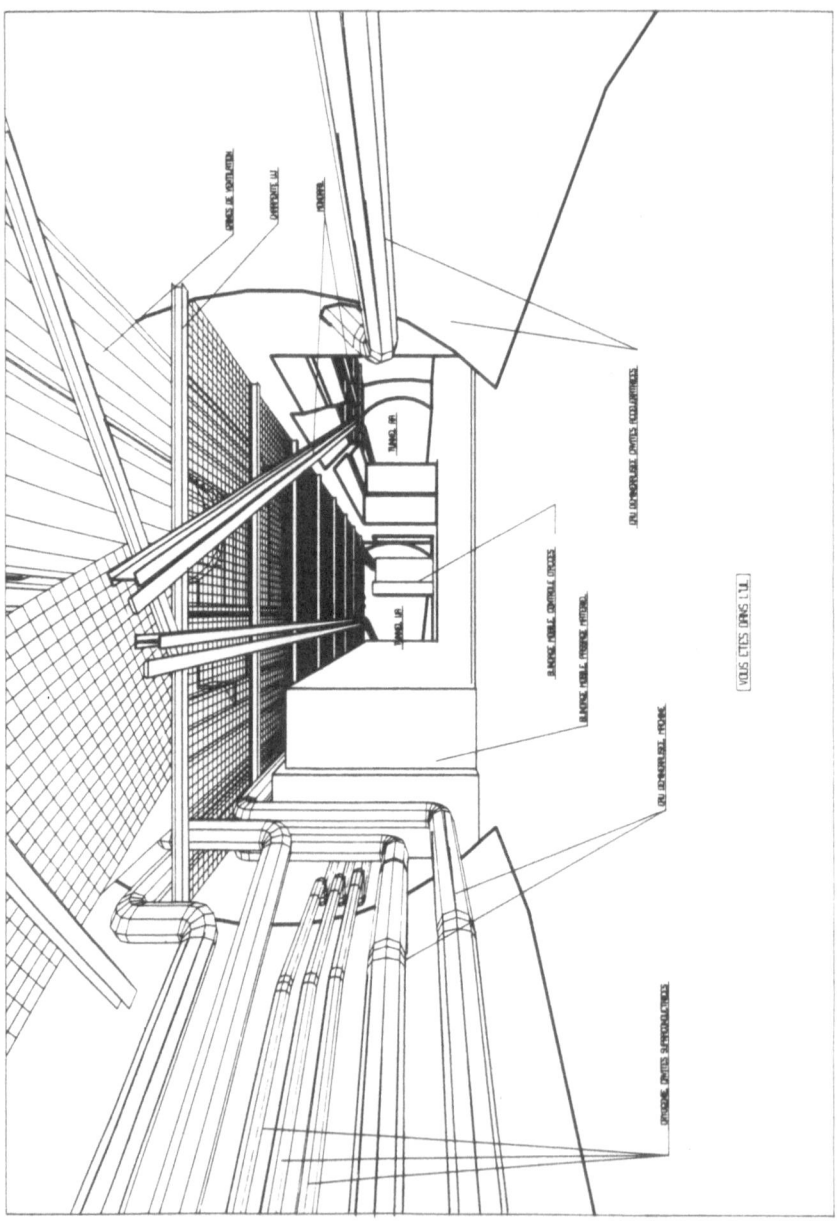

Figure 2. Large assemblies of 3D solids.
Part of the LEP tunnel.

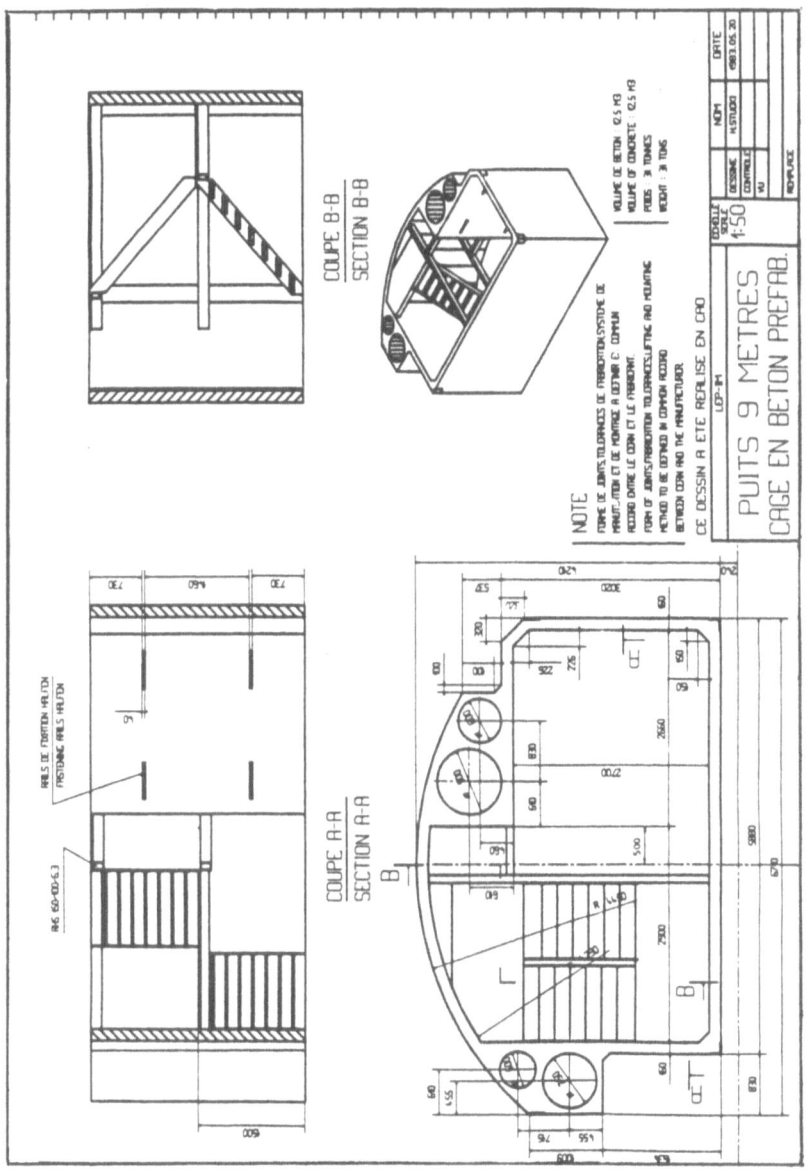

Figure 3. 3D enhancement of a 2D drawing.
A prefabricated access shaft.

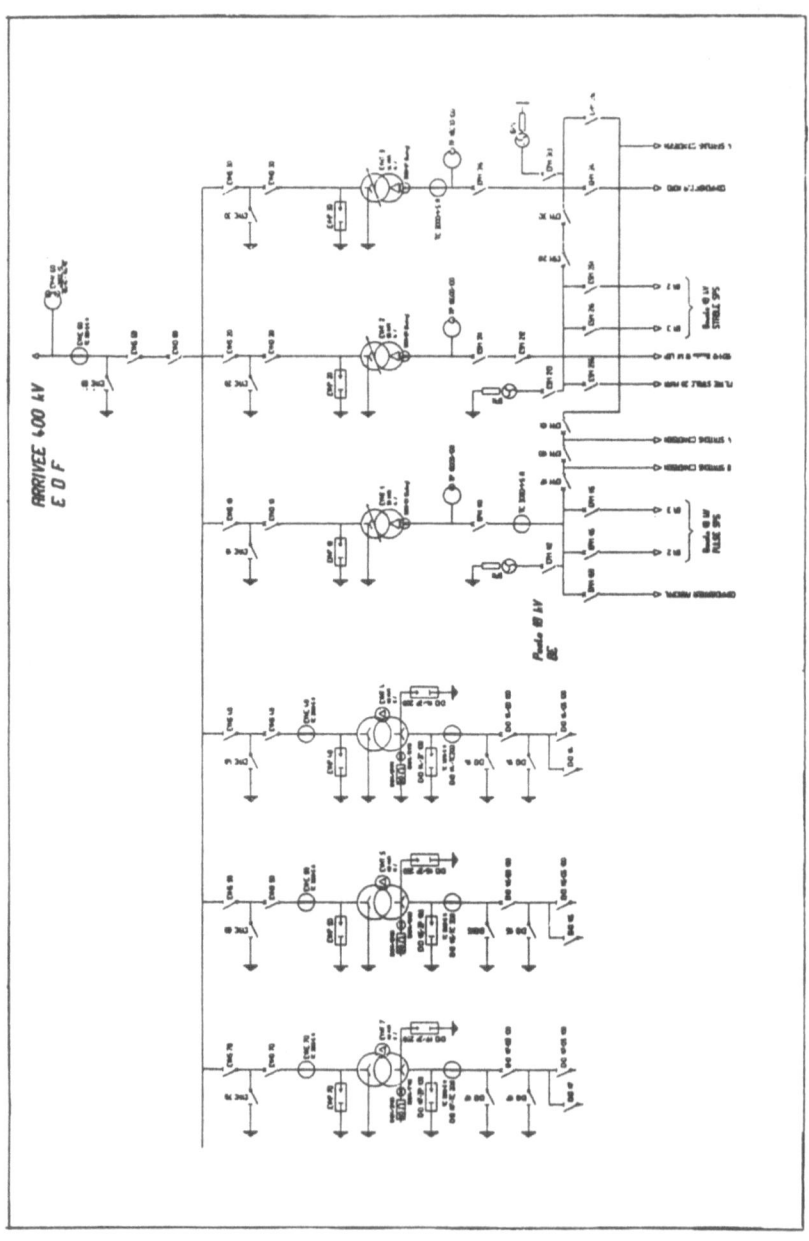

Figure 4. Pure 2D. An electrical schema.

5. INTERACTIVE EUCLID

Visualisation

Ten standard view types are available. The six standard 2D
views, front, back, top, bottom left and right, plus four 3D
views, axonometric (parallel lines remain parallel with
distance), perspective (parallel lines converge with distance),
panoramic (perspective with cylindrical lens) and fish-eye
(perspective with spherical lens).

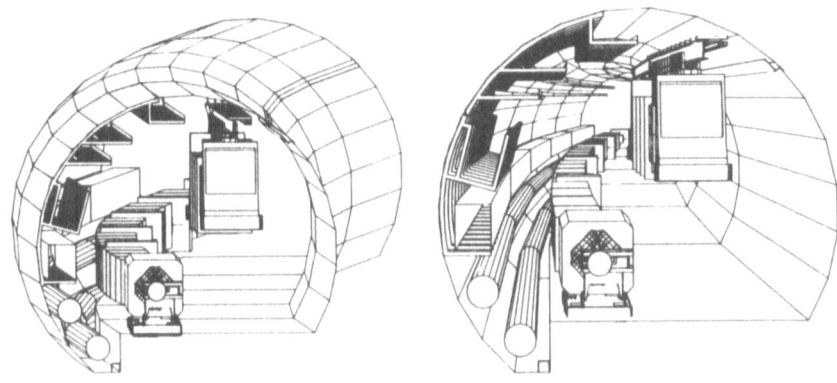

Figure 5. Axonometric and perspective views of the LEP tunnel.

All 3D views are calculated using the concept of an eye position
and focal point which define the line of sight. It is therefore
possible to modify the line of sight and look at an object from
various positions.

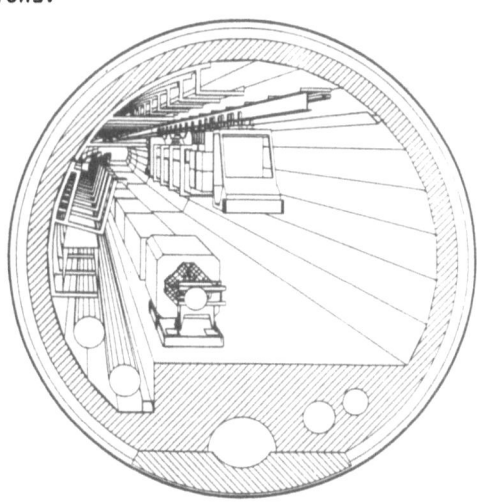

Figure 6. Line of sight change.

Measures

As 3D objects are represented as solids, and not as surfaces or
wire models, a full range of measuring utilities are available.
One is therefore able to obtain the volume, mass, surface area,
perimeter, centre of gravity and moments of inertia of objects.
It is also possible to check if two objects physically touch or
intersect.

Two dimensional work

Substantial possibilities to transform a 3D object into a form
presentable on a 2D plan also exist. This process consists
essentially of adding dimension lines, textual comments, hatching
and axes. One is able to work directly with the 3D object, so
dimension lines will automatically record the correct distance.
The following example shows such a drawing of a flange, for LEP.

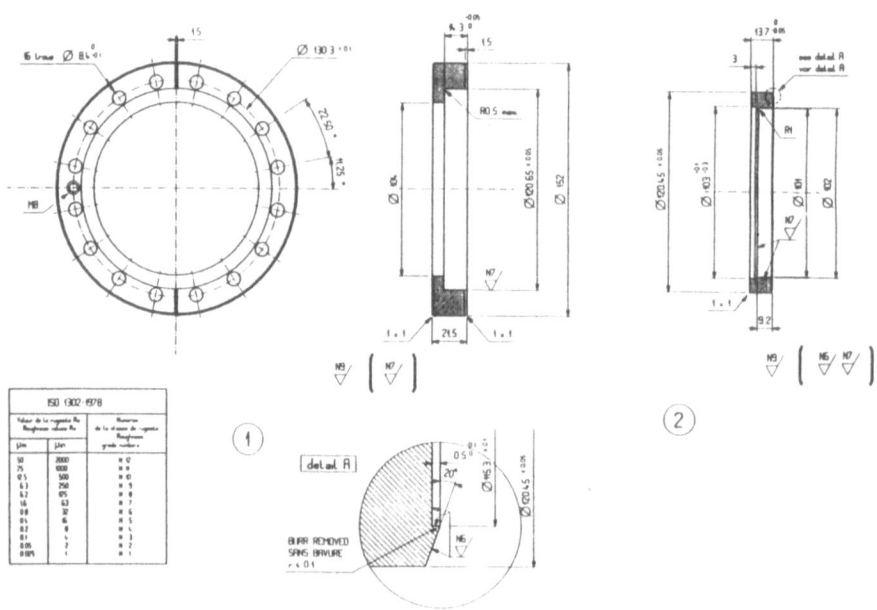

Figure 7. Dimensioning. A flange fabrication drawing.

6. EUCLID PROGRAMMING

One of the most powerful features of Euclid is the existance of an easy to use programming interface to its interior structure. Experience at CERN has shown that previously untrained draughtsmen can be creating Euclid extensions within a week, and those with prior programming experience can teach themselves from the manuals.

Because of its early origin, before the advent of many of today's modern programming languages, and for reasons of transportability, Euclid is written almost exclusively in Fortran IV. All user extensions therefore are generally also written in Fortran, but other languages may be used in certain circumstances.

Euclid extensions fall into two categories, being either additional tools for the interactive user (applications), or completely separate programs (batch).

Applications
LEP applications are classified under a number of headings.

Creation applications have allowed the user to create standard parts by providing their principal parameters. A hexagonal head bolt is defined by simply giving its diameter. More complicated examples allow the creation of large vacuum chambers and the assembly of existing elements.

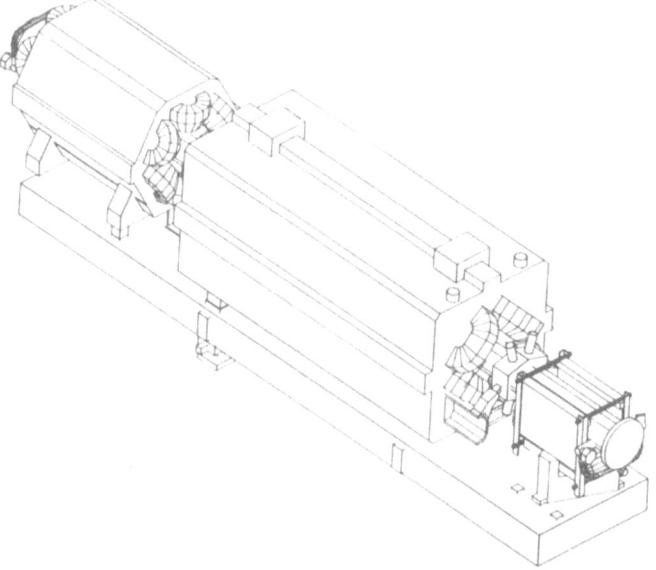

Figure 8. An assembly of LEP beam line elements.

Tool applications allow the user to perform specific operations on objects already created. This ranges from the automatic drilling of holes to the transportation of large magnets along an overhead monorail to test if they will hit any fixed objects.

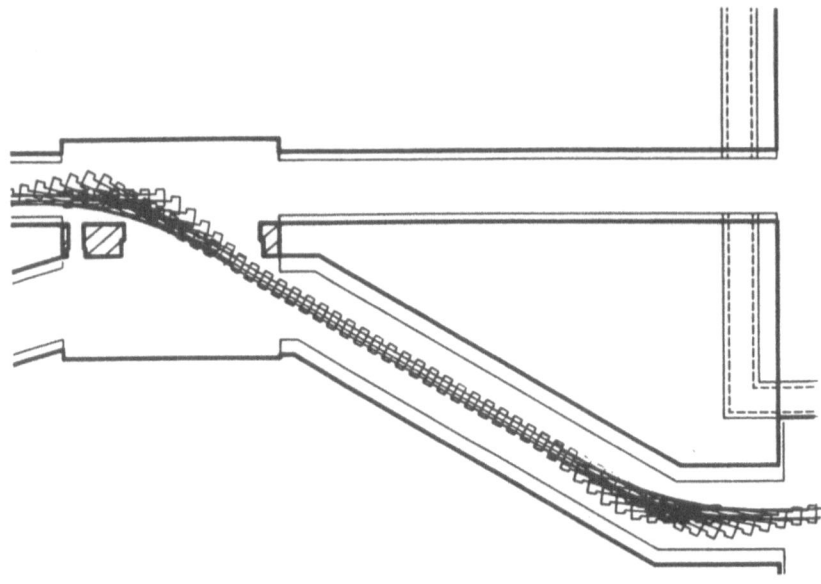

Figure 9. Passage of a dipole pair in the LEP tunnel.

Measuring applications allow the extraction of specific statistics about an object. One example decodes the physical construction of a vacuum chamber, which allows a calculation to be made of the thickness of solid material an atomic particle has to traverse to reach a detector. Another allows multiple cross sections to be made, and the centre of gravity and moments of inertia to be extracted from the resultant surfaces. These applications consist of two parts, the data extraction and analysis. The analysis step is normally performed outside the CAD framework by non-Euclid software. This type of application therefore provides a method of connection between Euclid and other specialist software.

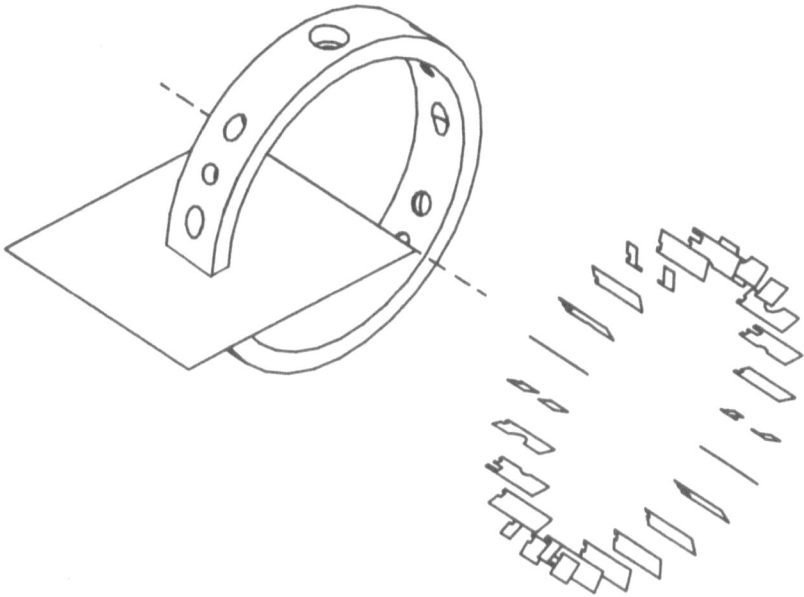

Figure 10. Multiple section measuring.

Batch

A batch program differs from an application in that it is not accessible from an interactive Euclid session. It performs a pre-defined repetitive task based upon user supplied data.

LEP, which contains more than 1,000 component types, with approximately 50,000 occurences of these components, is of such complexity that it is only feasible to define its structure by using computers. The dynamic nature of LEP also means that whenever an update is made, a complete new set of 24 layout drawings must be produced. This task can take up to 50 man hours, even with CAD.

During the day, Euclid is very heavily loaded with "normal" interactive work, and the task of creating layout drawings would bring the system to an complete halt. Such a situation is to be avoided, not only because of its economic insanity, but also as it has a very negative effect on users. Therefore, a specially developed batch program now allows this task to be performed overnight. It also eliminates all user errors that are inevitable in such a mammoth operation.

This batch program has a number of other advantages. It liberates work stations for interactive use, it reduces the day time load on the central computers, and by performing its task during off-peak hours, it allows resources to be used more economically.

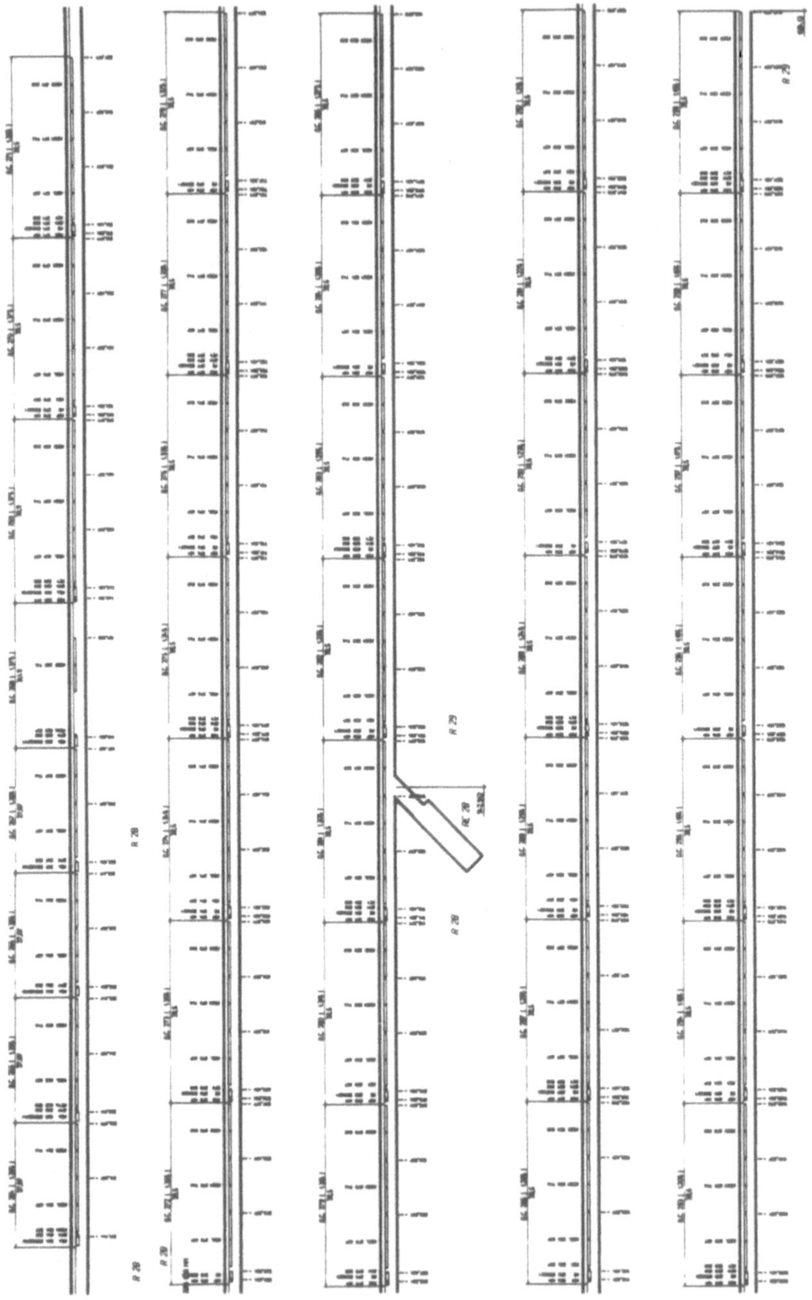

Figure 11. One of the 24 LEP layout drawings.

Other batch programs are used to perform pre-defined operations
on individual elements.   In some drawings of small regions of
LEP, only the relative positions of elements are of interest.
However the varing length of elements, ranging from 12 metres
down to a few centimetres prohibit a useful drawing being made.
The long elements must be drawn using a shortened representation.
A simple program performs this task, and ensures that the results
conform to ISO standards.

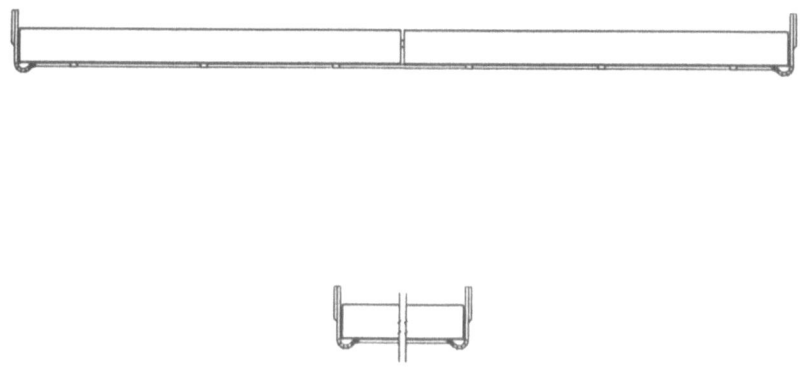

Figure 12. Shortened representation of a magnetic dipole pair.

More specialised programs can be used in a management context.
LEP machine elements that are considered to be in a final state
are stored in special databases known as standards.   A batch
program is used to take a copy of the original object from a user
"working" data base and transfer it to various standards. At the
same time it will translate the object to a pre-defined
position/orientation, optimise it (in terms of its data base
size), and produce a document describing it. The program must
also create two copies of the object, one in metres for the civil
engineers and one in millimetres for the mechanical engineers.
Such a task could take an hour interactively, but is done
overnight using batch.

The program supplied by Matra Datavision to plot drawings is
essentially a batch program. Because of this, LEP was able to
modify it for use with its Hewlett-Packard plotters and produce a
three fold improvement in plot time. Special pre-printed paper
can also be used; in this case the plotter will draw the title
block information directly into the appropiate box, which saves
time and increases quality.

Bug resolution

Like any advanced product in the field of information technology, Euclid contains a number of bugs. In general it is possible to avoid or work around known bugs, a task which has been aided by specially developed applications or batch programs. For example, the topological operation to remove a volume from an object will not always work if the object is too complex. An application specially developed in LEP, can perform this operation on small sub-units of the complex object, and then combine the results. The batch program mentioned above, to create metre and millimetre copies of an object, is another example of bug resolution.

7. PROGRAMMING PRINCIPLES

Any Euclid session , either interactive or batch is based on a data structure which contains the definition of all the objects currently available. This data structure consists essentially of three tables. The table of 3D points contains triples of real values. The table of liaisons contains connections between points, and defines open ended lines, closed loops, or plane surfaces. And finally, the table of elements defines logical assemblies of liaisons, and is used to represent continuous non-planar surfaces or solids.

Euclid entities are manipulated within a Fortran program as real variables. These variables, whose bit pattern have a special interpretation contain references to the data structure. The system used on VAX is illustrated below.

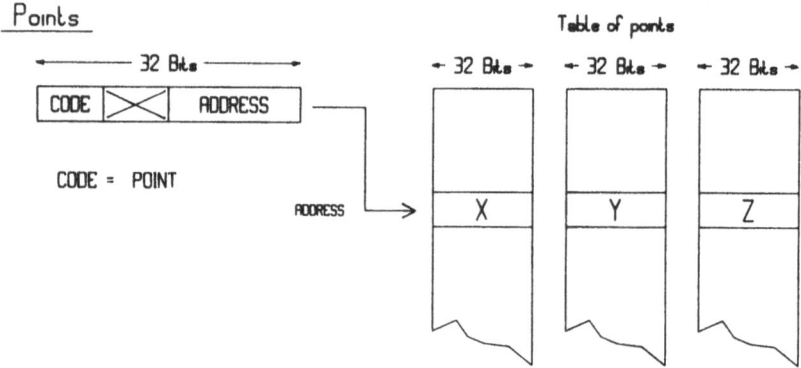

Figure 13. Representation of a point.

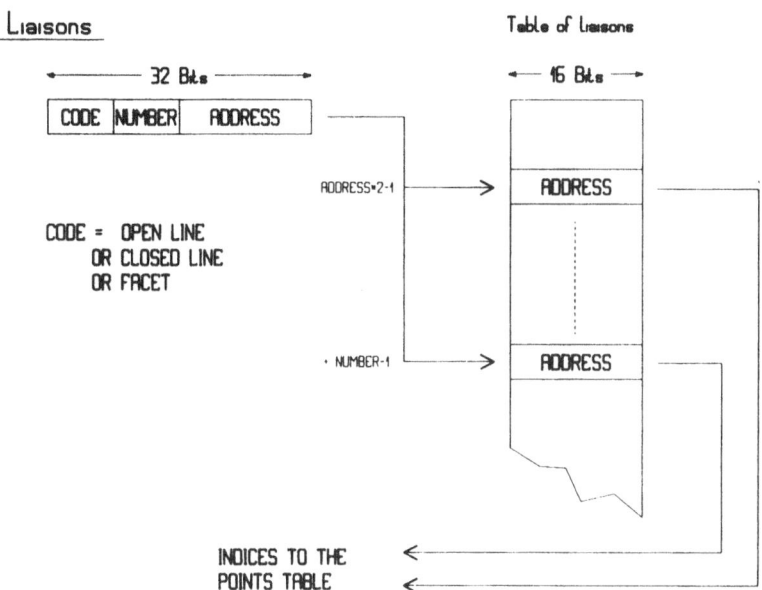

Figure 14. Representation of a liaison.

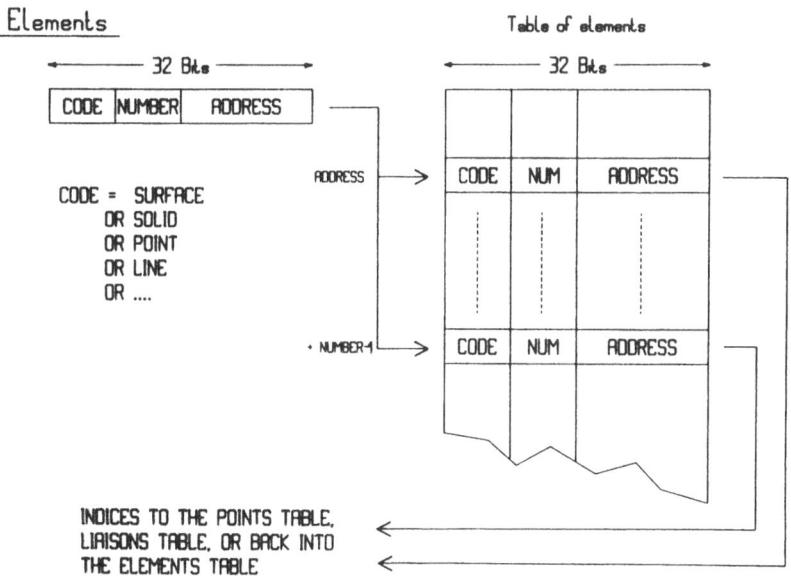

Figure 15. Representation of an element.

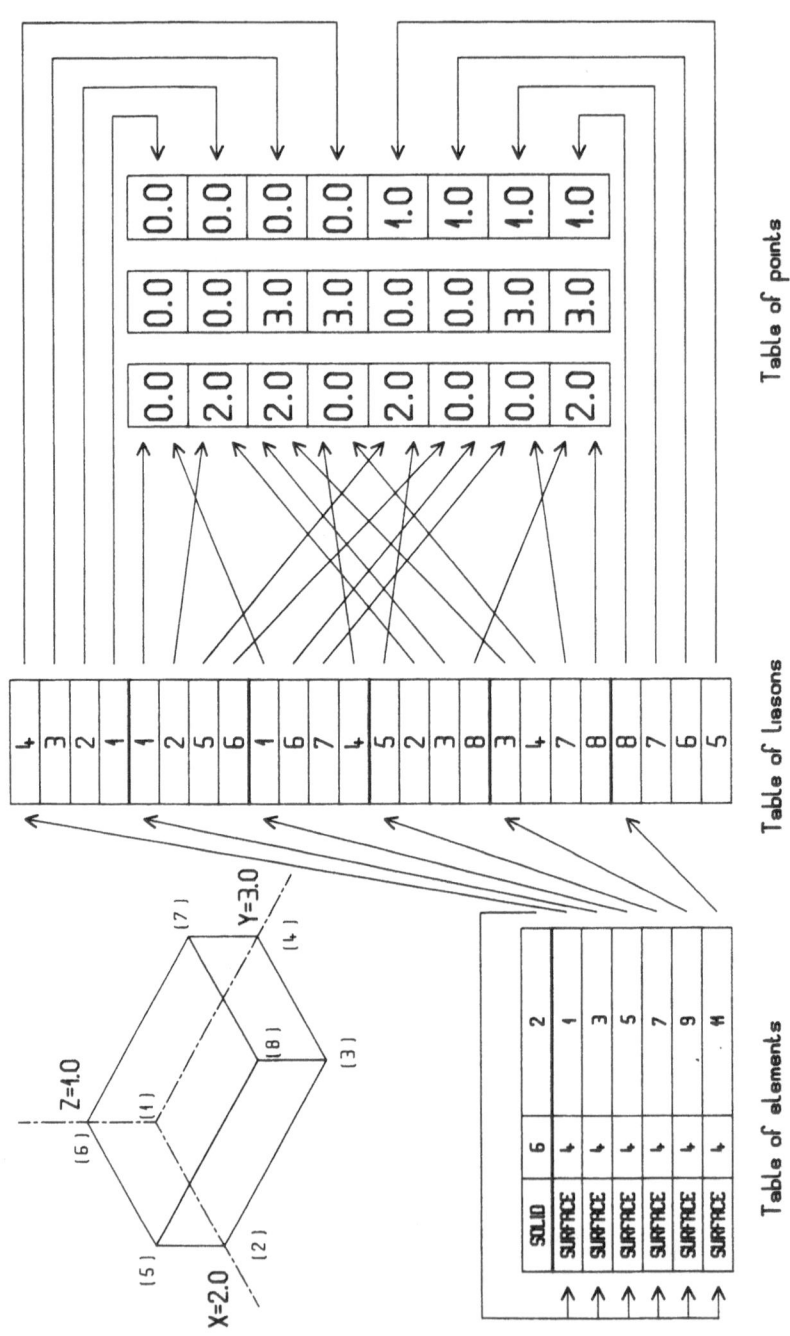

Figure 16. Complete data structure representation of a box.

However, knowledge of the data structure is unnecessary for the Euclid programmer as it is totally hidden from him by a pseudo high level language. To create a point he simply writes;

        P = POINT (5.9 , 10.4 , 7.24)

The function POINT is a Euclid resident routine. The local variable P will contain a reference to the 3D point 5.9,10.4,7.24. In a similar way a plane surface can be made.

        F1 = FACETE (P1 , P2 , P3 , ...)
    or F2 = FACETE (N , TABLE)

Where the P(i) are pre-defined points and TABLE is an array of N points. 3D objects are just as easily created.

        B = BOITE (X , Y , Z)

creates a box at the origin with side lengths of X, Y and Z.

        R = REVOLA (AXE , BASE , N , ANG)

creates a volume of revolution by rotating the surface BASE through an angle of ANG degrees, about the axis AXE. The resulting solid will have N exterior panels. Access to the topological functions could not be easier.

        F = FUSION (OBJ1 , OBJ2)

will fuse objects OBJ1 and OBJ2 together. Two subroutines

        RANGER and RAPPEL   (Store and recall)

provide direct access to the databases. Visualisation is trivial

        CALL VISION (PERS(OBJ), CADRER(50,50), VISE (0,0,0),
                      OEIL(500,500,500), ZOOM(10), CACHER(0))

will visualise the object OBJ in a perspective view with frame size 50 x 50. The focal point will be at the origin and the eye will be at the point 500,500,500. A zoom of 10 will be used and total hidden line removal will be employed.

8. CONNECTING EUCLID TO EXTERNAL SOFTWARE

The Euclid programming possibilities have provided the means by which interfaces have been created to other software, either supplied by other vendors or LEP developed. Some examples have already been cited.

These connections are in one of two directions. External data can be used to create Euclid objects, or previously created Euclid objects can be decoded and the data transmitted to other programs. This interface is normally achieved via an intermediate data file.

In LEP, use of this facility has been made with in-house programs. It is foreseen that Euclid will eventually be connected to the ORACLE data base management system, written by Tom Pederson International, and to a finite element system, yet to be selected. There is no reason why such work could not be extended into the field of numerical control.

9. LIMITATIONS AND PROBLEMS

Use of CAD in LEP has not only realised the advantages already cited, problems have of course also been encountered. Some of these problems are of a general nature and are applicable to any CAD system, while others are specific to the Euclid implementation.

Modifications

One of the best selling points of any CAD system is the speed with which modifications can be carried out. At the root of this feature there is usually a fully integrataed data base which contains one, and only one copy of each object. The idea is that when an object is modified, all other objects that make reference to it will automatically incorporate the modification.

This works well in a small user environment where everyone is aware of the work currently in progress, but on the LEP project, with more than 20 designers, the story is different. If a user leaves a modification in an unfinished state, or if it is not definitive, the cascade effect on other users making reference to the object can be catastrophic. In general, users have no exact knowledge of the work being carried out by their colleagues. Clearly this problem will apply to any CAD system using an integrated data base.

To bypass this, any standard LEP parts required by many users are copied into a special standard data base, and only this copy is referenced by other users. The users are therefore free to update their objects without fear of upsetting their colleagues. When a standard part does require updating, then it can be replaced in a controlled fashion, and all users making reference to it can be informed.

2D versus 3D

A common assumption made about 3D systems is that they should be able to automatically handle 2D. This is not in fact true as within the context of a 3D environment 2D requires special treatment.

For example, a 2D plane surface should, by its very nature, be visable in only one view. However, when combined with 3D, should it hide the 3D or should the 3D hide the 2D, or should there be no connection? Is it logical to treat texts and dimension lines in the way same as 2D objects? How should 2D be interpreted in the context of a topological operation? Many such questions can be posed and all need to be answered.

Data structure size

The Euclid data structure size is limited to 16K points, 32K liaisons and 32K elements. To significantly increase this maximum would require a total internal re-organisation of Euclid. Such an increase will however be counter productive as the visualisation response time is proportional to the square of the "complexity" of the object. This data structure size has proven to be a limitation at CERN.

Response time

No advanced CAD system responds instantly to user requests. The view of the LEP tunnel, presented in figure 2, took 2 hours real time (45 minutes of CPU) to be drawn. A design of the dipole wiggler magnet, presented in figure 1, takes 5 hours of real time to be drawn in front, top, left and axonometric views (during a normal working day).

This response time performance has been aggravated at CERN by an over-zealous training campaign that outstripped the capacity of the installation. However successive steps have been taken to rectify this by increasing the power of the CERN installation.

Matra Datavision are also taking steps to improve the response time by undertaking a massive consolidation of the program. This includes the re-writing of key modules in VAX assembly language. A new Euclid version, with a promised average performance increase of 3 is awaited with eager anticipation.

A second route to increased response under consideration at CERN, is the trend towards stand-alone systems with their own local micro VAX. This however is a very expensive option.

10. CONCLUSION

The Euclid system is a very powerful tool that is one of the leaders in its field. Its 3D capabilities have significantly improved the presentation of LEP drawings and designs. The programming possibilities have provided the means by which Euclid has been tuned to the LEP environment. Indeed, the programming side is now, in many ways, just as important as that of the interactive side.

However, using Euclid for LEP did not initially prove to be as easy as was hoped. A considerable ammount of groudwork needed to be completed before the true benifits, applicable to the LEP project, could be reaped. When evaluating the system, potential users must always keep in mind their own individual requirements and how Euclid will help realise them.

ACKNOWLEDGEMENTS

The preparation of this paper would not have been possible without the aid of many highly skilled Euclid users and programmers in the LEP division. Their work, presented here, is the foundation stone upon which CAD at CERN has been based.

REFERENCE

[1] K.Colman, C.Hauviller, M.Mottier - La CAO pour la construction et l'installation du LEP - Janvier 1984 - CERN-LEP-IM/84-05.

ON QUADRATIC SPLINES AND THEIR CAD-APPLICATION

Ferenc Fenyves — George Kovács

Computer and Automation Institute, Hungarian Academy of Sciences, Budapest, Hungary

1. INTRODUCTION

One of the important tasks of computer aided engineering design in geometric modeling. Because of theoretical and computational reasons most practical CAD-systems use parametric curves and surfaces (most commonly parametric cubic segments) in geometric modeling. Most basic ideas involved are presented in the survey paper [3] of Böhm, Farin and Kahmann, 1984. Our paper deals with quadratic splines and their application for curve and surface representation. The basic problem is to present a method for obtaining a smooth bivariate function which takes on certain prescribed values. The collection of these values is assumed to be on a rectangular grid, that is, for every point (x_i, y_j) on the rectangular grid $\{x_i\}_{i=0}^{n} \times \{y_j\}_{j=0}^{m}$ there is a given value z_{ij}.

The task is to construct a surface $S=S(x,y)$, (i.e. give an explicit expression for $S(x,y)$) that interpolates (passes through) the points z_{ij}, that is $S(x_i,y_j) = z_{ij}$, where $i=0,1,...,n$ and $j=0,1,...,m$. In Section 2 a short account of some results on quadratic (or parabolic) splines is presented using a selected form of the general quadratic splines, the so-called quadratic G-splines. The reason of introducing G-splines is that they will serve as a basic tool for a solution of the above mentioned problem, to construct interpolant surfaces, via bivariate functions, which are treated in Section 3. Thus Section 3 gives a new technique for representing continuous resp. smooth surfaces. The resulting piecewise bivariate polynomial functions are global representations and may be expressed in closed analytic form, allowing algebraic operations.

2. A REVIEW OF QUADRATIC SPLINES

Let us start by fixing some notations and by summarizing a few known facts for later reference. The reader is referred to [1] Ahlberg et al. 1967 and [2] de Boor, 1978 for background material and additional details.

Let $\{x_i\}_{i=0}^{n}$ be a given partition of a closed interval [a,b] of the real line, i.e.

$$a = x_0 < x_1 < \ldots < x_n = b \ .$$

A function $S(x)$ is called a quadratic spline on the interval $[a,b]$ if and only if it satisfies the following three conditions:

(i)   In each subinterval $[x_i, x_{i+1}]$, $i = 0,1,\ldots,n-1$, S is a quadratic polynomial denoted by P_i, say.

(ii) $P_i(x_{i+1}) = P_{i+1}(x_{i+1})$ for $i = 0,1,\ldots,n-2$.

(iii) $P_i'(x_{i+1}) = P_{i+1}'(x_{i+1})$ for $i = 0,1,\ldots,n-2$.

That is, $S(x)$ is a set of n quadratic polynomials with continuity properties up to the first derivative at the given inner points $x_i,\ldots,x_{n-1}$.

The points x_i are called the knots of the spline and the set $X = \left\{x_i\right\}_{i=0}^{n}$ is the knot vector. A section of the spline between two adjacent knots is referred to as a span, so P_i is the i-th span. A knot vector $X = \left\{x_i\right\}_{i=0}^{n}$ will always be assumed to   satisfy $x_i < x_{i+1}$ for $i = 0,1,\ldots,n-1$.

Assume that values $z_i = z(x_i)$; $i = 0,1,\ldots,n$, of a real function defined on $[a,b]$ are given and that $S(x)$ is a quadratic spline interpolant to $z(x)$, i.e.

$$S(x_i) = z_i \text{ for all } i \ .$$

Then a quadratic spline function $S(x)$ interpolating $z(x)$ at the points x_i and having the form

$$S(x) = P_i(x) = \frac{a_i}{h_i} (x - x_i)(x_{i+1} - x) + b_i (x - x_i) + z_i \ ;$$

where

$$x \in [x_i, x_{i+1}] \text{ and } h_i = x_{i+1} - x_i$$

is called a quadratic G-spline with coefficients $(a_i; b_i)$ for $i = 0,1,\ldots,n-1$.

So, the required expression $S(x)$ for the quadratic G-spline can be obtained by the following conditions:

(i)   The function $S(x)$ is piecewise polynomial of degree 2 such that

$$S(x) := P_i(x) := \frac{a_i}{h_i}(x-x_i)(x_{i+1}-x) + b_i(x-x_i) + z_i$$

   in each subinterval $x_i \leqslant x \leqslant x_{i+1}$ for $i = 0,1,\ldots,n-1$.

(ii) The function $S(x)$ passes through the values $\left\{z_i\right\}$ so that $S_i(x_{i+1}) = z_{i+1}$, i.e.

$$b_i = \frac{z_{i+1} - z_i}{h_i} \text{ for } i = 0,1,\ldots,n-1 \ . \tag{1}$$

(iii) The first derivative $S'(x)$ is continuous at x_i for $i = 1,2,\ldots,n-1$, which condition yields a set of linear equations for the coefficients a_i .

Namely

$$a_{i-1} + a_i = -B_{i-1} \quad \text{for} \quad i = 1,2,...,n-1. \tag{2}$$

where $B_{i-1} = b_i - b_{i-1}$.

From Eq.(2) it is obvious that we have a freedom to choose a_0 or a_{n-1}. Or, more generally, we may subject them to an additional equation

$$f_0 a_0 + f_1 a_{n-1} = f_2 \tag{3}$$

Where f_0, f_1 and f_2 are user specified with the requirement $f_0 \neq (-1)^n f_1$.

The resulting system of linear equations

$$\begin{cases} f_0 a_0 + f_1 a_{n-1} = f_2 \\ a_{i-1} + a_i = -B_{i-1} , \end{cases} \quad i = i, 2, ..., n-1 ; \tag{4}$$

has exactly one solution, and this can be found without any difficulty. Namely

$$\begin{cases} a_0 = \dfrac{(-1)^n f_2 + E_{n-2}}{(-1)^n f_0 - f_1} , \\ a_i = (-1)^i (a_0 + E_{i-1}) , \end{cases} \quad i = 1, 2, ..., n-1. \tag{5}$$

where $E_i = B_0 - B_1 + B_2 - B_3 + ... + (-1)^i B_i$, $\quad i = 0, 1, ..., n-2$.

We summarize and record our conclusions in

Theorem 1. If given the set of $n+1$ data points $\{(x_i, z_i)\}_{i=0}^n$ with distinct knots $x_0 < x_1 < ... < x_n$ then the quadratic interpolating G-spline $S(x)$ exists uniquely with additional requirement Eq.(3). The explicit expression for the coefficients $(a_i; b_i)$ of $S(x)$ are given by Eqs. (1) and (5).

Note that the given construction of a quadratic interpolating G-spline $S(x)$ is a global method, i.e. $S(x)$ is dependent on all data points, and addition or delation of a data point, or a change of one of the coordinates of a data point, will propagate throughout the interval of definition. We omit the proof of the following.

Theorem 2. If $z(x) \in C^1(a,b)$ then for a quadratic interpolating G-spline $S(x)$ we have the estimate

$$|z(x) - S(x)| \leqslant C_1 h \, \omega_1(h)$$

$$|z'(x) - S'(x)| \leqslant C_2 \omega_1(h)$$

with requirement (3), where C_1, C_2 are constants

$$h = \max\{h_i\}_{i=0}^{n-1} \quad \text{and} \quad \omega_1(h)$$

denotes the modulus of continuity of $z'(x)$.

3. SURFACE INTERPOLATION VIA QUADRATIC G-SPLINES ON RECTANGLES

In this section we deal with bivariate spline functions that are directly determined by a given rectangular grid configuration. Using G-splines and linear (resp. quadratic) blending functions we may obtain continuous (resp. smooth or C^1-continuous) interpolant surfaces.

Indeed, the Coons-Forrest methods (see [5] Forrest 1972) and the interpolation scheme by means of quadratic G-splines may also be used to construct bivariate splines interpolant to a function $z = z(x,y)$ given at the discrete values (x_i,y_j) i.e. $z_{ij} = z(x_i,y_j)$, with $0 \leqslant i \leqslant n$ and $0 \leqslant j \leqslant m$, on the rectangular grid $\{x_i\}_{i=0}^n \times \{y_j\}_{j=0}^m$ where $a = x_0 < x_1 < ..., < x_n = b$ and $c = y_0 < y_1 < ..., < y_m = d$, (see Figure 1.).

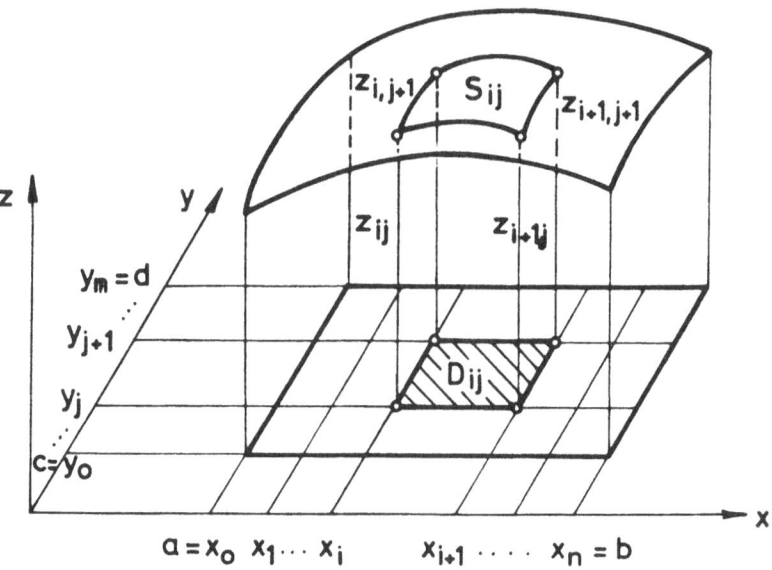

Figure 1. Rectangular grid configuration.

In what follows, let

$$S_j(x) := P_{ij}(x) := \frac{a_{ij}}{h_i}(x-x_i)(x_{i+1}-x) + b_{ij}(x-x_i) + z_{ij}$$

where

$$x \in [x_i, x_{i+1}], \quad h = x_{i+1} - x_i \quad \text{and} \quad 0 \leqslant i \leqslant n-1$$

be the quadratic G-spline interpolating z over the grid line $y=y_j$ (i.e. at the points $\{(x_i, y_j)\}_{i=0}^n$ for $j = 0,1,...,m$. This spline is produced by the algorithm described in Section 2, (see Theorem 1.).

Similarly, let

$$S_i^*(y) := Q_{ij}(y) := \frac{c_{ij}}{k_j}\,(y\text{-}y_j)(y_{j+1}\text{-}y) + d_{ij}(y\text{-}y_j) + z_{ij}$$

where

$$y \in [y_j,\, y_{j+1}]\,, \quad k_j = y_{j+1}\text{-}y_j \quad \text{and} \quad 0 \leqslant j \leqslant m\text{-}1$$

be the quadratic G-spline interpolating z over the grid line $x = x_i$ (i.e. at the grid points $\{(x_i,\, y_j)\}_{j=0}^{m}$) for $i = 0,1,...,n$.

Then by means of these splines, more precisely using the spans $P_{ij}(x)$ and $Q_{ij}(y)$ we can derive a surface $S(x,y)$ on the closed rectangular domain $D = [a,b] \times [c,d]$, as follows:

$$S(x,y) := S_{ij}(x,y) := P_{ij}(x)\,\frac{y_{j+1}\text{-}y}{k_j} + P_{i,j+1}(x)\,\frac{y\text{-}y_j}{k_j} + Q_{ij}(y)\,\frac{x_{i+1}\text{-}x}{h_i} +$$
$$+\, Q_{i+1,j}(y)\,\frac{x\text{-}x_i}{h_i} - z_{ij}\,\frac{(x_{i+1}\text{-}x)(y_{j+1}\text{-}y)}{h_i k_j} - z_{i+1,j}\,\frac{(x\text{-}x_i)(y_{j+1}\text{-}y)}{h_i k_j} -$$
$$-\, z_{i,j+1}\,\frac{(x_{i+1}\text{-}x)(y\text{-}y_j)}{k_i k_j} - z_{i+1,j+1}\,\frac{(x\text{-}x_i)(y\text{-}y_j)}{h_i k_j}\,,$$

where

$$(x,y) \in D_{ij} = [x_i,\, x_{i+1}] \times [y_j,\, y_{j+1}] \text{ for } i = 0,1,...,n\text{-}1 \text{ and } j = 0,1,...,m\text{-}1\,.$$

Then it is obvious that on any subrectangle D_{ij}, $S(x,y)$ is given by a bivariate algebraic polynomial $S_{ij}(x,y)$ of total degree 3 and of degree 2 in x and y separately. Next it is a straightforward matter to check that

$$S(x_i, y_j) = z_{ij} \tag{6}$$

$$S_{ij}(x_i,\, y) = Q_{i+1,j}(y) \text{ for } y \in [y_j, y_{j+1}] \tag{7}$$

where

$$0 \leqslant i \leqslant n\text{-}1 \quad \text{and} \quad 0 \leqslant j \leqslant m\text{-}1\,.$$

$$S_{ij}(x_{i+1}, y) = Q_{i+1,j}(y) \text{ for } y \in [y_j, y_{j+1}] \tag{8}$$

where

$$0 \leqslant i \leqslant n\text{-}2 \quad \text{and} \quad 0 \leqslant j \leqslant m\text{-}1\,.$$

$$S_{ij}(x,y_j) = P_{ij}(x) \qquad \text{for } x \in [x_i, x_{i+1}] \tag{9}$$

where

$$0 \leqslant i \leqslant n\text{-}1 \quad \text{and} \quad 0 \leqslant j \leqslant m\text{-}1\,.$$

$$S_{ij}(x,y_{j+1}) = P_{i,j+1}(x) \text{ for } x \in [x_i, x_{i+1}] \tag{10}$$

where

$$0 \leqslant i \leqslant n\text{-}1 \quad \text{and} \quad 0 \leqslant j \leqslant m\text{-}2\,.$$

To prove that $S(x,y)$ is well defined and continuous in D it is sufficient to show that

$$S_{ij}(x_{i+1},y) = S_{i+1,j}(x_{i+1},y) \quad \text{for} \quad 0 \leq i \leq n-2 \quad \text{and} \quad 0 \leq j \leq m-1 , \tag{11}$$

and

$$S_{ij}(x,y_{j+1}) = S_{i,j+1}(x,y_{j+1}) \quad \text{for} \quad 0 \leq i \leq n-1 \quad \text{and} \quad 0 \leq j \leq m-2 . \tag{12}$$

But Eq.(11) follows from Eqs.(7) and (8). Similarly Eq.(12) follows from Eqs.(9) and (10). Now Eq.(6) gives that $S(x,y)$ interpolates the data points z_{ij}. Consequently, $S(x,y)$ is a bivariate continuous spline function which interpolates the value z_{ij} at the grid point (x_i,y_j) for all i,j .

Also, it is evident by Eqs.(7)–(10) that the restriction of $S(x,y)$ to the grid line of type $y=y_j$ (resp. $x=x_i$) coincides with the quadratic G-spline $S_j(x)$ for all j (resp. $S_i^*(y)$ for all i). Moreover, any plane of type $x=g$ whose $g \epsilon [a,b]$ (resp. $y=g$ where $g \epsilon [c,d]$) intersects the surface $S(x,y)$ in such curve $S(g,y)$ (resp. $S(x,g)$) which is a piecewise quadratic polynomial over the line segment $x=g$ (resp. $y=g$) of the domain D.

Concluding our discussion of result on the considered bivariate spline $S(x,y)$ we mention – using the above notation – the following.

Theorem 3. If given the set of $(n+1) \times (m+1)$ data points $\{(x_i,y_j,z_{ij})\}$ where $0 \leq i \leq n, 0 \leq j \leq m$ with distinct knots $x_0 < x_1 < ... < x_n$ in the x-direction and $y_0 < y_1 < ... < y_m$ in the y-direction then a continuous bivariate interpolating spline function $S(x,y)$ of total degree 3 and of degree 2 in x and y separately always exists, and it depends on the choice of the quadratic G-spline function $S_j(x)$, $S_i^*(y)$.

In addition let

$$f_i(x) := \begin{cases} 1 - 2\left(\dfrac{x-x_i}{h_i}\right)^2 & \text{if } x_i \leq x \leq x_i + \dfrac{h_i}{2} \\[3mm] 2\left(\dfrac{x_{i+1}-x}{h_i}\right)^2 & \text{if } x_i + \dfrac{h_i}{2} \leq x \leq x_{i+1} , \end{cases}$$

where

$$h_i := x_{i+1} - x_i > 0 \quad \text{for} \quad 0 \leq i \leq n-1 , \text{(see Figure 2.).}$$

Similarly, let

$$g_j(y) := \begin{cases} 1 - 2\left(\dfrac{y-y_j}{k_j}\right)^2 & \text{if } y_j \leq y \leq y_j + \dfrac{k_j}{2} \\[3mm] 2\left(\dfrac{y_{j+1}-y}{k_j}\right)^2 & \text{if } y_j + \dfrac{k_j}{2} \leq y \leq y_{j+1} , \end{cases}$$

where

$$k_j := y_{j+1} - y_j > 0 \quad \text{for} \quad 0 \leq j \leq m-1 .$$

It is clear that $f_i(x)$ (resp. $g_j(y)$) is a piecewise quadratic (in other words twin-quadratic) and countinuously differentiable function on the interval $[x_i, x_{i+1}]$ (resp. $[y_j, y_{j+1}]$). Moreover these functions have the following properties:

$$f_i(x_i) = 1, \ f_i(x_{i+1}) = f_i'(x_i) = f_i'(x_{i+1}) = 0$$

for $0 \leqslant i \leqslant n-1$, and

$$g_j(y_j) = 1, \ g_j(y_{j+1}) = g_j'(y_j) = g_j'(y_{j+1}) = 0$$

for $0 \leqslant j \leqslant m-1$, (see Figure 2.).

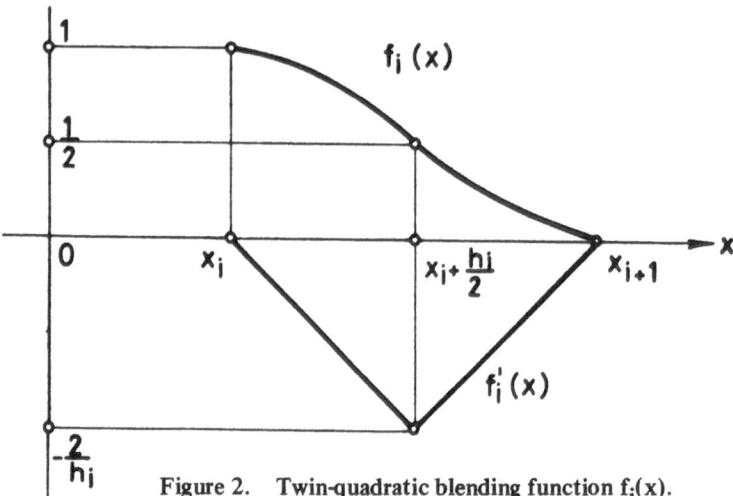

Figure 2. Twin-quadratic blending function $f_i(x)$.

Then using the G-splines S_j, S_i^* given as above and the twin-quadratic blending functions f_i, g_j, we define a surface $B(x,y)$ on the closed rectangular domain $D = [a,b] \times [c,d]$ as follows

$$B(x,y) := B_{ij}(x,y) := P_{ij}(x)g_j(y) + P_{i,j+1}(x)g_j(y_j + y_{j+1} - y) +$$

$$+ \ Q_{ij}(y)f_i(x) + Q_{i+1,j}(y)f_i(x_i + x_{i+1} - x) - z_{ij}f_i(x)g_j(y) -$$

$$- \ z_{i+1,j}f_i(x_i + x_{i+1} - x)g_j(y) - z_{i,j+1}f_i(x)g_j(y_j + y_{j+1} - y) -$$

$$- \ z_{i+1,j+1}f_i(x_i + x_{i+1} - x)g_j(y_j + y_{j+1} - y)$$

where

$a = x_0$, $b = x_n$, $c = y_0$, $d = y_m$ and $(x,y) \in D_{ij} := [x_i, x_{i+1}] \times [y_j, y_{j+1}]$

for $i = 0,1,...,n-1$, $j = 0,1,...,m-1$.

The properties of this surface $B(x,y)$ will now be investigated.

First of all, $B(x_i,y_j) = z_{ij}$ and it is obvious that on each of the four quarter parts of the arbitrary subrectangle D_{ij}, $B(x,y)$ is given by a bivariate algebraic polynomial of total degree 4 and of degree 2 in x and y separately, i.e. $B(x,y)$ is a biquadratic on D. Moreover it is clear that $B(x,y)$ is smooth in any D_{ij} for $0 \leqslant i \leqslant n-1$, $0 \leqslant j \leqslant m-1$.

Now we are going to show that $B(x,y)$ is smooth in the whole domain D, that is all derivatives

$$\frac{\partial^{k+\ell} B(x,y)}{\partial x^k \partial y^\ell}$$

for $0 \leqslant k, \ell \leqslant 1$ are continuous in D. In fact, it is a straightforward matter to verify by using the properties of the functions $P_{ij}(x)$, $Q_{ij}(y)$, $f_i(x)$ and $g_j(y)$ that

$$\frac{\partial^{k+\ell} B_{ij}(x,y)}{\partial x^k \partial y^\ell}\bigg|_{x=x_{i+1}} = \frac{\partial^{k+\ell} B_{i+1,j}(x,y)}{\partial x^k \partial y^\ell}\bigg|_{x=x_{i+1}}$$

for $0 \leqslant k, \ell \leqslant 1$ and $0 \leqslant i \leqslant n-2$, and

$$\frac{\partial^{k+\ell} B_{ij}(x,y)}{\partial x^k \partial y^\ell}\bigg|_{y=y_{j+1}} = \frac{\partial^{k+\ell} B_{ij+1}(x,y)}{\partial x^k \partial y^\ell}\bigg|_{y=y_{j+1}}$$

for $0 \leqslant k, \ell \leqslant 1$ and $0 \leqslant j \leqslant m-2$.

The validity of these relations yields the desired result. Consequently we obtain

Theorem 4. If given the set of $(n+1) \times (m+1)$ data points $\{(x_i, y_j, z_{ij})\}$ where $0 \leqslant i \leqslant n$, $0 \leqslant j \leqslant m$ with distinct knots $x_0, x_1, ..., x_n$ in the x-direction and $y_0, y_1, ..., y_m$ in the y-direction then a smooth bivariate interpolating biquadratic spline function $B(x,y)$ always exists and it depends on the choice of the quadratic G-spline functions $S_j(x)$, $S_i^*(y)$.

As a result, it seems that the class of considered bivariate spline functions $S(x,y)$, $B(x,y)$ can be useful to interpolate surfaces given on rectangular grids. Being a global method, it has the advantage of the simplicity of the algorithm, although it requires some more computation. It leads to continuous (resp. smooth) surfaces without excessive undulations.

4. CONCLUSION

Methods of constructing a continuous resp. a smooth interpolant to a surface given on a rectangular grid configuration has been described. The methods are based on a quadratic spline representation and may be expressed in closed analitic form, allowing algebraic operations.

The software package based on the above methods is now under test on a professional personal computer. The experiments show that the given surface modeling system can effectively be used to design complicated, free form 3D objects, as tools, moulds, dies, etc.

An earlier experiment concerned the so called translation-surfaces. Translation-surfaces are defined by means of two plane-curves: The generator is moved along the directrix; see [4], Fenyves−Licskó−Kovács, 1983. These plane-curves may be defined using quadratic G-splines. As tool-path calculations are necessary to generate control tapes for NC milling of surfaces defined as translation-surfaces, it is necessary to produce the equidistant offsets of the directrix curve.

One of the advantages of using quadratic G-splines for curve interpolation is the relatively simple way of offset generation (e.g. in comparison with cubic or other splines).

REFERENCES

[1]   Ahlberg, J.H. - Nilson, E.N. - Walsh, J.L., (1967). The theory of splines and their applications. Academic Press, New York.

[2]   de Boor, C., (1978). Apractical guide to splines. Springer-Verlag, New York-Heidelberg-Berlin.

[3]   Böhm, W. - Farin, G. - Kahmann, J., (1984). A survey of curve and surface methods in CAGD. Computer Aided Geometric Design, Vol.1. pp.1-60.

[4]   Fenyves, F. - Licskó, I. - Kovács, G., (1983). Translation-surfaces in the 3D subsystem of the COMECON NC programming system. Proceeding of ICED'83, Kobenhavn (WDK 10) Vol.1. pp.105-112.

[5]   Forrest, A.R., (1972). On Coons and other methods for the representation of curved surfaces. Computer Graphics and Image Processing. Vol.1. pp. 341-359.

ENGINEERING WITH SOFTWARE PROTOTYPES
Dr. M.P. Williamson, APPLICON (UK) Ltd, Stockport

The Traditional Approach

In a traditional engineering organisation the product
development cycle may be summarised as follows:-

* recognition of a marketing opportunity
* design of the product to satisfy that need
* production of detailed drawings
* construction of a prototype
* test
* preparation of manufacturing plans
* manufacture

However, before proceeding to the manufacturing of the
product it is necessary to feed back to the design part of
the cycle any faults detected in the test stage. Following
consequent modification of the design, the prototype must be
modified and retested. Thus, the design, detailed drawing
production, prototype construction, and test stages must be
repeated often enough to eliminate all the design errors
detected by testing the prototype.

It could be argued that this product development method has
been proved successful by frequent use in a wide variety of
engineering companies over many years. If it could be
ensured that all design faults could be detected in this way,
then perhaps the continued use of this development cycle
could be justified. However, this is by no means true.
How many times do design errors only become apparent after
extended use under normal working load? If the testing part

of the product development cycle can be exhaustive under real operating conditions, then all the faults will be detected. Unfortunately, there is usually not time for this kind of exhaustive testing.   In a company manufacturing parts, perhaps on a subcontract basis, then this sort of exhaustive testing under operating conditions is just not possible.

Yet another significant problem of this "design, test, build" technique results from the detection of faults only rather late in the development cycle.   The longer it takes to detect a fault the greater is the investment which has already been made in the product.   If the detected fault is a major one, then a total redesign of the part is called for.   The only alternative would be to degrade the quality of the product by resorting to a "botch-up".   Either solution is far from satisfactory, but the most important problem may well be the delay in the introduction of the product.   Deferring, taking the profit that this part was predicted to have made will have an effect on the business which can be quantified.   How can we predict accurately, however, the effect of allowing our competitors to steal a march on us and to introduce their equivalent product before our own?   In an increasingly competitive world this latter problem is likely to be the most damaging.

We have assumed that when this embryo product is introduced, all design faults have been eliminated.   Of course, this cannot be guaranteed, and the detrimental effect on future business of design inadequacies being discovered by customers must not be under estimated.

If there are so many potential problems associated with the traditional approach to engineering development, how can we overcome them? With the advance of the use of computers in engineering, more and more specific and powerful software is available to us. The questions which we have been answering experimentally with the prototypes discussed previously include:-

Is it big enough?
Where is the centre of gravity?
What is the weight of the part?
What does it look like?
Does it work as we predicted?
Is it strong enough?

Modern computer techniques can do much to provide answers to these questions. Making productivity gains with specific questions is useful, but a much greater impact can be seen by using a fully integrated CAE approach. The engineering - cycle now changes subtly:-

* recognition of a marketing opportunity
* design
* analysis of the design using computer techniques
* production of detailed drawings
* prototype construction
* test
* preparation of manufacturing plans
* manufacture

By carrying out analyses on the software prototype, the construction of the physical prototype is delayed. But, since the software prototype has been checked extensively by software analyses, we are more confident that the design is valid. This, in turn, ensures that the prototype is really only a final check and minimises the chances of having to return to the old " design, build, test" iterative loop.

The type of software analyses that we will use in the development cycle are exemplified by mass property analysis, mechanisms analysis and finite element analysis. Various suites of software are available from different vendors in order to effect these analyses. But, now we encounter another problem. Since these software programs are available from different sources, they all have quite different data input requirements. In pratical terms, then, the analyst will have to prepare (format) his data to satisfy the input requirements of each of the programs he will use. Inevitably this will take time and introduces potential errors caused by inaccurate transcription of the data. The latter errors are particularly dangerous since they may well lead to incorrect analyses, the results of which will be misleading and may well lead eventually to an incorrect design.

If our original design had been created within a computer system, then this need for entry of the input data for each of the analysis software suites is obviated. All that is now needed is that this data - already resident within the computer - is reformatted to meet the needs of any particular analysis software. In this way, transcription errors are eliminated. However, we still have a problem in that all (most probably) of the analysis suites have been written by different people, each with their own ideas of the best way to input commands to their programs. Thus, if a single designer is to perform his own analyses, he must learn several command input protocols - one for the Computer Aided Design System and one for each of the analysis programs he wishes to use.

We might refer to this scenario as the interfaced solution. Since each of the third-party analysis programs retains its own input protocols and is interfaced with the CAD system from where its input data is obtained.

If a way could be devised to ensure that a common command input protocol could be used for both computer aided design and for the analyses, then a far more user-friendly solution would result. We could refer to this as the integrated or fully integrated solution. This type of system is more difficult for the system vendor since more work is required to integrate the third-party analysis software into the CAD system. Since the interfaced solution is simpler (ie less costly) to the system vendor, this is the route most often offered.

If each of the analysis suites can operate in isolation, then perhaps an interfaced rather than integrated solution would be acceptable. Unfortunately this is not often the case. More frequently, data obtained from one analysis is needed as part of the input to another suite. This is best seen in an example.

Let us consider a connecting rod forming part of an engine. After design in the CAD system, a mass property analysis gives us weight, co-ordinates of the centre of gravity, moments of intertia, radii of gyration, etc. We will now need to satisfy ourselves that the mechanism, of which the connecting rod forms a part, will actually assemble. One of the pieces of input data needed by this software is the co-ordinates of the centre of gravity of each of the component parts and in particular of the connecting rod. This information was already calculated in the previous mass property analysis. With a fully integrated system this data is now available for the mechanisms analysis. With an interfaced system, it is much more difficult.

During the mechanism analysis, we check to ensure that the constituent parts will assemble. We can then check the mechanism kinematically to ensure that as it moves it does not foul anything. It is also possible to calculate the forces applied at both the little-end and at the big-end of the connecting rod.

We now know what the maximum forces applied to the connecting rod are as it operates. These can be used as an input to the finite-element analysis in which we satisfy ourselves that the connecting rod will withstand these applied forces.

If the design fails to pass any of the applied tests (or analyses) then a redesign is called for together with the appropriate retesting. Despite the fact that redesign and "re-test" may well be needed, these may be performed within our totally integrated Computer Aided Engineering system.

Throughout this whole process, we have been performing engineering with a software prototype. We have been adding to the knowledge of our design with all of the analyses we have performed. Not only has it been necessary to use a fully integrated CAE system but this will have had to be based on an efficient database manager in order to handle the (potentially) large amount of data we will have collected about each engineered part.

When we have satisfied ourselves, as far as possible, about the validity of the design, the same data that we have been analysing will be used to produce the detailed drawings. These are available directly, and the detailer will only need to decide which dimensions need to be displayed, what notes need to be added etc., and to compose the geometry into an aesthetically acceptable drawing. While these drawings may be generated from "wireframe" geometry, they are more easily created from solid models generated using Solids Modeling ^{TM} software. If this approach is adopted, hidden lines may be suppressed automatically. Other advantages which acrue from the use of Solids Modeling ^{TM} include more accurate and straightforward mass properties analyses, and the possibility of producing realistic colour shaded pictures ("photographs") of the object using only the computer system. It is often difficult to obtain accurate mass properties from wire-frame models. It may be equally difficult to visualise a 3D model constructed in wire-frame since all lines (both front and back) can be seen with equal clarity. Furthermore, the shaded image of the solid model may be examined in different colours and with illumination from any point in space. This enables the aesthetics of the part to be examined.

Once our geometry has been designed and validated by software analysis, this same data may be passed to Production Engineering. Decisions on fixing the job for machining can be made now with the security of knowing that the data is "as - designed" and "as - tested". Fixturing is performed with the same geometry and the latter also forms the starting point for any specific fixtures which need to be designed. In fact, with proper design and documentation, traditional drawings need play no part in transmitting design information to the production engineer.

Finally, for the whole philsophy of engineering with software prototypes to work, the computer which we will choose to use must satisfy certain criteria. It must have sufficient "computing horsepower" to run the programs needed in acceptable times. Today, this means that the computer must be based on 32-bit architecture. Secondly, the computer must be widely used so that there is a wide range of relevant engineering software available from third-parties for the system vendor to <u>integrate</u> within the CAE system. It is a well-known fact that the most widely used engineering computer is the VAX family from the Digital Equipment Company. This family also meets the criterion of 32 - bit architecture.

In summary then, we have seen that in order for Engineering with Software Prototypes to be viable we need:-

*   a <u>fully integrated</u> CAE system
*   based upon the DEC <u>VAX computer</u>
*   the foundation of a powerful <u>database manager</u>
*   the ability to design using <u>Solids Modeling</u> ^{TM}
    and to use this to create our computer model
*   an extensive fully integrated library of analysis
    programs
    to be used as necessary (eg mechanisms analysis,
    finite - element analysis etc)
*   the ability to create traditional, detailed drawings
    "automatically"
*   the possibility of allowing access to our verified model
    to other departments (other than Design and Analysis) as
    typified by Production Engineering.

15. SIMULATION

CAMAS, A COMPUTER AIDED MODELLING, ANALYSIS AND SIMULATION
ENVIRONMENT

Jan F. Broenink and G. Dick Nijen Twilhaar

Twente University of Technology, Department of Electrical
Engineering, Enschede, The Netherlands.

SUMMARY

Modelling and simulation of technical systems is rather error-
prone. A part of the errors may be due to the fact that the
commonly used CSSL.'s (Continuous Systems Simulation Languages),
have some important deficiencies. They have a good performance
in the computational sense but are not designed for structured
modelling. Furthermore often not much support for the man-
machine dialogue is offered, and the formalism they use for
model description is not related to physics.   In this
contribution a modelling and simulation aid especially designed
for structured physics-based model description is discussed.
This aid, CAMAS (Computer Aided Modelling, Analysis and
Simulation), may be used interactively.

CAMAS accepts the model description language SIDOPS (Structured
Inter-disciplinary Description Of Physical Systems) as input.
SIDOPS is based on a physical systems theory and defined in such
a way that structured description of models is encouraged. A
part of CAMAS is a SIDOPS interpreter, which, as an interactive
means for model input, can improve the man-machine dialog.
On-line checking of the input on syntactic errors and generating
proper error messages is one of its features. As SIDOPS is based
on a (multi-disciplinar) physical systems theory (bond graphs),
with use of a SIDOPS interpreter also *semantic* errors can be
detected.

1. INTRODUCTION

Modelling and simulation of technical/physical systems is an aid
with growing importance. Especially for large systems, often too
complex to be overseen by one person, computer based analysis
and simulation is important. The models of such complex systems
are bound to have errors, which may remain unrecognized. In
order to obtain correct and reliable models and simulation
outputs, it is important that modelling is performed in a
structured and verifiable way.

There are a lot of simulation languages (Cellier 1983), each
with its own applications. The CSSL-standard (Strauss 1967,
Crosbie 1982) is a valuable contribution to continuous systems
simulation, it provides a framework for simulation language
development. However, this framework does not encourage a

structured model description, due to its FORTRAN basis (Wirth 1983). Therefore, the CSSL standard is not used for the definition of SIDOPS.

Simular to the development in the field of high level general purpose programming languages, for instance Pascal as the succesor of FORTRAN, it can be argued that there is a need for the definition of a simulation tool which supports structured modelling. Although structured modelling is essential, it is not sufficient for the construction of reliable models. If a structured description does not have a clear link with physics, it still will be error-prone. Hence, a structured model description formalism based upon physics is needed.

Physical systems are often multi-discliplinar. This causes additional difficulties for the construction of correct models. The communication between modellers with different specialisms is sometimes difficult; every part of engineering has its own "language". A "meta-language" designed for the description of multi-disciplinar physical systems may improve this. The in this contribution presented physical systems theory may serve as such a language.

In order to meet the demands outlined above, a software tool named CAMAS (Computer Aided Modelling, Analysis and Simulation) is developed. CAMAS processes models described in SIDOPS (Structured Inter-disciplinary Desciption Of Physical Systems), a language which is based on a multi-disciplinar physical systems theory (bond graphs). Due to its "physics-base" semantic tests on described systems can be performed automatically to some extend.

In the next sections, a brief outline of the bond graph system theory is given. After that, SIDOPS and the SIDOPS interpreter, the latter being a part the software environment CAMAS, are outlined.

2. PHYSICAL MODELLING: AN EXAMPLE

The physical modelling method presented here is based on bond graphs (Paynter 1967, Karnopp 1975, van Dixhoorn 1982, Rosenberg 1983 and Breedveld 1984). The bond graph modelling theory is introduced by an example.

Let us consider an automatic gearbox which makes use of James Watt's speed governor (Kawase 1982), see fig. 1. The rotating balls are driven by an engine. A higher angular velocity of the engine results in a bigger deflection of the balls. This causes a higher ratio of ω_1 over ω_3 and hence a more or less constant ω_3

A SIDOPS description of this system is constructed in a number of top-down modelling steps. The first step is dividing the system (SpeedGovenor, fig. 2) into four subsystems, representing the axis, the govenor (balls and spring), the string and the transmission itself (transmission). See figure 3. The inter-connections between the subsystems are represented by so-called bonds. These bonds denote the energy exchange between the subsystems (see next section).

On its turn, each subsystem can be regarded as a system itself, which again may be divided into subsystems connected by bonds. This dividing pocess may be continued until all subsystems are so-called basic systems or systems which are specially defined. By doing so, a physical system can be modelled using a top-down approach.

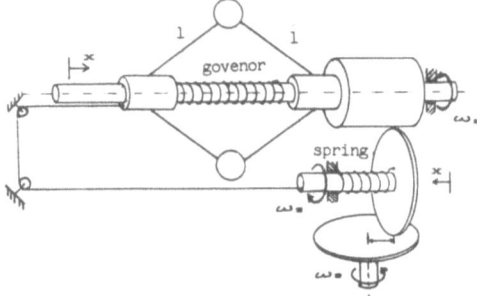

Figure 1. Speedgovenor

Figure 2. Speedgovenor as system

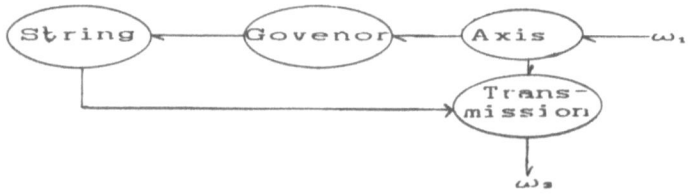

Figure 3. SpeedGovenor divided into 4 subsystems

3. BOND GRAPH THEORY IN GENERAL

As the example shows, there exists a close relation between the physical model with lumped elements, which is a generally used aid for abstraction, and the bond graph. In general, a bond graph is a number of systems which are inter-connected with bonds.

A bond represents transfer of energy (power). To each bond, a pair of conjugate physical variables is connected whose product is power (P). These variables have been given the generalized names of effort (e) and flow (f). There exists a close relation with thermodynamics. A effort is a intensive variable and a flow is the time derivative of an extensive variable. An extensive variable is (q) a conserved quantity. In many domains of

Physical domain	State variable q (dim)	Effort $e = \dfrac{\partial E_{tot}}{\partial q}$ (dim)	Flow $f = \dfrac{dq}{dt}$ (dim)
Translationally potential (elastic, gravitational)	Displacement x (m)	Force F (N)	$\dot{x}$ (m.s^{-1})
Special forms: hydraulic acoustic	Volume V (m^3)	Pressure p (Nm^{-2})	$\dot{V}$ (m^3.s^{-1})
Rotationally potential	Angular displacement ϕ (rad)	Moment M (Nm)	$\dot{\phi}$ (rad.s^{-1})
Translationally kinetic	Momentum of impulse p (Ns)	Velocity v (m.s^{-1})	$\dot{p}$ (N)
Special forms: hydraulic acoustic	Pressure momentum Γ (Nm^{-2}s)	Volume flow ϕ_V (m^{3}s^{-1})	$\dot{\Gamma}$ (Nm^{-2})
Rotationally kinetic	Angular momentum or moment of momentum b (Nms)	Angular velocity Ω (rad.s^{-1})	$\dot{b}$ (Nm)
Electric	Charge q (C)	Voltage u (V)	$\dot{q}$ (A)
Magnetic	Magnetic flux or flux linkage Φ (V.s)	Current or magnetomotive force $i(= HI)$ (A)	$\dot{\Phi}$ (V)
Material	Mole number N (mol)	Total material potential μ^{tot} (J.mol^{-1})	Molar flow $\dot{N} = f_N$ (mol.s^{-1})
Chemical	Species mole numbers N_i (mol)	Chemical potentials μ_i (J.mol^{-1})	$\dot{N_i} = f_{N_i}$ (mol.s^{-1})
Thermal ("thermodynamical").	Entropy S (J.K^{-1})	Temperature T (K)	Entropy flow $\dot{S} = f_S$ (J.s^{-1}.K^{-1})

Table 1. Bond graph variables (Breedveld 1982)

macro-physics an effort and flow variable (time rate of a conserved quantity) can be indicated. Examples of effort variables
are voltage (electrical domain), velocity (translational kinetic
domain) and pressure (hydraulic domain). The corresponding flow
variables are current, moment and fluid flow. The corresponding
conserved quantities are charge, momemtum of impulse and volume.
(See table 1.)

The relation P=e.f is derived from the Gibbs' equation (van
Dixhoorn 1982) which states that the rate of energy can be
written as a total differential with respect to all extensive
variables. An intensive variable (e) is the derivative of the
energy with respect to the corresponding extensive variable (q):

$$dU = \Sigma_i \frac{\delta U}{\delta q_i} dq_i$$

$$P = \frac{dU}{dt} = \Sigma_i \frac{\delta U}{\delta q_i} \frac{dq_i}{dt} = \Sigma_i e_i f_i$$

U: energy
q: conserved quantity of domain i

As mentioned before, with bond graphs a system is defined as a
number of elements which interact with each other. There are a
number of much used basic elements each representing one single
physical phenomenon. A classification of these elements is
grounded on thermodynamics. The basic elements are:

physical phenomenon	element	symbol
storage of energy	buffer	C
dissipation of energy	dissipator	R
non-reciprocal coupling	gyrator	GY
non-mixing coupling	junction	0, 1, TF
power discontinuous	source	S

Table 2. Basic elements (Breedveld 1982)

Buffer
A buffer (C) stores energy. This implies storage of a extensive
variable or conserved quantity (q). The constitutive relation
is:
    $e = q/C$ with $q = \int f dt$ and C is the capacity of the buffer
Examples: electric capacitor, translating body, spring.

Dissipator
A dissipator (R) represents the production of entropy: Energy
from a arbitrary domain of physics irreversibly flows to the
thermal domain. The constitutive relation is: $e=R.f$ (R is the
resistance of the dissipator). Examples: electrical resistor,
bearings friction, shock absorber. In terms of thermodynamics, a
process containing one or more dissipators is irreversible.

Gyrator
A gyrator (GY) couples two domains of physics in a non-dissipative manner. Often a gyrator represents one of the basic laws of physics. If a gyrator is essential (i.e. it cannot be eliminated), then the so-called Onsager-Casimir reciprocity relations (Breedveld 1982) do not hold. Hence, a gyrator is called a non-reciprocal element.

Junction
A junction (0, 1, TF) couples two or more subsystems. A junction can be characterised by a non-mixing relation of efforts and/or flows of the attached bonds (efforts are expressed in efforts and flows in flows). Examples: a series connection of some electrical components (1-juction, same flow, here electrical current).

Sources
Sources (S) describe the interaction of a system with its environment. Very often this interaction can be modelled with effort-sources (constant effort) or flow-sources (constant flow). Example: force of gravity, modelled as an effort-source, the electric mains modelled as an effort-source.

The above is a very brief outline of bond graphs as a systems theory for the modelling of physical/technical systems. The bond graph theory is the basis of the SIDOPS language, which is discussed in the next section.

4. DESCRIPTION OF SIDOPS

4.1. Two levels of modelling
SIDOPS (Structured Inter-disciplinary Description Of Physical Systems) is developed keeping three essential aims in mind:
- SIDOPS has a clear and direct link with physics:
  SIDOPS is based on a physical systems theory, it has the "sematics of physics". This is essential for the creation of correct and verifiable models. Due to the direct link with physics, SIDOPS offers the possibility to perform semantic tests automatically. It is expected that models described in SIDOPS will be less error-prone.
- Structured description of physical systems is encouraged:
  It is the natural way of using SIDOPS. A system can be described as the inter-connection of several subsystems. As a subsystem is again a system, this is a repeating process. A structured description of a complex system will enhance readability and understandability.
- Descriptions in SIDOPS are multi-disciplinary:
  As SIDOPS is based on bond graphs, multi-disciplinar systems can be modelled quite easily, so multi-disciplinar systems (like systems involving sensors and actuators) can be dealt with without additional problems.

With use of SIDOPS it is possible to distinguish the physical modelling process from the mathematical modelling process. The latter, with its numerical and algebraic problems, should be dealt with after the physical modelling process. Here the modelling process is divided into two levels:

First level: Physical modelling
This is the modelling process to create a SIDOPS model description. This means dividing the system into subsystems. This process is continued until basic systems have been isolated. These basic elements are represented by a bond graph elements, or described by formulas.

Second level: Mathematical modelling
The SIDOPS description is used to generate a computable code. This means actions like submodel expansion, performing (computationally) causal analysis and sorting of statements. After this a simulation run can be performed by evaluating the generated set of equations using an adequate integrating routine.

This second level will not be explained any further, as this is not the topic of this contribution.

4.2. SIDOPS
The SIDOPS interpreter is a part of the software environment CAMAS (Computer Aided Modelling, Analysis and Simulation).

A SIDOPS description of a physical system consists of a class description and a parameterset. A class description is a description in SIDOPS of a system, whereby all constants are formal parameters. This class description can be compared to the declaration with a record type in PASCAL. A parameterset is a list of parameter names, whereby numerical values are assigned to each parameter. (A parameterset can be compared with the assignment of values to a record type in PASCAL.)

A occurence of a system will be created by combining its class description and one of its parametersets. In these parametersets, the values of all parameters are stored. It is possible, of course, that there is more than one occurence of the same class. With the use of CAMAS, the declaration of models (class description), and the realisation of models are treated separately. This enhances the versitality of class descriptions. It is evident that a database is needed to store class descriptions, parametersets and realisations of models.

The description of a class consists of the following parts: heading, interface, subsystems and body. The body can either be a formulas body or a graph body. In the following these different parts are described.

Heading
The heading serves as an identification of the class, it consists of:
- a class name, the name which is used as identification.
- a version number, to distinguish different versions of implementation of the class. Changing a class description can have consequences for already defined systems which use this class description. Hence, one must refer to a class with its name *and* version number.
- a parameter, to parameterize the dimensions of the class description e.g. a mechanical model can be suitable for a description in 1, 2 or 3 dimensions.

Interface
The connections with other systems are described here. The described system can only have connections with other systems via the connections declared in this part (no global variables are permitted). The connections are described in terms of physical connections (bonds), and signals (inputs or outputs). This part is a comparable with the variable list of a procedure heading in PASCAL. These variable names are the formal idenfifiers of the connections. When a system is connected to other systems, only the name of the system needs to be given. The sequence of connections determines which actual connection maps onto which formal one. Only when a formulas body is following, these connections can be restricted. These restrictions are not discussed here.

Subsystems
In this part, the used subsystems are declared. The syntax is: Class name followed by the names of all occurences of that class.

Graph body
In this part, the connections of the subsystems with each other and with the interface are listed. If a model is defined with the use of the interpreter, the connections of every subsystem are given separately. Connections of the actual subsystem which were already typed in are shown, there is no need for the user to type redundant information.

There is a number of possible connections: bonds and signals, each bond or signal has a direction and a dimension. See table 3.

At the moment of completion of an input line, the interpreter checks if the connections made in this line, are permitted according to the class descriptions of the connected submodels, if not an error message is produced. After the model input is completed, an overall check is performed on the model consistency on a semantic level.

connection type	bond graph symbol	SIDOPS symbols
single signal	← ——	: -
	—— →	- :
single bond	← ——	< -
	—— →	- >
n signals	←⎯n⎯	: =n=
	⎯n⎯→	=n= :
n dimensional bond	⎯n⎯	< =n=
	⎯n⎯	=n= >
n * m * ... signals,	n / m	: =n,m,...=
	n / m	=n,m,...= :
n * m * ... dimensional bond	n / m	< =n,m,..=
	n / m	=n,m,...= >

Table 3. Connecting symbols of SIDOPS (Welleweerd 1985).

Formulas body
Here, the constitutive relations of a (basic) system are
provided by describing the relations using assignment
statements. The description is PASCAL like, so all (local)
variables must be declared. Besides 'normal' parameters, also
initial conditions for integrators can be declared, as a
special group of parameters.

The description has the following features:
- The statements may be written in an implicit or explicit
  form, in other words the computational causality need not to
  be known.
- Special functions as usual in continuous systems simulation
  languages are implemented in SIDOPS (e.g. integrator, pulse).
- Special loop-constructs have been designed in order to
  describe repetitive structures very compactly. This can be
  useful for the modelling of distributed systems.

The definiton of SIDOPS, as described here, is designed to
provide a formalism to perform modelling of physical systems in
a structured way and on a physical level. This enables

modellers to concentrate on physics exclusively. In order to simulate a model, it is of course necessary to describe it mathematically. But because this level is generated by CAMAS, this need not be concerned by the user.

5. EXAMPLE

The example, already discussed in section two, can now be described in SIDOPS. The submodels Axis and Transmission are decomposed into basic elements. The submodels String and Govenor are specially defined. Hence, the total system Speedgovenor consists of a bond graph with two non-standard subsystems (fig. 4).

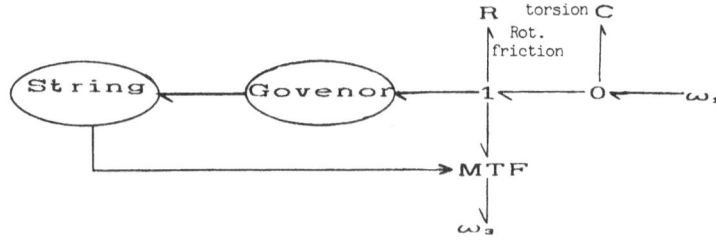

Figure 4. Bond graph of the system with two non standard subsystems

```
SYNTACTIC ANALYSIS OF A SIDOPS CLASS DESCRIPTION
             VERSION February 18, 1985

1    class SpeedGovenor version 1
2      interface
3        ports: SpeedIn, SpeedOut
4      subsystems
5        C         torsion
6        R         RotFriction
7        1J        1Jrot
8        0J        0Jrot
9        MTF       transmission
10       Govenor   govenor
11       String    string
12     structure
13       1J        1Jrot         <-0Jrot,->RotFriction,\
14                               ->transmission,->govenor
15       0J        0Jrot         --SpeedIn,->torsion,->1Jrot
16       Govenor   govenor       <-string,<-1Jrot
17       MTF       transmission  <-1Jrot,--SpeedOut,:-string
18       String    string        ->govenor,-:gearbox

0 error(s) detected.
```

Figure 5. Listing of class SpeedGovenor

The description of the bond graph elements is avaliable in the public part of the database of CAMAS. In fig. 5 a listing of the SIDOPS description of the speedgovenor is shown. The 4 parts (heading, interface, subsystems and graph body) are denoted by just these words, except for graph body. Here 'structure' is the key word.

Line number 1 of the listing denotes that a class of systems, named 'SpeedGovenor' is defined. The interface part describes the connections with the surrounding world. Here, there are two physical connections ('ports') called 'SpeedIn' and 'SpeedOut', which represent the velocities of the engine driven axis and the load axis respectively. In the subsystems part, the subsystems are declared, e.g. line 5 means that a subsystems named 'torsion' of the class 'C' will be used. The structure part describes the connections between the subsystems, e.g. on line 17 the connections of the subsystem 'transmission' of class 'MTF' are described: the two connected bonds denote a power from the 1-junction '1Jrot' to 'transmission' and an outgoing power to the output 'SpeedOut'. The signal input (':-string') is the signal that describes the relation between input and output as is derived in the subsystem 'string' of class 'String'.

In the class Speedgovenor two (sub)classes (govenor and string) are used which are not basic elements. They have to be defined.

Govenor
Here the coupling of the rotation and translation is modelled ('omega' is the connection in the rotational domain and 'speed' the connection in the translational domain, see figure 6). Using the energy function and the Gibbs' equation the relations between efforts and flows can be derived straightforwardly. Force F, velocity v and displacement x are the effort, flow and conserved quantity respectively of the translationally potential domain, while angular velocity o, torque T and rotational impulse momentum b are effort, flow and the conserve quantity of the kinetic rotational domain. The energy function is:

$$E = b^2/2J + x^2/2C$$

where J is the inertia of the balls and one over C the compliance of the spring. The non-tangential movements of the balls in the translation domain are neglected for simplicity. The inertia J is:

$$J = 2mr^2$$

As r is a function of x (see figure 1):

$$r^2 = L^2 - (2L-x)^2$$

resulting in

$$J = 2mLx - Lx^2/2$$

As $F = \delta E/\delta x$ and $\omega = \delta E/\delta b$,

$$F = -b^2/J^2 . \delta J/\delta x + x/C$$
$$\omega = b/J$$

SYNTACTIC ANALYSIS OF A SIDOPS CLASS DESCRIPTION
VERSION February 18, 1985

```
1     class Govenor version 1
2       interface
3         ports: speed, omega
4       parameters
5         real m,l,C1
6       initial conditions
7         real x0,b0
8       variables
9         real b,x,J,dJdx
10      dynamic
11        J:=2*m*l-m*x*x/2
12        dJdx:=2*m*l-m*x
13        speed.e:=-b*b*dJdx/(2*J*J)+x/C1
14        omega.f:=b/J
15        b:=int(b0,omega.e)
16        x:=int(x0,speed.f)
```

0 error(s) detected.

Figure 6. Listing of class Govenor

Substituting J yields the formulas as shown in fig. 6. In fig. 6 The keyword 'dynamic' denotes the beginning of the formulas body. The efforts and flows of the power bonds are denoted as 'speed.e' and 'speed.f', meaning the effort (F) and flow (v) of the power bond called 'speed' respectively.

String
Here the relation between the translation of the govenor and the transmission ratio is modelled. The standard system ATT (attenuator) is used. Here, also the spring and the translational friction are modelled. See fig. 7. The SIDOPS description is shown in figure 8.

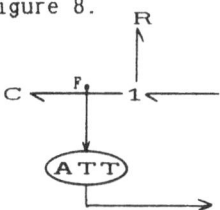

Figure 7. System String

SYNTACTIC ANALYSIS OF A SIDOPS CLASS DESCRIPTION
VERSION February 18, 1985

```
1    class String version 1
2      interface
3        ports: translation
4        outputs: ratio
5      subsystems
6        R          TransFriction
7        C          spring
8        ATT        relation
9        1J         1Jtrans
10     structure
11       1J         1Jtrans    --translation,->TransFriction,->spring
12       ATT        relation   :-spring.e,-:ratio
```

0 error(s) detected.

Figure 8. Listing of class Spring

With ths SIDOPS description of SpeedGovenor, Govenor and String
the description of the system of figure 1 is completed. With
use of CAMAS this system can now be simulated or analyzed.

6. CONCLUDING REMARKS

The SIDOPS language and the CAMAS environment do not have much in
common with commonly used simulation languages and programs. This
is an important disadvantage, while bringing them into practical
use. Still we hope that the offered advantages
    - structured
    - multi-disciplinary
    - 'hands-on' interactive
    - automatic semantic tests
    - based on physics
may show to be more important than the disadvantages.

REFERENCES

Breedveld, P.C., (1982) Thermodynamic bond graphs and the
problem of thermal inertance. J. of the Franklin Inst., 314,
1:15-40

Breedveld, P.C., (1984) Physical systems theory in terms of
bond graps. Ph.D. Thesis, Twente Univ. Of Techn., Enschede, The
netherlands.

Cellier, F.E., (1983) Simulation software: today and tomorrow.
In Mezencev, K., (ed.) Proceedings of IMACS international
symposium on simulation in engineering sciences, Nantes, France

Crosbie, R.E., Cellier, F.E., (1982) Progress in Simulation language standards. Proceedings 10th IMACS congress on systems simulation and scientific computation, 1:411-412, Montreal.

Dixhoorn, J.J. van, (1982) Bond Graphs and the Challange of a Unified Modelling Theory of Physical Systems, in "Progress in Modelling and Simulation", F.E. Cellier (ed.), 207-245, Academic Press, New York.

Karnopp, D.C., Rosenberg, R.C.,(1975) Systems Dynamics: A Unified Apprach, Wiley, New York.

Kawase, T., (1982) Modelling of dynamical systems. J. Japan soc precision engg. 48, 6:102-107

Paynter, H.M., (1961) Analysis and Design of Engineering Systems, MIT-press, Cambridge (Mass).

Rosenberg, R.C., Karnopp, D.C., (1983) Introduction to Physical Systems Dynamics", McGraw-Hill, New York.

Strauss, J.C., et al., (1967) The SCi continuous system simulation language. Simulation, 9, 6.

Welleweerd, A, (1985) Definition of the SIDOPS simulation language and realisation of the interpreter. M.Sc. thesis Twente Univ. of Techn., Enschede, The Netherlands.

Wirth, N., (1983) Program development by stepwise refinement. Comm. ACM, 26, 1:70-74. Reprint in the 25th anniversary issue of Comm. ACM. The original paper was published in (1967) Comm. ACM 14, 4:221-227.

ST.EX.OM - A STATISTICAL PACKAGE FOR THE PREDICTION OF
EXTREME VALUES

J. Labeyrie
IFREMER - Institut Français de Recherche pour l'Exploitation
de la Mer, Brest, France

1. INTRODUCTION

The estimation of extreme statistics of the environment
(wind, waves, current) and structure behaviour parameters
is a preliminary step which is proving crucial in a modern
approach of structural safety and reliability. For example,
design criteria for offshore structures are based on occurence
probabilities of extreme loads.

In a rational way, we have developed an asymptotic method
to infer extreme value estimates from data banks coming
from instrumented sites (Labeyrie, 1984). In particular,
we have shown how the theory of order statistics allows
to define the basic principles of the extrapolation techniques
and of the uncertainties estimates. We present here the
statistical package ST.EX.OM* which reflects these concepts
and where the data analysis capabilities include robust
methods. This package is also an user-oriented program which
can be extended to specific applications very conveniently
(failure ratio estimate, simulations, time series analysis).

2. STATISTICAL APPROACH

We are concerned with the inference of reliable extreme
values from instrumented measurements (Labeyrie - Lebas,
1985). If the period of record is long enough, we come to
fit an extreme model (see table 1 - modelization). Unfortunately this condition is not often satisfied. So, an extrapolation method is worked out as it follows.

Step 1 Let T be the period of record
  We qualify a data sample (size n) whose values all exceed
some prescribed threshold and which may be identified to
n realizations of independent identically distributed random
variables.
* STatistique EXtrême pour les Ouvrages en Mer

During a display procedure, informations about the underlying models may come back in memory as shown in table 3
and 4.

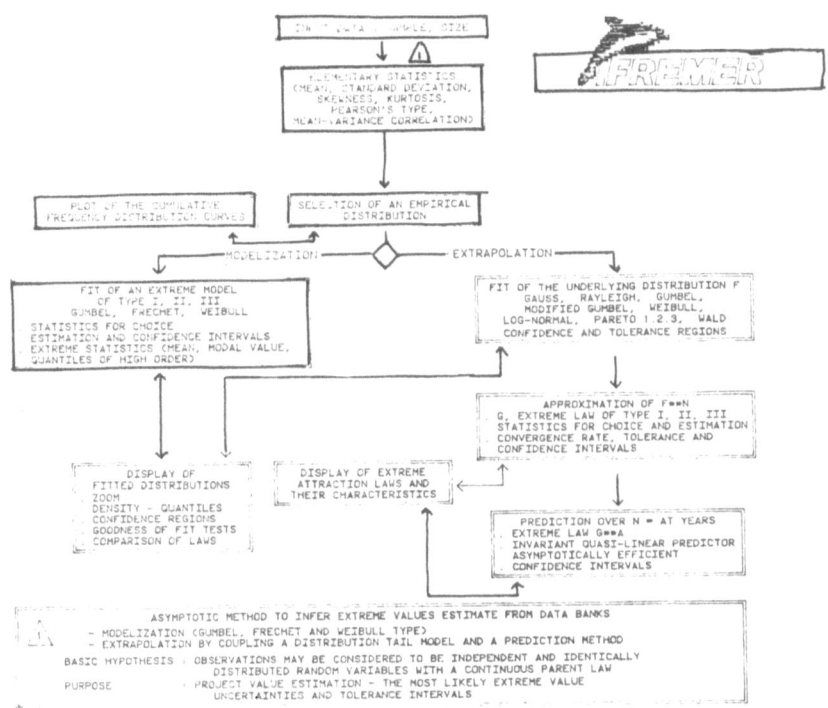

Table 1 : Bill display of ST.EX.OM

4. NUMERICAL OPTIMIZATION

Now we present some crucial aspects of the numerical optimization we have implemented.

When a distribution function F has been fitted to the
data sample, reliable statistical extrapolation may only
be expected if we have tested the goodness of fit of the
tail behaviour of F to the cumulative frequency distribution
one.

The cumulative frequency distribution curve is then fitted by a continuous distribution tail model F (see figure 1).

A central result to be used here, has been discovered by Fischer and Tippett (1928) and later proved in complete generality by Gnedenko (1943). It shows that after suitable standardizations the class of limiting laws for the distribution F^n of the largest order statistic contains only laws G of the three types, commonly called the three extreme value distributions.

Type I   : Gumbel $G(x) = \exp(-\exp-x)$ $-\infty < x < \infty$

Type II : Frechet
$$G(x) = \begin{cases} 0 & x \leq o \\ \exp(-x^{-\beta}) & \beta > o \quad\quad x > o \end{cases}$$

Type III : Weibull
$$G(x) = \begin{cases} 1 & x \geq o \\ \exp-(-x)^{\beta} & \beta > o \quad\quad x < o \end{cases}$$

So, we may obtain the following extreme parameters (see figure 4).

X_T - most likely extreme value over Tyears

$[X_{(1)}; X_{(2)}]_{\Theta,T}$ - confidence interval for X_T with coefficient Θ

$[S_{(1)}; S_{(2)}]_{\Xi,T}$ - tolerance interval for X_T with coefficient Ξ

Step 2 If the project life is N = a Tyears (a > 1)
A prediction method becomes necessary. As G is max-stable, the mode X_N of G^a is an invariant quasi-linear predictor over N years.

See (Labeyrie, 1985) for the mathematical and heuristical aspects of this method. Here we present the data analysis capabilities of the package ST.EX.OM we have developed to concretize this approach in the engineering context.

3. FUNCTIONAL ORGANIGRAM (table 1)

First, the bill display presents the application and during the investigation, gives informative outputs to let the user to select incrementally his path through the different phases (fit-approximation-prediction). For example the table 2 explains the interactive options available after the fitting phase (step 5).

Graphic outputs are used extensively to traduce the numerical results in a more comprehensible way.

The classical Chi-square and Kolmogorov tests are not very tractable. It is advisable to use a normality test based on a weighted deviation random variable.

$$(F^*(x) - F(x))\, \Psi\, (F(x))$$

where F^* is the empirical distribution and Ψ a function in the form,

$$\Psi\,(u) = \frac{u}{1 - u}$$

For example, if you consider the fitting investigation (see figure 1) we obtain,

Weighted deviation

	Mean	Variance
Log-normal	.05	.01
Rayleigh	.38	.75
Gauss	3.11	81.1

Accurate confidence regions for the distribution parameters are computed by a procedure we illustrate with the log-normal model.

(i) Maximum likelihood confidence ellipse for the mean and variance of the Gaussian random variable $Ln(x-x_o)$ (see figure 2)

(ii) Variation of the weighted deviation along the axis of the ellipse (see table 5)

(iii) Correction of the confidence region by taking into account the confidence level of the weighted deviation (see figure 3).

The distribution $F^n(a_n\, x_{(n)} + b_n)$ of the renormalised largest order statistic $x_{(n)}$ converges in distribution to an appropriate extreme value law. There is no "universal rate of convergence" as in central limit theory. So we have to specify the characteristics of the underlying distributions F encountered in our applications.

Let D_n be the maximum absolute difference defined by,

$$D_n = Sup|\, F^n(a_n\, x + b_n) - G(x)|$$

As the distribution tail of F is already close to exponential form, we obtain a fast convergence in the sense that D_n is proportional to an inverse power of n.

We used also tests statistics (Tiago de Oliveira, 1983) for selection between the three models to justify the applicability of the asymptotic efficiency to a finite n.

5. PACKAGE REPORT

ST.EX.OM is a modular package, practicable in a time-sharing environment with efficiency. The routines are written in FORTRAN IV with a care for a high portability. As we use IGL (Interactive Graphics Library) options 1 and 2, the host computer may be consistent with Tektronix series.

Inputs

Only one access to a sequential indexed file which contains the data sample is necessary for all the investigation. Data User's (name of file, treshold value, instrumentation duration, ...) are given interactively at the beginning of the application.

Outputs
  (i)   Bill displays to explain the planning of the application
  (ii) Plot of distribution curves (empirical, parametric model, density) and confidence region
  (iii) Tables (parameters, quantiles, variations of the weighted deviation, fitting control card (see table 6))
Interfaces with hard copy or desktop plotter (Tektronix and/or Benson) are also available.

6. SOME APPLICATIONS

Fruitful results on the tolerance and confidence intervals of extreme loads have been performed through ST.EX.OM over various data banks, coming from several instrumented sites among which large offshore platforms. These results have been obtained with reduced manpower and in very short delays.

   Computation and modelling have been realized in different domains, for example : the uncertainties connected with vibration frequencies estimated from an ARMA model (offshore platform behaviour), or the bacteria distribution along the walls of an exchanger tube (Ocean Thermal Energy Conversion).

   Now we study the feasibility to extend the package ST.EX.OM to multivariate extreme analysis. The practical purpose is to estimate a joint-probability of meteorological and oceanographic phenomena in offshore structural design. So, by taking into account more realistic combinations of extreme wind, wave and current loads we may in several cases avoid an increase in fabrication costs.

CONCLUSION

Instrumentations and then statistical softwares are proving necessary in order to evaluate reliable design factors in structural engineering.

Here, we have presented the interactive package ST.EX.OM, based on a probabilistic approach, which allows to estimate extreme values of observed phenomena.

This user-oriented program can be used very conveniently in reliability and modelling.

REFERENCES

Labeyrie, J. (1984) Statistique extrême des hauteurs de vagues de tempête. Rapport ARAE. Réf. IFP, Paris.
Labeyrie, J. and Lebas, G. (1985) Large excursions of extreme value estimates in offshore engineering. 4th Int. Cong. on Struct. Safe. and Rel., ICOSSAR'85, Japan.
Labeyrie, J. (1985) An asymptotic method to infer extreme value estimates from data banks. To appear.
Tiago de Oliveira, J. (1983) Univariate extremes, statistical choice. NATO ASI Série C, Holland, 91-107.
Deheuvels, P. (1980) Some applications of the dependence functions to statistical inference : non parametric estimates of extreme value distributions and a Kiefer type universal bound for the uniform test of independence. Coll. Math., Janos Bolyai 32, North Holland, 183-201.
Pickands, J. III (1975) Statistical inference using extreme order statistics. Ann. Statist., 3 : 119-131.

ST.EX.OM

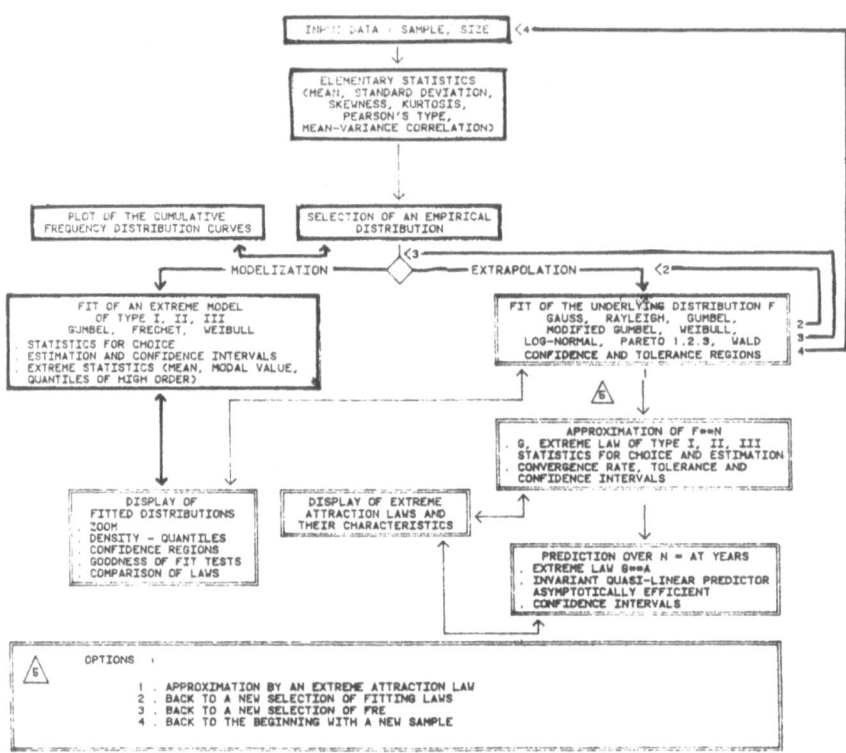

Table 2 : Definitions of interactive options
in STEP 5

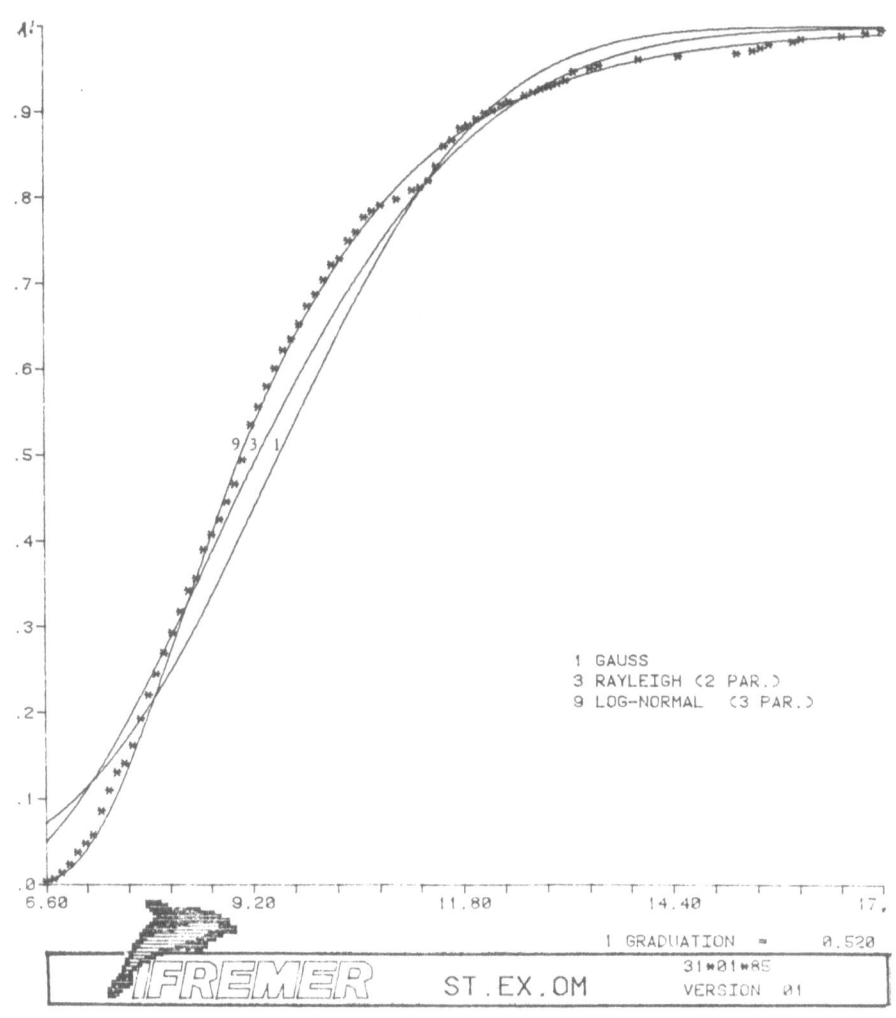

Figure 1 : Display of different fitted models

PARAMETERS AND QUANTILES

ORDER OF QUANTILES	LOG-NORMAL (3 PAR.)
0.1000	7.469
0.2000	7.903
0.3000	8.284
0.4000	8.663
0.5000	9.069
0.6000	9.533
0.7000	10.103
0.8000	10.882
0.9000	12.202
0.9500	13.546
0.9900	16.874
0.9950	18.422
0.9990	22.329
0.9995	24.159
0.9999	28.783

PARAMETERS				CONFIDENCE HALF-AXIS		WEIGHTED DEVIATION	
5.800	1.184	0.275		0.066	0.049	0.054	0.011

ST.EX.OM   31∗01∗85   VERSION 01

Table 3 : Statistical characteristics of the fitted model

TABLES OF QUANTILES

ORDER OF QUANTILES	GAUSS	RAYLEIGH (2 PAR.)	LOG-NORMAL (3 PAR.)
0.1000	6.966	7.035	7.469
0.2000	7.850	7.688	7.903
0.3000	8.488	8.240	8.284
0.4000	9.032	8.759	8.663
0.5000	9.540	9.280	9.069
0.6000	10.048	9.831	9.533
0.7000	10.592	10.450	10.103
0.8000	11.229	11.208	10.882
0.9000	12.113	12.308	12.202
0.9500	12.843	13.251	13.546
0.9900	14.212	15.086	16.874
0.9950	14.713	15.775	18.422
0.9990	15.745	17.218	22.329
0.9995	16.147	17.787	24.169
0.9999	17.007	19.015	28.783

Table 4 : Quantiles connected with different fitted models

CONFIDENCE REGION AT LEVEL 90.00 %

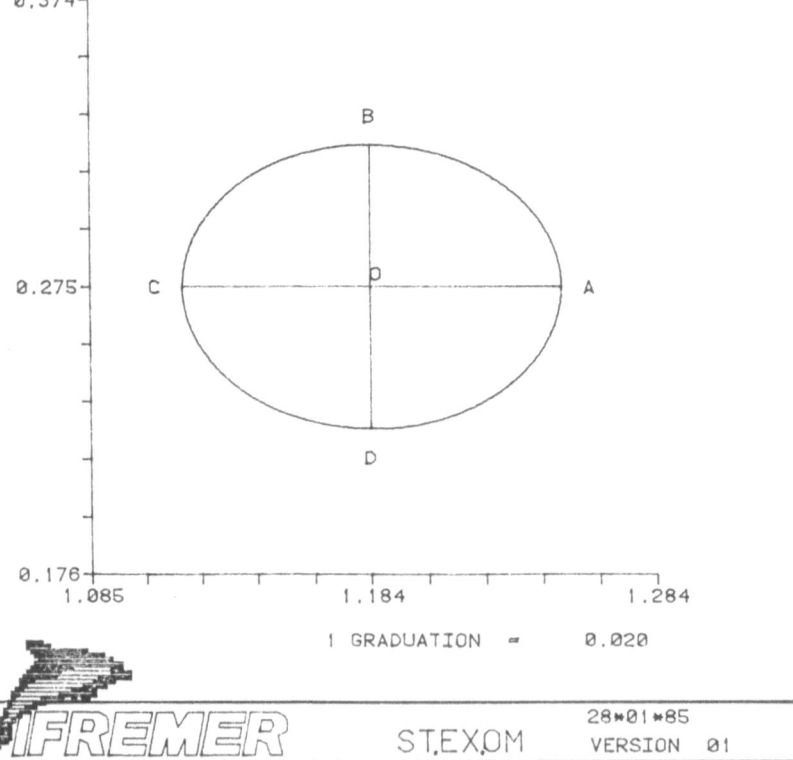

LOG-NORMAL (3 PAR.)		
HALF-AXIS LENGTH	0.066	0.049
CENTER O	1.184	0.275
A	1.251	0.275
B	1.184	0.324
C	1.118	0.275
D	1.184	0.226

1 GRADUATION = 0.020

IFREMER   ST.EX.OM   28*01*85 VERSION 01

Figure 2 : Confidence ellipse for the mean and variance of the Gaussian random variable $Ln(X-X_o)$

CONFIDENCE REGION AT LEVEL 90.00 %

LOG-NORMAL
3 PAR.

WEIGHTED DEVIATION ALONG
THE AXIS OF THE CONFIDENCE ELLIPSE

DISTRIBUTION PARAMETERS		WEIGHTED DEVIATION (MEAN	VARIANCE)
1.184	0.275	0.054	0.011
1.251	0.275	0.112	0.015
1.238	"	0.097	0.013
1.224	"	0.083	0.011
1.211	"	0.071	0.010
1.198	"	0.062	0.010
1.184	"	0.054	0.011
1.171	"	0.057	0.013
1.158	"	0.077	0.017
1.145	"	0.104	0.021
1.131	"	0.135	0.028
1.118	"	0.168	0.037
1.184	0.324	0.098	0.019
"	0.314	0.086	0.016
"	0.305	0.076	0.013
"	0.295	0.067	0.011
"	0.285	0.058	0.010
"	0.275	0.054	0.011
"	0.265	0.059	0.015
"	0.255	0.079	0.022
"	0.245	0.110	0.035
"	0.236	0.151	0.057
"	0.226	0.200	0.095

CONFIDENCE INTERVALS
FOR THE WEIGHTED DEVIATION

80 %	0.039	0.069
90 %	0.034	0.074
95 %	0.031	0.077
99 %	0.023	0.085

IFREMER   ST.EXOM   28*01*85   VERSION 01

Table 5 : Variation of the
weighted deviation

CONFIDENCE REGION AT LEVEL 90.00 %
(CORRECTIONS)

LOG-NORMAL
3 PAR.

WEIGHTED DEVIATION
(MEAN , VARIANCE)

U	0.0832	0.0111
V	0.0813	0.0145
W	0.0847	0.0178
R	0.0846	0.0239
K	0.0768	0.0189

CONFIDENCE INTERVALS
FOR THE WEIGHTED DEVIATION

80 %	0.0387	0.0693
90 %	0.0344	0.0736
95 %	0.0306	0.0774
99 %	0.0233	0.0847

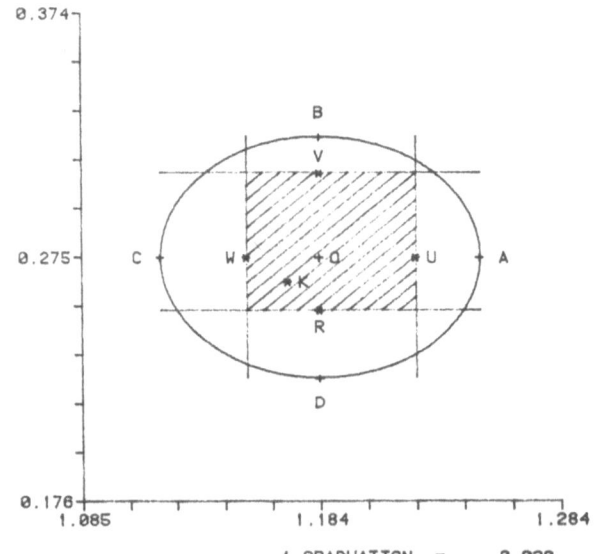

1 GRADUATION = 0.020

ST.EX.OM

04*02*85

VERSION 01

Figure 3 : Confidence region computed with a
normality test of the weighted deviation.

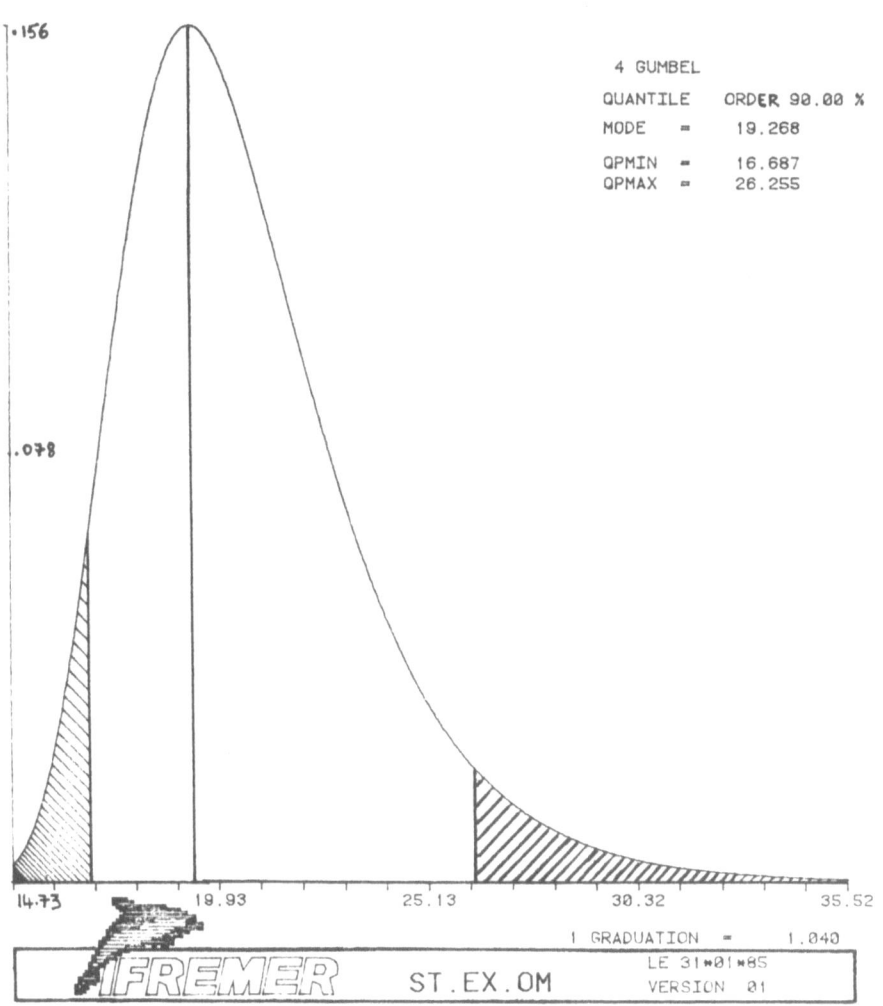

Figure 4 : Limiting extreme law over
T-YEARS (MODE AND TOLERANCE INTERVAL)

```
******************* EXTREME LAW TYPE I ***********************
*                    - GUMBEL -                              *
*                                                            *
*    ESTIMATION OPTION 1  ( MAXIMUM LIKELIHOOD )             *
*                                                            *
*    PARAMETERS :                      INITIAL VALUES        *
*                                      METHOD OF MOMENTS     *
*                                                            *
*        MODE   =    8.6618               8.6363             *
*        SCALE  =    1.4422               1.5655             *
*                                                            *
*    CONFIDENCE REGION AT LEVEL  (ALPHA) :                   *
*                                                            *
*       CENTERED IN :  8.6618    1.4422                      *
*                                                            *
*       ROTATION MATRIX (EIGENVECTORS) :,                    *
*                                                            *
*                          .92135   - .38873                *
*                          .38873     .92135                *
*                                                            *
*       LENGTH OF HALF-AXIS ( MODE) :   .0682 (50%)          *
*                                         .1415 (90%)        *
*       LENGTH OF HALF-AXIS (SCALE) :   .0437 (50%)          *
*                                         .0906 (90%)        *
*                                                            *
*    CALIBRATION TESTS :                                     *
*                                                            *
*       WEIGHTED DEVIATION :                                 *
*          FOLLOWS A NORMAL LAW :                            *
*          MEAN     =   .2290     VARIANCE =    .1777        *
*                                                            *
*       CHI-SQUARE TEST :                                    *
*          STATISTIC  :  5.6111   LEVEL  :    .4313          *
*                                                            *
*                          OVER THE       OVER THE HIGHEST   *
*                          WHOLE LAW        TEN PERCENT      *
*       ARITHMETIC DEVIATION                                 *
*          MEAN        :     .0032            - .0011        *
*          STD. DEVIATION:   .0213             .0037         *
*                                                            *
*       ABSOLUTE DEVIATION                                   *
*          MEAN        :     .0184             .0011         *
*          STD. DEVIATION:   .0110             .0037         *
*                                                            *
*       MAXIMUM  DEVIATION    .0413            .0192         *
*                                                            *
*                                                            *
*       KOLMOGOROV TEST   :                                  *
*                                                            *
*          2 SIDED - STATISTIC    .0413    LEVEL:   .2910    *
*                                          POWER:   .5000    *
*          1 SIDED - STATISTIC    .0413    LEVEL:   .2313    *
*                                                            *
*                                                            *
******************************************************************
```

Table 6 : *Fitting control card*

INTERACTION OF SUSPENSION DESIGN WITH VEHICLE RIDE

P J H Wormell

Royal Military College of Science, Cranfield, UK

1.   INTRODUCTION

The continual demand for improved mobility and agility of
military off-road vehicles has led to ever-increasing installed
power, with power/weight ratios currently over 2 kW/kN, even for
Main Battle Tanks. As a result of this, over a high
proportion of terrains the speed is now limited, not by the
power, but by the ability of the equipment and crew to sustain
the shock and vibration transmitted from the ground.

The way in which a vehicle rides, that is moves in bounce and
pitch, when driving over ground irregularities is principally
dependent upon the design of those parts of its suspension
which respond to motion with a force - the dynamic elements.
In the case of a passive suspension (one whose response
characteristics are predetermined), these elements are the
springs and dampers, generating forces that are a function
of displacement and velocity respectively. It is with the
choice of the characteristics for these components that this
paper is concerned.

Although, in the final event, there is no substitute for
full-scale prototype trials, these are extremely costly, and
the characteristics of suspension hardware are not easily
altered once installed. As a result the use of modelling to
investigate the interaction of suspension parameters with the
vehicle ride holds considerable attractions.

The areas involved in such modelling are indicated in Figure 1.
Evidently a terrain profile must be defined at the input end of
the chain, whilst at the other a means of quantifying the ride
experienced by the crew in the hull is essential in order to
provide a measure of the effectiveness of alterations to the
suspension.

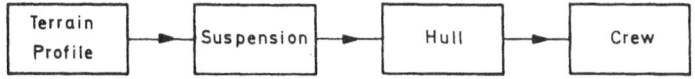

Figure 1   Flowpath for ride analysis

For any but the simplest of vehicle and disturbance situations
a general analytical solution of the ride is impracticable and
a step-by-step approach over short time intervals using a
digital computer becomes necessary. This paper describes a
programme intended to allow the designer rapidly to explore
the interaction of a wide range of vehicle and suspension
variables, over a variety of terrains, and to optimise his
design on the basis of the presentation of a number of
quantified ride criteria. The programme is written for
implementation on a Tektronic 4052 (32K) digital computer,
chosen for its high resolution graphics capability.

2.   VEHICLE MODEL

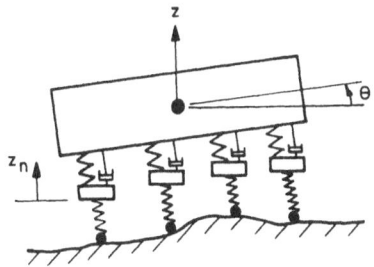

Figure 2   Degrees of freedom of vehicle model

The first priority of the programme has been to enhance the
interactive nature of design; this has meant that runs must be
kept as short as possible. Balanced against this is the depth
of complexity of modelling needed to produce valid results on
which to base sound conclusions. The result is a compromise,
with average run times kept down to a few minutes, and a
flexible model, having the degrees of freedom shown in Figure 2,
the scope of whose variables are discussed in the subsequent
Sections. In particular, however, it should be noted that no
attempt is made to include track effects; nonetheless the model
is found to give a good comparative assessment of types of
suspension for tracked as well as wheeled vehicles.

3. PROGRAMME STRUCTURE

The programme is divided into 8 separate segments. Variables
are grouped within these segments, each of which can be called
up on its own by means of one of the User Definable Keys (UDK)
1 to 8 (see Table 1). This arrangement makes for ease of
control during running, with any parameter alterable with a
minimum of disturbance to the remainder. Additionally input
can be made, either before or during, a run by calling up the
variable by name.

UDK	Segments
1	Master Menu (lists function of UDK 1 - 8)
2	Terrain (ground profile)
3	Vehicle (parameters other than suspension)
4	Suspension (springs)
5	Suspension (dampers)
6	Output presentation (selects parameters to be plotted during run)
7	Speed of run
8	Initiate run

Table 1 UDK functions

4. TERRAIN

The designer may choose from 3 types of ground profile, the
parameters of which he can specify, together with the length
of run (up to a maximum of 75 m).

Double Ramp
This profile is used to measure the ability of a suspension, in
particular the front units, to absorb without damage to themselves or to the crew. The results from tests over this
profile correlate with the ability of the vehicle to cross at
speed such large individual obstacles as banks and ditches.

A symmetrical double ramp is used, with specification needed
for slope (in practice 16°.7, 21°.8 and 26°.6 are used), and
height. The ramp can be made effectively single sided, if so
wished, by specifying a suitably large height in relation to
the length of the run.

Sine-wave
This profile is used to excite pitch in the vehicle and so to

test its resonant frequency and damping in this mode. Undue
pitch amplitudes, accompanied by frequent bottoming on the
idler or suspension, are found to correlate with poor cross-
country speed and performance.

The designer may specify wave-length and amplitude. In
practice military tracked vehicles are tested over profiles
ranging from 4.5 m x 50 mm for small vehicles, to 12 m x 100 mm
for large high speed vehicles.

Cross-country profiles

These profiles are derived from stave-and-level measurements
taken over a variety of terrains at Fort Knox, USA. The ground
heights, recorded at intervals of 1 ft, are stored in a series
of 11 data files on a magnetic tape from which the programme
can retrieve them, one at a time, as requested. The profile
between the defined points is taken to be a 4th order curve
with coefficients chosen to ensure continuity of height and
slope between successive 1 ft sections. This smoothing, which
actually occurs due to the finite wheel diameter, track
bridging effects, and to deformation of the ground, enables
simplification of the ground contact to be made so that it may
be considered as being at a single rigid point below the wheel
centre.

The nature and severity of the terrain can be judged from
spectral density analysis that has been made on them, or from
calculated values of coefficient of roughness. These range
from 1.7×10^{-4} over mild grassland to 12.8×10^{-4} over rocky
soil. For comparison, a value for a UK minor road might be
0.052×10^{-4} Labarre [1], and for a USA Highway 0.014×10^{-4}
Hedrick and Wormley [2].

As a further aid to selection of terrain the programme offers
the designer a visual display showing either a 15 m or 75 m
length of profile (Figure 3).

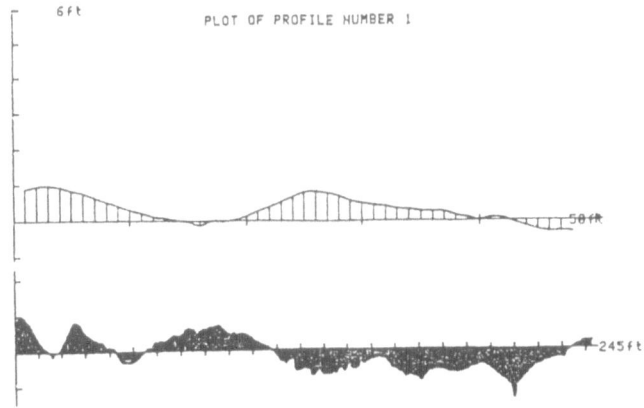

Figure 3 Terrain profile display

5. VEHICLE PARAMETERS

The following vehicle parameters need to be specified:

Sprung mass This is taken to be centrally placed with respect
to the wheel-base.

Unsprung mass This consists of the wheel and part of the
suspension supported by the tyre. Its importance lies in its
influence on maintainance of ground contact.

Wheel-base of a wheeled vehicle, or length of track on the
ground in the case of a tracked one. This influences the
pitch behaviour, but too high a value is precluded for a skid-
steered vehicle due to difficulty this produces in achieving
a turn.

Pitch radius of gyration of sprung mass. This again effects
the pitch response.

Number of wheel stations These are taken to be equi-spaced
and any number may be specified. For wheeled vehicles the
maximum practicable number is 4, but on tracks up to 7 will
be found.

6. SPRINGS

Military tracked vehicles nearly all use transverse torsion bar
springs, acting on trailing wheel arms. These offer an
essentially linear wheel rate with no built-in damping.
Alternatively there is a move towards the use of hydrogas
springs which, besides producing an inherent rising wheel rate,
are externally mounted and thus do not take up armoured volume
and raise the silhouette.

The designer is offered an initial choice of either torsion
bar or hydrogas springing. However, the former covers in
practice any system that gives an essentially linear wheel
rate. The wheels are assumed to move vertically, ie the
effect of any change of geometry is ignored.

The following parameters need to be specified:

Bump travel of wheel (static to bump stop contact). If no
bump stop is fitted then whatever practically limits upward
wheel movement is substituted eg hull contact.
This is the most important parameter for a suspension. It is
limited by space available, the width of the hull (in the case
of torsion bar), the hull/ground clearance, and problems with
track tension. Current values attained range from under 100 mm
for some logistic vehicles, up to nearly 400 mm for Main Battle
Tanks.

Rebound travel of wheel (static to rebound stop contact).
If no rebound stop is fitted some reasonable value eg
1/3 x bump travel may be substituted; results are not sensitive
to this variable. In the case of a tracked vehicle the 'catch-
ing' effect of the track beneath the wheel can be estimated
instead.

Spring reaction factor (wheel load at bump stop contact/static
load). This parameter is a measure of the ability to absorb
large impacts without bottoming; it is the number of g's at the
station before this occurs. How large it may be made is
limited by 2 factors. For a linear spring the implication of
increased stiffness will be accompanied by increased natural
frequency and greater force transmissibility over the range of
travel; a large reaction factor will also result in high stress
in the system, and the programme checks this out (see below).
A typical factor is 2.5 - 3 for transverse torsion bar and
4-5 for hydrogas.

Bump and rebound stop stiffnesses These are seldom known very
precisely and an estimate is usually sufficient. If too
high a value is chosen (eg to simulate metal-to-metal contact
with the hull), problems may be encountered with stability
of the model and the programme step length may need to be
reduced. If a 2 rate suspension is used (eg torsion bar plus
volute spring), then the characteristic of the helper spring
may be substituted for that of the bump stop.

Tyre static deflection This is a measure of the tyre radial
stiffness. The ride of tracked vehicles is not usually
sensitive to this parameter, although it may be expected to
play a significant part in the case of a wheeled one.

To enable the designer to review the implications of his choice
of spring characteristics the programme will now present a
number of calculated parameters. Figure 4 shows 3 typical
displays, covering the parameters of a light tracked vehicle,
its torsion bar springs and its dampers (see Section 7).

    a. Force vs Displacement characteristic of spring and
        bump stop.

    b. Principal mode frequencies of the hull. (ie in
        bounce and in pitch). These are around the static
        position in the case of the non-linear hydrogas
        spring. Frequencies of much more than 2 Hz are
        indicative of a hard suspension that is likely to
        transmit an uncomfortable degree of disturbance to
        the hull on irregular ground. A pitch frequency
        much below 1 Hz is likely to result in undue
        pitch amplitudes when braking and changing gear, or
        traversing critical wave-lengths.

‡‡‡ MENU OF VEHICLE PARAMETERS ‡‡‡

Input your values for each parameter in turn.

        Sprung mass (tonne) 6.75
        Pitch radius of gyration of sprung mass (m) 1.42
        Unsprung mass/wheel (tonne) .12
        Number of wheel stations 5
        Wheel-base (m) 2.49

‡‡‡ MENU OF SUSPENSION PARAMETERS (SPRINGS) ‡‡‡

Identical suspension units at each wheel station will be
assumed, with the wheels equi-spaced, and the centre of
sprung mass over the centre of the wheel-base.

Select type of spring:
        1. Torsion bar
        2. Hydrogas

Enter number selected: 1

Enter the following parameters:
        Bump travel (m) = .203
        Rebound travel (m) = .102
        Wheel load at bump / static load = 2.6
        Bump stop stiffness (kN/m) = 500
        Rebound stop stiffness (kN/m) = 500
        Tyre static deflection (m) = .01

Bar length is taken as (m) 1.249
Wheel arm is taken as (m) 0.386
   bar dia. (mm)   =32.986
   bar stress (MPa) = 940

Hull natural frequencies are:
   bounce(Hz) = 1.399
   pitch (Hz) = 0.868

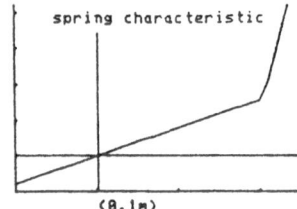

spring characteristic

(0.1m)

‡‡‡ MENU OF SUSPENSION PARAMETERS (DAMPERS) ‡‡‡

Dampers need not be fitted to all wheel stations.
Enter how many stations have dampers: 2
Enter station numbers to which dampers are fitted: 1,5

Enter the following parameters:
        Damper rate at wheel (kN s/m) = 4
        Damper force at blow-off / static load = 1.5

Hull damping ratios are:
        bounce 0.337
        pitch   0.209

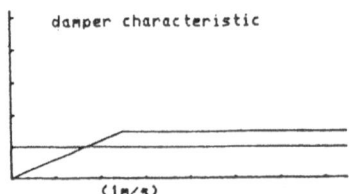

damper characteristic

(1m/s)

Figure 4   Displays for a light tracked vehicle and suspension

c. Limiting spring parameter values for:

(i) Transverse Torsion bar. The bar length is taken as half the wheel-base, and the length of trailing arm to be 0.62 x wheel centre distance. On this basis the maximum shear stress is computed. The limit for this, using the latest steel and manufacturing technology, is about 1.1 GPa.

(ii) Hydrogas. Here the limit is imposed by the pressure/cylinder bore combination. Current technology uses pressures up to about 0.5 k bar. On this basis, and assuming a typical leverage ratio between wheel and piston of 3.5, the cylinder bore is computed. This must be compact enough to fit into the space allocated for it.

7. DAMPERS

Initially the designer may specify those wheel stations that have dampers fitted. On tracked vehicles with torsion bar springs these are often limited to the first and last, where they have most effect in pitch; hydrogas units have intrinsic damping and thus are effectively damped at all stations.

Dampers are assumed to be viscous, ie offer a force at the wheel proportional to its velocity, up to a certain limiting velocity, whereafter the force remains constant. The characteristics in bump and in rebound are taken to be the same.

The user can specify 2 parameters to quantify this behaviour:

Damper rate This gives the force per unit wheel velocity. It needs to be high to control resonant amplitudes (especially in pitch), and to aid the spring in preventing bottoming over large bumps; however, if made too high there will be undue force transmission to the hull over normal terrain and a potential overheating problem. The effect that a given damper will produce will depend upon the size of vehicle to which it is fitted and the wheel rate. For this reason it is usual to quantify the amount of damping as the 'damping ratio' for the vehicle.

Damper reaction factor (wheel force at damper blow-off/static wheel load). This will be additive to the Spring Reaction Factor in giving the total number of g's that a unit can take without bottoming. It is limited by a pressure relief valve set to prevent damage to the damper. For a telescopic damper a figure of only 0.2 - 0.5 is usual; for a rotary damper or

one built into a hydrogas unit, values up to 1.6 have been
produced.

To enable the User to review the implications of his choice of
damper parameters the programme will now present (see Figure 4)

    a.   Force/Velocity characteristics.

    b.   Principal mode damping ratios (ie pitch and bounce).
          Values less than 0.3 are not generally found to be
          satisfactory.

8. OUTPUT PRESENTATION

To provide the basis for an objective assessment of the Ride of
the vehicle following a run, the User is presented with
collected data which may be in any or all of 3 forms. These
are: a plot showing variation of a chosen parameter with
distance, a table of critical values, and a Probability
Density Function (PDF) histogram of acceleration.

Initially a choice is offered from 6 parameters for the plot.
Depending upon which of these is selected, and upon which of
the 3 terrains has been chosen for the run, the remainder of
the output will follow.

Plot parameters
Whichever parameter is chosen, a plot of the ground profile
will be produced as well. As the run proceeds, and the
vehicle advances, the front wheel and the ground profile will
be plotted in the same position along the x-axis; the plot of a
measurement relating to any point on the vehicle will then be
displaced horizontally by the correct amount in relation to the
front wheel, eg the plot for the last wheel will appear behind
that for the first by the length of the wheel-base.

Displacement and Acceleration of road wheels will appear with
traces for successive wheels having their origins displaced
upwards by an arbitrary amount, simply to aid clarity. These
plots are generally used if a detailed study of the succession
of events over a single obstacle is required, perhaps to
identify the cause of a particular acceleration peak, or to
validate the programme against actual measurement. From the
displacement it will be clear when a wheel leaves the ground
and lands again; acceleration will be heavily dependent upon
tyre stiffness, and the amplitudes of its high frequency content
need to be treated with caution.

To further assist understanding of behaviour the designer may
choose to have the graphs annotated to show whenever the bump
(↑) or rebound (■)stops are inoperation, or the damper blow-off
pressure has been exceeded (↓).

Vertical displacement of centre of sprung mass (CG) This plot
also superimpose a line, corresponding in length to the wheel-
base, and rotated by the angle of pitch; this will appear
at intervals equal to the wheel-base.

Vertical acceleration of centre of sprung mass This is the
parameter most generally chosen for study. As with the wheels,
amplitudes of 'spikes' due to say bump stop contact, are
safer considered relatively rather than absolutely.

Acceleration components of other hull stations This enables
the acceleration at points on the hull, apart from the CG, to
be studied; eg the various crew positions (in particular the
driver and commander of an AFV receive very different motions).
The designer must specify the coordinates of the point chosen
relative to the pitch axis; this in practice is generally not
far from the centre of sprung mass. The programme will
compute the horizontal and vertical components of acceleration,
accounting for centripetal and pitch acceleration contributions
and for pitch angle. It assumes a constant horizontal velocity
of the CG. The trace for the vertical acceleration is
distinguished periodically by the appearance of an upward
pointing arrow.

Forces in suspension This quantifies the contribution made by
the spring, the damper and the stops to the total ground
reaction on the front unit. The principal use for this will be
to provide an input for the detailed design of the suspension
components eg wheels and linkages. The front unit is chosen
as being invariably the most highly stressed. The 3 forces at
the wheel may be referred back to find the corresponding forces
in the spring, damper and stops, and hence the loads on the
parts of the suspension on which these bear can be assessed.
The programme produces plots of the forces on the 3 components
concerned, annotated at intervals to distinguish them.

Table of critical values
Parameters that are included as appropriate:

Maximum forces in the front station spring, damper and stops
encountered during the run. These are needed for detailed
stressing of components.

Maximum acceleration at CG This is limited by human response
to shock. A working value for this limit is 2.5 g, but rate of
repetition will also play a part in other than single obstacle
cases.

RMS of acceleration at CG This is limited by human response to
vibration. Although ideally this measure of mean acceleration
should be weighted to account for variation in human
sensitivity with frequency, it will still provide a useful

criteria for comparative assessment of successive rungs over
the same piece of terrain in its unweighted form as here.

Maximum acceleration components at a selected hull station

RMS of acceleration components at a selected hull station

Mean damper power will give a measure of the heat dissipative
properties required of a damper, designed to have the
characteristics chosen in Segment 5, under the terrain and
speed conditions of the run.

PDF histogram
A histogram of the PDF of the vertical acceleration at any
chosen point on the hull may be presented. This takes the
form of an acceleration scale along the x-axis, divided into
0.1 g bands. The PDF, plotted along the y-axis, then gives
the probability of the acceleration at any instant lying
within a particular band. The area under the histogram will
evidently therefore be unity. Its interpretation should
give a measure as to how well the design has achieved the
compromise between having sufficient hardness to avoid
bottoming and shock over large bumps, and offering a soft
suspension to minimise transmitted disturbance over small
irregularities. This is illustrated in Figure 5.

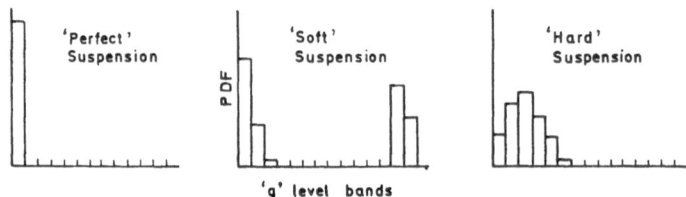

Figure 5 Examples of PDF histograms

9. RUNNING THE PROGRAMME

Run times depend upon the number of wheel stations and the
iteration time interval chosen. The latter defaults to 3 ms;
this is found to be a good general figure. If tyres and stops
are relatively soft, time may be saved by setting a somewhat
larger value; if taken to excess unstable oscillations on the
acceleration outputs will appear. A rough estimate of likely
run times on the Tektronix 4052 computer can be formed given
that, using a typical tracked vehicle, the computer model
speed is about equal to the actual speed/100 (eg if an actual
speed of 10 m/s is being simulated over a run of 20 m, then the
computer run time will be very approximately 200 s).

As an illustration of the running of the programme consider
again the light tracked vehicle whose parameters and suspension
characteristics were given in Figure 4. The vehicle is to be

subjected to a 20° ramp test at a speed of 30 km/h. Figure 6 is
a composite of 3 screen displays which together show the motion
of the wheels and hull, together with a table of maximum values
attained. It may be noted (from the top plot) that the fitting
of dampers with a high reaction factor to the end stations has
resulted in these absorbing this large obstacle by damper
blow-off (↓) rather than bump-stop contact (↑) such as has
occurred on the undamped second and fourth stations. This
would account for the relatively low double shock 1.27 g which
has occurred twice on the impact of the first and then the last
wheels with the ramp; between these times there has been a
brief period of weightlessness which has been repeated as the
front takes off over the top of the ramp , (see lowest plot).
The pitch attitude changes during this time are indicated on
the middle plot, which also shows that very little displacement
of the centre of sprung mass has occurred; this may well not be
the case at some other position on the hull, such as the
driver over the front wheels.

To illustrate this the acceleration components at this position
are shown in Figure 7. The driver is seen to suffer a vertical
shock which has increased to 1.69 g, although his position
low down in the hull has kept the horizontal acceleration level
down; this would not be so for his commander seated high up in
the turret. The advisability of wearing a restraint harness
is indicated by the negative values reached.

Figure 7 also shows the forces occurring in the suspension
components of the front wheel station. On this particular
run there has been no bump stop contact (b), the damper blow-
off on both bump and rebound is clearly indicated (d); whilst
the change in spring force(s) to compression (on impact with
ramp), expansion (on clearing apex), and recompression (on
landing again) may be noted.

Finally, in Figure 8, the vehicle has been run for a short
distance at 40 km/h over moderate cross-country terrain; the
road wheel displacement only being illustrated. The resulting
ride is at the extreme limit of tolerance, with the front half
of the vehicle off the ground for a considerable distance
(note the rebound stop contact (▪)), resulting in a 2.6 g peak
shock and an rms acceleration of 0.804 g. Of note is the
damper power dissipation of over 5 kW which might well
preclude any further increase in the rates of these components.

REFERENCES

1.   Labarre, R P et al   The Measurement and Analysis of Road
                           Roughness. MIRA Report 1970/5

2.   Hedrick, J K   Advanced Analytical Methods for
      Wormley, D N   Vehicle Analysis and Design, MIT 1981

Maximum values of:
spring force/static (#1)   = 2.55
bump stop force/static (#1) = 0
damper force/static (#1)   = 1.49
acceleration at G (g)   = 1.27

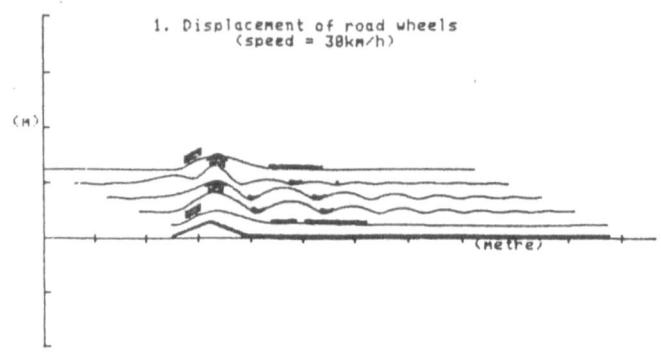

1. Displacement of road wheels
(speed = 30km/h)

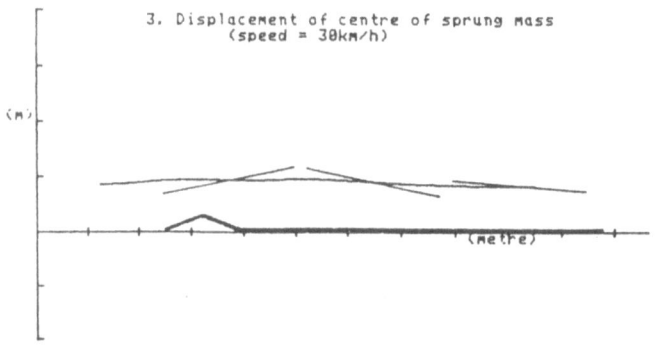

3. Displacement of centre of sprung mass
(speed = 30km/h)

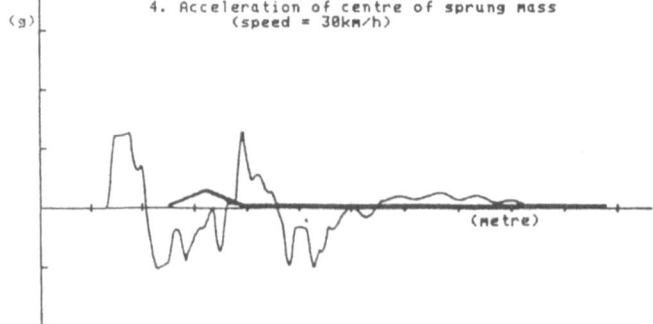

4. Acceleration of centre of sprung mass
(speed = 30km/h)

Figure 6 Motion of light tracked vehicle over 20° ramp

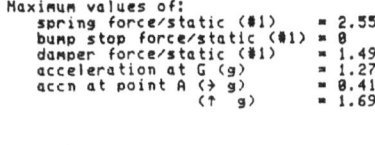

Maximum values of:
    spring force/static (#1)   = 2.55
    bump stop force/static (#1) = 0
    damper force/static (#1)   = 1.49
    acceleration at G (g)   = 1.27
    accn at point A (→ g)   = 0.41
                    (↑ g)   = 1.69

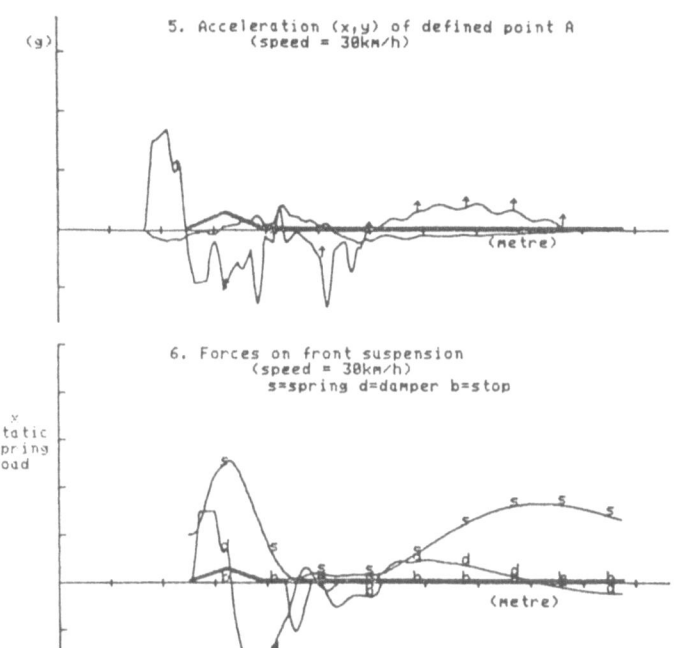

5. Acceleration (x,y) of defined point A
   (speed = 30km/h)

(g)

(metre)

6. Forces on front suspension
   (speed = 30km/h)
   s=spring d=damper b=stop

y
static
spring
load

(metre)

Figure 7   Driver acceleration and Suspension Forces

Maximum values of:
    spring force/static (#1)   = 3
    bump stop force/static (#1) = 3.06
    damper force/static (#1)   = 1.49
    acceleration at G (g)   = 2.6
RMS (g) values of:
    acceleration at G   = 0.804
Mean Power (#1 damper (kW))   = 5.44

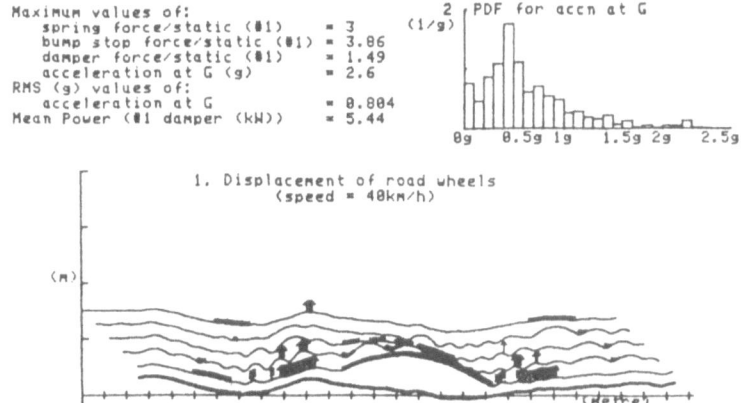

2 PDF for accn at G
(1/g)

0g   0.5g 1g   1.5g 2g   2.5g

1. Displacement of road wheels
   (speed = 40km/h)

(m)

(metre)

Figure 8   Movement over off-road profile

RANDOM SIGNALS ANALYSES APPLIED TO VEHICLE-TRACK INTERACTION.

C. Esveld

Netherlands Railways

The inter-relations between vehicle and track can be described in terms
of transfer functions, indicating how the various geometry components
contribute to a given response component of the vehicle. In conventional
approaches use is made of mathematical models, based on a schematisation
according to masses, springs and dampers. As a rule, the predictive merits
of such models turn out to be small, due to the impossibility of accurately determining the model parameters.
The method described in this article starts from measured geometry signals
(input) and a corresponding response signal of the vehicle (output), between which transfer functions are established with the aid of the MISO
method (Multiple Input, Single Output). This method is based on the theory
of random signal analysis and, as far as the application to the interaction
between vehicle and track is concerned, it was suggested for the first time
in [1].
In addition, this article presents a possible solution for real dynamic
measurement of vertical and lateral wheel/rail forces through measuring
wheelsets using MISO.

1. Introduction.

Track geometry forms an important aspect in the assessment of track quality. Tracks are laid according to a design geometry. Due to the wheel/
rail forces exerted under train traffic the geometry deteriorates and
consequently will gradually start deviating from the design values. This
leads to the requirement of corrective measures through track maintenance. The geometry components commonly considered as degrees of freedom of
a track are represented in figure 1.1.

MISO TRACK GEOMETRY COMPONENTS

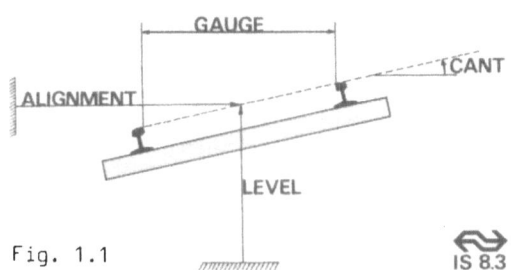

Fig. 1.1

About 10 years ago Netherlands Railways started on the digital analyses of track geometry signals, recorded by a track recording car.
At first the calculations were only aimed at computing mean values and
standard deviations. But before long, the Fast Fourier Transform (FFT)
was introduced, enabling power spectra to be determined. This was followed by an extension to the calculation of the inter-relations between
1 input and 1 output with applications especially in the field of assessment of track maintenance methods, comparing between situations prior to
and after maintenance operations. The theory and the relevant applications are described, amongst other things, in [1] and [2].
These references also describe the operations to be carried out
for digital data processing such as analog filtering to prevent alia-
sing, digital filtering [3], tapering, FFT procedure, ensemble averaging, frequency smoothing, etc. One of the problems in analysing track
geometry signals concerns the combination of individual results for
each component to an overall quality index. An obvious solution consists in choosing the car body acceleration, being a standard of ri-
ding comfort, as a starting point. The problem is then focused on the
question how the relationships between track geometry components on
the one hand and car body acceleration, defined in a given way, on
the other, can be plotted according to a practical method. In view of
the problems in applying mathematical vehicle models referred to before, NS have decided, by virtue of the proposals made in [1], that the
MISO method should be chosen as a starting point for further developments.

The Office of Research and Experiments (ORE), which co-ordinates railway
research in Europe, also adopted this method for a number of track/train
interaction studies carried out bij Committee C 152, and has meanwhile published the reports [4], [5] and [6], dealing with MISO-calculations applied to measurements taken by various Railways.

The present article aims at giving a survey of the MISO-theory, supplemented by a number of applications, dealing, above all, with measurements taken according to the new inertial track recording system BMS of NS. Finally, the article briefly reviews the procedure according to which the NS Permanent Way Department thinks to incorporate the MISO results into the BMS system and also how the Rolling Stock Department could use this method in assessing dynamic vehicle behaviour.

2. Basic principles for 1 input-1 output.

Since the primary object of this article consists in giving a survey of the main trends of the theory, without entering into all sorts of minor details, no derivations will be discussed. As far as details and more basic considerations with respect to the theory are concerned, reference is made in the first place, to the Standard work by Bendat and Piersol [7] and to references [1], [2] and [6], which, in addition to practical implementations, include information on rail applications.

The theory of random signal analyses distinguishes between the time domain (for dynamic processes) or spatial domain (for geometrical processes) on the one hand and the frequency domain on the other, the frequency being composed of the reciprocal time or the reciprocal distance for dynamic and geometrical processes respectively. Though in the following text, the magnitude t is used as a variable of time in the formulae, the distance may be imagined, too, for this. Likewise, it applies with respect to the frequency f that this may represent both the reciprocal time and the reciprocal distance. As a matter of fact, the variables time and distance are interlinked by the running speed.

If the signal x (t) denotes a magnitude in the time domain, the representation in the frequency domain is obtained via the so-called Fourier transform. provided that $\int_{-\infty}^{\infty} |x(t)|\, dt < \infty$ and consequently also $\int_{-\infty}^{\infty} |X(f)|\, df < \infty$. Both transforms from and to the time domain read as follows

$$X(f) = \int_{-\infty}^{\infty} x(t)\, e^{-i 2\pi f t}\, dt \qquad (2.1)$$

$$x(t) = \int_{-\infty}^{\infty} X(f)\, e^{i 2\pi f t}\, df \qquad (2.2)$$

If these transforms are made digitally this is done with the aid of the Fast Fourier Transform (FFT), which is, at present, becoming more easily available as hardware facility.

If the 1 input, 1 output model shown in Fig. 2.1, with transfer function H (f), is composed of linear physical realisable system, the transfer function H (f) can be explicitly determined on the basis of the system parameters. For a measured input x (t) with the corresponding X (f), an output value Y (f) can be calculated for any f as follows:

$$Y(f) = H(f) \cdot X(f) \qquad (2.3)$$

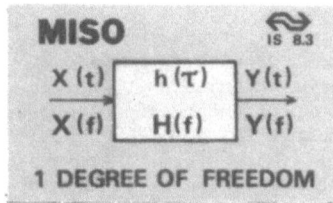

Fig. 2.1

If, however, both input and output are measured, it might be wrongly inferred from (2.3), that the transfer function would follow from the quotient of output and input; however:

$$H(f) \neq \frac{Y(f)}{X(f)} \qquad (2.4)$$

Both the real and the imaginary parts of the complex Fourier transforms have, as a rule, rather an irregular shape. Therefore, recourse must be had to quadratic spectral density functions, which must first be submitted to an averaging procedure, so as to obtain an acceptable statistical degree of reliability. Only after this may an estimate of H (f) be made, to be discussed later on in the chapter.

The relationship between input and output is described in the time domain as the convolution product of h and the input x according to (2.5), this complicated procedure being reduced, as already becoming apparent from (2.3), in the frequency domain to a simple multiplication.

$$y(t) = \int_{-\infty}^{\infty} h(\tau)\ x(t-\tau)\ d\tau \qquad \text{convolution} \qquad (2.5)$$

$$y(f) = H(f) \cdot X(f) \qquad \text{multiplication} \qquad (2.3)$$

in which h (τ) represents the unit impulse response and H (f) the transfer function, interrelated as follows;

$$H(f) = \int_{-\infty}^{\infty} h(\tau)\ e^{-i2\pi f\tau}\ d\tau \qquad (2.6)$$

$$h(\tau) = \int_{-\infty}^{\infty} H(f)\ e^{i2\pi f\tau}\ df \qquad (2.7)$$

From the Fourier transforms, spectral density functions may be deduced in multiplying the two function magnitudes with each other and by dividing them subsequently by the record length T. This leads, as a rule, to the (complex) cross spectrum S_{xy} (f) .If, however, y is replaced by x, a real valued auto-spectrum is obtained.
In (2.8), $\overline{X}$ stands for the complex conjugate of X.

$$S_{xy}(f) = \lim_{T \to \infty} \frac{1}{T}\ \overline{X}(f)\ y(f) \qquad (2.8)$$

$$R_{xy}(\tau) = \lim_{T \to \infty} \frac{1}{T} \int_{0}^{T} x(t)\ y(t+\tau)\ dt \qquad (2.9)$$

Here, too, an analog form of processing is valid in the time domain, which leads to the cross correlation fuction R_{xy} (τ), shown in (2.9). From the point of view of calculation technique, this expression is very similar to the convolution process discussed before. In an absolutely analog way as in (2.6) and (2.7), S_{xy} (f) and R_{xy} (τ) are interrelated via a Fourier transform. These expressions are known in literature as the Wiener-Kintchine relationships reading as follows:

$$S_{xy}(f) = \int_{-\infty}^{\infty} R_{xy}(\tau)\, e^{-i2\pi f\tau}\, d\tau \qquad (2.10)$$

$$R_{xy}(\tau) = \int_{-\infty}^{\infty} S_{xy}(f)\, e^{i2\pi f\tau}\, df \qquad (2.11)$$

An important feature of the auto-spectra is that they are symmetrical with respect to the line f = o and that the area equals the variance σ^2. This is illustrated in Fig. 2.2.

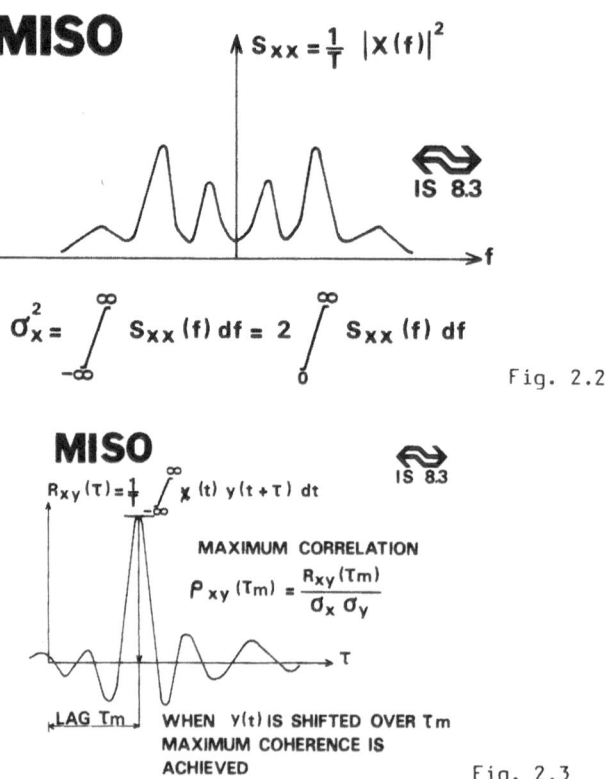

MISO $S_{xx} = \frac{1}{T}\, |X(f)|^2$

IS 8.3

$$\sigma_x^2 = \int_{-\infty}^{\infty} S_{xx}(f)\, df = 2\int_{0}^{\infty} S_{xx}(f)\, df$$

Fig. 2.2

MISO   IS 8.3

$R_{xy}(\tau) = \frac{1}{T}\int_{-\infty}^{\infty} x(t)\, y(t+\tau)\, dt$

MAXIMUM CORRELATION

$P_{xy}(Tm) = \dfrac{R_{xy}(Tm)}{\sigma_x\, \sigma_y}$

LAG Tm   WHEN Y(t) IS SHIFTED OVER Tm
MAXIMUM COHERENCE IS
ACHIEVED

Fig. 2.3

The cross correlation function is particularly frequently applied in plotting the shift between two signals. The displacement corresponds, in fact, to the place where the maximum correlation occurs. Fig. 2.3 shows the principle of this way of consideration.

In quantifying the correlation between 2 signals, once more a consideration in the frequency domain is applicable, designated as the coherence γ^2 xy (f) and a consideration in the time domain, resulting in the correlation fuction ρ^2 xy (τ). These expressions read as follows:

$$\gamma^2_{xy}(f) = \frac{|S_{xy}(f)|^2}{S_{xx}(f) \, S_{yy}(f)} \qquad 0 \leqslant \gamma^2_{xy}(f) \leqslant 1 \quad (2.12)$$

$$\rho^2_{xy}(\tau) = \frac{R^2_{xy}(\tau)}{R_{xx}(o) \, R_{yy}(o)} \qquad 0 \leqslant \rho^2_{xy}(\tau) \leqslant 1 \quad (2.13)$$

In (2.13) it applies: $R_{xx}(o) = \sigma^2_x$ and $R_{yy}(o) = \sigma^2_y$
In addition, it should already be observed here that the coherence according to (2.12) only furnishes sensible information, if the spectra S_{xx}, S_{yy} and S_{xy} have been averaged according to the rules to be discussed in chapter 4. This also applies to the formulae to be discussed now for estamating the transfer function H (f).
From the relationship between input and output, the following relation on a spectral level may be deduced:

$$S_{xy}(f) = H(f) \, S_{xx}(f) \qquad\qquad\qquad (2.14)$$

From this it follows for the transfer function H (f):

$$H(f) = \frac{S_{xy}(f)}{S_{xx}(f)} \qquad\qquad\qquad (2.15)$$

The coherence function γ^2xy (f) is obtained from:

$$\gamma^2_{xy}(f) = \frac{\overline{H}(f) \, S_{xy}(f)}{S_{yy}(f)} \qquad\qquad (2.16)$$

The NS have made a simulation study of the influence of non-correlated constributions to the output on the error in estimating the transfer function. This study, published in [5], has shown that about 10% of non-correlated data in the output leads to an error of about 10% in the transfer function estimate, the coherence being reduced to about 0.8.

Seen from this angle, whilst also allowing for possible other causses, NS now retain a limit value of γ^2 xy (f) = 0.85. Only estimates of H (f) to which applies that $\gamma^2_{xy}(f) \geqslant 0.85$, are accepted, estimates with a lower γ^2xy value being rejected.

3. Multiple input, single output (MISO).

The model shown in Fig. 3.1 shows how the q inputs x_i (t), produce, via q linear systems, q outputs y_i (t) together constituting the over-all output, according to:

$$y(t) = \sum_{i=1}^{q} y_i(t) \qquad\qquad\qquad (3.1)$$

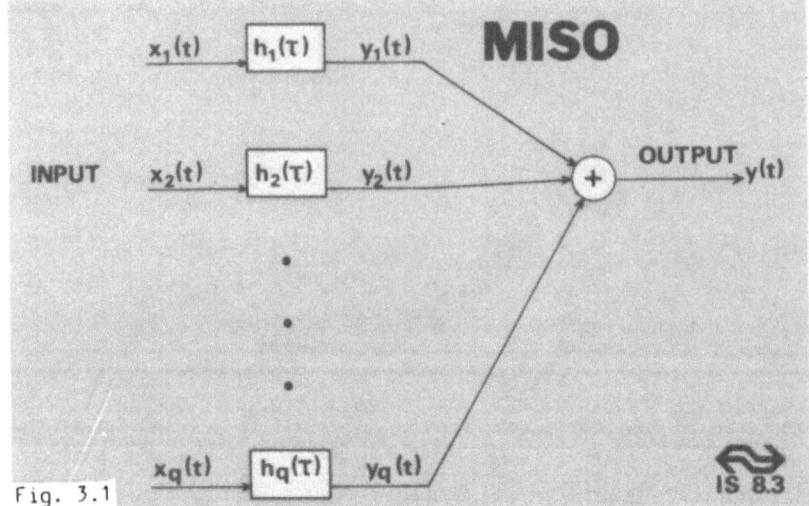

Fig. 3.1

De outputs y_i (t) follow from the inputs via the convolution products:

$$y_i(t) = \int_{-\infty}^{\infty} h_i(\tau) \; x(t-\tau) \; d\tau \qquad (3.2)$$

Assuming the process to be stationary, auto- and cross correlation functions may be deduced, furnishing after Fourier transformation, the following set of equations in the frequency domain:

$$S_{iy}(f) = \sum_{j=1}^{q} S_{ij}(f) \; H_j(f) \qquad (3.3)$$

in which according to (2.8):

$$S_{iy} = \frac{1}{T} \overline{X}_i(f) \; Y(f)$$

$$S_{ij} = \frac{1}{T} \overline{X}_i(f) \; X_j(f)$$

In matrix notation (3.3) reads as follows:

$$\{S_{xy}\} = [S_{xx}] \{H\} \qquad (3.4)$$

The generation and solution of these q complex equations is discussed in chapter 5. However, the formal solution can be provisionally written as follows:

$$\{H\} = [S_{xx}]^{-1} \{S_{xy}\} \qquad (3.5)$$

The reliability of the transfer functions thus estimated follows from the multiple coherence function $\gamma^2_{xy}(f)$, depicting the ratio between the output spectrum calculated on the basis of (3.5) and the measured output spectrum according to:

$$\gamma^2_{x.y}(f) = \frac{S_{yy} \text{ calculated}}{S_{yy} \text{ measured}} \qquad 0 \leq \gamma^2_{x.y}(f) \leq 1 \quad (3.6)$$

After substitution this leads to:

$$\gamma^2_{x.y}(f) = \frac{\sum\limits_{i=1}^{q} \bar{H}_i(f)\, S_{iy}(f)}{S_{yy}(f)} = \frac{\{\bar{H}\}^T \{S_{xy}\}}{S_{yy}(f)} \qquad (3.7)$$

Summation in the numerator indicates the contribution of the various inputs to the output spectrum.
In this case, too, it applies that for a reliable estimate $\gamma^2_{xy}(f)$ must lie in the interval

$$0.85 \leq \gamma^2_{x.y}(f) \leq 1 \qquad (3.8)$$

4. Statistical reliability.

4.1 General consideration.

In this chapter some attention is paid to random errors E_r and systematic errors (bias errors) E_b.
Systematic errors can be compensated for by correction or calibration, according to:

$$X = \tilde{X}(1 - E_b) \qquad -1 \leq E_b \leq 1 \qquad (4.1)$$

Where X denotes the true value and $\tilde{X}$ the estimator. If the systematic error is negligibly small ($E_b < 0.02$) and the random error is small ($E_r < 0.10$), the 95% confidence interval for X is given approximately by:

$$\tilde{X}(1 - 2E_r) \leq X \leq \tilde{X}(1 + 2E_r) \qquad (4.2)$$

In this chapter, only some main trends are given. For a more detailed discussion, see references [7], [8], [9] and [10].

4.2 Random errors.

As stated before, spectral density functions must be averaged, so as to keep the random error at an acceptably low value. This averaging procedure can be made in two different ways, i.e. either by averaging over the records (ensemble averaging), this number amounting to NSEC, or by combining a number of frequency components, so-called frequency smoothing; this number amounts to NA. The overall number of averaging operations n thus is;

$$n = NA * NSEC \qquad (4.3)$$

The random error in spectral density functions is inversely proportional to $\sqrt{n}$ and it follows from:

$$S_{xx} \rightarrow \quad \varepsilon_r = \frac{1}{\sqrt{n}} \qquad (4.4)$$

$$S_{xy} \rightarrow \quad \varepsilon_r = \frac{1}{|\gamma_{xy}|\sqrt{n}} \qquad (4.5)$$

The random error in the modulus of the transfer functions H_i, determined via MISO, can be approximated according to:

$$|H_i| \rightarrow \quad \varepsilon_r = \sqrt{\frac{q}{n-q} F_{n_1, n_2, \alpha} \frac{1-\gamma_{x.y}^2}{1-\gamma_{i.x}^2} \frac{S_{yy}}{S_{xx}}} \qquad (4.6)$$

where: q = number of inputs
         n = number of averaging operations
         F = 100 α percentage point of the F-distribution ,
           with $n_1 = 2_q$ en $n_2 = 2$ (n-q)
$\gamma_{i.x}^2$ = multiple coherence function between input x_i and the other
           inputs.

From the first term under the square root sign it becomes apparent that the number of averaging operations n must at least be equal to q + 1; in practice, the value of n will certainly have to be one order of magnitude higher.
The random error in the argument of H_i follows from:

$$arg\, H_i \rightarrow \quad \varepsilon_r = arcsin \frac{\varepsilon_{r|H_i|}}{|H_i|} \qquad (4.7)$$

4.3 Bias errors.

On the subject of estimating bias errors relatively little is known. For auto-spectra, a Taylor-series expansion for approximating the bias error may be applied. This leads to the following formula:

$$\varepsilon_b \simeq \frac{B_e^2}{24} \frac{S_{xx}''}{S_{xx}} \simeq -\frac{1}{3} \frac{B_e^2}{B_r^2} \qquad (4.8)$$

where: B_e = resolution band width

$$= \Delta f = \frac{NA}{L}$$

   L = record length
   NA = number of frequency smoothing operations
   B_r ="half power point" band width $\simeq$ 0.03 m^{-1} for track geometry
       spectra.

4.4 Summary of random and bias errors.

Obviously, the requirement made so as to keep the random error and the bias error small,are sometimes incompatible, since:

$$\varepsilon_b \therefore \frac{NA^2}{L^2} \tag{4.9}$$

$$\varepsilon_r \therefore \frac{1}{\sqrt{NA * NSEC}} \tag{4.10}$$

Averaging over frequencies soon leads to large bias errors, whereas long records and a high value of NSEC are attractive, but soon lead to measuring problems.
NS are attempting to adhere to the following maximum values of random and bias errors:

$$\varepsilon_r < 0.10 \tag{4.11}$$
$$\varepsilon_b < 0.025$$

With $B_r \approx 0.03$ m^{-1}, this leads to the following choice of parameters

$$
\begin{aligned}
NA &= 4 \\
NSEC &= 25 \\
L &= 500 \text{ m}
\end{aligned} \tag{4.12}
$$

The distance to be measured in this case is therefore, 12.5 km. At present, NS are adhering to a minimum length of 10 km for their measurements.

5. Numerical aspects.

The majority of numerical questions forms the subject of a detailed description in references [1], [2] and [6]. Despite this, some aspects will be explained in greater detail here.

5.1 Frequency smoothing.

Smoothing of frequencies should be made symmetrically about the line $f = 0$, so as to preserve the even and odd character of the real and the imaginary parts respectively. This averaging is made separately for the real part and the imaginary one. In case of odd NA, NA values are always summated and divided by NA, whereas in case of even NA, NA + 1 points are considered, of which the central NA-1 points are included entirely and the 2 boundary points only for 50%; the sum is once more divided by NA.

6. Applications.

6.1 General remarks.

On NS, the applications of the MISO method are primarily confined to the field of interaction between vehicle and track. The method is aimed at determining the relationships between track geometry components, serving as inputs and a vehicle response magnitude, representing the output.

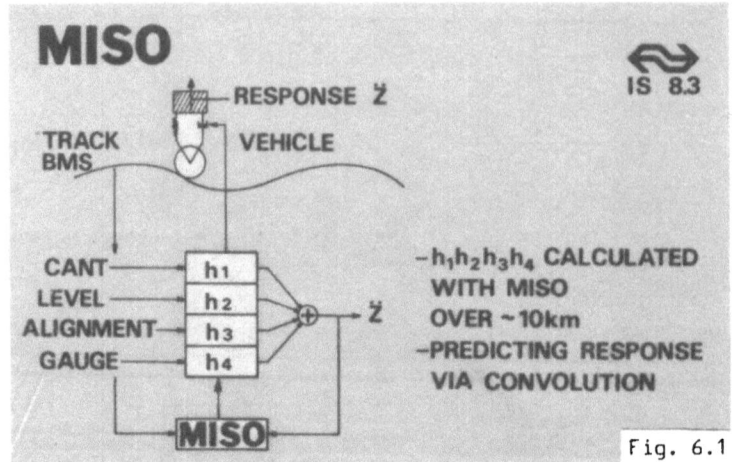

Fig. 6.1

The model describing the approach via MISO is depicted in Fig. 6.1.
The geometry components cant, level, alignment and gauge constitute
the input, whereas so far only car body accelerations have been considered as output.

6.2 Some examples.

The new track recording car of NS.

Early in 1983, the BMS track recording system was completed in the
universal measuring car of NS. This system enables non-
contact measurements of the track geometry to be made and is based
on the principle of the stabilized platform technique.
Further details on BMS are given in [11]

BMS measures the track geometry in wave bands between 0.5 and 70 m
and for this purpose uses, inter alia, two accelerometers for measuring the vertical and horizontal car body accelerations and a rate
gyro, measuring the derivative of the car body rotation. The accelerometers are located centrically above the measuring bogie. The car
itself is fitted with the modern Y-32 boqies, assessing a very linear
charasteristic. Within the scope of the C 152 studies, a MISO analysis
has been applied to this recording car, where the vertical body acceleration furnished by the BMS system is considered as vehicle reaction. The results are given in Figures 6.2 - 6.7.

Figure 6.2 shows the power spectrum of the track geometry component
level as a function of the spatial frequency f [m^{-1}] and the wave
length λ [m]. The response spectra at 80 and 120 km/h are shown in
Fig. 6.3. These spectra were calculated from 20 records of 500 m
length and a frequency smoothing factor NA = 4, so that the bias and
random errors remain within the limits mentioned in (4.11).

As an example, Fig. 6.4 shows the transfer functions H_2 between level
and vertical car body response; the other transfer functions are negligibly small. These are, in fact, the primary results of the MISO
analysis according to (3.5). In agreement with formula (2.7), these
functions have been Fourier transformed, so as to obtain the unit
impulse respons functions h_i of which h_2 is illustrated in Fig. 6.5.

In this case, too, all other h functions can be neglected, implying that only the level contributes to the vertical acceleration.

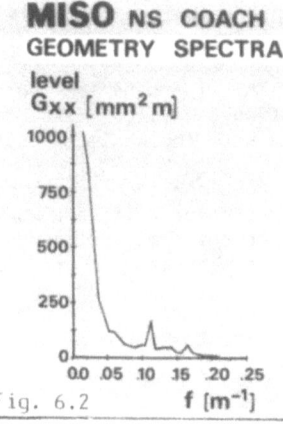

Fig. 6.2

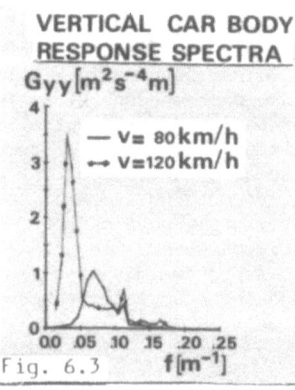

Fig. 6.3

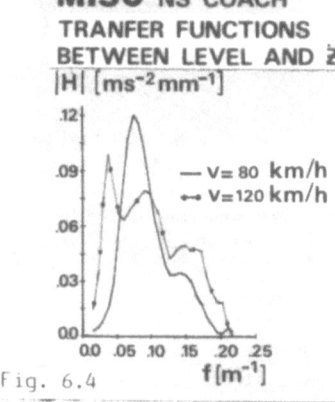

Fig. 6.4

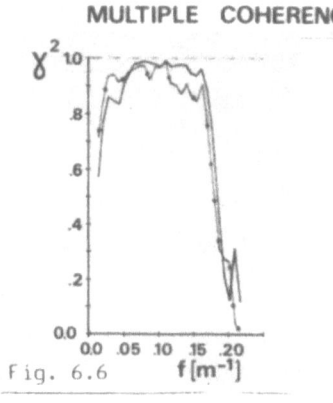

Fig. 6.6

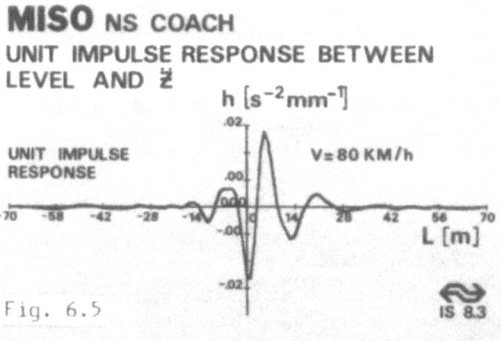

Fig. 6.5

The degree to which the transfer function values are reliable is shown by the multiple coherence function $\gamma^2_{x.y}(f)$, depicted in Fig. 6.6. As stated before, the $\gamma^2_{x.y}$ (f) value should be higher than 0.85 if practical applications are to be made possible. On further analysis of the results, it appears that the γ^2_{xy} (f) value only meets this demand in the frequency bands where the measured basis signals contain enough energy. This is rather obvious and it also explains, perhaps in a different way, why long measuring sections should be chosen, by preference with the maximum possible variations in the geometry spectra.

To complete the sequence of computations, the response is calculated once more as a function of the distance covered via the convolution principle according to (2.5), using the unit impulse response functions previously obtained, and compared with the response signal originally measured. Fig. 6.7 contains a graphical representation of the calculated and measured signals; the similarity between the measured and the calculated response appears to be remarkable.

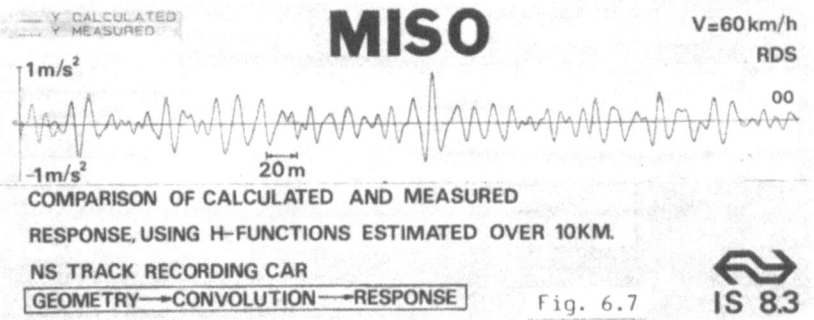

COMPARISON OF CALCULATED AND MEASURED
RESPONSE, USING H-FUNCTIONS ESTIMATED OVER 10KM.
NS TRACK RECORDING CAR
GEOMETRY→CONVOLUTION→RESPONSE   Fig. 6.7   IS 8.3

So as to quantify the deviation between the two signals, the RDS value (relative difference between standard diviations) is determined for each 200 m sub-section and defined as:

$$RDS = \frac{\sigma_{measured} - \sigma_{calculated}}{\sigma_{measured}} \quad (6.1)$$

This value roughly conforms to the value $1 - \gamma_{x.y}$, with respect to which the mean value of $\gamma_{x.y}$ must be imagined over the area in which the energy in the signal is concentrated.

The RDS value denotes, in fact, the error in standard deviation, if the latter has been calculated, using the transfer functions obtained from MISO. The quality indices, determined by NS for each 200 m section for assessing track geometry are also based on standard deviations. The difficulty of an overall assessment lies in the way in which the various parameters have to be combined to an overall figure. The examples given here show the possibility of the direct calculation of a response signal from the geometry signals. Such signals provide a much more realistic basis for assessing the quality of the track in relation to the vehicle response.

Results of other measurements.

The last example concerns measurements on two-axled freight wagons of the KS design, taken by the Hungarian State Railways (MAV). Despite the rather non-linear spring characteristic of this type of rolling stock, the coherence turns out to be fairly high. Fig. 6.10 shows the coherence vertically, with a peak value of 0.95. The response calculated according to Fig. 6.11 conforms, in consequence, very well to the measured vehicle response.

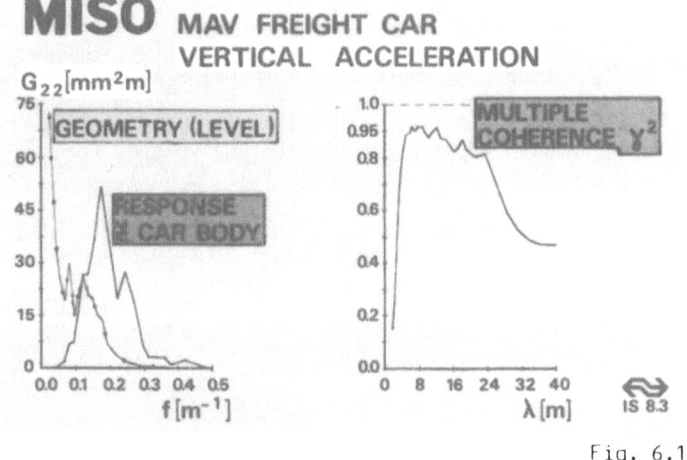

Fig. 6.10

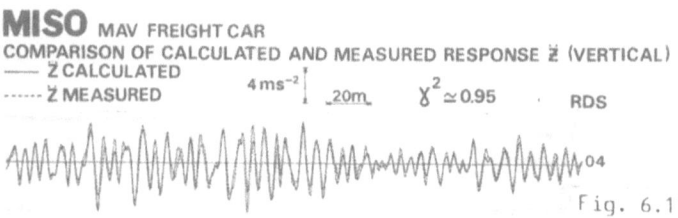

Fig. 6.11

7. Final remarks.

The experience obtained from MISO so far show that the results gained from this method lend themselves very well for practical applications, provided that strict rules are adhered to during measurements and dataprocessing. The NS Permanent Way Department will use the MISO method in the first place to obtain transfer functions from a representative cross-sectional view of the rolling stock park at various line speeds. These will be implemented in the BMS track recording system, so as to enable the on line computation of the vehicle response during measuring runs, for assessing the track quality. The main problem in this development is the on-line processing for which ultra-rapid microprocessors will be required.

However, there is also a number of applications in the field of rolling stock assesment. Thus, MISO could be an extremely practical resource for verifying the transfer characteristics of new rolling stock categories. This method also offers possibilities in studies on the vehicle behaviour. The overall performance can be determined with the aid of MISO, after which a more detailed study of components could be undertaken with the aid of mathematical models, whose overall behaviour should be chosen such as to agree with the MISO results.

References.

[1] Esveld, C: "Comparison between Theoretical and Actual Transfer
            Function of Track Maintenance Machines", Doctoral Thesis
            1978.

[2] Esveld, C: "Spectral Analysis of Track Geometry for Assessing the
            Performance of Maintenace Machines", ORE DT 77, 1978.

[3] Esveld, C:"A Numerical Butterworth Filter for Digital Data Proces-
            sing with Applications to the Analysis of Track Geo-
            metry", Rail International, Vol 11, No 1, p.35, 1980.

[4] ORE C 152, Rp 1: "Introductory study to the problem of assessing
            track geometry on the basis of vehicle response", 1981.

[5] ORE C 152, Rp 2: "Preliminary study concerning the application of
            the mathematical methods for characterising the vehicle/
            track interaction", 1983.

[6] Esveld, C. Kosieradzki, W. and Peerboom, A:
            "Theory and Application of the Multiple Input Single Out-
            put method". ORE DT 136, 1982.

[7] Bendat, J.S. and Piersol. A.G.:
            "Random Data: Analysis and Measurement Procedures", Wiley-
            Interscience, New York, 1971.

[8] Bendat, J.S. and Piersol, A.G.:
            "Engineering Applications of Correlation and Spectral
            Analysis", Wiley-Interscience, New York, 1980.

[9] Bendat, J.S.: "Statistical Errors in Measurement of Coherence Func-
            tions and Input/Output Quantities", Journal of Sound and
            Vibration, Vol.59, No 3, p. 405, 1978.

[10] Halvorson, W.G. and Bendat, J.S.: " Noise Source Identifaction
            Using Coherent Output Power Spectra", Sound and Vibra-
            ting, Vol. 9, No 8, p.15, 1975.

[11] Esveld, C.:"NS adopts contact-free measurement of track geometry",
            Railway Gazette, November 1983.

[12] Peerboom, A., e.a.: "A method to measure wheel-rail forces" (in Dutch),
            NS-report CTO/6/10.235/41, April 1981.

[13] Proceedings "International conference on wheel/rail load en displa-
            cement measurement techniques", September 1983.

2D FINITE ELEMENT ANALYSIS PACKAGE FOR SENSIBILITY ANALYSIS IN ELECTROTECHNIQUE

J.L. Coulomb   A. Kouyoumdjian

Laboratoire D'Electrotechnique   DE GRENOBLE (LA 355 CNRS)
BP 46 38402 SAINT MARTIN D'HERES FRANCE

0 EPITOME

Techniques calling up CAE are developping in electrotechnique. And now the results expected from this tool are becoming more and more general. In order to preserve the possibilities of creativity required in engineering, we are here presenting the means to carry out directly a study of sensitivity in relation to a characteristical parameter of the studied object.

From the finite element formulation, uncertainties and the direction of variation of a global quantity resulting from mistake or modification of a parameter are directly evaluated. A few examples illustrate and prove the method.

1 REVIEW : from CAD to CAX

Computer intended for scientific use became common only after 1960, CAD (Computer Aided Design) just followed, and was linked to the appearance of the first interactive systems. The CAD term itself was created around 1965.

At the same time, after mechanics, the application field of that technique of study was extended to other sciences such as electrotechnique thanks to finite element formulation generalization.

After 1975, incredible progresses of graphic consoles possibilities and performances were made ; electronics became one of the main users of what is tranformed in CAX. Indeed computer aided design, as well as drawing, making, production management engineering, ... are included in that designation.

2 PHILOSOPHY OF CAE : a source of creativity

At its early stages, the future CAE only knew packages consisting of a gathering of programs written without relation between them. Then it evoluated towards a more and more reinforced integration. Even if we do not go as far as to envisage a complete CAX system going from the theoretical study of a product to try it out when coming out of the production line passing through the numerical commands machines which made the necessary tools to realize the product, even if we limit ourselves to C.A.E., the same remark prevails.

It means that the quantity of calculations remaining to do by the conceiver is seriously reduced ; that also means that everything concerning the management of the various files necessary to calculation-datas, intervening and final results files- is totally occulted. Relieved from all these burdens, the operator can devote himself to his only work, using for that his vocational experience, his theoretical knowledge, and his creativity (SABONNADIERE JC,1982).

Thus, when there is not yet scheme of a product, this one may have already been the object of numerous testings, and some non dimensioning characteristics may have been determined or chosen. That practice knew such a development because it allows to reduce to one, exceptionnaly two, scale models, the testings made with real materials, resulting in important savings. As to the necessary delays to try various alternatives out, they are considerably shorten ed, which allows possibly to study unusual configurations, a source of technical progress.

3 HELP FOR DESIGN : sensitivity to a parameter

3 - 1 - The advantages of such a calculation:

We have just seen all the freedom owned by the conceiver when he defines a product. Nevertheless, if the numerical results, by which he estimates the quality and appropriateness of his creation in relation with what he wants to obtain, must be easily understood by him -for instance, it may be an energy, a force, a torque- the calculations which allowed to obtain them are very abstruse- solving of partial derivative equations-. Anyway the operator does not have to know in detail the algorithms leading to the results.

Consequently, for a complicated thing, and despite all his common sense, the conceiver will evaluate with difficulties the consequences- the direction of variation and the intensity of that variation- of a modification of the datas, or of a gap between the accepted hypotheses and the real specifications of the materials.

The advantage of computing sensibility lies in obtaining this information directly for every main option, the conceiver has no more to study a large number of variants separately.

3 - 2 - Necessity of a parametrization :

Thus, that is why we set up a tool of parametrized 2D topological description on a package using the finite element method. In that way, the new possibilities of results, i.e. senstibility analysis, are refering directly the parameter by which the operator is interested.

That package, is written in FØRTRAN, according to the ANSI 77 norm, in order to ensure its transportable character, because it is realized simultaneously on two mega-mini 32 bits computers (Norsk - Data 500 and Apollo DN420)

3 - 3 - Preprocessor : description of the parametrization

Even before starting to describe the geometry which will make a representation of the product, the characteristic parameters by which the geometry and physical properties are built, are formally described by numbers -other parameters, but recursivity is forbidden, or constants-, by arithmetics operators (+, -, *, /,^) and by usual functions (log, sin, min,..), which are fifteen at present, but it is possible to add others.

The formule itself is then memorized, and the numerical value of the result is computed. In case of mistake such as a forgotten parenthesis, or of impossibility for instance, the argument of an Arccos higher than 1. - it is pointed out a message, and the interactivity of the package allows to correct the improperty or to give up to carry on the operation.

If the new parameter is acceptable, a name, left to the user's choice is given to it, for instance DUMMY, TRY, NU, THICKNESS. That name is accompanied by a short comment in order to specify the exact part of that parameter, for the name only may be insufficient to ensure that mnemonic service.

Once a parameter has been created, the user refers to it by using its name. Afterwards, for convenience reasons the interactivity of the programme allows to modify formula, name and comment, and even to suppress the parameter if it is of no use, having the conceiver changed his mind in the meantime, and being the parameter useless.

The modification of a parameter is immediately and automatically (figure 1-1 and 1-2) transfered to the mesh. That mesh is constituted

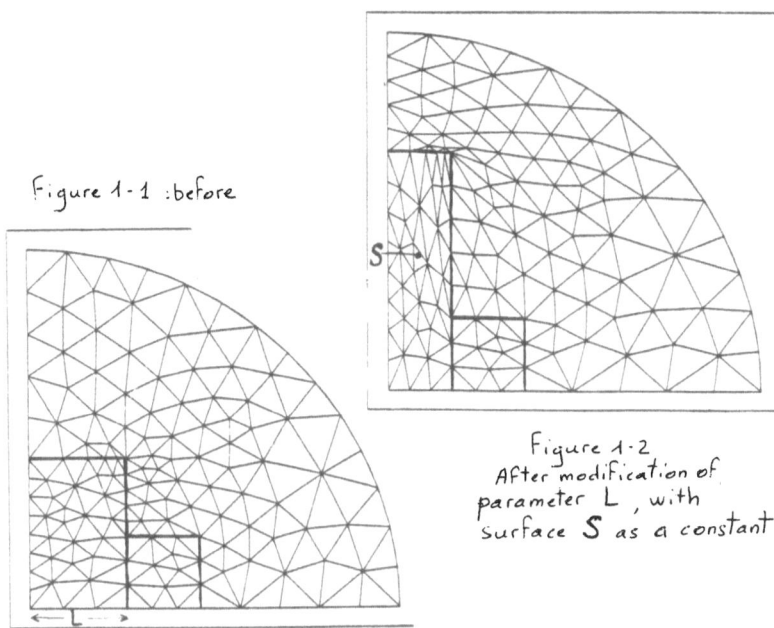

Figure 1-1 :before

Figure 1-2
After modification of
parameter L , with
surface S as a constant

of triangles of 1st or 2nd order, at user's will, which are automatically built for them to verify the property of being a Delaunay triangulation (COULOMB J.L., DUTERRAIL Y., MEUNIER G., 1984). Then, mesh regularisation and parameters modifications can destroy that property which is only important when are created the inside points of the meshed domains.

3 - 4 - Computing processor : Poisson's equation
  That step where is solved, in electromagnetism, on the field taken in account, with given boundary conditions, Poisson's equation
$$\nabla \times (\nu \nabla \times A) = J \qquad (1)$$
A magnetic potentiel
ν magnetic reluctivity (inverse of permeability)
J density of source currents

is entirely automatic. Only the requested precision for the solution is left to the user's assessment.
  The solving of that equation which allows to compute the values of A at the nodes of the mesh is done by a variant of the conjugate gradient method.

3 - 5 - Post processor : exploitation of the results

In the domain where our experimental package is applied, viz. electromagnetism, magnetic potential has been given to us. And the question is thus, for a given structure, described by mean of parameters, being numerically known these parameters, to precise to which parameter we particularly want to deal with. That choice is done in the list of the parameters used in the model, a list put up by the program to the user.

After having computed the derivative of the potentiel versus the selected parameter (COULOMB J.L.,KOUYOUMDJIAN A, 1985),the possible exploitations are numerous. One of the easiest to start is the derivative of the whole energy. The sign of the result gives the direction of variation of the energy should the parameters value tend to increase, the numeric value of the result compared with the value of energy allows us to determine influence in percentage of a small variation of the parameter. Indeed, by a first order Taylor's development, only according to the p parameter, we can write for the quantity energy W :

$$W(p) = W(po) + \frac{\partial W}{\partial p} (p-po)$$

i.e. $$100 \times \frac{W(p) - W(po)}{W(po)} = 100 \times \frac{p-po}{po} \frac{po}{W(po)} \frac{\partial W}{\partial p}$$

or if po is nil

$$100 \times \frac{W(p) - W(po)}{W(po)} = 100 \times \frac{p}{W(po)} \frac{\partial W}{\partial p}$$

po is given at the beginning or the process, W(po) is computed after the solving of the partial differential equation 1.

$\frac{\partial W}{\partial p}$ has been just computed. It is the conceiver who introduce directly the parameter variation interesting him.

If the user wishes the same type of result about an other parameter, he selects it after having came back at the beginning of the exploitation.

3 - 6 - Mathematical formulation : derivative versus a parameter

Nodal values of A have been computed by the finite element method , i.e. by the extremalization of a functionnal. First we compute the derivatives of A versus the selected parameter at each node of the mesh remarking the functionnal is also extremal regarding both A and p, the parameter. After that, such quantities as B, the magnetic flux can be derivated versus p by using the values of A, $\frac{\partial A}{\partial p}$ and the operators ∇ and ∇_p, i.e. the influence of the variation of the parameter upon the way to compute a derivative.

Numerically, even in the case of non linear properties, the system to solve to compute $\frac{\partial A}{\partial p}$ is alwyas linear. And from $\frac{\partial A}{\partial p}$ all results concerning a global quantity can be computed, being applied the virtual works principle.

3 - 7 - <u>Non dimensionnal parameter : magnetic reluctivity</u>
  3 - 7 - 1 - <u>Formulation</u> :
        For simplicity reasons in the formulation, reasons that appears afterwards, reluctivity has been just a follow
$$v(p) = v_o \, v_r (B) \cdot (1 + p)$$
In the way, giving the value .005 to p, at the end of the computing phase, is encreasing v by .5 % whatever the B magnetic flux is, and the v characteristic may correspond to a linear material as well as to a saturable one. The derivative of v versus the parameter is then
$$\frac{\partial v}{\partial p} = v_o \, v_r(B)$$
where v(o) is recognized.

In these conditions, on each finite element where v depends from the p parameter, energy and its derivative computed by the same formula
$$\frac{dWe l}{dp} = W_{el} = \int_{el} (\int_0^B v(b) . b . db) \, d\Omega$$
on the others, $\frac{dWe l}{dp}$ is nil of course.

By applying the § 3-5 formula, we thus ends in the value of imprecision on the global quantity energy held in the system that follows from imprecision about v, which is altogether useful for the manufacturers give only a few points of measurement, spoilt as a consequence by some mistake we can estimate, and then we deduce a continuous characteristic curve only by interpolation.
    3 - 7 - 2 - <u>Example</u> :
        That simplistic example

has no other goal than to show a coherent result appears at the end, surpriseless for it can be obtained without computer.

We create a 1m x 1m square, and as we are working in 2.D. Space, we find the results for 1m depth when extensive quantities – it has no importance when result is a pourcentage.

We give it a 1 reluctivity (the air's one), and divide it artificially into two regions (figure 2).

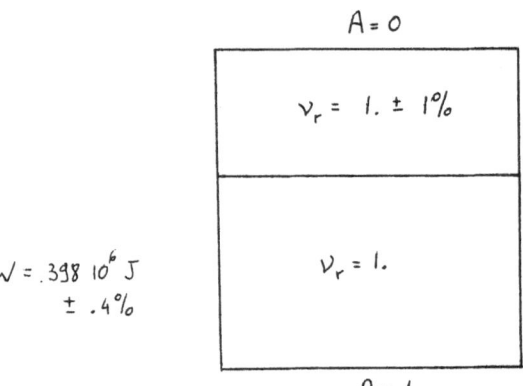

$A = 0$

$\nu_r = 1. \pm 1\%$

$\nu_r = 1.$

$W = .398 \ 10^6 \ J$
$\pm .4\%$

$A = 1.$

Figure 2. Uncertainty

One represents 40 % of the surface, the other 60 %. The first one will be assumed to be constituted by a material of uncertain reluctivity. The result is the one hoped ; a 1 % uncertainty on ν_r leads to a .4 % uncertainty on the global energy value.

3 - 8 – Dimentionnal parameter : an other application :
3 - 8 - 1 - Formulation :
The calculation of the energy remains unchanged, on the contrary its derivative formulation is a little more complicated for we have to take into account the finite elements dimensions modification influence.

As a consequence, we are induced to evaluate the following integral, assuming that characteristics of materials are linear.

$$I = \frac{1}{2} \int_\Omega (\frac{\partial H}{\partial p} B + H \frac{\partial B}{\partial p} + HB \frac{\partial /G/}{\partial p}) d\Omega$$

Where the last term is the expression of the deformation of the finite element. Afterthat, the same treatment previously pointed out leads to the result.

3 - 8 - 2 - Example :

        We have chosen, to illustrate
this paragraph, the same 1m × 1m square,
assumed to have a 1m depth.
We study the variation of the energy pulled
by the displacement to the top of the bottom
side limit (figure 3)

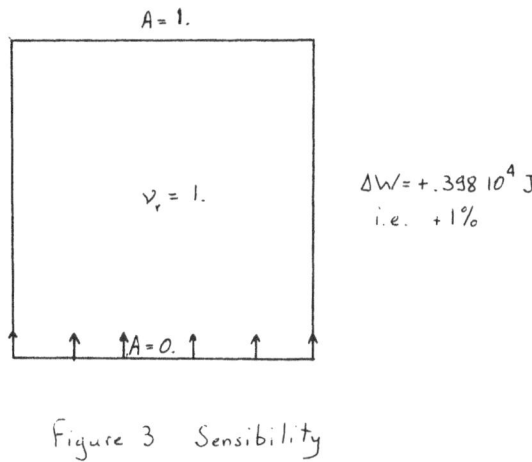

$A = 1.$

$\nu_r = 1.$

$\Delta W = +.398 \; 10^4 \; J$

i.e. $+1\%$

$A = 0.$

Figure 3   Sensibility

We find that if the material is homogenous,
with a 1 relative reluctivity, a .01 m displacement, i.e 1 %, leads to a + 1 % energy variation, being identical the boundary conditions.

3 - 9 - Extention to other quantities

        That exploitation we have just pointed
out for energy, can obviously be extended
to mechanical, electromechanical and electrical
quantities deduced from energy always using finite
element formulation and leading to formulas
indicated elsewhere (particularly in COULOMB
JL, 1983) The principles to obtain the derivatives corresponding to force, stiffness induc-
tance,... are the same as that we used previously in 3.7.1 and 3.8.1.

3 - 10 - Practical example :

        We have chosen a compensed DC motor,
whose motor is 20 cm diameter and air-gap
1 cm, meshed with 1100 finite elements. With

second order triangles that leads to about
2500 nodes (figure 5)

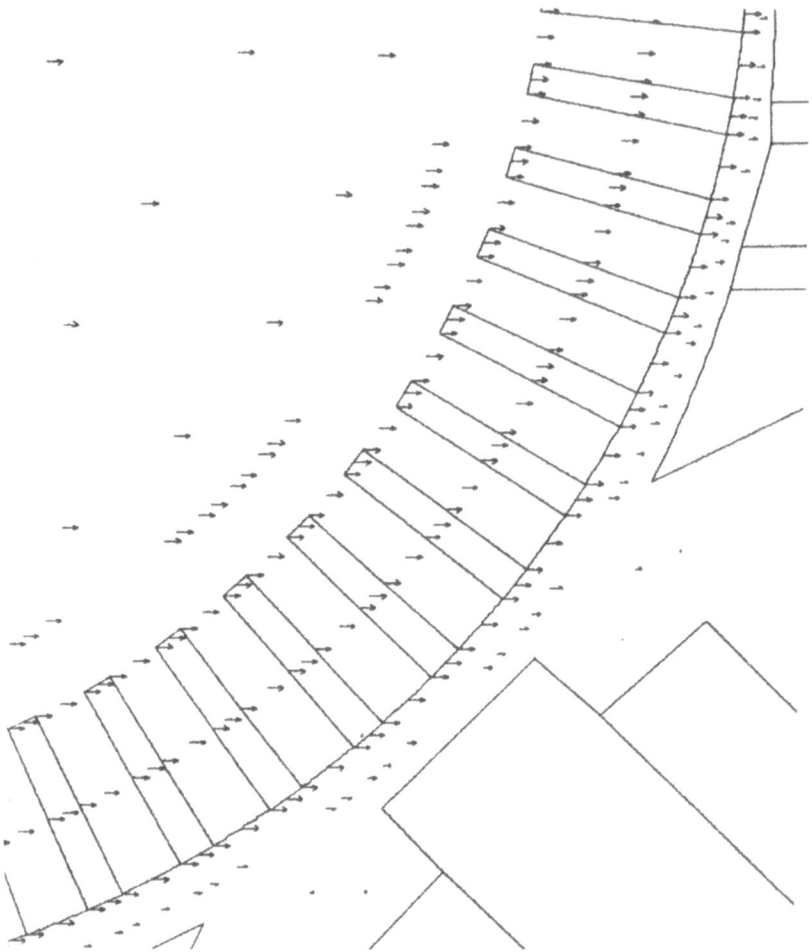

figure 4 Field of virtual nodal
displacements

3 - 10 - 1 - Translation   of the rotor :
         We parametrize the coordinates
of the points describing the geometry to provoke
an horizontal translation by modifying a
unique parameter. Once the field of displacements computed (figure 4, 5 et 6 for details),
we solve an always linear system. The results
are presented in table I. By an other method

we should solve generally non linear systems
to obtain less precise results for we obtain
them by substraction. The coherence between
the two approches is as good as it could
be hoped.

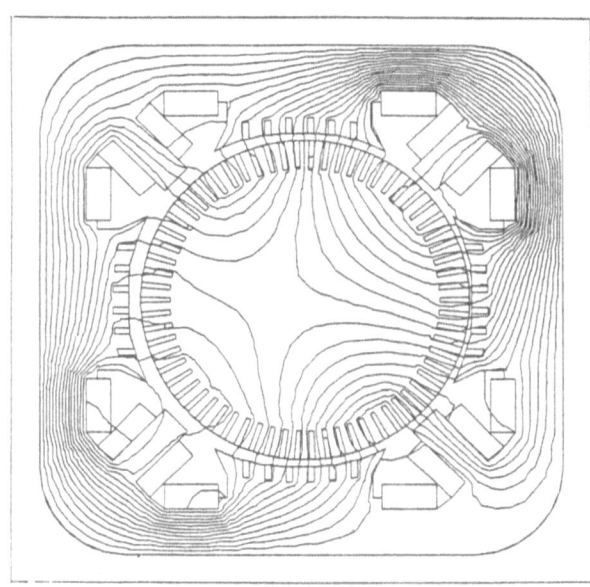

Copier
Retracer
Agrandir
Recadrer
?
Quitter

figure 5

Magnetic equipotential lines of the
modelized example ; with "menu"
for interactivity

     - Copy
    - Draw again
    - Zoom
    - Return to initial drawing
     - ? (what is possible to do?)
    - Quit

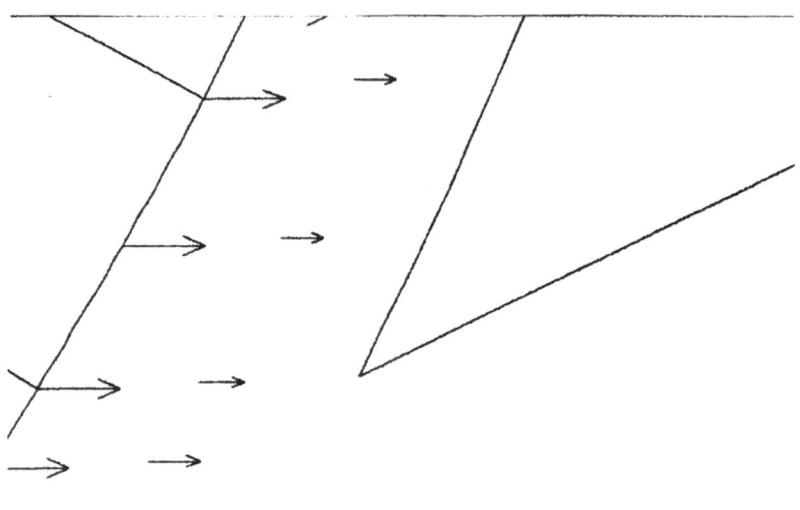

Figure 6 . Detail of fig. 4

Table I : Excentrage ; numerical results

Value of p the excentrage: (mm)	: Energy (10⁴J)	Difference (10⁴J/mm)	Mean of 2 differences (10⁴J/mm)	: : :	Calculated derivative (10²/mm)
0	: · 504	. 0015		:	
2	: · 507	. ·0045	· 003	:	· 278
4	: · 516	. ·0075	·006	:	· 581
6	: · 531			:	

3 – 10 – 2 – <u>Physical uncertainty</u> :

We have chosen for the stator a relative magnetic reluctivity $v_r = 6.10^{-4}$ and for the rotor $v_r = 3.5. \ 10^{-4}$ .The same process as described before in 3.7.1. has given the results presented in the table II. We have also

computed the results by a direct way, giving
to the reluctivities, assumed to be linear,
their extreme possible values. We see that
both results give quite the same tendency
and numerical value.

Table II Physical uncertainties : numerical results

Reluctivity	Energy (10^4J)	Difference (%)	Direct computation
- 15 %	. 510		
		+ 1.19	⎧ sign negative
basis	. 504		⎨
		- .992	⎩ ± 1.12 %
+ 15 %	. 499		

4 - CONCLUSION : AN ADDITIONAL TOOL FOR THE CONCEIVER

We perfectly see the great number of results which can be
obtained by this way.Our aim, here, was only to show the usefulness and feasability of such computations. Anyway, these
results may lead us to foresee the transformation ofthe initial
2D package (SABONNADIERE J.C., MEUNIER G, MOREL B , 1982) allowing calculation into one allowing conception. It is to say
that it should be able to point out the best solution of the
problem, as it has been represented. This approach allows us
not only to understand some phenomena by a new technique, but
also its development leads to original results which render
ineluctable, we hope, the use of the tool of a main help to
C.A.E. And all that opens new prospects in mixing irksome
computations and expert systems.

Bibliography :

1 - COULOMB J.L. - (1983) "A methodology for the determination
                   of global electromechanical quantities from
                   a finite element analysis and its applica-
                   tion to the evaluation of magnetic forces,
                   torques and stiffness.
                   I.E.E.E. Trans. on Mag, (MAG-19),6,pp 2514
                   -19.

2 - COULOMB J.L. - DU TERRAIL Y., MEUNIER G (1984) "Two 3D pa-
                   rametered mesh generators for the magnetic
                   field computation.
                   I.E.E.E. Trans.on Mag, (MAG-20), 5, pp 1900-
                   02.

3 - COULOMB J.L. - KOUYOUMDJIAN A. (1985) "An original approach

to compute global electrical quantities by
the finite element method and sensibility to
a parameter.
COMPUMAG'85, to be published.

4 - SABONNADIERE J.C. (1982)- "Modèles, méthodes et logiciels
        d'aide à la conception de systèmes électro-
        mécaniques".
        Revue Générale de l'Electricité, 10, pp645-57

5 - SABONNADIERE J.C., MEUNIER G, MOREL B. (1982) "Flux, a ge-
        neral interactive finite element package for
        2D electromagnetic field.
        I.E.E.E. Trans. on MAG, (MAG-18), 2, pp624-26

SYNTRA - AN INTERACTIVE PROGRAM TO DESIGN GRASHOF PLANAR
FOUR BAR MECHANISMS

C. R. Barker

University of Missouri, Rolla, MO, U.S.A.

1. INTRODUCTION

Linkage synthesis has always been the most challenging task in
kinematics. It is therefore natural that a computer can be of
great assistance to the mechanism designer in arriving at a
successful solution. Beginning around 1970 several synthesis
programs appeared which were implemented on mainframe computers.
Recently the advent of powerful personal computers has made the
cost of the hardware less of a barrier to a large number of small
design firms. As a result, several of the larger programs have
been rewritten to run on small computers. Krouse [1] describes
the evolution of Micro-Kinsyn [2] and Micro-Linkages [3] both of
which are currently available for relatively inexpensive
computers. These programs are based upon Burmester's graphical
solution for four precision points and use the computer to
expedite the complex calculations required.

Shoup [4] evaluates the use of microcomputers for mechanism
and machine theory problems and concludes that the future is very
bright. He presents an example of a synthesis of a function
generator using a spatial slider crank mechanism.

Wilson and Soni [5] used a microcomputer to synthesize four
bar linkages for three and four precision points. Their
assessment of the capabilities is also positive, and they suggest
extending the applications to include rigid body motion and path
generation.

2. THEORETICAL BASIS FOR SYNTRA

Recently, Barker [6,7,8] has proposed a new method of classifying
planar four bar mechanisms based upon a solution space concept.
A typical planar four bar mechanism has a fixed distance between
the frame pivots designated as R_1. The input bar has a length
R_2, the coupler length is R_3, and the output bar length is R_4.
In order to eliminate the physical size of the mechanism from the
discussion, all of the lengths can be divided by R_2 to define

non-dimensional parameters where

$$\begin{aligned}
\lambda_1 &= R_1/R_2 \\
\lambda_3 &= R_3/R_2 \\
\lambda_4 &= R_4/R_2.
\end{aligned} \qquad (1)$$

The effect of normalizing the link lengths is to reduce a four dimensional space to a three dimensional space which will be more convenient for what follows. Only finite positive values of the λ's need to be considered, and therefore the positive octant of a $\lambda_3\lambda_4\lambda_1$ coordinate system would represent every possible combination of link lengths without regard to the scale of the mechanism.

Zero Mobility Planes

Certain regions of this positive octant can be eliminated from consideration because of the zero mobility condition. This means that a particular bar length is sufficiently large that the mechanism can be assembled, but can not move after assembly. For example, if the distance between the fixed pivots is equal to the sum of the lengths of the input bar, the coupler, and the output bar, then the zero mobility condition would exist. Any of the four bar lengths may be the longest, and this leads to the four conditions

$$\begin{aligned}
1 &= \lambda_1 + \lambda_3 + \lambda_4 \qquad &(2) \\
\lambda_3 &= \lambda_1 + 1 + \lambda_4 \qquad &(3) \\
\lambda_4 &= \lambda_1 + 1 + \lambda_3 \qquad &(4) \\
\lambda_1 &= 1 + \lambda_3 + \lambda_4 \qquad &(5)
\end{aligned}$$

which define four zero mobility planes (#1 through #4 in the order given which is consistent with Ref. 6). The zero mobility planes reduce the volume of the positive octant which must be considered because there are no mechanisms which can be assembled on the outside of the zero mobility planes. When the volumes outside of the zero mobility planes are subtracted from the positive octant, what remains is the solution space.

Change Point Planes

The solution space is then partitioned into eight volumes by the three change point planes. These planes are defined by the condition that the shortest bar length when added to the longest bar length is equal to the sum of the other two bar lengths. The equations describing the three change point planes are

$$\begin{aligned}
1 + \lambda_1 &= \lambda_3 + \lambda_4 \qquad &(6) \\
1 + \lambda_3 &= \lambda_1 + \lambda_4 \qquad &(7) \\
1 + \lambda_4 &= \lambda_1 + \lambda_3. \qquad &(8)
\end{aligned}$$

Complete Classification

Each of the eight volumes of the solution space is unique because all of the mechanisms represented by points on the interior possess a common characteristic in regard to the lengths of the bars. It is therefore possible to say that a certain bar is a crank, which can rotate through a full revolution, or a rocker, which can not complete a full revolution, and have these statements remain valid for every mechanism within the volume. This leads to the classification system given in Ref. 6. Fig. 1 shows an exploded view of the solution space and the volumes 1-4 which represent Grashof mechanisms while the volumes 5-8 represent non-Grashof mechanisms. The three dimensional depiction of the solution space can be represented in two dimensions if the viewing orientation is chosen to be toward the origin along the line $\lambda_1 = \lambda_3 = \lambda_4$. There are only three typical cross-sections of the solution space when viewed in this manner and these are shown in Figs. 2-4.

When λ, the perpendicular distance from the origin to the cross-section, is in the range $\sqrt{3}/3$ to $\sqrt{3}$, the shape of the cross-section will appear as shown in Fig. 2. When the distance is $\sqrt{3}$, Fig. 3 will be the cross-section, and when the distance is greater than $\sqrt{3}$, Fig. 4 is the correct shape.

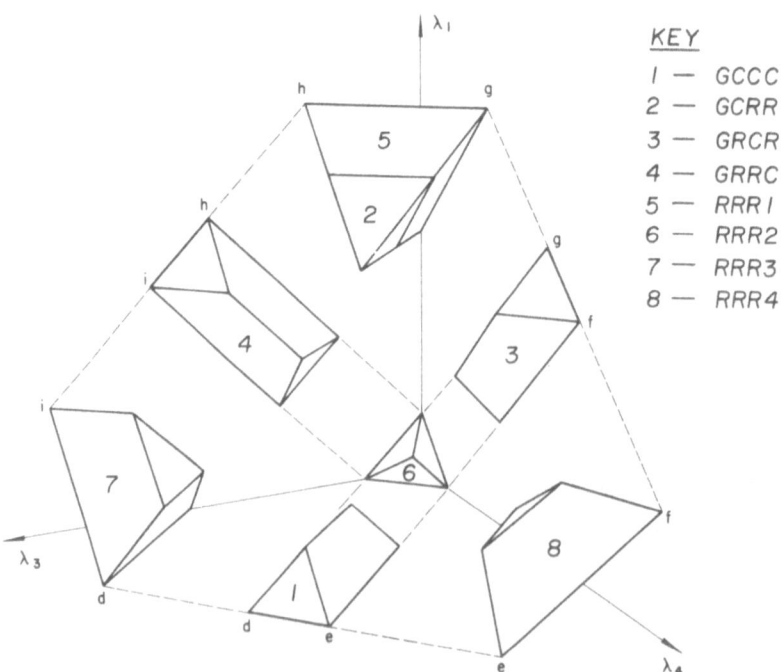

KEY

1 — GCCC
2 — GCRR
3 — GRCR
4 — GRRC
5 — RRR1
6 — RRR2
7 — RRR3
8 — RRR4

Fig. 1. Exploded View of Solution Space

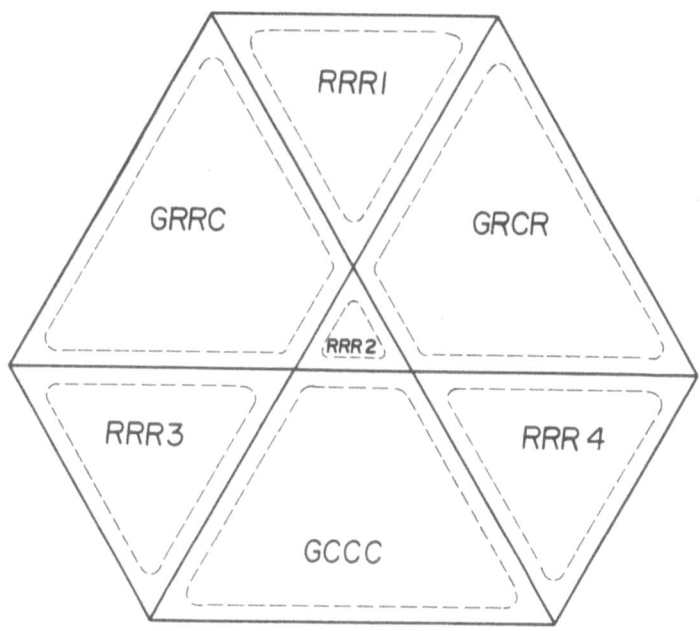

Fig. 2 Cross-section Where $\sqrt{3}/3 < \lambda < \sqrt{3}$

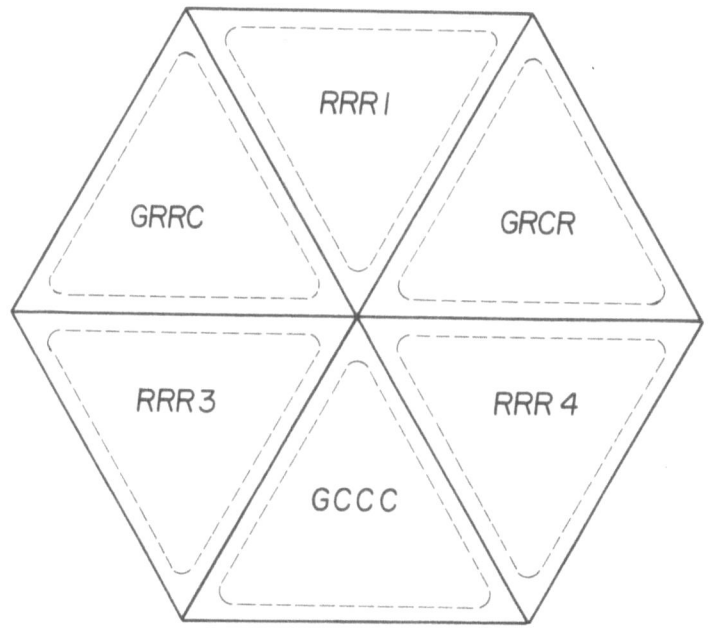

Fig. 3 Cross-section Where $\lambda = \sqrt{3}$

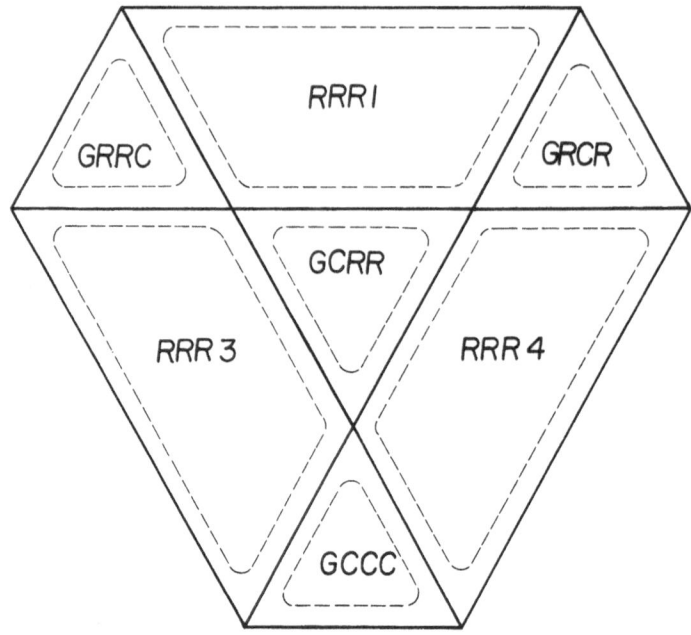

Fig. 4 Cross-section Where $\lambda > \sqrt{3}$

3. COMPUTER IMPLEMENTATION

The computer program SYNTRA (acronym for Synthesis using Traces)
is an interactive program which can be used to design Grashof
planar four bar mechanisms. The program is written in Advanced
Basic and runs on an IBM Personal Computer with 128K memory, one
disk drive, and Color/Graphics Adapter. SYNTRA is based upon the
fact that for every property of the four bar mechanism a surface
exists which joins together all the mechanisms which share that
characteristic property. The surfaces are three dimensional, but
their trace in a plane chosen by the user is two dimensional,
i.e. planes such as those shown in Figs. 2-4. Therefore it is
possible to synthesize a mechanism with certain desired
characteristics by searching for an intersection of the desired
traces in a plane whose location within the solution space is
controlled by the user. In practice, the user inputs the desired
properties such as maximum and minimum transmission angle,
maximum and minimum output angle, and extreme value of the
angular velocity ratio.

When the desired traces are entered, the computer displays
the traces for each successive plane chosen by the user. The
user observes the various trace intersections and guides the
computer to a location where all of the desired traces intersect
at a common point. The final step is for the user to position a

cross at the common point which then allows the computer to display the required link lengths for the solution. Successful designs can be achieved quickly, and an extensive knowledge of theory is not required to use the program.

Definition of Desired Properties

The user defines the characteristic he wishes the mechanism to have by assigning numerical values to a list of properties provided for each of the four types of Grashof mechanisms, GCCC, GCRR, GRCR, and GRRC. Each letter "G" stands for Grashof, each "C" for crank, and each "R" for rocker. The order of naming the links is input, coupler, and output link. Thus a GCCC is a Grashof crank-crank-crank or the familiar drag linkage.

Once the user has input the numerical values, the computer displays the trace of the desired properties. The user then interacts with the computer to search for a plane where all of the desired traces intersect at a common point. The common point represents a solution to the problem if one does in fact exist. The locations where the traces nearly all intersect at a common point represents a mechanism group which comes close to satisfying the design requirements.

As the number of design specifications increase, the number of potential solutions decrease. Experience using SYNTRA suggests that the maximum number of independent properties which can be exactly achieved is three. If more than three properties are specified, then all solutions found will be a compromise solution. Many compromise solutions can be found which are completely satisfactory solutions for design purposes.

Analysis Capabilities of SYNTRA

After a mechanism has been synthesized using the interactive features of SYNTRA, the user may wish to analyze the motion of this particular configuration. The program has this capability and for any Grashof mechanism it can compute position angles, angular velocities, and angular accelerations for a full cycle of motion. These results are displayed on the screen and it is possible to obtain a hard copy of the results using the graphics printer. In addition, the analysis portion of SYNTRA allows the user to generate coupler curves for up to nine chosen points in the coupler. The coupler curves are displayed on the screen as well as an animation of the mechanism itself. The analysis capabilities of SYNTRA were described in detail in Ref. [11].

4. EXAMPLES

The following examples illustrate the program capabilities. It is difficult to explain the exact sequence of steps followed to reach the final design because of the interactive nature of the program. Also in most of the examples, the solution obtained is not a unique solution so that there are other mechanisms which would satisfy the design requirements. The examples focus on the

GCCC and GCRR types of Grashof mechanisms, since these are by far the largest classes used in industrial practice.

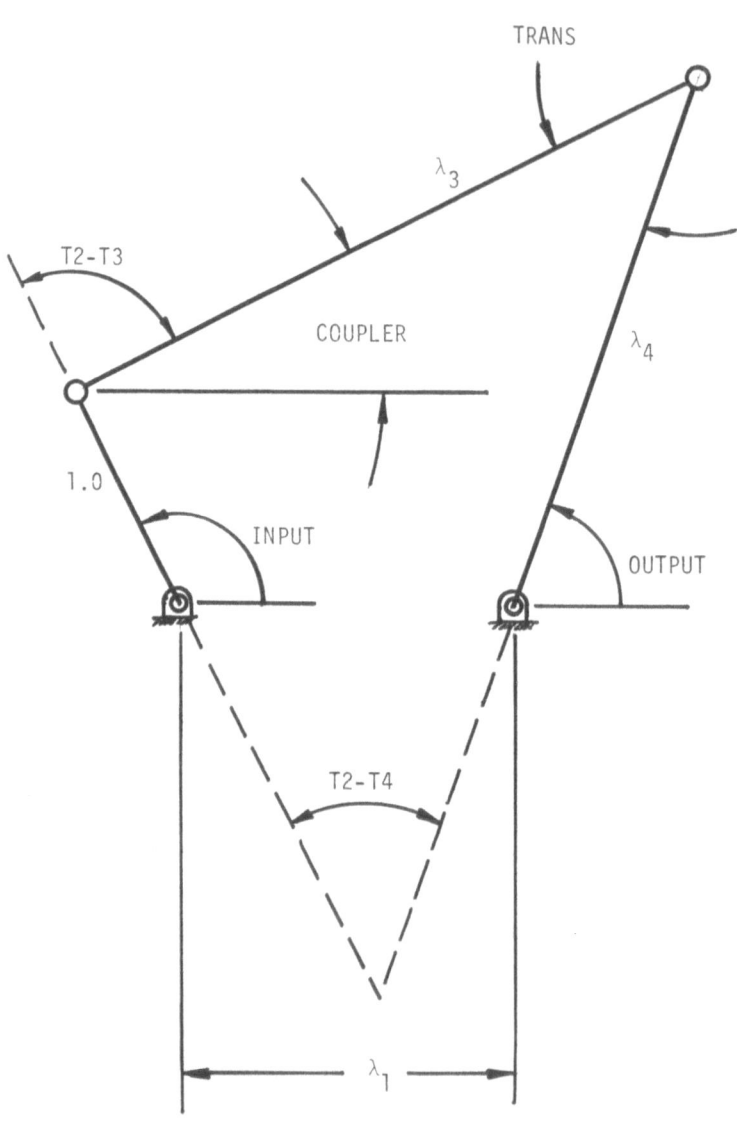

Fig. 5 Terminology

Example One

The terminology to be used in the examples is shown in Figure 5.
Six angles are shown in this figure and labelled INPUT, COUPLER,
OUTPUT, TRANS, T2-T3, and T2-T4. For Grashof mechanisms, the
designer has no control over the range of three of the angles
since they will always range from 0 to 2π in a full cycle of
motion. The other three angles can be controlled by the
appropriate choice of the link lengths λ_1, λ_3, and λ_4.

For example, for a GCCC the angles T2-T3, T2-T4, and TRANS
can be controlled. Suppose it is desired that the lag angle
T2-T4 be a maximum of 74 degrees and a minimum of 34 degrees [9].
The program automatically sets $\lambda = 5$ and presents a menu of the
four types of Grashof mechanism for which the user would choose a
GCCC in this case. Then, as shown in Figure 6 the list of
maximum and minimum angles appears for the GCCC which contains
the six values that can be controlled for this type. Notice in
the figure that a cross is positioned in the center of the
diagram and that the link ratio data shows that for this location
the mechanism is a GCRR with $\lambda_1 = \lambda_3 = \lambda_4 = 2.887$.

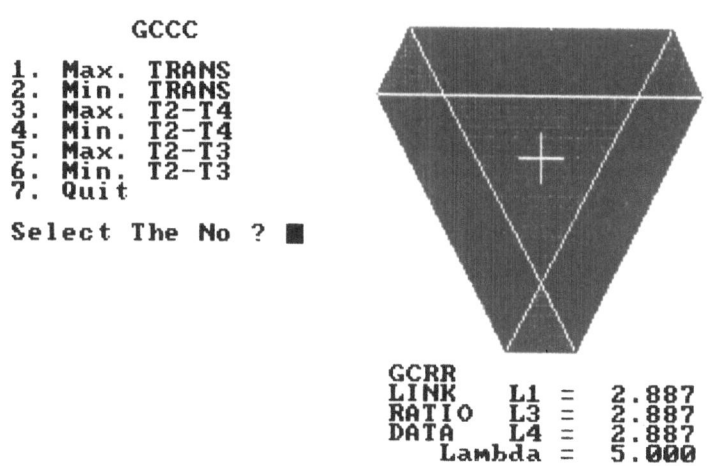

GCCC

```
1. Max. TRANS
2. Min. TRANS
3. Max. T2-T4
4. Min. T2-T4
5. Max. T2-T3
6. Min. T2-T3
7. Quit

Select The No ? ■
```

```
GCRR
LINK    L1 =  2.887
RATIO   L3 =  2.887
DATA    L4 =  2.887
      Lambda = 5.000
```

Fig. 6 List of Angles for GCCC

When #3 is selected and the value 74 is entered, the straight
line trace shown in Figure 7 appears which extends from the
bottom upwards and to the right. All of the mechanisms along this
trace have the property that angle T2-T4 maximum is 74 degrees.
The trace shown begins in the GCCC zone (the bottom triangle in
the diagram) extends into the RRR4 region and ends in the GRCR
zone (the upper triangle on the right side of the diagram). When
#4 is selected and the value 34 is entered, two traces appear as
shown in Figure 7. They both start at the upper left of the
diagram. Of the two branches only the lower is of any interest
in this example because it intersects the straight line trace
inside the GCCC zone.

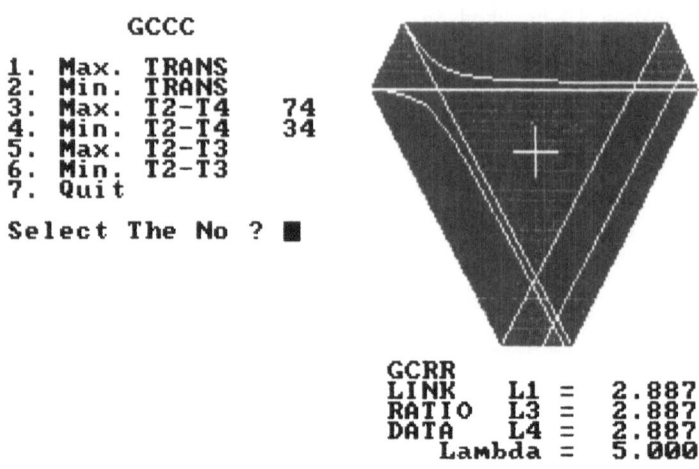

Fig. 7 Traces for Example One

Once the intersection has been defined, the next step is for
the user to position the cross at the location where the traces
intersect. This is accomplished using the cursor control keys on
the keyboard and an increment size which the user selects. While
this process is occurring the screen display has changed to the
format shown in Figure 8. The left half of the screen now has
six circles shaded to represent the range of the six angles
defined in Figure 5. Notice that λ is different in Figure 8 than
it was in Figure 7. The reason for this is that the range of the
transmission angle for the solution represented by Figure 7 was
very unacceptable. By trying different λ's the final design
shown in Figure 8 was selected where λ_1 = 0.234, λ_3 = 0.811, and
λ_4 = 0.687. For these dimensions the transmission angle will
range from 60.8 to 110.8 degrees.

Example Two
In this example additional constraints will be placed upon the
synthesis as imposed by Tsai [9]. He specified that the lag
would be the same as for Example One and in addition the maximum
transmission angle would be 115 degrees while the minimum
transmission angle would be 65 degrees. When these values are
entered into the program, and several different values of λ are
chosen, the user will find that a solution does in fact exist as
shown in Figure 9 where λ_1 = 0.214, λ_3 = .809, and λ_4 = 0.625. In
the process of locating this solution several alternate solutions
are found which do not satisfy the design requirements exactly,
but which would be perfectly acceptable solutions. By examining
the neighborhood around the solution given by Tsai, it appears
that his solution is a unique solution to the problem.

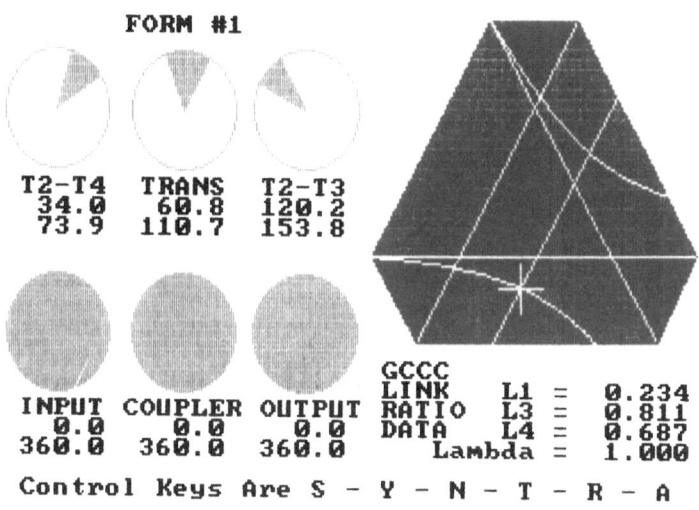

Fig. 8   Range of Angles For Example One

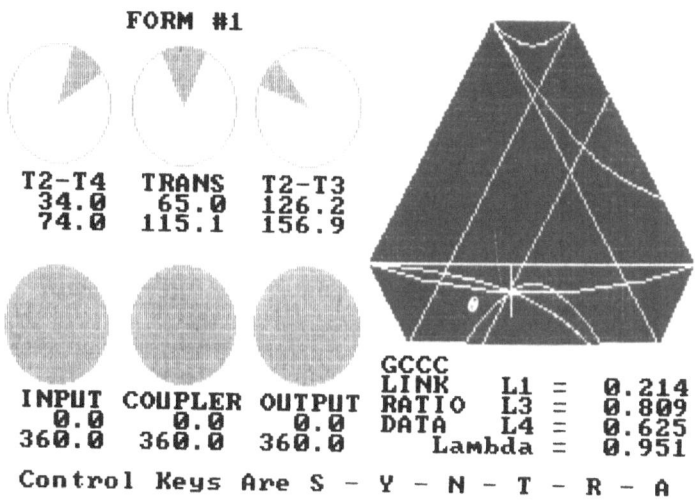

Fig. 9   Screen Display For Example Two

Example Three

Suppose a GCRR mechanism is desired where the maximum anglular
velocity ratio of the output link to the input link is 0.5. The
minimum angular velocity ratio is to be -0.5 and the transmission
angle should be in the range 45 to 135 degrees. Figure 10 shows
the list of traces which are possible for a GCRR. The maximum
and minimum transmission angle traces are shown in the figure and
they define a zone within which potential solutions will be
located. All the mechanisms within this zone possess the
required range of transmission angle.

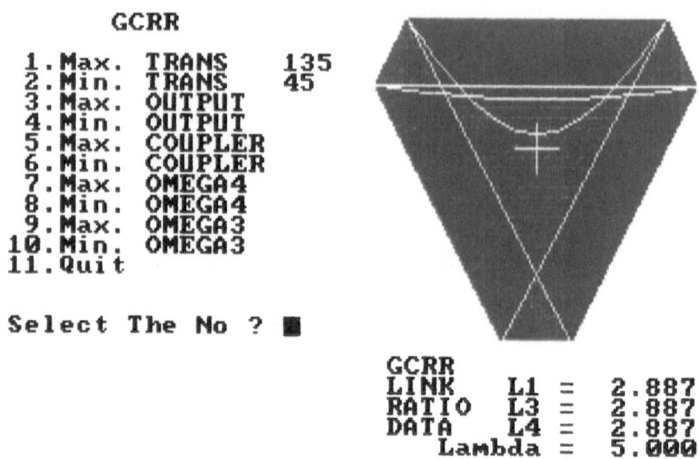

Fig. 10   First Display For Example Three

Figure 11 shows two additional traces for the maximum and minimum angular velocity ratio. The intersection of these traces is within the required zone and hence a solution has been found. The cross is positioned in Figure 11 at the intersection of the angular velocity traces and the link dimensions are read as λ_1 = 5.226, λ_3 = 4.773, and λ_4 = 2.086. Using the analysis portion of SYNTRA shows that the maximum angular velocity ratio is 0.491 and the minimum is -0.494. These values are very close to the desired values. In addition, the transmission angle is in the range 63.3 to 127.5 degrees.

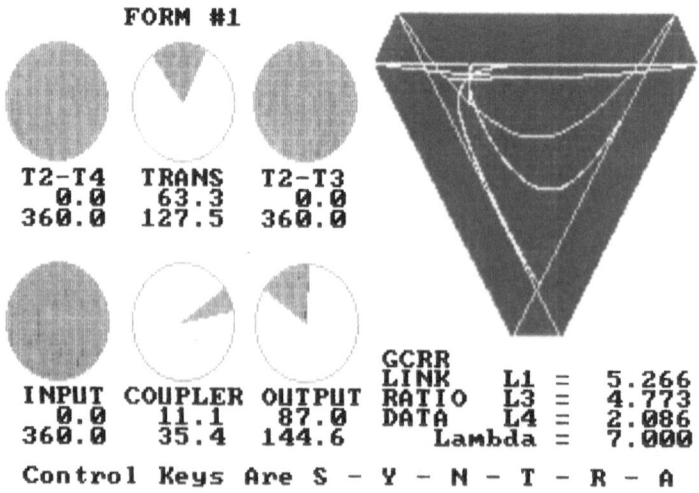

Fig. 11   Second Display For Example Three

Example Four

SYNTRA can be used to perform a kinematic analysis of any of the
Grashof mechanisms which are synthesized. The analysis
capability can also be used to study mechanisms obtained in other
ways. For example Savage & Suchora [10] present an example where
they synthesized a GCRR to have a maximum angular velocity ratio
of 0.3 and a minimum of -0.5. In their approach the optimum
minimum transmission angle for this case would be approximately
40 degrees and this leads to link dimensions of 0.49, 1.05, 1.61,
and 1.5 for respectively the input, coupler, output, and frame.
These are equivalent to λ_1 = 3.061, λ_3 = 2.143, and λ_4 = 3.286.
When these values are entered into the program, the display shown
in Figure 12 results.

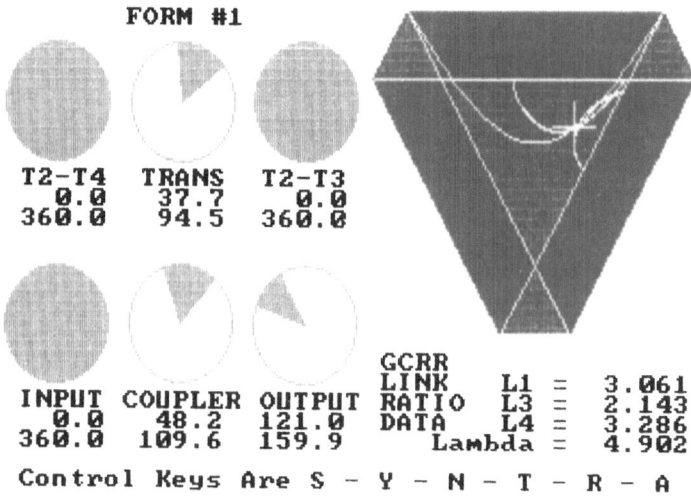

Fig. 12 Screen Display For Example Four

This figure shows that the transmission angle will have a
range from 37.7 to 94.5 degrees. The coupler angle will vary
between 48.2 and 109.6 degrees, and the output angle will cover
the range 121.0 to 159.9 degrees. The traces for the maximum and
minimum angular velocity ratio and the minimum transmission angle
all intersect very near to the location of the cross. The
results of Savage and Suchora have therefore been shown to be in
agreement with those obtained here.

In addition, the program will analyze this GCRR for a full
cycle of motion and display the angles COUPLER (Theta 3) and
OUTPUT (Theta 4) versus INPUT (Theta 2) for a full cycle of
motion as shown in Figure 13. The angular velocity and angular
acceleration of the coupler and output links are also computed
and displayed as shown in Figures 14 & 15 normalized with respect

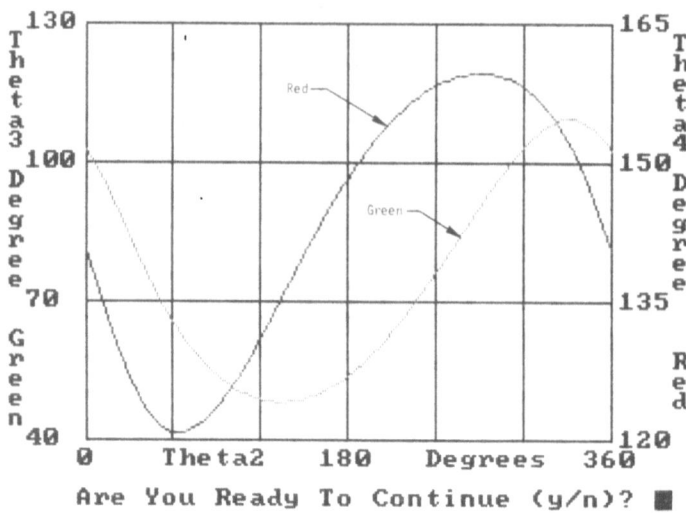

Fig. 13 Example Four - Position Analysis

to the constant rate of rotation of the input link. From Figure
14 the maximum and minimum angular velocity ratio are very close
to the values 0.3 and -0.5 as given in Ref. [10].

Figure 16 shows the result of the user selecting four points
in the coupler of the GCRR of this example. SYNTRA allows up to
nine points to be selected at any desired location in the
coupler. Here the four points were chosen along the axis of the
coupler. Points A and B were taken 4 and 5 units in one
direction away from the pin joint between the coupler and input
link. Points C and D were taken 3 and 4 units in the opposite
direction away from the same location.

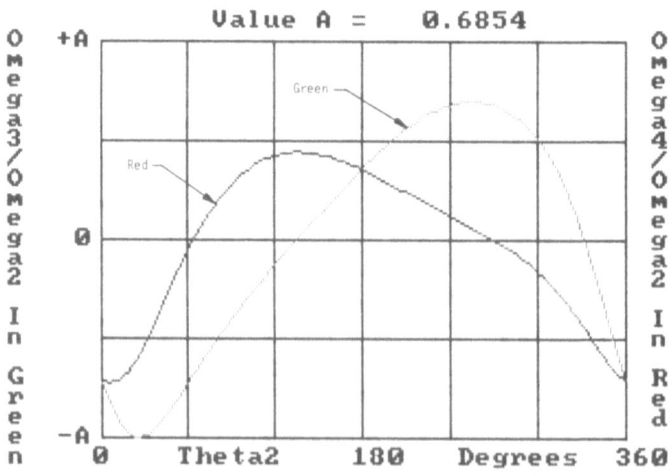

Fig. 14 Example Four - Angular Velocity Analysis

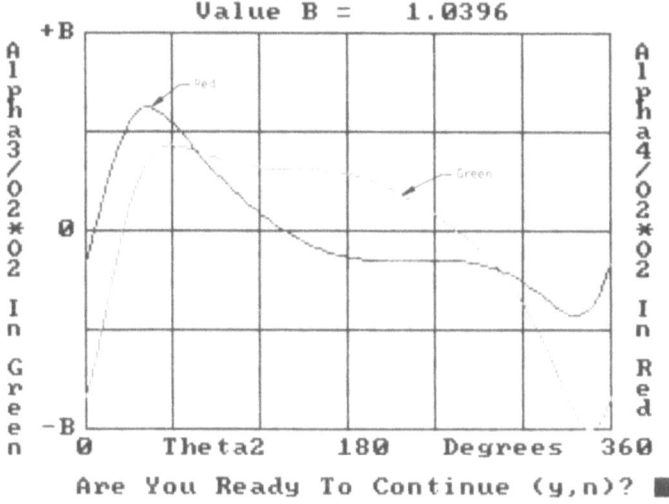

Fig. 15   Example Four - Angular Acceleration Analysis

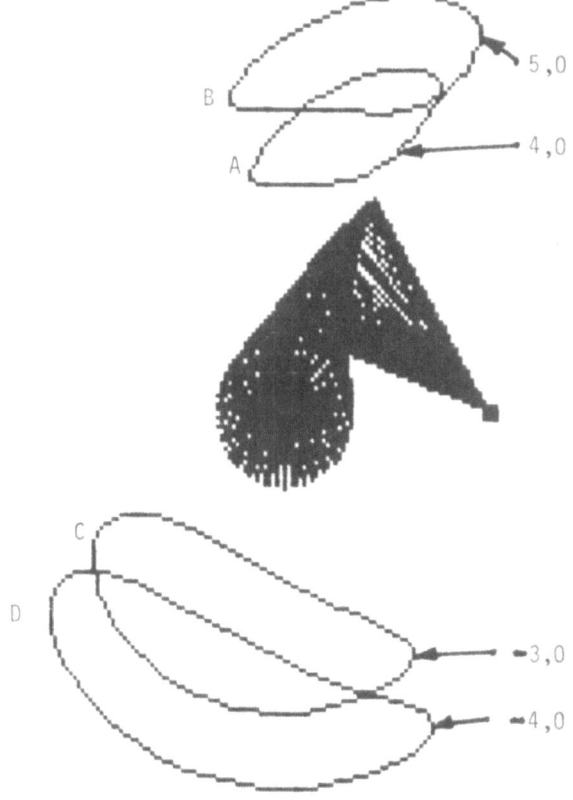

Fig. 16   Example Four - Coupler Curves

5. CONCLUSION

The program SYNTRA described here is a useful tool for the dimensional synthesis of planar four bar mechanisms. The organization of the program allows the user to quickly arrive at an acceptable solution to his design problem if such a solution does in fact exist. The user does not have to be an expert in the theory behind the program to use it successfully.

The utility of the program is being enhanced by the addition of more traces. For example traces for time ratio will be added to the GCRR list. Maximum and minimum angular velocity ratio will be added to the GCCC list. Numerical results show that traces also exist for extreme values of angular acceleration. When these enhancements are completed, the program will represent an extremely useful design package for kinematic synthesis.

6. REFERENCES

1. Krouse, J. K., "Designing Mechanisms On a Personal Computer", Machine Design, March 24, 1983, pp. 94-99.

2. Kaufman, R., sales literature from Kintech, Inc., 1441 Springvale Ave., McLean, VA, 22101.

3. Lysdahl, H., Barris, W., Riley, D., and Erdman, A., "Micro-Linkages - A Mechanism Design Package", 8th OSU Applied Mechanisms Conference, St. Louis, MO., Sept., 1983, Paper No. 23.

4. Shoup, T., "The Use of the Personal Computer As An Aid To the Kinematic Synthesis of Linkages", 8th OSU Applied Mechanisms Conference, St. Louis, MO, Sept., 1983, Paper No. 63.

5. Wilson, P., and Soni, A., "Micro-Computer, Pascal, and Linkage Synthesis", 8th OSU Applied Mechanisms Conference, St. Louis, MO, Sept., 1983, Paper No. 59.

6. Barker, C., "Complete Classification of Planar Four Bar Linkages", 8th OSU Applied Mechanisms Conference, St. Louis, MO, Sept., 1983, Paper No. 32. Accepted for publication in Mechanism and Machine Theory.

7. Barker, C., and Jeng, Y., "Range of the Six Fundamental
    Position Angles of a Planar Four Bar Mechanism", accepted for
    publication in Mechanism and Machine Theory, 1984.

8. Barker, C., "Characteristic Surfaces in the Solution Space of
    Planar Four Bar Mechanisms, Part 1 & 2", 18th ASME Mechanisms
    Conference, Cambridge, MA, Oct. 1984, Paper 84-DET-126 &
    84-DET-127.

9. Tsai, L. "Design of Drag-Link Mechanisms with Optimum
    Transmission Angle", ASME Journal of Mechanisms,
    Transmissions, and Automation in Design, Vol. 105, No. 2,
    June 1983, pp. 254-259.

10. Savage, M., and Suchora, D., "Optimum Design of Four Bar
    Crank Rocker Mechanisms with Prescribed Extreme Velocity
    Ratios", Journal of Engineering for Industry, Trans. ASME,
    Series B, Vol. 96, 1974 pp. 1314-1321.

11. Barker, C. R., "Analysing Mechanisms With The IBM PC,"
    Computers in Mechanical Engineering, Vol. 2, No., 1, July
    1983, pp. 37-44.

THE USE OF THE FLUID FLOW PROGRAM PHOENICS IN
ENGINEERING DESIGN

N Rhodes, S A Al-Sanea and K A Pericleous

Concentration Heat and Momentum Ltd, 40 High Street,
Wimbledon, London SW19 5AU

INTRODUCTION

In recent years the use of computational methods to solve fluid flow
problems has increased enormously. The problem-solving needs of
the aerospace and nuclear industries, and the availability of
suitable computers, led to the development of the numerical
techniques which are required to solve the governing equations.
There use, however, has now spread much wider, and includes
environmental flows, process equipment, car aerodynamics, glass
and metal flows and many others.

In their infancy, computer codes for fluid flow analysis generally
required the insight of their originator in order to obtain useful
results. This situation reflects the degree of expert knowledge
required to write such a program and, since early codes usually
simulated a specific process or equipment, the solution method was
oriented towards the particular flow situation. As experience
increased, however, it became apparent that a more unified
strategy was required, and a single, general-purpose code was
desirable. The main specification for such a code is that it should
be adaptable to any of the problems normally addressed using
computational techniques. It therefore needs the capability of
providing solutions for steady or transient forms of the
Navier-Stokes Equations in one, two or three space dimensions, to
account for the effects of turbulence and chemical reaction, to
allow consideration of additional fluid or solid components having
different velocities to the continuous fluid and so on.

The PHOENICS code, Rosten et al (1983), has been designed with
this specification in mind. It removes the burden of adapting an
existing mathematical model, or worse, writing one from scratch,
to simulate a required process or item of equipment. It allows
engineers, who need not necessarily be well versed in
computational fluid flow methods, to develop models of a particular
process and to investigate, for example, the performance and
effects of design changes on the process. The general framework

which the program provides has thus led to the development of more design-oriented models and a movement away from the specialised and sometimes difficult to use single-purpose models.

The purpose of this paper is to illustrate the alternative levels of modelling which can be taken using PHOENICS by way of three examples of recent areas of application. The first application is to a steam condenser. This represents the use of a fluid flow model to investigate the detailed three-dimensional flow and associated heat transfer processes taking place in a condenser. In the development of this model, the attention to the physics of the processes and geometrical details has led to computational tool which gives the designer enormous scope for investigating the performance of different designs, and the influence of design changes. The second example is an air dryer. The purpose of the study was to determine whether the existing designs could be adapted to provide higher volume through-put without increasing the overall size of the equipment. In this case, a simple one-dimensional model sufficed, and all that was required was the incorporation of suitable heat-transfer and pressure drop correlations. The final example concerns the study of hydrocyclone performance. In this case PHOENICS is being used to unravel the very complex flow and separation mechanisms which occur in a hydrocyclone by developing hypothetical models of the physical behaviour, expressing these in mathematical terms, incorporating them in PHOENICS and assessing the performance of the mathematical model by comparison with experimental data.

STRUCTURE OF PHOENICS

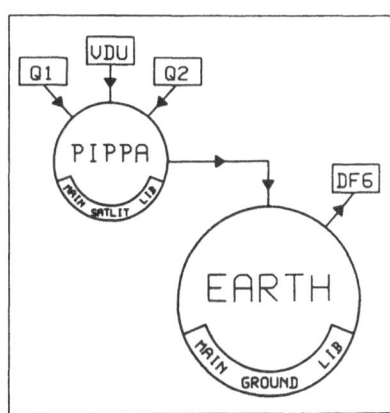

The PHOENICS code achieves its generality by separating the equation-solving parts of the program from those which describe the particular process under consideration. Figure 1 illustrates the main components. A central core program, named EARTH, is the main equation solver and is used in all flow simulations. To ensure that generality and reliability can be maintained, it is not accessible to the user.

Figure 1: Main Components of PHOENICS '84

A model of a particular flow situation is built up from a user 'satellite' program which activates the capabilities of the EARTH program. The satellite itself comprises two main sections, SATLIT and GROUND. In SATLIT, the user inserts problem-specific information such as the equations to be solved, the geometry, invariant boundary conditions, relaxation parameters and printout specification.

In the GROUND section, any additional information which requires interaction with the iterative calculation is inserted. Inputs provided through GROUND might include, for example, particular physical property formulations, alternative turbulence modelling data, flow-regime selection criteria, or correlations for interphase heat and mass transfer.

In setting up a new problem, a user can create a SATLIT – GROUND combination either through the creation of a "Quick-Input File", a Q1 file in Figure 1, or interactively through the PIPPA facility. The latter method is ideal for new users, as explanation of the required items to be set can be studied as one proceeds. Either way, the basic model is developed in the first instance by setting appropriate 'switches' for the EARTH program. The independent variables are x, y and z (or θ) in a cartesian or polar coordinate system. A general body-fitted-coordinate facility is also available for modelling the flow over streamlined bodies such as aerofoils, cars etc. For transient flows, time is also an independent variable.

The user can specify an unlimited number of dependent variables. The major ones which might apply to a three-dimensional, two-phase problem are:

* pressure, p, and pressure correction p';

* three first-phase velocities, u, v, and w;

* three second-phase velocities, u_2, v_2, and w_2;

* phase volume fractions, r_1 and r_2 (and r_2^*);

* turbulence energy and dissipation rate, k and ϵ;

* first and second phase enthalpies, h_1 and h_2;

* enthalpies of a third fluid, h_3;

* radiation-flux sums, Rx, Ry, Rz; and

* any concentration equations required to simulate chemical reactions, C1, C2.

The equations for all variables have the common form:

$$\frac{\partial}{\partial t} (r\rho\phi) + \text{div} (r\rho\vec{v}\phi - r\,\Gamma_\phi \text{grad } \phi) = rS\phi$$

where:

r   = phase volume fraction;

ρ = density;

ϕ = dependent variable;

v   = velocity vector;

$\Gamma\phi$ = exchange coefficient (laminar or turbulent); and

$S\phi$ = source or sink term.

Integration of these equations over a staggered grid results in a set of finite-domain equations which are solved in the EARTH program.

It is not appropriate in this paper to attempt to describe the mechanics of setting up a PHOENICS model, save to say that the procedures are fully documented, Rosten & Spalding, (1985 a,b,c,) and many coding examples are available in the form of PHOENICS Demonstration Reports, eg Glynn and Tatchell (1984).

Summarising, the program user has at his disposal a means of:

*   activating the solution of equations governing fluid motion

*   inserting boundary values, and sources or sinks of momentum, heat, species etc.

*   inserting special laws which might be specific to a particular problem, and

*   directly manipulating the equations to suit his particular problem.

The remainder of this paper will describe three applications of PHOENICS which illustrate quite different requirements of the user.

APPLICATIONS OF PHOENICS

(a) Steam Condenser

Steam condenser models using PHOENICS have undergone rigorous development. Initial studies, Al-Sanea et al (1983), concentrated on suitable representations for heat-transfer and turbulence and a two-dimensional, single phase model was validated against experimental information. This model has been used for design studies and has been extended to analyse three-dimensional flows.

The main features of the three-dimensional model can be summarised as follows:

* Equations are solved for three velocity components of a steam/air mixture, the concentration of steam in air, and pressure.

* Source terms are included for heat transfer, condensation and tube friction based on local conditions.

* The condensate is assumed to disappear, although its effect on heat transfer is included through an inundation term in the heat-transfer model.

* The effect of local pressure on condensation is included.

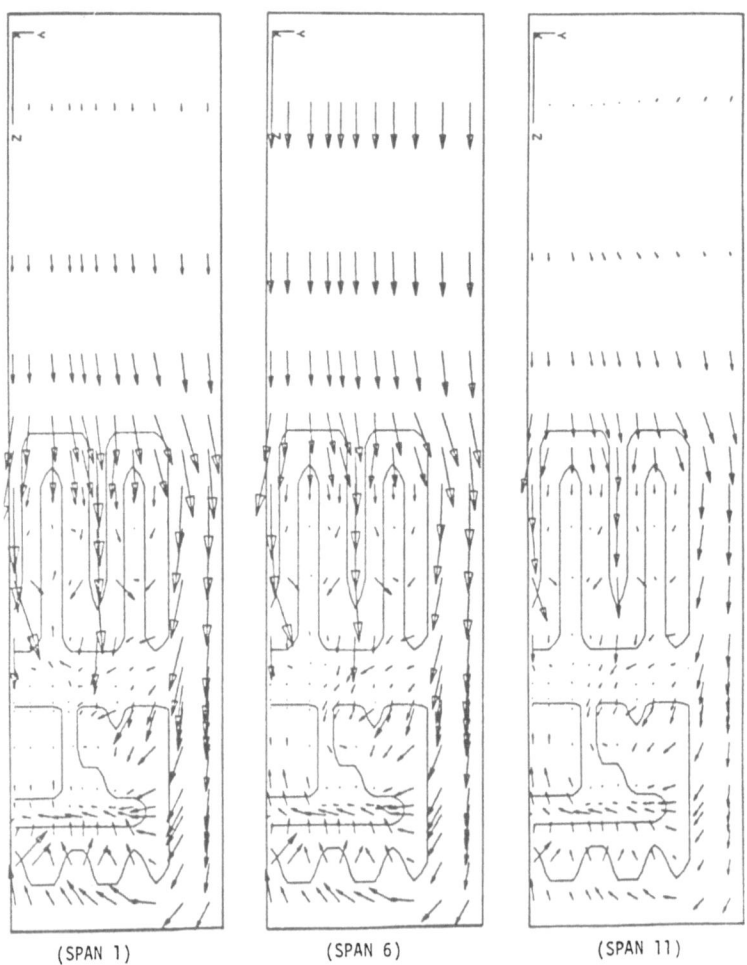

(SPAN 1)   (SPAN 6)   (SPAN 11)

Figure 2:   Velocity Vectors at Three Planes in a Steam Condenser

* The internal features of the condenser, ie tubes, baffles, drip-trays and support structures are represented by way of wholly or partially blocking cells or cell faces.

In parallel with the application of the "single-phase" model described above, the extension of the condenser model to include the effect of the condensate phase is being developed. This requires additional equations for the motion and enthalpy of the condensate and incorporation of appropriate mathematics to describe, for example the droplet/liquid film interactions. The status of this model developed so far is described in Al-Sanea et al (1985).

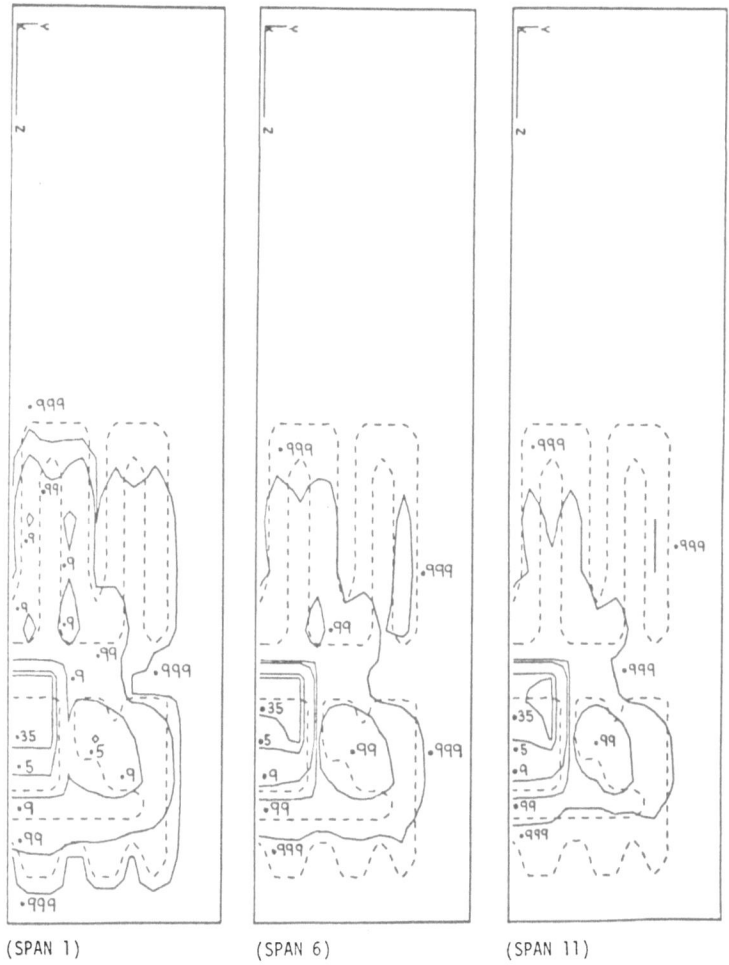

Figure 3: Concentration Contours in a Steam Condenser

A study was performed with the single-phase model to determine the effect on condenser performance of both uniform and non-uniform distributions of steam flowing into it. It should be said that the steam flow from the turbine exhaust into the condenser is itself very complex, and the designer has no certain knowledge of the inlet distribution to the condenser.

Figure 2 and 3 show velocity vectors and steam concentration contours at three locations in the condenser for the case with uniform steam inflow. Span 1 is at the cooling water inlet end, Span 6 in the middle and Span 11 at the cooling water outlet end.

The tube nest in these cross-sections is also illustrated. It can be seen from Figure 2 that the flow velocity into the tube nest is highest in the centre, rather low at the cooling water outlet end, and somewhat higher at the inlet end. The flow features within the tube nest also show differences according to the location. Figure 3 shows a high concentration of steam at spans 6 and 11, but at span 1 there is a higher air concentration within the nest. This is due to the higher rate of condensation in this region since the cooling water is at its lowest temperature.

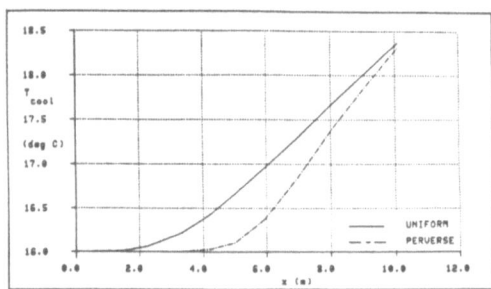

Figure 4: Variation of Cooling Water in the Air-
Cooling Section of a Condenser

Figure 4 shows the variation in cooling water temperature along the air-cooling section for both uniform and non-uniform distributions of steam inlet. The differences in the two cases are quite clear, indicating that in this particular region there is little heat transfer taking place at the inlet end of the condenser for the perverse distribution.

These Figures show only a very small fraction of the results which are obtained from the model, but give an indication of how a clear insight into the flow behaviour can be obtained.

(b) Compressed Air Dryer

Compressed air is used in a wide range of industrial applications. Moisture in compressed air is undesirable and is frequently removed by using a refrigeration drying system.

An important aim in the design of an air dryer, as indeed in all kinds of heat exchanger, is to increase the efficiency of the unit by increasing the heat-transfer rate, decreasing the pressure drop and reducing the heat exchanger size. In addition, certain constraints are to be met regarding the temperature of the air/water mixture at the end of the refrigeration chiller cycle: the mixture should be cooled to the dewpoint temperature but not low enough to freeze-up the condensed water within the chiller.

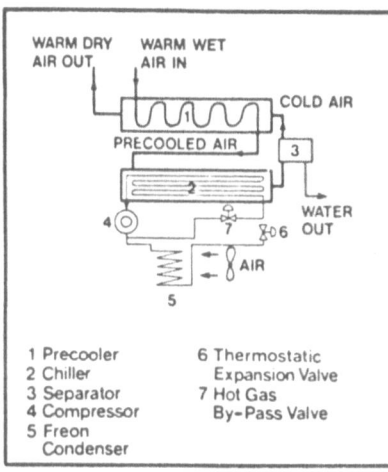

1 Precooler
2 Chiller
3 Separator
4 Compressor
5 Freon
   Condenser

6 Thermostatic
   Expansion Valve
7 Hot Gas
   By-Pass Valve

Figure 5: Schematic of an Air-Dryer

An air-dryer which PHOENICS has been used to model is shown in Figure 5. The dryer is basically made up of three parts:

(i)   The air-to-air heat exchanger loops for precooling the incoming wet air in the annuli, and reheating the outgoing dry air in the inner tubes. This reduces the energy requirement to power the refrigeration system.

(ii) The air-to-refrigerant heat exchanger loops which cool the air to the dewpoint temperature of about 3°C to enable water vapour to condense. The flow of air takes place inside the inner tubes: that of the refrigerant takes place in the annuli.

(iii) The condensate separator which removes the water from the chilled air.

A model was set up for this device in order to perform parametric studies of the effect of design changes and operating conditions on performance. The main features of the model are as follows:

*   Steady-state, one-dimensional

*   Only one loop of each heat exchanger was considered. Other parts of the unit, eg. separator and headers, were accounted for by adjustments to the friction factors.

*   The pressure-drop and heat-transfer correlations and the geometry specification were implemented in a general way to facilitate the parametric studies.

The model was set up and the empirical data governing the heat transfer and pressure drop were modified by factors which accounted for their surface roughness. The factors were adjusted so that the model predictions agreed with known data for given operating conditions. The model was then used to study various design changes. Figure 6 and 7 show the effect on pressure drop and air temperature of increasing the diameter of the outer pipe by 15%, the inner pipe remaining at the same diameter.

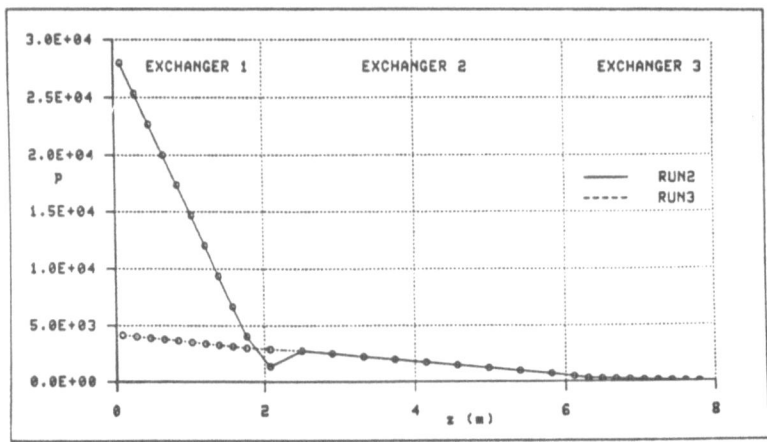

Figure 6: Variation of Pressure through the Air Dryer

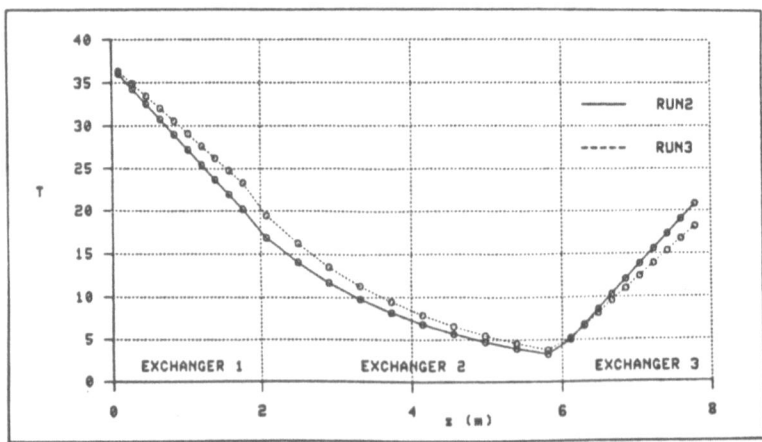

Figure 7: Variation of Temperature through the Air Dryer

It can be seen from Figure 6 that the major pressure drop occurs in the air-to-air heat exchanger, and that this has been substantially reduced by increased the outer annulus diameter.

The effect on air temperature, Figure 7, is quite noticeable in the air-to-air heat exchanger, but the overall drop is only 0.5°C higher than the standard case. The air temperature could be reduced further by employing a slightly longer air-to-refrigerant exchanger.

(c) Hydrocyclone

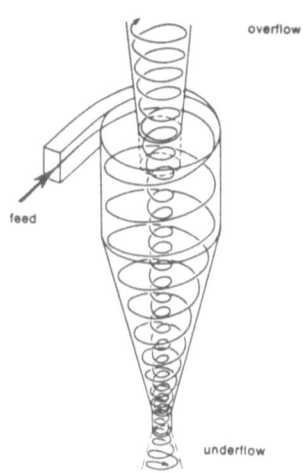

overflow

feed

underflow

A hydrocyclone is a device used for classification of solids. A typical geometry is shown in Figure 8. The slurry enters tangentially and the subsequenct rotation causes heavier or larger particles to move outwards, depending on the balance of centrifugal and drag forces. Although the device itself is quite simple, the flow mechanisms are very complex. A lack of basic understanding of the flow mechanisms within a hydrocyclone has resulted in an empirical approach to their design.

Figure 8: Hydrocyclone

Experimental measurements of pressure drop, flow split (ie. ratio of flow exiting at underflow and overflow), the particle size which reports in equal proportions to the overflow and underflow, and the sharpness of separation are expressed in empirical equations for design use. Unfortunately, there are, literally, hundreds of equations available for each parameter, eg Bednarski et al (1984), so there is no universal set of design rules.

A recent study of hydrocyclone performance, Pericleous et al (1984), has centred on modelling the flow using PHOENICS. The basic outline of the model is as simple as the device itself, being steady-state two-dimensional with swirl. The shape of the hydrocyclone is represented using porosities. Here the simplicity ends, and the multi-phase nature of the flow needs to be taken into account. The modelling practices adopted so far have been to use separate concentration equations to represent the air-core, which forms when the device is operated with atmospheric outlet, and each particle size range. The motion of each component is affected by the centrifugal and drag forces, and special source terms are introduced to compute the separation velocities of each component.

The effects of turbulence, particle/fluid interactions, and particle crowding have to be considered. During the course of the study mentioned in 'refs() many hypotheses have been examined using PHOENICS. Some have been found to be adequate, and others rejected along the way. Several representations of turbulence, for example, were investigated. The emphasis of the study is not so much on the fundamentals of numerical modelling, but more the physical modelling involved. PHOENICS has proved to be an excellent means of testing ways of representing the physical mechanisms.

Figures 9 and 10 show the success of this application so far. Results have been compared with the flow in a given hydrocyclone which has been studied experimentally at Warren Spring Laboratory. The predicted position and size of the air-core, the pressure drop and flow split are all in agreement with the experimental data. The swirl distribution in a hydrocyclone is shown in Figure 9, and compared with the experiments of Kelsall. Figure 10 compares the separation characteristics of a cyclone with an empirical equation. This too shows good agreement – the predicted case was selected from the experimental data from which the empirical equation was deduced.

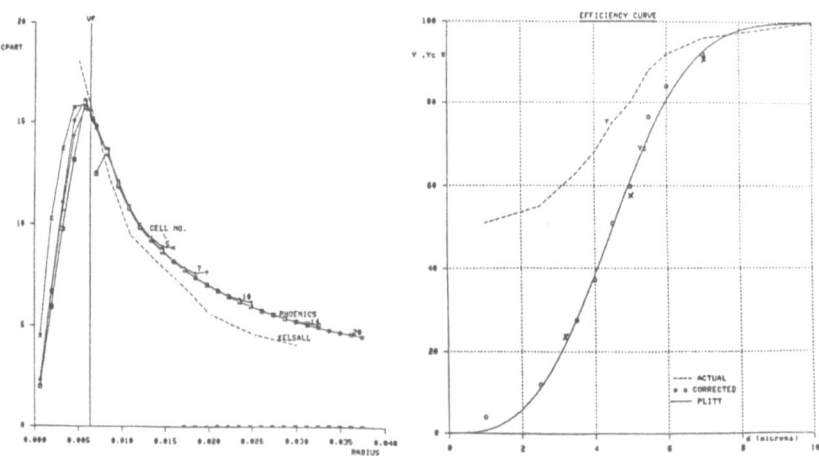

Figure 9: Variation of Swirl Velocity with Radius in a Hydrocyclone

Figure 10: Predicted Efficiency Curve for a Hydrocyclone

CONCLUDING REMARKS

The foregoing sections have provided a brief description of the general-purpose fluid-flow program PHOENICS and its application to a variety of engineering design problems. The use of the program merely as a tool for solving flow problems may not appeal to the engineers whose interests lie in the mechanics of computational techniques. For him, the dedicated program with access to all levels of the model, both physical and computational, is still the ideal. It must be recognised, of course, that the development of such dedicated models can take very much time and effort.

For the engineer concerned with answering practical problems on a shorter time scale, the use of PHOENICS can be a viable proposition. The air-dryer model mentioned above, for example, was developed in only a few days and gave the user immediate insight into the possibilities of design improvement. The condenser models are far more complex than the air-dryer and consequently take longer to establish. Much depends on the degree of detail required in the geometry specification. Using the condenser model described above, the user can set up a new design, perform preliminary calculations and begin iterating on design improvement within one or two months. The hydrocyclone problem is in the nature of a research exercise where the PHOENICS user must expect that some of his ideas will not lead to successful representations of the flow. A change of ideas in one area may have unsuspected consequences elsewhere. Nevertheless, removing the need to write dedicated numerical schemes leaves the researcher free to experiment with physical assumptions without interruptions caused by numerical aspects.

ACKNOWLEDGMENT

The authors wish to express their thanks to NEI Parson Ltd for permission to publish the results of the condenser studies and to Deltech Engineering for permission to publish the air-dryer results.

REFERENCES

Al-Sanea S A, Rhodes N, Tatchell D G & Wilkinson T S (1983) A Computer Model for Detailed Calculation of the Flow in Power Station Condensers. I Chem E Symposium Series No. 75.

Al-Sanea S A, Rhodes N & Wilkinson T S (1985) Mathematical Modelling of Two-Phase Condenser Flows. BHRA Conference on Multiphase Flows, London, 1985.

Bednarski S & Wiechowski A (1984) A Review of Hydrocyclone Performance Correlation. BHRA Conference on Hydrocyclones, Bath.

Glynn D R & Tatchell D G (1984) Flow in a Gas Turbine Engine Transition Duct, CHAM PDR/23.

Kelsall D F (1952) A Study of the Motion of Solid Particles in a Hydraulic Cyclone Trans. Inst. Chem. Eng., 30, p. 87.

Pericleous K A, Rhodes N, Cutting G W (1984) A Mathematical Model for Predicting the Flow-Field in a Hydrocyclone Classifier. BHRA Conference on Hydrocyclones, Bath, 1984.

Rosten H I, Spalding D B & Tatchell D G (1983) PHOENICS: A General-Purpose Program for Fluid-Flow, Heat-Transfer and Chemical Reaction Processes. Proc. of 3rd Int. Conf. on Engineering Software, London, 1983, pp 639-655.

Rosten H I & Spalding D B (1985 a) Beginners Guide to PHOENICS-84. CHAM Technical Report TR/99.

Rosten H I & Spalding D B (1985 b) PHOENICS-84 Reference Handbook CHAM Technical Report TR/100.

Rosten H I & Spalding D B (1985 c) PHOENICS-84 Input Examples CHAM Technical Report TR/101.

SIMULATION OF FUEL INJECTION IN DIESEL ENGINES

G. Chiatti, R. Ruscitti

Dipartimento di Meccanica e Aeronautica, Università degli Studi di Roma "La Sapienza"

1. INTRODUCTION

The numerical simulation of fuel injection in diesel engines has received much attention in recent years according to the diesel powered cars increase in number in Europe [1]. The aim of these studies is the improvement of the engine drivability by the control of the power output fluctuations.
The analysis of unsteady phenomena in   fuel injection equipments has been generally carried out by application of the method of characteristics to the waterhammer governing equations [2], improved in order to take account of non-linear effects due to parameter variations and liquid cavitation [3,4].

Here is proposed a new physical modelling of the system based on the thermodynamic analysis of the state equation of the liquid and the phase change during cavitation.

The numerical method is a finite difference procedure based on an explicit 2nd order accurate λ-scheme [5]. The Authors have modified the original method, developed for the study of ideal gas flow [6,7], to take account of liquid state equation, parameter variations, and new kind of discontinuities due to cavitation.

2. PHYSICAL MODEL

The physical model of the injection system is represented in figs. 1 and 2. The first element is a volumetric fuel pump feeded at constant pressure by a reservoir (R) and consisting of: a piston (P) directly moved by a camshaft, working in a pumping chamber (PC); a valve (V) connecting pumping and delivery chamber (DC); and a spring (SP) acting on the valve.

The second element is a duct (D) of variable section connecting the pump to the injector (I).

The third element, I, consists of: an injection chamber (IC); an injection delivery chamber (ID); a needle (N) loaded by a spring (SI); and a chamber (C) where the fuel is injected, representing the engine cylinder.

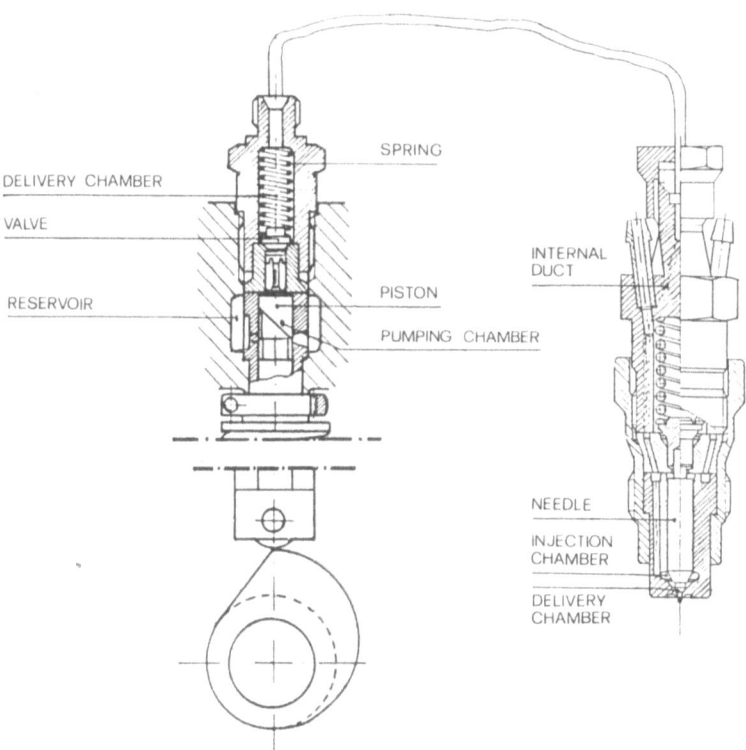

Fig. 1 - Fuel injection equipment.

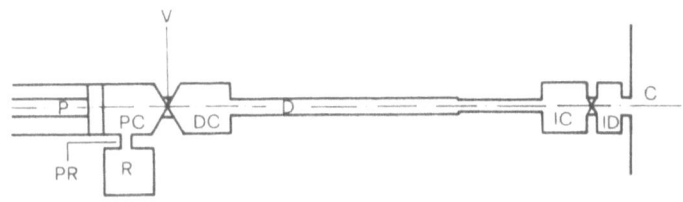

Fig. 2 - Physical model.

3. MATHEMATICAL MODEL

The thermodynamic state equation for the liquid is

$$\frac{\partial \rho}{\rho} = kdP - \beta dT \tag{1}$$

where ρ, P, T are the volumic mass, the pressure and the temperature; k is the isothermal compressibility; and β the volume coefficient of expansion.

The cavitation is controlled by the volumic vapor ratio, x_v, evaluated following the phase change process on the thermodynamic plane. A distributed bubble model is considered just to a limiting value of x_v, vapor is then lumped and a column separation is produced [8,9].

The behaviour of the liquid in the pump is described by a lumped parameter model; friction effects are introduced only by the consideration of discharge coefficients. Mass flow rates are calculated from the pressure differences between chambers.

The piston displacement, h_p, is imposed by a camshaft; fuel at constant pressure is feeded to the pumping chamber through the port (PR); back flow to the reservoir is allowed according to engine torque requirements. The valve movement, h_v, against the spring, is caused by the pressure difference between pumping and delivery chamber. Fuel flows from the delivery chamber to the duct through the port (AT). The continuity equation for the liquid in the pumping chamber is:

$$V_p \rho_p k \frac{dP_p}{dt} + \rho_p (-A_p \frac{dh_p}{dt} + A_v \frac{dh_v}{dt}) - \dot{m}_{in} + \dot{m}_v = 0 \tag{2}$$

where V_p is the instantaneous volume of the pumping chamber, A_p is the cross-section of the piston, A_v the section of delivery valve, $\dot{m}_{in}$ the mass flow rate from the reservoir and $\dot{m}_v$ the mass flow rate to the delivery chamber. The mass flow rates are related to the pressure difference between reservoir and pumping chamber and between pumping and delivery chamber:

$$\dot{m}_{in} = \pm A_{PR}C_{RR} \sqrt{(P_R - P_P)\rho_R} \tag{3}$$

$$\dot{m}_v = \pm A_{FV}C_V \sqrt{(P_P - P_D)\rho_P} \tag{4}$$

where A_{PR} and A_{FV} are the flow sections and C_{PR} and C_V the corresponding discharge coefficients, variable with the piston and valve displacements.

The continuity equation for the delivery chamber is:

$$V_D \rho_D k \frac{dP_D}{dt} + \rho_D (- A_{FV} \frac{dh_V}{dt}) - \dot{m}_V + \dot{m}_T = 0 \tag{5}$$

where $\dot{m}_T$ is the mass flow rate to the duct:

$$\dot{m}_T = \pm A_T C_T \sqrt{(P_D - P_T) \rho_D} \tag{6}$$

being A_T the duct section at pump outlet and C_T the corresponding discharge coefficient.

The valve equation of motion is:

$$m_V \ddot{h}_V + k_V h_V + K_V + A_V P_D - A_V P_P = 0 \tag{7}$$

k_V and K_V are the stiffness and the control load of the valve
spring; friction forces on the valve are neglected.

The entropies in the pumping and delivery chambers are:

$$(\dot{\rho V s})_P - \dot{m}_{in} s_{in} + \dot{m}_V s_P = \Delta S_{iP} \tag{8}$$

$$(\dot{\rho V s})_D - \dot{m}_V s_V + \dot{m}_T s_D = \Delta S_{iV} \tag{9}$$

where s is the entropy; s_{in} the entropy of the fluid coming into the pumping chamber; s_V the entropy of the fluid coming into
the delivery chamber taking account of the loss through the
ports. ΔS_{iP} and ΔS_{iV} are the entropy increase for inversibili-
ties.

Just like the pump, the continuity equations for the injector
are:

$$V_{IC} \rho_{IC} K \dot{P}_{IC} + \rho_{IC} (A_{NC} \dot{h}_N) - \dot{m}_{IC} + \dot{m}_{ID} = 0 \tag{10}$$

$$V_{ID} \rho_{ID} K \dot{P}_{ID} - \dot{m}_{ID} + \dot{m}_C = 0 \tag{11}$$

where A_{NC} is the needle area subjected to pressure of fuel; $\dot{m}_{IC}$,
$\dot{m}_{ID}$ and $\dot{m}_C$ are the mass flow rate between the end of the duct,
the injection chamber and the delivery chamber; h_N the needle
displacement.

$$\dot{m}_{ID} = A_{ID} C_{ID} \sqrt{(P_{IC} - P_{ID}) \rho_{IC}} \tag{12}$$

$$\dot{m}_C = A_C C_C \sqrt{(P_{ID} - P_C)\rho_{ID}} \qquad (13)$$

being A_{ID}, A_C and C_{ID}, C_C the flow sections and the corresponding discharge coefficients. For the entropies we have:

$$(\rho Vs)_{IC} - \dot{m}_{IC} s_{mC} + \dot{m}_{ID} s_{IC} = \Delta S_{iIC} \qquad (14)$$

$$(\rho Vs)_{ID} - \dot{m}_{ID} s_{mD} + \dot{m}_C s_{ID} = \Delta S_{iID} \qquad (15)$$

Variables play the same role as in the equations (8) and (9).
As the flow in the duct connecting pump and injector is conside
red one-dimensional, the equations of momentum, mass and energy
conservation are:

$$\frac{Du}{Dt} + \frac{P_x}{\rho} + \frac{f}{2H} u^2 = 0 \qquad (16)$$

$$\frac{D}{Dt}(\rho A) + \rho A u_x = 0 \qquad (17)$$

$$\frac{Ds}{Dt} - f\frac{u^3}{2HT} = 0 \qquad (18)$$

where D indicates the total differential and subscript, x or t,
the partial derivative with respect to space or time variable.
A is the cross section of the duct, f the friction coefficient
and H the hydraulic diameter.

Rearranging the above equations in order to facilitate the nume
rical computation, the following system is obtained:

$$\frac{Du}{Dt} + \frac{P_x}{\rho} + \frac{f}{2H} u^2 = 0 \qquad (19)$$

$$\frac{DP}{Dt} + \frac{a^2}{1 + (d/Ek\sigma)} \rho u_x = 0 \qquad (20)$$

$$\frac{Ds}{Dt} - \frac{fu^3}{2HT} = 0 \qquad (21)$$

being d the diameter and σ the thickness of the tube, E the
Young modulus, and a the adiabatic sonic speed.

4. NUMERICAL METHOD

The set of equations (19), (20) and (21) may be written in the
form

$$\frac{Du}{Dt} + \frac{\alpha}{\delta} P_x + \frac{f}{2H} u^2 = 0 \qquad (22)$$

$$\frac{1}{\delta} \frac{DP}{Dt} + \alpha u_x = 0 \qquad (23)$$

$$\frac{Ds}{Dt} - \frac{f}{2H} u^3 = 0 \qquad (24)$$

being

$$\alpha^2 = \frac{a^2}{1 + (d/Ek\delta)}$$

$$\delta = \alpha\rho$$

By addition and subtraction, equations (22) and (23) transform themselves in:

$$(\frac{P_t}{\delta} + u_t) + (\alpha + u)(\frac{P_x}{\delta} + u_x) = -\frac{f}{2}\frac{u^2}{H} \qquad (25)$$

$$(\frac{P_t}{\delta} - u_t) + (-\alpha + u)(\frac{P_x}{\delta} - u_x) = \frac{f}{2}\frac{u^2}{H} \qquad (26)$$

The integration of the above equations is carried out by an explicit scheme, so parameters α and δ maintain for each time step the initial values, and the system assumes the form:

$$(\frac{P}{\delta} + u)_t = -(\alpha + u)(\frac{P_x}{\delta} + u_x) - \frac{f}{2}\frac{u^2}{H} \qquad (27)$$

$$(\frac{P}{\delta} - u)_t = -(-\alpha + u)(\frac{P_x}{\delta} - u_x) + \frac{f}{2}\frac{u^2}{H} \qquad (28)$$

$$s_t = -us_x + \frac{f}{2}\frac{u^3}{HT} \qquad (29)$$

The computation is performed over a grid of points and the spatial derivatives are approximated by finite differences taking account of the domain of dependence. To improve the accuracy, the time step is subdivided in two partial ones, the code remaining formally the same as in a first order difference scheme.

Following the integration scheme, in each boundary one relationship between pressure and velocity is imposed and one more thermodynamic variable is assigned if the flow is going into the tube. Pump and injector differential equations are integrated by an explicit numerical procedure.

5. RESULTS

At first the simulation of the fuel pump was performed imposing a constant pressure discharge. In figures 3.a, 3.b, 3.c, results are reported for a 5 MPa discharge reservoir pressure and a 1500 rpm cam revolution speed.

The injector model was tested imposing pressure impulses of dif ferent amplitude at inlet. Results for 15 MPa pressure amplitude and $4 \cdot 10^{-3}$ s time duration impulse are presented in figures 4.a, 4.b, 4.c.

The λ-scheme treatment of unsteady flow in the connecting duct has been implemented by considering two limiting cases of impul ses generated at one end and reflected at the opposite one by a zero flux condition or a constant pressure condition; results are shown in figs. 5 and 6.

Subsystems were then matched in the complete one that simulates the whole injection apparatus. Results for the condition of maximum fuel delivery are reported in fig. 7.

6. CONCLUSIONS

The model proposed is an original application of 2nd order accu rate λ-scheme integration method to the computation of unsteady phenomena in fuel ducts, thus reducing the computer burden and allowing to perform numerical simulations on a PC computer.

The thermodynamic approach introduced for cavitation is coherent with physical treatment of unsteady flow as performed by λ-scheme methods.

Results have confirmed the possibilities of the method proposed in order to find the working limits of injection equipments for different operating conditions.

REFERENCES

1. Becchi, G.A. (1983) Osservazioni sulla natura dei fenomeni di instabilità dei motori Diesel. ATA, vol. 36, 97-105.

2. Evangelisti, G. (1974) Waterhammer Analysis by the Method of Characteristics. Energia Elettrica, 46, 673-692.

3. Weyler, M.E., Streeter, V.L., Larsen, P.S. (1971) An Investigation of the Effect of Cavitation Bubbles on the Momentum Loss in Transient Pipe Flow. J. Basic Eng., 1-10.

4. De Bernardinis, B., Federici, G., Siccardi F. (1975) Transient with Liquid Column Separation: Numerical Evaluation and Comparison with Experimental Results. Energia Elettrica, 9, 471-477.

5. Moretti, G. (1979) The λ-scheme. Computers and Fluids, 7, 191-205.

6. Moretti, G., Di Piano, M.T. (1983) An Improved λ-scheme for One-Dimensional Flows. Contract NAS 1-16946.

7. Zannetti, L. Colasurdo, G. (1981) Unsteady Compressible Flow: A Computation Method Consistent with the Physical Phenomena. AIAA Journal, 19, 851-856.

8. Wiggert, D.C., Sundquist, M.J. (1977) The Effect of Released Gases on Hydraulic Transients. Final Research Report. Michigan State University.

9. Raiteri, E., Siccardi, F. (1975) Transient in Conduits Conveying a Two-Phase Bubbly Flow: Experimental Measurements of Celerity. Energia Elettrica, 5, 256-261.

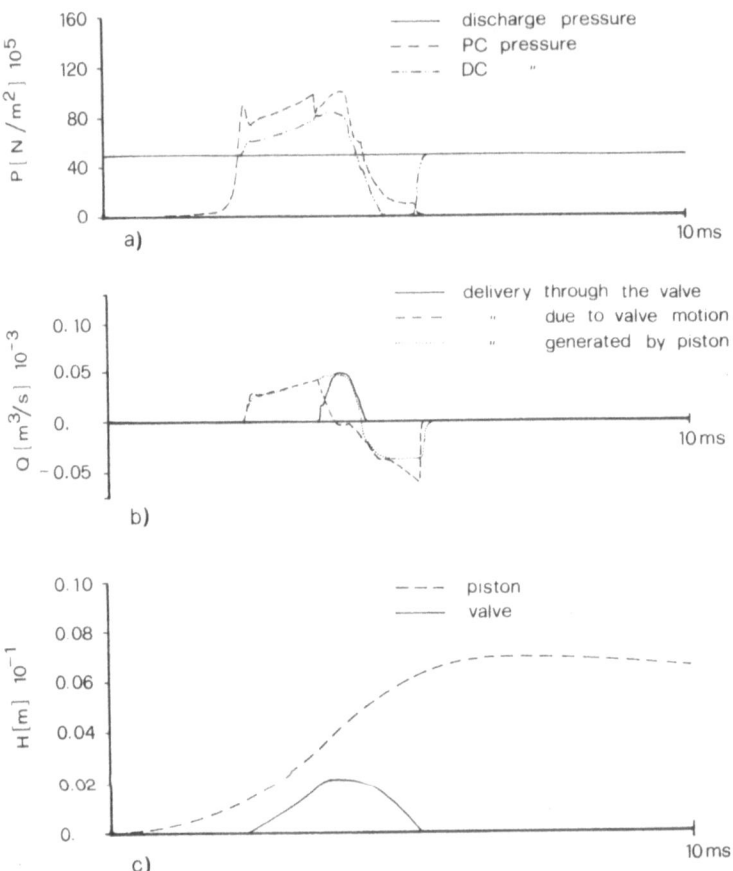

Fig. 3 - Fuel pump simulation: constant pressure discharge.

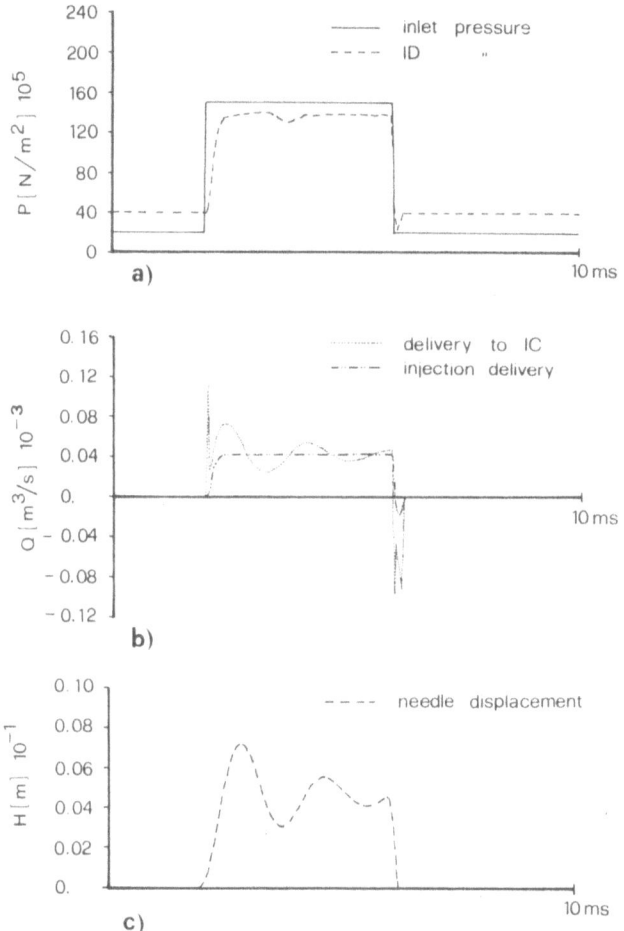

Fig. 4 - Injector simulation.

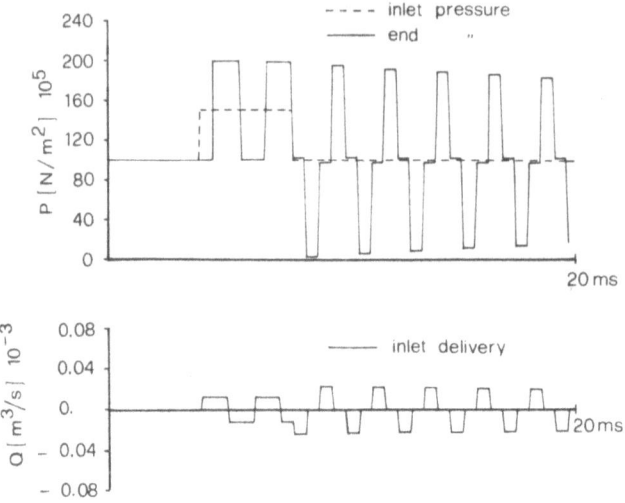

Fig. 5 - Duct simulation: end zero flux condition.

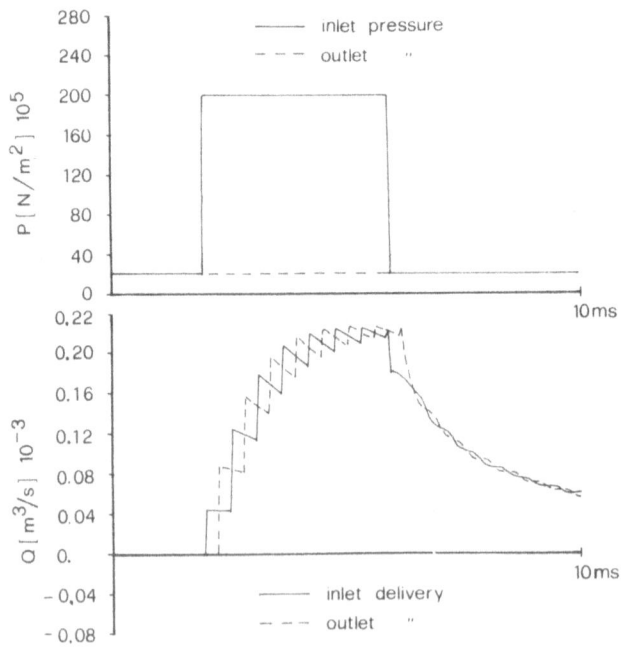

Fig. 6 - Duct simulation: end constant pressure condition.

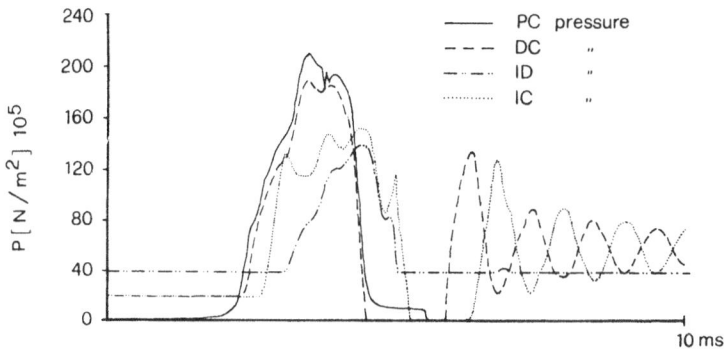

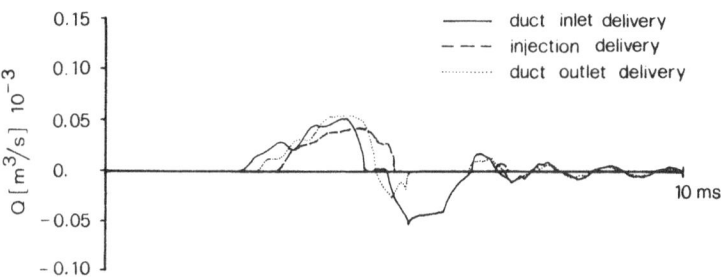

Fig. 7 - Complete system simulation.

COMPUTER AIDED DESIGN FOR THE SELECTION PROCESS OF HAZARDOUS AND NUCLEAR WASTES SITES

G.V. Abi-Ghanem and V. Nguyen

ARDI Corporation, P.O. Box 27113, Minneapolis, MN
55427 U.S.A.

I. INTRODUCTION

The long-term performance evaluation of a
hazardous/nuclear waste site is a scientific analysis
process that quantifies how well the system achieves
its basic design objective, i.e. the isolation of the
hazardous/nuclear waste from the accessible
environment. Application of performance in site
selection analysis is to predict the long-term
behavior of the candidate site system (in the order
of decades for a hazardous waste site and centuries
for a nuclear waste repository) under normal and
disruptive man-induced/geologic events. The
forecasting of contaminant/radionuclide movement
under changing site conditions can only be achieved
using both deterministic and stochastic computer
modeling techniques. The analysis must consider
uncertainties in (1) field and laboratory data, (2)
theoretical completeness of the predictive models,
and (3) predictions of future conditions.
 In this communication, a SITECAD package of
softwares is introduced to assist in the
identification of model parameters using field and
laboratory data (e.g. from well logs, pump test,
tracer test, geomechanics, geochemistry, thermal and
radiation enhanced diffusion testing, etc.), and to
predict the fate of hazardous chemicals/radionuclides
in the environment. This package is interactive and
has graphic design and statistical analysis
capability as supporting features. It is mounted on a
network of microcomputers and can easily interface
with mainframe systems.

1

II. FEASIBILITY AND PERFORMANCE ASSESSMENT

A. Conceptual Approach
To achieve the long-term performance objectives, the following is required:
1) Identification of the fundamental physical processes that significantly affect the quantity, direction, and rate of radionuclide/hazardous waste movement;
2) Adequate understanding of the conditions and processes pertinent to the site under investigation;
3) Development of a systems model that relates the most important processes and conditions to the waste isolation criteria;
4) Assemblage and adaptation of the evaluation tools (predictive models and data analysis codes) with appropriate hardware equipment to allow best networking and interfacing system configuration;
5) Identification of data needs in terms of each model requirements and collection of data;
6) Quantitative or probabilistic description of scenarios of anticipated and unanticipated disruptive events (catastrophic or not) that can be used by the models for global and local site performance assessment.

The systems approach, inherent in the master code SITECAD, provides the optimal assemblage of the data analysis and predictive models (i.e. computer codes) into an overall systems model. This master code selects the best networking path to test one by one the design alternative of the engineered barrier in light of the anticipated site conditions.

B. Scientific Method of Analysis
Predicting waste disposal system performance over large spatial scales and long time frames indicates the need for mathematical models to adequately simulate the observed physical and chemical/nuclear processes, and to extrapolate the measured system behavior. Construction of conceptual models is a necessary condition to the development of more detailed mathematical models to assess the effect of key system parameters on system performance or design.

Because the critical processes in many instances are scale-dependent, the overall long-term performance analysis problem is best broken down into three subregions: (1) very near fields, i.e. between the waste and the backfill material boundaries for a

2

nuclear repository, and in the zone extending beneath the waste to the natural/engineered barrier for a hazardous wastes site, (2) near field, i.e. from the natural or engineered-barrier to the controlled-area (site boundary), and (3) far-field, i.e. from the controlled-area to the accessible environmental (rivers and springs, water supply wells, populated areas, etc.). This subdivision is developed from a recognition that some processes are more important at one scale than at another scale. Using this approach, mathematical models for each subregion can be developed that realistically portray the dominant physical processes, without neglecting less important effects. Thus, the objective of adequate predictive analysis is achieved in a practical manner (see ▪2▪).

C.Model Classification

Once the important chemical/nuclear and physical processes and associated parameters are identified, the model types can be identified. Not all model types require implementation in the form of sophisticated, numerical codes. Grosso modo, the models can be classified as follows:

  a) Groundwater flow in a porous medium (in general, suited for far field modeling of a nuclear waste repository (NWR), and for a hazardous waste site (HWS));

  b) Groundwater flow in a fractured system (used for near field simulation of NWR) ▪3▪;

  c) Thermal analysis (performed on waste package and engineered subsystem for a NWR);

  d) Waste package degradation in terms of probability of failure due to corrosion, radiation damage, and cracking of canister and hazardous waste container, dessication of natural clay, dissolution of artificial liner, etc.;

  e) Statistical quantitative analysis for direct release rates from nuclear waste canister or hazardous waste container or lining system;

  f) Deterministic and statistical analysis of radionuclide or hazardous waste release rate across the engineered or natural barrier;

  g) Mass transport study in the very-near field applied to radionuclides for NWR, and hazardous chemicals for HWS.

  h) Structural analysis for testing at various depth, design and construction;

  i) Groundwater velocity and path orientation in the far-field (both deterministic and stochastic). Examples of the relationships among the various processes, characteristics, and related models are discussed in ▪1▪ and ▪2▪. The computer codes are

3

assembled into a systems model based upon the
processes modeled and the appropriate coupling of
theses processes. Probabilistic analysis techniques
have been programmed into several of the computer
codes to permit a statistical analysis of system
performance to be evaluated. Combination of the two
types of models yields an interactive network of
SITECAD softwares easy to adapt to problem-specific
objectives. A schematic description of a SITECAD
application is given in Figure 1.

D. Selection and Characterization of Long-Term Scenarios

Identification, selection, and characterization of
postclosure disruptive scenarios (e.g., continuous,
intermittent, or sudden releases) define the credible
events, processes, and resultant hazardous/
radionuclide chemicals releases and pathways that
must be considered as the basis for site-suitability
selection and engineered system design. The basic
objective of this activity is to identify and
characterize, in detail, scenarios having a
reasonable probability of occurrence and sufficient
consequences to represent a meaningful design
concern.

The framework developed for disruption scenario
analyses uses guidelines, when existing, set forth by
regulatory agencies , or considerations provided by
experts and system estimation methods for
categorizing occurrence probabilities of site-
specific scenarios.

Implementation of the selection process is a
structured process. For purposes of selection of
´disruption scenarios for detailed parametric
characterization and release-risk analysis, a step-
by-step process of disruption scenario has been
defined and adopted. Seven steps have been
systematically addressed:

1) assembling of a comprehensive list of credible
site-specific disruption,
2) adoption of scenario selection criteria,
3) assessment of scenario occurrence probability, and
4) selection of an initial group of scenarios to be
analyzed in detail,
5) determination of specific parameters required for
credible scenario characterization,
6) collection and analysis of additional data,
7) completion of scenario consequence analysis, with
assessment of repository performance under scenario
conditions as measured by regulatory criteria.

E.Development, Documentation, and Validation of SITECAD

The preliminary development of computer models provides a description of the unique site characteristics, and information on key parameters. Because of the uncertainty in establishing an initial baseline for site performance analysis, the adopted method of analysis incorporates the use of both deterministic and probabilistic models. With the aid of currently known site characteristics, major deficiencies in system and subsystem performance are easily identified. The updating of postclosure performance assessment baseline will be performed on-line with major incoming information.

To check the reliability of the above codes, a three-step trouble-shooting approach is required:
 1) Stage of verification to ensure that the software product is free of errors by reviewing and/or testing the code;
2) Benchmarking to compare the numerical results from different codes.
3) Validation to demonstrate that the simulator represents physical reality, i.e. data recorded from laboratory and field tests. For stochastic models, more care must be taken to validate the prediction. SITECAD has a built-in software management plan which can be used to categorize, select, test and control computer codes for a HWS or NWR problem.

III.CONCLUSION

A description of the development and implementation procedure of SITECAD to hazardous waste sites has been presented. Examples of its application can be found in ∎4∎. Testing on nuclear waste repository problem is being documented. Quantitative analysis and results will be reported elsewhere in a future communication.
SITECAD is available for inquiry by writing to the authors.

VI.References

1) G.V. Abi-Ghanem, V.V. Nguyen, and H.O. Pfannkuch (1983) Practical Solution to Chemical Spillages and Groundwater Contamination, ARDI Publications, 425pp..
2) V.V. Nguyen and L. Lehman (1984) Interscale Transfer of Information in Nuclear Waste Repository Multibarrier Systems, NWWA 1985 Western Regional Groundwater Conference, Reno, Nevada, March 1985, 15pp..
3) V.V. Nguyen and G.V. Abi-Ghanem (1984) Adaptive Collocation Method for the Transport Problem Induced by Irregular Well Patterns, 5th International Conference on Finite Elements in Water Resources,Burlington, Vermont, June 1984, 9pp..

5

4) G.V. Abi-Ghanem and V.V. Nguyen (1983) Numerical Simulation and Performance Assessment of a Hazardous Waste Site Using Adaptive Optimization Filter, EOS Transaction, AGU Proceedings of 1983 Fall meeting, San Francisco, 64 (15): 714.

Figure 1. AN APPLICATION SCHEMATIC OF SITECAD TO SITE SELECTION

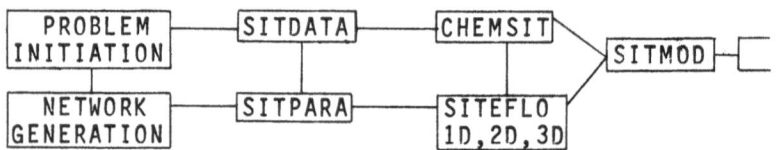

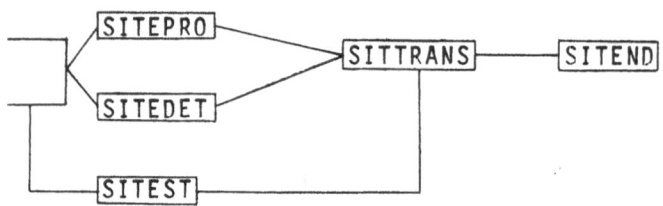

```
DATA  : data input
PARA  : parameter identification
CHEM  : chemical species indexing
FLO   : flow codes (FE,FD,BIEM,SPECTRAL ELEMENT,
        MONTE-CARLO, 2nd ORDER)
MOD   : deterministic (DET) or probabilistic (PRO)
        classification of simulation
TRANS: transport codes (FE,FD,BIEM,SPECTRAL ELEMENT,
        MONTE-CARLO, 2nd ORDER)
TEST  : inverse problem or verification package
        iteration
END   : digital output
```

AN APPLICATION OF SYSTEMS MODELLING IN STEEL PRODUCTION SCHEDULING

Jeff R. Wright and Mark H. Houck

School of Civil Engineering, Purdue University, U.S.A.

1. INTRODUCTION

Modern steel production is a complex operation with literally hundreds of separate, though operationally related processes. As a production industry with a high volume, total profits depend heavily on efficient operation on the margin. A key factor in efficient operation is the ability to schedule individual operations toward a common goal of optimal production.

Proponents and practitioners of systems engineering point proudly to a great many successful applications of analytical models to problems of production scheduling. These systems are typically well defined and governed by a fairly tight set of evaluative criteria. It seems only natural that production scheduling in a modern steel-making facility would profit from a systems perspective.

The focus of this research is the development of a systems methodology for improving one small process within the large number of activities. The process being studied is the Hot Strip Mill Operation and though the process is small relative to the magnitude of the total steel manufacturing process, its smooth and efficient operation is an important factor in overall plant profitability.

The function of the hot strip mill is to transform large steel bars into long coils of product. The coils are produced in a wide variety of lengths, widths and thicknesses, and have potentially different metallurgical properties. Coils are produced according to pre-defined specifications requested by the purchaser of the product. Each coil must meet, or nearly meet those specifications or be used as surplus and thus represent lost revenue.

The scheduling of bars through the hot strip mill is complicated by several factors. First, several different product types are produced in the mill. Each product type may require a specific type of rolling equipment, and may have specific quality requirements. Furthermore, the physical process of reducing the dimensions of bars of different product types may affect the quality of products differently. Second, the sequence of bars through the mill has a direct impact on the level and frequency of maintenance on the mill. This is significant because maintenance requires halting production. The greater the frequency of such maintenance, the higher the overall maintenance costs as well. Lastly, only a limited amount of storage exists at the mill. It is therefore important that production is scheduled in such a way that products move through the mill as rapidly as possible. Since each product has a specific target delivery date associated with it, this also must be considered in the scheduling of production.

The goal of this research was the development of an analytical capability that will help to improve the scheduling of production through a hot strip mill. Toward this end, two models have been developed; one an optimization model formulation which is extremely large and costly to solve but which captures many of the concerns of scheduling hot mill production; the other, an interactive heuristic model designed to evaluate existing production schedules, to attempt to make improvements to those schedules, and to provide a variety of graphical displays of schedule information. The heuristic scheduling model is the subject of this paper.

2. CONSIDERATIONS IN HOT STRIP MILL PRODUCTION SCHEDULING

On any given day, a steel mill is faced with the task of fulfilling a wide assortment of customer orders for steel products. Many of these orders require processing in the hot mill, an operation that, in essence, transforms large steel slabs or bars into coils of relatively thin steel sheets. A hot mill scheduling system is employed to expedite the completion of customer orders and to coordinate the various milling operations in an economical manner. The objective of scheduling is to take each order and assign it to a certain time slot in the milling process such that it is most economically efficient for the company.

Over the course of a campaign, —defined as 24 8-hour consecutive work shifts--each steel bar will be assigned to a specific location within the sequence of production necessary to fulfill all available orders. To accommodate the fact that different types of orders require different types of milling equipment, the campaign is split into smaller sections known as schedules. Each customer order will typically fall into one of six production categories. A schedule is a group of customer orders of the same product category, with the restriction that the total footage length of a schedule (as measured by the length of the final products) cannot exceed some maximum value. This limit is typically described by the total footage of steel that can be run through the hot mill without having to stop operations to change mill equipment.

Once a tentative campaign has been developed, the first step in fulfilling a customers order is to select a set of steel bars from the stock yard. These slabs of steel, which come in varying dimensions, weights, and chemical composition, are then heated to a temperature of several thousand degrees to prepare the bars for milling. After a critical temperature is reached, each bar is ejected onto a conveyer platform that leads to the rolling stands. The rolling stands perform the "squeezing" necessary to reduce a steel slab 12" thick into a sheet less than 1/2" thick. After proceeding through the roller stands, the bar emerges as a long sheet of steel that is then cooled by water spray and wound into a coil. The coils are then stored temporarily at the mill for either further processing or shipment to customers.

The primary goal when scheduling steel bars through the hot mill is to achieve economic efficiency. The attainment of this goal is directly related to the degree to which three distinct, but interrelated production objectives can be achieved: (1) the maximization of product quality; (2) the maximization of production rate; and (3) the optimization of product handling.

The first objective is necessary to ensure the reputation of the company's product. Clients may reject orders of poor quality, or they may not place orders in the future if the final product does not meet quality standards. Both of these actions will result in direct loss of revenue for the company.

The second objective is based on the premise that the hot mill operation is the limiting factor in the production process (as opposed to marketing, ingot production, warehousing, transportation etc.). An increase in the rate of production through the hot mill would translate directly into an increase in the overall plant production rate, which in turn would have immediate consequences on profits.

Objective (3) is concerned with the on time delivery of the finished product. If deliveries to a client are consistently tardy, there is a great likelihood that business will suffer, as the client may seek to obtain resources from another source. While late delivery of product may have serious consequences for the client, products delivered too early may also cause problems. Early delivery may impose significant storage and handling costs on the client, particularly if the client's production activity requires many inputs.

Product quality
The mill is made up of a long conveyor that moves the red-hot

steel slabs through five <u>roughing</u> <u>stands</u> and seven <u>finishing</u> <u>stands</u> (see Fig. 1). Each of these stands is comprised of a series of heavy rollers, capable of applying thousands of pounds of force to squeeze the steel into successively smaller and smaller gauges from one stand to the next. Because of the forces and temperatures involved, the rollers at each stand are subjected to tremendous stresses. This stress can cause the pitting, gouging and general wear that can be observed on roller as they are removed from service.

Wear on the rollers affects the quality of the product being processed through the mill. Uneven wear on any of the rollers, for example, can create abnormalities in the surface or shape of the finished steel product. For products that require critical surface specifications, even small scratches in any of rollers can translate into poor product quality and lost profits.

Although the rollers undergo tremendous stresses throughout any given schedule, these stresses, and the concomitant wear of the rollers, are not independent of the order in which steel bars are scheduled through the stands. One can imagine that if several thousand feet of steel were to be rolled to a finished width of 50" on a roller 100" wide, the inner 50" of the roller would receive a disproportionate amount of wear compared to the outer portions of the roller. Likewise, a schedule that seeks to distribute the product widths, beginning with the largest width and progressing to smaller widths from one bar to the next, will result in reduced wear on the roller at any given width. Therefore, by considering the width of the finished product when scheduling bars through the mill, it would appear that wear on the rollers could be reduced, the frequency of roll changes could be reduced, the quality of the product could be improved, and thus profits could be increased.

If width were the only consideration in reducing roller wear, then scheduling bars through the hot mill would be greatly simplified. An optimal schedule would consist of a series of steel coils of one product type progressing from the largest finished width to the smallest finished width. This schedule would lengthen the life of the rollers by reducing wear. The attainment of economic efficiency would be close at hand. Unfortunately, width is not the only parameter that affects the extent of wear on the stand rollers. As one would expect, the hardness of the steel being milled greatly influences the stresses and wear that rollers undergo. It requires greater force to squeeze a "hard" grade of steel to the same dimensions as it does to squeeze a "soft" grade of steel. This differential force translates to differential wear on the rollers. Likewise, the finished gauge of the product determines, to a great extent, the amount of pressure that must be applied to each steel bar. Since the number of rolling stands is constant, producing .061" gauge steel requires greater total force than producing .301" gauge steel. One should not expect the wear on the rollers to be

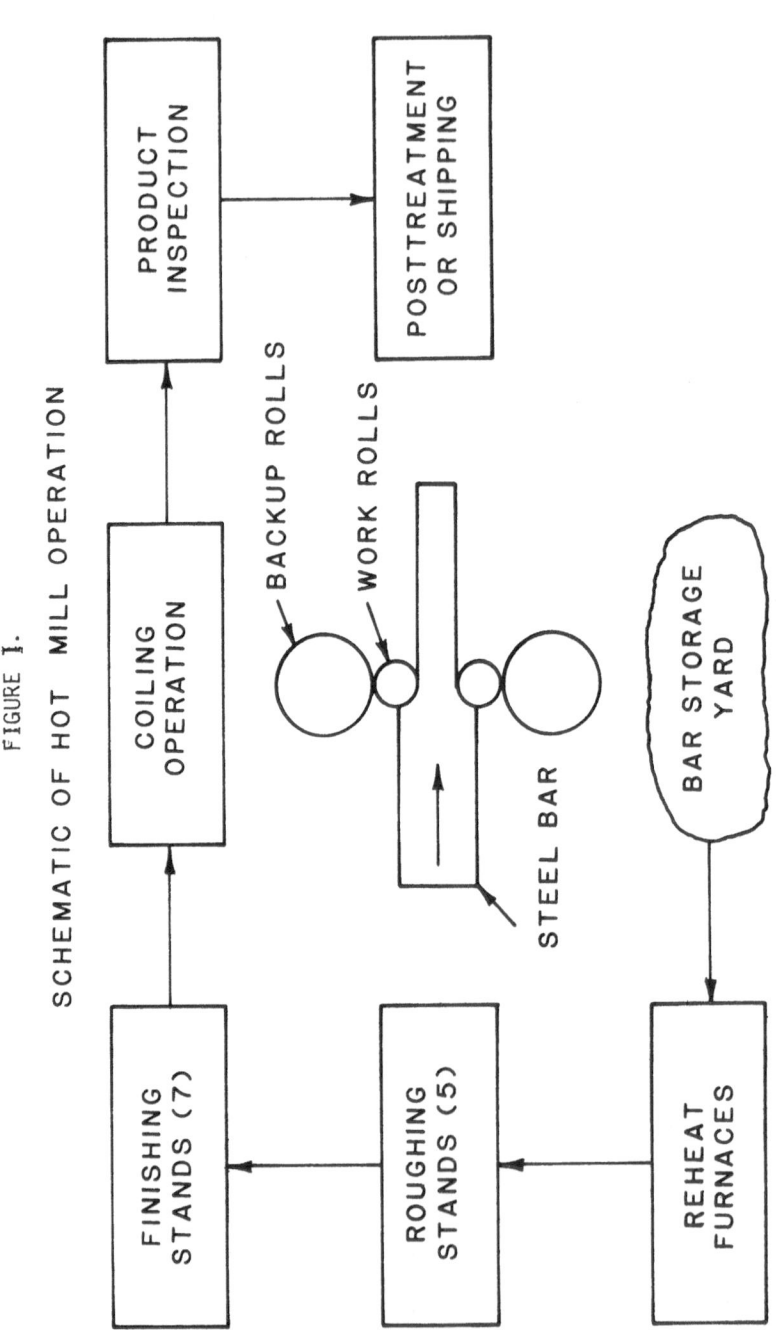

FIGURE 1.

SCHEMATIC OF HOT MILL OPERATION

equivalent for these processes.

Production rate

The attainment of economic efficiency is a difficult task at best. Each of the three objectives necessary to obtain economic efficiency conflict with each other to some degree. The maximization of product quality requires frequent changes of the hot mill finishing rollers. On the other hand, the attainment of a high production rate is facilitated by minimizing the number of roller changes. To deliver products on time, it may be necessary to schedule orders for rolling during times that would be less than ideal from the standpoint of product quality. In effect, the attainment of economic efficiency involves various tradeoffs between conflicting objectives, and the final scheduling decision must be based on a accurate assessment of these tradeoffs and the benefits and disbenefits associated with any given schedule.

To ensure product quality, it becomes necessary to change the rollers frequently, before wear translates into the loss of a quality product. Because the finishing rollers are critical in obtaining the exact product specifications, the finishing rollers are changed more frequently than the roughing rollers. It is at this point when the tradeoff between product quality and production rate becomes evident. In order to obtain a high quality product, it is necessary to change the finishing rollers frequently, but these frequent changes necessitate that production come to a halt for some period of time. To further complicate the situation, it becomes necessary to change rollers whenever there is to be a change in product type. Clearly, achieving high product quality comes at the expense of achieving a high production rate. If the assumption is made that the hot mill is the "bottleneck" or limiting factor in the production of steel, it becomes of paramount importance to speed up hot mill processing as much as possible. Thousands of dollars of profits are at stake every time the mill must stop operations to change rollers.

Product handling

Thus far, it has been argued that by observing certain scheduling rules, progress can be made towards improving product quality. By implementing a modified roll change policy, progress can be made towards improving production rate. Progress in both areas is desirable and consistent with the goals of achieving economic efficiency, although, as we have seen, the pursuit of one objective may conflict with the pursuit of another. In the case of product quality versus production rate, production rate frequently suffers from the necessity of making roll changes to ensure product quality. In the case of optimizing the handling of the final product, there may be a conflict with one or both of the other objectives.

When orders for a steel product are placed, there is typically a delivery date associated with that order. In order to maintain favorable client relations, it is assumed that the delivery of the final product on time is an important issue in scheduling orders through the hot mill. If deliveries to a client are consistently late, this will undoubtedly damage relations with that client and may result in the loss of a customer.

To compensate for this problem, orders for a customer could be filled many days before the actual delivery date, so that there would be plenty of time to deliver the product to the customer in case there were any unforeseen delays. On the other hand, if the order for a client were processed far in advance, there may be some problem with storage space, since warehouse space is limited. Plant managers would like to avoid taking up valuable floor space for orders that will not be delivered for several months. Therefore, plant managers must contend with another tradeoff, that of completing an order sufficiently early so as to avoid tardy deliveries, and completing the order as near the target date as possible to avoid occupying scarce floor space.

But the issue is even more complex. If a client had a unique order, perhaps requiring a special work roller, then scheduling the completion of that order close to the delivery date may require that the normal scheduling pattern be preempted temporarily just to fill that special order. If plant managers did not have to concern themselves with delivery dates, then several small orders of a special product could be clumped together into one complete schedule. However, if these same orders had delivery dates that were several months apart, attempting to process these orders all at the same time would result in part of the orders either taking up floor space while waiting for delivery, or being delivered to the client several months late.

Since delivery dates are an important factor in maintaining client relations, the scheduling of small and specialized orders at inconvenient times is a necessary evil. That is, it may considered more important to deliver a product on time than it is to delay the production of that product until additional orders become available, at which time an entire schedule of similar orders could be processed at once. In order to fulfill small specialized orders, it becomes necessary to make frequent and inefficient roller changes, all of which result in a decrease in the overall hot mill production rate.

Penalty Functions
Attempting to achieve economic efficiency through the pursuit of the three objectives outlined above is anything but straightforward. The tradeoffs that exist between each of the objectives makes it nearly impossible to find any one optimal schedule or scheduling algorithm. A decision is made only through the careful analysis of the tradeoffs involved in pursuing one course of action versus another. Defining the tradeoffs that exist between alternatives is an equally difficult task, but the more accurate the information regarding these tradeoffs, the better will be the

resulting decision.

One way of assessing the tradeoffs that exist between the various objectives is to assign penalties to those actions or parameters which are considered important in scheduling bars through the hot mill. These penalties can then be used to evaluate the relative desirability of one scheduling configuration versus another. For instance, tardiness penalties can be assigned to orders of steel that are not, or might not, be delivered to the client on time. Gauge penalties can be assigned to a sequence of coils whose gauge jump from one coil to the next is extremely damaging to the finishing rollers and which will result in reduced product quality. Roll change penalties can be assigned to those schedules which require frequent roll changes and that slow down the production rate. All of these penalties could be used in a model to evaluate which schedule produces the smallest penalty and contributes the most toward achieving economic efficiency.

Aside from width, gauge, and hardness, there are numerous other parameters of the finished product which determine the amount of wear on the rollers. Each parameter interacts with every other in such a complex fashion that predicting the amount of wear that a particular bar sequence will produce is an art in itself. A penalty structure has been developed, based on roller wear, that will assess the relative desirability of one steel schedule from the next. There are penalties associated with jumps (from one bar to the next) in width, gauge, and hardness. For example, a jump of five and one half inches in width from one bar to the next would be assessed a penalty of 20 points. Similar penalties would be assessed for jumps in gauge and hardness. For each schedule in a campaign, and for each campaign, these penalties are accumulated and will provide an overall indication of the relative desirability of that schedule and that campaign.

This discussion has served to describe the major issues and concerns with respect to scheduling production through a hot strip mill. It was shown that the order in which coils are produced can have a profound influence on product quality, cost of production and customer satisfaction. Each of these factors relates directly to overall plant profitability. And yet, the development of a scheduling policy which incorporates these considerations in a meaningful way is far from simple.

3. USING MINIMUM PENALTIES TO IMPROVE ROLLING EFFICIENCY

Recall from the quality of any particular schedule may be measured in terms of a penalty structure specified for various parameters. Thus, a procedure for scheduling coils through the hot strip mill may be evaluated in terms of this penalty structure. Assuming that the penalty structure is an accurate representation of "actual" concerns, these procedures or algorithms for scheduling production can be evaluated in a quantitative way. One scheduling algorithm is "better" than another if, on average, it schedules bars in such a way that total penalty is less.

Suppose that a schedule has been generated by a particular algorithm. That schedule can be represented by a two dimensional array consisting of a bar identification number and a number indicating the relative position of that bar in a schedule. With a given sequence of bars specified, and with the existing penalty structure, a "score" may be computed for that schedule as the simple sum of all penalties associated with that ordering. If a bar can be removed from its position in the schedule and moved to another position in the schedule in such a way that total penalty is reduced, one might argue that the original schedule has been improved. Since complete information exists about the penalty structure, this process can be accomplished using a trial-and-error process until no further improvement can be made.

Clearly, the quality of the resulting schedule, being driven by the penalty specification, assumes that the penalties are accurate and meaningful. If, for example, the penalty structure would be changed and the improvement process repeated, the resulting schedule might look considerably different from the one generated under the original penalty scheme. If some individual knowledgeable about he hot mill process were to compare the two schedules and make a judgement about which one is superior, he/she would in effect be making some statement about which penalty structure is better as well.

The heuristic model developed in this research implements the logic of the previous scenario. It attempts to improve a given schedule consistent with a given penalty structure to a degree specified by the user. The details of the algorithm which is called UMPIRE (Using Minimum Penalties to Improve Rolling Efficiency) will be discussed in the next section.

The UMPIRE algorithm

Referring to the flowchart of Fig. 2, a more detailed description of UMPIRE can be given. The first major step involves calling the EVALUATOR. This subroutine adds up the penalties associated with the scheduling of each bar. EVALUATOR will return either the penalties associated with each individual bar or the total penalty for a turn depending on the value of a parameter that is passed to it. The first time EVALUATOR is called, it returns a whole array of penalties -- width, gauge, hardness, etc. penalties -- for each bar.

These bars are then sorted in decreasing order according to their associated total penalty. This process is accomplished by a bubble sort subroutine. The array that contains the sorted bars consists of the bar number and its penalty. The bars are stored in this order so that the bar causing the most damage (i.e. the bar with the largest penalty) will be considered for

FIGURE 2.

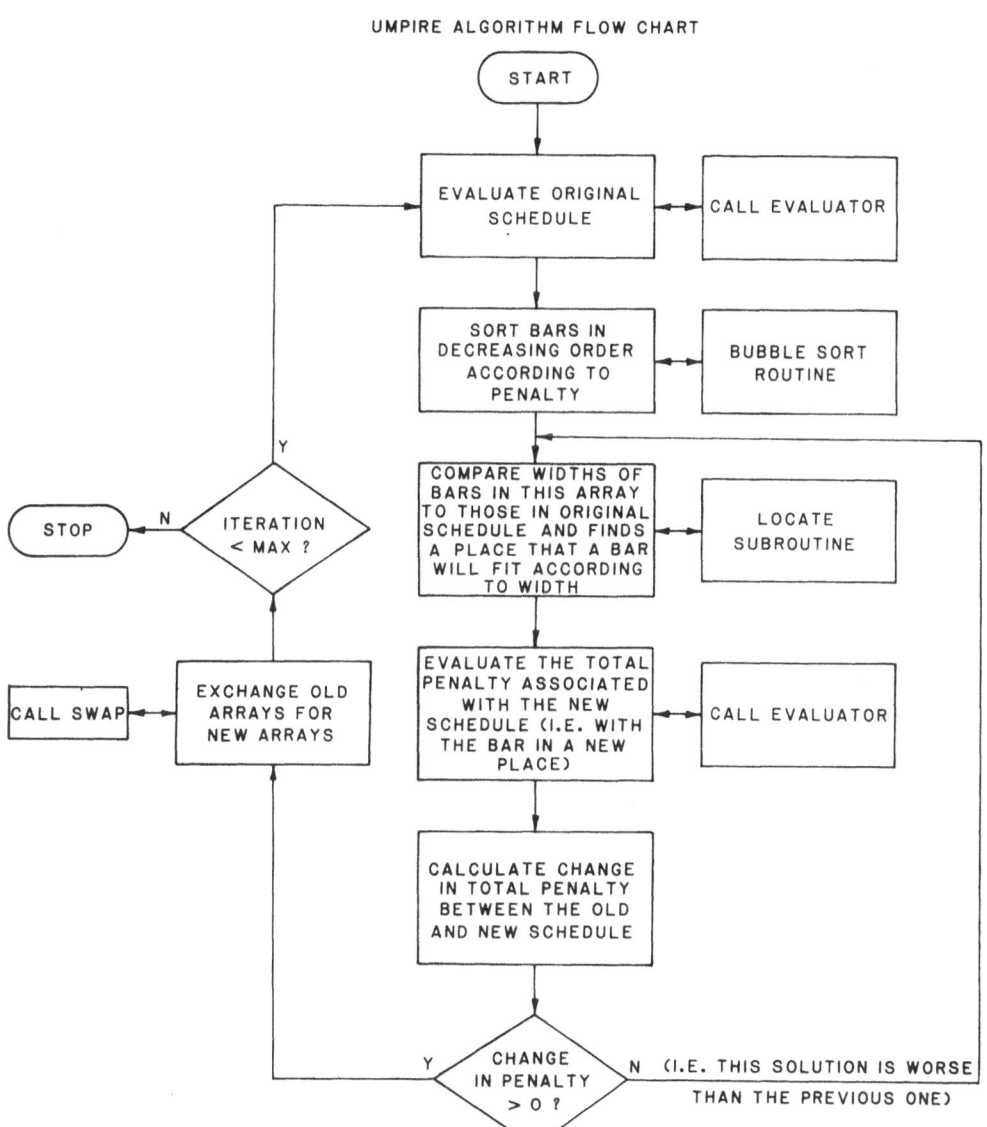

UMPIRE ALGORITHM FLOW CHART

moving first. The bars could be checked randomly or in the scheduled order, but the objective of the algorithm is to decrease the total penalty; therefore, the bar with the largest penalty is moved first.

Because it requires extra computer time, the evaluator is called as little as possible. In order to achieve this goal, each bar is not evaluated in every position. The total penalty is not calculated for the bar in a new position unless that bar meets the width requirement. These stipulations require that a bar not be placed in any position if its width is not between the widths of the preceding and the following bars or equal to one or both of their widths. This process of checking whether the bar being considered for moving meets the width specifications is carried out in the subroutine named LOCATE.

After LOCATE has found an acceptable position, the bar is placed in the new location temporarily. The new schedule is evaluated and the total penalty is computed. If the new penalty is less than the existing penalty, the new schedule replaces the old one. However, if the new schedule is not better, two possibilities arise. First, if there are any other possible positions for the bar, they will be checked to determine if an improvement in the schedule is possible. If, however, no suitable position is found, the bar with the next highest penalty is checked in all positions. This process continues until all the bars have been checked, or until a limit set by the user is reached.

In order to set a limit on the time that UMPIRE will run, the number of iterations that the heuristic will go through can be changed. An iteration is defined as one attempt to move one bar. After the desired number of iterations has been performed, a final schedule is produced and stored.

Because of the nature of UMPIRE, the final schedule will never be worse than the original schedule. The heuristic only moves a bar if it decreases the total penalty. Although this is a major advantage of UMPIRE, it also is a limitation. It is possible that two or more bars must be moved simultaneously to achieve a decrease in total penalty.

The heuristic could be revised so that two bars could be moved at once or so that an initial increase in penalty for one bar is tolerated in order to reduce the total penalty eventually. However, these revisions induce an excessive amount of computer time and space, which are exactly the original reasons for implementing this heuristic rather than an optimization model.

The sequence of schedules within a campaign

Suppose that all schedules within a campaign are passed through the improvement algorithm described in the previous section and that the penalty structure used for this task is "correct". Knowing the width of the widest coil within each schedule one could sequence the schedules through the campaign in such a way that the widest coil in a given schedule is no wider than the widest coil in the previous schedule. This would comply with the desired campaign width profile discussed. However, it is possible that a reduction in total campaign penalty could be further reduced by moving a bar out of one schedule and into another; again, driven by the penalty structure. Thus, the allocation of bars to schedules in a satisfactory manner is a key element in a comprehensive scheduling algorithm.

The allocation of bars to schedules has not been addressed in the development of UMPIRE. The primary reason for this is that there is not a clear penalty structure specified for inter-schedule parameters. In addition, since a heuristic approach is the focus of this research, the two problems are considered separable. That is, the problem of allocating bars to schedules is considered separate from the one of allocation bars to positions within a schedule. The inter-campaign allocation of coils to schedules is, however, an important aspect of overall production efficiency and should be addressed in future research.

4. MODEL IMPLEMENTATION

The improvement algorithm described in the previous chapter is the centerpiece of a rich set to FORTRAN subroutines which make up the UMPIRE model. The model is designed to be used in an interactive decision-making environment and as such, is properly viewed as a tool which can be used by steel-making professionals toward improving hot strip mill production. In this section, the implementation of the model will be discussed focusing on the overall structure and function of the computer code.

Model Structure and Function

The UMPIRE model can be thought of as a series of processes and files systematically linked through an interactive structure. Files are data representations, either on storage media or in memory which may be used, modified or displayed as the model is being executed. Processes are activities which act on those data representations according to a rigid set of rules.

Two of the processes (the pilot read filter and the penalty reader) provide an input function when the model is first executed. Raw data about the bars to be produced during the current schedule is passed through the read filter where those data items not pertaining to the scheduling activity are stripped away. At the same time, the penalty structure is read from a separate file and all data arrays are initialized. The result of this input phase of the model is a representation of the original data (called PU TURN).

Each data record in PU TURN contains information about each bar or coil which is to be produced during the current schedule. The position of the record within the file corresponds to the position of that bar in the schedule. In theory, the ordering may be random, such as the order in which the bars were allocated to the current schedule. In actuality, the order should be the result of some prior scheduling algorithm, such as the one currently in use at the mill. In either case, this ordering is considered to be an original schedule.

Three separate processes may be applied to this reduced data representation. First, the schedule may be evaluated which means that the ordering of the bars will be compared to the penalty structure and width, gauge, hardness, and total penalties computed. Here information about each bar is displayed adjacent to the number indicating its position within the schedule. Between each record, penalty points are displayed. These are the points in each penalty category resulting from assigning the following bar to its current position. The column totals thus indicate the total penalty in each category for the current schedule.

A second option is to apply the improvement algorithm to the current schedule as described in the previous section. Since the improvement algorithm makes heavy use of the evaluation process, the two are shown linked in the schematic. If improvement is made, a new schedule is produced.

The third process that may be elected is to request a graphical representation of the current schedule. Several options for graphical display are available.

While these three processes have been described as separate activities, actual use of UMPIRE allows each to be used as needed. The original schedule may be graphed prior to evaluation or improvement, for example. At present, graphs are stored in one of 10 different files for later examination.

The use of UMPIRE as discussed above presumes that the penalty structure has been pre-specified and is consistent with the current wisdom of hot mill scheduling. A major feature of the model, however, is the ability to change the penalty structure.

The quality of the schedule produced using the UMPIRE depends on the validity of the penalty structure. Since, the production objectives of minimizing width, hardness and gauge jumps are in conflict to some degree, one might question the accuracy of the specified penalty structure. What if the penalty structure does not accurately represent the objectives of the scheduling process? Perhaps the most important feature of the UMPIRE model is the ability to modify any given penalty structure and repeat the analysis described above.

Suppose, for example, one decides that the actual impact of hardness jumps is not well understood and that, say, hardness jumps at the low end of the hardness scale are relatively worse than previously thought.

The penalty changing process within UMPIRE provides for this kind of analysis. Penalties for all parameters may be changed and new penalties may be added in an iterative fashion. The improvement algorithm may be re-applied to the schedule, and the changes may be compared graphically as described previously.

Thoughts on Model Use

More research needs to be done as to the use of UMPIRE towards the development of an efficient hot strip mill scheduling procedure. The model is best viewed as a tool which will not dictate how production should be scheduled, but will let steel production experts test their ideas and intuitions about the relationships between physical and metallurgical factors contributing to overall product quality.

One way in which this might be accomplished it to iteratively schedule production according to a specified penalty structure such as the one currently being used. The graphics capability of UMPIRE can be used to display schedule information explicitly. By changing the penalty structure, new production profiles can be produced in this manner. If production personnel are able to judge whether one profile is superior to another, they are implicitly making a judgement about the accuracy of the penalty designations. By using such a procedure systematically and over a period of time, there might be a collective convergence on a best penalty structure. Once this is identified, the improvement algorithm might actually be used to develop production schedules.

REFERENCES

Fabian, Tibor, (1958) "A Linear Programming Model of Integrated iron and Steel Production, Management Science, Vol. 4, No. 4, pp. 415-449., July.

Mackulak, Gerald T., et al., (1977) "Advanced Scheduling in Steel Production," Report No. 92; Purdue laboratory for Applied Industrial Control, Purdue University Schools of Engineering, September.

Maggio, Ralph A., (1983) "Simulation and Geographical Determination of an Integrated Steel Mill," Modeling and Simulation, Vol. 14, Part 4, pp. 1099-1107, April.

Sato, Shuzo, et al., (1977) "Development of Integrated Production Scheduling System for Iron and Steel Works," International Journal of Production Research, Vol. 15, No. 6, pp. 539-552, July.

BEASY Boundary Element Analysis SYstem

C.A. Brebbia, D.J. Danson and J.M.W. Baynham

CM BEASY Ltd.

ABSTRACT

The BEASY code is based on the Boundary Element Method (BEM) of
Analysis which offers important advantages over the classical
finite element approach. In BEM only the boundary needs to be
discretized thus reducing by one dimension the data required to
run a problem. Direct boundary element solution of the type
used in BEASY gives displacements and stresses (or temperature
and fluxes, etc.) with the same degree of accuracy, which makes
the technique well suited to problems - for example involving
stress concentration - where reliable and highly accurate results
are required. Boundary elements are also well adapted to
problems with infinite domains.

    BEASY can solve potential - including temperature - and
elasticity problems. The code has its own pre and postprocessing facilities and in addition interfaces with well known
modelling systems such as PATRAN.

INTRODUCTION

Over the past decade the finite element method (FEM) has become
established as a tool for the solution of a wide variety of
problems in engineering.

    The FEM may be seen as a method of solving problems for
which the phenomenon under consideration obeys known differential
equations within the domain. In the FEM the domain is discretized
into a number of elements in each of which the solution of the
governing equations is approximated by some functions which are
required to satisfy the boundary conditions. A set of equations
is then created which forces the solution at the various (nodal)
points to be such that the overall solution is the best approximation allowed by the chosen functions.

    An alternative approach is to use functions which satisfy

the differential equations in the domain but not the boundary conditions. The boundary may be divided into elements and the solution is assumed to vary in some manner within these elements. A set of equations may then be formulated in terms of nodal values, with the nodes this time only on the boundaries. The solution of these equations forms the best solution compatible with the assumptions of boundary value variation along the elements on the boundary. The attractions of such an approach are obvious. Only the boundary needs to be discretized thus reducing by one dimension the long list of nodal coordinates and connectivity tables which make the finite element method so tedious. The idea is the basis of the method known as boundary elements [1][2]. Because of the type of influence functions used, boundary elements are also well adapted to problems with infinite domains.

Because of its advantages the mathematical formulation of the boundary element method (BEM) is more complicated than in the FEM. Nevertheless the BEM has been applied successfully to problems in potential theory (including temperature), elasticity, plasticity and time dependent problems such as those governed by the diffusion and the wave equations.

2. GENERAL ARCHITECTURE OF BEASY

The general architecture of BEASY is shown in figure 1. BEASY includes pre and post processing modules and six independent modules for the solution of potential problems - including temperature - and linear isotropic stress analysis. The box labelled OPTIONS covers various more advanced applications of the BEM such as: elastoplasticity, time dependent problems and wave propagation.

BEASY can be used to solve two dimensional, axisymmetric and three dimensional problems, and in the latter case one dimensional elements may be used for a certain range of problems. The elements are illustrated in figure 2.

The constant and linear elements have linear geometry described by linear shape functions and the quadratic elements have fully quadratic geometry described by quadratic shape functions. The elements are called discontinuous because the nodes for the unknowns are situated within the element body rather than on the edge of the element. The reason for choosing this type of element is mostly practical. It is very easy to mix discontinuous elements putting quadratic elements in areas of rapidly varying stress (or potential) and linear or constant elements elsewhere. It is also not necessary for the points defining the boundary element mesh (called mesh or geometry description points - and which would be nodes if continuous elements were used) to be common to several elements. Figure 3 shows a typical discretization for a bearing cap which is a perfectly valid mesh for BEASY but which would be difficult to

handle if continuous elements were used. Figure 3 was produced using PATRAN. Using this technique there are no problems
of "fanning out" elements from areas of high element density.

The 3D quadratic element, which has numerous practical
applications, is a complete bi-quadratic element containing nine
nodes as opposed to the more common incomplete 8-node quadrilateral. The reasons for choosing a 9-node element are not only
that it contains all the terms for a full bi-quadratic expansion
but also because the classical BEM is a collocation technique
with the nodes as collocation points and the use of a 9-node
element gives rise to an even pattern of collocation points.

Options include some bar elements for potential problems
with quadratic variation of unknowns.

Loadings and Boundary Conditions

The boundary conditions available in the potential modules are
prescribed potential or flux density, or a linear relation
between the two (generally called a heat transfer boundary
condition because it is commonly used in thermal analysis).
Non-linear boundary conditions as specified by clients for a
particular application are easily modelled and this has been
done successfully in the field of cathodic protection where the
potential and flux (current) density on the cathode are related
by the polarization curve. The 2D and 3D modules can also model
concentrated point and line sources which is a useful feature
for many applications.

The stress analysis modules allow for prescribed displacement or loads or spring boundary conditions. The boundary
conditions may be entered either in the global coordinate system
or in a local system, one of whose axes always coincides with
the normal to the boundary surface. This local system is not
only useful when applying boundary conditions of a single type
(e.g. when specifying an internal pressure in a spherical
pressure vessel) but is absolutely essential when specifying a
mixed type of boundary condition such as sliding, where the
displacement is prescribed normal to the boundary and the loading is prescribed tangential to the boundary. Problems with
gravitational or rotational loading or problems where the
stresses are due to steady state thermal loading may also be
analysed. These problems involving body forces may be analysed
quite simply [3]. In the case of thermal analyses it is
necessary first to solve the potential problem to obtain the
temperature and fluxes at the boundary nodes and the temperatures
at any internal points. This information is then fed into the
stress analysis module which then calculates the thermal stresses.
Exactly the same data file may be used for both analyses.

Analysis Steps

BEASY carries out a typical Boundary Element analysis in six
distinct steps (see Table 1). The analysis may be started or
stopped at each step.

Step	Input Required	Computation	Output Generated
1	Data file	Check on data	
2	Data file	Formation of influence matrices	Influence matrices
3	Data file Influence matrices	Application of the type of boundary conditions to form the system matrix	System matrix $\underset{\sim}{A}$ (Part of system of equations $\underset{\sim}{A}\ \underset{\sim}{x} = \underset{\sim}{d}$)
4	Data file System matrix	Reduction of the left hand side of the system of equations	Reduced left hand side
5	Data file Reduced left hand side	Application of the magnitude of the boundary conditions to form the right hand vector $\underset{\sim}{d}$ Reduction and back-substitution to obtain the boundary solution	Boundary solution
6	Data file Boundary solution	Computation of results at internal points	Results at internal points

Table 1   BEM Steps

BEASY carries out some comprehensive checks on the data. If clients request the addition of more sophisticated checks then these can be provided.

The boundary element method results from applying the techniques of finite element discretization to the boundary integral formulation of the problem. This results in the formation of influence matrices (usually $\underset{\sim}{H}$ and $\underset{\sim}{G}$) which describe the behaviour at each node due to unit excitation at each node (Fig. 4). The influence matrices $\underset{\sim}{H}$ and $\underset{\sim}{G}$ are related by the equation

$$\underset{\sim}{H}\ \underset{\sim}{u} = \underset{\sim}{G}\ \underset{\sim}{p} + \underset{\sim}{b}$$

where $\underset{\sim}{u}$ is the vector of nodal boundary potentials or displacements
       $\underset{\sim}{p}$ is the vector of nodal boundary fluxes or loadings
and $\underset{\sim}{b}$ is a vector which results from sources within the problem or from body forces.

The H and G matrices are analogous to the stiffness matrix which one õbtains when using the Finite Element Method (FEM). The fact that there are two matrices and not one is a consequence of the mixed character of BEM solutions, i.e., one works with potential and fluxes, displacements and stresses, etc. This mixed

approach makes BEM results generally more accurate than FEM
solutions.

Application of the type of boundary conditions enables the
elements of the matrices to be rearranged so that we may write

$$\underset{\sim}{A} \underset{\sim}{x} = \underset{\sim}{B} \underset{\sim}{f} + \underset{\sim}{b}$$

where $\underset{\sim}{A}$ is the system matrix
$\qquad \underset{\sim}{x}$ is the vector of unknowns
$\qquad \underset{\sim}{B}$ is the complementary matrix
and $\underset{\sim}{f}$ is a vector of prescribed boundary values.

The system matrix $\underset{\sim}{A}$ and the complementary matrix $\underset{\sim}{B}$ are
stored on disc.

Gauss elimination is now applied to the system matrix $\underset{\sim}{A}$.
This reduction of the left hand side also consumes a large proportion of the total run time for a typical problem. The reduced
matrices are stored on disc.

Not until this stage is the vector b due to sources/body
forces evaluated and the right hand side vector $\underset{\sim}{d}$ calculated
from the equation

$$\underset{\sim}{d} = \underset{\sim}{B} \underset{\sim}{f} + \underset{\sim}{b}$$

The final equations may be solved for the vector of unknowns x.
BEASY uses an out of core solver when necessary to enable large
problems to be run efficiently on quite small machines.

The final step calculates the values at internal points
using the boundary solution just obtained.

The analysis in the manner described above minimizes the
run times required for repeated analyses. As the influence
matrices are dependent only on the mesh geometry and material
properties, then having designed a satisfactory mesh it is only
necessary to create the influence matrices once. The analysis
for subsequent boundary conditions need only be restarted at
step 3 if the type of boundary condition has been altered, e.g.
a prescribed displacement is specified where previously the
loading was specified. If only the magnitude of the boundary
condition is altered, or new sources/body forces added then it
is not necessary to repeat steps 3 and 4 but the run may be
restarted at step 5. Thus the time consuming steps of forming
the influence matrices and of decomposing the system matrix are
avoided.

Once the boundary solution is obtained the solution at internal points is calculated by a fairly simple procedure. No
further equation solution is required. If after inspection of
results the user decides more information is required within the
boundary then only step 6 need be repeated.

Symmetry

Symmetry is handled by the simple expedient of reflecting the boundary about the plane of symmetry and continuing the boundary integration around the reflected part of the structure. By making use of the fact that not only the geometry but also the potentials, fluxes, displacements and loadings are symmetric the number of equations is not increased. Indeed it is reduced by the fact that no elements are required on the plane of symmetry. The time taken to compute the influence matrices is increased but the extra cost is nearly always outweighed by the reduced number of man-hours taken up in data preparation.

Zones or Subregions

Although BEASY does not enable the user to model problems with continuously varying material properties such as Young's modulus or conductivity piecewise constant properties may be handled by considering each part of the problem as a boundary element zone or subregion. This is a very useful facility as it may also be used in certain types of problems to reduce the total run time as its effect is to start to band the influence matrices.

BEASY Options

The options in BEASY cover a number of more advanced BEM applications. They each have only one element type and cannot handle zoned problems. They are:-

BETA2D   Solves the diffusion equation in two dimensions. Its main application is in time-dependent thermal calculations. Uses a conforming linear element.

BETAAX   As BETA2D but for axisymmetric geometries. [4]

BEPLAS   Solves problems in elasto-plasticity in two dimensions.

BEPLAX   As BEPLAS but for axisymmetric geometries. [5]

BERPOT   Solves the scalar/potential wave equation in three dimensions. Has been used to analyse the transient response of liquids due to an explosion at a point [6]. Uses a non-conforming complete biquadratic quadrilateral element.

3. APPLICATIONS

i) Connecting Rod

Figure 5 shows the BE mesh for stress analysis of a connecting rod when subjected to a load from a crankshaft. Notice that no elements are required on the plane of symmetry. This example is included to illustrate how easy it is to create the necessary data using the BEASYG preprocessing facilities.

The necessary commands to create the mesh and loadings are given below.

```
LE 3                                 BC 3,3,4,15,1
SX                                   BL 4,4,5,5
BP 1,0,0                             BC 5,5,6,16,2
BP 2,35,0                            BL 6,6,7,1
BP 3,35,47.803                       BL 7,7,8,3
BP 4,37,56.75                        BL 8,8,9,2
BP 5,55,95                           BC 9,9,10,17,2
BP 6,63.759,100.56                   BC 10,10,11,14,6
BP 7,67,100.56                       BC 11,12,13,14,11
BP 8,67,144.29                       ZE 2100
BP 9,65,754,155                      ZP 0.28
BP 10,52.846,159.64                  ZI 1,4.135,66
BP 11,0,187.5                        ZI 9,37.218,66
BP 12,0,167.5                        ZI 10,42.723,144.29
BP 13,0,82.5                         ZI 14,32.145,144.29
BP 14,0,125                          EP 29,-1,-7.802,-7.705,-7.775
BP 15,56.1,47.803                    EP 30,-1,-7.705,-6.947,-7.424
BP 16,63.759,90.88                   EP 31,-1,-6.947,-5.533,-6.293
BP 17,65.754,175.7                   EP 32,-1,-5.533,-3.522,-4.586
BL 1,1,2,2                           EP 33,-1,-3.522,-1.231,-2.442
BL 2,2,3,4                           PD 1,-1,0,0
```

ii) Corrosion Protection System for an Offshore Platform

Fig. 6 shows the boundary element mesh required for the analysis of the impressed current corrosion protection system for an off-shore platform to be operated in the North Sea. The requirement is to solve Laplace's equation in the infinite domain of the seawater and the designer wishes to know the current density on the cathode, which is the hull of the platform. Since the boundary of the seawater is the platform itself this example is ideally suited to the BEM. There are three planes of symmetry in the problem. Two of them are genuine planes of symmetry. The mesh shown in Fig. 6 represents only one quarter of the total problem. The third plane of symmetry is put on the sea surface to enforce the boundary condition $\partial u/\partial n = 0$ on the sea surface, which saves having to put elements there. There are also some elements (not shown in Fig. 6) a large distance from the platform. These elements were used to enforce the boundary condition required by the designer of $\partial u/\partial n = 0$ at infinity. Without these elements the BEM would automatically enforce a boundary condition of $u = 0$ at infinity which is not what the designer desired. It should be noted that this mesh at "infinity" is not connected to the mesh on the structure itself. The smallest elements used in the problem had dimensions of 460×45 mm (on the anodes) and the largest had dimensions of 30×30 m (on the "infinite" boundary). The problem was run on an early version of BEASY which only had constant elements, and was analysed using 653 elements. If it were to be re-analysed today the number of elements could be considerably reduced by using quadratic elements. Several analyses were performed as the designer not only wanted results for the complete system but also wanted to know what would happen if various combinations of anodes were switched off. The ability to restart the analysis at various intermediate points was very useful here. Current intensity contours near a typical K-joint with protective anodes are shown in Figure 7.

iii) Crankshaft Analysis

Figure 8 describes the mesh used to analyse a typical crankshaft for an automotive manufacturer. The object of the analysis was to calculate the stiffness of the shaft section. Notice that two planes of symmetry have been taken into consideration. The rest of the external surface has been discretized into 51 quadrilateral elements with 1377 unknowns. The module BE3DTE was used in this analysis, the data plots were done using PATRAN for which an interface with BEASY is now available. A plot of the displaced shape (figure 9) was also obtained using PATRAN.

Because of the way the example was run CPU times were not available. More recently another example was run in a CRAY-1 machine consisting of 67 elements, i.e. 1809 unknowns and its solution took 15' of CPU time.

iv) Bearing Cap Analysis

Figure 3 shows a mesh used to investigate behaviour of a bearing cap under load from the bearing. Again BE3DTE was used and figure 10 (produced using PATRAN) shows contours of direct stress.

v) Turbine Disc

Figure 11 shows the analysis of transient temperature distributions in a turbine disc. The initial temperature of the turbine disc is 295.1°K and the values of the thermal conductivity, density and specific heat are $5W/m^{-1}{}^{\circ}K^{-1}$, 8221 kgm^{-3} and 550 JKg$^{-1}{}^{\circ}K^{-1}$ respectively. There are 18 different zones along the boundary each with a different set of prescribed values for the heat transfer coefficient and the temperature of the surrounding gas and their time variation at one of such boundary zones is shown in Fig. 11. Note that the mathematical representation of the heat transfer coefficient implies the use of mixed boundary conditions of the type $\alpha u + \beta q = \gamma$. No special difficulty is associated with this implementation.

For comparison purposes a FEM analysis employing 71 quadratic isoparametric elements and 278 nodes was also carried out. The BEM discretization employed 90 linear elements and 106 nodes (there are 16 double nodes to allow for the discontinuities on the boundary data at the intersection of boundary zones). A stepwise linear variation was prescribed for the boundary temperature. For the boundary flux it was assumed to be linear or quasi-quadratic according to the variation of the heat transfer coefficient and external temperature within each step.

BEASY results (isothermals) at a typical time are plotted in Fig. 11. The BEM and FEM (not shown) results were in excellent agreement.

It is important to point out that the BEM results were obtained using a novel approach of referring the integral equation always to the initial conditions. As the initial conditions

are usually everywhere zero this means that one is always
solving a boundary problem, i.e., only boundary integrals need
be computed. This technique presents important advantages for
unbounded domains. The time dependent option of the BEASY
program can use internal cells within the domain and the above
technique of referring all variables to the boundary. The time
dependent codes can be used for the solution of two-dimensional
and axisymmetric potential problems.

4. CONCLUSIONS

BEASY is a comprehensive commercial application package of the
BEM. It is being used by many large industrial companies who
are finding it a useful tool in their design offices.

Its chief advantages over FEM packages are

i)   Only having to discretize the boundary greatly eases data
      preparation. This is in the authors' and users' opinion
      the greatest advantage of the method. With computing costs
      still declining and engineers' time becoming more expensive
      the saving in engineers' time is far more significant than
      savings in machine time. Also, engineers welcome anything
      which relieves them of the dreary chore of data preparation
      and leaves them free to concentrate on more important tasks.
      Even more fundamental is the fact that analysis invariably
      lies on the "critical path" in the design and production
      process and any tool which can shorten the "turn-around"
      time through the analysis office can bring forward the date
      of completion of the project. This in turn has very sig-
      nificant economic effects, particularly important in a
      competitive world. It is often said that modern mesh
      generators can make FEM data as easy to prepare as BEM data.
      However, it is the authors' opinion, based on our constant
      visits to industrial companies many of which have invested
      heavily in FEM packages, that mesh generation is still a
      major problem.
ii)   The ability to handle infinite domains. A surprising
      number of problems fall into this category and the diffic-
      ulty of using FEM for this type of problem is obvious.
iii) The reduced number of degrees of freedom needed to analyse
      a typical problem means that large problems can be handled
      on small minicomputers.
iv)   Results at internal points are obtained only at points
      requested by the user. As the user is rarely interested
      in the full field solution (and is in fact often interested
      only in the boundary solution) this is a distinct advantage.

All these advantages point to the importance of boundary
element methods and the need of providing industry with adequate
software. BEASY is the first comprehensive boundary element
package available to the practising engineer.

REFERENCES

1. BREBBIA, C.A. "The Boundary Element Method for Engineers",
   Pentech Press, London, Halstead Press, New York, 1978:
   second printing 1980.

2. BREBBIA, C.A., J. TELLES and L. WROBEL "Boundary Element
   Techniques - Theory and Applications in Engineering"
   Springer Verlag, Berlin, NY, 1984.

3. DANSON, D.J. A Boundary Element Formulation of Problems in
   Linear Isotropic Elasticity with Body Forces. Boundary
   Element Methods. Proceedings of the Third International
   Seminar, Irvine, California, July 1981. C.A. Brebbia
   (Editor).

4. TELLES, J.C.F. and C.A. BREBBIA "Elasto-Plastic Boundary
   Element Analysis", in "Non Linear Finite Element Analysis
   in Structural Mechanics", Wunderlich, W., Stein, E. and
   Bathe, K.J. (Editors), Springer Verlag, 1981.

5. WROBEL, L. and C.A. BREBBIA "Time Dependent Potential
   Problems", Chapter in "Progress in Boundary Elements,
   Volume I" Pentech Press, London, Halstead Press, New York,
   1981.

6. GROENENBOOM, P.H.L. The Application of Boundary Elements
   to Steady and Unsteady Potential Fluid Flow Problems in
   Two and Three Dimensions. Boundary Element Methods.
   Proceedings of the Third International Seminar, Irvine,
   California, July 1981. C.A. Brebbia (Editor).

This paper has also been published in the Finite Element
Systems Handbook.

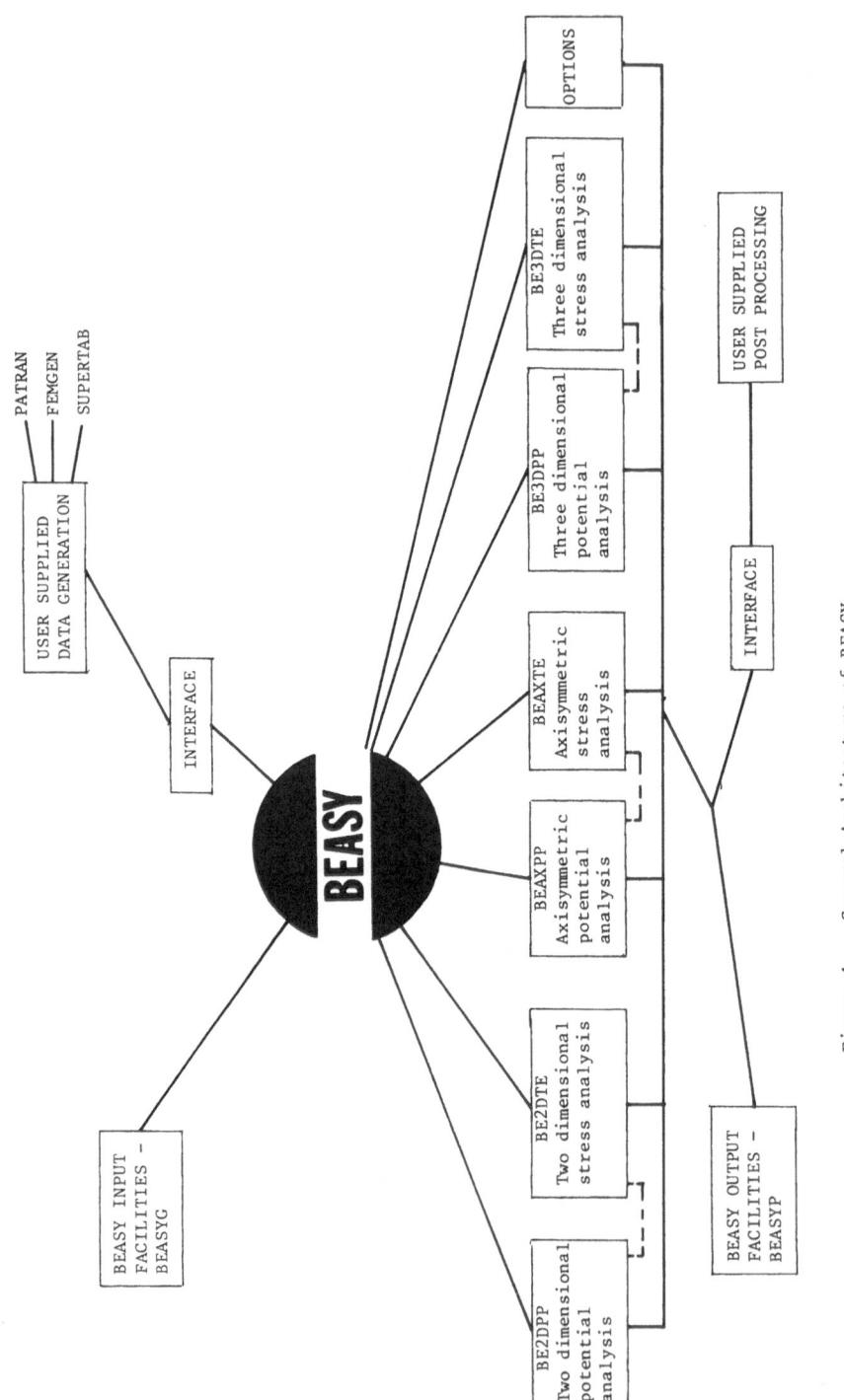

Figure 1   General Architecture of BEASY

Problem Type / Element Type	Two Dimensional and Axisymmetric problems		Three Dimensional Problems	
	Geometry Description	Nodal Unknowns	Geometry Description	Nodal Unknowns
CONSTANT				
LINEAR				
QUADRATIC				

Figure 2 Main Element Types

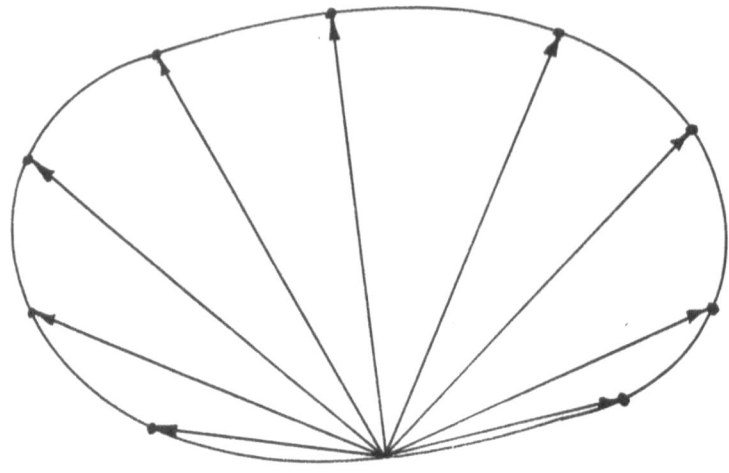

Figure 4   Influence Function

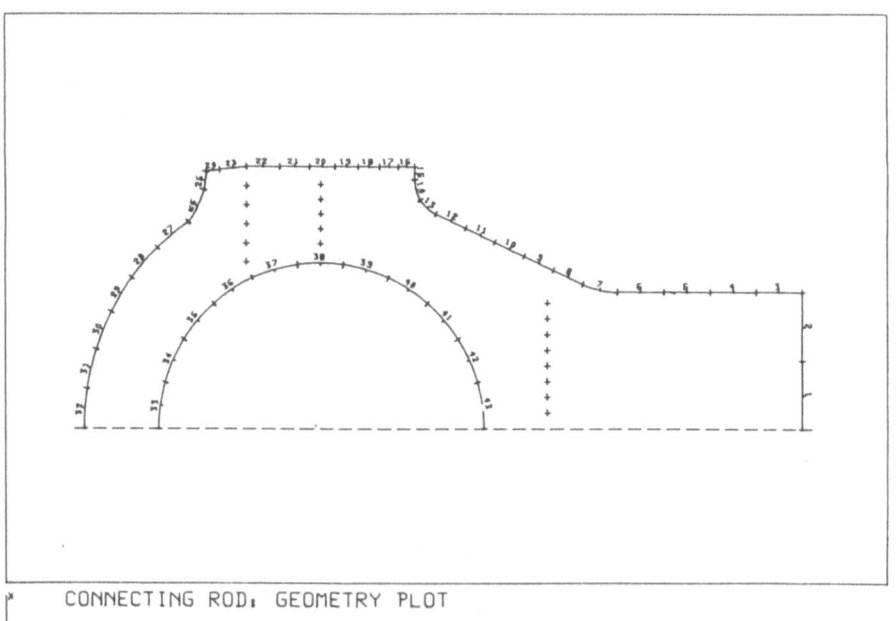

CONNECTING ROD: GEOMETRY PLOT

PLOT   1.20 ,1   CMC/BEASY

Figure 5   Connecting Rod: Discretization

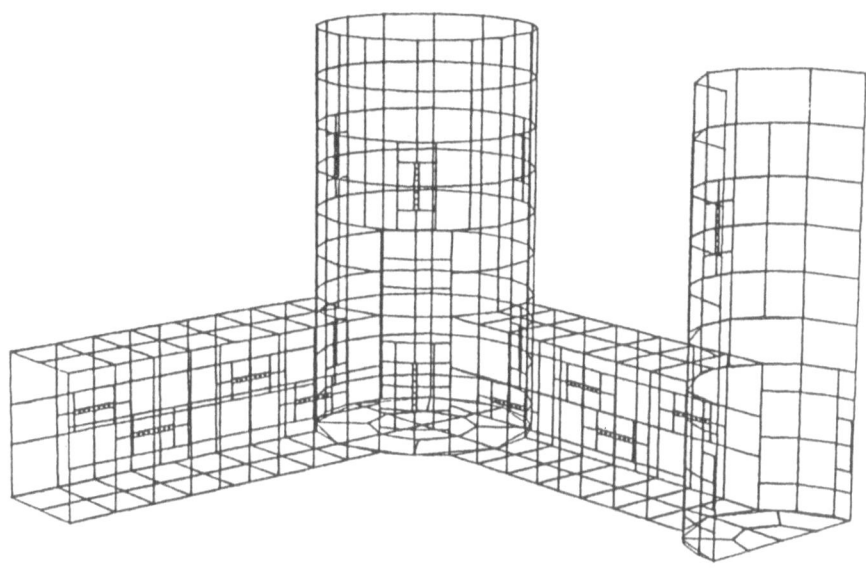

Figure 6   BE Mesh for an Offshore Platform

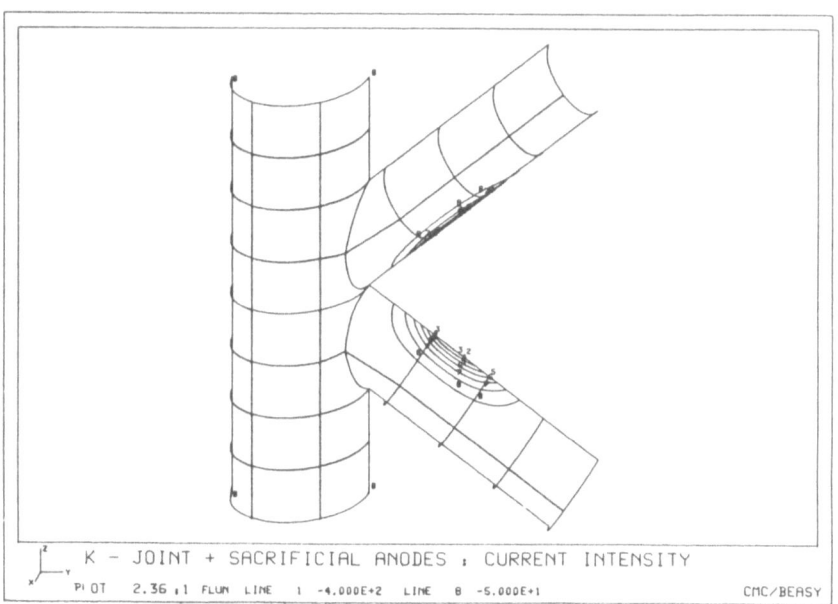

Figure 7   Contours of Current Intensity on K-Joint
            with Anodes

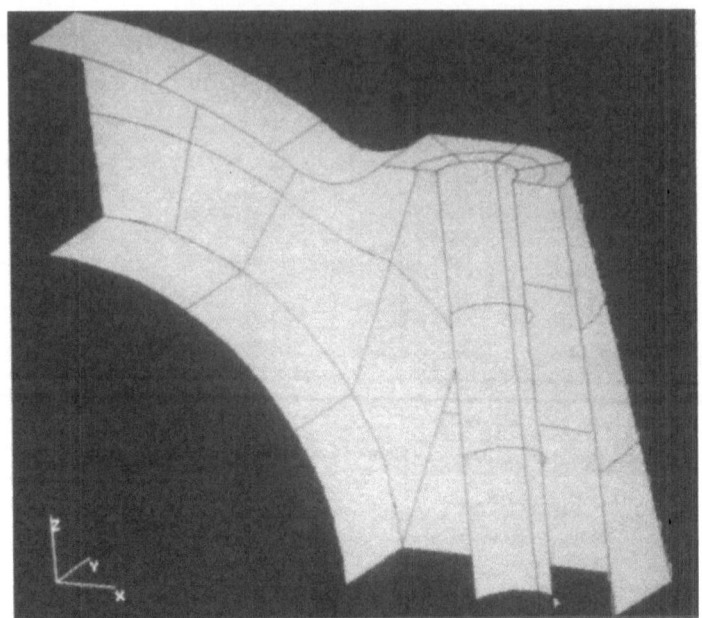

Figure 3 Typical Discretization of a Bearing Cap (notice the use of discontinuous elements throughout)

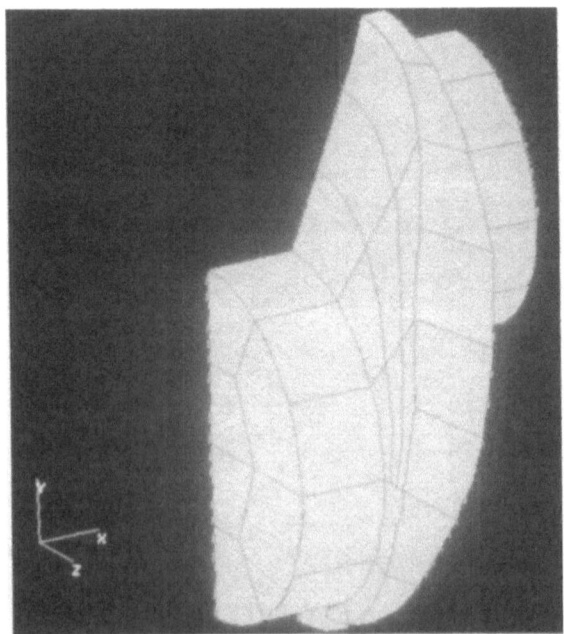

Figure 8 Crankshaft: Discretization

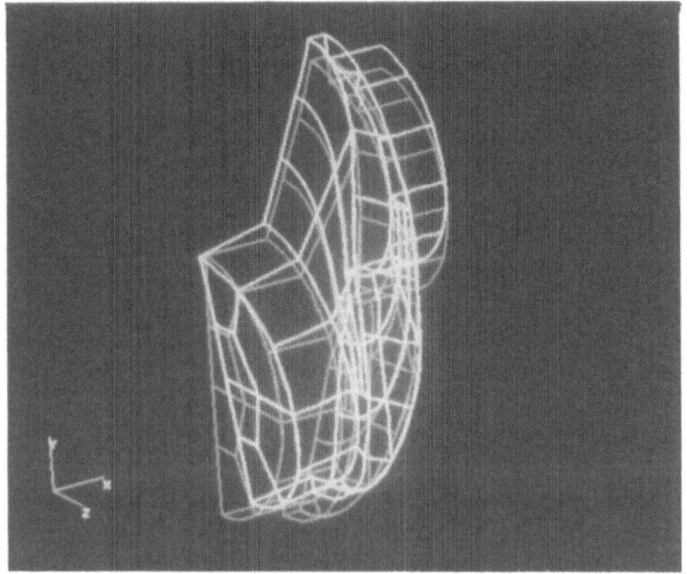

Figure 9   Crankshaft: Displaced Shape

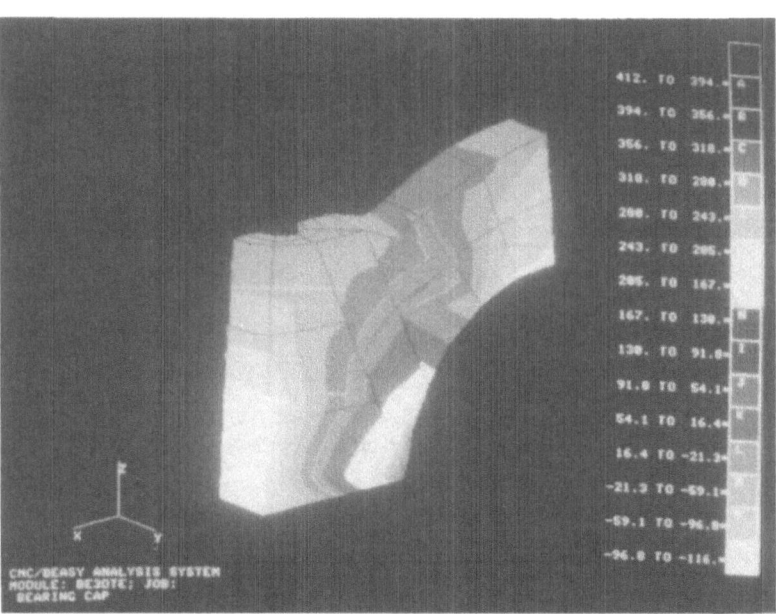

Figure 10 Bearing Cap: Contours of Direct Stress

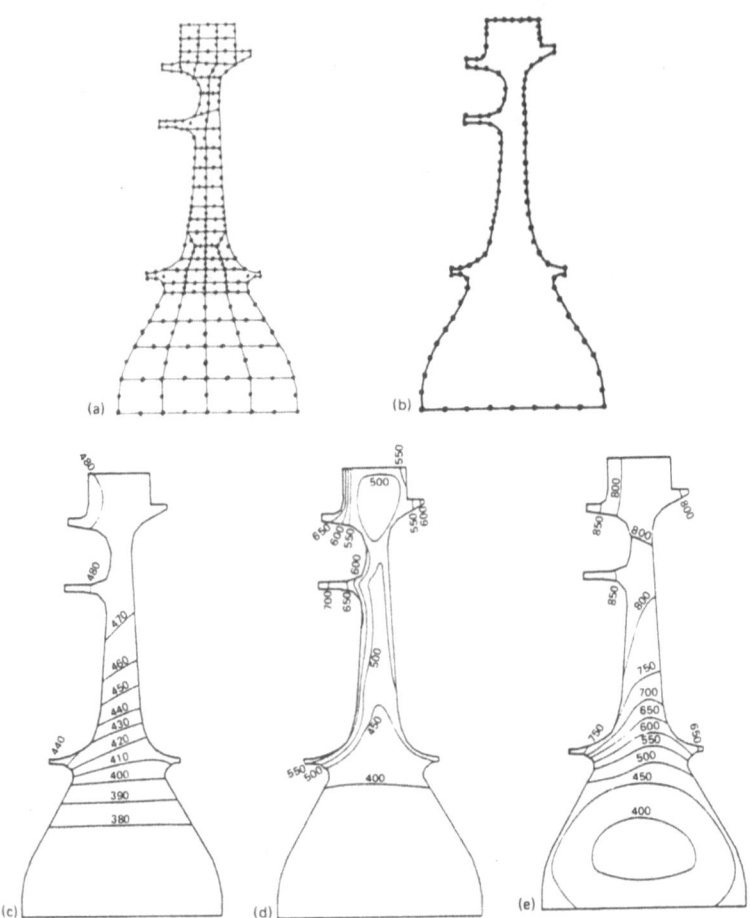

Figure 11   Turbine disc: (a) FEM mesh   (b) BEM discretization
(c-e) boundary element results

16. SIMULATION POSTER SESSION

COMPAMM - A PROGRAM FOR THE DYNAMIC ANALYSIS OF MULTI-RIGID-BODY
SYSTEMS

J. Unda, A. Avello, J.M. Jimenez, J. García de Jalón

Department of Applied Mechanics, University of Navarra, San Se-
bastián, Spain.
Centro de Estudios e Investigaciones Tecnicas de Guipuzcoa,
Spain

INTRODUCTION

Many mechanical systems can be effectively modelled by systems of
interconnected rigid bodies. The dynamic analysis of such systems
can be done by analytical or numerical methods. The analytical me
thods as those described for three-dimensional systems by Duffy
(1980), have considerably extended the field of application in
the last few years, but the theoretical and practical difficulties
are so great that the only alternative to general analysis are
the numerical methods. These methods have opened up the possibility of developing general purpose computer programs, like the
well known IMP, ADAMS and DADS-3D packages, described by Sheth
et al. (1972), Orlandea et al. (1977) and Wehage et al. (1982),
respectively.

In this paper the distinctive features of a general pro
gram for the dynamic analysis of multi-rigid-body systems are des
cribed. The program COMPAMM (COMputer Analysis of Mechanisms and
Machines) carries out the dynamic analysis of any planar mechanism made up of a set of rigid bodies linked by revolute and/or
prismatic pairs, which can contain linear or non-linear springs
and dampers between its elements and also rigid and radially deformable wheels running over a road profile. The analysis can be
performed under any kind of external forces, previously defined
by the user, including the forces of viscous friction in the kinematic pairs. The situations of impact between the mechanism
elements and wheels, and the road profile, can be analyzed by
the program under any situation of restitution and friction at
the point of contact.

The program COMPAMM can also perform the dynamic analysis of any three-dimensional mechanism formed by rigid bodies
linked by spherical (S), revolute (R), cylindrical (C) or prisma
tic (P) pairs as well as universal joints (U), and containing
springs and dampers, and also under any set of external forces.
The program capabilities are now being developed and the next

features to be added in the near future is the posibility of ana
lyzing three-dimensional wheels and three-dimensional impact situations.

The program COMPAMM is based on the method of natural
coordinates developed by the authors in the reference Avello et
al. (1984), García de Jalón et al. (1981) and (1983), Serna et
al. (1982), Unda et al. (1983), (1983) and Villalonga et al.
(1984), which uses as coordinates of the mechanism the cartesian
coordinates of some of its points and the cartesian components
of some unitary vectors. These "natural coordinates" provide a
simple and easy way of considering the kinematic pairs naturally
associated with a particular direction like R, C,P and U pairs.
This method is caracterized by its conceptual simplicity and its
easy implementation on a computer.

In this paper the natural coordinates method will be
briefly presented, with also some examples of application which
show the capabilities of the COMPAMM program.

THE "NATURAL COORDINATES" METHOD

Mechanism   coordinates
The central point in any numerical method for the kinematic and
dynamic analysis of mechanisms is the definition of the mechanism
"coordinates". These coordinates are a set of non-independent pa
rameters which unequivocally define the position of each element
of the mechanism. They are not independent because any set of pa
rameters greater in number than the number of degrees of freedom
of the mechanism, must satisfy some geometric compatility equations, known as "constraint equations". The constraint equations
play a fundamental role in the analysis of these rigid-body systems and are closely related to the type of mechanism coordinates chosen.

On the other hand, the "generalized coordinates" of a
mechanism are defined as a set of independent coordinates whose
number coincides with the number of degrees of freedom. The generalized coordinates do not directly determine the position of
each element, and it is necessary to solve the non-linear position problem to obtain the mechanism coordinates from the genera
lized ones. For this reason, the generalized coordinates can not
be used to determine the position of the mechanism. They can be
used to define the velocities and accelerations of the input ele
ments, or in the numerical integration of the differential equations of motion.

Figure 1 shows an RSCR three-dimensional mechanism whose position is defined using natural coordinates.

Constraint equations
The constraint equations associated to the natural coordinates
are obtained in two ways; from the rigid body conditions between

points and vectors belonging to the same element, and from the constraints imposed on the relative motion of two rigid bodies by the kinematic pair that links them.

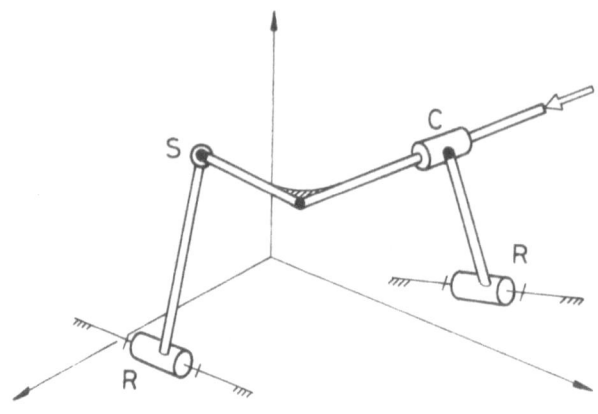

Figure 1

In the planar case the rigid body condition for a binary element is established through a constant distance condition between the two points that belong to the element, as is explained by García de Jalón et at. (1981) and (1983). Sometimes, in more complicated elements, it is necessary to impose several constant distance conditions, and even a constant area constraint between three points, that belonging to the same element, are on a straight line. This is necessary because the three constant distance conditions that could be imposed are not linearly independent (Figure 2).

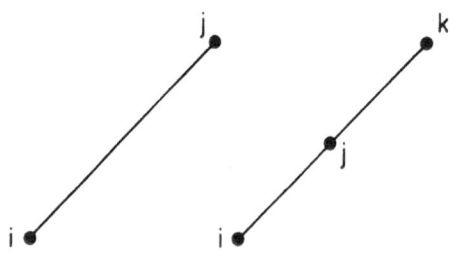

Figure 2

In the three-dimensional case the constraint equations associated to the rigid body condition of an element are also formulated very easily, establishing constant angle and constant distance conditions between the points and unitary vectors that belong to the same element. These constraint equations are formulated as scalar products between unitary vectors, as scalar products between a vector formed by two points and a unitary vector, and as constant distance conditions between points belonging of the same element.

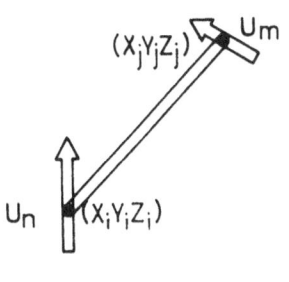

Figure 3

Figure 3 shows an element with two points and two unitary vectors. Its element or rigid body constraints are the constant distance condition between points i and j, the constant angle conditions between the vector {rij} and the unitary vectors {Un} and {Um} respectively, the constant angle between the unitary {Um} and {Un} and the conditions of unitary norm for vectors {Un} and {Um}.

Once the rigid body conditions have been established for every element, the constraint equations due to the kinematic pairs must be formulated.

In the planar case, the revolute pair (R) shown in figure 4a, does not imply the introduction of any new constraint equation, because this is automatically considered when the two elements share the point located at the pair. The prismatic pair (P) shown in figure 4b determines two constant area constraint equations between points i, j, k and between points i, j, l.

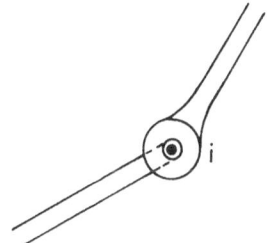

Figure 4a

In the three dimensional case the natural coordinates allow again a very simple formulation of the kinematic constraints. In a spherical pair (S) (figure 5a), the constraint imposed on the relative motion of the elements is automatically considered, as in the planar case for the revolute pair, when the two elements share the point located in the pair.

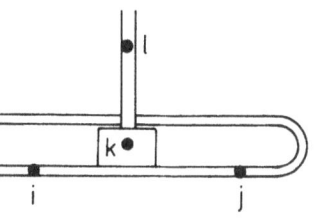

Figure 4b

A three-dimensional revolute pair (R) (figure 5b) does not need any new constraint equation either, when the two elements joined by it share a point located in the pair and a unitary vector associated with its revolution axis.

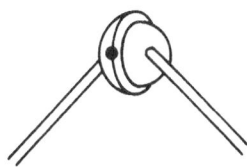

Figure 5a

A cylindrical pair (C) (figure 5c) is considered in the analysis when both elements share a unitary vector aligned with the revolution axis of the pair, and taking into account the condition of alignement between points i, j and that unitary vector. This condition is imposed through a vector product of the vector determined by the two points.

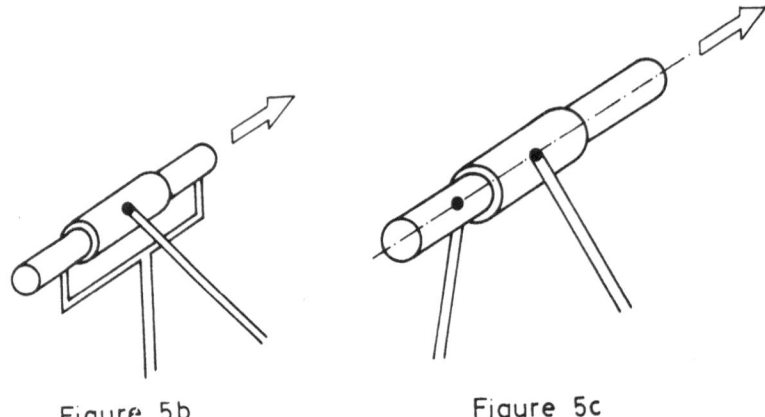

Figure 5b   Figure 5c

It is also possible to consider in the analysis, in a similarly easy way, other pairs as the prismatics pair, the universal joint, etc.

As has been shown, the natural coordinates allow the definition of the mechanism position with a small number of coordinates, an important fact for the efficiency of the method. They also permit the easy establishment of the constraint equations, which are quadratic, and therefore can be easily differentiated and numerically treated.

KINEMATIC ANALYSIS

The kinematic analysis of mechanism/García de Jalón et al.(1981), (1983) and Villalonga et al. (1984)/, can be done through the differentiation in respect to the time of the element and pair constraint equations defined in the previous section. The following system of linear equations can be obtained

$$|A(\{x\})| \ \{\dot{x}\} = \{0\} \tag{1}$$

where $\{x\}$ is the vector of natural coordinates of the mechanism, $\{\dot{x}\}$ is the vector of natural velocities and $|A(\{x\})|$ is the "kinematic matrix", which is a sparse rectangular non-symmetric matrix, which depends on the mechanism position. The rows of this matrix are formed by the derivatives of the constraint equations with respect to the $\{x\}$ coordinates. As a consequence of the constraint equations being quadratic equations, matrix $|A|$ elements are obtained directly from the mechanism position vector $\{x\}$. Completing the system with as many known velocities as the number of degrees of freedom of the mechanism, and solving the resultant linear system of equations, the vector of velocities of the natural coordinates $\{\dot{x}\}$ is obtained.

Differentiating again expression (1) with respect to time, the following result is obtained

$$|A(\{x\})| \ \{\ddot{x}\} = -|A(\{x\},\{\dot{x}\})| \ \{\dot{x}\}$$

where $\{\ddot{x}\}$ is the vector of accelerations of the natural coordinates, and the right hand term of this expression is easily obtained knowing the velocities and the position of the mechanism. Completing the resulting system as in the velocity analysis, the vector of mechanism accelerations is obtained.

DYNAMIC ANALYSIS

Inertia Forces. Mass Matrix
It has been demonstrated by Avello et al. (1984), that for every element it is possible to express the virtual power of its inertia forces using an inertia matrix $|M|_e$, square and symmetric, and using the acceleration vector of the element, in the following way

$$W_{ie} = \{\dot{\underline{x}}\}_e^T \, (-|M|_e \, \{\ddot{x}\}_e) \tag{3}$$

where $\{\dot{x}\}_e$ is the virtual velocity. Matrix $|M|_e$ can be computed very easily from the inertia properties of the element.

The virtual power of the inertia forces of the whole mechanism can be expressed using a global mass matrix $|M|$, formed by the assembly of the inertia matrices of each element, in the following way

$$W_i = \{\dot{\underline{x}}\}^T \, (-|M| \, \{\ddot{x}\}) \tag{4}$$

Natural Coordinates and Generalized Coordinates
As has been stated previously, the natural coordinates are not independent, because they are interrelated through the constraint equations. Nevertheless it is always possible to select an independent subset of natural coordinates whose number coincides with the number of degrees of freedom of the mechanism, and which are capable of defining unequivocally the mechanism motion. This set of natural coordinates has been called previously generalized coordinates.

The relation between the velocities of the natural coordinates and the generalized velocities can be expressed in the following matrix form

$$\{\dot{x}\} = |R| \, \{\dot{z}\} \tag{5}$$

where matrix $|R|$ is a rectangular matrix with as many columns as the number of degrees of freedom of the mechanism. This matrix can be easily calculated solving the velocity problem as many times as the existing number of degrees of freedom, giving alternately value one to a generalized velocity and zero to the others. The columns of matrix $|R|$ constitute a basis for the space of allowable motions. Differentiating expression (5) with respect to time

$$\{\ddot{x}\} = |\dot{R}| \, \{\dot{z}\} + |R| \, \{\ddot{z}\} \tag{6}$$

is obtained.

In this expression, matrix $|\dot{R}|$ can not be easily calculated, but the term $|\dot{R}|\{\dot{z}\}$ has a physical sense which permits its easy calculation; this physical sense can be seen in expression (6), the term $|\dot{R}|\{\dot{z}\}$ being the natural accelerations when the generalized ones are zero.

Differential equations of motion in generalized coordinates

The differential equations of motion can be established via the Theorem Virtual Power/Serna et al. (1982), Unda et al. (1983), Avello et al. (1984)/. The virtual power generated by the forces acting on the mechanism for a given virtual velocity compatible with the mechanism constraints can be expressed as

$$\{\dot{x}\}^t \; (\{F_{ex}\} + \{F_{in}\}) = \{0\} \tag{7}$$

where $\{\dot{x}\}^t \{F_{ex}\}$ is the virtual power due to the external forces and $\{\dot{x}\}^t \{F_{in}\}$ the virtual power due to the inertia forces which can be substitued by the expression (4), resulting in

$$\{\dot{x}\}^t \; (\{F_{ex}\} - |M| \; \{\ddot{x}\} \;) = \{0\} \tag{8}$$

Substituting in this expression the virtual velocity in terms of the generalized velocities $\{\dot{z}\}$ (5)

$$\{\dot{z}\}^t |R|^t \; (\{F_{ex}\} - |M| \; \{\ddot{x}\}) = \{0\} \tag{9}$$

in this expression the components of $\{\dot{z}\}$ are independent and arbitrary, and therefore it can be established that

$$|R|^t \; (\{F_{ex}\} - |M| \; \{\ddot{x}\}) = \{0\} \tag{10}$$

Substituting in this expression the vector of natural accelerations in terms of generalized velocities and accelerations using expression (6)

$$|R|^t \; (\{F_{ex}\} - |M||R|\{\ddot{z}\} - |M||\dot{R}|\{\dot{z}\}) = \{0\} \tag{11}$$

is obtained, which is a system of F differential equations (as many as the number of degrees of freedom of mechanism).

The external forces acting on the mechanism can be taken into account through the vector $\{F_{ex}\}$. This vector can be obtained reducing the external forces to the direction of the natural coordinates in such a way that these reduced forces generate the same virtual power $\{\dot{x}\}^t \{F_{ex}\}$ as the external forces. The program can take into account any kind of external forces through a simple subroutine provided by the user.

Springs and Dampers. Viscous Friction

The COMPAMM program can study the effect of springs and dampers acting between any two points of the mechanisms. The spring or damper acting on them can be substituted by its effect: the force acting between these two points which can be formulated as a function of the mechanism coordinates and velocites

$$\{F_k\} \;\; = \{F_k \; (\{x\})\} \tag{12}$$

$$\{F_d\} \;\; = \{F_d \; (\{x\}, \; \{\dot{x}\})\} \tag{13}$$

these effects are then taken into account integrating these expressions into the vector of external forces $\{F_{ex}\}$ of expression (11).

The viscous friction forces acting on the kinematic pairs can be studied in a very similar way, as they are velocity dependent forces that after being reduced to the direction of the natural coordinates, can also be integrated into the external forces vector $\{F_{ex}\}$.

$$\{F_f\} = \{F_f (\{\dot{x}\})\} \tag{14}$$

Impacts

Impacts, i.e. very large forces actuating on very small time intervals, can play an important role in the dynamic behaviour of many mechanical systems.

In rigid body mechanics, impacts are usually studied by means of the equation of momentum conservation, taking the form

$$|M| (\{\dot{x}\}_a - \{\dot{x}\}_b) = k\{p\} \tag{15}$$

where $\{\dot{x}\}_b$ and $\{\dot{x}\}_a$ are the velocities before and after the impact, $\{p\}$ is a unitary vector in the impact direction (comprising external, internal and reaction impacts), and k is a scale factor. Applying the Theorem of Virtual Power to equation (15), the following result can be obtained

$$|R|^t (|M| (\{\dot{x}\}_a - \{\dot{x}\}_b)) = k|R|^t \{p\}_e \tag{16}$$

where $\{p\}_e$ is an unitary vector in the direction of external impacts only, because the others do not give resultant power. Ordinarily $\{p\}_e$ is a known vector.

In order to compute $\{\dot{x}\}_a$ and k it is necessary to add new equations to system (16). These equations arise from the compatibility of velocities after the impact

$$|A| \{\dot{x}\} = \{0\} \tag{17}$$

and from Newton's equation

$$\varepsilon = \frac{(\{\dot{x}\}_a)_r}{(\{\dot{x}\}_b)_r} \tag{18}$$

where ε is the Newton coefficient of restitution ($\varepsilon=1$ for perfectly elastic impacts, and $\varepsilon=0$ for perfectly plastic ones),and subscript r denotes the component in the direction perpendicular to both surfaces at the point of contact.

The Wheel-Road Subsystem

A subsystem composed of a rigid or radially deformable planar wheel, and a road of a general profile has been developed,/Unda et al. (1983)/ in order to broaden the range of applications of the program.

The road profile can be modelized in the form of several spans. Each span is defined by the user, by the position of several points and by the tangent angles at the beginning and at the end of the span; the use of cubic splines defined in local axes allows continuity in first and second derivatives and non single valued functions in global coordinates. It also permits slope discontinuity between contiguous spans. As a particular case straight spans can also be included.

The wheel is mode lized with two points, as can be seen in figure 6, one in the center and ano ther on the extreme of a radius. These points are related by a constant dis tance constraint equation. In order to consider the wheel inertia forces in the analysis, its total mass and moment of iner tia are substitued by its equivalent masses, two punctual masses on both points, and a dis tributed mass in the line between them. Sometimes negative va lues for the equivalent masses are obtained, but this is not a problem for the numerical solution.

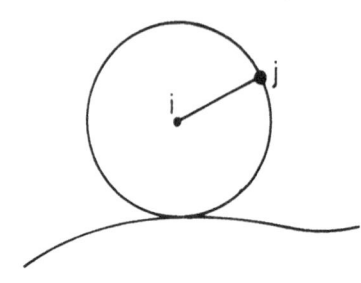

Figure 6

The rolling condition of the rigid wheel over the road profile is equivalent to constant distance conditions between points A and B and their centers of path curvature O_a and O_b. These conditions imposed through Lagrange Multipliers, also pro vide the means to compute directly the contact force, and then the rolling condition can be checked.

In the radially deformable wheel the contact force is a function of the wheel deformation and can be introduced into the analysis through the term of external forces in a similar way as in the analysis of springs and dampers. The rolling condition in a deformable wheel needs a new constraint equation which imposes a relation between the velocity parallel to the road profile of the center of the wheel, and its angular velocity.

When the contact force angle exceeds the allowable va lue determined by the coefficient of friction, a sliding condi tion occurs.

In this case, for the rigid wheel, there is only one ki nematic constraint that corresponds to the constant distance condition between the wheel center and its center of path curva ture. The second condition is a dynamical one, where a tangent friction force must be imposed on to the wheel whose value is proportional to the normal force. If the kinematical constraint is imposed with a Lagrangian Multiplier, the normal force appears as an unknown and then the friction force can be introduced di rectly into the equilibrium equations, resulting in an extremely simple way of considering Coulomb friction.

In the radially deformable wheel the sliding condition
is only a dynamical one, similar to that of the rigid wheel. The
normal force is now known because it is a function of the wheel
deformation and then the friction force can be easily introduced
into the equilibrium equations.

The sliding condition must also be checked continuously
in order to detect a possible change to rolling. This check can
be carried out by following the sign of the contact point veloci
ty.

The loss of contact of the wheel can be detected easily
checking the change of sign of the normal contact force. When
this happens, none of the aforesaid kinematic or dynamic constraints must be imposed. The system evolves until a geometrical
interference between wheel and road is detected. At this moment
in the case of rigid wheel, an impact occurs and equations (16-
18) must be applied to obtain the velocities after this impact.
For a deformable wheel, the loss of contact is detected in a si
milar way, but there is no need to use the impact equations when
a new contact is reached.In this situation the contact forces
between road and wheel are applied again and the sliding, rolling and loss of contact checks are now activated.

Numerical integration of the motion differential equations

The resulting system of differential equations (11) can be numerically integrated, beginning from the initial position and velo
city, to obtain the evolution of the mechanism.

For this numerical integration, first order differential equation integration methods can be used, as Runge Kutta, methods of polynomical interpolation, Adams predictor /corrector,
etc. ... The subroutines of application of these methods published by Gear (1971), Forsythe et al. (1977) and AERE Harwell
(1981) are very interesting. These subroutines can estimate the
integration error and can change the step length to mantain this
error under a maximum value established previously by the user.

The system of second order differential equations (11)
can be easily reduced to a system of first order differential
equations with twice the number of variables, in order to use the
subroutines referred to previously.

The COMPAMM Data Input

The program data can be prepared very easily. First of all, it
is necessary to choose the natural coordinates of the mechanism,
numbering them and giving their initial value, in order to define the initial position of the mechanism. Then, the elements
must be defined, numbering them and indicating the natural coordiantes belonging to the same element. The kinematic pairs must
also be defined in a similar way.

The rest of the data needed, depends on the problem to
be analyzed. In a kinematic analysis it is necessary to define
the velocities and accelerations of the input elements. In a dynamic analysis, a FORTRAN subroutine of external forces must be

supplied by the user , also the inertial characteristics of the elements, the initial velocity of the mechanism, the duration in seconds of the simulation and the maximum error permitted to the numerical integration are to be supplied.

Once the program has been provided with this data, the dynamic simulation can be started.

The COMPAMM Results Output

The easy interpretation of the natural coordinates, velocities and accelerations, eliminates the need of the postprocessing of the dynamic simulation results. The output of the program is the position or velocity or acceleration histories of any natural coordinate of the mechanism.

The program can also represent the motion of the mechanism in a plotter as a sequence of static positions separated by the same interval of time.

The motion can also be represented dynamically in a refresh graphics display, where the mechanism can be seen in movement during the time that the dynamic simulation lasts. The motion can be slowed or accelerated on the display as the user wishes. This allows a quick interpretation of the results.

EXAMPLES

Three different examples are presented. The first example shows the planar dynamic simulation of the behaviour of a single bar under the gravitational force, falling and hitting with a static obstacle. Figure 7 shows the motion of the bar during 5 seconds, representing a position each 0.04 seconds. The impact situations have been analyzed with a Newton coefficient value of 0.9 and with a coefficient of friction of 0.2.

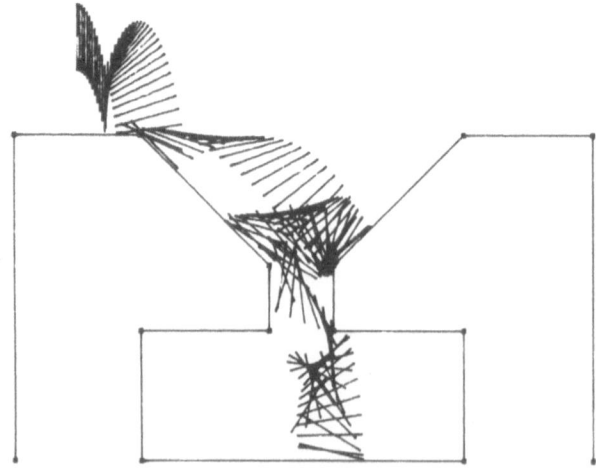

Figure 7

Figures 8 to 11 show the planar dynamic simulation of a two wheel vehicle with rigid wheels over a stair shaped road at a speed of 90 Km/h. Each figure shows the motion of the vehicle for 0.3 seconds of simulation, figure 8 from 0.0 to 0.3 seconds, figure 9 from 0.3 to 0.6, figure 10 from 0.6 to 0.9 and figure 11 from 0.9 to 1.2 seconds.

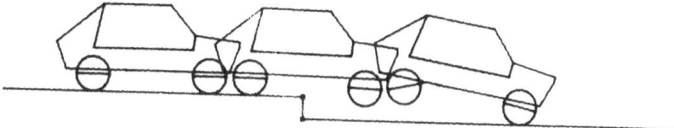

Figure 8

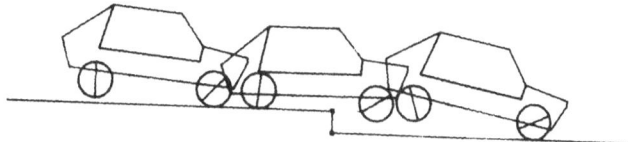

Figure 9

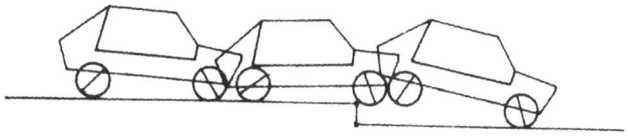

Figure 10

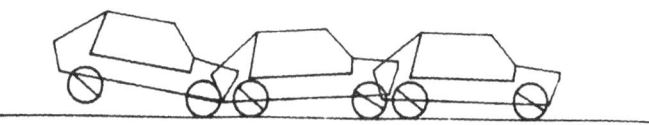

Figure 11

Finally figure 12 shows the evolution of a SSC three dimensional mechanism under gravitational force. The motion is represented for 1.2 seconds each 0.03 seconds.

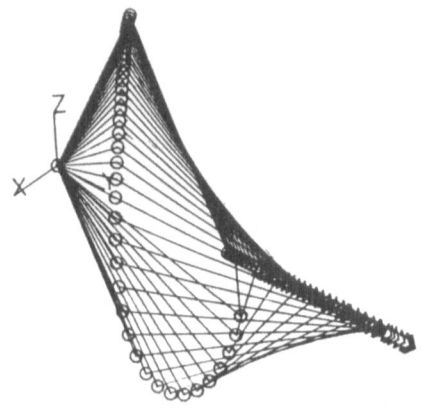

Figure 12

CONCLUSIONS

This paper presents the features of a general program for the dynamic analysis of multi-rigid-body systems. The program COMPAMM (COMputer Analysis of Mechanisms and Machines) is based on the natural coordinates method of numerical analysis of mechanisms, developed by the authors, which uses as coordinates of the mechanism the cartesian coordinates of some of its points and the cartesian components of unitary vectors. The conceptual simplicity of the method, the easy formulation of the constraint equations, which are always linear or quadratic, and the direct interpretation of the natural coordinates, velocities and accelerations, allow an easy implementation of the program.

The program capabilities are now being developed in order to incorporate the possibility of analyzing three dimensional wheels and three-dimensional impact situations.

REFERENCES

AERE Harwell (1981). Harwell Subroutine Library. AERE Harwell. Oxfordshire. U.K.

Avello, A., Unda, J. and García de Jalón, J.(1984). Analisis Dinámico de Mecanismos Tridimensionales en Coordenadas Naturales. Proceedings of the III Congreso Nacional de Ingeniería Mecánica, Gijón, Spain.

Duffy, J. (1980). Analysis of Mechanisms and Robot Manipulators. Edward Arnold.

Forsythe, G.E., Malcolm, M.A. and Moler, C.B. (1977). Computer Methods for Mathematical Computations. Prentice-Hall, Englewood Cliffs, New Jersey.

García de Jalón, J., Serna, M.A. and Avilés, R. (1981). Computer Method for Kinematic Analysis of Lower-Pair Mechanisms - Part I Velocities and Accelerations. Part II Position Problem. Mechanism and Machine Theory, 16, 543-566.

García de Jalón, J. and Unda, J. (1983). Kinematic Analysis of Lower-Pair Planar Mechanisms in Basic Coordiantes. Proceeding of the VIth IFToMM World Congress on the Theory of Machines and Mechanisms, New Delhi, 183-186.

Gear, C.W. (1971). Numerical Initial Value Problems in Ordinary Differential Equations. Prentice-Hall, Englewood Cliffs, New Jersey.

Orlandea, N., Chace, M.A. and Calahan, D.A. (1977) A Sparsity Oriented Approach to the Dynamic Analysis and Design of Mechanical Systems -Part I and Part 2, Journal of Engineering for Industry, 773-784.

Serna, M.A., Avilés, R. and García de Jalón, J. (1982). Dynamic Analysis of Plane Mechanisms with Lower Pairs in Basic Coordinates. Mechanism and Machine Theory, 17, 397-403.

Sheth,P.N. and Uicker, J.J. (1972) IMP (Integrated Mechanisms Program); A Computer-Aided Design Analysis System for Mechanisms and Linkages, Journal of Engineering for Industry, 454-464.

Unda, J., Gimenez, J.G. and García de Jalón, J.,(1983). Computer Simulation of the Dynamic Behaviour of Vehicles, in: J. K. Hedrick, ed., The Dynamics of Vehicles on Roads and on Railway Tracks, (Swets & Zeitlinger, Lisse, 1984)

Unda, J. and García de Jalón, J. (1983). Dynamic Analysis of Lower-Pair Planar Mechanisms in Basic Coordinates. Proceeding of the VIth IFToMM World Congress on the Theory of Machines and Mechanisms, New Delhi, 391-395.

Villalonga, G., Unda, J. and García de Jalón, J., (1984). Numerical Kinematic Analysis of Three-Dimensional Mechanisms Using a "Natural" System of Lagrangian Coordinates, ASME Paper 84-DET-199.

Wehage, R.A. and Haug, E.J., (1982). Dynamic Analysis of Mechanical Systems with Intermittent Motion, Journal of Mechanical Design, 778-784.

Wehage, R.A., and Haug. E.J., (1982). Generalized Coordinate Partitioning for Dimension Reduction in Analysis of Constrained Dynamic Systems, Journal of Mechanical Design, 247-255.

COMPUTER AIDED METHODS FOR THE SYNTHESIS OF MECHANISM KINEMATIC MODELS: PLANAR AND SPATIAL CASES

P. Fanghella, C. Galletti

Istituto di Meccanica applicata, University of Genoa

1. INTRODUCTION

In the past few years, many works on computer-aided mechanism analysis and synthesis have appeared in the scientific literature. This is a complex and diffi‑ cult subject which requires specific expertise, rang- ing from a good knowledge of mechanical hardware and the use of modelling techniques and computer programs. When the term 'mechanism model' is used, some ambigu- ity may arise. Model is, in fact, a word with differ- ent meanings depending on the context and environment in which it is used. For the purpose of mechanism a- nalysis, we shall follow classical definitions (Cannon) of 'physical model' and 'mathematical model'; moreo- ver, we shall regard a 'functional mathematical model' as a set of differential and/or trascendental equa‑ tions which give the position, velocity, acceleration, reactions, and internal forces of any mechanism part. In most applications, very simple physical models can be used for mechanism design: a model based on rigid links and kinematic pairs can often be accepted, and and we shall adopt it in this paper.
Given such a simple physical model, the procedure for building the corresponding mathematical model can be regarded as a straightforward process: from a theoret- ic point of view this may even be true, but several difficulties usually arise in actual operations. In fact, many alternatives to the same model can be ob- tained, all expressing a unique mathematical reality (though from different points of view), with various

types of notation and variables, which can be more or less useful in solving different design problems.

A recent work of Galletti (1985) shows that a model of mechanism information can be represented by a multilevel hierarchical structure whose lower levels correspond to kinematic information; such information must be accessible in different ways, depending on the design requirements, and consequently alternative mathematical models should be used.

A remark on common functional models: the majority of such models (see Paul for references) are of the numerical type and are utilized to carry out analysis. Minor attention is given to formal models and synthesis problems (Angeles). Moreover important design objectives, like number synthesis, are not considered within the same framework as other design phases, thereby reducing the degrees of flexibility and power of many computer-aided design programs.

Machine theory provides effective tools for deriving kinematic mechanism models which can be very useful in computer-aided design applications. In section 2 we outline an approach (which is based on topology and compatibility considerations) to kinematic model building; it can meet both analysis and synthesis requirements and yield both analytical and numerical results.

Clearly, many software problems may arise, depending on the formal methodology we want to implement by computer programs. Section 3 present a topological approach to planar linkages, using graph handling. Section 4 faces the more complex problem of 3-D mechanisms (like robot and manipulators). Section 5 discusses the problem of computer integration, which is a new subject of high practical interest.

2. KINEMATICS AND KINEMATIC MODELS

Kinematic models of planar and spatial mechanisms can be partioned into various classes, depending on the hypotheses used to define such models.

The most important is undoubtedly the class of rigid-link and ideal (no clearance) joint linkages. The excellent and practical results that have been obtained by models based on such hypotheses make this class almost a standard one. Indeed, the best known and most powerful computer programs for mechanism analysis (Sheth; Smith; Rossi) and synthesis were implemented according to this type of model.

Over the years, several different approaches have been proposed in order to obtain compatibility equations which give the kinematic mechanism model in accordance with the related physical

hypotheses.
Such approaches differ in the usage of topological data, in the choice of geometric invariants for stating the compatibility, and in the types of mathematical tools used.
Many Authors (see Galletti - 1982 - for references) demonstrated the possibility and the advantages of using topological information in order to decompose a planar linkage into special substructures. Since the relevant compatibility equations are independent of one another, this approach can better describe the structure and complexity of a model, and allows the implementation of very effective analysis and synthesis procedures.
When the decomposition step has been performed, there arise the problem of solving kinematic relations for the resulting substructures. A solution to this problem, which is common to every approach based on linkage compatibility, requires the use of geometric invariants. The choice can be made either from a general point of view, as proposed by Paul and by Smith, or by taking into account the knowledge of the mechanism structure.
To sum up, it is worth noting that:
-as the complexity of an algebric compatibility problem only depends on the mechanism topology and structure, all the approaches adopted can lead to the same results through an appropriate formal elaboration procedure;
-taking into account topological and structural data, before formulating the compatibility relations, directly yields models which are easier to handle and solve;
-this structural approach brings into clear evidence the geometric nature of the kinematic problem: this fact can be a great advantage in understanding the role of mechanism parameters, especially in synthesis applications.

3. MODEL SYNTHESIS FOR PLANAR LINKAGES

The main idea of this section is that a topological mechanism scheme can be obtained by a constructive approach, starting from a given basic structure and adding to it   certain link groups according to some strategic procedure (see Rossi). If we give a graphical representation of any possible process of this kind, we obtain several trees for which the roots are the given basic structures; the nodes are the structures made up; the connections represent the path followed by the construction process.
A tree data structure can easily be implemented using a high-level programming language (e.g. Pascal). At each node we store the required information about topological data for every struc-

ture: the joint matrix, the number of links, the last added
group, and the relationships with other nodes.
Addition of new links to a basic structure can be performed in
several different ways, and an implemented tree has the structure shown in Figure 1 where nodes are displayed.

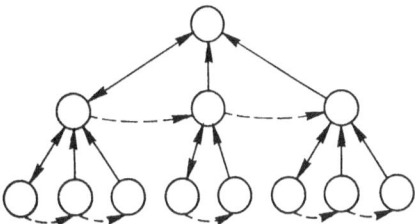

Figure 1 Mechanism tree for constructive process

Pointers are also provided (marked with arrows in Figure 1) to
allow effective searching strategies.
Moreover,we can observe that the number of mechanisms generated
through this process is 'a priori' unknown; therefore, dynamic
variables must be used.
The above process of model building is straightforward, but some difficulties may arise: e.g. a simple notation for identifying kinematic chains must be determined; identical structures
found by following different paths of the same tree are to be
detected,and so on. All these problems can be handled by making
minor changes in the basic data structures, and by storing additional information at each node (Figure 2).

```
const   maxgroup=   ; maxmem=   ;
type    numem=1..maxmem;
        grtipo=0..maxgroup;
        strucptr=^strtype;
        polinomio=array numem  of integer;
        matrixnm=array numem,numem  of integer;
        infotype=record
                admat:matrixnm;
                last:grtipo;
                mem:numem;
                coeff:polinomio
                end;
        strctype=record
                father:strucptr;
                info:ifotype;
                frstson:strucptr;
                nxtbrthr:strucptr
                end;
```

Figure 2 Data definition for mechanism tree

Auxiliary data structures (like lists, with pointers to and
from a given tree) can be useful in speeding up searching operations (see Pedemonte).
Clearly, the approach outlined in this section is best suited
to buiding a structural model of planar mechanism, and is useful in several design processes,like number synthesis, in the
field of mechanics.
However, the model can easily be augmented to perform other
types of functional simulation. In fact, by adding kinematic
or dynamic algorithms the basic structures and the link groups,
we can derive, from the same tree, additional paths for solving
the whole mechanism.
In other words, we can consider the tree in Figure 1 from a
twofold point of view: as a constructive process for mechanism
generation and as a simulation strategy for mechanism analysis
and synthesis.

4. MODEL SYNTHESIS FOR SPATIAL LINKAGES

Hervé and Fanghella demonstrated that the theory of finite dimensional transformation groups can be successfully used to
deal with both mobility and compatibility problems of spatial
single loop mechanisms.
A computer package based on this methodology was designed and
implemented by Russo in order to solve the kinematic inverse
problem of spatial 6 degree-of-freedom manipulators. The package, which was developed in Pascal using a personal computer,
can be defined 'mechanism compiler' and is capable to treat
single loop mechanisms and manipulators whose positions can be
determined in closed form.
Figure 3 describes the procedure which is to be used in order
to generate an executable program for the analysis of a mechanism. The developed software consists of two distinct parts
(inside the dashed boxes in Figure 3).
The mechanism compiler program performs three basic sequential
actions. Mobility analysis is first fulfilled: this module takes from the input file (Figure 4) the kinematic description
of the considered mechanism.
Taking account of this description, which must be made in terms
of revolute and prismatic pairs (more complex pairs can be treated through kinematic equivalence), and verifying a set of geometric conditions, the program creates a matrix containing all
possible constraints and the related invariantive parameters.
This mobility analyser, which is also suited for recognizing
passive degree-of-fredom, is based on a table of geometric rules whose entries are the pairs to be joined and whose output

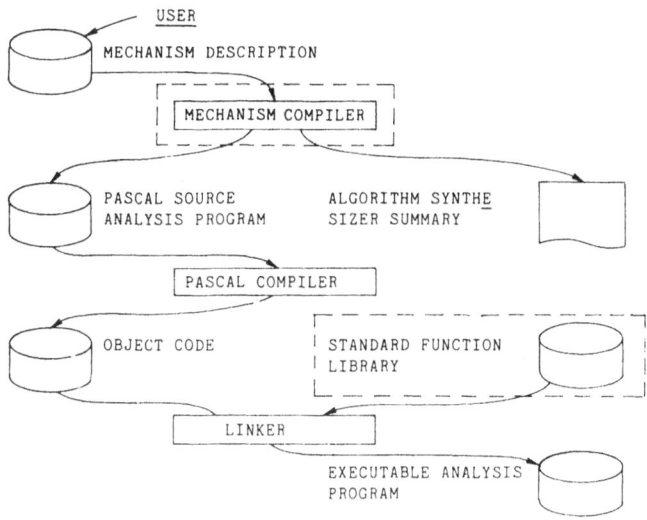

Figure 3   Procedure for mechanism model building

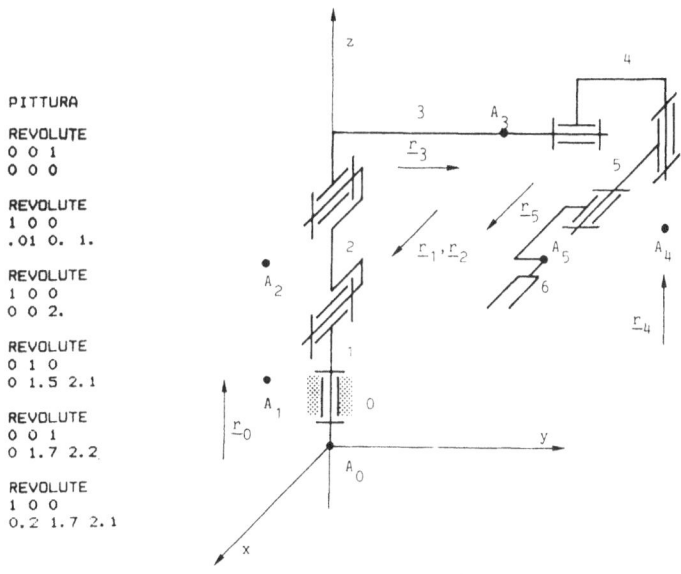

Figures 4   Example mechanism

is, if the related rule is satisfied, the resulting composed constraint.
The second part of the program attempts to generate a set of uncoupled and explicitly solvable compatibility equations.
These synthesized equations are displayed in a table (Figure 5) containing:
-the pair to be solved;
-the invariantive entity and the invariantive relation used to build up the equations;
-the structure of the actual kinematic chain and the related solved transormations to be used in order to evaluate the coefficients of equations.
Two nested algorithms make up the equation synthetizer: a tree data structure scanner explores all the kinematic chains that can be deduced from the constraint matrix until a complete set of equations is obtained or the scansion fails. At each tree node an attempt is made in order to single an equation out.
Since the kind of the available invariantive quantities and relations only depends on the type of the pairs in the actual kinematic chain, the algorithm is based on consulting a set of tables which supplies the required geometric invariants and tests their functional dependence upon the unknown pair variable.
The last part of the program is a sort of symbolic computation module: starting from the synthetized equations,it produces a Pascal source program for the analysis of the considered mechanism.
First the type and the coefficients of each equation are symbolically computed by means of the properties of the scalar and vector products, the operative definition of coordinate transformations and the chain data. The program is generated tailoring the grammatical representation of all possible analysis programs to the actual case. Such resulting program contains data structures and procedures that are both present in all other analysis programs and typical of the actual mechanism: the former are supplied by the relocatable library of standard functions,the latter are explicitly enclosed in the program.

5. A DISCUSSION ON PROGRAM INTEGRATION

As discussed by Galletti (1985), the use of a mechanism model for design applications involves complex procedures for the following reasons:
-the level of information required by the user varies according to the type of application;

```
numero equazioni=6
EQUAZIONI SINTETIZZATE:

coppia in esame: pair12
segno della coordinata di coppia: -
parametro incognito:       TIPO=  punto
                    COORDS=  0.00  1.70  2.10
                         solidale al liaison di indice 3
relazione di invarianza: (P,e)
                    e=   1.00  0.00  0.00
                         solidali al liaison di indice 2


coppia in esame: pair23
segno della coordinata di coppia: +
parametro incognito:       TIPO=  punto
                    COORDS=  0.00  0.00  2.00
                         solidale al liaison di indice 3
relazione di invarianza: (P-A,P-A)
                    A=   0.00  1.70  2.10
                         solidali al liaison di indice 4


coppia in esame: pair34
segno della coordinata di coppia: +
parametro incognito:       TIPO=  punto
                    COORDS=  0.00  1.50  2.10
                         solidale al liaison di indice 4
relazione di invarianza: (P-A,P-A)
                    A=   0.00  1.70  2.10
                         solidali al liaison di indice 5


coppia in esame: pair45
segno della coordinata di coppia: +
parametro incognito:       TIPO=versore
                    COORDS=  0.00  0.00  1.00
                         solidale al liaison di indice 5
relazione di invarianza: (u,e)
                    e=   1.00  0.00  0.00
                         solidali al liaison di indice 6


coppia in esame: pair56
segno della coordinata di coppia: +
parametro incognito:       TIPO=versore
                    COORDS=  1.00  0.00  0.00
                         solidale a link6
relazione di invarianza: (u,e)
                    e=   1.00  0.00  0.00
                         solidali al liaison di indice 6


coppia in esame: pair67
segno della coordinata di coppia: +
parametro incognito:       TIPO=versore
                    COORDS=  0.00  1.00  0.00
                         solidale a link7
relazione di invarianza: (u,w)
                    w=   0.00  1.00  0.00
                         solidali a link6
```

Figure 5 Output of equation synthetizer

-a mechanism can be installed in very different machine assemblies, and simulation can require that its model be embedded into a complete machine model;

-users often want to run a mechanism model using externally provided software, (e.g. a general-purpose simulator or an optimization package), or to utilize its output in further computations, (e.g. data base storing and retrieval operations, graphic display, and so on);

-model users may have very different skills and knowledge about programming and data processing; we can regard them as designers without any previous CAD experience, technicians with some computer background, or M.E. graduates who are practising mechanism analysis and design.

Therefore, it is necessary, and today possible, to implement computer programs, working with mechanism models which can meet different needs (e.g., integration of the various phases of model building and running flexibility; modularity; interfacing and so on).

Unfortunately, computer programs so far implemented exibit many drawbacks, in particular when the designer wants to use them in real CAD applications. In fact three main requirements should be pointed out in order to tailor computer programs to the user's needs (see Parnas).

-some users often require only a subset of the features needed by other users: consequently, these less exacting users do not want to pay for computation resources relating to unneeded features;

-programs which allow changes must.be implemented;

-tools for high-level abstraction must be given, thus providing the user with 'software instructions' that can be followed using suitable types of data, if desired.

Several approaches to developing computer programs for some of the above-mentioned tasks are available: two of them have aredy been adopted for the development of existing programs, while the others have not yet been fully implemented.

1)-Stand alone packages: they represent the earliest,most used implementation strategy; the most widely used programs have in fact been developed according to this approach. In general, users can have access to these packages from the outside only: any interfacing with other programs would be difficult. No changes or tailoring operations can be performed.

2)-A set of independent routines (i.e. Subroutine) can also be offered to users to 'assemble' their mechanism models. They allow great flexibility but require that the user understand the assembly process and   know the actions performed by eve-

ry routine. Usually,this approach is implemented in Fortran, so
that little data-structuring can be carried out by standard
users: then lower levels of model complexity are to be considered (see Suh).Little or no attention is given to abstraction,
and the great flexibility of this approach cannot be utilized
at high level of model complexity without considerable programming effort.
3)-In a recent work (Galletti, 1984), a different solution is
suggested: to provide the user with a kind of environment made
up of procedures, modules, data structures at high abstraction
levels, and I/O facilities. This approach does not exhibit any
drawbacks and can strongly facilitate the user's tasks.
4)-A last approach, which was adopted to develop the software
described in section 4, is to realize an expansible set of pre-
compilers which give output programs for specific design applications.
This approach can be very helpful to the trained user, who can
see and modify a precompiled program at any instruction or data
level, making it very flexible. However, at present this option does not seem to be a very important one for mechanical
CAD applications.
In conclusion, our experience suggests the third strategy as
the most stimulating and promising one; if a suitable programming language is chosen, this approach allows the user to reach
very high degrees of abstraction and flexibility.

6. CONCLUSION

The strategy we suggested in the previous sections for building
the kinematic model of mechanical devices, is aimed to obtain
rather a formal treatment then a numerical one. Specially for
synthesis purposes this approach appears the most convenient.
Moreover faster and more precise algorithms can be deduced by
this constructive method. As a final result more simplex codes
are obtainable.
While for the analysis purpose good results were achieved, a lot
of work must be made in order to solve synthesis problems and to
develop integrated CAD packages as outlined in section 5.

7. ACKNOWLEDGMENT

Section 3 and 4 of this paper make references to the works developed by Mr. Pedemonte and Mr. Russo in their M.S. degree the-
sys.

8. REFERENCES

Angeles J. (1982) Spatial kinematic chains. Springer-Verlag.

Cannon R. (1967) Dynamics of physical systems. McGraw-Hill

Fanghella P. (1984) Sull'analisi di compatibilità dei meccanismi in moto rigido generale. VII Congr. AIMETA. Trieste.

Galletti C. (1982) Osservazioni sui gruppi di Assur nei meccanismi piani. VI Congresso AIMETA. Genova.

Galletti C. Giannotti E. (1984) A personal computer approach to Computer-aided mechanism design. IASTED Conference CAD.

Galletti C. Giannotti E. (1985) CAD of mechanical components: a conceptual point of view. III IASTED Symp. Appl.Informatics.

Gupta V.K. (1973) Kinematic analysis of plane and spatial mechanism. Trans. ASME J.Eng.Ind.,95,481-486

Hervé J. (1978) Analyse structurelle des mécanismes par groupe des déplacements. Mech.mach.theory,13,437-450.

Litvin F. (1975) Simplification of the matrix method of linkage analysis by division of a mechanism into unclosed kinematic chains. Mech.mach.theory,10,315-326.

Paul B. (1979) Kinematics and dynamics of planar machinery. Prentice-Hall.

Parnas D. (1979) Designing software for ease of extension and contraction. IEEE Trans.Soft.Eng.,SE-5,128-137.

Pedemonte D. (1983) Analisi automatica dei meccanismi piani: metodi diretti e indiretti di scomposizione. Tesi. Genova.

Rossi A. Fanghella P. Giannotti E. (1981) An interactive computer package for planar mechanism analysis. II Conference Engineering software, London.

Russo G. (1984) Analisi di posizione dei meccanismi spaziali: applicazioni all'analisi dei manipolatori. Tesi. Genova.

Sheth P. Uicker J. (1972) IMP: Integrated mechanism program. Trans. ASME J.Eng.Ind.,94,454-464.

Smith D. Chace M. Rubens A. (1972) The automatic generation of a mathematical model for machinery systems. ASME paper 72mech31

Suh C. Radcliffe C. (1978) Kinematics and mechanisms design. Wiley & sons.

THE SPREAD OF POLLUTANTS EMITTED FROM LONG AND LARGE OBSTACLES
IN ATMOSPHERE

M.A.Serag-Eldin

IBM Cairo Scientific Center, Egypt

ABSTRACT

The paper presents a computational model for the prediction of
the pollutant spread from a source close to a relatively large
object in the atmosphere. The mathematical model comprises a
quasi three-dimensional steady flow model, and a fully three-
dimensional, unsteady concentration model which employs the results of the computed flow field.

The computational model is applied to predict the spread of
pollutants for a typical illustration case, and the results
are displayed and demonstrated to be plausible and informative

INTRODUCTION

The present paper is concerned with the computation of the
spread of pollutant gases emitted from a source in the close
surroundings of relatively large objects. Typical applications
might be the exhaust from the chimney of a factory, long building or ship-tanker.

The main interest lies in the prediction of the pollutant concentration in the neighbourhood of the flow obstruction. In
this region, the concentration field would depend mainly on
the local flow field and turbulent mixing generated by the obstacle, which would generally be very different from that generated by the free atmospheric flow.

Many "Gaussian plume" models exist in the literature for the
prediction of the spread of pollutants over flat open terrains
of large areas (see for e.g.VKI Series; Pollutant Dispersal
Feb.1985). These models are emperical to a large extent, and
assume the spread to be controlled by free atmospheric conditions only. They are thus unsuitable for the prediction of the
pollutant dispersion in the neighbourhood of flow obstacles.

A computational model is presented here which introduces the effect of the obstacle disturbed flow on the spread of pollutants.

The problem considered

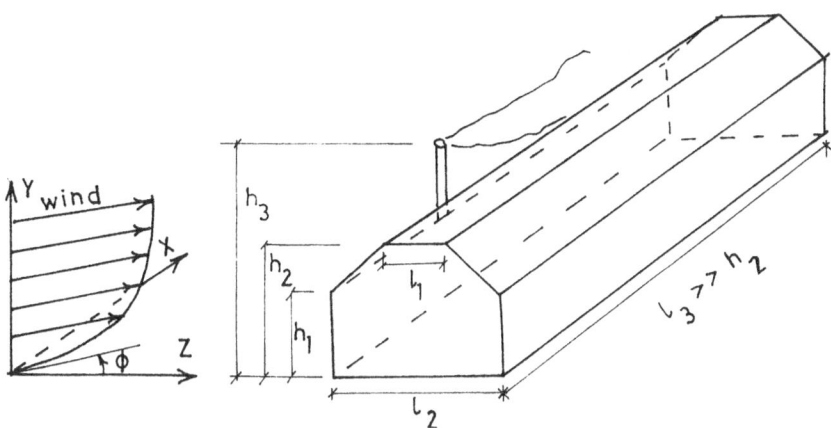

Figure 1. Problem Considered

The problem considered here is the prediction of the pollutant spread from a source neighbouring a long obstacle of uniform cross-section placed on the ground and within the atmospheric surface layer. The object must be long enough for end effects to be negligible.

Figure 1 displays a typical obstacle together with an associated cartesian coordinate system. The X-direction is parrallel to the longitudenal axis of the object, the Z-direction is the horizontal normal to it, and the y-direction gives the vertical direction and its coordinate is measured outwards from the ground.

The free atmospheric wind may approach the object at any arbitrary angle ϕ with the Z-direction, and may display any steady velocity profile which is independent of X. Pollutants are emitted from a given source at a specified rate which may vary with time.

The computational domain engulfs the object and extends upto the regions where the object no longer influences the rate of pollutant spread, or else the concentration levels are so low that they are no longer of practical importance. If there is still interest in predicting the concentrations in the undisturbed far off flow field, the integration domain may still be expanded to include these; however it is recommended in this case to resort to simpler models outside the disturbed flow domain.

MATHEMATICAL MODEL

The mathematical model contains a flow field model and a concentration field model. The flow field influences the concentration field through the convection and diffusion terms; however buoyancy forces are neglected and therefore there is no feedback from the concentration field.

Flow Model

The flow model comprises the continuity equation, the momentum equations in the x,y and Z- directions, and the transport equations for the kinetic energy of turbulence k and its rate of dissipation ε. The latter two equations are introduced by the adopted turbulence model which was presented by Launder and Spalding (1974). In this model the eddy diffusivity μ_t is derived from k and ε according to :

$$\mu_t = \rho_a \, C_\mu \, k^2/\varepsilon \tag{1}$$

where ρ_a is the atmospheric air density and C_μ is a turbulence model constant whose value is displayed with other model constants in Table 1.

The effective viscosity μ_{eff} is thus given by :

$$\mu_{eff} = \mu_t + \mu_a \tag{2}$$

where μ_a is the molecular viscosity of air.

Governing equations : An outcome of neglecting end effects, assuming the cross-section of the obstacle to be uniform, and considering the free wind profile to be independent of x, is that gradients in the x-direction may be omitted from the governing equations. Moreover since the flow field is considered steady, the governing equations may all be cast in the following concise form :

$$\frac{\partial}{\partial y}(\rho v\phi) + \frac{\partial}{\partial z}(\rho w\phi) = \frac{\partial}{\partial y}(\Gamma_\phi \frac{\partial \phi}{\partial y}) + \frac{\partial}{\partial z}(\Gamma_\phi \frac{\partial \phi}{\partial z}) + S_\phi \tag{3}$$

where $\phi=1$ expresses the continuity equation, otherwise ϕ represents any of the dependent variables listed in Table 2.; v and w are the velocity components in the y and Z directions, respectively; Γ_ϕ is the exchange coefficient for the variable ϕ, and S_ϕ is its source term. The expressions for Γ_ϕ and S_ϕ are displayed in Table 2., where u is the velocity component in the x-direction, and C_1, C_2, σ_k and σ_ε are turbulence model constants whose values are revealed in Table 1.

Boundary Conditions : The integration domain covers a single x-plane and its boundaries may be classified into inflow boundaries (Z=0), outflow boundaries (furthest Z-plane), top boundary (highest y-plane) and bottom boundary (ground and external surface of obstacle). The inflow and top boundaries are specified atmospheric conditions; the outflow boundary is one of zero outward gradient; whereas the bottom boundary is treated with the

Table 1. Turbulence Model Constants

C_μ	C_1	C_2	σ_k	σ_ε
.09	1.44	1.92	1.0	1.3

Table 2. Flow Field Governing Equations.

ϕ	Γ_ϕ	S_ϕ
1	0	0
u	μ_{eff}	0
v	μ_{eff}	$\frac{\partial}{\partial y}(\mu_{eff}\frac{\partial v}{\partial y}) + \frac{\partial}{\partial z}(\mu_{eff}\frac{\partial w}{\partial y}) - \frac{\partial P}{\partial y}$
w	μ_{eff}	$\frac{\partial}{\partial z}(\mu_{eff}\frac{\partial w}{\partial z}) + \frac{\partial}{\partial y}(\mu_{eff}\frac{\partial v}{\partial z}) - \frac{\partial P}{\partial z}$
k	$\frac{\mu_t}{\sigma_k}$	$G_k^* - \rho\varepsilon$
ε	$\frac{\mu_t}{\sigma_\varepsilon}$	$C_1\frac{\varepsilon}{k}G_k^* - C_2\rho\frac{\varepsilon^2}{k}$

$$*G_k = \mu_t\left[2(\frac{\partial w}{\partial z})^2 + 2(\frac{\partial v}{\partial y})^2 + (\frac{\partial w}{\partial y} + \frac{\partial v}{\partial z})^2 + (\frac{\partial u}{\partial z})^2 + (\frac{\partial u}{\partial y})^2\right]$$

aid of standard wall functions. The boundary conditions are summarized in Table.3; for details of atmospheric conditions, Serag-Eldin (1984b) is to be consulted, whereas Patankar and Spalding (1970), Launder and Spalding (1972) are to be consulted for wall functions.

Concentration Model

The concentration field is derived from the full three-dimensional, unsteady form of the advection-diffusion equation, expressed by :

Table 3. Flow Field Boundary Conditions

Variable	BOUNDARY			
	$Z = 0$	$Z=Z_{max}$	$y=y_{max}$	Bottom
u	$\sqrt{\dfrac{\tau_w}{\rho}} \cdot \dfrac{\sin\phi}{K} \ell n(\dfrac{y}{y_o})$	$\dfrac{\partial u}{\partial z} = 0$	same as Z=0 with $y=y_{max}$	$U_w = 0$
v	0	$\dfrac{\partial v}{\partial z} = 0$	" "	$\dfrac{U_w}{\sqrt{\tau_w/\rho}} = \dfrac{1}{K}\ell n(\dfrac{y_n}{y_o})$
w	$\sqrt{\dfrac{\tau_w}{\rho}} \cdot \dfrac{\cos\phi}{K} \ell n(\dfrac{y}{y_o})$	$\dfrac{\partial w}{\partial z} = 0$	" "	$V_w=0, \dfrac{\partial v_w}{\partial y_n} = 0$
k	$\dfrac{\tau_w}{\rho\sqrt{C_\mu}}$	$\dfrac{\partial k}{\partial z} = 0$	" "	$\partial k/\partial y_n = 0$ $G_{k,w} = \tau_w \cdot \dfrac{\partial U_w}{\partial y_n}$
ε	$\dfrac{1}{Ky} (\dfrac{\tau_w}{\rho})^{3/2}$	$\dfrac{\partial \varepsilon}{\partial z} = 0$	" "	$\varepsilon_w = \dfrac{C_\mu^{3/4} k^{3/2}}{K y_n}$

where:

y_{max}, Z_{max} = maximum value of y and Z coordinates, respectively.

τ_w = wall shear stress , y_o = surface roughness height

K = 0.435 is von Karman constant

y_n = distance of near wall node from wall

$G_{k,w}$ = G_k near wall

U_w, V_w = near wall velocity components, parrallel and normal to wall, respectively

ε_w = value of ε near wall.

$$\frac{\partial c}{\partial t} + \frac{u\partial c}{\partial x} + \frac{v\partial c}{\partial y} + \frac{w\partial c}{\partial z} = \frac{\partial}{\partial x}\left(\frac{\Gamma_{c,eff}}{\rho_a}\frac{\partial c}{\partial x}\right) + \frac{\partial}{\partial y}\left(\frac{\Gamma_{c,eff}}{\rho_a}\frac{\partial c}{\partial y}\right)$$

$$+ \frac{\partial}{\partial z}\left(\frac{\Gamma_{c,eff}}{\rho_a}\frac{\partial c}{\partial z}\right) + S_c \qquad (4)$$

where C is the local mass concentration of the pollutant defined as the mass of pollutant per unit volume of the mixture, S_c is rate of c per unit volume and time and $\Gamma_{c,eff}$ is the effective turbulence diffusion coefficient expressed in terms of μ_{eff} according to :

$$\Gamma_{c,eff} = \frac{\mu_{eff}}{\sigma_{c,eff}} \qquad (5)$$

where $\sigma_{c,eff}$ is a model constant ascribed the value $\sigma_{c,eff}=0.7$. The local values of u,v and w are derived from the computation of the aerodynamic field.

Boundary Conditions : For this field the boundary values/flux of c have to be specified at all six boundaries of the three-dimensional integration domain, and for all time. The particular boundary conditions to be employed are problem dependent. However, a widely applicable set would be to set : C=0 at the two inflow boundaries (x=0, Z=0),zero normal gradients at the remaining four boundaries, and the specification of the emitted flux of c at the source location as a source term S_c.

If the pollutants emitted from the source are lighter than air, partial correction for buoyancy effects may be introduced through an effective height correction, as is commonly the practice in Gaussian plume models, see for e.g. Perkins (1974).

SOLUTION PROCEDURE

The solution is obtained in two steps employing two separate computer programs. The first step produces the flow field and the second step employs the results of the computed flow field to produce the concentration field.

Solution of Flow Field

As result of the assumptions made, the solution of the flow model equations need be performed over a single X-plane only, and hence it is obtained with the aid of a general computer code for the solution of two-dimensional recirculating flows presented by Pun and Spalding (1977), after introducing the improvements reported by Serag-Eldin (1984a). This code adopts the "SIMPLE" algorithm described by Caretto et al (1975) and Patankar (1980), which is an iterative finite difference solution procedure that employs a "pressure correction" equation to derive the pressure field, and satisfy the continuity equation indirectly.

Within each iteration, the solution sequence is as follows :

   i)the finite-difference equations for w are solved.
  ii)the finite-difference equations for v are solved.
 iii)the finite-difference equations for u are solved.
  iv)the "pressure-correction" equation is solved for y and Z
     directions only, and the p, v and w fields are updated.
   v)the finite-difference equations for k are solved.
  vi)the finite-difference equations for ϵ are solved.

Solution of Concentration Field

Once the flow field is computed, the resulting velocity and effective diffusivity fields are stored on disc and the flow code
is released. A specially developed concentration code is then
loaded,and made to read the above fields offdisc and generate
the required three-dimensional   coefficients for the finite-
difference form of equation (4). The finite-difference formulation is of the conservative type employing the control volume
approach. The upwind-differencing scheme is employed for the
evaluation of the coefficients.

The solution starts from a given set of initial values and progresses semi-implicitly step by step in time, until either a
steady state solution is obtained or the end of the specified
time duration is reached.

Extension to Unsteady Flow

Although the present solution procedure assumes the flow field
to be steady, this need not be the case. If the flow field is
variant with time, then the two-step solution (flow field then
concentration field) would merely have to be repeated each time
interval. Of course, the unsteady terms would also have to be
introduced in the flow model equations.

DEMONSTRATION OF APPLICATION

Demonstration Case

In order to demonstrate the application of the model, the mathematical model and solution procedure are applied to predict the
flow over a long factory building, of the shape sketched in
Figure 1, and of the following dimensions : $\ell_1 = 3.5$m, $\ell_2 = 20$ m,
$h_1 = 6.5$ m, $h_2 = 9.0$ m, $h_3 = 9.9$ m.

The free wind approaches at an angle $\phi = \tan^{-1} 0.5$, and for the
terrain considered $\tau_w = 0.18$ N/m^2 and $y_o = .02$ m. The concentration
field is assumed to be initially zero everywhere, and then a
sudden continuous steady release of 0.6 kg/sec of pollutant is

emitted from the chimney exit.

Computed Flow Field

The computed flow field is quasi three-dimensional i.e. the streamlines are three-dimensional, but invariant with x. Thus the resultant streamsurfaces give the same projections over any x-plane.

Figure 2 displays the projection of some stream-surfaces on a x-plane. The projection reveals a large recirculation zone downstream the building and a relatively small one at the forward corner. Due to the u component of velocity, the flow in the recirculation zone follows a helical path, as sketched in Figure 3. This motion is responsible for convecting properties (such as c) along the building length (X-direction).

The streamlines traced over any streamsurface exhibit varying curvature, i.e. w/u ratios which vary from one Z-location to another on the same stream-surface, and from one stream surface to another at the same Z location. However, since the flow gradients are all independent of x, each streamsurface displays an infinite set of parrallel streamlines.

Computed Concentration Field

Since the computed concentration field is three-dimensional and time dependent, a full display of the field requires more space than available here. Thus only sample results are presented for the purpose of illustration.

Figures 4-11 display the computed contour lines 2.1 seconds after the start of pollutant emission, at the Z-plane passing through the plume source $Z=Z_S$, and at Z-planes 3.5, 8.5, 12, 17.5, 20, 28.5 and 43.5 m downstream of it, respectively. It may be worth noting that $Z-Z_S=20$ m gives the first computational plane downwind the building, and that the furthest Z-plane, $Z-Z_S= 45.5$ m is 7 m before the edge of the downwind recirculation zone. The plots display contour lines $C=10^{-2}$, $10^{-3}, 10^{-4}$, 10^{-5}, 10^{-6} and 10^{-7} kg/m^3. The absence of any of these contour lines on a plot, indicates the absence of this level of C at the pertaining Z location.

Examination of Figures 4-11 reveals that :

i) the concentration of the core of the plume decreases monotonically in the Z-direction, which is expected due to the diffusion processes. The latter also causes the external contour lines to expand, so that the peak-to-average concentration falls rapidly in the downstream direction.

ii) there is a lateral shift of contour lines in the X-direction

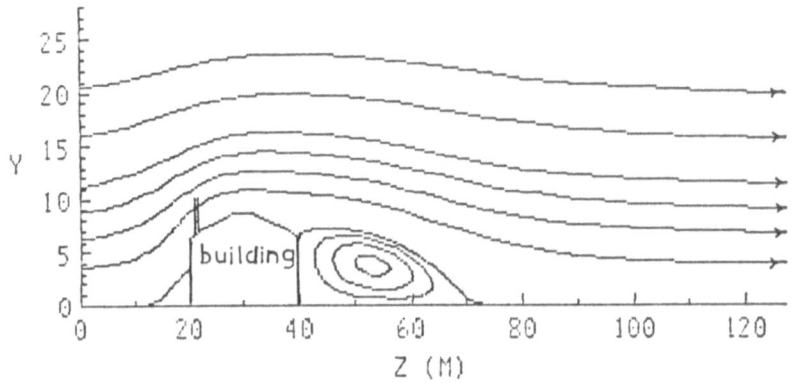

FIGURE.2. PROJECTION OF STREAMLINES IN X-PLANE

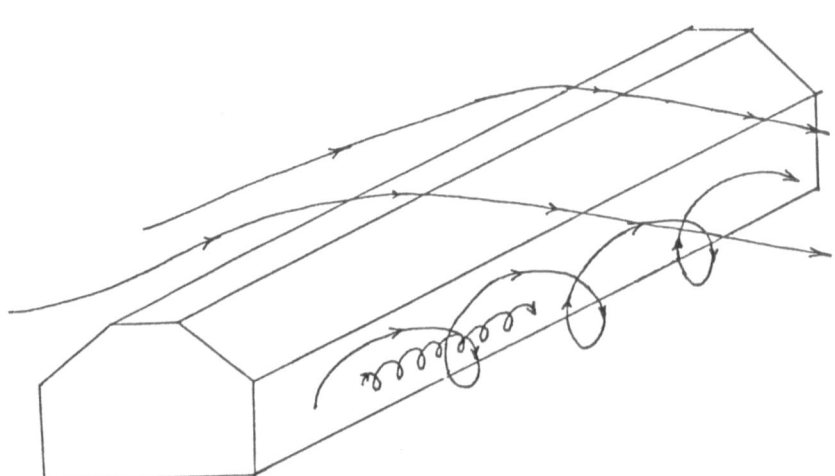

Figure 3. Sketch of Streamlines.

as the plume flows downstream, due to the convective effect of u.

iii) after $Z-ZS=28.5$ m, the mass of pollutant present over the downstream planes is much reduced, due to the limited downstream propagation of the plume in 2.1 seconds.

After 6.1 seconds from the start of emission the same C contour lines show only minor changes in profile over the ones after 2.1 seconds, for all Z-planes upto $Z-ZS=12$ m; hence they will not be presented here. The contour lines for $Z-ZS=17.5,20,29$ and 43.5 m are displayed in Figures 12-15. Comparison against the corresponding profiles after 2.1 seconds, reveals an increase in the core concentration and a lateral and vertical expansion of the plume, particularly so for the furthest downstream section. Figures 16-19 display the contour lines 16 seconds after emission. They reveal more expansion of the plume in the forward-direction with time, particularly so for the further downstream sections.

After 32 seconds, the contour lines have practically stablized for $(Z-Z_S)$ of atleast 20 m; thus only the profiles at $Z-ZS=28.5$ m and 43.5 m are displayed in Figures 20,21, respectively. The corresponding profiles after 182 seconds are revealed in Figures 22,23, respectively. Comparison together and with previous time steps reveals that, at large values of x within the recirculation zone, there is still considerable expansion of the contour lines in the x-direction, even though most of the remaining distribution has fairly well stabilized. This is because the expansion of the plume in·this region is due mainly to the helical motion of the fluid in the recirculation zone, and since this motion is slower than the forward moving motion, the local spread in the x-direction takes longer to stabilize than in the other regions.

Figures 24-27 display the variations of ground concentrations downwind the building, after 6.1, 16,32 and 182 seconds respectively.

It is apparent that after 6.1 seconds there is a relatively small area with concentration above $C=10^{-7}$, but that the ground concentration rises to above 10^{-4} after 32 seconds. After 182 seconds, Figure 27 reveals that far away from the building the concentration profiles assume a cigar like shape pointed in the direction of $\phi=\tan^{-}1/2$, which is the free-wind direction, as expected.

DISCUSSION AND CONCLUSION

The model presented here introduces certain assumptions regarding body dimensions and buoyancy effects, hence it is not universally applicable; however, it does find many useful applications. For these applications the model offers a relatively

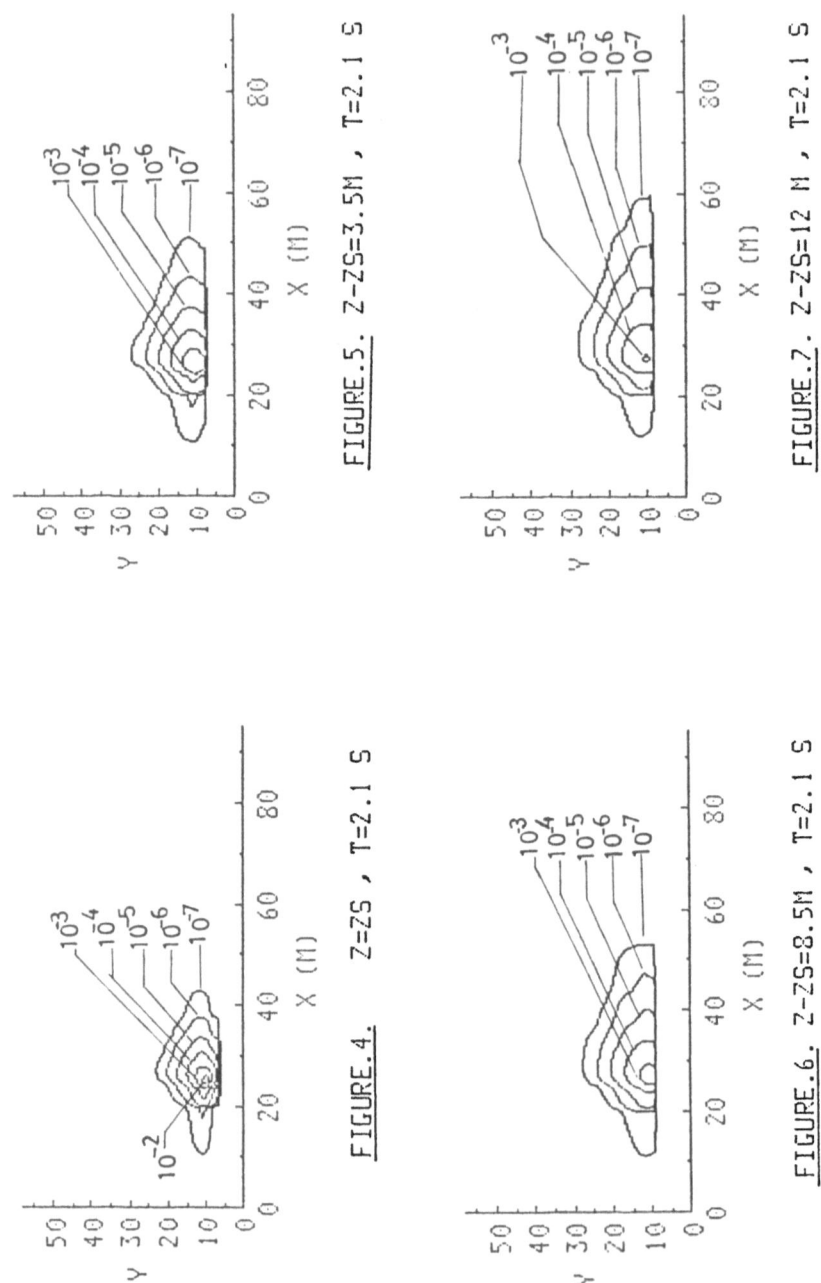

FIGURE.4.   Z=ZS , T=2.1 S

FIGURE.5. Z-ZS=3.5M , T=2.1 S

FIGURE.6. Z-ZS=8.5M , T=2.1 S

FIGURE.7. Z-ZS=12 M , T=2.1 S

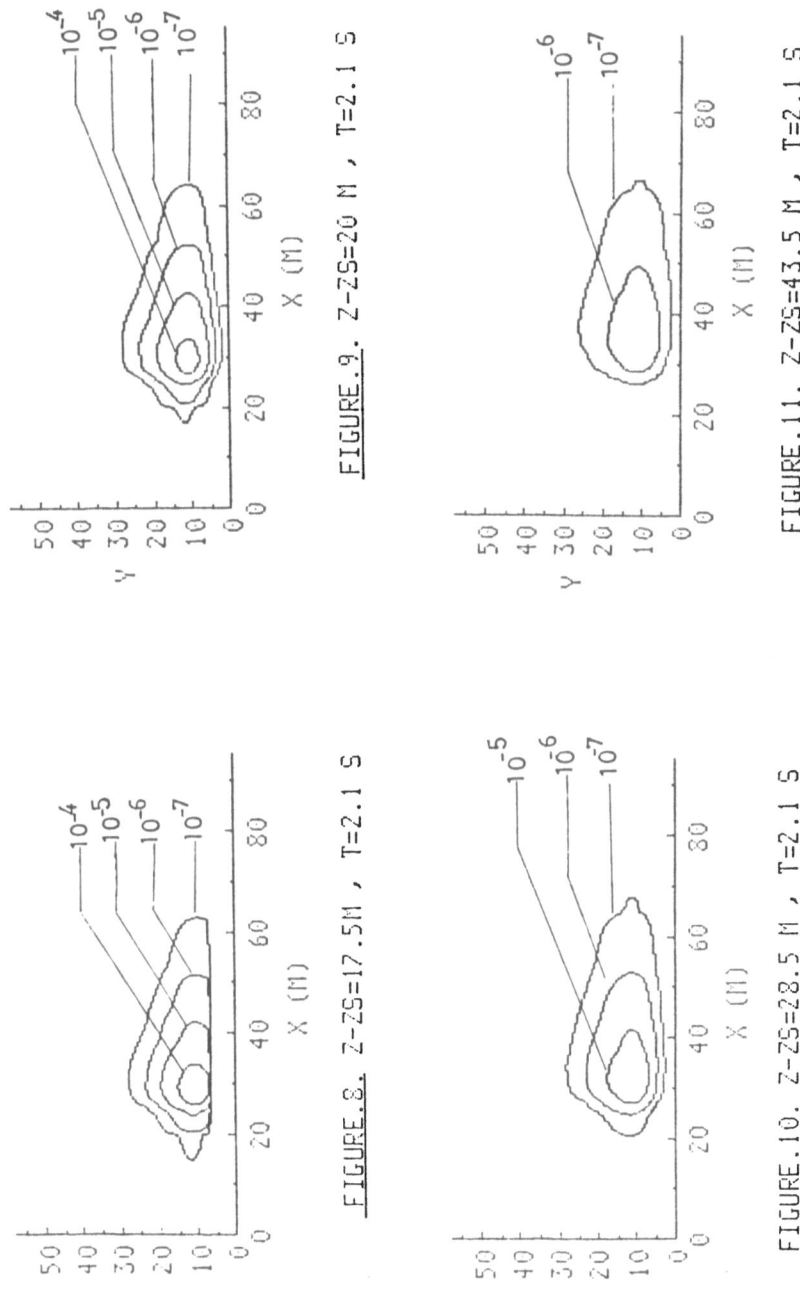

FIGURE.8. Z-ZS=17.5M , T=2.1 S

FIGURE.9. Z-ZS=20 M , T=2.1 S

FIGURE.10. Z-ZS=28.5 M , T=2.1 S

FIGURE.11. Z-ZS=43.5 M , T=2.1 S

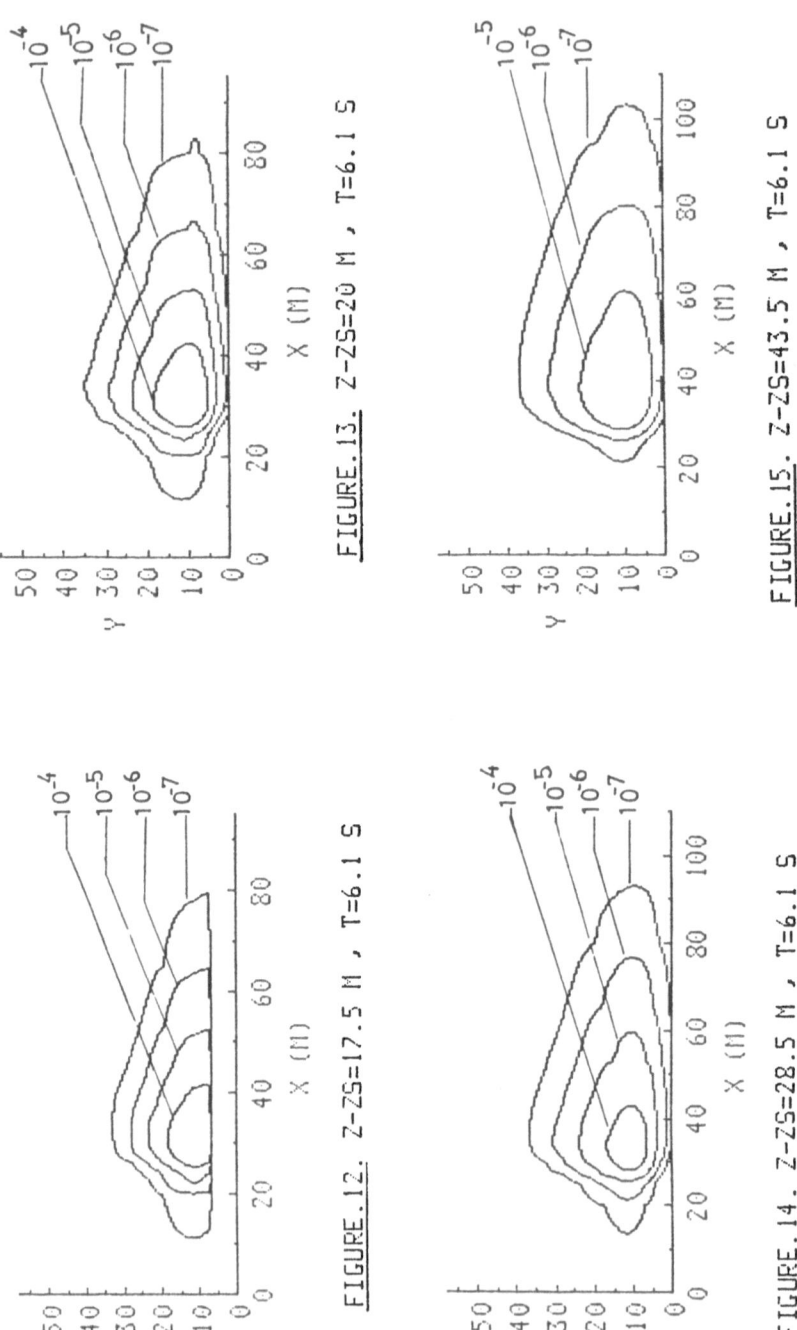

FIGURE. 13. Z-ZS=20 M , T=6.1 S

FIGURE. 15. Z-ZS=43.5 M , T=6.1 S

FIGURE. 12. Z-ZS=17.5 M , T=6.1 S

FIGURE. 14. Z-ZS=28.5 M , T=6.1 S

16-42

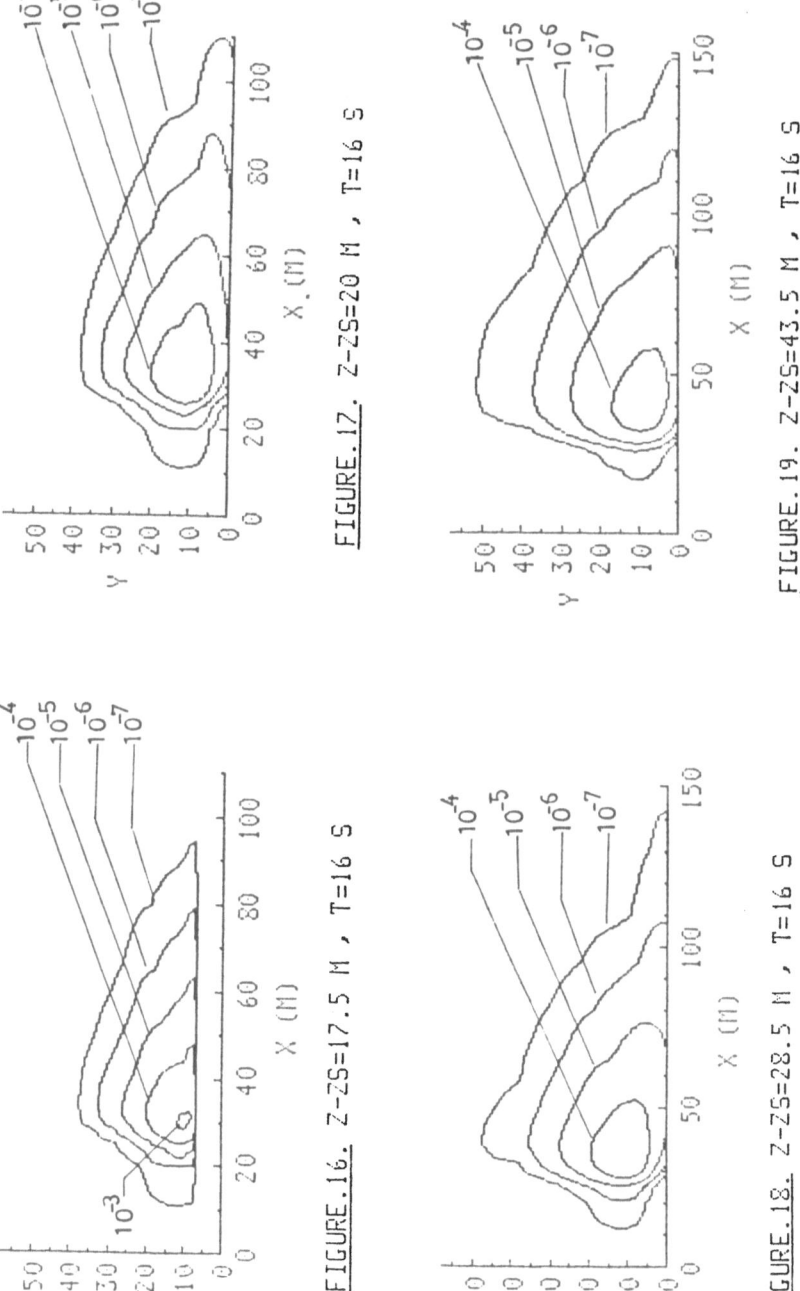

FIGURE.16. Z-ZS=17.5 M , T=16 S

FIGURE.17. Z-ZS=20 M , T=16 S

FIGURE.18. Z-ZS=28.5 M , T=16 S

FIGURE.19. Z-ZS=43.5 M , T=16 S

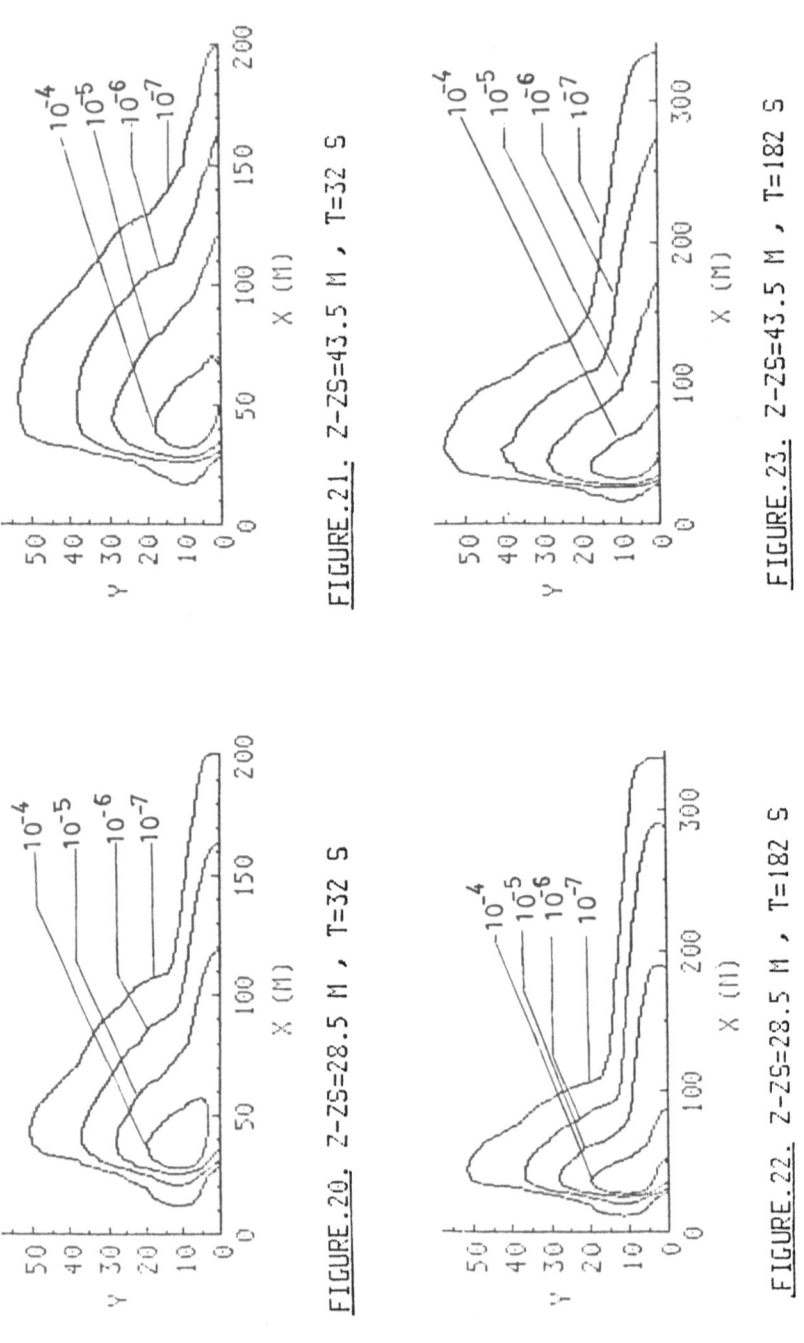

FIGURE.20. Z-ZS=28.5 M , T=32 S

FIGURE.21. Z-ZS=43.5 M , T=32 S

FIGURE.22. Z-ZS=28.5 M , T=182 S

FIGURE.23. Z-ZS=43.5 M , T=182 S

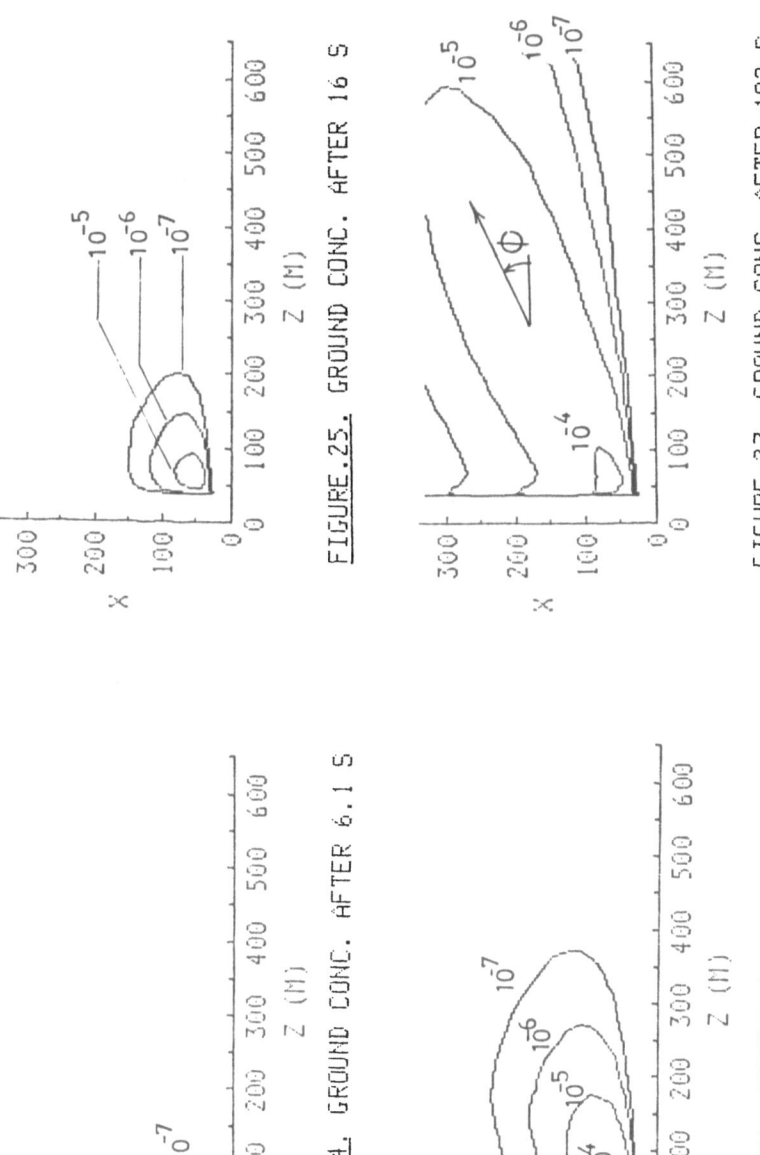

FIGURE.24. GROUND CONC. AFTER 6.1 S

FIGURE.25. GROUND CONC. AFTER 16 S

FIGURE.26. GROUND CONC. AFTER 32 S

FIGURE.27. GROUND CONC. AFTER 182 S

economic and reliable means of predicting the fully three-dimensional, time dependent concentration field in a complex flow pattern created by an obstacle in the atmosphere. Universal application would require the solution of fully three-dimensional, unsteady and coupled equations of the flow and concentration models, and would necessitate several orders of magnitude more computational requirements than the present model.

REFERENCES

Carretto,L.S.,Gosman, A.D., Patankar, S.V. and Spalding D.B. (1975). Two Calculation Procedures for Steady, Three-Dimensional Flows with Recirculation. Proceedings of the 3rd Int .Conf. on Num. Methods in Fluid Mechs.,Vol.II.

Launder, B.E. and Spalding, D.B. (1972). Mathematical Models of turbulence, Academic Press, London and N.Y.

Launder, B.E. and Spalding, D.B. (1974). The Numerical Computation of Turbulent Flows. J. Computer Methods in Applied Mechanics and Eng.,3, p. 269-289.

Patankar, S.V. and Spalding, D.B. (1970). Heat & Mass Transfer in Boundary Layers. Intertext Books, London 2nd edition.

Patankar, S.V. (1980). Numerical Heat Transfer and Fluid Flow. Hemisphere Publishing Corporation, Washington, N.Y., London.

Perkins, H.C. (1974) Air Pollution McGraw-Hill, London, Tokyo, Sydney.

Pun, W.M. and Spalding, D.B. (1977). A General Computer Program for Two Dimensional Elliptic Flows. Mech.Eng.Dept.,Report No. HTS/76/2, Imperial College, London.

Serag-Eldin, M.A. (1984a). Computation of the Flow Field Over Long Two-Dimensional Obstacles in the Atmosphere. IBM Cairo Scientific Center, Tech.Rep. No. 002.

Serag-Eldin, M.A. (1984b). Computation of the Atmospheric Surface-Layer Flow Over a Wide, Inclined, Backward Facing Step. Proceedings of the 11th IASTED Int.Conf. Applied Modelling and Simulation, Nice 1984, p.69-73. ACTA Press, Anaheim, Calgary and Zurich.

VKI Lecture Series: Pollutant Dispersal (Feb.1985) Von Karman Institute, Belgium.

17. DATABASES IN ENGINEERING

ESTSOL - A SUBROUTINE LIBRARY FOR ESTABLISHMENT AND SOLUTION
OF LINEAR EQUATION SYSTEMS

Harald Tägnfors, Nils-Erik Wiberg, and Lars Bernspång

Chalmers University of Technology, Göteborg, Sweden

1. INTRODUCTION

One of the crucial steps in the solution of problems with Finite
Element Methods is establishment and solution of systems of
equations. A numerous number of papers have been presented
concerning different techniques of solution of equation systems
such as direct methods, Mondkar and Powell (1974), Wilson (1978),
Kamel and McCabe (1978), Argyris and Brönlund (1975), and iterative methods such as conjugate gradient, Axelsson and Gustafsson
(1980), Gustafsson and Lindskog (1984), and viscous relaxations,
Hughes et al. (1983), Zienkiewicz (1984). However, the user of
these methods normally has to write a number of subroutines for
the establishment and solution which are different for each
method. The presented subroutine library will establish and
solve a linear system of equations for a number of right hand
sides using almost the same data for all methods. It provides
the user of an application program with the possibility to
select the solution method which is best suited concerning
efficiency and accuracy for the problem at hand.

The subroutine library presented here covers both the
establishment and the solution and is supported by a data base
management system (DBMS). Several direct and iterative methods
may be chosen. Independent of which method is to be used at the
solution the required data is the same. The subroutine library
is a part of the programming system SITU, Tägnfors et al. (1985),
which is both a tool for program development and an assemblage
of engineering programs.

A program developer only has to calculate required data in
a space provided by the data base management system, which in
this version is PRISEC, Tägnfors et al. (1981). A program developer first has to assure that some arrays containing information about topology are available and then call subroutine PRESOL
for analysis of variable topology, which means calculation of
global variable numbers which will be coupled. Arrays produced

in this subroutine will facilate the establishment of the array for specification of prescribed variables and the right hand side array.

After the establishment of element matrices in a user made routine any of the routines for establishment and solution may be called.

The direct methods are Crout elimination and factorization based on element factorization. If the available size of primary memory is small compared to the space required for the coefficient matrix the Crout solution can be made in steps with use of secondary memory. The iterative method is the conjugate gradient method with different types of preconditioning and viscous relaxation with splitting.

It is possible to specify and utilize previously made work such as an already factorized coefficient matrix, or that a starting vector is available in an iterative procedure.

When a subroutine in ESTSOL is called the data base management system will assure that data to be used are secured to be in primary memory by calls to DBMS-routines and that space in primary memory for required arrays is available. The result of the solution is one array for each right hand side containing the variable values. These arrays will be stored in the data base.

In the paper a comparison of required time and accuracy for some model problems is made.

2. SYSTEM EQUATION

The solution of differential equations by the finite element method gives as a result a system of equations

$$S\tilde{u} = \tilde{f} \tag{1}$$

which has to be solved. For a problem in elasticity solved by the displacement method S is the stiffness matrix, $\tilde{u}$ the displacements in the nodes and $\tilde{f}$ the corresponding load vector. For finite element problems S is large and sparse, which has to be taken advantage of in order to obtain an economic solution. The stiffness matrix S in Equation 1 is obtained from element matrices S^e by use of coupling data. Some of the variables $\tilde{u}$ are prescribed, $\tilde{u}_p$, so we can formally partition the equation system Equation 1 into

$$\begin{bmatrix} S_f & S_{fp} \\ S_{pf} & S_p \end{bmatrix} \begin{bmatrix} \tilde{u}_f \\ \tilde{u}_p \end{bmatrix} = \begin{bmatrix} \tilde{f}_f \\ \tilde{f}_p \end{bmatrix} \tag{2}$$

Using the subroutine library ESTSOL this equation system can be solved by different methods, Figure 1.

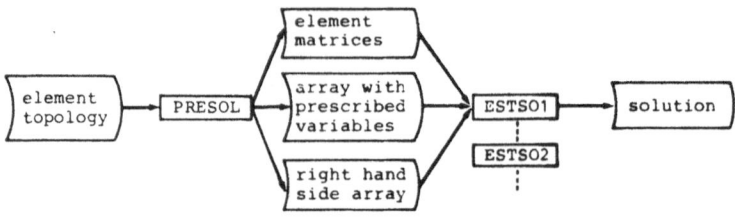

Figure 1.   Subroutines and data for solution

The direct methods can be used for non-definite systems of equations but the iterative methods only for positive definite systems.

3. DATA MANAGEMENT

The data management is a crusial matter in a finite element analysis program. In this subroutine library it is managed by a DBMS (Data Base Management System) called PRISEC, Tägnfors et al. (1981). The data management can be handled by any other suitable DBMS-system.

4. SOLUTION METHODS

In this chapter the different methods of solution will shortly be described.

4.1   Direct solution with Crout method. In-core solution

The symmetric system of equation, Equation 1, will be solved by the Crout method. Only the unknown variables $\tilde{u}_f$ will be solved so the system to establish and solve is

$$S_f \tilde{u}_f = \tilde{\underline{f}}_f; \quad \tilde{\underline{f}}_f = \tilde{f}_f - S_{fp} \tilde{u}_p \tag{3}$$

The stiffness matrix S_f is stored within a skyline profile, which is utilized in the Crout solution process, Figure 2a. The Crout method is used here because it may give less rounding off errors than straight-forward Gauss elimination. A similar routine is earlier published, Wiberg and Tägnfors (1974). It is also possible to make static condensation with the routine.

4.2 Direct method with Crout method. Out-of-core solution

For an out-of-core solution the same method as above is used.
Only a part of the system is within this core memory, Figure 2b.
For each move of the working space is moved as far as possible
in order to minimize the administration. The procedure was originally produced for the program FEMWT, Wiberg and Tägnfors
(1972).

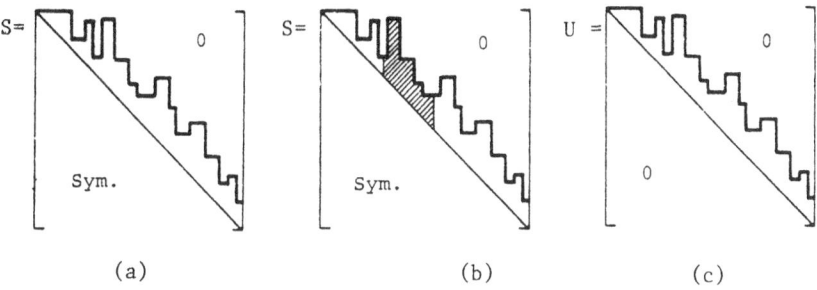

Figure 2.   (a) Symmetric system with skyline profile
            (b) Working space for out-of-core solution
            (c) Triangular factor U for the system.

4.3 Direct method, element factorization

The factorization of the assembled stiffness matrix $S = U^t U$ in
the standard displacement method may give large numerical error.
It has been shown that a direct factorization of the differential stiffness matrix $S_d = U_d^t U_d$ (or of the non-singular element
stiffness matrix $S_n = U_n^t U_n$) reduces the numerical errors,
Argyris and Brönlund (1975):

$$S = U^t U = B^t U_d^t U_d B \qquad (4)$$

The triangular factor U, Figure 2c, may be obtained by a
QR-transformation of the rectangular matrix $U_d B$. The error
produced by a Crout elimination of standard displacement method
equations is proportional to the spectral condition number
$\kappa(S) = \lambda_n/\lambda_1$, where λ_1 and λ_n are the smallest and largest
eigenvalues of the stiffness matrix S. The numerical error of
this procedure is reduced by the factor $1/\sqrt{\kappa(S)}$ compared to the
standard displacement method. It is to be noted that the triangular factor U based on element factorization is the Cholesky
factor of S and thus needs exactly the same storage space as
the symmetric matrix S, as seen in Figure 2c.

4.4 Iterative methods; conjugate gradient with preconditioning

Solution of $S\tilde{u} = \tilde{f}$ iteratively by the preconditioned conjugate
gradient method is made according to

$$\tilde{r}^{(i)} = S\tilde{u}^{(i)} - \tilde{f} \qquad\qquad (5)$$

$$\tilde{d}^{(i)} = -C^{-1}\tilde{r}^{(i)} + \beta_i\tilde{d}^{(i-1)}$$

$$\tilde{u}^{(i+1)} = \tilde{u}^{(i)} + \lambda_i\tilde{d}^{(i)}, \quad i = 0,1,2,\ldots$$

where β_i and λ_i are determined from the condition of conjugate vectors $\tilde{d}^{(i)}$ and the minimization of potential energy, and C is a well chosen preconditioning matrix. The convergence depends on the condition number $\kappa = \kappa(C^{-1}S)$ which is less than $\kappa(S)$, for a suitable choice of C.

Many proposals for selection of an efficient C have been made. In Axelsson and Gustafsson (1980) a modified incomplete Cholesky factorization (MIC) in a multigrid manner with kept row sums is applied and in Hughes et al (1984) and Gustafsson and Lindskog (1984) element-by-element factorization techniques are utilized. In Gustafsson and Lindskog (1984) it is pointed out that the element-by-element preconditioning method is not very successful for structural elements, such as beams, plates, and shells.

The preconditioned matrix can in problem with hierarchical formulation be chosen as

$$C = \begin{bmatrix} S_b & 0 \\ 0 & D \end{bmatrix}$$

where S_b belongs to variables from the basic mesh and D the diagonal stiffness elements corresponding to the hierarchical variables, see Wiberg et al. (1984a) and (1984b). The greatest advantage of iterative methods is that much less core storage is needed.

4.5 Iterative method, viscous relaxation with splitting

Viscous relaxation with line search - that is Equation 5 with $\beta_i = 0$, see Hughes et al. (1983), Zienkiewicz (1984) and Wiberg et al. (1984a) has been successfully applied to many problems, also in a split element-by-element version.

We have observed that in all iterative procedures it is essential to calculate the residual with a minimum of rounding errors. A simple and efficient way is to calculate it from relative displacements, see Axelsson and Gustafsson (1980).

5. SUBROUTINE LIBRARY - ESTSOL

5.1 Structure of subroutines

The subroutine library ESTSOL contains one subroutine for each of the methods to be called in an applications program. The library contains the following subroutines:

ESTS01 – Direct. CROUT–method, in core
ESTS02 – Direct. CROUT–method, out of core
ESTS03 – Direct. Element factorization, in core
ESTS04 – Iterative. Conjugate gradient with MIC–
      preconditioning
ESTS05 – Iterative. Conjugate gradient with weighted
      elementfactor – preconditioning
ESTS06 – Iterative. Jacobi viscous relaxation, element
      by element

Within each of these routines the following calculations are performed:

. Analysis of equation topology, e.g. equation numbers are calculated for non-prescribed variables and a number of work arrays facilitating establishment and solution are calculated.

. Establishment and solution

In ESTS04 C is calculated according to MIC(0) as proposed by Axelsson and Gustafsson (1980) and for ESTS05 C is assembled from element factors calculated as for ESTS03 but weighted. As C and S for ESTS05 are assembled as sums over all elements this method can be done element by element, but at present assemblage is adopted for ESTS05. in ESTS04 C is calculated from S and can not be performed element-by-element. Both these methods store only non-zero earlier in S and C so each of them is smaller than S or U in other methods.

In ESTS06 C is calculated according to Hughes et al (1983) as a product of element contributions that are never assembled. Iterations are made in such a way that each element contribution is multiplied by an array calculated from the residuals and thus the storage requirement is kept to a minimum. Line search is used because it is essential for good convergence. It is also possible to calculate residuals from global or relative displacements. In the problems below relative displacements are used.

Independent of the choice of method for the solution, a subroutine PRESOL shall be called before any of the solution routines is used. In this routine the variable topology is analysed, e.g. variable numbers are calculated from a specification of variables and node incidences for elements. Arrays calculated at this step are normally used when the right hand side array is established.

5.2 Input to the routines
All solution routines need almost identical input data. Scalars required in the routines are transferred as arguments. Arrays are supposed to be stored with specified names in the data base. The data management is here described in connection with the data base management system PRISEC.

By scalars information are given concerning problem size, calculations to be performed and accuracy to be reached in iterative methods. Calculation can be controlled so that data in work arrays, for example triangular factors which have been calculated at a previous call, need not to be recalculated.

Data stored in arrays are either element data or data for the whole structure. The arrays shall be stored in a hierarchic data structure commencing in an arbitrary data node XXXX.

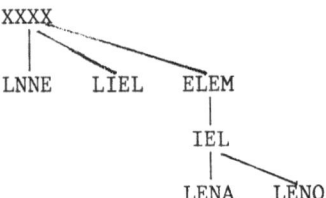

Figure 3. Required arrays for PRESOL

The required arrays for PRESOL are given in Figure 3:

LENA contains name of element variables (ordered locally).
LENO contains element nodes for element variables (ordered locally).

If these arrays are equal for all elements, the arrays may be inserted in data node XXXX instead of in node IEL for each element.

LNNE contains numbers of incidenting nodes for all elements.
LIEL is an index defining for each element where in LNNE data is stored.

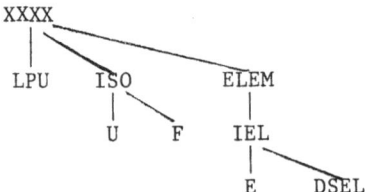

Figure 4. Required arrays for solution routines

The required arrays for solution routines are given in Figure 4:

LPU contains information for each variable whether it is prescribed or not. This arry is used for the analysis of equation topology, e.g. calculation of equation numbers for variables.

U and F   are left and right hand side arrays for equation
          system for solution case ISO. U may for conveniency
          contain prescribed values but they are not used at
          the solution. If only one right hand side array is
          solved for U and F may be inserted in node XXXX.

E   contains for each element IEL the stiffness matrix

DSEL   contains constitutive matrix for each element IEL.
          It is only needed for ESTSO3 and ESTSO5 and then
          an external function also has to be supplied for
          calculation of derivatives.

5.3 Output from routines

Output from the solution is the solution stored in the same
array U which was used at the input stage. Triangular factors
are stored for later use on request.

6. NUMERICAL EXAMPLES

In order to compare the different methods that are available in
ESTSOL we study three different structures (2D-problems) shown
in Figure 5. The comparison is made concerning efficiency, and
accuracy. Elements based on a bilinear displacement approximation are used.

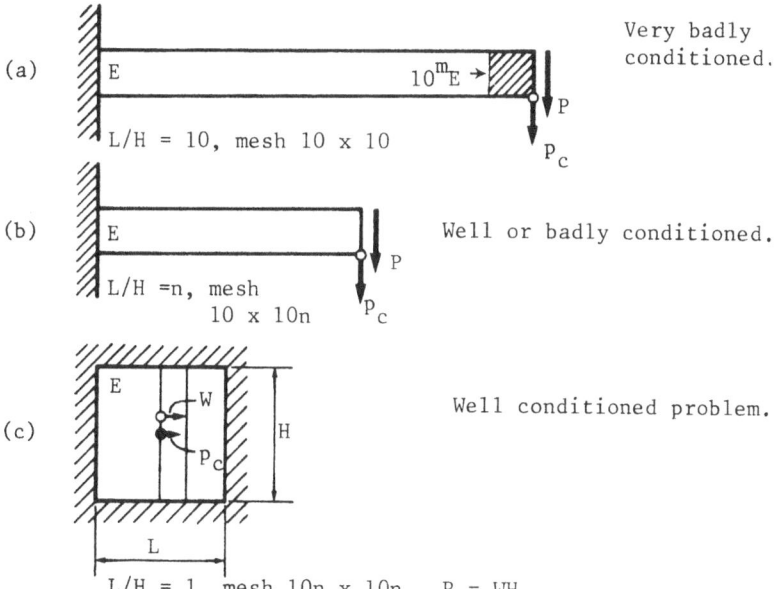

(a) $L/H = 10$, mesh 10 x 10   Very badly conditioned.

(b) $L/H = n$, mesh 10 x 10n   Well or badly conditioned.

(c) $L/H = 1$, mesh 10n x 10n, $P = WH$ Well conditioned problem.

Figure 5. Studied structures

6.1 Very badly conditioned problem

The cantilever in Figure 5a is given a large stiffness at the top by increasing the E-modulus to 10^mE. For increasing m-values the stiffness matrix becomes more and more badly conditioned, which means that the condition number becomes large and so the numerical errors. Figure 6 shows calculated top displacement p_c divided by displacement p_b according to technical beam theory for different methods when the stiffness at the top is increased by the factor 10^m, for m = 0,1,2,...

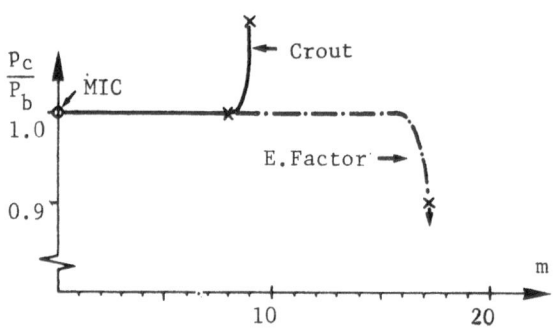

Figure 6. Top displacement of a cantilever structure

We see that for this badly conditioned problem the direct method based on element factorization is surprisingly good and much better than the ordinary Crout-elimination method. All iterative methods fail except the conjugate gradient with MIC-preconditioning for m = 0.

6.2 Well or badly conditioned problem

A cantilever, Figure 5b, with different L/H ratios and meshes with square elements loaded with a point load at the top end, is calculated with the different methods.

The calculated top displacements p_c compared to the displacements according to technical beam theory p_b are shown in Table I. It also gives the CPU-time in seconds and the number of iterations for the iterative methods. The last iterative method is suited for element-by-element treatment. The iterative methods gave the same displacements as the direct methods with 4 digits accuracy in almost all cases when the iterations is stopped when the square root of the sum of squares is less than 10^{-3}P.

When the L/H-ratio increases the stiffness matrix becomes more and more badly conditioned and the iterative methods fails due to very slow convergence. Assembled preconditioned matrix combined with the conjugate gradient method gives a much faster convergence than the element-by-element Jacobi viscous relaxation.

Table I. Calculation of uniform beams, Figure 5b, with different lengths and meshes with square bilinear elements. Calculation time in seconds and number of iterations for iterative methods within parentheses. Iterations stopped when the square root of the sum of squares of residuals is less than $10^{-3}P$. Displacements are then obtained with 4 digits accuracy.

| | | | Direct | | Iterative | | |
| | | | | | with assemblage | | element-by-element |
L/H	Mesh	Displ. P_c/P_b	Crout	Nat.fact.	Conj. grad. MIC-precond.	Conj. grad. Nat. fact.	Jacobi. Viscous relax.
1	10×10	1.646	0.5	2.4	5.9(53)	6.5(68)	15.3(100)
2	10×20	L.161	0.9	5.1	21.6(90)	22.5(123)	68.9(226)
3	10×30	1.069	1.4	8.0	47.0(127)	50.0(189)	352.7(765)
4	10×40	1.036	1.7	11.2	85.5(178)	89.1(263)	908.5(1472)
5	10×50	1.021	2.2	14.6	139.5(242)	152.3(377)	*
10	10×100	1.002	5.7	35.6	585.0(523)	1041.9(1493)	
20	10×200	0.997	10.0	93.1	*	*	

* CPU-time > 20 min

Table II. Calculation of square plate, Figure 5c, with different meshes of square bilinear elements. Calculation time in seconds and number of iterations for iterative methods within parentheses. Iterations stopped when the square root of the sum of square of residuals is less than $10^{-3}P$. Displacements are then obtained about 3 digits accuracy.

| | | Direct | | Iterative | | |
| | | | | with assemblage | | element-by-element |
Mesh	Displ. P_c Crout	Crout	Nat.fact.	Conj. grad. MIC-precond.	Conj. grad. Nat. fact.	Jacobi. Viscous relax.
10×10	0.2157	0.2	1.8	2.4(16)	3.0(21)	1.3(9)
20×20	0.2151	3.1	18.4	37.4(31)	31.4(40)	12.4(20)
30×30	0.2150	19.8	90.1	191.7(45)	139.5(57)	41.4(29)
40×40	0.2150	48.8	270.9	606.0(58)	413.4(76)	105.6(41)
50×50	0.2150	135.5	*	*	*	178.7(44)

* Not calculated (out of core procedure needed)

6.3 Well-conditioned problem

The plate in Figure 5c with all sides changed is a well conditioned problem. The midpoint displacement p_C due to a line
load along the midsection for different square meshes is calculated. The result is given in Table II.

This problem is fairly well-conditioned, and from Table II
we find that the iterative methods are competitive to the direct
ones, only when a few digits (3) accuracy is needed. If a good
start vector is available and thus a few iterations are needed
the iterative methods will take shorter time than the direct
ones.

7. REFERENCES

Argyris, J.H. and Brönlund, O.E. (1975) The natural factor formulation of the stiffness for the matrix displacement method,
Computer Methods in Applied Mechanics and Engineering 5, 97-119

Axelsson, O. and Gustafsson, I. (1980) A Preconditioned Conjugate Gradient Method for Finite Element Equations, Which is
Stable for Rounding Errors. Information Processing, 80.

Gustafsson, I. and Lindskog, G. (1984) A Preconditioning Technique Based on Element Matrix Factorization. Chalmers Univ. of
Techn., Dept. Comp. Sci., Report 84.01R.

Hughes, T., Levit, I., and Winget, J. (1983) Element-By-Element
Solution Algorithm for Problems of Structural and Solid Mechanics. Comp. Meth. Appl. Mech and Engng. 36, 241-254.

Hughes, T., Raefsky, A., Muller, A., Winget, J., and Levit, I.
(1984) A Progress Report on EBE Solution Procedures in Solid
Mechanics. 2nd Int. Conf. on Numerical Methods for Nonlinear
Problems, Barcelona, 18-26

Kamel, H.A. and McCabe, M.W. (1978) Direct Solution of Large Sets
of Simultaneous Equations. Computers and Structures, 9, 113-123.

Mondkar, D. and Powell, G. (1974) Toward Optimal In-Core Equation Solving. Computers and Structures, 4, 531-548.

Tägnfors, H., Runesson, K., and Wiberg, N-E. (1981) PRISEC -
Version 1, A Tool for the Data Management in FORTRAN-Programs.
Chalmers Univ. of Techn., Dept. Struct. Mech., Göteborg.

Tägnfors, H., Runesson, K., and Wiberg, N-E. (1985) SITU -
General Manual for a Programming System which is SImple To Use.
Chalmers Univ. of Techn., Dept. Struct. Mech., Göteborg (In
Preparation).

Wiberg, N-E. and Tägnfors, H. (1972) FEMWT - A General Purpose Program for Solving Problems by the Finite Element Method. Chalmers Univ. of Techn., Dept. Struct. Mech., Publ. 72:2, Göteborg.

Wiberg, N-E. and Tägnfors, H. (1974) General Solution Routines for Symmetric Equation Systems. MAFELAP 2, London, 499-509.

Wiberg, N-E., Samuelsson, A., and Bernspång, L. (1984a) On the Improvement of the Numerical Accuracy of FEM-Solutions. MAFELAP 5, London.

Wiberg, N-E., Möller, P., and Samuelsson, A. (1984b) Use of Trigonometric Functions for Hierarchical Improvement of Finite Element Solutions of 2D-Static Problems. Proc. NUMETA 85, Swansea 9-11 Jan. 1985, 77-86.

Wilson, E.L. (1978) Solution of Sparse Stiffness Matrices for Structural Systems. SIAM Symposium on Sparse Matrix Computations, Nov. 2-3, Knoxvill, Tennessee.

Zienkiewicz, O.C. (1984) Iterative Methods and Hierarchical Approaches, A Prospect for the Future of the Finite Element Method. ASEA Centenniel Lecture.

MANAGEMENT METHOD OF A DATABASE SYSTEM FOR SCIENTIFIC COMPUTER CODES

Martine Paolillo

Electricité de France, Direction des Etudes et Recherches

1, avenue du Général de Gaulle

92141 - CLAMART Cédex

INTRODUCTION

This paper presents a software Database System created for the development of industrial engineering computer codes.

Mathematical objects of the Database which are in core memory or on file can be accessed automatically.

This software package, written in FORTRAN 77, has the following aims :
- To increase the capabilities of the FORTRAN 77 language ;
- To obtain a reliable code with portability on any machine ;
- To acquire programming flexibility ;
- To create and use a complete, documented and self-describing database ;
- To make a database recovery ;
- To work in a dynamic core memory context and manage file access as well as possible to minimize the costs.

Considering all these objectives, an important and efficient aid is brought to the package maker of scientific computer codes.

SPECIFICATION OF OBJECTS

Description

There are different types of objects :
- single objects (vectors, elements, matrices,...) of various
  types (INTEGER, REAL, COMPLEX, CHARACTER, LOGICAL,...) ;
- families : sets of single objects with varying sizes.

An object can be partitionned (vector divided into sub-vectors
or matrix into sub-matrices) or not.

Example :

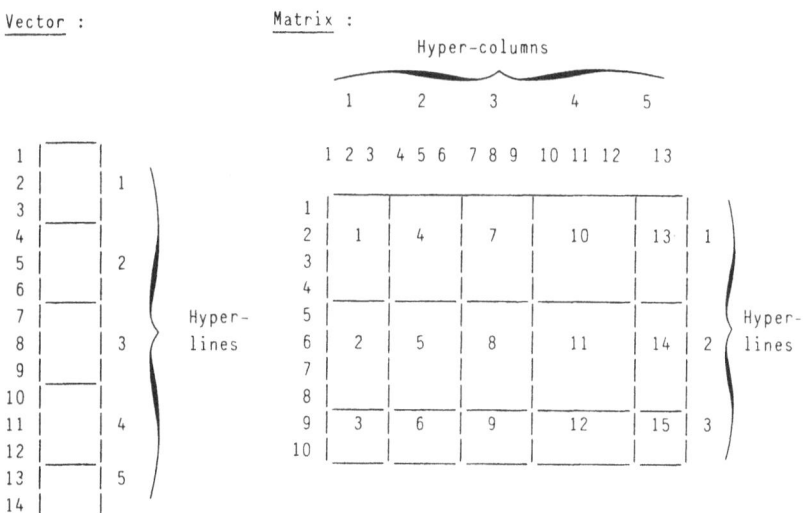

Note : The blocks, sub-parts of a matrix are numbered
       according to a hyper-column.

Justification of the objects structure choice

- Partitionned ones :

  Some objects are big and cannot stay as a whole in main
  storage, so a partitionned structure is built up, well
  adapted to the current problem. Each partition can be
  retrieved and stored in core independently of the others.
  Thus, matrices can be computed with very large dimensions on
  machines which do not have a great memory size.

- Family :

Some objects have particular forms : they can be triangular matrices, matrices structured in variable profile used for instance in mechanical calculus, their profile are called skyline and over the skyline, there are many zero values. So they can be considered as families of vectors (part of non zero columns of the matrices) with varying dimensions so as to use only on core memory or store on file the non-zero values. Moreover, if these ones are symetrical, only the upper (or lower) band is kept.

Here can be seen the schema of a stiffness matrix in profile with a largest band of 8048 terms.

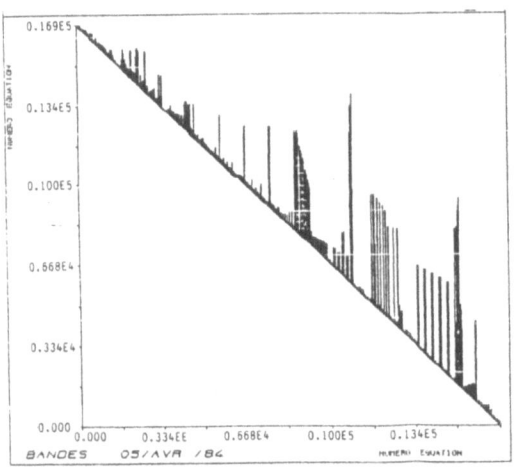

This stiffness matrix is a square symetrical one with 279 millions of elements, 6 millions of which are useful operands.

Thus, two per cent of the totality of the matrix is stored in core and mass memory.

The introduction of families simplifies programming. Indeed, in any case, it only writes a programming loop, and the value of the iteration count is the number of objects of the family. Thus the routines become shorter.

A family can have pointers of length if its objects'sizes vary. Any single object is named ; a family can be named or numbered.

A step of creating objects is necessary to constitute the database. Their caracteristics are stored in a set of attributes. Their names are inserted in a repertory with a "hash-coding" method.

Just a specific CALL to the names is sufficient to access directly to the values if they are in core memory or to retrieve them from the file.

Syntactic description

Attributes of objects are clustered in the following way :

- Essential attributes :
  < CLASS > :: G/L/V
      with G : Global database ;
           L : Local database ;
           V : Volatile database.

There are three kinds of direct access files, we shall see later on the significance of these expressions.

< KIND > :: E/V/RM/SQ/SM
    with E : Single Element (a real, complex, integer,..);
         V : Vector ;
         RM : Rectangular Matrix ;
         SQ : SQuare matrix ;
         SM : Symetrical Matrix.

< TYPE > :: I/R/C/R8/L/K [n]
    with I   : Integer ;
         K [n] : Character$_\star$n ;
         R   : Real ;
         C   : Complex ;
         R8   : Real$_\star$8 ;
         L   : Logical.

- Optional attributes :
  < PARTITION > :: PB/PC/PV
      with PB : Bloc Partitionned matrix ;
           PC : matrix partitionned in columns ;
           PV : vector partitionned in sub-vectors.

< LENGTH POINTER> :: L/C/LC/L.NAMEFX/C.NAMEFY/LC.NAMEFZ

L | When these letters are specified, pointers of
C } lines and/or columns are created, attached to
LC | the family named in the syntax ;

L.NAMEFX | We access to the length pointers of the
C.NAMEFX } families whose names are NAMEFX, NAMEFY, NAMEFZ.
LC.NAMEFZ |

So each lengths' pointer can be owned by one family or shared between several families.

- Creation of attributes of a single object :

First its name and references are generated :

CALL JCONOM ('list of arguments')

with 'list of arguments' :

< object name > < class > < kind > < type > [< partition >]

then the dimensions have to be issued :

CALL JCRNOL ('Name of object',NOL) ← Creation of the number of lines.

CALL JCRNOC ('Name of object',NOC) ← number of columns.

Creation of the possible partitions :

CALL JCRLSV ('name',LSV)   ← Creation of the Length of Sub-Vector.

CALL JCRLPB ('name',LSM,CSM) ← Creation of the number of Lines and Columns of Sub-Matrix.

Note : A partitionned object has an internal pointer of partitions that indexes the set of addresses of each partition in core and file.

- Creation of attributes of a family :
  Name and its reference :

  $\qquad$ CALL JCFNOM('List of arguments',NOB).

  NOB : number of objects of the family.

  'List of arguments' =

  <family name[.pointer name]> <class> <kind> <type>

  [<lengths pointers>] [<partition>].

  The family can have either named objects and then a pointer
  of names is created, or the family is numbered.

The dimensions of objects are fixed as the same way as a
single object with different calls.

Note : A criterion is added : when each object can be read
       and written from the file, it is said to be
       "Modifiable", if it is only read, it is said to be
       "Not Modifiable".

Functions exist to retrieve attributes :

example : NOL = JRVNOL ('Name') the number of lines is
          retrieved.

DATABASE INTERNAL MANAGEMENT

Description of the core memory

A work area is defined in core, dynamically managed by the
package routine JEVEUX.

For this, an internal COMMON /ZONE/ is used, its workspace is
created in only one subroutine of the code.

It is accessed by a dummy array TAB(1), with a subroutine
written in Assembly Language that localises the address of the
common.

User Program

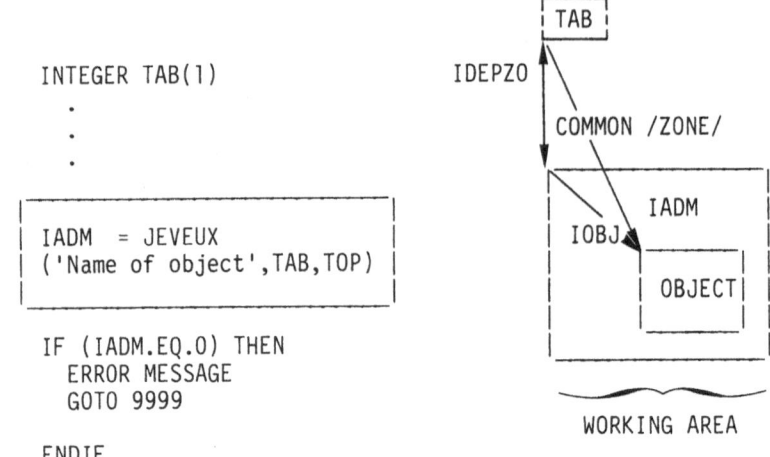

```
        INTEGER TAB(1)                 IDEPZO
              .
              .
              .
       |----------------------------|
       |                            |
       | IADM  = JEVEUX             |
       | ('Name of object',TAB,TOP) |
       |                            |
       |----------------------------|

        IF (IADM.EQ.0) THEN
           ERROR MESSAGE
           GOTO 9999

        ENDIF

  9999 RETURN
        END
```

SCHEMA 2

IDEPZO : displacement between TAB and /ZONE/ :

IOBJ   : address of the required object in core memory, with
         regard to the beginning of common /ZONE/ :

$$IADM = IDEPZO + IOBJ$$

Thus, the system "JEVEUX" dynamically gives the address of the object.

Note : Some machines have an OVERLAY structure, and the area
       memory can be arborescent like this :

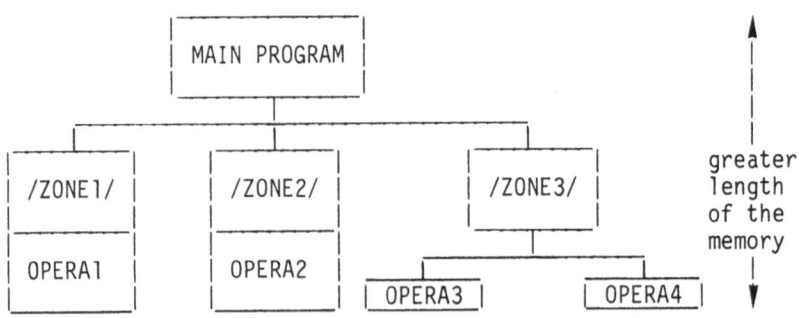

SCHEMA 3

The memory is divided into two dynamic areas :
- The System Memory : this is a control field which groups all
  the informations of objects and database organization.
- The Values Memory : this is the working storage.

The Control Field is divided into pages. The dimension of
pages is determined by the Page Turning System of the machine
(for instance, on IBM 30xx, 1 page = 1024 words of 4 bytes).
And the pages of common /ZONE/ are adjusted with the pages of
the system.
It contains :
- an Address Directory of named objects indexed by a
  "hashcoding" method ; it is constituated by contiguous
  pages ;
- a set of Attributes ;
- a chained list to manage free spaces in the workspace ;
- buffer areas to load in core or store on file Data ;
- a page called "zero" owned by the system "JEVEUX", its
  function is to keep informations of loading or unloading of
  the whole database.

The attributes and chained lists are stored in chained pages
of the System Memory.
The memory locations of objects are read in the attributes.

Here can be seen the schemas describing the transfers between
the Control Field and the Working Area :

1/ Schema describing a single object :

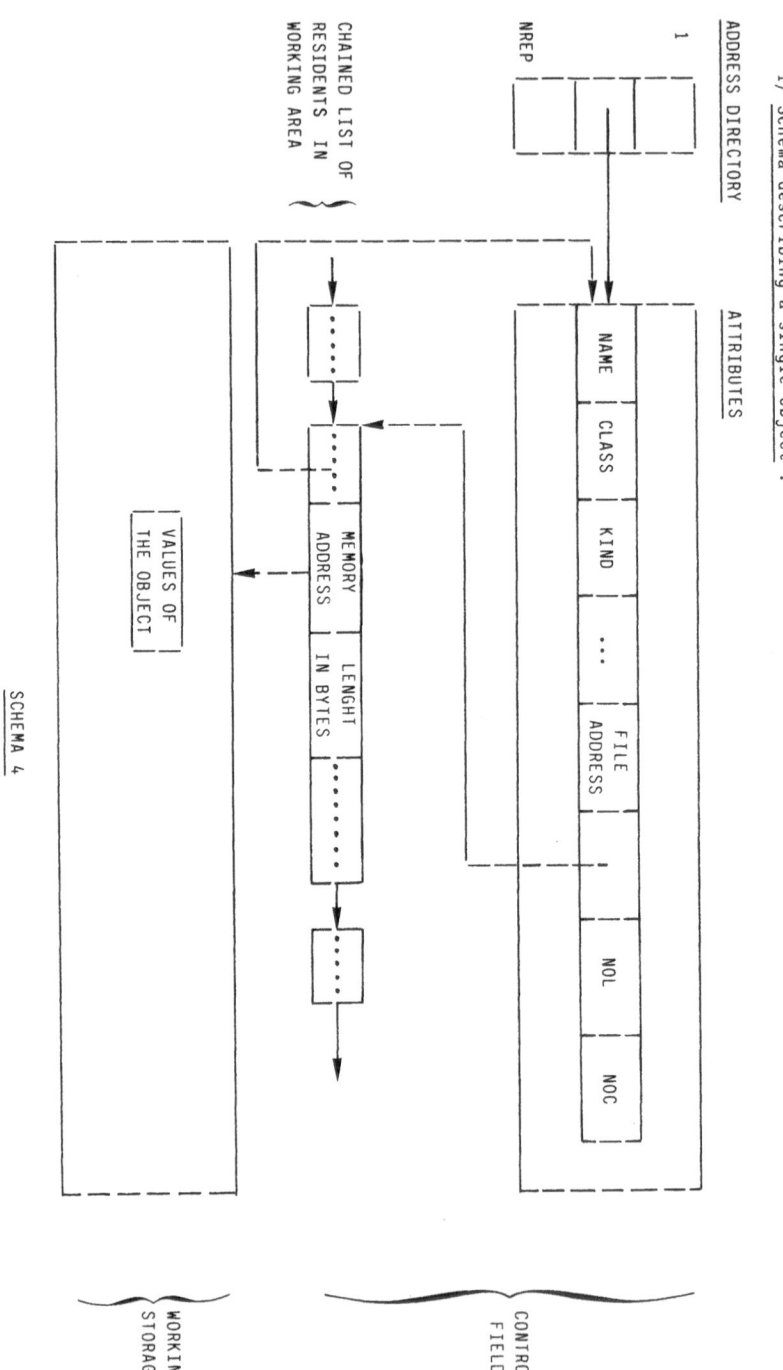

SCHEMA 4

2/ Schema describing a numbered family :

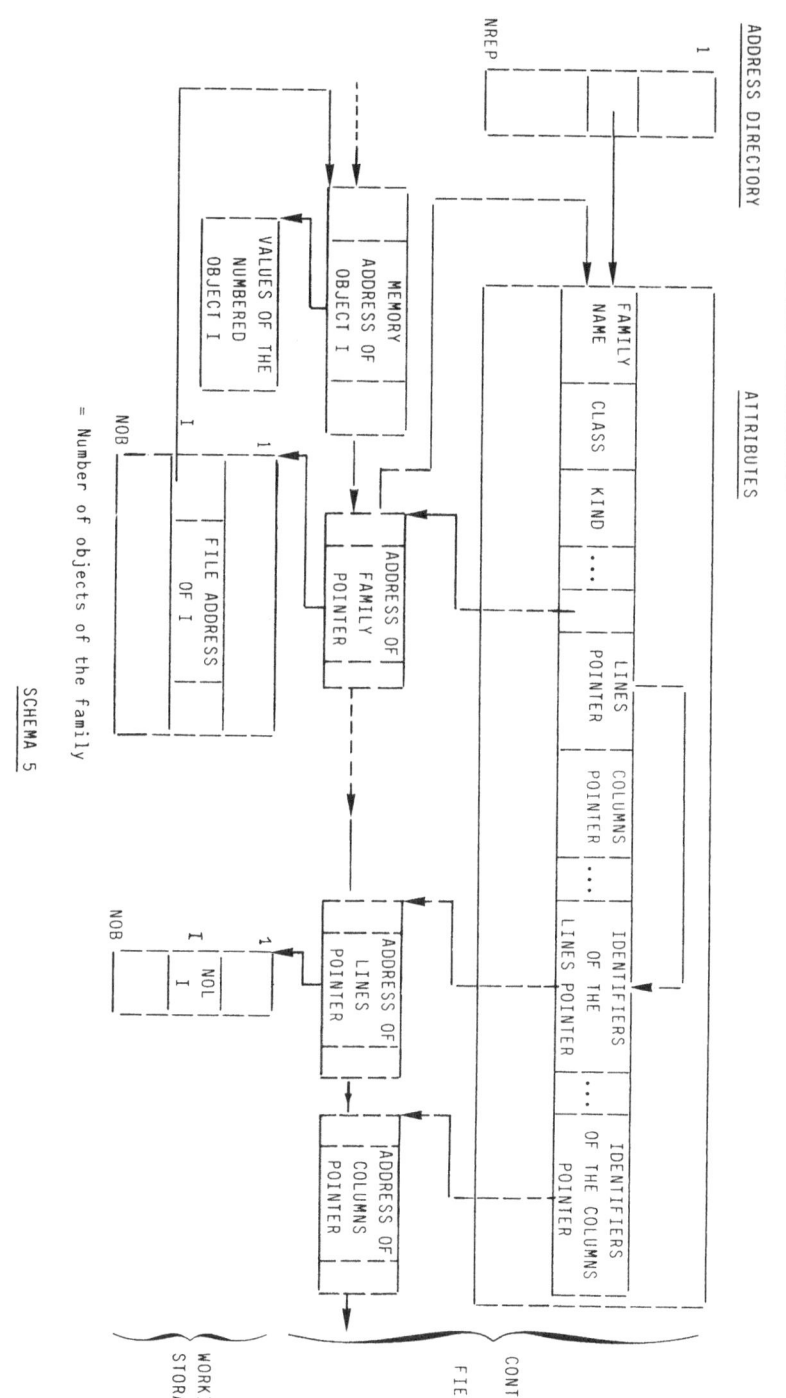

SCHEMA 5

= Number of objects of the family

Note : A named family has a more complex diagram because additionaly a named pointer of family's objects is introduced. There is an alternate area between the control field and the working memory which is used when one of them is completely filled.
The working area, composed of allocated and free spaces, is entirely maintained by the Control Field.

Management of allocated objects in core area :

The list of allocated objects has two chained parts : the first one is built up of temporary objects and the second one of permanent stored objects.
This criterion is introduced to assure the calculus the permanency of some objects in core memory and avoid that the system unloads them to the bulk storage unit and reloads them immediately after this.
Thus the transfer time is minimized with regard to other managerial tools.
A temporary object can be transformed into a permanent object and conversely.

When the working storage is completely filled, the built-in check of the system makes the release of an allocated object in searching into chained list of allocated fields (ranked in occuring order in core memory). The oldest temporary object with convenient dimension to store in core the replacing one, is either unloaded if it is "modifiable", or left if it has already been unloaded and is classed "not modifiable", taking into account the eventual neighbouring free spaces.

Structure of the virtual memory

The Virtual Memory of the system "JEVEUX" is a set of Direct Access files divided into logical records. Record lengh is calculated in computer words, parameterized, and fixed at the beginning of the running in the case of a file creation. The mass storage is word addressable.

There are three classes of direct-access file :

- The "Global database" : it is a permanent archival file. It gathers objects in class 'G', address directory and attributes, backed up at the end of running. Thus a complete and documented rerun of the database system can be made. This global file permits eventual graphic processings and editings of output data.

- The "Local database" : this file can be introduced when the machines have an overlay structure . Their objects are shared into several tasks and disregarded at the end of the module.

- The "Volatile database" : this file is a set of working objects not saved at the end of running.
  In an overlay structure, the working arrays are internal to a task and disregarded at the end of the task.

An object can be moved from one class to another.

Transfers between core memory and file

Two kinds of objects are considered for the transfer :

- The small objects : they are smaller than a logical record ; on the contrary there are big objects. Buffers are used for transfers and stored in the Control Field of the system "JEVEUX". They have the same length as a file record.
  The criterion of objects modifiability minimizes disk access and transfers become efficient.

- Big object treatment : There is a READ/WRITE Buffer, RWB, for the big objects.It is only used if the started or ended address of the object does not coincide with the start or end of the file record.

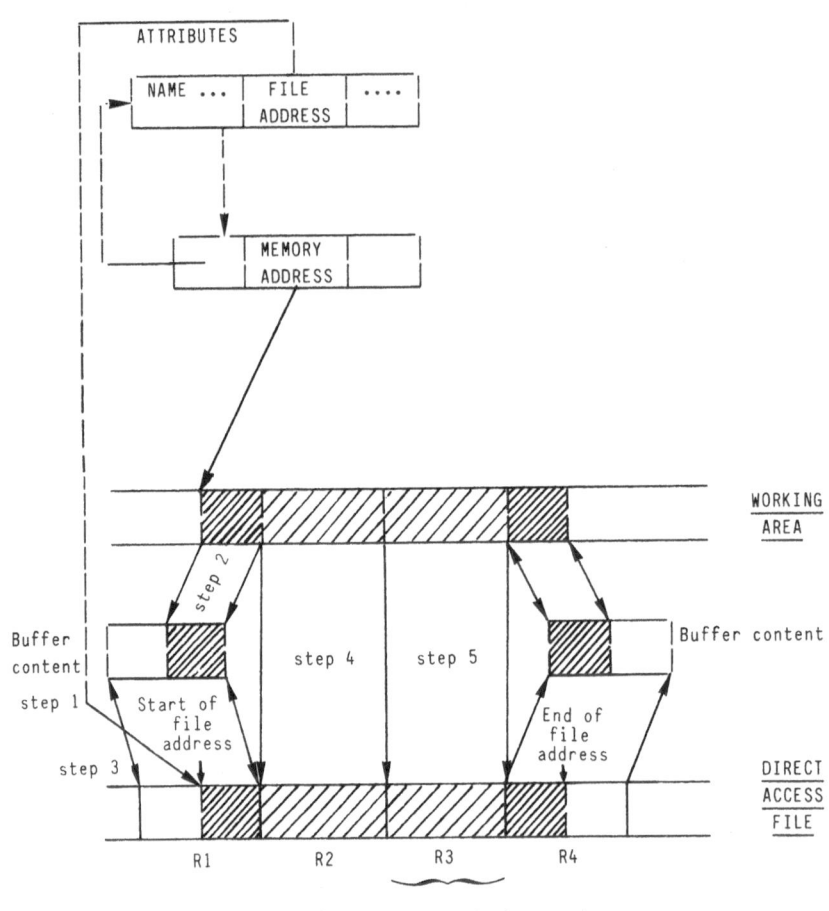

SCHEMA 6 : BIG OBJECT UNLOADING

Schema 6 : Here, can be seen an object that occupies four
records of the Direct Access File.
Treatment of the first record :

Step 1 : the record R1 is retrieved from the file in the
buffer RWB.

Step 2 : the object values are written from the core memory
into the buffer without modifying the preceeding
others.

Step 3 : the buffer content is unloaded in record R1.

The two other blocks are directly unloaded (Step 4 and 5) ;
the last block is treated as the same way as the first one.

Small object treatment :
Small objects created at the same moment in core area will be
simultaneously used in every case.

Thus, to minimize the cost of the Input/Output system, they
are clustered at their first unloading in the Unloading Buffer
of Small Objects called UBSO. This buffer is filled in
proportion as request of automatic swapping out onto secondary
storage.
As the same way, there is a Loading Buffer of Small Objects,
LBSO.
In a small object loading, its neighbourings are also
retrieved in core memory.
Thus, many Input/Output accesses are avoided.

Note : One object can be modified into buffers.

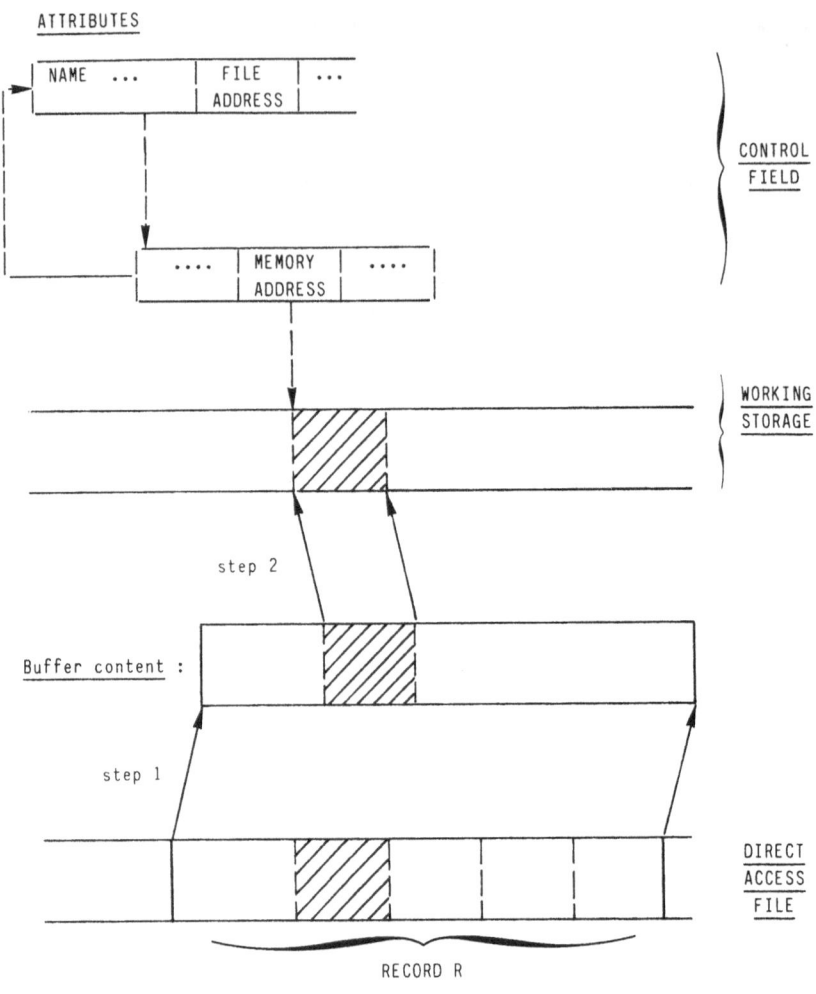

SCHEMA 7 : SMALL OBJECT LOADING

Schema 7 : Here, can be seen the loading of a small object.

Step 1   : The content of record R, group of small objects is
          loaded into the buffer.

Step 2   : The object values are transfered from the buffer in
          core area.

CONCLUSION

The processing database system "JEVEUX" can be used in any scientific computer codes where problems using mathematical objects with the easiest to the most complicated structures are resolved.

This database is documented and can be entirely used again. An internal software system permits the database system to run on any machine (IBM, BULL, CRAY,...).

At the moment, the system "JEVEUX" is operating on IBM 3081 and CRAY-1S computers. It will be implemented later on a Hewlett-Packard A700 computer.

This database management system is operational in a finite element computer code called **SIVA**, for static and dynamic analysis of industrial mechanical structures developed and maintened at "Electricity of France".

REFERENCES

C.A. FELIPPA (1982)
Fortran 77 simulation of word-addressable files.

L.A. LOPEZ
Polo II.

DATABASE ADMINISTRATOR FACILITIES
FOR
ENGINEERING DATA MANAGEMENT

H. Steenbergen and F.J. Heerema

National Aerospace Laboratory NLR, The Netherlands

SUMMARY

Databases in engineering are operated in a multi-disciplinary
environment by users with different skills. Database administrator facilities have to be available to keep control over a
database and to keep track on its contents. This paper deals
with the database administration task and with the requirements for the facilities to support this task. As an example,
the facilities for database administration in EDIPAS are used.
EDIPAS (Engineering Data Interactive Presentation and Analysis
System) has been designed and made operational at the National
Aerospace Laboratory NLR (The Netherlands). This system provides a standard framework for the data management and the
data evaluation and presentation tasks for a variety of engineering applications. It provides database administrator facilities to create a database, to inspect and to maintain the
contents of the database, including the chosen user's views on
the data as stored in the database.

1. INTRODUCTION

To date, the application of database technology in engineering
is generally accepted. The database technology is to be distinguished from the usage of a file manager, by the application of a database management system (DBMS) and related
schemas. These schemas contain the description of the contents
of a database and are used by the application programs to
access the data in the database.
Engineering and design optimization are multi-disciplinary
environments by nature. The database technology provides
eminent means to transfer data between the various engineering
groups involved and between the processes they perform.
Although there are differences between the application of

databases in engineering and in business applications, some lessons can be learned from the latter with respect to the control over the database contents.

As distinct from business database applications, in engineering the data derived from the basic (input) or primary data is also stored in the database. The derived data in engineering is often the result of some creative design activity or of rather time consuming algorithmic processes, as for instance analysis programs for strength or aerodynamic properties. Given a set of basic data, various sets of derived data can be obtained. The derived data depends strongly on the choice of the designer (or the engineer) and on the processes applied to generate the derived data. For instance, the drag of an already known shape of an aircraft or a car can be estimated roughly, or can be calculated by means of high quality and very accurate numerical simulation methods, or can be obtained by means of experiments with models. Both the primary and the derived data are acquired against considerable costs, which already points to requirements for facilities in the data management software to control the contents.

An issue in the application of databases is the internal consistency of the database, which means that the data is in accordance with the descriptions in the schemas. Application of a qualified existing DBMS solves the internal consistency problems. In an engineering application of the DBMS, it still will be possible to store in and to obtain from the database consistent data, that however do not reflect the reality of the application, i.e. the integrity is not maintained. As the DBMS software used in different engineering applications will not cope with these integrity problems, they have to be solved in the application dependent parts of the information system that uses the DBMS. This DBMS software has to provide facilities with which the contents of the database can be kept in consistent order. These facilities control the access of groups or persons to the database, provide a means to maintain the data in the database, and are an aid to spot the necessary data. Such facilities support the database administration task, that was explicit in business applications already for years. This paper discusses the database administration task in engineering applications. Requirements are given for facilities to support that task.

As an example, the database administrator support of the EDIPAS-package will be given. This package has been developed and is in operational use at NLR. The acronym stands for "Engineering Data Interactive Presentation and Analysis System". The package has grown to a set of facilities for both engineering data management and engineering data evaluation and presentation. The data management facilities are built around a commercially available DBMS, that provides internal consistency and controlled redundancy. The EDIPAS-package comprises facilities to support the database administration task. With these facilities the user is able to start a database for an application and to control the contents of an existing database.

2. DATA MANAGEMENT IN ENGINEERING

Data management tools become generally accepted in engineering. Distinct from the more traditional file organization, a database is controlled by means of a database management system (DBMS). A DBMS applies schemas in which the contents of the database is described. Application programs obtain access to the database using the information in these schemas. The generally accepted three-schema approach gives the conceptual, the external and the internal schema. The conceptual schema describes the structure of all types of data that the database deals with, and the relationships among these types of data. The description reflects the possible context in which the database can be used. An external schema selects a particular view of the data in the database, in accordance with the needs of an application program or user. The external schema controls to what extent the data may be seen, as well as the way in which it may be accessed, when permitted. Several external schemas can exist for the same database, providing different views for the different users and programs. The internal schema provides the information how the types of data are mapped to physical records on files. As the information in this schema determines the speed of update and retrieval processes, it is a means for tuning the performance of a database. A DBMS provides a data definition language (DDL) to specify the schemas. The DDL is used during the design phase of a database, and later on when modifications are necessary. Manipulation of the database requires a specialized language, called a data manipulation language (DML). The DML is used for the storage, retrieval, modification, and deletion of data in the database. One may say, that the DDL is the language of the database designer and that the DML is the language of the database user, in general an application programmer.

Database technology has been applied in business applications for about two decades. As the contents and the use of databases in engineering differ from business applications, some attention has to be paid to the differences and the consequences.
More than in business applications, the database in engineering contains the derived data as well as the basic or primary data. Whereas in business applications a selection is made from the primary data and the results are presented to the user after rather simple processing, in engineering the derived data is often obtained by means of complex (numerical) processing. Another aspect of databases in engineering that

strikes the eye is the dynamic character of the data and datastructure, i.e. the way in which the data is organized. This dynamic character is found in the variable amount of data related to an object (e.g. the order of a matrix) and in the modification of the type of data that describe an object when the object changes. When the description of the data in a database changes, this leads to modifications in the schemas. Application of techniques that provide facilities for the management of dynamic data and datastructures, bring the flexibility back to the users and the application programs that access the database. As a consequence users do need facilities to inform them about the actual contents of the database.

Engineering is a multi-disciplinary activity, where many users share a common information set, stored in a database (Fig. 1). In the engineering environment the database is seen as the point where data can be transferred from one discipline or group to another. The various processes communicate via the database (Loeve, 1982). The data management facilities will be used in various projects with different goals. In this situation the data management facilities should have additional provisions to:
- distinguish the different databases used for various projects;
- grant access to the database of a project to the members of the project-team, with different levels of authorization with respect to the data manipulation possibilities (store, delete, modify, etc.);
- maintain the data in the database in accordance with the logical views of the users on the data and with the requirements with respect to performance of the processes.
To accommodate these overall requirements, the introduction of the database administration function in engineering applications is necessary, together with utilities in the data management software that support that function.

3. THE DATABASE ADMINISTRATION TASK

In the previous section, the database administration function in engineering is indicated. This function is performed by the database administrator (DBA). The DBA is the person responsible for overall control of the database. He or she has to have already expertise in database applications in an engineering environment and the ability to understand and interpret management requirements at a senior level. In business database applications the database administrator task is sometimes performed by more than one professional. In the engineering field the benefit of database administration has been recognized too. Some organizations appoint in their projects a person who is responsible for the operational shape of the project database. In projects of limited size, the principal

user will take care of the database administration task. In
the multi-user environment of engineering projects, at least
one person has to be assigned to the database administration
task. Only in that situation the project team will be able to
keep control over the database and its contents. The responsi-
bilities of a project database administrator are:
- to advise the project team on the definition of the database
  contents;
- to open the database for the project and store the initial
  data, such as naming conventions for the data items as
  agreed upon in the project team;
- to grant access to the database for users within the project
  team. The various users will get different levels of
  authorization. For instance, modification of basic data
  could be granted only to the project-leader or the DBA.
- to define and to restructure access paths to the data. The
  access paths can be defined according to specific views of
  project members, using their own terminology. Furthermore a
  well chosen access path may increase the performance considerably.
- to maintain the basic and derived data;
- to inform the project members that use the database about
  the status of the database and its contents, on a regular
  basis;
- to make full or partial copies of the database contents for
  back-up and performance purposes. Partial copies of the
  database contents will be made for instance if the data
  involved will not be used during some time, but may become
  of interest again at a later time.
- to maintain the archive of (partial) database copies in
  order to have them ready for re-use on a selective basis.

In larger organizations several project databases may be
maintained at the same time on central computer facilities.
The project database administrators will optimize the overall
performance of the database, they are responsible for. The
management of the computer centre will benefit themselves when
they assign a system database administrator. This person will
direct his/her attention to the system and its performance.
The system database administrator (SDBA), a database management professional, will be responsible for the maintenance of
the data management software and the schemas that are used for
the various applications. Requirements from the user's community that affect the schemas will be handled using the DDL
facilities of the DBMS. The SDBA will gather statistics about
the performance of the system given the running database
applications and about resource usage of the individual
project databases. Of course the SDBA can provide information
that will be of great help to the project database administrators.

4. SURVEY OF EDIPAS FACILITIES

In order to give the scope for the database administrator
facilities in EDIPAS, a short survey of the capabilities of
the system is presented. The system provides facilities for
engineering data management and for engineering data evaluation and presentation. With the EDIPAS data management
facilities it is possible to build an information system for a
specific purpose (Troost, et al, 1984) with a project database
that is the transfer-point of data between processes applied
in the various disciplines involved in the project (Fig. 2).
The data in the database can be inspected and evaluated for
interpretation purposes by means of the data analysis and
presentation facilities. These facilities are command driven
and can be seen as engineering query and report facilities for
an EDIPAS database. The commands can be grouped into procedures which can be applied for complex evaluation tasks
(V.d. Draai, 1984). The procedures can be stored in the
database for later usage.

The data model applied in EDIPAS allows users to define (and
to change at will) naming conventions for the data according
to the terminology they want to apply. The database is used to
store primary data, derived data, data supporting the operations with the database, including procedures to execute the
analysis and presentation facilities. To support the operational use of the database, time tags are added to the
entities.

The EDIPAS facilities are grouped into three subsystems
(Fig. 3):
- data management,
- database administrator support,
- data analysis and presentation.

The subsystems interface to the database by means of a set of
database communication modules. These modules on their turn
interface to the commercial database management system, that
has been used as the data management kernel. This data
management kernel is completely hidden for the user's community by the EDIPAS facilities.

The main components of EDIPAS, contained in the subsystems are
(Fig. 4):
a) for data management:
   - facilities to load data from auxiliary datafiles,
   - interface modules to give auxiliary (user-)programs or
     program packages access to the database;

b) for database administrator support:
   - a facility to initialize a database for a project,
   - a facility to audit the contents of a database (i.e. to
     generate an overview of the contents),
   - a facility to manipulate the database contents for
     maintenance purposes,
   - facilities to dump and to reload the database;

c) for data analysis and presentation:
   - facilities that comprise various functions for the selec-
     tion, interpolation, computation with and graphical and
     tabular presentation of the data. The facilities are
     controlled with commands, that can be grouped upon user
     request into procedures that can be executed in batch or
     interactively.
   - a facility to generate the command sequences for a batch
     run of the analysis and presentation facilities, applying
     prepared procedures.

In (Heerema and Kreijkamp, 1984) the capabilities of EDIPAS
are described in more detail.

5. EDIPAS DATA MANAGEMENT AND THE DATABASE ADMINISTRATION TASK

The facilities in EDIPAS for database administration depend on
the properties of the applied data model of the database and
the related data management facilities. The concept of the
EDIPAS data model (Kreijkamp, et al, 1985) is based on basic
data entities with very general properties. The data entities
applied in EDIPAS are translated into the physical records in
the database by means of the DBMS used in the kernel. This
translation is defined once for all applications by means of
the DDL of the DBMS. The data management facilities of EDIPAS
provide the interface to the end-users, the database administrator and the application programmers. For all these users
there is no need to learn details about the DDL and DML of the
DBMS, they "operate" with the EDIPAS database entities.

A basic entity of EDIPAS is the datablock. One datablock
contains engineering data that is interrelated in a sense
defined by the user. Within a datablock, single values
(scalars) and matrices (a row or a set of values) can be
recognized, each of them having a user supplied name. A
datablock is uniquely identified with one or more so-called
characteristic values, again with user supplied names. There
is no restriction on the number of elements in a datablock, on
the length of a matrix or on the number of characteristic
values. The various datablocks may contain a different number
of elements with different names. Other entities in EDIPAS
concern the datastructures, the derived data and the operational use of the database.
Time tags are added to the entities, providing information on

date of storage and/or date of latest usage. The datablocks
are given the date of initial load. The remaining data
entities are given the date of creation and the date of latest
usage. The users, introduced to the database, get the date of
latest access to the database tagged to their user name.

The datablocks are organized in one or more multilevel
hierarchical structures, with a characteristic value name on
each level. Datablocks are logically placed in a structure at
the lowest level, according to the characteristic value names
belonging to the datablocks (Fig. 5). As a consequence of the
possibility of having more than one structure in a database,
each user or group of users may apply a structure to access
the datablocks in accordance with a view related to the
specific application. EDIPAS allows to (re-)define the contents of datablocks and the structures in an operational
database. It will be clear from this, that for the EDIPAS
user's community the data definition and data manipulation
phase coincide. For data definition and manipulation, the
EDIPAS data management facilities provide DDL and DML functions, called EDDL (EDIPAS-DDL) and EDML (EDIPAS-DML), respectively. The EDDL and EDML are the interface to the users,
that utilize the EDIPAS data evaluation and presentation
functions, the DBA-support functions and user programs interfaced to the database (Fig. 6). The data management facilities
of EDIPAS translate the EDDL and EDML to the DML of the DBMS.
The DBMS takes care of the storage of the physical records.

The database administrator assigned to a project using EDIPAS,
will advise the users how to define the contents of the
datablocks. Knowing the contents of a database and the way it
is used, he or she is able to increase the performance for the
users, by means of (a) well chosen structure(s). The DBA will
advise users and programmers how to interface application
programs to the project database. The project DBA utilizes the
EDIPAS database administration facilities, without knowledge
of the internals of the DBMS. The time tags of the EDIPAS
database entities are used by the DBA to get information about
operational use of the database.

A system database administrator (SDBA) is responsible for the
maintenance of the schemas as applied by the DBMS (Fig. 6).
The person or the group of persons that perform the task of
the SDBA has detailed knowledge of the DDL and DML of the
DBMS. The SDBA is able to increase the performance of the
project databases utilizing EDIPAS, by tuning the internal
schema of the DBMS. Required changes or enhancements of the
available EDIPAS facilities, that imply updates of the schemas
are the responsibility of the SDBA.

6. FACILITIES IN EDIPAS FOR PROJECT DATABASE ADMINISTRATION

The database administrator subsystem of EDIPAS consists of
facilities to initialize a database, to get an overview of the
contents of a database, to maintain the database contents and
to perform a dump or reload of the database. These facilities,
to be used by the project database administrator, are described below in more detail.

6.1 Initialization of a database

When a project team utilizes EDIPAS, the data to be kept in
the database are to be analysed in order to agree upon the
contents of the datablocks and the names of the characteristic
values and data elements. A strategy may be developed, how to
organize the datablocks into one or more structures. The DBA
(being an experienced person or the prime user, but not a DBMS
professional) will contribute to this process into a great
extent.

To open the database, EDIPAS provides an initialization
utility. With this utility the DBA identifies a database for
the project. The database gets a name, and the project and the
responsible person's name (of the project-leader or the DBA)
are entered. During initialization, the DBA may specify:
- the names and initial passwords of the intended users of the
  database and their authorization to access the database,
- if already known, the name(s) of one (or more) structure(s)
  and the names of the levels,
- the default values of a number of system parameters chosen
  for the project.
The initialization process is depicted in figure 7.

6.2 Audit of the database contents

When a project database is in operational use, the audit
utility of EDIPAS can be used to obtain an overview of the
contents of the database. The options of the audit utility
allow to obtain the overview of the complete contents of the
database or of a part of it.

The DBA audits a part of the basic and derived data by making
a choice from the available structures and the levels in a
chosen structure. Within this window, the data can be listed
according to their entity type (basic and derived) or according to their location in the window. Auditing according to
locations gives all the level occurrences within the specified
window and the data entities connected to it. The engineering
data contained in the datablocks is listed by giving the names
(and values, if asked for) of the single values and the
matrices.
Without specifying a window, the audit facility provides a
means to list those datablocks, that can not be accessed via
structures.

These audit options can be used to check the result of a data load process.

By using the time tags given to the database entities, the DBA gets information about their operational usage. All EDIPAS database entities can be selected on their time tag(s). The backup facility for (selective) load/dump shall have to provide means to dump the data not used recently.

A project-member can get an overview of his or her own supporting data, for example his or her own procedures used to run evaluation and presentation tasks. An overview of a procedure contains the commands of which the procedure consists and the parameters of that procedure and their default values.

With the audit facility, the DBA can provide a short overview of the database contents for the project team members on a regular basis. Such an overview contains the descriptive information of the project, the names of the current users (optional) and structures, and total amounts of EDIPAS entities, currently available in the database. See figure 8 for an example of such an overview.
During interactive evaluation and presentation, the user can consult the database in order to get an overview of the actual contents of the datablocks, and the data entities and their location.

6.3 Maintenance of the database contents
The project members utilizing the project database may change their view on the data when the database has been in operation for some time. Using the database has as effect, that the data kept in it will not reflect the "reality" as seen by the project group, although the database is still consistent. With the maintenance utility the project DBA has the possibility to bring the contents of the database in accordance with the view of the project team, and to update the database with respect to the basic data, the structures and the supporting data.

The operations on the basic data and structures include the following:
- structures, being the access paths to the datablocks, can be added or deleted. To the levels of a structure, aliasses can be added.
- datablocks, containing the basic data elements, can be accessed by means of an access path as determined by a structure. Characteristic values may be added.
  Characteristic values, not being a level of a structure, may be deleted or modified. Basic data elements (single values, matrices, and matrix values) can be added, deleted or modified.

The operations on supporting data include the following:
- text can be entered or deleted, that serves documentation purposes for the project database. The DBA can use this option to keep a project database history.
- new users can be granted access to the database, existing users can be detached from the database or be given another level of authorization,
- the presentation format and the name of the engineering unit of a data item can be specified or modified,
- aliasses can be introduced for the default EDDL and EDML commands and command parameters as available in the project database. With this possibility the DBA can bring the default EDIPAS "language" in accordance with adopted project terminology.
- help information that will be given upon user request can be adapted.

The usage of the maintenance subsystem is restricted to the DBA and to those members of a project team that have the same (=highest possible) authorization level as the DBA.

6.4 Backup and archive of the database contents
The backup facilities are only directed to the database as a whole. The project DBA is able to copy a project database to off-line storage (such as magnetic tape). In this way it is possible to keep different versions of the project database. So that from a certain moment a copy of the project database can be used for experimental purposes, without destroying the contents of the original one.

A selective dump of the contents of a project database is defined but not yet implemented. With this facility the data not relevant for on-line access can be stored on off-line media. With the re-load utility it is possible to restore the data again in the originating database or to transfer data to the database of another project. The latter possibility, implemented in a previous operational version, has proven to be of great help to the users of different project databases that contain partly the same data. With this facility the names of the data items of the original data can be replaced by names to be applied in the receiving database.

A future development will be that the partial database dumps can be administrated by means of an archive utility. In that situation the project database can be seen as the combination of the on-line and off-line stored data. Access to the off-line stored data however will be in an indirect way, i.e. by (automatically) restoring the required data into the on-line database.

7. CONCLUDING REMARKS

The utilization of database management tools in engineering applications is generally accepted. In a multi-disciplinary environment the database administrator function is necessary to keep control over the access to the database and its contents. The database administration function consists of tasks to be performed by a database administrator for a project and tasks to be performed by a system database administrator. The project database administrator is not a database management professional. The activities of this person are directed to the engineering aspects of the project databases. The system database administrator covers the internal affairs of the data management tools. He/she is able to adapt the data management facilities for all project databases with respect to new functional requirements and system performance.

The EDIPAS package, made operational at National Aerospace Laboratory NLR provides various facilities that support the project database administrator tasks. A database can be created for a project, the contents of a database can be audited and be maintained. Some facilities are available for backup purposes.

The available facilities for database administration have proven to be of great help to keep control over the contents of a project database.

8. REFERENCES

Draai, R.K. van der (1984). Postprocessing NLR-HST Windtunnel data by means of EDIPAS. The 62nd STA-meeting, October 10-12, 1984, Goettingen, Federal Republic of Germany, (NLR MP 84083 U).

Heerema, F.J. and Kreijkamp, H.A. (1984). A multi-discipline and integrated facility for management and evaluation of engineering data. The 6th International Conference and Exhibition on Computers in Design Engineering, CAD84, April 3-5, 1984, Brighton, United Kingdom (NLR MP 83062 U).

Kreijkamp, H.A., Heerema, F.J., Hedel, H. van (1985). Flexible and dynamic data model in an Engineering Data Management System. Conference and Exhibition on Computer Applications in Production and Engineering, CAPE'85, May 1-3, 1985, Amsterdam, The Netherlands (NLR MP 85011 U) (to be published in Dutch).

Loeve, W. (1982). An Infra-structure for Computational Fluid Dynamics to serve Computer Aided Design. Symposium for users of finite element methods, Delft Technical University, The Netherlands (NLR MP 82046 U) (Paper in Dutch).

Troost, G.K., Kemper, C.A.L, Verbruggen, T.W. (FDO) and Wijngaart, R.F. van der, Lindhout, J.P.F., Draai, R.K. van der (NLR) (1984). Channel Design for Closed Cycle MHD. The 22nd symposium on engineering aspects of magnetohydrodynamics, June 1984, Starkville, USA (NLR MP 84053 U and FDO TPV 84-014).

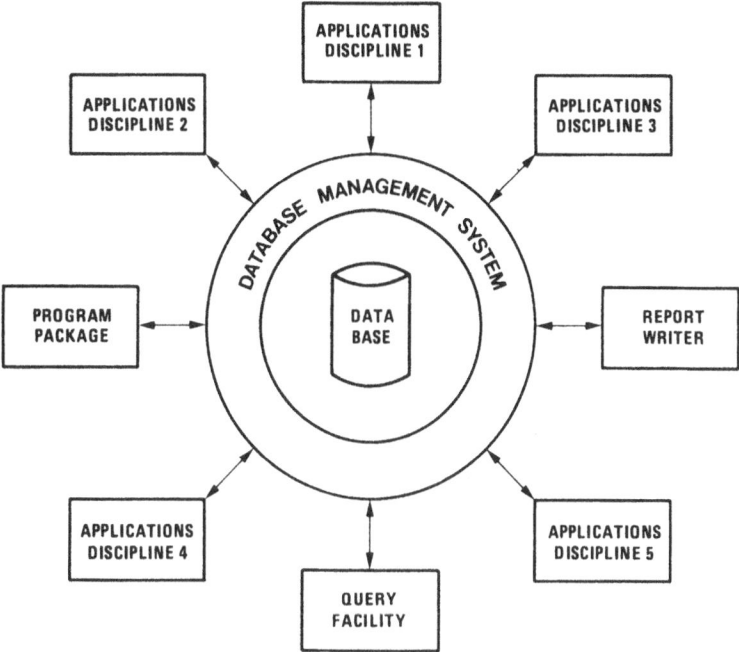

Figure 1 Database management in a multi-disciplinary
environment

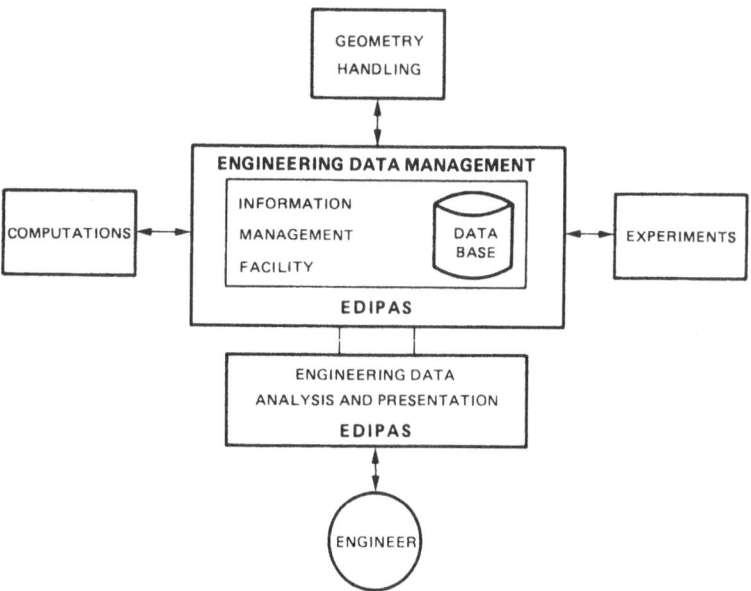

Figure 2 System architecture using EDIPAS

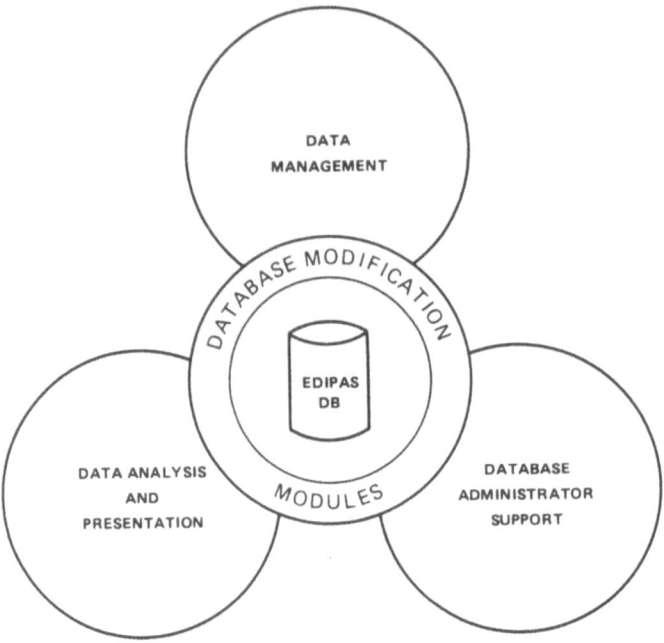

Figure 3   Subsystems of EDIPAS

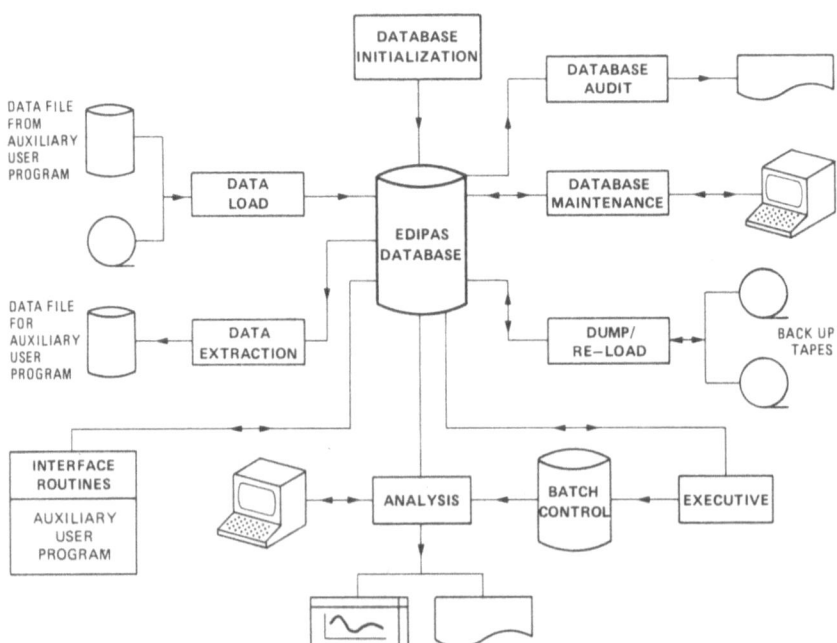

Figure 4   EDIPAS system components

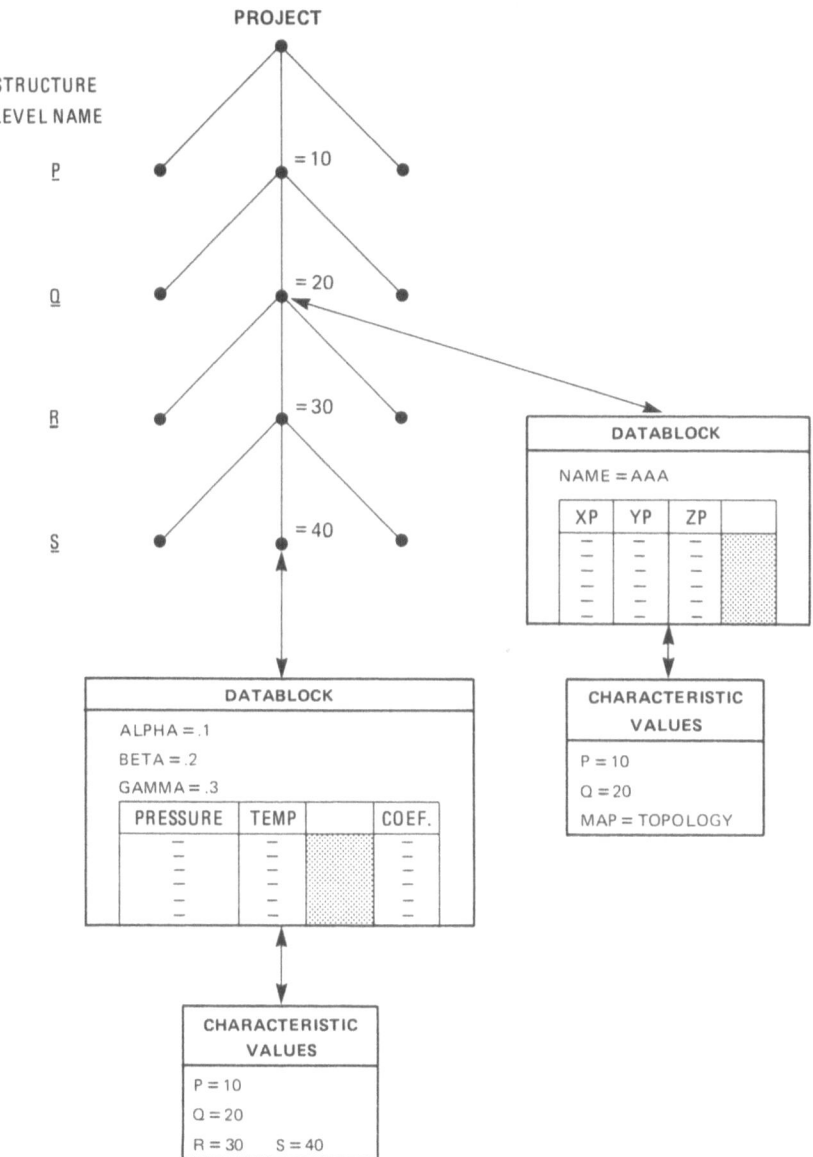

Figure 5   Datablocks at different levels of a structure in an
EDIPAS database

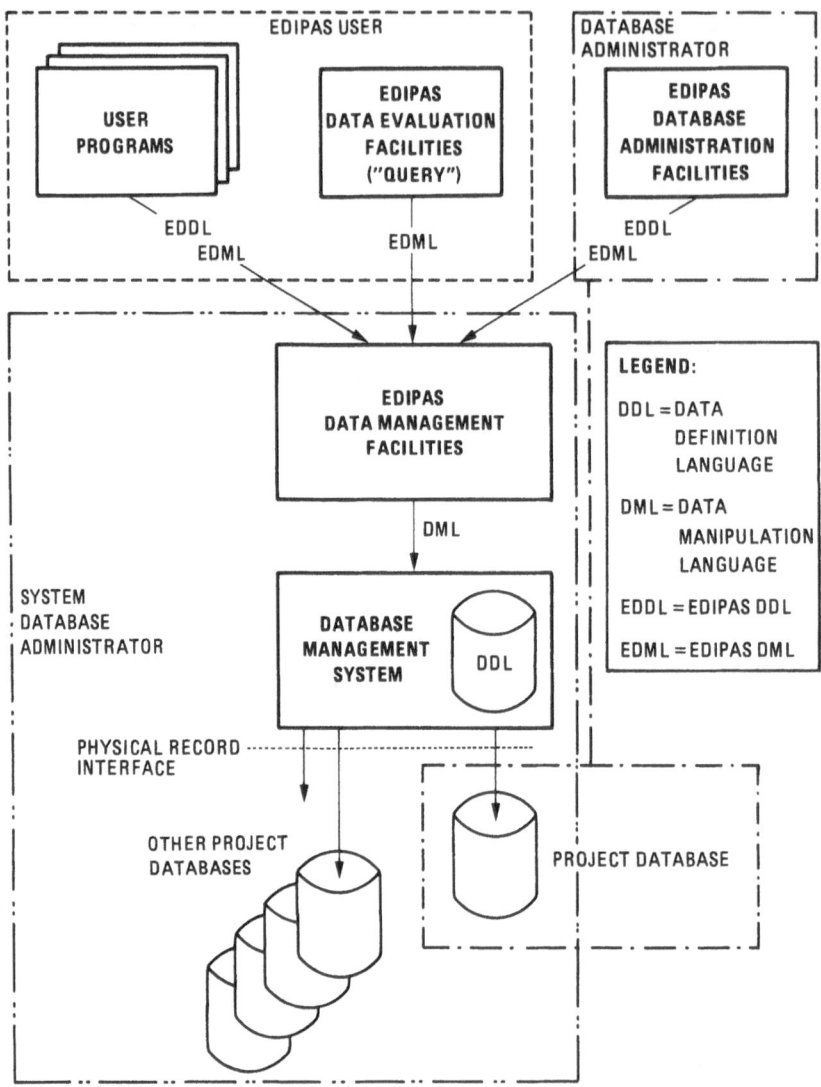

Figure 6   Scheme of applied data management language interfaces
in EDIPAS

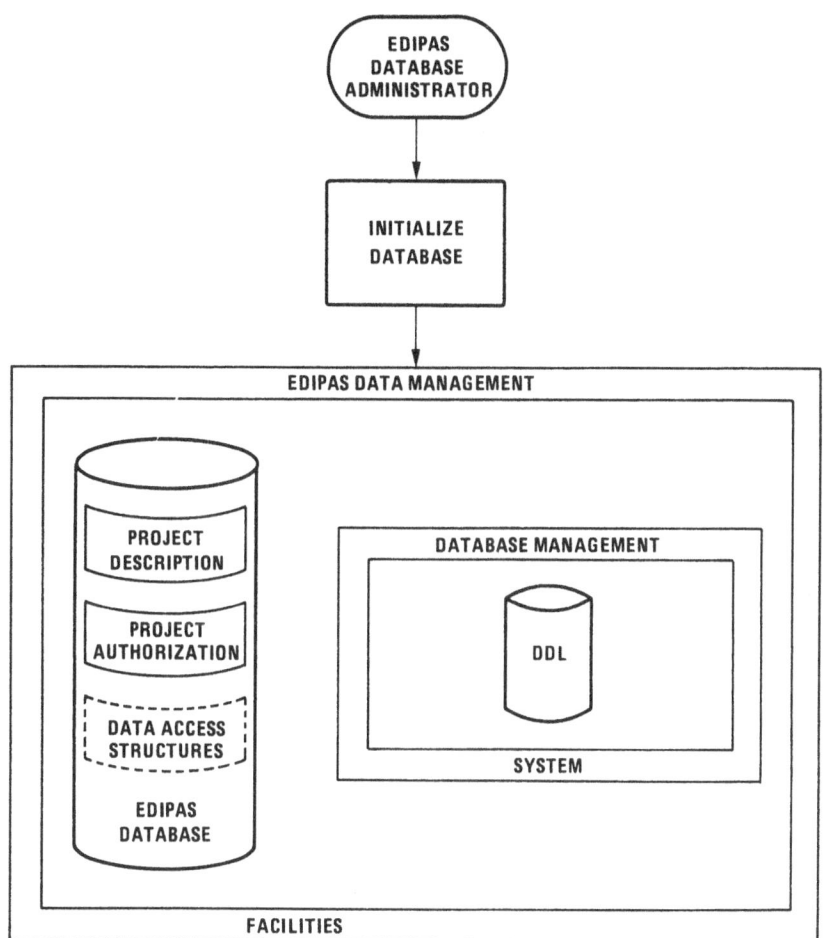

Figure 7   Initialization process of an EDIPAS database

```
EDIPAS    AUDIT  VERSION 4.002   84/12/19.        PROJECT INFORMATION


PROJECT : TPS

TURBO POWERED SIMULATORS

JANUARY 1984 :
CALIBRATION OF TPS 2400,
CONFIGURATION AB6,
LOW SPEED CONDITIONS.

DATA OF THE FOLLOWING CONFIGURATIONS ARE AVAILABLE :

TEST 4901 - SERIE 530 : TPS 6343, SHORT CORE NOZZLE, PORT ENGINE
          - SERIE 540 : TPS 6344, SHORT CORE NOZZLE, STARBOARD ENGINE
          - SERIE 541 : TPS 6344, LONG CORE NOZZLE,  STARBOARD ENGINE


     CLASSIFICATION :
     CUSTOMER      : NLR
     PROJECT-LEADER : JANSEN
     ALL STRING    : ALL
     EMPTY VALUE   :     .1000000000+321
     EMPTY STRING  :


     STRUCTURE(S) :

     NAME

     AERO
     CONF

     TOTAL NUMBER OF STRUCTURES :         2



     TOTAL NUMBER OF USERS      :         7

     TOTAL NUMBER OF DATABLOCKS :       464
     TOTAL NUMBER OF CURVES     :       231
     TOTAL NUMBER OF ARRAYS     :        17
     TOTAL NUMBER OF ISOPLOTS   :         6
     TOTAL NUMBER OF TABLES     :         6
     TOTAL NUMBER OF SESSIONS   :        12
     TOTAL NUMBER OF PROCEDURES :        37
     TOTAL NUMBER OF PICTURES   :        10
     TOTAL NUMBER OF MACROS     :         0
```

Figure 8 Example of audit result

A DATA BASE MANAGER FOR ENGINEERING COMPUTATIONS

E. Backx

Catholic University of Louvain

SUMMARY

During the last conference on engineering software in 1983 [Backx, 1983], a data base manager was presented. This manager was written in basic and was implemented in a finite element program. The aim was to be able to solve large problems in a computer with a small core size (64K). This implementation turned out to be very successful because, not only the primary aim was reached but it turned out that coding additions to the program becomes much simpler and faster.

Since that time, the same ideas were used to write a data manager for other computers and in other languages. (MS-basic, Fortran 77, Pascal).

INTRODUCTION

Writing large programs for scientific computations such as finite element and boundary element programs and pre- and post-processors is a time demanding job. Every effort has to be made to decrease the number of dependencies of one module to another module because these links are often the reason of bugs. Storage allocation both incore and out of core is one of these problem areas that oblige the programmer to be inventive and very disciplined. Dynamic incore storage allocation in FORTRAN programs is sometimes solved by using a one dimensional array in COMMON.
During execution, data arrays are given the required space in this array and the calls to subroutines is done by passing the appropriate pointers. The space used by arrays which are not needed anymore, can be reused by other arrays if the size is smaller than the previous array. To avoid errors in reusing the space of deleted arrays, the data needed have to be well structured before programming is started. Additions to the programs become tedious.

When the programs data needs become too large to justify a complete incore solution, the data have to be transferred to disk files for intermediate or permanent storage. Some programs need as much as 20 different external files to solve this problem.

This is a very unsatisfactory solution for the programmer (and the user).

A remedy for these problems is to use a data manager. Data base managers are well known in administrative applications but not so much used in scientific applications. The reasons are that these data base managers are very heavy programs in their own and due to their capabilities, they tend to decrease drastically the execution of the application. Moreover, these data base managers are expensive.

This paper proposes the use of a scientific data manager to solve the problems mentioned above. This scientific data manager is by no means comparable in size and capabilities to a commercial data base manager but it is shown from experience that it is a valuable tool in program development.

DESCRIPTION OF THE DATA MANAGER

The intention is to free the programmer from managing memory and disk space organisation by giving him easy to use commands for storing and retrieving data. Both incore and out of core data management is taken care off. From the application programmers view, the use is very simple. There are only five different functions he needs to know:
1. Open the database
2. Close the database
3. Put a data item in the database
4. Get a data item from the database
5. Delete an array in the database.

The way to execute one of these functions is different in the different implementations. At this moment, five different versions have been written:
1. BASIC-2 on Wang
2. MS Basic
3. Fortran 77 on Vax
4. Pascal on Vax
5. MS PASCAL

The first version was described earlier [Backx, 1983]. This first version was designed because primary memory was the most important problem to implement finite element programs in BASIC. This primary memory is generally limited to 64K on these machines. After the implementation, the other advantages such as decreasing program development time and managing large dynamic arrays became predominant. The advantage of such a tool in environments with virtual memory and higher level languages is obvious. It was however not possible to translate the original version in these higher level languages directly.
The original version will now be described functionally. Afterwards, the FORTRAN and PASCAL versions will be discussed.

BASIC-2 ON WANG

Constant lenght blocks are used in primary and secondary memory. The programmer has to dimension several arrays in each program and has to load the database functions as part of his program. To put a data item in the database, the following CALL is send to the data base routines:

GOSUB'70(N$, W∅, S4, W3, W7)

input
- N$ 3-character name of array
- W∅ index
- S4 dummy
- W3 length of data item

output
- W7 starting address of required information in memory array C$()

Next, the programmer should put the data item in C$() starting at the address W7. This is done with a PACK instruction.
To get a data item from the database, the same CALL is send and the data is extracted by an UNPACK statement.

To delete an array in the data base, the following CALL is issued

GOSUB'74(N$)
N$ 3-character name of array.

The blocks which are deleted by this call are freed and become available to hold newly created arrays.
The implementation is very system dependent because it uses special possibilities of BASIC-2 such as:

```
MAT SEARCH
PACK
UNPACK
STR
VAL
CONVERT
INIT
```

The interaction between primary and secondary memory is done in direct access-mode.

FORTRAN 77 on VAX

A first version has been written (Ickroth-Maes, 1984) and was used in a boundary element program. The PACK and UNPACK instructions were replaced by WRITE and READ to and from an array. The resulting code was too long and too slow.

A second version was then made by the author. This version was developed from the point of view that the application programmer should be able to use the routines as simple as possible. Now, the conversion from real data to character data (PACK-UNPACK) is obtained using the EQUIVALENCE statement and the substring capabilities of FORTRAN 77.

There are 5 user callable routines which cover the functions described before. These are:

1. CALL DBOPENEN (ICODE)
   ICODE = $\emptyset$: new database
   = 1: existing database

2. CALL DBCLOSE

3. CALL DBPUT(A, NR, NAME)
   A : variable to add to the data base
   NAME: name of array to which variable A has to be added
   NR : index of variable A in array NAME

4. A = DBGET (NR,NAME)
   A is the variable in array NAME with index NR

5. CALL DBDELETE (NAME)
   Delete the array NAME in the database.

Only real variables can be handled.

The size of the data base is controlled by a BLOCK DATA subprogram. The programmer has to specify:

1. NBLIN. The size in number of blocks of the primary memory that he wants to reserve for incore data (minimum 2)

2. NBLOUT. The size in number of blocks of the secondary memory that he wants to manage for out of core data

3. NBLSIZE. The size of the block in bytes.

Example 1: NBLIN=10
           NBLOUT=150
           NBLSIZE=1024

           The total amount of data that can be handled is 150K (150 x 1024). The total amount of memory required is 11K.

Example 2: NBLIN=10
           NBLOUT=1000
           NBLSIZE=5120

           The total amount of data that can be handled is 5MB (1000 x 5120). The total amount of memory required is 57K.

Chosing NBLIN too small will lead to a large amount of input-output with the disk. Choosing NBLSIZE too large will lead to an uneconomic use of disk space. The choice therefore depends on the problem. If there are only few different but large arrays, increase NBLSIZE. If there are a large number of different arrays of smaller size, decrease NBLSIZE.

PASCAL ON VAX

The first PASCAL implementation was written in MS-Pascal. After-wards, the same implementation was used on VAX. Only the disk I-O routines had to be rewritten.

The PASCAL version is somewhat longer in code and is still not as flexible as the FORTRAN implementation. However,integer, real and character data can be handled.

APPLICATIONS

The BASIC-2 version is implemented in the programs of SCIA*.
These are finite element and steel structure design programs.
The FORTRAN version is used in our university in a boundary ele-ment program (Ickroth-Maes, 1984) and is also used to couple FEMGEN (a finite element preprocessor) with other programs (Van Eynde-Verheyden, 1985). The finite element mesh generated by FEMGEN is stored in our database. The input for the finite element program is generated from these data, the bandwith is minimised, the structure is computed and the results are stored in the same database. The post-processor program only needs this database as an input to pre-sent the results in a most convenient way.

CONCLUSIONS

A data manager for scientific data is designed and implemented in BASIC, FORTRAN and PASCAL. The advantages of such a manager are:

- the programmer does not need to organise internal and external memory
- the size of primary memory needed is minimal
- improved use of secondary storage space
- development time is decreased
- additions to programs are much easier.

REFERENCES

BACKX E., RAMMANT J.P.: "A virtual storage data management system for finite element analysis on micro's", Engineering Software III, April 1983, pp.819-831.

ICKROTH , MAES D.: "Computerprogram with data management for two-dimensional boundary elements", E 1983-1984, K.U.Lv. (In Dutch).

VAN EYNDE F., VERHEYDEN P.: "Coupling of CDM300 with fini-te element programs", E 1984-1985, K.U.Lv. (In Dutch).

---

* SCIA sv, Steenweg 108, Herk de Stad, Belgium

ENGINEERING DATABASE - WEIGHTS AND CENTRES.

S M Fraser

University of Strathclyde, Glasgow, U.K.

1. INTRODUCTION

In the engineering design of a complete installation it is important to keep the total cost and the total weight under constant review and in many applications, such as aircraft and ship design, the nett centre of gravity of the overall structure must be estimated as accurately as possible.

It is essential that the cost estimate and the weight estimate are based on the same component list and therefore a common database should be used by both; a suitable component numbering system should be employed which simplifies database interrogation by computer programs.

The philosophy behind the development of this program was that it should incorporate the following features:-

a) program to run on a desk-top micro - APPLE IIE.

b) user friendly - for infrequent use by designer/ draughtmen.

c) input/ output file handling to be controlled efficiently by the program to maximise use of disc space and reduce access time.

d) program to feature a date-lock facility such that a time history at various stages of the project would be available for reviewing the design procedure.

Because of the limited RAM capacity of the APPLE II the cost estimate program is run separately from the weights and centres program and only the latter is described here, however the cost estimate program is similar in structure but with features specific to the financial aspect.

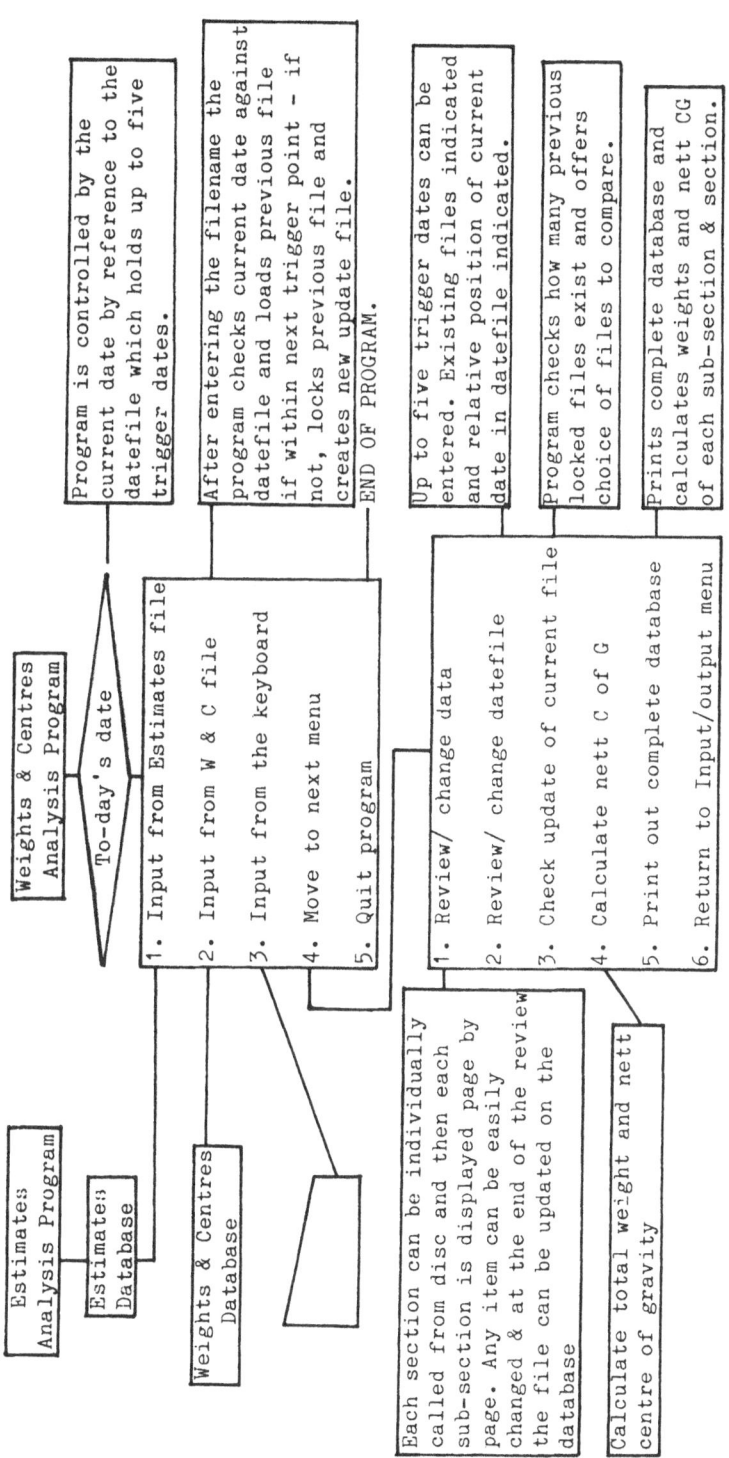

17-58

Estimates
Analysis Program

Estimates
Database

Weights & Centres
Analysis Program

To-day's date

Weights & Centres
Database

Program is controlled by the current date by reference to the datefile which holds up to five trigger dates.

1. Input from Estimates file
2. Input from W & C file
3. Input from the keyboard
4. Move to next menu
5. Quit program

After entering the filename the program checks current date against datefile and loads previous file if within next trigger point – if not, locks previous file and creates new update file.
END OF PROGRAM.

1. Review/ change data
2. Review/ change datefile
3. Check update of current file
4. Calculate nett C of G
5. Print out complete database
6. Return to Input/output menu

Up to five trigger dates can be entered. Existing files indicated and relative position of current date in datefile indicated.

Program checks how many previous locked files exist and offers choice of files to compare.

Prints complete database and calculates weights and nett CG of each sub-section & section.

Each section can be individually called from disc and then each sub-section is displayed page by page. Any item can be easily changed & at the end of the review the file can be updated on the database

Calculate total weight and nett centre of gravity

FIGURE 1 SCHEMATIC PROGRAM LAYOUT.

2. LIMITATIONS OF THE DESK-TOP MICRO

The computer system consists of an APPLE IIE with 64K RAM and
80-column card, twin 140K floppy disc drives, monitor and
printer. The disc operating system is capable of handling
sequential or random access files but random access files are
wasteful of space on a large database and the limited memory
capacity will not allow the full database to be read in from
a sequential file. For this reason the data is split into
nine sequential files which are accessed individually by the
program; this gives a good compromise between speed of
accessing and loading data relative to the efficient use of
disc space.

Once the program is operating the processing speed of the
micro is not found to be a limitation since the operator is
the slowest element in the time taken to read, enter or
change data. Printer speed limitation can be adequately
covered by using a buffered interface.

3. PROGRAM STRUCTURE

The program operates from two option menus, the first dealing
with input/ output options and the second menu offers the
various updating, data checking and calculating options; the
schematic program structure is shown in Figure 1.

The input/ output menu indicates the following options:-

        1.   Input data from estimates file.
        2.   Input data from weights & centres file.
        3.   Input data from keyboard.
        4.   Move to next menu.
        5.   Quit program.

It is usual for a first quotation to be drawn up based on a
cost estimate/ quotation procedure and hence the part
description of individual components will be available in the
estimates database. Therefore option 1 in the menu allows
the program to draw in the relevant information required by
the weights and centres program. This information is then
automatically stored as a set of weights and centres
sequential files which will, in future, be called back by
selecting option 2. Option 3 allows a database to be
generated from scratch although this option will not
generally be used. Options 4 and 5 allow transfer to the
main menu or exit from the program.

As stated previously the database is stored as a set of nine
sequential files ie the project is divided into nine specific
section headings; each section is sub- divided into nine
sub-sections and each sub-section has a maximum of fifteen

DATA INPUT   SECTION 1/2   LINE 4   TYPE ? FOR HELP

	PART DESCRIPTION	X M	Y M	Z M	WEIGHT TONNES
1201	Part Number 1	12.5	0.65	0.15	12.75 U1
1202	Part Number 2	28.75	-0.898	7.65	1.23 U1
1203	Part Number 3	2.15	1.115	2.35	0.65 U1
1204					
1205					
1206					
1207					
1208					
1209					
1210					
1211					
1212					
1213					
1214					
1215					

Total Weight = 14.63

Enter new data Part Number 4

Figure 2 SCREEN EDITING DISPLAY.

items. Thus a component numbering system is used whereby the first two digits represent the section and sub-section numbers, the next two numbers represent the item number and the remainder of a 40 character string is available for the rest of the part description eg

2512 Hydraulic motor sub-assembly

Section 2   Sub-section 5   Item 12

With the above limitations the maximum number of items is 1215 but this is sufficient for many small design applications; however the theoretical limit for this numbering system is 10000.

In addition to the part description each item record stores a value for the weight of the item and the x,y and z co-ordinates of its centre of gravity relative to some reference datum point.

The main menu indicates the following options:-

        1. Review/ change data.
        2. Review/ change datefile.
        3. Check update of current file.
        4. Calculate nett centre of gravity.
        5. Print out complete database.
        6. Return to input/ output menu.

Selecting option 1 allows the further option to select any of the nine sections for screen editing. The screen display is shown in Figure 2 and the layout and editing functions have been designed to be as user friendly as possible.

As shown in Figure 2 the reason for choosing 15 items per sub-section was to allow additional information to be displayed and still give a readable information display. The screen editing cursor control, using the arrow keys, restricts the cursor movement to the 15 lines of data and moves immediately to the beginning of each field in the horizontal direction; automatic wrap-around is provided in all directions. When the cursor is positioned at the required location, new data can be entered and is immediately over- written at the chosen location. The total weight of the items in the sub- section is displayed and is automatically re-calculated whenever a new weight is entered.

    Since the user does not access the program on a daily basis and frequently has difficulty remembering the editing keys, a help page facility is a pre-requisite for users of this program. At any time whilst in cursor screen editing mode, entering ?   will throw up the help page, Figure 3.

```
                UP                Use arrow keys to
        LEFT        RIGHT         Move cursor to
              DOWN                required position

   Type    /              to enter new data
   Press   RETURN         after data entered.

   Type   C(ontinue)      to move to next sub-section.

   Type   Q(uit)          to quit entry & ask to record data.
   Press  ESC             to store data automatically
                          & return to menu.
   Type   U(pdate)        to accept current values as update.
```

Press any key to return to screen editing

Figure 3 SCREEN EDITING HELP PAGE.

Option 2 allows the user to review/ change the datefile; this
is a set of up to five dates which are used to control the
historical sequence of file generation. The first date is
automatically set as the date on which the database was
created and up to four other "trigger" dates can be entered
into the datefile. The datefile screen display is shown in
Figure 4.

```
                        File
                        exists

   Update file ONE starts      Y     1/10/84

   Update file TWO starts      Y     1/1/85
                                              1/2/85 - To-day's
   Update file THREE starts           1/6/85          date

   Update file FOUR starts            1/10/85

   Update file FIVE starts            1/3/86
```

Figure 4 DATEFILE SCREEN DISPLAY

When the program is run, the current date is entered and the datefile is scanned to find whether the current date has passed a trigger date; if it has then the existing file is locked as a historical record and a new set of files is created with the appropriate update number appended to the filename.

When there is more than one set of files, option 3 checks the current file against any selected previous file and lists, to the screen or printer, any part number whose current update number has not changed since the previous file. As shown in Figure 2 the update number is displayed at the extreme right of each line and this number is changed to the current update number when new data is entered or if the cursor is placed on that line and the U(pdate) key is pressed. Using this option immediately highlights items which have not been changed in the current file and therefore indicates to the designer what items should be checked with respect to the current design of the project.

Option 4 is the main output requirement from the program since it calculates the total weight and nett centre of gravity of the structure.

Option 5 lists to the printer the complete database section by section and for each section and sub- section calculates the total weight and nett centre of gravity of that element.

4. CONCLUSIONS

A user- friendly program has been created which makes effective use of the limited memory and disc handling capability of a desk-top micro- computer. It provides the designer with a facility for quickly updating the database as the project design changes and, in addition, provides a historical record which charts the changes which have been made between initial design concept and final weight of the manufactured structure.

DATABASE USED FOR OPTIMAL CONTROL OF THE DJERDAP
HYDROPOWER SYSTEM

Serafim Opricović
Faculty of Civil Engineering, Belgrade - Yugoslavia

Milorad Miloradov
"Jaroslav Černi" Institute for Development of Water Resources

SUMMARY
The optimisation model is realised as a program package with
appropriate database. Some of the files contain: morphological
data (of the river system), characteristics of the units of power plants, spillways, and extreme values of controlable variables. Other files contain the discrete values of the dynamic
vairables during the given time horizon as: required power,
predicted hydrological inflows, and water levels and flows.
    All the programs from the package use the database by
the particular subroutines. The original input data are entered
by user of the program package according to the specified formats.

1. INTRODUCTION

To achieve the optimal control of complex dynamic systems, it
is essential to have the support of a developed software program package, and of an information system with the relevant
database. The structure of the information system and database
as well as how it should be developed (estabilished) and used,
all largely depend on the character and complexity of the system
that should be managed on one hand, and on the model of optimal
control on the other hand. This is why it is essential to study
and develop the information system and its database when developing the optimal control model. The authors of this paper had
faced this problem developing the optimal control model for the
hydropower stations Iron Gate I and II on the Danube. The database was then developed as a very important component part of
the software support required for the optimal control of the
system.

2. DESCRIPTION OF THE SYSTEM

The Iron Gate system included in the mathematical model consists of power plants I and II (both the Yugoslav and the Rumanian sides), of reservoirs I and II and of the downstream reach of Danube up to the junction with the river Timok. The inflow into reservoir I consists of the inflows (UQ) at the following input profiles:Ilok on the Danube, N.Bečej on the river Tisza, Sremska Mitrovica on the Sava and Ljubičevski Most on the Morava, and the junction of smaller tributaries (the Mlava, Pek, Nera, etc). The system also includes the spillways at dams I and II (BI and BII).

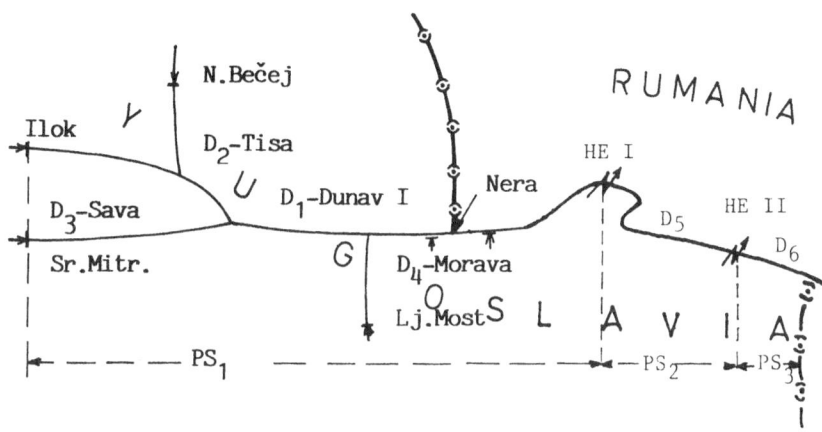

Fig. 1. Scheme of the Iron Gate System

The system can be decomposed into three subsystems: PS_1, PS_2 and PS_3, presented on the Fig. 1.

Subsystem PS_1 consists of the hydroelectric power plant (HE I) and the dam (BI) of Iron Gate I, the reservoir that spreads into the reach of the Danube I (D_1) and the reaches of the tributaries of the rivers Tisza (D_2), Sava (D_3) and Morava (D_4) as well as the inflows at the junctions of smaller tributaries (UQ)

$$PS_1 = (HEI \times BI) \times (UQ) \times D_1 \times D_2 \times D_3 \times D_4$$

System PS_2 consists of the hydroelectric power plant (HE II) and the dam (BII) of the Iron Gate II as well as of the reach of the Danube (D_5) between dams I and II.

$$PS_2 = (HEII \times BII) \times D_5$$

Subsystem PS_3 consists of the reach of the Danube (D_6) starting from dam II and the reach up to the border between Yugoslavia and Bulgaria, including the conditions and limitations (G_t) at this border.

$$PS_3 = D_6 \times (G_t)$$

From the aspect of control, there are three alternative configurations of the system:

$$S_o = PS_1 \times PS_2 \times PS_3$$
$$S_1 = PS_1 \times (D_5U\ PS_3)$$
$$S_2 = PS_2 \times PS_3$$

The alternative S_o represents the entire Iron Gate System. Alternative S_1 represents the Iron Gate system without the Iron Gate II while alternative S_2 is the part of the system that refers only to the Iron Gate II.

The Iron Gate system produces the energy for the power supply systems of Yugoslavia and Rumania. Due to the different loading diagrams, the operating regimes of the following four power plants are considered within the Iron Gate System: HEIJ, HEIR, HEIIJ, HEIIR (J- Yugoslav and R- Rumanian). Since each side controls its own spilways, the following four groups of spillways are considered within the system BIJ, BIR, BIIJ, BIIR.

3. OPTIMAL EXPLOATATION MODEL

The optimal exploatation model should enable the dispatchers at the Iron Gate to realize the optimal control of this complex system. The basic objectives of the optimal operation are:

a) Maximization of the total energy production

$$E_{max}(T) = \max_{\{Q^I, Q^{II}\}} \int_o^T \left[P^I(t) + P^{II}(t) \right] \cdot dt$$

where

   T - time horizon (seven days or less)
   Q - outflow through the turbines
   P - power
   I and II - denote the hydropower plants

b) Maximization of the production benefit

$$D_{max}(T) = \max_{\{Q^I, Q^{II}\}} \int_o^T D(P^I(t),\ P^{II}(t)) \cdot dt$$

where

   D - benefit from the produced power, taking into consideration the differentiated prices of the peak and basic energy

c) Production according to the diagram of the required power $P^R(t)$ (or energy) with an optimal distribution of the loading to both the power plants and utits.

$$\min_{\{Q^I, Q^{II}\}} \int_0^T \left[P^R(t) - P^I(t) - P^{II}(t) \right]^2 \times dt$$

The solution of these optimization problems must be such that
it meets all the constraints on the flow and the water level in
the reservoirs. These constraints have been defined by the Convention and Agreement on the Use of Hydroelectric Power Plants
on the Danube.

An optimization done for the purpose of achieving a maximum
benefit better suits interests of the Iron Gate System. However,
there is always the question whether it is possible to include
this type of energy production into the power systems of the
two countries involved. This is actually the reason why it is
important to have an optimal production according to the given
power diagram, i.e. simulate certain given manner of use of
hydropower plants.

Besides solving these already mentioned problems, the model
should also meet other additional requirements such as these
given bellow, that could arise during the actual implementation
of the model:

- production according to a temporary control strategy
- operation of only one hydro-power plant
- change (or replacement) of a constraint
- change of the required power at any moment of time

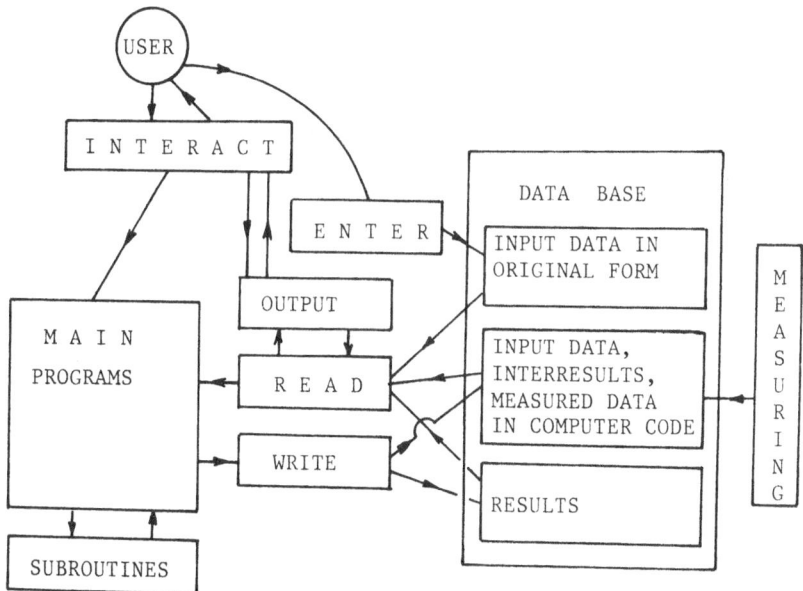

Fig. 2. Scheme of program package and data base

Besides planning the production, the model should also register
the produced power and the conditions in the reservoirs in real
time. It should also compute the difference between the planned
and actual production as well as the distribution of energy
between the two national power systems.

The optimization model was realized as a program package with
the corresponding database. (Fig. 2). It consists of an interactive part and of several basic programs and of several subprograms. The structure of the program package is a result of
the required structure of the optimization model. The input
data is introduced by the user of the program package either
directly or by using corresponding routines. The results are
obtained from the database using output routines. The connection
with the base can also be achieved through the interactive
program. The basic role of this interactive part is to enable
the user to start the desired computational procedure within
the program package. The data from the real system (that are
included in the model) are measured and automatically stored
in the database with a fixed sampling period.

The maximum time horizon of the optimization model is seven
days. The beginning and duration (t_0,T) of the time horizon are
given as the initial hour and day in a week for the required
computation time interval.

4. DATABASE

According to the characteristics of the hydropower plants as
well as the control objectives and the model itself, the database consisting of three types of data was developed. So, there
are three types of files:

I. The files with original input data given by the user directly
on the terminal (manually). The data is given in the decade
numbering system in accordance with the input formats. These
files are of sequential type.

II. The files used for computation purposes as well as some
with intermediate results (binary). The measured data also appear in the binary code. Most of these files are of the direct
access type.

III.The files with the final results in the form as given by
the output formats.

The users of the program package and database introduce the
input data in their original form which enables the changes in
the input data if this is necessary. The routines for entering
the data "read" the input data and transmit them to the basic
programs. The basic programs are not direct connected with the
database. The results of the basic programs are stored in the
database by the output routine. Users can obtain the results

by the interactive program as well as certain subprograms.

The values of the dynamic variables (time series) as well as the data without the time dimension (static values) are in the database. The relationships given in the studies as diagrams are entered as pairs of values for the points on the diagram. The number of the points for which the pairs of values are read depends on the curve of diagram. The number of pairs should not be greater than 30 for the part of the diagram between the two extremes. The pairs of values should be read with a maximum precision. The multi-dimensional relationships are entered in the form of matrices whose structure should be the result of the analysis of a given relationship.

The values of the dynamic variable are entered into the files as discrete values measured per hour for a period of seven days (168 numbers). The first number in the time series refers to the value of the dynamic variable at 7.00 a.m. on Monday or 7.00 and 8.00 a.m. for the mean hour values. The files with the data in the decade system have records of 72 Bytes. One record contains the data in the format 12F6.0 (or if necessary (8F9.0). The part of the file containing the values of a dynamic variable in decade form is presented in Fig. 3.

Hours	00	01	02	03	04	05	06	07	08	09	10	11	12
Days	h	d	m					x	x	x	x	x	
I	12	13	14	15	16	17	18	19	20	21	22	23	24
		x	x	x	x	x	x	x	x	x	x	x	x
II	00	01	02	03	04	05	06	07	08	09	10	11	12
		x	x	x	x	x	x	x	x	x	x	x	x
	12	13	14	15	16	17	18	19	20	21	22	23	24
		x	x	x	x	x	x	x	x	x	x	x	x
.					.			.					
.					.			.					
.					.			.					
VII	00	01	02	03	04	05	06	07	08	09	10	11	12
		x	x	x	x	x	x	x	x	x	x	x	x
	12	13	14	15	16	17	18	19	20	21	22	23	24
		x	x	x	x	x	x	x	x	x	x	x	x
	00	01	02	03	04	05	06	07	08	09	10	11	12
I		x	x	x	x	x	x	x					

Fig. 3. The records for the dynamic variable

The files with the data in binary form have a record of 171 * 4
Bytes so that one record contains all the values of a dynamic
variable for a period of seven days, while the last three numbers
represent the hour, day and month when the last change of the data
in the record was entered.

The following is a list of all the files within the database
grouped according to the nature of the data they contain. The
numbers in this list are used as symbolic units of files in the
input and output routines.

1. The files with the parameter values

11. IMEPROG - file with the names of the programs
12. PARAMNEST - parameters for computing the unsteady flow
13. PINTERAK - parameters and messages for interactive program

2. The files containing data of the system

21. MORFO - file with morphological data
22. RAPAVO - file with data of the roughness coefficient
23. GRANICE - file of contraints
25. PSTANJE - flow and water levels at the cross-sections within
             the system for the initial moment of optimization
26. AGREGATI - characteristics of the units
27. MORFOBIN - contains the data from the MORFO and RAPAVO in the
              binary code
28. PRELIVK - characteristics of a spillway
29. PRELIVP - contains the summed up characteristics of a spillway

3. The files with time series

31. SNAGAZ - required power (loading diagram)
32. DIJSNAGE - typical loading diagrams
33. DOTOK - inflows at the input profiles into the HE Iron Gate I
           (binary)
34. GRUSLOV 1 - boundary conditions for the reach above dam I
              (binary)
35. GRUSLOV 3 - downstream boundary contidion for reservoir II
              (junction of the river Timok)
37. GORUSI - data of the upstream boundary conditions for the
            HE Iron Gate I (given by the user if the data are
            not those from DOTOK)
38. PRDOTOK - forecasted inflows at the input profiles of the
             HE Iron Gate I

4. The files with the results

40. REZODP - intermediate results of dynamic programming
41. REZNEST - results of the unsteady flow computation
42. REZAHTE - computation results using a given loading diagram
43. REZEMAKS - results of the total energy maximization
44. REZOPTI - results of the optimization of energy production
45. REZUPR - results of simulation with a given control
46. RNADZOR - results of the program for the supervision of the
             operation of the system

47. RKONGRAN - results of the program for testing whether the
    constraints are met
48. REZPREL 1 - computation results of the overflow for the HE
    Iron Gate I
49. REZPREL 2 - computation results of the overflow for the HE
    Iron Gate II

5. The files with control data

51. ZADATUPR - trajectories at control profiles
52. PLANVRED - planned values of the energy, water levels and
    flows within the system.

6. The files with measured data

61. MERENO - produced energy, measured water level at control
    profiles and measured (derived) discharge values
62. MEREPREL - measured data of the overflow

Depending on the enter of data, some of the files will be grouped
as given below:

a) The files with the data which the user enter directly, but
   less often than once in seven days

11. IMEPROG, 12. PARAMNEST, 13. PINTERAK, 21. MORFO, 22. RAPAVO,
23. GRANICE, 26. AGREGATI, 28. PRELIVK, 32. DIJSNAGE,

b) The files with the input data entered by the user at least
   once over a seven days period during his interactive work with
   the package:

25. PSTANJA, 31, SNAGAZ, 38 PROTOK, 51. ZATATUPR,

The input data from the following files should be adopted and a
brought into accordance with the Rumanian partner:

21. MORFO, 23. GRANICE, 26. AGREGATI, 28. PRELIVK, 51. ZADATUPR

while the same should be done at least once in 7 daysfor files:
31. SNAGAZ, 38 PRDOTOK .

The measured data stored into the database are transferred each
Monday by 7.00 a.m. fromthe files of the operational base into
the archival database.This should be defined within the project
of the Iron Gate information system.

ACKNOWLEDGEMENT

The research reported herein is supported by the Serbian Community of Sciences - Beograd, and by the work organization HE
Djerdap, Kladovo.

REFERENCES

Jakovljević P. (1979) Power Plants in the Yugoslav-Rumanian Reach of the Danube (in Serbo-Croatian), Niro Export Press, Beograd.

Miloradov M., Opricović S. and Tanasković P. (1978) "The Mathematical Model of Unsteady Flow and Its Application to Water Reservoir Control with Daily Regulation", Proc. 2nd International Conference on Applied Numerical Modelling, Madrid.

Miloradov M. and Opricović S. (1983) "Program Package Used for the Optimal Control of the Djerdap Hydropower System", Proc. of the 3rd International Conference on Engineering Software, London

Preliminary project "The Mathematical Model of Optimal Exploatation of the Hydropower Plants Djerdap I and II" (in Serbo-Croatian), "Jaroslav Černi" Institute, Beograd, Dec. 1981.